T0406876

# Handbook of Geometric Constraint Systems Principles

## Discrete Mathematics and Its Applications



**Handbook of Product Graphs, Second Edition**
*Richard Hammack, Wilfried Imrich, and Sandi Klavžar*

**Handbook of Mathematical Induction:**
**Theory and Applications**
*David S. Gunderson*

**Graph Polynomials**
*Yongtang Shi, Matthias Dehmer, Xueliang Li, and Ivan Gutman*

**Introduction to Combinatorics, Second Edition**
*W. D. Wallis and J. C. George*

**Representation Theory of Symmetric Groups**
*Pierre-Loïc Méliot*

**Advanced Number Theory with Applications**
*Richard A. Mollin*

**A Multidisciplinary Introduction to Information Security**
*Stig F. Mjølsnes*

**Combinatorics of Compositions and Words**
*Silvia Heubach and Toufik Mansour*

**Handbook of Linear Algebra, Second Edition**
*Leslie Hogben*

**Combinatorics, Second Edition**
*Nicholas A. Loehr*

**Handbook of Discrete and Computational Geometry, Third Edition**
*C. Toth, Jacob E. Goodman and Joseph O'Rourke*

**Handbook of Discrete and Combinatorial Mathematics, Second Edition**
*Kenneth H. Rosen*

**Crossing Numbers of Graphs**
*Marcus Schaefer*

**Graph Searching Games and Probabilistic Methods**
*Anthony Bonato and Paweł Prałat*

**Handbook of Geometric Constraint Systems Principles**
*Meera Sitharam, Audrey St. John, and Jessica Sidman,*

https://www.crcpress.com/Discrete-Mathematics-and-Its-Applications/book-series/CHDISMTHAPP?page=1
&order=dtitle&size=12&view=list&status=published,forthcoming

# Handbook of Geometric Constraint Systems Principles

Meera Sitharam
Audrey St. John
Jessica Sidman

CRC Press is an imprint of the
Taylor & Francis Group, an **informa** business

CRC Press
Taylor & Francis Group
6000 Broken Sound Parkway NW, Suite 300
Boca Raton, FL 33487-2742



Printed on acid-free paper
Version Date: 20180528

International Standard Book Number-13: 978-1-4987-3891-0 (Hardback)



**Visit the Taylor & Francis Web site at**
**http://www.taylorandfrancis.com**

**and the CRC Press Web site at**
**http://www.crcpress.com**

*To the works of:*

*Hilda Polaczek Geiringer 1893–1973*

*Neil White 1945–2014*

*Michel Marie Deza 1939–2016*

*Wenjun Wu 1919–2017*

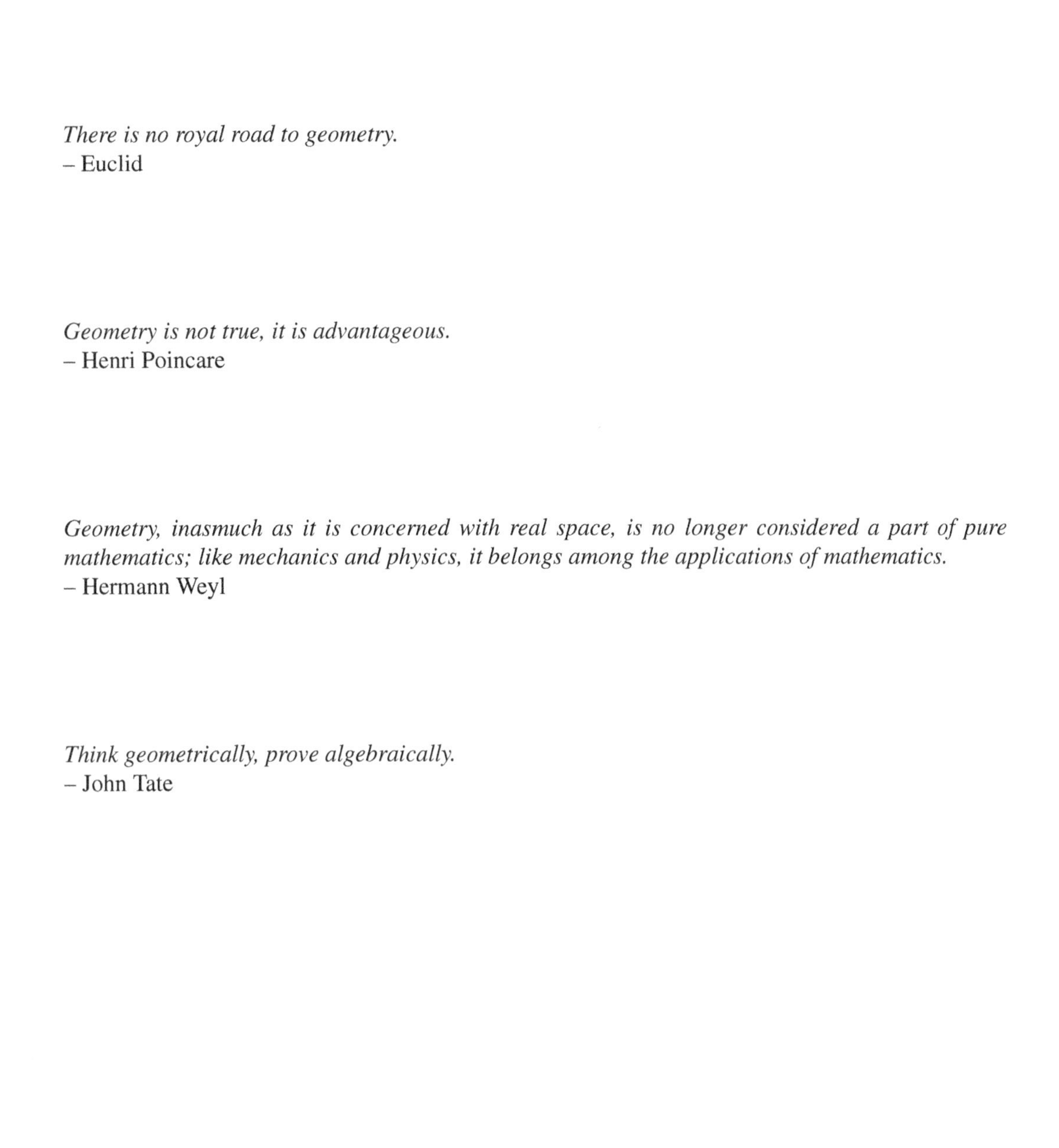

*There is no royal road to geometry.*
– Euclid

*Geometry is not true, it is advantageous.*
– Henri Poincare

*Geometry, inasmuch as it is concerned with real space, is no longer considered a part of pure mathematics; like mechanics and physics, it belongs among the applications of mathematics.*
– Hermann Weyl

*Think geometrically, prove algebraically.*
– John Tate

# Contents

# *Foreword*

Geometric constraint systems arise in a diverse range of applications including: computer aided engineering and architectural design, molecular and materials modeling, robotics and animation, sensor networks, machine learning, and dimension reduction. Broadly, a geometric constraint system (GCS) is defined on a set of geometric primitives (e.g., points, lines, rigid bodies) by specifying geometric relationships (such as distances, angles, or incidences). The core GCS foundations come from at least four interwoven topic areas and research communities: (i) combinatorial and geometric rigidity, (ii) automated geometric theorem proving, (iii) geometrically constrained configuration spaces and, (iv) distance geometry. Indeed, the principles, tools, and techniques rely on invariant theory, combinatorial and discrete geometry, algebraic geometry and topology, convex/semidefinite analysis, with algorithmic foundations and complexity going back to Cauchy, Cayley, Hilbert, Klein, and Maxwell.

With such a rich array of communities working on GCS research, this handbook is intended as an entry point to the principal mathematical and computational tools, techniques and results currently in use. It was born out of continued requests for a single source containing the core principles and results that would be accessible to beginners and experts alike (from the graduate student starting research to the algebraic geometer interested in applications to the roboticist seeking to engineer a swarm of autonomous agents). Recognizing that readers may come from a wide variety of backgrounds, we hope that this book will be a useful tool for navigating the concepts, approaches, and results found in GCS research.

We are grateful to the authors of the chapters that follow; their expertise provides the roadmap for developing a unified view of the varied perspectives. We pledge any royalties toward supporting the activities of the four research communities represented by the four parts of this handbook, especially the activities of young researchers. We would like to thank Louis Theran for his feedback. We thank Rahul Prabhu for his kind and timely expert help with LaTeX. And, finally, we cannot put into words the debt of gratitude owed to our families for their unconditional support and patience during this process.

Meera Sitharam
Audrey St. John
Jessica Sidman

# Preface

The goal of this book is to provide a resource for those aiming to become acquainted with the fundamentals as well as experts looking to pinpoint specific results or approaches in the broad landscape. The flow of the handbook is intended to take readers from the general algebro-geometric approaches to more specialized contexts permitting combinatorial analysis and efficient algorithms. Chapters are grouped by the main techniques being deployed, in the hopes that readers can find the material best-suited to their expertise. Of course, the overlapping nature of the material being presented prevents a neat partitioning of chapters by topic area, but we hope the juxtapositioning of the chapters helps the reader to see how the subject is connected.

Chapter 1 provides an overview of the book as a more detailed starting point and is expected to help the reader navigate the book effectively. It includes a basic introduction, some preliminaries, and an overview of the various topics and methods. We also give an alternative pathway through the book, intended to help a newcomer become acquainted with the domain. We hope this is a first step toward a unifying foundation for the rich set of GCS problems.

# *Contributors*

**Abdo Y. Alfakih**
University of Windsor
Windsor Ontario, Canada

**Troy Baker**
University of Florida
Gainesville, Florida

**Robert Connelly**
Cornell University
Ithaca, New York

**Michel-Marie Deza**
École Normale Supérieure
Paris, France

**Elena I. Deza**
Moscow State Pedagogical University
Moscow, Russia

**Mathieu Dutour Sikirić**
Rudjer Bosković Institute
Zagreb, Croatia

**Jacques Fleuriot**
University of Edinburgh
Scotland, United Kingdom

**Ioannis Fudos**
University of Ioannina
Ioannina, Greece

**Steven J. Gortler**
Harvard University
Cambridge, Massachusetts

**Timothy F. Havel**
Energy Compression Inc.
Boston, Massachusetts

**Christoph M. Hoffmann**
Purdue University
West Lafayette, Indiana

**Bill Jackson**
Queen Mary University of London
London, England

**Predrag Janičić**
University of Belgrade
Belgrade, Serbia

**Robert Joan-Arinyo**
Universitat Politècnica de Catalunya
Barcelona, Catalonia

**Tibor Jordán**
ELTE Eötvös Loránd University and
MTA-ELTE Egerváry Research Group on
Combinatorial Optimization
Budapest, Hungary

**Oleg Karpenkov**
University of Liverpool
Liverpool, United Kingdom

**Csaba Király**
ELTE Eötvös Loránd University and
MTA-ELTE Egerváry Research Group on
Combinatorial Optimization
Budapest, Hungary

**Derek Kitson**
Lancaster University
Lancaster, United Kingdom

**Hongbo Li**
Chinese Academy of Sciences
Beijing, China

**Julien Narboux**
University of Strasbourg
Strasbourg, France

**Anthony Nixon**
Lancaster University
Lancaster, United Kingdom

**John C. Owen**
Siemens

Cambridge, England

**Martin Pfurner**
University of Innsbruck
Innsbruck, Austria

**Rahul Prabhu**
University of Florida
Gainesville, Florida

**Jürgen Richter-Gebert**
Technical University of Munich
Munich, Germany

**Elissa Ross**
MESH Consultants Inc., Fields Institute
Toronto, Canada

**Josef Schadlbauer**
University of Innsbruck
Innsbruck, Austria

**Hans-Peter Schröcker**
University of Innsbruck
Innsbruck, Austria

**Bernd Schulze**
Lancaster University
Lancaster, United Kingdom

**Brigitte Servatius**
Worcester Polytechnic Institute
Worcester, Massachusetts

**Herman Servatius**
Worcester Polytechnic Institute
Worcester, Massachusetts

**Jessica Sidman**
Mount Holyoke College
South Hadley, Massachusetts

**Meera Sitharam**
University of Florida
Gainesville, Florida

**Frank Sottile**
Texas A&M University
College Station, Texas

**Audrey St. John**
Mount Holyoke College
South Hadley, Massachusetts

**Shin-ichi Tanigawa**
The University of Tokyo
Tokyo, Japan

**William Traves**
US Naval Academy
Annapolis, Maryland

**Menghan Wang**
University of Florida
Gainesville, Florida

**Walter Whiteley**
York University
Toronto, Canada

**Joel Willoughby**
University of Florida
Gainesville, Florida

# Chapter 1

## *Overview and Preliminaries*

**Meera Sitharam**
*University of Florida*

**Troy Baker**
*University of Florida*

**CONTENTS**



In this chapter, we begin with a generalized introduction to geometric constraint systems before giving an overview of the book's contents. We conclude with a section reconciling terminology and concepts that have arisen in different communities and as well as an alternative pathway through the book especially for the novice reader.

## 1.1 Introduction

A *geometric constraint system* is generally defined on a finite collection of *geometric primitives* (e.g., 0-dimensional points, 1-dimensional lines, general $d$-dimensional hyperplanes, $d$-dimensional rigid bodies, conics, cylinders). The setting is a given Euclidean or non-Euclidean geometry over the reals, generally of fixed dimension, and constraints specify geometric $n$-ary relationships among the primitives. These constraints can be logical (e.g., incidence, perpendicularity, tangency) or metric (e.g., distance, angle, orientation) and may be either equalities or inequalities. Typically, a constraint can be expressed as a set of quadratic polynomials with real (often rational or even integer) coefficients. The combinatorics of a GCS are usually captured separately in a *constraint graph*: a (hyper)graph where each vertex represents a geometric primitive and each (hyper)edge represents a constraint on the corresponding primitives.

A *realization* (or *solution*) of a GCS is a placement (or configuration) of the geometric primitives that satisfies the constraints. The realizations of a GCS can be found algebraically by solving a system of polynomial equations corresponding to the GCS, where the variables are the coordinates of the geometric primitives. Thus, the set of realizations of such a system consists of the solutions to a finite collection of polynomial equations and is hence a variety. Typically, our primary interest is in the real points of this variety, but if we consider solutions over $\mathbb{C}$, then the full power of algebro-geometric methods may be brought to bear.

In the geometric setting it is generally implied (as it is implied throughout this chapter, unless explicitly stated otherwise) that we are concerned with the realization space modulo some group of *trivial motions* that is designated a priori. For example, in Euclidean space the trivial motions are comprised of translation and rotation; in $d$-dimensional Euclidean space, there is a $d$-dimensional space of translations and $\binom{d}{2}$-dimensional space of rotations giving a $\binom{d+1}{2}$-dimensional space of trivial motions. Embedding $\mathbb{R}^d$ into projective space can help us to see unifying principles in incidence and other constraint systems. Sometimes, the realization is *pinned* (or *grounded*), i.e., the trivial group is chosen to be the empty group.

### 1.1.1 Specifying a GCS

To illustrate these core concepts, consider specifying the most common GCS for classical rigidity theory: the *Euclidean bar-and-joint*, or *Euclidean distance constraint* system. The geometric primitives are 0-dimensional points (called "joints"), the constraints are specified distances between points (called "bars") and the ambient space is $\mathbb{R}^d$. $G=(V,E)$ associates a vertex to each joint and an edge $(u,v)$ to each bar constraining the joints represented by vertices $u$ and $v$. Then a bar-and-joint constraint system of $G$ can be defined as a tuple $(G,\delta)$ where $\delta : E \to \mathbb{R}$ assigns distance values to the bars. A bar-and-joint constraint system is also called a *linkage*. A configuration of the joints in $\mathbb{R}^d$ is given by a map $p : V \to \mathbb{R}^d$ and is a realization of $(G,\delta)$ if the distance between $p(u)$ and $p(v)$ is $\delta(u,v)$ for all $(u,v) \in E$. For example, let $(G=(\{1,2\},\{(1,2)\}),\delta)$ be a Euclidean bar-and-joint system consisting of two joints with one bar between them specifying a distance of 4; i.e., $\delta((1,2))=4$. Then, if $(x_1,y_1)$ and $(x_2,y_2)$ are the variables for the coordinates of joints 1 and 2, respectively, realizations of this linkage are the solutions to the single constraint equation

$$(x_1-x_2)^2+(y_1-y_2)^2=\delta^2.$$

Notice that some geometric constraints may lead to multiple equations; if we were to place an incidence constraint between two points, it would give the equations:

$$x_1=x_2 \text{ and } y_1=y_2.$$

Now consider specifying a GCS whose geometric primitives are rigid bodies in Euclidean space; a "bar" constraint can be placed between two bodies by picking a point on each and constraining the distance between them. Such a system describes a *Euclidean body-and-bar constraint system.* Realizations are solutions to the quadratic system of distance equations, i.e., placements of the bodies (e.g., by assigning elements of the special Euclidean group $SE(d)$) that satisfy the bar lengths.

### 1.1.2 Fundamental GCS Questions

Given a GCS $C$ with $n$ equality constraints, we seek approaches for finding realizations and/or giving structural characterization of $C$ based on properties of the resulting *set* $S$ of geometrically constrained configurations or realizations. That is, when the solution space $S$ has co-dimension $n$ (independent $C$); is finite (locally rigid $C$); has the singleton property (globally rigid $C$). Other properties of $S$ such as dimension (degrees of freedom), connectedness, singularities (deformation paths and extreme configurations) are also of interest. More generally, many of these properties can be deduced by deriving dependent (often inconsistent) constraints that are locally or globally implied by the given GCS, or by ascertaining its independence.

### 1.1.3 Tractability and Computational Complexity

In its full generality, the fundamental questions encompass the first order theory of the real closed fields. This theory is complete, and automated theorem proving over the reals (RCF) is decidable as shown by Tarski [36] (i.e., does not suffer from Gödel's incompleteness of Peano's first order theory of natural numbers). That said, its algorithmic complexity is essentially that of polynomial ideal membership, commonly using Gröbner bases or cylindrical algebraic decomposition [4, 2], which is prohibitive, being complete for the class EXPSPACE [25]. Even the existential theory of the reals is NP-hard, with the best-known algorithms requiring doubly exponential time [25].

## 1.2 Parts and Chapters of the Handbook

This handbook contains a sampling of a wide range of theories and methods that attempt to circumvent the above intractability by taking advantage of properties specific to various types of geometric constraint systems. We have organized it into parts based on the main techniques underlying each chapter, starting with the algebro-geometric techniques and concluding with combinatorial approaches. This essentially provides a flow from approaches that address the general (generic and non-generic) GCS setting to those that work under restricted (generic) settings. Indeed, assumptions of genericity appear throughout Part IV, while Part I contains approaches to address non-generic situations.

Part I **Geometric Reasoning Techniques:** Chapters 2–7 address more general geometric constraint systems, with many of the approaches based on algebraic methods.

Part II **Distance Geometry, Configuration Space, and Real Algebraic Geometry Techniques:** Chapters 8–12 span the underlying topics of distance geometry, configuration spaces, and (real) algebraic geometry.

Part III **Geometric Rigidity Techniques:** Chapters 13–17 (mostly based in Rigidity Theory), while often restricted to generic assumptions, require geometric analyses.

Part IV **Combinatorial Rigidity Techniques:** Chapters 18–25 conclude the book with the Rigidity Theory settings that permit combinatorial approaches.

### 1.2.1 Part I: Geometric Reasoning Techniques

For the specific case of deriving dependent geometric constraints from a GCS, i.e., automated geometry theorem proving, an algorithm that is significantly more efficient than either of the previously mentioned (exponential time and polynomial space) approaches is the Wu-Ritt characteristic set or triangle decomposition method [9, 35]. This and other techniques for automated geometric theorem proving are discussed in Chapter 2.

Since we are restricting ourselves to geometric constraints, the relevant polynomials are typically invariants of transformation or trivial motion groups that define the underlying geometry (Euclidean, Projective, etc.). Invariant polynomials permit synthetic, coordinate-free, and even metric-free computational approaches to deriving dependent constraints, e.g., using bracket algebras. For example, the Grassman-Cayley algebra [40] yields a synthetic computational approach to deriving dependent constraints in projective and incidence geometry. In fact, such invariant theoretic methods even extend to finite geometries [33, 32].

Chapter 5 introduces the bracket algebra and Grassmann-Cayley algebra for the plane with a view toward proving theorems in Projective and Euclidean geometry. The bracket algebra and Grassmann-Cayley algebra appear again in Chapter 4, this time in the context of nongeneric or special realizations of a GCS. The goal is to provide a geometric meaning to the algebraic condition (so-called *pure condition*) that makes the realization special, and this involves the technique of *multilinear Cayley factorization*. The chapter includes introductions to projective space, homogeneous coordinates, the Grassmannian, and Plücker coordinates and ends with examples applying the theory to body-and-bar GCS.

As another example of a similar approach, Chapter 3 develops the theory and applications of a variety of types of Euclidean invariants in deriving constraint dependences. The $\binom{n}{2}$ pairwise distance polynomials between $n$ points in real Euclidean space are Euclidean invariants that are related by the Cayley-Menger syzygies. These are described in the following classical theorem on Euclidean distance matrices.

A *Euclidean distance matrix (EDM)* for $\mathbb{R}^d$ is an $n \times n$ square matrix of pairwise (squared) distances between $n$ points in $\mathbb{R}^d$. It is denoted $\Delta_{[n]}$ with distance entries $\delta_{ij}$ for $1 \leq i, j \leq n$. For $S \subseteq [n]$, the submatrix $\Delta_S$ has entries $\delta_{ij}$ for $i, j \in S$. The volume matrix $\hat{\Delta}_S$ is the $|S|+1 \times |S|+1$ matrix obtained from $\Delta_S$ by bordering $\Delta_S$ with a top row $(0, 1, \ldots, 1)$ and a left column $(0, 1, \ldots, 1)^T$. Now $det(\hat{\Delta}_S)$ computes the volume of the simplex with points in $S$, and is called a Cayley-Menger determinant [34]. The next theorem effectively says that the volumes of simplices formed by $d+2$ points in $\mathbb{R}^d$ is 0.

**Theorem 1.1 (Cayley-Menger Relations)** *A real symmetric matrix $\Delta_{[n]}$ with 0 diagonal and positive entries is a Euclidean distance matrix for $\mathbb{R}^d$ only if $det(\hat{\Delta}_S) = 0$ for all $S \subseteq [n]$ with $|S| \geq d+2$.*

A more direct approach for both Euclidean distance or Projective constraint systems is to simply solve for the set of realizations of the constraint system to determine (in)dependent constraints. For example, some GCS permit a ruler and compass type construction of the (finite set of) solutions, which is equivalent to solving a recursive triangular block decomposition of the constraint system. A broad class of GCS occurring in computer aided design are of this type, i.e., their underlying combinatorial structure, or constraint graph is *triangle-decomposable* or *tree-decomposable* [7].

It is known that such GCS are Quadratically Radically Solvable (QRS), i.e., the coordinates of the solutions are in nested quadratic extensions of the coefficient field. As to whether all *generic*

QRS systems have triangle-decomposable constraint graphs is still an open problem, with the equivalence being shown only when the constraint graph is topologically planar [26]. Solving these systems entails two stages: the recursive triangle decomposition stage, a combinatorial procedure on the constraint graph, and a solution or realization stage, that obtains the solution of the corresponding recursively decomposed system through a bottom-up process of assembling or recombining the (generically finitely many) solutions of subsystems.

Chapter 6 gives many natural examples of triangle-decomposable constraint systems and formal algorithms for recursive decomposition and recombination. In general, the time complexity of solving or realization is bottle-necked by the largest subsystem that must be solved simultaneously and a polynomial time preprocessing algorithm to identify the subsystems can be beneficial. For non-triangle-decomposable GCS, the above method can be generalized to obtain recursive decompositions into subgraphs that approximate generically rigid subsystems that have finitely many isolated solutions or realizations (see Section 1.3 for Terminology and Basic Concepts).

The process of identifying these subsystems and finding a partial order in which to solve them is called *decomposition-recombination (DR-) planning*, the topic of Chapter 7. Such algorithms vary in their generality but often leverage geometric properties of specific constraint and primitive types or use a priori knowledge of patterns in their arrangement. However, in general, at some point an algebraic system must be solved for recombination of the decomposed subsystem solutions. Chapter 7 surveys several such methods for decomposition and recombination of more general constraint systems.

However, most of the above-mentioned approaches (including those that rely on generic rigidity or finiteness of the solution set for decomposition) do not differentiate between real and complex solution spaces. In particular to specialize the invariant-theoretic approach to the reals requires imposing additional inequality constraints beyond the Cayley-Menger conditions in the previous theorem, by asserting that all simplices have positive volumes. For 1- and 2-dimensional simplices (line segments and triangles), this gives exactly the metric condition on real Euclidean distances.

**Theorem 1.2 (Cayley-Menger Inequalities)** *A real symmetric matrix with 0 diagonal and positive entries $\Delta_{[n]}$ is a Euclidean distance matrix for $\mathbb{R}^d$ if and only if $det(\hat{\Delta}_S) \geq 0$ for all $S \subseteq [n], |S| \geq 2$; and $det(\hat{\Delta}_S) = 0$ for all $S \subseteq [n]$ with $|S| \geq d+2$.*

Note that the Euclidean invariant approaches described in Chapter 5 use all the Cayley-Menger conditions including the above inequalities.

### 1.2.2 Part II: Distance Geometry, Configuration Space, and Real Algebraic Geometry Techniques

The inequalitites in Theorem 1.2 can be viewed as partly arising from the metric property of real Euclidean space. This leads to another tool for dealing with geometric constraints with an underlying metric, namely distance or metric geometry. The classical theorem of Schoenberg [30, 31] (which generalizes to infinite dimensional Hilbert spaces) is stated for finite dimensional real Euclidean distance matrices below. It is equivalent to the conjunction of the two Cayley-Menger theorems above.

**Theorem 1.3 (Schoenberg's Theorem)** *A real symmetric matrix with 0 diagonal and positive entries $\Delta_{[n]}$ is a Euclidean distance matrix for $\mathbb{R}^d$ if and only if it is negative semidefinite on the subspace of all vectors orthogonal to the all 1's vector and $rank(\Delta_{[n]}) \leq d+1$.*

The convexity and face structure of the Euclidean distance cone yield powerful techniques for understanding distance constraint systems, including implied or dependent constraints, and different types of rigidity. Chapter 8 surveys some of these techniques. Chapter 12 introduces the tools of real algebraic geometry, specifically semialgebraic sets that involve polynomial equalities and inequalities (such as the Cayley-Menger determinantal equalities and inequalities above) and the positivenstellensatz as tools for defining generic realizations and dealing with distance constraint systems. It starts with a brief introduction to the correspondence between ideals and varieties over $\mathbb{C}$, and then turns to a discussion of varieties defined by polynomials with real coefficients, which may be viewed as either real or complex varieties. It culminates with a view of the projection of the $d$-dimensional stratum of the Euclidean (squared) distance cone as a semialgebraic set and its application to rigidity.

Chapter 9 explores the structure of general metric cones. Chapter 10 employs properties of projections and fibers of rank $d$ strata of the Euclidean distance cones to characterize distance constraint systems (their underlying graphs) whose configuration spaces generically map finitely-many-to-one to a convex set and whose singular configurations have a simple description. These characterizations extend to when the distance constraints are inequalities and the distances are $l_p^p$ norms. The techniques yield interesting configuration space properties of a common class of plane linkage mechanisms (Euclidean distance constraint systems with one degree of freedom in $\mathbb{R}^2$) arising from the QRS or triangle decomposable constraint graphs mentioned above.

However, questions about linkage mechanisms in general are inherently difficult: Kempe's universality theorem [20] states that the space of configurations can trace out any desired algebraic curve. Chapter 11 explores constraint varieties of mechanisms and describes how so-called study parameters and dual quaternions are used in kinematics.

### 1.2.3 Part III: Geometric Rigidity Techniques

The question of local and global uniqueness of polyhedra whose faces have a given combinatorics has been studied by a long line of researchers starting from Cauchy in the early 1800s to Alexandrov in the 1950s. These results concern the geometric rigidity of *polyhedral frameworks*, or just polyhedra, namely 3-dimensional polytopes that are composed of planar rigid panel faces; face panels can rotate about the edge or hinge on which they are incident with another panel. A *triangulated* polyhedron's faces are all triangular.

Although these results concern the rigidity-related properties of *frameworks*, i.e., specific realizations of a combinatorial structure of constraints (from which the GCS can be extracted), as opposed to rigidity-related properties of GCSs (as discussed so far), in the latter part of this chapter we will reconcile the slight differences in these two ways of thinking.

Cauchy [5] showed that all convex polyhedra are *rigid* and in fact globally rigid if convexity is stipulated. (Mistakes in Cauchy's proof were fixed and the result extended by a series of subsequent researchers.) Despite a long standing conjecture that the result extended to all triangulated polyhedra (convex or not), verified for many subclasses [10], the general statement was disproven by counterexample [6]. These results and further developments are discussed in Chapter 13.

Let $G = (V,E)$ be the graph associated to the edge skeleton (or 1-skeleton) of a triangulated polyhedron and let $p : V \to \mathbb{R}^3$ be the map that assigns the coordinates of each vertex of the polyhedron to a vertex of $G$. Then $(G,p)$ is a *bar-joint framework* that is a realization of a bar-joint constraint system in 3D, with a distance constraint graph $G$. Thus, Cauchy's theorem shows that bar-joint frameworks arising from convex triangulated polyhedra are rigid (and globally rigid if convexity of the framework is stipulated).

Chapter 16 discusses geometric conditions for global rigidity of *generic* bar-joint frameworks in arbitrary dimensions [11]. Recall that a framework is globally rigid if it is the unique realization of its underlying GCS. A framework $(G,p)$ is *generic w.r.t. a property P* (such as global rigidity) if for some neighborhood $\mathcal{N}(p)$, for all frameworks $(G,q)$ with $q \in \mathcal{N}(p)$ $(G,q)$ satisfies $P$ if and only if $(G,p)$ satisfies $P$. When the context, namely the property $P$ is clear, we simply say the framework is generic.

The result employs a feature of the framework's equilibrium self-stress (defined in Section 1.3) and further shows that global rigidity is in fact a *generic property*, i.e., either all generic frameworks $(G, p)$ of a graph $G$ are globally rigid or none are. In other words, the property of the framework only depends on the graph $G$, but is given a geometric characterization in Chapter 16. Chapters that additionally give combinatorial as opposed to geometric characterizations of such properties that depend only on the constraint graph are described in the next section on combinatorial rigidity.

Chapter 14 considers *tensegrity frameworks* [28] in which the underlying GCS involves inequality as opposed to equality constraints. Some edges of the constraint graph, called struts, have distance lower bounds and others, called ties have distance upper bounds which restrict the sign of the equilibrium self-stress they can carry. Tensegrity frameworks that represent packed incompressible spheres contain only struts. Bar-joint frameworks are special cases of tensegrity frameworks where all edges have both distance upper and lower bounds (fixed distances).

Chapter 14 gives geometric, equilibrium self-stress based characterizations for rigidity and other related properties of tensegrity frameworks both in general and generic settings, and connects them to rigidity properties of bar frameworks.

Chapter 15 specifically considers nongeneric tensegrity frameworks and uses extended Cayley algebra (discussed earlier under the Geometric Reasoning Section of the handbook) to give geometric conditions for rigidity related properties.

Chapter 17 deals with properties related to rigidity that are invariant under various transformations (beyond the trivial motion group of the underlying geometry). Using characterizations of properties related to rigidity theory, techniques (such as Coning or Maxwell-Cremona diagrams for understanding stresses) for GCSs and frameworks in one geometry can be extended to another. For example, techniques and characterizations from Euclidean geometry can be extended to say Affine, Projective, Spherical, Minkowski, and Hyperbolic geometries that are defined using Cayley-Klein metrics or using the trivial motion groups under which the metrics are invariant.

### 1.2.4 Part IV: Combinatorial Rigidity Techniques

As mentioned earlier, (in)dependence and other properties related to rigidity are often generic (under appropriate, careful definitions of genericity), i.e., they hold for all generic frameworks and/or GCSs with a given constraint graph $G$, or for none of them. They depend only on the underlying constraint graph $G$; (in)dependence is captured by a geometric or algebraic matroid such as the (generic) rigidity matroid defined formally in the second part of this chapter, whose ground set is related to the algebraic constraints represented by the edges (and nonedges) of $G$.

This section of the handbook deals with such (in)dependence properties that are equivalently characterized by purely combinatorial sparsity or graphic matroids, which do not use the algebraic structure of the constraint polynomials or brackets over the reals (or of the coefficient field of the constraint polynomials). The book [12] has so far been the trusted source on combinatorial rigidity, but significant progress has been made since it was published.

A classic example of a purely combinatorial rigidity characterization is a celebrated result of Laman [21] published in 1970, though recently it has also been found in a forgotten work of Hilda Pollaczek-Beiringer [27] from 1927.

**Theorem 1.4 ([27, 21])** *A 2D bar-joint graph* $G = (V, E)$ *is rigid (all its generic frameworks/GCSs are rigid) if and only if* $|E| = 2|V| - 3$ *and for any subsystem* $G' = (V', E')$ *where* $|V'| > 1$, $|E'| \leq 2|V'| - 3$.

Graphs satisfying such counting conditions – which keep track of the internal degrees of freedom (dof) of the system – are often referred to as *Laman graphs*. This combinatorial characterization of bar-joint rigidity in 2D led to a series of increasingly refined algorithms to detect rigidity, as well as maximal rigid subgraphs in flexible constraint graphs. Chronologically, there was a network flow based algorithm [16], a matroid sums algorithm [8], a bipartite matching algorithm [14], and finally what is known as the pebble game [18]. A version of the pebble game (a special case of network flow) is used for recursive decomposition into approximately rigid subgraphs in the DR-planning algorithms mentioned earlier. The idea of pebble games has since been extended to the class $(k,l)$-sparse graphs [22] where $l < 2k$, as discussed in Section 1.3.4.

It turns out that the latter condition in the above theorem defines independence in a sparsity matroid on $V \times V$. The analogous condition $d|V| - \binom{d+1}{2}$ formulated by Maxwell in the nineteenth century [24] is necessary but not sufficient for $d \geq 3$ as discussed later in this chapter, see the famous "double banana" graph, Figure 1.4. Typically, the so-called "Maxwell direction" is showing that independence in the combinatorial matroid is necessary for independence in the algebraic rigidity matroid, and is the easier one. The converse direction – that completes the equivalence of the two types of matroids – is the challenging one.

However, there are combinatorial characterizations of independence and local rigidity of bar-joint frameworks in 2D which extend to other types of 2D frameworks such as body-bar, body-hyperpin, 2D bar-joint on the sphere (or 3D line-angle), 2D point-line-incidence-direction, etc. Often there are more than one equivalent characterization. For example, Lovász and Yemini [23] found an alternate characterization of 2D bar-joint rigidity that is superficially quite different from Laman-Pollaczek-Beiringer's characterization. Chapter 18 and Chapter 21 discuss, respectively, such local and global rigidity characterizations. The latter chapter begins with the first combinatorial characterization of generic global rigidity, namely that of 2D bar-joint frameworks [17]. Chapter 22 gives a combinatorial matroid that captures the rigidity matroid of more challenging frameworks in 2D involving angles between lines with distances between points and lines.

#### 1.2.4.1 Inductive Constructions

In 1911, Henneberg [15] gave the following constructive definition of the class of what later became Laman-Pollaczek-Beiringer graphs.

**Definition 1.1 Henneberg Construction** A *Henneberg construction* of a graph $G$ is a sequence of the following operations which, beginning with a single edge, results in $G$.

(a) Add a new vertex and two edges connecting it to two existing vertices.

(b) Subdivide an existing edge and add an additional edge from the new vertex to another existing vertex.

Laman [21] used this class in the proof of Theorem 1.4. The basic structure of the proof was to show that the class described in the theorem was exactly the class of graphs with Henneberg constructions. Then he proved that for any Henneberg construction there is a positioning of the vertices that will have no infinitesimal motions. The earlier proof of Pollaczek-Beiringer was stronger, pointing out that almost all (or generic positionings) that avoid certain algebraic conditions will have no infinitesimal motions.

Inductive constructions are one of the mainstays of results that show equivalence between an algebraic rigidity matroid and a combinatorial or graphic matroid, specifically for the difficult direction, i.e., showing that independence in the combinatorial matroid implies independence in the algebraic rigidity matroid. Chapter 19 systematically surveys such inductive constructions in many proofs of combinatorial rigidity characterizations.

#### 1.2.4.2 Body Frameworks

A *body* geometric primitive is a finite $n$-dimensional rigid object; this is rather general, including points, line segments, plane segments, etc. but also any other rigid free-form shape of the same dimension as the space. A *body-hinge* system has body primitives and *hinge* constraints, which are incidences between two primitives in $d$-dimensions where $d$ is less than the dimension of either object about which the primitives can rotate. In 2D, this must be a bar-joint system. A *body-bar* system also has body primitives (as discussed in Section 1.2.4.2), but the constraints are distances between generic points on the body. Unlike the higher dimensional bar-joint systems, body-bar-hinge systems have a combinatorial characterization of independence and local rigidity in arbitrary dimensions, first proved in [38, 39] who pioneered the use of so-called pure conditions (where certain determinants vanish) to describe genericity. The characterizations extend to special classes of bar-joint systems that can be cast as body-bar-hinge systems. Combinatorial characterizations of global rigidity also exist.

Polyhedra (see Section 1.2.3) are a subclass of body-hinge structures. The bodies are panels (polygonal faces) and the hinges connect the panels; moreover, the system must completely enclose a volume. The so-called *molecular conjecture* [37] (referring to the ability to model protein backbones as body-hinge structures) stated that the rigidity of coplanar hinge and panel hinge frameworks obeyed the same combinatorial characterization as generic body-hinge structures. It was proven for general dimension in [19].

Chapter 20 surveys combinatorial characterizations of both local and global rigidity for body-hinge structures in arbitrary dimensions.

#### 1.2.4.3 Body-Cad, and Point-Line Frameworks

The set of body-cad frameworks is a catch-all category for 3D constraint systems [13]. Motivated by CAD design software, it includes many of the common constraints and primitives seen in the industry. The above-mentioned categories are in fact specializations of this class. Primitives include points, lines, and planes, and constraints include **c**oincidence, **a**ngular (parallel, perpendicular, or arbitrary fixed angles), and **d**istance (cad). Chapter 23 discusses a combinatorial characterization of (infinitesimal) rigidity for such systems.

#### 1.2.4.4 Symmetric and Periodic Frameworks and Frameworks under Polyhedral Norms

Chapter 25 develops the set up and techniques for extending combinatorial rigidity characterizations to symmetric and periodic frameworks (for different symmetry groups). Chapter 24 does the same for extending from Euclidean distance to polyhedral norms.

### 1.2.5 Missing Topics and Chapters

The sampling of topics in this handbook would have been more comprehensive with chapters on (a) Wu-Ritt's characteristic set method for automated geometry theorem proving mentioned earlier, (b) the topology (homology and cohomology) of linkage configuration spaces, related to Walker's problem, (c) combinatorial and algorithmic studies on expansive motions of linkages and origami related to the Carpenter's rule problem, (d) sphere-packing rigidity with results arising from analytic perspectives besides the tensegrity perspective, (e) conjectures and progress on characterizing generic bar-joint rigidity in 3D, and (f) the exploration of genericity, rigidity, and configuration spaces of periodic and infinite frameworks. Of these, there is extensive expository literature on (a),(b),(c). We hope that the next revision of the handbook may include chapters condensing the substantial amount of work on (d), (e), and (f). Finally the area is rich in subtopics and directions arising from numerous applications, from computer aided engineering and architectural design, molecular and materials modeling, machine learning and complexity. These however would be outside the scope of this handbook on geometric constraint principles, being more appropriate for a handbook on geometric constraints applications.

## 1.3 Terminology Reconciliation and Basic Concepts

In this section we clarify slightly different types of terminology used to talk about GCS and frameworks, and their relationship. We additionally introduce the overall program of combinatorial characterizations of GCS and frameworks: define notions of *genericity,* introduce the concept of a rigidity matrix and the notion of *infinitesimal rigidity*, which is a linearization of local rigidity that is generically equivalent and is used to define the so-called (generic) *rigidity matroid*.

### 1.3.1 Constrainedness

The notion of *constrainedness* exists to discuss the characteristics of the solution space of a GCS. An *over-constrained* system is one that has no solutions. A *well-constrained* system is one that has a finite number of solutions. An *under-constrained* system is one that has infinitely many solutions. Each variable in a system has an *approximate degree-of-freedom* (dof). Each constraint contributes at least one equation.

The definitions above apply to a system of equations with sufficiently general parameters. However, some specific assignments of values to parameters (e.g., length and angle measures) in the equations corresponding to the same underlying constraint graph may result in a different classification. That is, these properties are not structural (or combinatorial) A system is called *generic* over a ground field $K$ (usually $\mathbb{Q}$ or $\mathbb{R}$ in our setting) if the designated constraint parameters are *algebraically independent* over $K$, i.e., they are not the solutions of any nontrivial polynomial equation with coefficients in $K$. Weaker definitions of genericity of GCS exist, for example, stating that the parameters are not the zeroes of a given set of polynomials, or some finite but unspecified set of polynomials. In all these cases, the set of nongeneric parameter values is measure-zero, i.e., choosing the parameters at random will result in a generic system with probability of one. Unless otherwise specified, the strongest definition of algebraic independence of parameter values is implied.

A generic constraint system being under-, well-, or over-constrained implies that all generic parameter assignments to the same underlying constraint graph result in an under-, well-, or over-constrained system, respectively. Therefore, being generically $*$-constrained is a combinatorial property. Thus, generic constrainedness terminology can be extended to the underlying constraint graph. Generic $*$-constrainedness does not imply $*$-constrainedness of a nongeneric system, and similarly the converse does not hold either. Therefore, classifying nongeneric constraint systems is an intractable problem.

However, in the *generically over-constrained* setting, a specific class of non-generic systems is of interest. Such a system with a generic assignment to the parameters will have no solution, but those non-generic systems that do have a solution are called *consistently over-constrained.* The sets of *generically well-over-constrained* and *generically under-over-constrained* systems are disjoint, complementary subsets of generically over-constrained. A generically well-over-constrained system is one with a spanning generically well-constrained subsystem. A generically under-over-constrained is any other over-constrained system.

Note that traditionally generically under-constrained was taken to mean the union of generically under-constrained and generically under-over-constrained as defined here. The current definition is cleaner since it ensures that the sets of generically under-, well-, and over-constrained are disjoint.

### 1.3.2 Rigidity of Frameworks

Two frameworks $(G,p)$ and $(G,q)$ are congruent if $p$ and $q$ are congruent modulo trivial motions. They are equivalent if the underlying GCS of $(G,p)$ is identical to that of $(G,q)$.

A framework $(G,p)$ is *rigid* if there exists a nonempty neighborhood $\mathcal{N}(p)$ (in the Euclidean

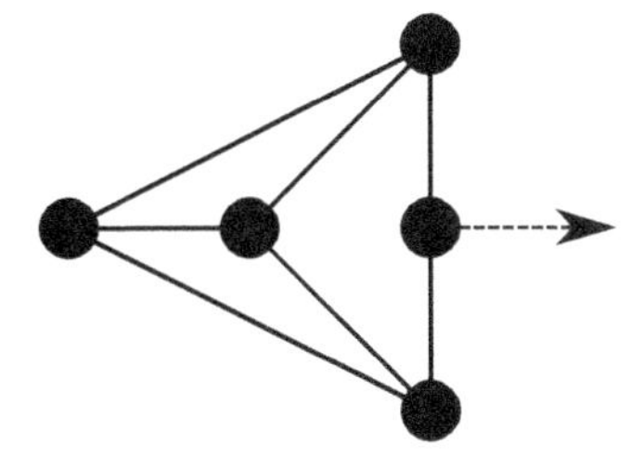

**Figure 1.1**
A 2D bar-joint framework that is rigid but not infinitesimally rigid. A nontrivial infinitesimal motion is indicated by the dashed arrow.

topology) such that for all $p' \in \mathcal{N}(p)$, congruence of frameworks $(G,p)$ and $(G,p')$ implies equivalence. If the framework is not rigid, it is *flexible*. A rigid framework is *minimally rigid* (or *isostatic*) if the removal of any hyperedge of $G$ results in a flexible framework. The framework is *globally rigid* (or *strongly rigid*) if, for all $p'$ for which the framework $(G,p')$ is equivalent to $(G,p)$, it also holds that they are congruent.

Given a constraint graph $G$, there is a system of polynomial equations $F = \{f_1,\ldots,f_m\}$ and variables $X = \{x_1,\ldots,x_n\}$ (as explained in Section 1.1.) The *rigidity matrix* $R(G)$ is the Jacobian of this system with respect to $X$, i.e., the $m \times n$ matrix with element $(i,j)$ equal to $\partial f_i/\partial x_j$. The rigidity matrix of a framework $(G,p)$ (written as $R(G,p)$) is $R(G)$ with all variables $x_i$ replaced by $p(x_i)$. When working with most constraint systems, the equations are quadratic and this process is often referred to as *linearization* of the polynomial system.

As explained earlier, the solution space of a geometric constraint system is a variety in $K^n$ (where we may let $K = \mathbb{C}$ to make full use of the algebraic theory); it is the set of zeros of the corresponding polynomial system. The row span of the rigidity matrix $R(G,p)$ is the space of normals to this variety at point $(p(x_1),p(x_2),\ldots,p(x_n))$. Any infinitesimal movement on the tangent space, orthogonal to the space of normals (i.e., along the variety), will give another solution. The tangent space is the right nullspace (or kernel) of $R(G,p)$. Geometrically, any infinitesimal vector in the right nullspace represents an infinitesimal change to each primitive such that the resulting framework still satisfies all of the constraints.

The left nullspace of the rigidity matrix also has a geometric interpretation. The left nullspace only has nonzero vectors if there is a linear dependency in the rigidity matrix, which corresponds to an *equilibrium self-stress* of the system.

The *degree-of-freedom* (*dof*) of the framework is the nullity of $R(G,p)$ (i.e., the rank of the right nullspace.) The framework is *infinitesimally rigid* if the dof is equal to the number of trivial motions. Geometrically, this means that all infinitesimal motions arise from trivial motions of the space. By the rank-nullity theorem, a framework is also infinitesimally rigid exactly when the rank of $R(G,p)$ is $n-k$, where $n$ is the number of variables and $k$ is the number of trivial motions. In the case of $d$-dimensional Euclidean space, the space of trivial motions has dimension $\binom{d+1}{2}$, which are the $d$ translations plus $\binom{d}{2}$ rotations. If the right nullspace of $R(G,p)$ has dimension greater than $\binom{d+1}{2}$, then the framework $(G,p)$ is said to have *infinitesimal motions* (or *infinitesimal flexes*).

It is clear that infinitesimal rigidity implies rigidity, and we give an example of a bar-joint framework that is rigid but not infinitesimally rigid to show that the converse is false. A comprehensive introductory treatment of the rigidity of graphs (bar-joint systems) can be found in Ref. [12].

To prove that rigidity does not imply infinitesimal rigidity, consider Figure 1.1. This is a rigid bar-joint graph that has an infinitesimal motion that is zero at all places except at the single vertex in the direction of the arrow. A rigid framework with a nontrivial infinitesimal motion is called *degenerate* and must be in a nongeneric realization. A combinatorial characterization of rigidity is

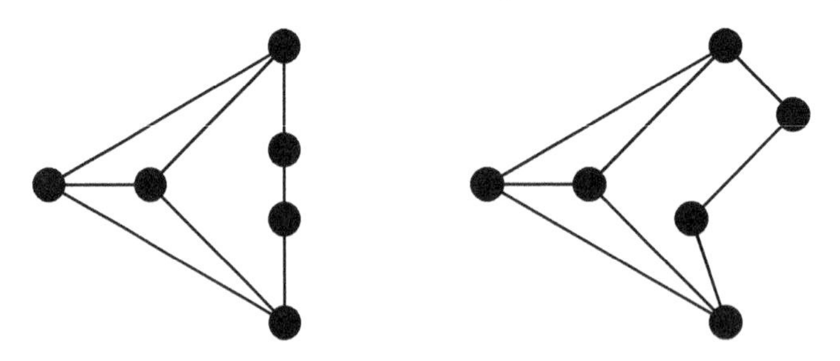

**Figure 1.2**
Two different 2D frameworks of the same generically flexible bar-joint constraint graph. The degenerate framework on the left is rigid, whereas the generic configuration on the right is flexible.

only guaranteed to hold for a certain generic class of realizations, and we formalize what we mean by genericity in the next section.

The rigidity matrix has a natural notion of dependence, based on the linear dependence of the rows (or columns) of the matrix. As such, the *rigidity matroid* of a framework [12] is simply the linear matroid of the rows of its rigidity matrix. That is, the row vectors of the matrix comprise the ground set, and the linearly independent subsets of rows comprise the family of independent sets. Therefore, the matroid rank (the maximum cardinality of an element in the matroid) is exactly the rank of the matrix. Since each row corresponds to some constraint, a dependent row corresponds to a dependent constraint. The framework as a whole is *independent* if there are no dependent constraints and is *dependent* otherwise.

### 1.3.3 Generic Rigidity of Frameworks

Determining rigidity of frameworks is difficult; deciding global [29] and local rigidity [1] are both strongly NP-hard for bar-joint systems. However, if a certain measure-zero set of primitive arrangements is excluded, determining rigidity can become much easier for certain constraints and primitives. This is the set of degenerate frameworks, which will be discussed in this section.

In Section 1.3.1, the notion of genericity was discussed in the context of geometric constraint systems. When considering frameworks, genericity has a different meaning. A framework $(G,p)$ is said to be *generic with respect to property* $\mathcal{P}$ $(G,p)$ satisfies $\mathcal{NP}$ if and only if there exists some neighborhood $\mathcal{N}(p)$ around $p$ such that for all points $p' \in \mathcal{N}(p)$ the framework $(G,p')$ satisfies $\mathcal{P}$.

A property $\mathcal{P}$ is said to be *generic* when, for all constraint graphs $G$, either all generic frameworks of $G$ (w.r.t. $\mathcal{P}$) satisfy $\mathcal{P}$ or none satisfy $\mathcal{P}$. If a property of frameworks $(G,p)$ is generic, then it is a combinatorial property of the underlying constraint graph alone. Intuitively, a generic property of a framework is one that is maintained if primitives were "wiggled" by small amounts in any direction. For example, the independence of the rigidity matrix is a generic property of frameworks.

To illustrate the importance of considering generic frameworks, consider the following examples. See Figure 1.2 which depicts two frameworks of the same constraint graph. The nongeneric framework on the left is rigid while being generically flexible. The three bar "chain" is taut, disallowing finite flexes; with only slightly different lengths, this would no longer be rigid. See also Figure 1.3 which also depicts two frameworks corresponding to a single constraint graph. The nongeneric framework on the left is flexible while being generically rigid. The three vertical bars are the same length, permitting a vertical shear; with different lengths, there would be no infinitesimal motions.

Whereas rigidity does not imply infinitesimal rigidity of a framework, in the generic case it does. The Implicit Function Theorem from multivariate calculus, as in Ref. [3], shows that when a framework is generic w.r.t. infinitesimal rigidity, every infinitesimal flex can be converted into a

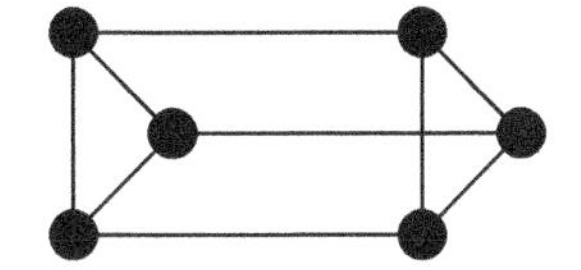
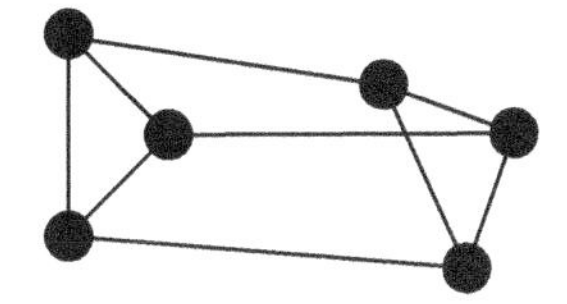

**Figure 1.3**
Two different 2D frameworks of the same generically rigid bar-joint constraint graph (known as $C_2 \times C_3$). The degenerate framework on the left is flexible, whereas the generic configuration on the right is rigid.

**Table 1.1**
Correspondence between constrainedness terminology when used in the context of generic systems and generic frameworks.

| Generic Systems | Realizations | Generic Framework |
|---|---|---|
| Under-constrained | Infinite solutions | Independent and flexible |
| Well-constrained | Finite solutions | Independent and rigid |
| Over-constrained | No solutions | Dependent |
| Under-over-constrained | No solutions | Dependent and flexible |
| Well-over-constrained | No solutions | Dependent and rigid |

finite flex, i.e., rigidity in fact implies infinitesimal rigidity. Since infinitesimal rigidity is a generic property, this shows that rigidity is also a generic property. As mentioned, this property can be thought of as a combinatorial property of the underlying constraint graph. Therefore, a constraint graph is called rigid if some generic framework of the graph is infinitesimally rigid. The notions of flexibility and minimal rigidity have obvious meanings in the generic sense as well.

The *generic rigidity matroid* of a geometric constraint system with constraint graph $G$ can be defined in two ways: (1) the rigidity matroid of any generic framework $(G, p)$, or (2) the rigidity matroid formed by the rows of the rigidity matrix $R(G)$ of indeterminates. This leads to combinatorial notions of independence among constraints.

Table 1.1 establishes a rough correspondence between the genericity of a GCS, a framework, and rigidity. Traditionally, generically under-constrained was used to describe all flexible frameworks and was therefore a union of under- and under-over-constrained as defined here. The symmetry of the new definitions displayed above is another argument for the new terminology.

### 1.3.4 Approximate Degree-of-Freedom and Sparsity

There is a real, algebraic notion of degree-of-freedom that is discussed in Section 1.3.2. This section introduces a different, combinatorial idea of degree-of-freedom that can be used to approximate rigidity. This was briefly mentioned in Section 1.3.1, and has an obvious relationship with constrainedness. The following methods are very general, working for any type of primitive or constraint. However, a positive result for rigidity by this characterization is often only a necessary condition for true rigidity. Many systems require additional considerations, if there is a combinatorial characterization at all.

In the constraint graph, each primitive has some *degrees-of-freedom (dofs)* and each constraint eliminates some dofs between participating primitives. In some of the literature, particularly that which uses network flow based algorithms, dof corresponds to the negation of the *density* of the constraint graph. Given a constraint graph $G = (P,C)$, with primitives $P$ and constraints $C$, and a

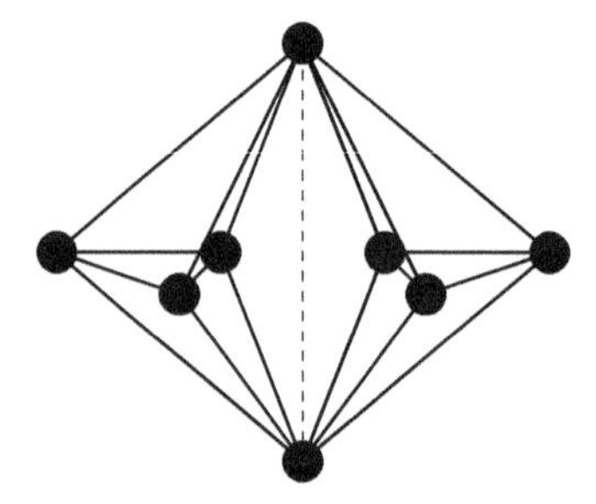

**Figure 1.4**
A 3D bar-joint framework, known as the double-banana. It is flexible due to a rotation about the dashed line, despite being dof-rigid (i.e., $(3,6)$-tight).

weight function $w$ on $P$ and $C$, the density of $G$ is

$$d(G) = \sum_{c \in C} w(c) - \sum_{p \in P} w(p).$$

Given some constant $k$ equal to the number of trivial motions of the underlying geometry, a constraint graph is *minimally dof-rigid* if it has $k$ dofs and every subgraph has at least $k$ dofs. In terms of density, the graph is minimally dof-rigid if $d(G) = -k$ and for all subgraphs $G'$, $d(G') \leq -k$. A graph is *dof-rigid* if it contains some minimally dof-rigid subgraph.

For example, consider 2D bar-joint systems: points can be thought of as having 2 dofs (translation but not rotation) and an edge between two points destroys 1 dof, leaving a system with 3 dof (translation and rotation.) In Euclidean 2D space, there are 3 trivial motions and therefore a single edge would be dof-rigid. In fact, as mentioned earlier, this notion of dof-rigid exactly captures generic rigidity of 2D bar-joint systems. We follow the convention of referring to methods using dof analysis as *Laman counts*.

We can understand the rigidity of some systems combinatorially using only dof analysis. However, this combinatorial notion of dof-rigid does not usually imply generic rigidity. Consider 3D bar-joint systems: points instead have 3 dofs, edges still eliminate 1 dof, and the space has 6 trivial motions. By the counts, the famous "double banana" graph in Figure 1.4 is dof-rigid; however, the "bananas" can clearly swivel about the dashed hinge.

As mentioned earlier, the Laman-Pollaczek-Beiringer theorem gives a purely combinatorial property (no algebra, simply counting) to capture the properies of the rigidity matrix and therefore the matroid. These subsystems are exactly the independent sets of the 2D bar-joint rigidity matroid.

This theorem motivated the notion of *sparsity* and *sparsity counts* [22]. This terminology is used to discuss constraint graphs where all primitives have $k$ dofs and all constraints eliminate one dof and are binary; however, the theory does allow loops, effectively permitting unary constraints, and allows for multiedges, so constraints that eliminate $n$ dofs can be represented in as many edges. A graph $G = (V,E)$ is $(k,l)$-*sparse*, for every induced subgraph $G' = (V',E')$, the inequality $|E'| \leq k|V'| - l$. The graph is $(k,l)$-*tight* if it additionally satisfies $|E| = k|V| - l$.

For example, Laman graphs would be the set of $(2,3)$-tight graphs. A 2D system of 2D rigid bodies and distance constraints would use $k = 3$ and $l = 3$; in fact, $(3,3)$-tight graphs are exactly the class of 2D rigid body-bar graphs. For a fixed $k$ and $l$ there is often a natural interpretation of a $(k,l)$-tight graph as a constraint system in which geometric primitives have $k$ dof.

For all $l < 2k$, these counts define a *sparsity matroid* where the basis is the set of edges and the independent sets are the edges in the $(k,l)$-sparse subgraphs. This allows for an efficient class of algorithms, called *pebble games*, which can detect $(k,l)$-sparse graphs in polynomial time, if $l < 2k$.

## 1.4 Alternative Pathway through the Book

Given the interconnected nature of the chapters of this book there are many logical ways of ordering the material. Here, we describe an alternative navigation that might be more accessible to a newcomer.

In this suggested pathway, Group I contains chapters highlighting the different perspectives from developing solvers for general geometric constraint systems to analyzing the rigidity of bar-and-joint structures to constructing purely combinatorial objects arising in rigidity theory. Groups II and III focus on the problems studied in Rigidity Theory, introducing concepts such as local and global rigidity and considering models initially restricted to points before generalizing to other types of geometric primitives (e.g., rigid bodies or lines). Groups IV and V shift the structure to partition by the underlying approaches (metric geometry and algebraic methods).

Group I **Getting Started: Chapters 1, 6, 7, 18, and 19**
To get the lay of the land, start with two sets of chapters highlighting the perspectives of historically distinct communities. Chapters 6 and 7 focus on decomposition-recombination approaches used in computing realizations of a general GCS. In contrast, Chapter 18 restricts its content to the classical structure studied in Rigidity Theory of 2-dimensional bar-and-joint frameworks (introduced in Section 1.1.1); the combinatorial property characterizing generic bar-and-joint rigidity is studied in a generalized setting in Chapter 19.

Group II **Rigidity Theory for Point Primitives: Chapters 21, 16, 8, 14, 15, and 25**
Building upon the fundamentals introduced in Group I, continue the Rigidity Theory perspective, with topics posed in the setting of GCS with points as the geometric primitives.

Group III **Rigidity Theory for Other Primitives: Chapters 20, 13, 22, and 23**
Next, move to the Rigidity Theory for systems defined on geometric primitives beyond simply points (e.g., rigid bodies, lines).

Group IV **Metric Geometry: Chapters 24, 17, 9, and 10**
Shifting to an organization based on underlying techniques, start with approaches relying on distance and metric geometry.

Group V **Algebraic Methods: Chapters 12, 11, 3, 2, 5, and 4**
Finally, conclude with chapters based on algebraic methods.

The intention of this pathway is to help to get readers started with the varying perspectives on GCS formulation (Group I). Then, start with topics in Rigidity Theory (Groups II and III) before shifting to an order partitioned more by the underlying machinery (Groups IV and V).

**Acknowledgment**

We thank Jessica Sidman and Rahul Prabhu for a careful reading, and Audrey St. John for providing the alternative pathway.

## References

[1] Timothy Good Abbott. *Generalizations of Kempe's universality theorem.* PhD thesis, Massachusetts Institute of Technology, 2008.

[2] Dennis S Arnon, George E Collins, and Scott McCallum. Cylindrical algebraic decomposition I: The basic algorithm. *SIAM Journal on Computing*, 13(4):865–877, 1984.

[3] L. Asimow and B. Roth. The rigidity of graphs. *Transactions of the American Mathematical Society*, 245:279–289, November 1978.

[4] B. Buchberger. *A note on the complexity of constructing Gröbner-bases*, pages 137–145. Springer Berlin Heidelberg, Berlin, Heidelberg, 1983.

[5] Augustin-Louis Cauchy. Sur les polygones et les polyèdres: second mémoire. *Journal de l'École Polytechnique*, 9:87, 1813.

[6] Robert Connelly. The rigidity of polyhedral surfaces. *Mathematics Magazine*, 52(5):275–283, November 1979.

[7] Ioannis Fudos and Christoph M. Hoffmann. Correctness proof of a geometric constraint solver. *International Journal of Computational Geometry & Applications*, 06(04):405–420, 1996.

[8] Harold N Gabow and Herbert H Westermann. Forests, frames, and games: algorithms for matroid sums and applications. *Algorithmica*, 7(1-6):465–497, 1992.

[9] Giovanni Gallo and Bud Mishra. Wu-Ritt characteristic sets and their complexity. *Discrete and Computational Geometry: Papers from the DIMACS Special Year*, 6:111–136, 1991.

[10] Herman Gluck. Almost all simply connected closed surfaces are rigid. In *Geometric topology*, volume 438 of *Lecture Notes in Mathematics*, pages 225–239. Springer, 1975.

[11] Steven J Gortler, Alexander D Healy, and Dylan P Thurston. Characterizing generic global rigidity. *American Journal of Mathematics*, 132(4):897–939, 2010.

[12] Jack E. Graver, Brigitte Servatius, and Herman Servatius. *Combinatorial Rigidity*, volume 2. Graduate Studies in Math., AMS, 1993.

[13] Kirk Haller, Audrey Lee-St.John, Meera Sitharam, Ileana Streinu, and Neil White. Body-and-cad geometric constraint systems. *Comput. Geom. Theory Appl.*, 45(8):385–405, October 2012.

[14] Bruce Hendrickson. Conditions for unique graph realizations. *SIAM J. Comput.*, 21(1):65–84, February 1992.

[15] Lebrecht Henneberg. *Die graphische Statik der starren Systeme*, volume 31. BG Teubner, Leipzig, 1911.

[16] Hiroshi Imai. On combinatorial structures of line drawings of polyhedra. *Discrete Applied Mathematics*, 10(1):79 – 92, 1985.

[17] B. Jackson and T. Jordán. Connected rigidity matroids and unique realizations of graphs. *J. Combinatorial Theory Ser B*, 94:1–29, 2005.

[18] Donald J. Jacobs and Bruce Hendrickson. An algorithm for two-dimensional rigidity percolation: The pebble game. *Journal of Computational Physics*, 137(2):346 – 365, 1997.

[19] Naoki Katoh and Shin-ichi Tanigawa. A proof of the molecular conjecture. *Discrete & Computational Geometry*, 45(4):647–700, 2011.

[20] Alfred B Kempe. On a general method of describing plane curves of the nth degree by linkwork. *Proceedings of the London Mathematical Society*, 1(1):213–216, 1875.

[21] G. Laman. On graphs and rigidity of plane skeletal structures. *Journal of Engineering Mathematics*, 4(4):331–340, October 1970.

[22] Audrey Lee and Ileana Streinu. Pebble game algorithms and sparse graphs. *Discrete Mathematics*, 308(8):1425–1437, 2008. Third European Conference on Combinatorics, Graph Theory, and Applications.

[23] L. Lovász and Y. Yemini. On generic rigidity in the plane. *SIAM Journal on Algebraic Discrete Methods*, 3(1):91–98, 1982.

[24] James Clerk Maxwell. On reciprocal figures and diagrams of forces. *Philosophical Magazine*, 27:250–261, 1864.

[25] Ernst Mayr and R. Cori. *Membership in polynomial ideals over Q is exponential space complete*, pages 400–406. Springer Berlin Heidelberg, Berlin, Heidelberg, 1989.

[26] J Owen and S Power. The non-solvability by radicals of generic 3-connected planar laman graphs. *Transactions of the American Mathematical Society*, 359(5):2269–2303, 2007.

[27] H. Pollaczek-Geiringer. über die gliederung ebener fachwerk. *Zeitschrift fur Angewandte Mathematik und Mechanik (ZAMM)*, 7:58–72, 1927.

[28] B. Roth and W. Whiteley. Tensegrity frameworks. *Transactions of the American Mathematical Society*, 265(2):419–446, June 1981.

[29] James B Saxe. *Embeddability of weighted graphs in k-space is strongly NP-hard.* Carnegie-Mellon University, Department of Computer Science, 1980.

[30] I. J. Schoenberg. On certain metric spaces arising from euclidean spaces by a change of metric and their imbedding in hilbert space. *Annals of Mathematics*, 38(4):787–793, December 1936.

[31] I. J. Schoenberg. Metric spaces and positive definite functions. *Transactions of the American Mathematical Society*, 44(3):522–526, November 1938.

[32] Neil JA Sloane. Error-correcting codes and invariant theory: new applications of a nineteenth-century technique. *American Mathematical Monthly*, pages 82–107, 1977.

[33] Larry Smith. Polynomial invariants of finite groups. a survey of recent developments. *Bulletin of the American Mathematical Society*, 34(3):211–250, 1997.

[34] D. M. Y. Sommerville. *An Introduction to the Geometry of n Dimensions.* 1958.

[35] Ágnes Szántó. Complexity of the Wu-Ritt decomposition. In *Proceedings of the second international symposium on parallel symbolic computation*, pages 139–149. ACM, 1997.

[36] Alfred Tarski. A decision method for elementary algebra and geometry. In *Quantifier elimination and cylindrical algebraic decomposition*, pages 24–84. Springer, 1998.

[37] T.-S. Tay and W. Whiteley. Recent advances in the generic rigidity of structures. *Structural Topology*, 9:31–38, 1984.

[38] Tiong-Seng Tay. Rigidity of multi-graphs. i. linking rigid bodies in n-space. *Journal of Combinatorial Theory, Series B*, 36(1):95 - 112, 1984.

[39] Neil White and Walter Whiteley. The algebraic geometry of motions of bar-and-body frameworks. *SIAM J. Algebraic Discrete Methods*, 8(1):1–32, 1987.

[40] Neil L. White. *A Tutorial on Grassmann-Cayley Algebra*, pages 93–106. Springer Netherlands, Dordrecht, 1995.

# Part I

# Geometric Reasoning, Factorization and Decomposition

# Chapter 2

## *Computer-Assisted Theorem Proving in Synthetic Geometry*

**Julien Narboux**
*University of Strasbourg*

**Predrag Janičić**
*University of Belgrade*

**Jacques Fleuriot**
*University of Edinburgh*

### CONTENTS

## 2.1 Introduction

Computer-assisted proof in mathematics has been underway since the pioneering times of computers in the 1950s. Starting from early systems with very limited capability, computer-assisted theorem proving has evolved to demonstrate theorems never proved before by humans [156] and assist with monumental efforts spanning several man-years [116]. In this endeavour, geometry plays an important part, just as it has throughout the history of mathematics. This is due to its pervasive role: it is a paradigmatic form of reasoning, with applications to education, mathematical and physical research, but also to many applied areas such as robotics, computer vision, and CAD [48]. Moreover, many of the search techniques and other algorithmic features developed for geometric reasoning have influenced other areas of artificial intelligence.

As for other subfields of computer-assisted proof, the mechanization of geometry spans both *automated* and *interactive* theorem proving. In the former, computers aim to prove theorems completely automatically, while in the latter, the role of the system is to act as a *proof assistant* that verifies the reasoning steps of the user, guides the proving process, and provides some limited automation. These two branches are often connected through methods that can produce geometric proofs automatically, where either the proofs or the methods themselves are fully verified.

Just as for its pen-and-paper counterpart, computer-assisted proof in geometry is subject to different foundations: algebraic or synthetic (axiomatic) ones. This chapter aims to provide a comprehensive survey of the latter and of its applications. In particular, it will mostly deal with *planar Euclidean geometry* [6], which in this case generally means a theory consisting of geometric statements true in $\mathbb{R}^2$. There are several formal systems that aim to axiomatize this theory or its subtheories, including those due to Euclid, Hilbert, and Tarski.

In what follows, to ensure a coherent exposition, we will use a uniform notation (which may differ at times from that used by the original authors). In particular, we will denote points by uppercase letters, lines by lowercase ones, the strict notion of *betweenness* by $A \dot{-} B \dot{-} C$ (i.e., $B$ belongs to segment $AC$ and is different from $A$ and $C$) while its nonstrict version will be denoted by $A–B–C$, collinearity by $\mathrm{Col}\,ABC$, perpendicularity by $\perp$, cyclicity of points, i.e., points lying on the same circle, by $cyclic(A,B,C,D)$, the angle between half-lines $AB$ and $AC$ by $\angle ABC$, the full-angle between *lines* $AB$ and $CD$ by $\angle[AB,CD]$, a triangle with vertices $A$, $B$, $C$ by $\triangle ABC$, congruence between segments, triangles, or between angles by $\cong$, and equality over measures of angles or over full-angles by $=$.

## 2.2 Automated Theorem Proving

Automated theorem proving in geometry is often considered a "classical AI domain." Its methods can be split into three major families or styles: algebraic, synthetic, and semisynthetic.

The algebraic methods deal with the algebraized formulation of geometric statements and usually involve dealing with the membership of polynomial ideals [45, 217, 224], quantifier eliminations [64, 212], or use coordinate-free approaches based on bracket and Cayley algebras, described in Chapter 3. Although they are powerful, they cannot produce human-readable proofs and, generally, consist of steps that do not have any obvious meaning in synthetic geometry. The second and the third groups of methods, the subjects of this review, focus on proving theorems via geometric axioms (or higher-level geometric lemmas) and often try to automate the traditional theorem proving approaches, while attempting to generate human-readable proofs.

Automated theorem proving is used in various contexts, e.g., for mathematical education [19], in the simplification of geometric axiom systems [69] and in the study of incidence geometry using term rewriting techniques [5].

Unless otherwise stated, the methods presented next only deal with planar geometries assuming the parallel postulate.

Note also that the examples will be presented in a uniform way, although we have taken care to preserve the essence of the methods being applied. Finally, we remark that there are other surveys [48, 90, 97, 131, 218] that cover some of the approaches that we will discuss next.

### 2.2.1 Foundations

Aside from the algorithmic techniques used, automated theorem proving methods rely on various choices with regard to the underlying logic and geometric knowledge. One issue relates to a set of axioms to be used. Some methods use well-known geometric axioms sets, but most use custom (finite sets of first-order) axioms. In the latter case, the axioms are actually simple theorems of Euclidean geometry whose proofs are not considered and are asserted as facts belonging to common geometric knowledge (hence they are often called "lemmas" or "rules"). The set of axioms is not necessarily minimal as they are often selected to ease the automatic proof of more complex theorems. For many methods, choosing an appropriate set and level of axioms is one of most critical issues when it comes to power and efficiency.

### 2.2.2 Nondegenerate Conditions

The notion of nondegenerate (NDG) conditions arises for each style of automated geometric reasoning. Namely, it is often the case that the goal (denoted by $G$) is implied by the configuration (denoted by $C$) plus some *additional* conditions [46, 48, 55]. For such conditions (denoted by $ndg$), the following formulae (where $\forall*$ and $\exists*$ denote universal and existential closure) have to be valid: $\exists * (C \wedge ndg)$ and $\forall * (C \wedge ndg \Rightarrow G)$. In many cases, the methods can automatically produce such NDG conditions, although not necessarily the "weakest" ones. There are, however, algorithms for computing the weakest NDG conditions [44].

### 2.2.3 Purely Synthetic Methods

Purely synthetic methods, or simply synthetic methods do not use coordinates and algebraic forms for the geometric statements. Many of these techniques add auxiliary elements to the geometric configuration under consideration, so that certain axioms can be applied. This usually leads to a combinatorial explosion in the search space. The challenge then rests in controlling this explosion and in developing suitable heuristics in order to avoid unnecessary construction steps. Due to the nature of these problems, such synthetic proof techniques are sometimes called Artificial Intelligence (AI) methods for automated theorem proving in geometry.

The very first AI method was developed by Gelernter et al. Quoting the authors: "In early spring, 1959, an IBM 704 computer, with the assistance of a program comprising some 20000 individual instructions, proved its first theorem in elementary Euclidean plane geometry" [101]. Their program, called the Geometry Machine and written in FORTRAN [99, 100], was not only the first automated theorem prover for geometry, but also one of the very first automated reasoning systems for mathematics. Although its power, from a modern point of view, was very limited, this system is important both for historical reasons and for introducing a number of ideas and techniques that were used by many subsequent reasoning systems. At the time, geometry was viewed as a typical, paradigmatic AI domain but also as a potentially easy domain where simple *ad-hoc* rules, basic forward and backward chaining* applied exhaustively, and simple heuristics could be used to bear easy fruit e.g. such

*Forward and backward chaining are two important forms of inferences within reasoning systems. The former can be viewed as a sequence of applications of modus ponens that derives new facts from existing premises in order to prove the goal, while

as cracking the whole area of high school geometry. In the description of the approaches, logical representation, algorithms, implementation details, and even the description of features of the programming languages were often intermixed. Over time, more sophisticated and mature approaches have emerged, with many still using some of the early techniques and ideas though.

#### 2.2.3.1 Early Systems

Gelernter's Geometry Machine implemented several reasoning techniques. These included the use of "diagrams" (concrete models in the Cartesian plane) in an attempt to reject false subgoals before any attempt to prove them, dealing with symmetries, and the use of simplification rules (akin to modern rewrite rules). The system worked by backward reasoning – starting with the given goal and trying to decompose it to simpler provable subgoals. Its basic rules were based on axioms about the congruence of triangles. Thus, proving that two segments were congruent could be done by showing that these two segments were the corresponding edges of two congruent triangles.

**Example.** Proving the following theorem, in less than 20s, was one of the big triumphs of Geometry Machine: if $\angle ABD \cong \angle DBC$, $AD \perp AB$, $DC \perp BC$, then $AD \cong CD$ (Figure 2.1):

| | |
|---|---|
| $\angle ABD \cong \angle DBC$ | (Premise) |
| $\angle DAB$ is right angle | (by Definition of perpendicular) |
| $\angle DCB$ is right angle | (by Definition of perpendicular) |
| $\angle BAD \cong \angle BCD$ | (by All right angles are congruent) |
| $BD \cong BD$ | (by Reflexivity of congruence) |
| $\triangle ADB \cong \triangle CDB$ | (by Congruence of triangles, rule Side-Angle-Angle) |
| $AD \cong CD$ | (by Corresponding elements of congruent triangles are congruent) |

There were several subsequent systems improving on Gelernter's ideas, e.g., by combining backward and forward chaining, by trying to model the human solving process more faithfully, or by being designed to serve as support for tutoring systems [2, 109, 56, 57, 89, 103, 105, 136, 166]. However, despite all these efforts, these early systems had a very limited scope and were only able to prove geometric problems of small or moderate complexity. They didn't treat NDG conditions and were not able (or were able only to a limited extent) to add new, "auxiliary" points, necessary in many proofs. So, they typically dealt only with axioms and conjectures of the following form (universal closure is assumed): $A_1(\vec{x}) \wedge \ldots \wedge A_n(\vec{x}) \Rightarrow B(\vec{x})$, where $\vec{x}$ denotes a sequence of variables, $A_i$ and $B$ are atomic formulae or their negations.

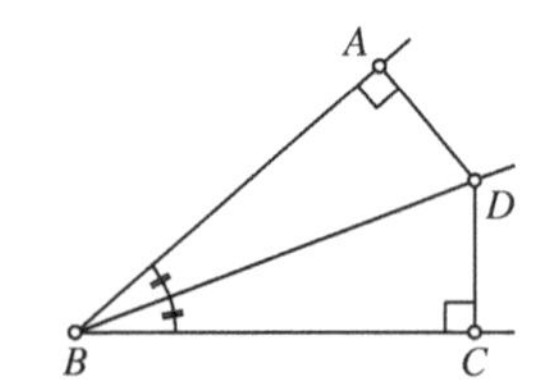


**Figure 2.1**
Geometry machine diagram.

#### 2.2.3.2 Deductive Database Method, GRAMY, and iGeoTutor

There are several theorem proving methods, including the deductive database (DD) [54] and those used by the systems GRAMY [155] and iGeoTutor [220], that deal with "rules" and conjectures of the form (universal closure is assumed): $A_1(\vec{x}) \wedge \ldots \wedge A_n(\vec{x}) \Rightarrow B_1(\vec{x})$ and also rules of the form: $A_1(\vec{x}) \wedge \ldots \wedge A_n(\vec{x}) \Rightarrow \exists \vec{y}(B_1(\vec{x},\vec{y}) \wedge \ldots \wedge B_m(\vec{x},\vec{y}))$ where $\vec{x}$ and $\vec{y}$ denote sequences of variables, $B_i$ are atomic formulae and $A_j$ are atomic formulae or their negations. There are no disjunctions either in the rules or in the conjectures. Hence, these methods cannot prove conjectures involving existential quantifiers, but can use new, auxiliary points (or segments), while searching for a proof.[†]

the latter can be viewed as applying modus ponens backward to refine the goal into subgoals that can hopefully be proven from the premises.

[†]Using auxiliary points makes a substantial change compared to the early methods (Section 2.2.3.1), a change that enabled the proof of a wider set of complex theorems.

One of the main challenges, though, lies in controlling the introduction of additional objects since these can lead to a combinatorial explosion.

Unlike for algebraic methods, a common motivation here is the generation of human-readable synthetic proofs that are as close as possible to those taught in schools. Moreover, in the case of GRAMY, the generation of several proofs is attempted, making it suitable for some forms of tutoring.

**Scope.** The methods deal with formulae containing no function symbols and with fixed sets of predicates symbols. For instance, the DD method uses predicate symbols corresponding to geometric relations (over points), such as Col, $\perp$, $cyclic()$, and equality over full-angles (see Section 2.2.4.2). GRAMY deals not only with points, but also with segments, angles and triangles, and with a set of predicate symbols that includes $\cong$, $\perp$, $\|$, membership, etc. Each system uses a fixed set of axioms, for instance, the DD method uses around 75 axioms of the first form, including:

D41: $cyclic(A,B,P,Q) \Rightarrow \angle[PA,PB] = \angle[QA,QB]$
D42: $\angle[PA,PB] = \angle[QA,QB] \wedge \neg \mathrm{Col}\, PQAB \Rightarrow cyclic(A,B,P,Q)$
D74: $\angle[AB,CD] = \angle[PQ,UV] \wedge PQ \perp UV \Rightarrow AB \perp CD$

and around 20 rules of the second form, including, for example:

X1: $OM \perp MA \wedge \angle[XO,MO] = \angle[MO,AO] \Rightarrow \exists B\, (\mathrm{Col}\, BAM \wedge \mathrm{Col}\, BOX)$

The DD method was reported as being able to prove, in a matter of seconds and via hundreds of derived facts, 160 out of the 600 theorems in the authors' collection of results proved by Wu's method. GRAMY and iGeoTutor were applied on smaller benchmark sets gathered from different sources.

**Theorem proving mechanisms.** For a given conjecture, atomic formulae from the hypotheses are considered as "facts." All three methods use forward chaining for deriving new facts using the available axioms (the DD method takes some ideas from deductive database theory [96]). There is a number of techniques (e.g., based on symmetries) used for keeping the number of stored derived facts low and the proving process efficient.

Auxiliary points are introduced only in a controlled manner, determined by various strategies, such as introducing new points only if new facts cannot be derived using forward chaining, or introducing points only through a very limited number of *templates*, specific geometry configurations.

GRAMY and iGeoTutor do not handle NDG conditions, while the DD method treats them using a form of negation as failure augmented with some basic diagrammatic model checking.

iGeoTutor can deal with conjectures involving some arithmetical constraints and for that purpose uses external provers that are specialized for theories like linear arithmetic and are available within the SMT (Satisfiability Modulo Theory) solver Z3 [160].

**Example.** The system iGeoTutor can prove the following theorem: Given a square $ABCD$, a point in its interior such that $AP \cong PD$ and $\angle PAD = 15°$, prove that the triangle $\triangle PBC$ is equilateral. The system, using the fact $AB \cong AD$, decides to use the *congruent triangles template*, and adds a points $Q$ such that $\angle BAQ \cong \angle PAD$ and $AQ \cong AP$ (such that $\triangle AQB \cong \triangle APD$ holds). Later, the system decides to use another template and introduces the segment $QP$. By the initial forward chaining, the system deduces $BP \cong CP$, $\angle APD = 150°$ (among others facts). The rest of the proof is as follows:

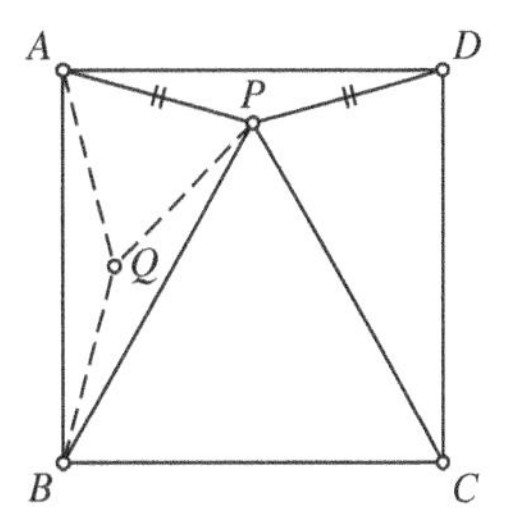


**Figure 2.2**
iGeoTutor diagram.

$\triangle AQB \cong \triangle APD$ (by Congruence of triangles, rule Side-Angle-Side)
$\angle BAQ = 15°$, $\angle BQA = 150°$ (by Corresponding parts of congruent triangles are congruent)
$\angle QAP = 60°$ ($\angle QAP = 90° - \angle BAQ - \angle PAD$)
$\angle AQP = 60°$, $\angle APQ = 60°$ (by $AQ \cong AP$, Isosceles triangle)
$AQ \cong PQ$ (by $\triangle AQP$ is equilateral)
$\angle BQP = 150°$ ($\angle BQP = 360° - \angle BQA - \angle AQP$)
$\triangle AQB \cong \triangle PQB$ (by $BQ \cong BQ$, $\angle BQA \cong \angle BQP$ $AQ \cong PQ$, Side-Angle-Side)
$AB \cong BP$ (by Corresponding parts of congruent triangles are congruent)
$BC \cong BP$, $BC \cong CP$ (by $AB \cong BC$, $AB \cong BP$, $BP \cong CP$, Transitivity of congruence)

**Properties.** These methods are complete, with respect to the set of rules being used, for proofs that do not require the introduction of auxiliary points since the search space is then finite. This feature enables not only proof, but also the deduction of additional facts from a given configuration and hence the discovery of new theorems. If a proof requires auxiliary elements, completeness can be proved only under specific conditions.

### 2.2.3.3 Logic-Based Approaches

Geometric theorems can be proved not only by dedicated systems but also by general ones, typically by theorem provers for first order logic or some of its fragments such as *Coherent logic* (CL). Such automated provers usually cover rich sets of formulae that include existential quantification. We review some of these approaches next.

Coherent logic consists of formulae of the following form (universal closure is assumed) [15, 85]: $A_1(\vec{x}) \wedge \ldots \wedge A_n(\vec{x}) \Rightarrow \exists \vec{y}(B_1(\vec{x},\vec{y}) \vee \ldots \vee B_m(\vec{x},\vec{y}))$, $\vec{x}$ and $\vec{y}$ denote sequences of variables, $A_i$ denotes an atomic formula, and $B_j$ denotes a conjunction of atomic formulae. There are no function symbols with arity greater than zero and there is no negation. *Resolution logic* (RL) deals with *clauses*, i.e., formulae of the following form (universal closure is assumed): $A_1(\vec{x}) \vee \ldots \vee A_n(\vec{x})$, where $\vec{x}$ denotes sequences of variables, and $A_i$ denotes an atomic formula or a negation of an atomic formula.

CL conveys a wider range of formulae compared to the DD method, for instance (Section 2.2.3.2) – there can be existential quantification over variables, not only in axioms but also in conjectures and, also, there can be disjunctions. CL can also be considered as an extension of RL, but in contrast to the resolution-based proving, the CL conjecture can be proved directly and unchanged (refutation, Skolemization and transformation to clausal form are not used). The domain of procedures for CL actually covers first-order logic because every first-order theory can be translated into coherent logic, possibly with additional predicate symbols [185, 85]. Checking validity of an arbitrary first-order formula can be replaced by checking unsatisfiability of a corresponding set of clauses (after refutation, Skolemization and transformation to clausal form).

Provability in CL and unsatisfiability in RL are semi-decidable and there is a number of methods and provers for coherent logic (some of them are based on simple forward reasoning and iterative deepening, with a number of techniques for narrowing the search space [15, 208, 209], while some use more advanced techniques, like lemma learning and back-jumping [167, 169]) and much more for RL [196]. CL admits a simple sequent-calculus style proof system, and any corresponding CL proof has a simple structure [208]. Readable proofs in a forward reasoning style can be easily obtained in CL [15]. The existing theorem proving methods for CL and RL do not deal with NDG conditions but can use the same heuristics as for the DD method (Section 2.2.3.2).

CL provers have been used in a variety of settings and domains. In particular, they have been applied to Euclidean geometry using an axiom system similar to Borsuk's [18, 129] and to Hilbert's and Tarski's axiomatics, with proofs exported to Isabelle, Coq, and natural language [208, 209]. They have also been used to prove the correctness of solutions to ruler and compass construction problems [153]. They have been used for projective plane geometry, where a proof of Hessenberg's theorem was carried out with Coq proof objects generated [16]. Recent work has also seen them combined with resolution theorem provers such as Vampire (for filtering relevant axioms) [210].

**Example.** The following theorem from Tarski's geometry [201] can be proved in CL (using the theorem prover ArgoCLP [209]): Assuming that $A$–$B$–$C$, $(A, B) \cong (A, D)$, and $(C, B) \cong (C, D)$, show that $B = D$. The presented proof is obtained by simplifying and transforming the generated proof, so it reintroduces negation and uses *reductio ad absurdum* [152].

1. It holds that $B$–$A$–$A$ (by th_3_1).
2. From the facts that $A$–$B$–$C$, it holds that Col $CAB$ (by ax_4_10_3).
3. From the facts that $(A, B) \cong (A, D)$, it holds that $(A, D) \cong (A, B)$ (by th_2_2).
4. It holds that $A = B$ or $A \neq B$ (by ax_g1).
   5. Assume that $A = B$.
      6. From the facts that $(A, D) \cong (A, B)$ and $A = B$ it holds that $(A, D) \cong (A, A)$.
      7. From the facts that $(A, D) \cong (A, A)$, it holds that $A = D$ (by ax_3).
      8. From the facts that $A = B$ and $A = D$ it holds that $B = D$.

      This proves the conjecture.
   9. Assume that $A \neq B$.

      Let us prove that $A \neq C$ by reductio ad absurdum.
      10. Assume that $A = C$.
          11. From the facts that $A$–$B$–$C$ and $A = C$ it holds that $A$–$B$–$A$.
          12. From the facts that $A$–$B$–$A$, and $B$–$A$–$A$, it holds that $A = B$ (by th_3_4).
          13. From the facts that $A \neq B$, and $A = B$ we get a contradiction.

          Contradiction.

      Therefore, it holds that $A \neq C$.
      14. From the fact that $A \neq C$, it holds that $C \neq A$ (by the equality axioms).
      15. From the facts that $C \neq A$, Col $CAB$, $(C, B) \cong (C, D)$, and $(A, B) \cong (A, D)$, it holds that $B = D$ (by th_4_18).

      This proves the conjecture.

Quaife used the resolution theorem prover OTTER to prove several non-trivial theorems in Tarski's geometry [201], with a slightly modified axiom system [189]. During theorem proving, a number of techniques were employed to guide resolution and, upon success, some post-processing used to translate the resolution proofs into a more readable form. More recently, Beeson and Wos carried out some similar work, but with a newer version of OTTER and with much more success thanks to a number of techniques and strategies that have become available in the meantime [10, 11]. They proved around 200 theorems from the book by Schwabhäuser et al. [201]. Of these theorems, 76% were proved automatically using different custom heuristics and strategies, while for the others heavy human support (in a form of lemmas and hints) was required. This latest work did not involve the production of readable proofs from the resolution ones. Even more recently, other resolution provers with state-of-the-art techniques have led to an even higher percentage of theorems from the same corpus being proved completely automatically, without any guidance by humans [216].

**Example.** The following proof, slightly reformulated for the sake of uniformity, is generated by Quaife's approach: if $C$ is between $B$ and $D$, each of which is between $A$ and $E$, then $C$ is between $A$ and $E$.

| | | |
|---|---|---|
| 37 | $U\text{–}V\text{–}W \Rightarrow W\text{–}V\text{–}U$ | (by Axiom) |
| 45 | $U\text{–}V\text{–}X, V\text{–}W\text{–}X \Rightarrow U\text{–}W\text{–}X$ | (by Lemma) |
| 46 | $U\text{–}V\text{–}W, U\text{–}W\text{–}X \Rightarrow U\text{–}V\text{–}X$ | (by Lemma) |
| 74 | $U\text{–}V\text{–}X, U\text{–}W\text{–}X \Rightarrow U\text{–}V\text{–}W, U\text{–}W\text{–}V$ | (by Lemma) |
| 77 | $A\text{–}B\text{–}E$ | (by Hyp) |
| 78 | $B\text{–}C\text{–}D$ | (by Hyp) |
| 79 | $A\text{–}D\text{–}E$ | (by Hyp) |
| 80 | $\neg A\text{–}C\text{–}E$ | (by negated goal) |
| 91 | $\neg A\text{–}C\text{–}D$ | (by 80, 46, 79) |
| 92 | $\neg A\text{–}C\text{–}B$ | (by 80, 46, 77) |
| 109 | $\neg A\text{–}B\text{–}D$ | (by 91, 45, 78) |
| 127 | $\neg B\text{–}C\text{–}A$ | (by 92, 37) |
| 184 | $A\text{–}D\text{–}B$ | (by 109, 74, 79, 77) |
| 253 | $\neg B\text{–}D\text{–}A$ | (by 127, 46, 78) |
| 309 | Contradiction! | (by 184, 37, 253) |

### 2.2.4 Semisynthetic Methods

Semisynthetic methods, sometimes also called coordinate-free methods or geometric invariant methods, do not use algebraic formulation of geometry problems, but express conjectures in terms of certain *geometric quantities* and prove them by manipulating equalities over expressions in these quantities. This approach can also lead to combinatorial explosion, but in many cases can give short and readable proofs.

#### 2.2.4.1 Area Method

The area method is a procedure for a fragment of Euclidean plane geometry [50, 51, 130, 229]. It uses suitably chosen geometry quantities, such as area of triangle, and can efficiently prove many non-trivial theorems and produces proofs that are often very concise and human-readable. The method had been extended to solid Euclidean geometry [52], to non-Euclidean geometries [225, 226] and, in conjunction with Collins algorithm [63], to a system for proving geometry inequalities [197].

**Scope.** A conjecture consists of a construction and a goal, where the construction is expressed in terms of (five basic) specific construction primitives (or constructions composed of the primitive ones), and the goal is an equality over expressions given in terms of (three basic) specific primitive geometry quantities. Both the construction and the goal are expressed only in terms of points (i.e., cannot involve lines or circles explicitly).

An example of a construction primitive is INTER $Y\ U\ V\ P\ Q$, which indicates that point $Y$ is the intersection of lines $UV$ and $PQ$. For a construction step to be well-defined, certain NDG conditions may be required. The above construction step has a NDG condition $U \neq V \wedge P \neq Q \wedge UV \nparallel PQ$. Intersections of two circles and intersections of a line and a circle are supported by construction primitives only in some special cases. Additional construction steps can be expressed in terms of the basic ones.

The geometric quantities used are: the signed ratio of parallel directed segments, denoted $\frac{\overline{AB}}{\overline{CD}}$, the signed area for a triangle $ABC$; denoted $\mathcal{S}_{ABC}$ (negated for the triangle with the opposite orientation); the Pythagoras difference, denoted $\mathcal{P}_{ABC}$ (for the points $A$, $B$, $C$, defined as $\mathcal{P}_{ABC} = \overline{AB}^2 + \overline{CB}^2 - \overline{AC}^2$). Using these quantities, a number of geometric predicates can be simply expressed, for instance: $A = B$ iff $\mathcal{P}_{ABA} = 0$; Col $ABC$ iff $\mathcal{S}_{ABC} = 0$; $AB \perp CD$ iff $\mathcal{P}_{ABA} \neq 0 \wedge \mathcal{P}_{CDC} \neq 0 \wedge \mathcal{P}_{ACD} = \mathcal{P}_{BCD}$; $AB \parallel CD$ iff $\mathcal{P}_{ABA} \neq 0 \wedge \mathcal{P}_{CDC} \neq 0 \wedge \mathcal{S}_{ACD} = \mathcal{S}_{BCD}$, etc.

The method implemented by its authors proved 500 theorems from their collection [51].

**Theorem proving mechanism.** The method works by the elimination of constructed points in reverse order, using a set of specific elimination lemmas.[‡] All the lemmas used by the method can be proved by an elegant, custom axiom system (Section 2.3.1.4).

There is an elimination lemma for each pair of construction step and geometric quantity. For instance, the following lemma is used for eliminating a point constructed on a line at a given ratio from a signed area:

*If $Y$ is a point constructed on line $PQ$, such that $\frac{\overline{PY}}{\overline{PQ}} = \lambda$ then for any points $A$ and $B$*

$$S_{ABY} = \lambda S_{ABQ} + (1-\lambda) S_{ABP}.$$

The combined NDG conditions of the conjecture is the conjunction of those for the corresponding construction steps, of the conditions that the denominators of the ratios of parallel directed segments in the goal equality are not equal to zero, and of the conditions that lines appearing in ratios of segments in the goal are parallel. It is then proved that the goal equality follows from the construction specification and the combined NDG conditions.[§]

Apart from the basic NDG conditions, there are also side conditions in some of the elimination lemmas having two cases — positive (always of the form "$A$ is on $PQ$") and negative (always of the form "$A$ is not on $PQ$"). If one side condition can be proved, then that case is applied. Otherwise, in one variation of the method, the proof process branches into two cases, and in another, the negative case is assumed and added to the NDG conditions [130].

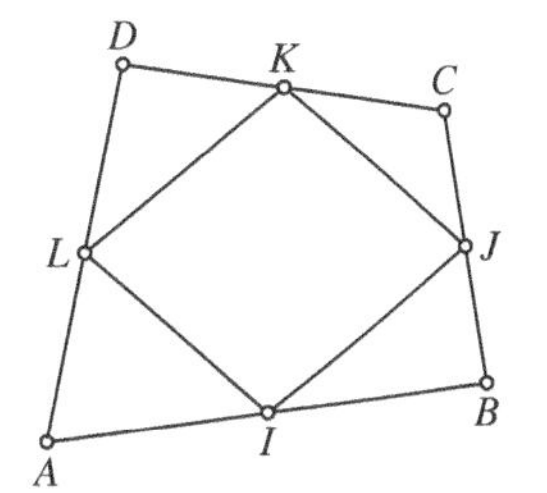


**Figure 2.3**
Varignon's theorem.

**Example.** As an example, we give the proof of Varignon's theorem: Given a quadrilateral $ABCD$, let $I$, $J$, $K$ and $L$ be the midpoints of $AB$, $BC$, $CD$, $DA$, then $IJKL$ is a parallelogram. We give below the proof that $IJ \parallel KL$, the proof that $JK \parallel IL$ is similar. Note that a synthetic proof within Coq is given in Figure 2.9.

$$\begin{array}{lll}
 & S_{KIJ} - S_{LIJ} & \\
= & \frac{S_{KIB}}{2} + \frac{S_{KIC}}{2} - \frac{S_{LIB}}{2} - \frac{S_{LIC}}{2} & J \text{ Eliminated} \\
= & \frac{S_{BKA}}{2} + \frac{S_{BKB}}{2} + \frac{S_{CKA}}{2} + \frac{S_{CKB}}{2} - \frac{S_{BLA}}{2} - \frac{S_{BLB}}{2} - \frac{S_{CLA}}{2} - \frac{S_{CLB}}{2} & I \text{ Eliminated} \\
= & \frac{1}{2}(S_{BKA} + S_{CKA} + S_{CKB} - S_{BLA} - S_{CLA} - S_{CLB}) & \text{Simplification} \\
= & \frac{1}{2}(\frac{S_{ABC}}{2} + \frac{S_{ABD}}{2} + \frac{S_{ACC}}{2} + \frac{S_{ACD}}{2} + \frac{S_{BCC}}{2} + \frac{S_{BCD}}{2} - S_{BLA} - S_{CLA} - S_{CLB}) & K \text{ Eliminated} \\
= & \frac{1}{2}(\frac{S_{ABC}}{2} + \frac{S_{ABD}}{2} + \frac{S_{ACC}}{2} + \frac{S_{ACD}}{2} + \frac{S_{BCC}}{2} + \frac{S_{BCD}}{2} - \frac{S_{ABA}}{2} - \frac{S_{ABD}}{2} - \frac{S_{ACA}}{2} - \frac{S_{ACD}}{2} - \frac{S_{BCA}}{2} - \frac{S_{BCD}}{2}) & L \text{ Eliminated} \\
= & \frac{1}{4}(S_{ABC} + S_{BCA}) & \text{Simplification} \\
= & 0 & \text{Simplification}
\end{array}$$

**Properties.** The method is terminating, sound, and complete: for each geometric statement in its scope, it can decide whether it is a theorem, i.e., it is a decision procedure for this fragment of geometry. Its complexity is exponential in the number of points involved [229].

[‡] A later variant of the method also deals with nonconstructive statements, described in terms of various geometric predicates [53].

[§] If the negation of some NDG condition of a geometric statement is implied by the remaining construction steps, the left-hand side of the implication is inconsistent and the statement is trivially valid.

#### 2.2.4.2 Full-Angle Method

The full-angle method [47] is, in spirit, closely related to the area method (Section 2.2.4.1) and can also produce elegant proofs for a number of complex theorems. The idea of eliminating points is extended to eliminating lines. The main motivation of the full-angle method is the fact that using "traditional angles" in geometrical proofs typically leads to considering a number of cases. For instance, for four distinct cyclic points $A$, $B$, $C$, $D$, one can claim that the angles $\angle ABC$ and $\angle ADC$ are congruent if $B$ and $D$ are on the same side of line $AC$ *or* complementary if they are on opposite sides (Figure 2.4). On the other hand, with full-angles one can simply (without using order relation or orientations of plane) state $\angle[AD,CD] = \angle[AB,CB]$. Namely, it holds $\angle[AB,BC] = \angle[DE,EF]$ iff $\angle ABC \cong \angle DEF$ and the two angles have the same orientation or $\angle ABC = 180° - \angle DEF$ and the two angles have opposite orientations.

A full-angle is defined to be an ordered pair $\angle[m,n]$ of two intersecting lines $m$ and $n$, such that $\angle[m,n]$ is equal to another full-angle $\angle[u,v]$ if there is a rotation $\mathcal{R}$ such that $\mathcal{R}(m) \parallel u$ and $\mathcal{R}(n) \parallel v$ (therefore, any full-angle can be considered as an equivalence class) [51]. The sum of two full-angles is defined as follows: given four lines $m$, $n$, $u$, and $v$, and a rotation $\mathcal{R}$ such that $\mathcal{R}(u) \parallel n$, then $\angle[m,n] + \angle[u,v] = \angle[m,\mathcal{R}(v)]$. For arbitrary line $m$, $\angle[m,m]$ is denoted by $\mathbf{0}$. For arbitrary lines $m$ and $n$, $\angle[m,n]$ can be denoted also by $-\angle[n,m]$. It can be proved that full-angles form an Abelian group with the operation $+$, the neutral element $\mathbf{0}$, and with inverse element corresponding to the (unary) operator $-$. In addition, $\angle[m,n] + -\angle[u,v]$ is abbreviated by $\angle[m,n] - \angle[u,v]$, and for arbitrary perpendicular lines $m$ and $n$, $\angle[m,n]$ is denoted by $\mathbf{1}$.

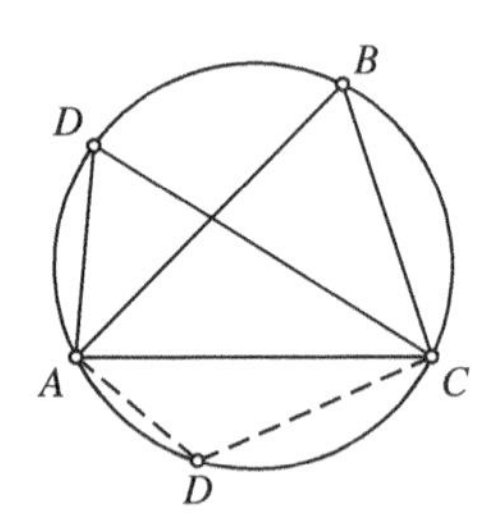


**Figure 2.4**
Cyclic points and peripheral angles.

It can be proved that full-angles satisfy around 20 properties useful for transforming goals, including the following ones (where, for each full-angle $\angle[AB,CD]$, it is assumed that $A \neq B$ and $C \neq D$):

R4: $\mathbf{1} + \mathbf{1} = \mathbf{0}$
R6: if Col $PQX$ then $\angle[AB,PX] = \angle[AB,PQ]$
R10: if $cyclic(A,B,C,D)$ then $\angle[AD,CD] = \angle[AB,CB]$
R13: $\angle[AB,CD] = -\angle[CD,AB]$
R14: for any line $UV$, $\angle[AB,CD] = \angle[AB,UV] + \angle[UV,CD]$

**Scope.** The full-angle method deals with conjectures consisting of hypotheses, expressed in terms of relevant construction steps (cf. the area method in Section 2.2.4.1) or in terms of other geometric predicates (as in the later variation of the area method), and of a goal that is an equality over full-angles.

**Theorem proving mechanism.** The proof method uses forward chaining for exhaustively deducing new facts from the existing ones, using lemmas (rules) like:

F1: if $m \parallel n$ and $m \parallel l$, then $n \parallel l$
F5: if $PA \perp PB$ then $QA \perp QB$ iff $cyclic(A,B,P,Q)$
F8: if $AB \parallel AC$, then Col $ABC$
K2: if $m \perp n$ and $u \perp v$, then $\angle[m,u] = \angle[n,v]$

Some rules have NDG conditions attached and they can be treated as for the area method (see Section 2.2.4.1).

Using derived facts, rules like the ones listed above are used for elimination of points (R6) or lines (R10) from the goal – this is again analogous to the area method although the expressions are now simpler since there are no multiplications or divisions over full-angles.

The configuration is not necessarily expressed in terms of constructive statements and so (unlike in the basic version of the area method) there is no implicit order in which points can be eliminated. So, an ordering has to be imposed over points, which extends to full-angles, in order to control the application of the rules. For instance, the rule: *For any line UV,* $\angle[AB,CD] = \angle[AB,UV] + \angle[UV,CD]$ is used only if the two new full-angles can be further reduced to full-angles less than $\angle[AB,CD]$ (in the ordering).

**Example.** For Simson's theorem (Figure 2.5) the hypotheses are $cyclic(A,B,C,D)$, $E$ is the foot from $D$ to $BC$ (i.e., Col $BCE$ and $DE \perp BC$), $F$ is the foot from $D$ to $AC$ (i.e., Col $ACF$ and $DF \perp AC$), $G$ is the foot from $D$ to $AB$ (i.e., Col $ABG$ and $DG \perp AB$), and the goal is that $G$, $F$, $E$ are collinear, i.e., $\angle[GF,GE] = \mathbf{0}$.

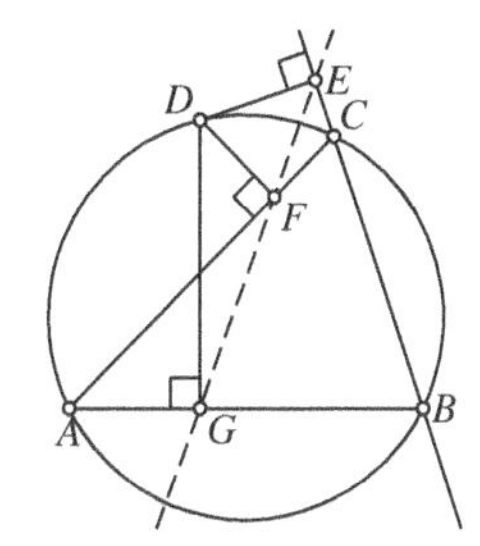


**Figure 2.5**
Simson's theorem.

The proof assumes that for each full-angle $\angle[XY,UV]$ used it holds that $X \neq U$ and $Y \neq V$. The following point order is used $O,A,B,C,D,E,F,G$, and the following facts (among others) can be derived from the hypotheses: $cyclic(F,A,D,G)$ (because $FA \perp FD, GA \perp GD$), $cyclic(E,B,D,G)$ (because $EB \perp ED, GB \perp GD$). In the following proof outline, applications of rules related to the symmetry properties of relations are not shown.

$$\begin{aligned}
\angle[GF,GE] &= \\
&= \angle[GF,GD] + \angle[GD,GE] && \text{(by R14)} \\
&= \angle[AF,AD] + \angle[GD,GE] && \text{(by R10, } cyclic(F,A,D,G)) \\
&= \angle[AF,AD] + \angle[BD,BE] && \text{(by R10, } cyclic(E,B,D,G)) \\
&= \angle[AF,AD] - \angle[BE,BD] && \text{(by R13)} \\
&= \angle[AC,AD] - \angle[BE,BD] && \text{(by R6, Col } AFC) \\
&= \angle[AC,AD] - \angle[BC,BD] && \text{(by R6, Col } BCE) \\
&= \angle[AC,AD] - \angle[AC,AD] && \text{(by R10, } cyclic(A,B,C,D)) \\
&= \mathbf{0}
\end{aligned}$$

**Properties.** The method is not complete, but can be used as a complement to the area method. When applied to a conjecture in its scope, if it succeeds, the generated proof is typically short and readable. Otherwise, the goal is transformed into a goal for the area method: an equality $\alpha = \beta$ is transformed[¶] into $\tan(\alpha) = \tan(\beta)$ and then further using the following equations (where $\mathcal{P}_{ABCD} = \mathcal{P}_{ABD} - \mathcal{P}_{CBD}$):

$$\tan(\angle[AB,CD] + \angle[PQ,UV]) = \frac{\tan(\angle[AB,CD]) + \tan(\angle[PQ,UV])}{1 - \tan(\angle[AB,CD])\tan(\angle[PQ,UV])}$$

$$\tan(\angle[AB,CD]) = \frac{4\mathcal{S}_{ACBD}}{\mathcal{P}_{ADBC}}$$

Since the area method is complete, the above gives a decision procedure for formulae belonging to the scope of the full-angle method [47].

#### 2.2.4.3 Vector-Based Method

The idea of using vectors for automating geometric proofs has been proposed by several authors, but probably the most important work in the area is due to Chou, Gao, and Zhang [49]. Their method

¶The function tan for the full-angle (corresponding to the usual trigonometric function), $\tan(\angle[AB,CD]) = \frac{4\mathscr{S}_{ACBD}}{\mathscr{P}_{ADBC}}$, is well-defined, thanks to the fact that $\angle[AB,CD] = \angle[PQ,UV]$ iff $\mathscr{S}_{ACBD}\mathscr{P}_{PUQV} = \mathscr{S}_{PUQV}\mathscr{P}_{ACBD}$.

is, in spirit, close to the area method (Section 2.2.4.1), in the way the hypotheses are described constructively and the constructed points are eliminated from the goal one by one using appropriate lemmas.

**Scope.** The hypotheses are expressed in terms of (four) specific construction primitives. For instance, PRATIO $A\ W\ U\ V\ r$ denotes the construction of a point $A$ such that $\overrightarrow{WA} = r\overrightarrow{UV}$, where $r$ is a rational number, an expression over geometric quantities, or a parameter (the NDG condition is $U \neq V$). Additional construction steps can be suitably expressed in terms of the basic ones. For instance, the construction of the midpoint, denoted MIDPOINT $M\ A\ B$, can be expressed as PRATIO $M\ A\ A\ B$ 1/2.

A goal is either an equality over vectors or an equality involving the inner products ($\langle\overrightarrow{AB},\overrightarrow{CD}\rangle$) and exterior products ($[\overrightarrow{AB},\overrightarrow{CD}]$) of vectors over constructed points.

There are two kinds of NDG conditions: those induced by the construction steps and those necessary for the goal to be defined (denominators are not zero). Then, the conjecture is changed by augmenting the hypotheses with these NDG conditions.

**Theorem proving mechanism.** The method works by the elimination of constructed points in reverse order, using a set of specific elimination lemmas, like for the area method. There are elimination lemmas for each pair (construction step, geometry quantity). For instance, the following lemma is used for the elimination of a point $Y$ constructed by the PRATIO step from the *linear* quantity $G(Y)$ satisfying $G(\alpha Y_1 + \beta Y_2) = \alpha G(Y_1) + \beta G(Y_2)$, for any real numbers $\alpha$ and $\beta$:

*If $Y$ is introduced by* PRATIO $Y\ W\ U\ V\ r$*, then* $G(Y) = G(W) + r(G(V) - G(U))$.

**Properties.** The method is terminating, sound, and complete: for each geometry statement in its domain, it can decide whether it is a theorem, i.e., the method is a decision procedure for its fragment of geometry. The complexity of the method is exponential in the number of involved points [49].

#### 2.2.4.4 Mass-Point Method

Barycentric coordinates and mass points have been used in geometry at least since 1969 [68], and were introduced in automated theorem proving for geometry by Zou and Zhang [230]. In the non-complex case, the method is similar to the method used by Kimberling for studying triangle centers (Section 2.3.2). A *mass point* is $mP$, where $m$ ("mass") is a positive real number, and $P$ is a point in a plane. Two mass points $mP$ and $nQ$ are equal iff $m = n$ and $P = Q$.

**Scope.** The conjecture consists of hypotheses (in the form of a construction) and a goal. Hypotheses are expressed in terms of three free (arbitrary) points and subsequent points are obtained by five basic geometric constructions and some compound ones (that enable a constructed point to be expressed as a linear combination of the three basis points), including:

C3 LRATIO $X\ A\ B\ r$, that gives a point $X$ on the line $AB$ such that $\overrightarrow{AX} = r\overrightarrow{AB}$, where $r$ is a rational number, a rational expression, or a variable. Specially, MIDPOINT $X\ A\ B$ denotes LRATIO $X\ A\ B$ $1/2$. Constructing a point $X$ such that $\overrightarrow{AX} = r\overrightarrow{XB}$ (or $(1+r)X = A + rB$) is denoted by MRATIO $X$ $A\ B\ r$.

C5 INTER $X\ U\ V\ A\ B$, that gives the intersection point $X$ of lines $UV$ and $AB$ (the NDG condition is that $X$ is not equal to some of the points $U, V, A, B$, and that $UV$ and $AB$ are not parallel; otherwise, the prover fails).

The goal is a predicate over constructed points, one from a set that includes, for instance, Col $ABC$. For this predicate, it can be proved [228]: if $P$, $Q$, and $R$ are points of the plane $ABC$ ($A$, $B$, $C$ are noncollinear points), and $P = a_pA + b_pB + c_pC$, $Q = a_qA + b_qB + c_qC$, $R = a_rA + b_rB + c_rC$, then $P$, $Q$ and $R$ are collinear iff

$$\begin{vmatrix} a_p & b_p & c_p \\ a_q & b_q & c_q \\ a_r & b_r & c_r \end{vmatrix} = 0.$$

The method is extended to deal with additional constructions (such as a construction of a circle) and uses complex numbers for convenience. This extended version has a scope strictly wider than the basic version.

The authors of the implementation successfully used it for proving hundreds of nontrivial theorems. Although the generated proofs are understandable, they are still not human-like proofs.

**Theorem proving mechanism.** The mass point method works by expressing all constructed points as a linear combination, of three (or two, for some simple conjectures) free points, then reformulating the goal the same way and finally proving it as a goal over real numbers.

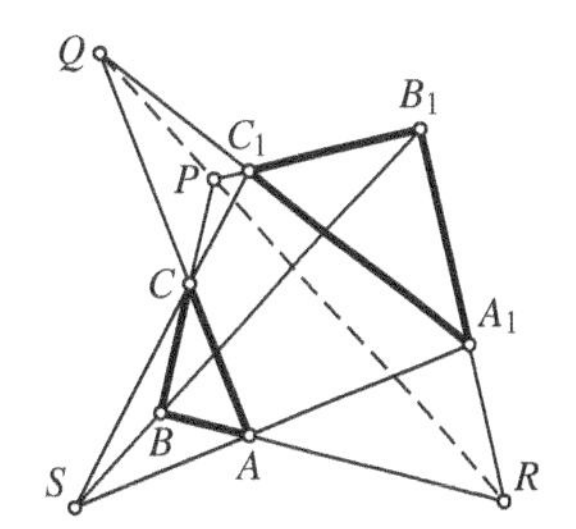


**Figure 2.6**
Desargues's theorem.

**Example.** Desargues's theorem (see also Section 2.3.2) states that, given two triangles $\triangle ABC$ and $\triangle A_1B_1C_1$, if the lines $AA_1$, $BB_1$, $CC_1$ intersect in a point $S$, then the intersection points $P$ of $BC$ and $B_1C_1$, $Q$ of $CA$ and $C_1A_1$, and $R$ of $AB$ and $A_1B_1$ are collinear (Figure 2.6). The theorem has to be slightly reformulated in order to use the mass point method. The construction is as follows:

Let $A$, $B$, and $C$ be three free points.

Let $S$ be an arbitrary point of the plane $ABC$, hence $S = aA + bB + cC$ for some real numbers $a$, $b$, $c$, such that $a+b+c=1$.

MRATIO $A_1$ $S$ $A$ $x$, for some $x$ (then it holds $A_1 = \frac{1}{1+x}S + \frac{x}{1+x}A = \frac{1}{1+x}(aA+bB+cC) + \frac{x}{1+x}A = \frac{a+x}{1+x}A + \frac{b}{1+x}B + \frac{c}{1+x}C$).

MRATIO $B_1$ $S$ $B$ $y$, for some $y$ (then it holds $B_1 = \frac{a}{1+y}A + \frac{b+y}{1+y}B + \frac{c}{1+y}C$).

MRATIO $C_1$ $S$ $C$ $z$, for some $z$ (then it holds $C_1 = \frac{a}{1+z}A + \frac{b}{1+z}B + \frac{c+z}{1+z}C$).

INTER $P$ $B$ $C$ $B_1$ $C_1$ (then it holds $yB - zC = (1+y)B_1 - (1+z)C_1 = (y-z)P$, i.e., $P = \frac{y}{y-z}B - \frac{z}{y-z}C$).

INTER $Q$ $A$ $C$ $A_1$ $C_1$ (then it holds $Q = \frac{x}{x-z}A - \frac{z}{x-z}C$).

INTER $R$ $A$ $B$ $A_1$ $B_1$ (then it holds $R = \frac{x}{x-y}A - \frac{y}{x-y}B$).

The goal is to prove that $P$, $Q$, $R$ are collinear, which is done by showing that:

$$\begin{vmatrix} 0 & \frac{y}{y-z} & -\frac{z}{y-z} \\ \frac{x}{x-z} & 0 & -\frac{z}{x-z} \\ \frac{x}{x-y} & -\frac{y}{x-y} & 0 \end{vmatrix} = 0$$

**Properties.** The mass point method provides a decision procedure for conjectures within its scope [230].

### 2.2.5 Provers Implementations and Repositories of Theorems

There are a number of tools, typically providing dynamic geometry functionalities, that have support for the automated proof of geometry theorems. We mention the most notable ones next.

GEX/jGEX/MMP/Geometer is a family of systems equipped with provers based on algebraic approaches, the DD method, the area method, the vector method, and the full-angle method [98, 227]. GeoGebra [19], also equipped with several algebraic-based provers and tools based on the area method, is aimed at education. It can also work with the Coq proof assistant to support interactive proofs [180]. GeoProof is another tool linked to Coq that can use provers based on the area method, Wu's method, and the Gröbner basis method for generating machine verifiable proofs [163]. GCLC is a system that supports two algebraic methods and the area method [128]. Theorema [36], built on top of Mathematica, is a general mathematical tool with support for several theorem proving approaches, including the area method. OpenGeoProver is a library with several algebraic provers and one based on the area method [151]. Geometry Explorer uses the full-angle method [223] and provides means of visualizing geometric proofs as graphs.

There are ongoing efforts toward linking dynamic geometry systems with automated theorem proving and also with automated discovery, intelligent management of geometry knowledge, tutoring, eLearning, and so on [43, 142, 191, 219].

Finally, aside from collections of theorems available within the above tools, we note the existence of a dedicated repository of geometry theorems known as TGTP [190].

## 2.3 Interactive Theorem Proving

A proof assistant is a piece of software that can check mathematical assertions interactively. The main ones that have been used for the formalization of geometry are Coq [14, 66, 67], Isabelle [168, 175, 221], HOL4 [207], HOL-Light [119], and Mizar [215, 222]. They differ in their mathematical foundations (e.g., type theory, higher order logic [HOL], or set theory) and their proof language. In procedural style proof assistants (e.g., Coq and HOL Light), proofs are described as sequences of commands that modify the proof state whereas in proof assistants that use a declarative language (e.g., Mizar and Isabelle), the proofs are structured and contain the intermediate assertions that were given by the user and justified by the system.

### 2.3.1 Formalization of Foundations of Geometry

There are several ways in which the foundations of geometry can be laid.

In the *synthetic* approach, the geometry theory is built from axioms, with non-logical symbols corresponding to geometric predicates, and sorts corresponding to geometric objects.

The best-known *modern* axiomatic systems along these lines are those of Hilbert [124] and Tarski [214], which we will examine in detail next.

In the *analytic* approach, a field $\mathbb{F}$ is assumed (usually $\mathbb{R}$, the reals), the space is defined as $\mathbb{F}^n$, and the geometric objects and predicates are *defined.*

In the mixed analytic/synthetic approaches, one assumes both the existence of a field and also some geometric axioms. For example, the axiomatic systems for geometry proposed for education in North America by the School Mathematics Study Group in the 1960s are based on Birkhoff's axiomatic system [17] in which the underlying field ($\mathbb{R}$) serves to measure distances and angles. This approach, known as the metric approach, is developed in a number of modern sources [154, 159]. A similar approach is used by Chou, Gao, and Zhang for the foundations of the area method [229] (Section 2.2.4.1), where the underlying field is used to express ratios of signed distances and areas. The axioms and properties of the area method have been formalized in Coq [130]. Geometry can also be defined as a space of objects and a group of transformations acting on it (Erlangen program [135]), and several axiom systems based on this approach have been proposed [86, 170].

Axiom systems based on intuitionistic logic have also been proposed for geometry. Von Plato, for instance, uses the concept of apartness of points and convergence of lines to study plane geometry [183, 184]. Beeson, for his part, introduces a constructive version of Tarski's axiom system [7, 8].

Axiomatics exist for a variety of other geometries (hyperbolic, elliptic, projective, etc.) that will not be examined in this chapter. Details about such axiomatizations can be found in surveys by Pambuccian [171, 172, 173], for instance. In what follows, we mainly concentrate on Hilbert and Tarski's axiomatic systems and their formalizations.

#### 2.3.1.1 Hilbert's Geometry

**Hilbert's axiom system** Hilbert stated his axioms in natural language, but left various aspects implicit and imprecise [124]. One formulation of the first four groups of planar axioms (the fifth group of axioms is discussed in Section 2.3.1.3) in terms of first-order logic (FOL) is given next, although there are other possible formalizations [39, 123, 199, 200].

Hilbert's axiom system can be formulated in a three-sorted first-order language, with individual variables interpreted as points, lines, and planes (the third sort is only needed when dealing with space geometry). It involves four nonlogical predicate symbols: one for the incidence of a point on a line (denoted by $P \in l$ in what follows), one for strict betweenness (denoted by $A \dot{-} B \dot{-} C$), one for the congruence of segments (denoted by $AB \cong CD$), and one for the congruence of angles (denoted by $\angle ABC \cong \angle DEF$).

The axioms are stated by Hilbert using notions from set theory, but the use of set theory is inessential and the formulation of Hilbert's axioms given next does not rely on it (the use of $\in$ is merely as a syntactically convenient predicate).

The first group of Hilbert's axioms deals with incidence: for any two points there is one and only one line through the two points; on every line there are two points; there is a line and point not incident to it (or alternatively: there are three noncollinear points). The axioms, given in FOL form, are as follows:

Incidence Axioms

I 1 $A \neq B \Rightarrow \exists l\, A \in l \wedge B \in l$

I 2 $A \neq B \wedge A \in l \wedge B \in l \wedge A \in m \wedge B \in m \Rightarrow l = m$

I 3 $\exists AB\, B \in l \wedge A \in l \wedge A \neq B$

I 4 $\exists P_0 l_0\, \neg P_0 \in l_0$

The second group of axioms provides properties of betweenness. The betweenness predicate of Hilbert is strict (as opposed to that of Tarski) as ensured by the first axiom (II 1). The fourth axiom is the axiom of Pasch: if a line intersects one side of a triangle, then it also intersects one of the two other sides.

Order axioms

II 1 $A \dot{-} B \dot{-} C \Rightarrow \text{Col}\, ABC$

$A \dot{-} B \dot{-} C \Rightarrow A \neq C$

$A \dot{-} B \dot{-} C \Rightarrow C \dot{-} B \dot{-} A$

II 2 $A \neq B \Rightarrow \exists C\, A \dot{-} B \dot{-} C$

II 3 $A \dot{-} B \dot{-} C \Rightarrow \neg B \dot{-} C \dot{-} A$

Def cut l A B $:= \neg A \in l \wedge \neg B \in l \wedge \exists I\, (I \in l \wedge A \dot{-} I \dot{-} B)$

II 4 $\neg \text{Col}\, ABC \wedge \neg C \in l \wedge \text{cut}\, lAB \Rightarrow (\text{cut}\, lAC \vee \text{cut}\, lBC)$

The third group of axioms provides congruence properties. The aspects that are implicit in Hilbert's text or are consequences of the set theoretical formulation of the axioms are suffixed with lowercase letters. For example, in Hilbert's text, the segment $CD$ is the same as segment $DC$, a fact captured by axiom IV-a in our case. Axiom IV 2 is the transitivity property of the congruence predicate. Axiom IV 3 is the segment addition axiom. Axiom IV 4 is the angle construction axiom: given an angle $\angle BAC$ and a side of a half line $\overrightarrow{OX}$, there is a unique half line $\overrightarrow{OY}$ such that the angle between $\overrightarrow{OX}$ and $\overrightarrow{OY}$ is $\angle BAC$. Axiom IV 5 is the side-angle-side property (two triangles are congruent if two pairs of their corresponding sides and angles determined by them are congruent).

Congruence Axioms

| | |
|---|---|
| IV-a | $AB \cong CD \Rightarrow AB \cong DC$ |
| IV 1 | $A \neq B \wedge M \in l \Rightarrow \exists A'B'\ (A' \in l \wedge B' \in l \wedge A' \dot{-} M \dot{-} B' \wedge MA' \cong AB \wedge MB' \cong AB)$ |
| IV 2 | $AB \cong CD \wedge AB \cong EF \Rightarrow CD \cong EF$ |
| Def | disjoint A B C D := $\neg \exists P\ (A \dot{-} P \dot{-} B \wedge C \dot{-} P \dot{-} D)$ |
| IV 3 | $\text{Col}\, ABC \wedge \text{Col}\, A'B'C' \wedge \text{disjoint}\, ABBC \wedge \text{disjoint}\, A'B'B'C' \wedge AB \cong A'B' \wedge BC \cong B'C' \Rightarrow AC \cong A'C'$ |
| IV-b | $\neg \text{Col}\, ABC \Rightarrow \angle ABC \cong \angle ABC$ |
| IV-c | $\neg \text{Col}\, ABC \Rightarrow \angle ABC \cong \angle CBA$ |
| IV-d | $\angle ABC \cong \angle DEF \Rightarrow \angle CBA \cong \angle FED$ |
| Def | same side A B l := $\exists P$ cut $lAP \wedge$ cut $lBP$ |
| Def | same side' A B X Y := $X \neq Y \wedge \forall l\ X \in l \wedge Y \in l \Rightarrow$ same side $ABl$ |
| Def | $B \in \overrightarrow{PA} := P \dot{-} A \dot{-} B \vee P \dot{-} B \dot{-} A \vee (P \neq A \wedge A = B)$ |
| IV-e | $\angle ABC \cong \angle DEF \wedge A' \in \overrightarrow{BA} \wedge C' \in \overrightarrow{BC} \wedge D' \in \overrightarrow{ED} \wedge F' \in \overrightarrow{EF} \Rightarrow \angle A'BC' \cong \angle D'EF'$ |
| IV 4.1 | $\neg \text{Col}\, POX \wedge \neg \text{Col}\, ABC \Rightarrow \exists Y (\angle ABC \cong \angle XOY \wedge$ same side' $PYOX)$ |
| IV 4.2 | $\neg \text{Col}\, POX \wedge \neg \text{Col}\, ABC \wedge \angle ABC \cong \angle XOY \wedge \angle ABC \cong \angle XOY' \wedge$ same side' $PYOX \wedge$ same side' $PY'OX \Rightarrow Y' \in \overrightarrow{OY}$ |
| IV 5 | $\neg \text{Col}\, ABC \wedge \neg \text{Col}\, A'B'C' \wedge AB \cong A'B' \wedge AC \cong A'C' \wedge \angle BAC \cong \angle B'A'C' \Rightarrow \angle ABC \cong \angle A'B'C'$ |

The fourth group of axioms consists of a single axiom, Playfair's version of Euclid's fifth postulate which states that there is at most one parallel line to a given line through a given point.

Parallels Axiom

| | |
|---|---|
| Def | $l \parallel m := \neg \exists X\ (X \in l \wedge X \in m)$ |
| IV I | $\forall l P m_1 m_2\ \neg P \in l \wedge l \parallel m_1 \wedge P \in m_1 \wedge l \parallel m_2 \wedge P \in m_2 \Rightarrow m_1 = m_2$ |

**Formalization of Hilbert's *Gründlagen der Geometrie*** Dehlinger, Dufourd, and Schreck have studied the formalization of Hilbert's foundations of geometry in the intuitionist setting of Coq [71]. In this context, they highlighted that Hilbert's proofs require the excluded middle axiom applied to the equality between points. They focus on the first two groups of axioms and prove some betweenness properties. Meikle and Fleuriot formalized the first three groups of axioms in their three-dimensional version within the Isabelle/HOL proof assistant [157]. They went up to the twelfth theorem of Hilbert's book[‖]. Scott continued that formalization and revised it [202]. He corrected some "subtle errors in the formalization of Group III" such as the definition of a point being inside an angle, which led to the set of points inside an angle being always empty. But Scott's definition, which states that a point $P$ is inside an angle if there are two points $A$ and $B$ on the sides of the angle such that $P$ belongs to segment $AB$, is also flawed in the sense that it is only valid in Euclidean geometry since this property is equivalent to the parallel postulate [23]. A more general definition is used by Schwabäuser et al. [201] (see Section 2.3.1.1). In further work, this time using HOL Light, Scott and Fleuriot developed a discovery system that attempted to fill gaps in Hilbert's incidence-related proofs automatically [203].

Independently, Richter also formalized a substantial number of results based on Hilbert's axioms and a metric axiom system within HOL Light [193] while Braun et al. have formalized the fragment of [124] necessary to prove Tarski's axioms [32].

[‖]We use the numbering of theorems from the tenth edition.

#### 2.3.1.2 Tarski's Geometry

Szmielew provided a rigorous development of geometry based on Tarski's axioms, which was later extended by Schwabäuser et al. [201]. The proofs presented there include the important and nontrivial proofs obtained by Gupta that the midpoint can be constructed without any continuity axiom [113]. Tarski's axiom system has evolved over time and a comprehensive description of its history can be found elsewhere [214].

**Tarski's axiom system** Tarski's axiom system can be formulated in an unsorted first-order language (with all individual variables interpreted as points). It involves only two nonlogical predicate symbols: one for nonstrict betweenness (denoted by $A–B–C$) and one for congruence (denoted by $AB \cong CD$). The axioms are formulated as follows:

| | | |
|---|---|---|
| $A_1$ | Symmetry | $AB \cong BA$ |
| $A_2$ | Pseudo-Transitivity | $AB \cong CD \wedge AB \cong EF \Rightarrow CD \cong EF$ |
| $A_3$ | Cong Identity | $AB \cong CC \Rightarrow A = B$ |
| $A_4$ | Segment construction | $\exists E\ A–B–E \wedge BE \cong CD$ |
| $A_5$ | Five-segment | $AB \cong A'B' \wedge BC \cong B'C' \wedge$ <br> $AD \cong A'D' \wedge BD \cong B'D' \wedge$ <br> $A–B–C \wedge A'–B'–C' \wedge A \neq B \Rightarrow CD \cong C'D'$ |
| $A_6$ | Between Identity | $A–B–A \Rightarrow A = B$ |
| $A_7$ | Inner Pasch | $A–P–C \wedge B–Q–C \Rightarrow \exists X\ P–X–B \wedge Q–X–A$ |
| $A_8$ | Lower Dimension | $\exists ABC\ \neg A–B–C \wedge \neg B–C–A \wedge \neg C–A–B$ |
| $A_9$ | Upper Dimension | $AP \cong AQ \wedge BP \cong BQ \wedge CP \cong CQ \wedge P \neq Q$ <br> $\Rightarrow A–B–C \vee B–C–A \vee C–A–B.$ |
| $A_{10}$ | Parallel postulate | $\exists XY\ (A–D–T \wedge B–D–C \wedge A \neq D \Rightarrow$ <br> $A–B–X \wedge A–C–Y \wedge X–T–Y)$ |

The symmetry and transitivity axioms, i.e., $A_1$ and $A_2$, imply that equidistance is an equivalence relation. The identity axiom for equidistance ($A_3$) ensures that only degenerate line segments can be congruent to a degenerate line segment. The axiom of segment construction ($A_4$) allows the extension of a line segment by a given length. The five-segments axiom ($A_5$) is similar to the side-angle-side property, but expressed without mentioning angles, using only the betweenness and congruence relations (Figure 2.7(a)). The lengths of $\overline{AB}$, $\overline{AD}$, and $\overline{BD}$ determine the angle $\angle CBD$. The identity axiom for betweenness expresses that the only possibility to have $B$ between $A$ and $A$ is to have $A$ and $B$ equal. The inner form of Pasch's axiom ($A_7$, Figure 2.7(b)) is a variant of the axiom that Pasch introduced to repair the defects of Euclid [174]. This version of Pasch's axiom was introduced by Peano [176]. Inner Pasch's axiom is a form of the axiom that holds even in space, i.e., it does not assume a dimension axiom. The lower 2-dimensional axiom ($A_8$) asserts the existence of three noncollinear points. The upper 2-dimensional axiom ($A_9$) implies that all the points are coplanar. The version of the parallel postulate ($A_{10}$) is a statement which can be expressed easily in the language of Tarski's geometry (Figure 2.7(c)). It is equivalent to the uniqueness of parallels (Playfair's postulate) that Hilbert uses or to Euclid's fifth postulate.

**Formalization of *Metamathematische Methoden in Der Geometrie*** Narboux formalized the first eight chapters of Schwabhäuser et al.'s axiomatic development using the Coq proof assistant [162, 164]. Braun and Narboux extended this formalization up to the twelfth chapter and provided the formal proof that every model of Tarski's axiom $A_1$–$A_{10}$ is a model of Hilbert's first four groups of axioms [32]. Makarios formally verified that the Cartesian plane over $\mathbb{R}$ is a model of Tarski's axioms using the Isabelle/HOL proof assistant. He also proved that the Klein-Beltrami plane is a model of Tarski's axioms (except for the parallel postulate), thereby demonstrating the independence of the parallel postulate [149]. Furthermore, Makarios proposed a simplification to

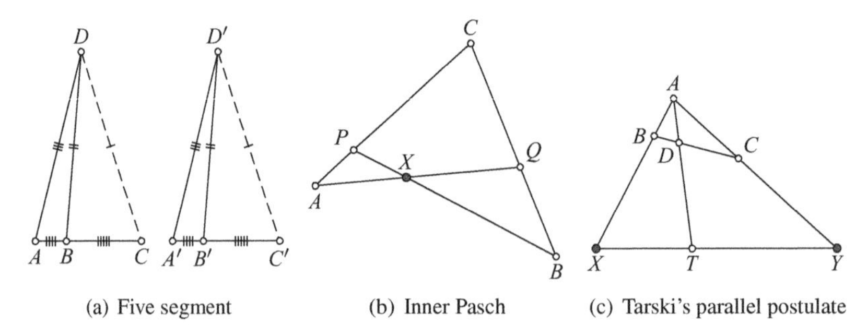

(a) Five segment (b) Inner Pasch (c) Tarski's parallel postulate

**Figure 2.7**
Illustration for some of Tarski's axioms.

Tarski's axioms: by changing the order of the points in the five-segment axiom, the pseudo reflexivity of congruence was formally derived from the other axioms [149]. In Coq, Braun and Narboux formalized two synthetic proofs of Pappus's theorem [33]: the usual version that assumes the parallel postulate and another one in neutral geometry (i.e., without assuming the parallel postulate). The proofs are formalized versions of those presented by Hilbert [124] and described in detail by Schwabhäuser et al. [201]. Having a synthetic proof is necessary because Pappus's theorem is used in the arithmetization of geometry that justifies analytic proofs. Boutry et al. have studied Tarski's geometry in the context of intuitionist logic (without the excluded middle axiom) but assuming the decidability of equality of points. They showed that the decidability of equality of points is equivalent to that of congruence and betweenness. They also proved that the decidability of the other predicates, with the exception of line intersection, can be derived from that of equality [25]. Richter et al. [108, 194] ported some of the earlier Coq proofs by Narboux to Mizar and then extended the formalization. They proved that the real line is a model of Tarskis axioms $A_1$–$A_7$ and formalized about 50 theorems by Schwabhäuser et al. [201]. Also in Mizar, Coghetto and Grabowski proved that the Cartesian plane over $\mathbb{R}$ is a model of Tarski's axiom [62], as was done by Harrison using HOL-Light.

In Isabelle/HOL, Marić and Petrović proved that the Cartesian plane over $\mathbb{R}$ is a model of both Tarski's axioms and Hilbert's axioms [177, 151]. Boutry and Cohen demonstrated that Cartesian planes over Pythagorean ordered fields (i.e., ones in which sums of squares are squares) are models of Tarski's axioms $A_1$–$A_{10}$ [20]. Moreover, Boutry et al. formalized the arithmetization of Tarski's geometry, i.e., that from any model of its planar axioms, it is possible to define a Pythagorean ordered field and associate pairs of coordinates to points [21]. This is the culminating result of both the development by Schwabhäuser et al. [201] and Hilbert's book [124]. Boutry et al. also proved the usual algebraic characterization of geometric predicates, which allows the use of algebraic methods for automatic theorem proving in a synthetic setting.

#### 2.3.1.3 Axiom Systems and Continuity Properties

Although the synthetic and analytic approaches to geometry seem quite different, Descartes proved that the second one can be derived from the first one. By defining addition, multiplication, and square roots geometrically, he introduced the arithmetization and coordinatization of geometry [74]. However, despite this analytic recasting, there are geometries that are too weak for arithmetization and, moreover, not all geometries can be reduced to algebra. Thus, having a synthetic approach to both the foundations of geometry and to automated reasoning in geometry is still vital.

Hilbert's axioms (Groups I–IV) as presented in Section 2.3.1.1 are bi-interpretable with Tarski's axioms $A_1$–$A_{10}$. The claim that there exists at least one line parallel to $l$ through $P$ is not part of these axioms: it can be derived from the other axioms. This implies that a different set of axioms is needed for the foundations of elliptic geometry.

Neutral geometry is the set of theorems which are valid both in hyperbolic and Euclidean geometry, it can defined by the axioms $A_1$–$A_9$ of Tarski or Hilbert's Groups I–III.

Our earlier presentation of Tarski and Hilbert's axiomatics did not include the continuity axioms as a large part of geometry can be built without them. We now describe some continuity properties starting with a class (Figure 2.8) that states the existence of a point of intersection between circles/lines/segments. The following three properties are equivalent assuming Tarski's axioms $A_1$–$A_9$ [122, 211]:

**Circle–Circle continuity** Given two circles $\mathscr{C}_1$ and $\mathscr{C}_2$, if there is a point on $\mathscr{C}_1$ which is inside $\mathscr{C}_2$ and vice-versa, then there is a point of intersection of the two circles.

**Line–Circle continuity** Given a line $l$ which contains a point inside the circle $\mathscr{C}$, there is a point on $l$ which is also on $\mathscr{C}$.

**Segment–Circle continuity** Given a segment $PQ$ and a circle $\mathscr{C}$, if $P$ is inside $\mathscr{C}$, and $Q$ outside then there is a point of the segment $PQ$ which is on $\mathscr{C}$.

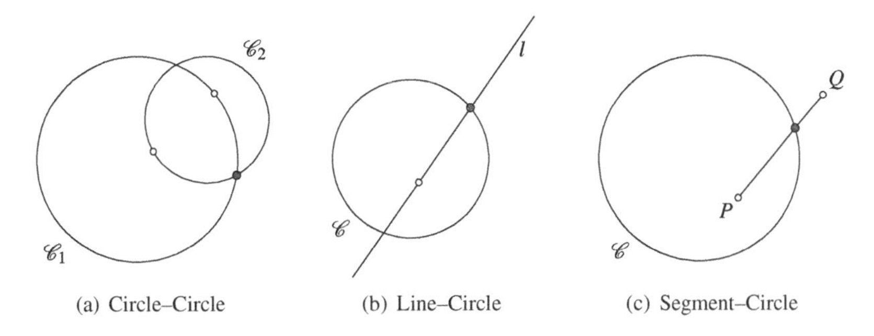


**Figure 2.8**
Elementary continuity.

$\mathscr{M}$ is a model of $A_1$–$A_{10}$, if, as mentioned in the previous section, it is isomorphic to Cartesian planes over Pythagorean fields. By adding the Circle–Circle continuity axiom, an axiomatization of ruler and compass geometry is obtained. $\mathscr{M}$ is a model of $A_1$–$A_{10}$ plus Circle–Circle if $\mathscr{M}$ is isomorphic to the Cartesian plane over a Euclidean field (positives are squares).

The continuity property ($A_{11}$) introduced by Tarski is a first-order axiom schema which restricts the Dedekind cut axiom to first-order definable cuts:

**First-Order Dedekind's Schema** $A_{11}$

$$(\exists a\, \forall xy\, (X \wedge Y \Rightarrow a\text{–}x\text{–}y)) \Rightarrow (\exists b\, \forall xy\, (X \wedge Y \Rightarrow x\text{–}b\text{–}y))$$

where $X$ and $Y$ stand for any first-order formula in the language of Tarski's axiom system which does not contain any free occurrence of $a, b, x$ in $X$ and $a, b, y$ in $Y$.

Tarski showed [213] that $\mathscr{M}$ is a model of $A_1$–$A_{11}$ if and only if it is isomorphic to the Cartesian plane over some real closed field.

The full-continuity axiom is the second-order axiom.

**Dedekind's Axiom** $A_{11}^*$

$$\forall XY\, (\exists a\, \forall xy\, (x \in X \wedge y \in Y \Rightarrow a\text{–}x\text{–}y)) \Rightarrow$$
$$(\exists b\, \forall xy\, (x \in X \wedge y \in Y \Rightarrow x\text{–}b\text{–}y))$$

The theory $A_1$–$A_{10}, A^*_{11}$ is categorical. $\mathscr{M}$ is a model of $A_1$–$A_{10}$,$A^*_{11}$ if and only if it is isomorphic to the real Cartesian plane.

Hilbert has two continuity axioms. The first one is:

**V 1: Archimedes's Axiom** Given two line segments *AB* and *CD*, there exists a number *n* such that *n* copies of segment *CD* is greater than *AB*.

This can be derived from $A^*_{11}$ but not from $A_{11}$ as it is a second order property. If $\mathscr{M}$ is a model of $A_1$–$A_{10}$ in which the geometric version of Archimedes's axiom holds then $\mathscr{M}$ is isomorphic to the Cartesian plane over an Archimedean and Pythagorean field.

Hilbert's second continuity axiom is a property about the models of the other axioms:

**V 2: Line completeness** An extension of a set of points on a line with its order and congruence relations that would preserve the relations existing among the original elements as well as the fundamental properties of line order and congruence that follows from Axioms I–III and from V 1 is impossible.

Giovanni argues that Hilbert was aware of the Dedekind continuity axiom $A^*_{11}$ and chose Axiom V 2 because it does not imply Archimedes's axiom [104]. The role of this axiom and its relationship with Archimedes's axiom is discussed in [88].

**Formalization of continuity properties and ruler and compass geometry** Duprat has proposed the formalization of several axiom systems for Euclidean plane geometry in Coq. His aim was to axiomatize a geometry where objects correspond to ruler and compass constructible ones [81, 82, 83]. He proved the corresponding axioms of Hilbert (except continuity) from his set of 20 axioms [84] with two nonlogical predicate symbols: one for ternary left-turn and one for quaternary congruence. Tarski's and Hilbert's systems can also be used to axiomatize ruler and compass geometry. Gries et al. have formalized (as part of the GeoCoq library) the equivalence between different versions of the Circle–Circle intersection axiom and also the equivalence between Circle–Line and Circle–Segment axioms. They have shown that the Circle–Line intersection can be derived from the Circle–Circle intersection and that the Circle–Circle intersection can be derived from $A_{11}$. This proof relies on the nontrivial fact that the triangular inequality can be derived without the circle continuity axiom.

### 2.3.1.4 Other Axiom Systems and Geometries

**Euclid's geometry** Avigad et al. have proposed a formal system for Euclid's Elements [4]. This includes Tarski's axioms, Circle/Line continuity, and some other axioms related to the relative position of points in the plane that can be derived from Tarski's axioms. Beeson et al. have given another formalization of Euclid's axioms and produced proofs in a forward reasoning style for all the propositions of the first book of Euclid's Elements [9]. Their proofs are checked using both HOL Light and Coq.

**Foundations of the area method** The lemmas used by the area method (Section 2.2.4.1) can be proved within, say, Hilbert's geometry (i.e., within its fragment for plane geometry), but the proofs are involved because in order to define the signed area and Pythagoras difference, arithmetization of geometry is needed. As an alternative, Chou et al. proposed an elegant higher-level axiomatization [50], suitable for the area method, later further extended and improved by Narboux, yielding an axiomatic system that includes axioms for fields and only 13 axioms (such as $\mathcal{S}_{ABC} = -\mathcal{S}_{BAC}$). Using this system, Narboux proved (formally, within Coq) all properties used by the area method, thus demonstrating the correctness of the method [130, 161]. It was also proved in a machine verifiable way that the axioms can be derived from either Guilhot's formal development [179, 181], Hilbert's or Tarski's axioms [22].

**Projective geometry** Magaud et al. have proposed a Coq formalization of projective plane geometry based on the notion of rank [147]. The correspondence between the rank axiom system and traditional axiom system was then completed and extended by Braun et al. [30, 31]. In Mizar, Coghetto has formalized a proof of Pascal's theorem [61] that is based on several articles by Leończuk and Prażmowskin [140], and by Skaba [206].

**Complex plane** Marić and Petrović have formalized the extended complex plane using the Isabelle/HOL proof assistant [150]. They defined homogeneous coordinates, stereographic projection, Möbius transformations, generalized circles and proved their properties.

#### 2.3.1.5 Meta-Theory

There has also been some mechanization efforts targeted at proving meta-theoretical results. Makarios proved the independence of the parallel postulate and the equivalence between different variants of Tarski's axiom system in Isabelle/HOL [149]. Boutry et al. formalized the equivalence of 34 versions of the parallel postulate and studied the impact of intuitionist logic on it [23]. Romanos and Paulson formalized a proof of the impossibility of trisecting the angle and doubling the cube [198] in Isabelle/HOL as did Harrison in HOL Light.

### 2.3.2 Higher Level Results

In the section, we present the formalizations of some of the best known theorems in synthetic geometry.

**Desargues' theorem** Desargues' theorem is an important result in geometry. It states that two triangles are in perspective axially if and only if they are in perspective centrally (see also Section 2.2.4.4). The theorem can be proved without any assumption in space but there are non-Desarguesian planes such as Moulton's plane. Kusak formalized the proof of Desargues' theorem in the projective space using Mizar [138]. A formalization of the same result and of Moulton's plane using the concept of rank was obtained using the Coq proof assistant by Magaud et al. [146, 147, 148]. Braun has formalized Hessenberg's proof of Desargues' theorem in the context of Tarski's geometry [201]. Desargues' theorem can be also verified by the Coq's implementation of the area method [130]. Harrison mechanized the proof of Desargues' theorem using a technique based on the computation of determinants [195].

**Morley's theorem** Morley's theorem states that the three points of intersection of the adjacent angle trisectors form an equilateral triangle. Coghetto formalized the proof of Letac [141] using the Mizar proof assistant [58]. The proof is based on trigonometry, the law of sines and the law of cosines. Harrison mechanized Connes' proof [65] in HOL Light. Guilhot formalized a proof based on a set of lemmas about trigonometric functions, while Ida et al. proposed a computational proof using origami [126, 127].

**Triangle centers** Triangle centers such as the center of gravity, the circumcenter, the orthocenter have been studied for centuries by geometers. The existence and uniqueness of the circumcenter, incenter, orthocenter, and gravity center have been formalized in Coq by Boutry et al. In particular, they examined the difference between the computer checked proof of the existence of the center of gravity and the usual proof published in textbooks [26]. The equivalence of the existence of the circumcenter with the parallel postulate was also formalized [23]. In Mizar, Coghetto has formalized the existence and uniqueness of circumcenter, orthocenter, and centroid [59, 60].

Kimberling has curated an electronic encyclopedia of triangle centers (ETC) that contains over 12,000 centers and many of their properties [134]. Recently, Narboux developed a certified version of (parts of) the ETC such that some of the properties are accompanied by a Coq-checked proof of their validity [165].

**Sum of angles of a triangle** The fact that the sum of angles of a triangle is congruent to the sum of two right angles has also been studied. Boutry et al. formalized in Coq its equivalence with the parallel postulate [23]. This required the sum of angles to be defined geometrically without using the Archimedean axiom, hence without a means of measuring angles as real numbers [111]. Guilhot proved in Coq that the measure of the sum of angles of a triangle is $\pi$ using her axiom system based on mass points and by assuming axioms similar to the protractor postulate [112]. Harrison formalized the same theorem in HOL Light using the law of sines and cosines. Eberl has ported Harrison's proof to Isabelle/HOL [87]. Mizar's formalization by Chang et al. uses complex numbers [42].

**Pythagorean theorem** The Pythagorean theorem is an important property in Euclidean geometry. Guilhot has formalized an analytic proof using Coq and the ring tactic [112]. A formal proof can also be obtained in one line using algebraic techniques such as Gröbner basis or Wu's method as implemented in Coq, Isabelle/HOL and HOL Light. The Mizar proof by Kubo and Nakamura uses the complex numbers [42]. Braun et al. have formalized a synthetic proof starting from Tarski's axioms [22]. Having a synthetic proof of the Pythagorean theorem is necessary as the latter is used in the justification of the analytic approach that relies on the arithmetization of geometry.

**Heron's formula** The formalization in Mizar [192] uses a coordinate-based definition of the area of a triangle. The HOL Light formalization by Harrison uses a more general definition of area based on measure theory. There are (at least) two Coq's formalization. The one by Narboux is based on the axioms of the area method (see Section 2.2.4.1) while Guilhot has one derived from her axioms about mass points.

**Thales' circle theorem** Braun and Magaud have formalized several analytic and synthetic proofs of this theorem and of its reciprocal in Coq [29]. Boutry et al. have formalized its equivalence to the parallel postulate [23] and the proof can also be obtained using algebraic methods in HOL Light and in Coq (see Section 2.3.4).

**Intersecting chord and Ptolemy's theorems** Pham proposed a formalization in Coq based on Guilhot's axiom system [178]. Bulwahn formalized those theorems in the real Cartesian plane using Isabelle/HOL [37, 38]. Pottier et al. proved a version of these theorems not involving betweenness and square of distances using Gröbner basis [110]. Harrison verified these theorems using algebraic techniques in HOL Light and, as for the Pythagorean theorem, the Mizar formalization used complex numbers [192].

### 2.3.3 Other Formalizations Related to Geometry

In this section, we briefly describe some important formalization efforts which are not in the field of synthetic geometry, but are related to geometry.

**$\pi$ and the area of the circle** The number $\pi$ can be given by many different definitions. Bertot and Allais prove formally using the Coq proof assistant that several definitions are equivalent [13]. The cosine function is defined as a power series and $\pi$ is then defined as twice the unique value $\alpha$ between 1 and 2 such that $\cos(\alpha) = 0$. They show that this definition is equivalent to a definition based on the area of a unit circle using Riemann integral. They also proved that Archimedes's process to compute $\pi$ is also equivalent. The perimeter of the circle can be seen as the limit of a sequence of inscribed regular polygons. Finally, they also provide means to actually compute approximations of $\pi$. The area of a circle has been derived by Eberl in Isabelle/HOL (via integration by substitution) and Harrison in HOL Light. In both cases, this involves a definition of area based on measure theory.

**Euler's formula** Euler's polyhedron formula asserts that $V - E + F = 2$ where $V$, $E$, and $F$ are the number of vertices, edges, and faces of polyhedron, respectively. This formula is interesting from the formalization point of view because of its long history of proofs and refutations [139]. Alama formalized Poincaré's proof using a definition of polyhedrons based on incidence matrices [1]. Dufourd formalized an intuitionist proof in Coq involving polyhedron defined using hyper-maps [78].

**The Jordan Curve Theorem** The Jordan Curve Theorem is a well-known example of a statement that looks trivial but is indeed very hard to prove rigorously. The Mizar formalization started in 1992 and was completed in September 2005 [137]. Hales completed the formalization of the theorem in HOL Light in January 2005 [117]. Scott and Fleuriot formalized a version of the theorem for polygons using only Hilbert's axioms Group I and II, i.e., with only the weak notions of incidence and order [205]. Dufourd formalized a discrete version using hyper-maps [79].

**Four color theorem** The incorrect proof by Kempe [133] and the fact that Appel and Haken's proof [3] involved ad-hoc computer programs made the four color theorem a good candidate for formalization. Gonthier produced a formal proof using hyper-maps [106, 107]. He used a version of Euler's formula to define planar maps and proved a combinatorial version of the Jordan Curve Theorem equivalent to his version the of Euler formula.

**The Kepler conjecture** The largest formalization effort related to geometry is that of Ferguson and Hales' proof of the Kepler conjecture. Their pen-and-paper proof was completed in 1998 but not published in full until 2006, with the editor adding as a side note that the reviewers were only "99% certain" of the correctness of the proof. This compelled Hales to start the Flyspeck formalization project [116], with the aim of providing a fully verified proof in HOL Light. During the formalization, which involved a large international collaboration and the formalization of some early results in Isabelle/HOL, a few minor errors and incomplete arguments were uncovered [118]. Hales published a book describing the formalized version of the proof [114], while the final, fully checked result was finally achieved using HOL Light in August 2014 [115].

**Nonstandard analysis** Fleuriot carried out the formalization of geometric proofs from Newton's Principia using nonstandard analysis in Isabelle/HOL [92]. By combining synthetic proofs based on the area method (see Section 2.2.4.1) and full-angles (see Section 2.2.4.2) with a rigorous treatment of infinitesimals [91], he formalized Kepler's law of Equal Areas and discovered a flaw in Newton's famous Kepler Proposition [94]. Magaud, Cholet, and Fuchs carried out formalization of the discrete model of the continuum introduced by Harthong and Reeb in Coq [145] and Fleuriot performed a similar study using Isabelle/HOL [93].

**Origami** Origami, the art of paper folding, provides geometrical tools allowing some constructions that cannot be realized by ruler and compass. Ida et al. have studied the mathematical and computational aspect of origami. Kaliszyk and Ida have also investigated how the Mathematica proofs based on Gröbner basis and Cylindrical Algebraic Decomposition can be reproduced in Isabelle/HOL, HOL Light, and Coq [132]. Despite the limitations of proof assistants when it comes to algebraic reasoning, such combinations of interactive and automatic theorem proving can be useful for the mechanized verification of theorems about origami constructions [125].

**Computational and combinatorial geometry** Dufourd and Puitg were the first to use a proof assistant to formalize geometric data structures [187, 188]. Pichardie and Bertot proved a convex hull algorithm using the Coq proof assistant [182]. They show that Knuth's counterclockwise systems (CCS) hold in $\mathbb{R}^2$ and then verify an incremental algorithm for computing convex hulls. Dehlinger and Dufourd defined a formal specification of generalized maps in Coq and proved the trading theorem [70]. Dufourd also developed a Coq formalization of hyper-maps [77] and applied it to the verification of an image segmentation algorithm [75, 76]. Meikle and Fleuriot verified in Isabelle/HOL the Graham Scan convex hull algorithm using an embedded While language, Hoare logic, and the properties of signed areas formalized via Knuth's CCS [158]. Brun et al. formalized the same algorithm using hyper-maps [35]. They also have derived an imperative program [34]. Dufourd and Bertot applied these formalizations to the proof of a plane Delaunay triangulation algorithm [80]. Brandt proposed the use of three-valued logic to simplify the formalization of degenerated conditions and apply its method to formalize a convex-hull algorithm [27, 28]. Fuchs and Théry have formalized Grassmann-Cayley algebra in Coq [95]. Ma et al. have used geometric algebra to formalize robotic manipulation algorithms [144]. Liu et al. have used Coq to verify an aircraft proximity property [143].

### 2.3.4 Verified Automated Reasoning

Proof assistants are designed for checking proofs generated by the user, but it is also possible to check proofs generated by automated theorem provers. The latter can be used to prove theorems wholesale or, at a smaller scale, to automate parts of proofs which are produced interactively. There are several ways to integrate an ATP in a proof assistant. It is possible to define the ATP inside the proof assistant and prove its correctness; this is called the reflexive approach [12]. It is also possible to rely on external tools that generate a proof which is then checked by the proof assistant. Instead of exporting the whole proof it is also possible to export a certificate, i.e., a piece of information that is sufficient to reconstruct a proof efficiently inside the proof assistant. For example, to prove that a polynomial is not irreducible, an oracle can use a complex factorization algorithm to compute the factors. The proof assistant can then use a simple algorithm to check that the factorization is correct [73].

Narboux has used the tactic language of Coq [72] to implement the area method in Coq (see Section 2.2.4.1) [130, 161]. Grégoire et al. have used this approach to bring the power of Gröbner basis algorithms to the Coq proof assistant [110, 186]. Genevaux et al. then extended their work to an implementation of Wu's method [102]. Chaieb et al. integrated quantifier elimination and Gröbner basis within Isabelle/HOL [40, 41] and applied these to some geometric examples. Petrović proposed an Isabelle/HOL formalization of the concept of geometric construction, with an algebraic semantics [177]. Harrison used algebraic methods to check some examples in geometry using HOL Light [120, 121].

At a small scale level, Boutry et al. proposed a reflexive approach to prove some transitivity properties and deals with some symmetries implicitly [24, 26] generalizing the theory $E_L$ that Balbiani et al. studied using term rewriting techniques [5]. The example shown in Figure 2.9 uses the tactics described by Boutry et al. [26]. Scott and Fleuriot introduced a generic combinator language that was then instantiated to give automatic forward proof strategies that could deal with incidence reasoning in Hilbert's proof, especially in cases involving Pasch's axiom [204]. As pointed out in Section 2.2.3.2, Stojanović et al. generated readable proofs (of theorems from Tarski's geometry) which are checked by Coq and Isabelle/HOL [208].

**Acknowledgment** The authors are grateful to Michael Beeson, Roland Cogheto, Pedro Quaresma, Predrag Tanović, Dongming Wang, and the anonymous reviewer for their comments on preliminary versions of this chapter. Janičić was partly supported by grant ON174021 of the Ministry of Science of Serbia. Fleuriot was supported by EPSRC grants EP/L011794/1 and EP/N014758/1.

```
Lemma varignon : forall A B C D I J K L, A<>C -> B<>D -> ~ Col I J K ->
 Midpoint I A B -> Midpoint J B C -> Midpoint K C D -> Midpoint L A D ->
 Parallelogram I J K L.
Proof.
intros.
assert_diffs. (* Deduce some inequalities *)
assert (Par I L B D) by perm_apply (triangle_mid_par B D A L I).
(* Applying the midpoint theorem in the triangle BDA. *)
assert (Par J K B D) by perm_apply (triangle_mid_par B D C K J).
(* Applying the midpoint theorem in the triangle BDC. *)
assert (Par I L J K) by (apply par_trans with B D;finish).
(* Transitivity of parallelism *)
assert (Par I J A C) by perm_apply (triangle_mid_par A C B J I).
(* Applying the midpoint theorem in the triangle ACB. *)
assert (Par L K A C) by perm_apply (triangle_mid_par A C D K L).
(* Applying the midpoint theorem in the triangle ACD. *)
assert (Par I J K L) by (apply par_trans with A C;finish).
(* Transitivity of parallelism *)
apply par_2_plg;finish.
(* If in the opposite sides of a quadrilatral are parallel and
two opposite side are distinct then it is a parallelogram. *)
Qed.
```

**Figure 2.9**
Coq script of the proof of Varignon's theorem based on some tactics for partial automation.

# References


[1] Jesse Alama. Euler's polyhedron formula. *Formalized Mathematics*, 16(1):7–17, 2008.

[2] J. R. Anderson, C. F. Boyle, and G. Yost. The geometry tutor. *The Journal of Mathematical Behavior*, pages 5–20, 1986.

[3] K. Appel and W. Haken. Every planar map is four colorable. Part I: Discharging. *Illinois Journal of Mathematics*, 21(3):429–490, 1977.

[4] Jeremy Avigad, Edward Dean, and John Mumma. A Formal System for Euclid's Elements. *The Review of Symbolic Logic*, 2:700–768, 2009.

[5] Philippe Balbiani and Luis del Cerro. Affine geometry of collinearity and conditional term rewriting. In Hubert Comon and Jean-Pierre Jounnaud, editors, *Term Rewriting*, volume 909 of *Lecture Notes in Computer Science*, pages 196–213. Springer Berlin / Heidelberg, 1995. 10.1007/3-540-59340-3_14.

[6] Michael Beeson. Foreword to Metamathematische Methoden in der Geometrie. Ishi Press, 2011.

[7] Michael Beeson. A constructive version of Tarski's geometry. *Annals of Pure and Applied Logic*, 166(11):1199–1273, 2015.

[8] Michael Beeson. Constructive geometry and the parallel postulate. *Bulletin of Symbolic Logic*, 22(1):1–104, 2016.

[9] Michael Beeson, Julien Narboux, and Freek Wiedijk. Proof-checking Euclid. 2017.

[10] Michael Beeson and Larry Wos. OTTER proofs of theorems in Tarskian geometry. In Stéphane Demri, Deepak Kapur, and Christoph Weidenbach, editors, *7th International Joint Conference, IJCAR 2014, Held as Part of the Vienna Summer of Logic, Vienna, Austria, July 19-22, 2014, Proceedings*, volume 8562 of *Lecture Notes in Computer Science*, pages 495–510. Springer, 2014.

[11] Michael Beeson and Larry Wos. Finding Proofs in Tarskian Geometry. *Journal of Automated Reasoning*, submitted, 2015.

[12] Yves Bertot. Coq in a Hurry. May 2010.

[13] Yves Bertot and Guillaume Allais. Views of PI: Definition and computation. *Journal of Formalized Reasoning*, 7(1):105–129, 2014.

[14] Yves Bertot and Pierre Castéran. *Interactive Theorem Proving and Program Development, Coq'Art: The Calculus of Inductive Constructions*. Texts in Theoretical Computer Science. An EATCS Series. Springer, 2004.

[15] Marc Bezem and Thierry Coquand. Automating Coherent Logic. In Geoff Sutcliffe and Andrei Voronkov, editors, *12th International Conference on Logic for Programming, Artificial Intelligence, and Reasoning — LPAR 2005*, volume 3835 of *Lecture Notes in Computer Science*, pages 246–260. Springer-Verlag, 2005.

[16] Marc Bezem and Dimitri Hendriks. On the Mechanization of the Proof of Hessenberg's Theorem in Coherent Logic. *Journal of Automated Reasoning*, 40(1):61–85, 2008.

[17] George D. Birkhoff. A set of postulates for plane geometry, based on scale and protractor. *Annals of Mathematics*, 33:329–345, 1932.

[18] Karol Borsuk and Wanda Szmielew. *Foundations of Geometry: Euclidean and Bolyai-Lobachevskian geometry, Projective Geometry*. North-Holland Pub Co, 1960.

[19] Francisco Botana, Markus Hohenwarter, Predrag Janičić, Zoltán Kovács, Ivan Petrović, Tomás Recio, and Simon Weitzhofer. Automated Theorem Proving in GeoGebra: Current Achievements. *Journal of Automated Reasoning*, 55(1):39–59, 2015.

[20] Pierre Boutry. *On the Formalization of Foundations of Geometry*. PhD thesis, University of Strasbourg. in preparation.

[21] Pierre Boutry, Gabriel Braun, and Julien Narboux. From Tarski to Descartes: Formalization of the Arithmetization of Euclidean Geometry. In *The 7th International Symposium on Symbolic Computation in Software (SCSS 2016)*, EasyChair Proceedings in Computing, page 15, Tokyo, Japan, March 2016.

[22] Pierre Boutry, Gabriel Braun, and Julien Narboux. Formalization of the Arithmetization of Euclidean Plane Geometry and Applications. *Journal of Symbolic Computation*, page 23, 2017. In Press.

[23] Pierre Boutry, Charly Gries, Julien Narboux, and Pascal Schreck. Parallel postulates and continuity axioms: a mechanized study in intuitionistic logic using Coq. *Journal of Automated Reasoning*, page 72, 2017. to appear.

[24] Pierre Boutry, Julien Narboux, and Pascal Schreck. A reflexive tactic for automated generation of proofs of incidence to an affine variety. October 2015.

[25] Pierre Boutry, Julien Narboux, Pascal Schreck, and Gabriel Braun. A short note about case distinctions in Tarski's geometry. In Francisco Botana and Pedro Quaresma, editors, *Proceedings of the 10th Int. Workshop on Automated Deduction in Geometry*, volume TR 2014/01 of *Proceedings of ADG 2014*, pages 51–65, Coimbra, Portugal, July 2014. University of Coimbra.

[26] Pierre Boutry, Julien Narboux, Pascal Schreck, and Gabriel Braun. Using small scale automation to improve both accessibility and readability of formal proofs in geometry. In Francisco Botana and Pedro Quaresma, editors, *Proceedings of the 10th Int. Workshop on*

*Automated Deduction in Geometry*, volume TR 2014/01 of *Proceedings of ADG 2014*, pages 31–49, Coimbra, Portugal, July 2014. University of Coimbra.

[27] Jens Brandt. *A Layered Approach to Polygon Processing for Safety-Critical Embedded Systems*. PhD thesis, Universität Kaiserslautern, 2007.

[28] Jens Brandt and Klaus Schneider. Dependable Polygon-Processing Algorithms for Safety-Critical Embedded Systems. In Laurence T. Yang, Makoto Amamiya, Zhen Liu, Minyi Guo, and Franz J. Rammig, editors, *Embedded and Ubiquitous Computing – EUC 2005*, number 3824 in Lecture Notes in Computer Science, pages 405–417. Springer Berlin Heidelberg, December 2005. DOI: 10.1007/11596356_42.

[29] David Braun and Nicolas Magaud. Des preuves formelles en Coq du théorème de Thalès pour les cercles. In David Baelde and Jade Alglave, editors, *Vingt-sixièmes Journées Francophones des Langages Applicatifs (JFLA 2015)*, Le Val d'Ajol, France, January 2015.

[30] David Braun, Nicolas Magaud, and Pascal Schreck. An Equivalence Proof Between Rank Theory and Incidence Projective Geometry. In *Proceedings of Automated Deduction in Geometry*, pages 62–77, Strasbourg, 2016.

[31] David Braun, Nicolas Magaud, and Pascal Schreck. Formalizing Finite Models of Projective Geometry in Coq: a Benchmark for Proof Automation. 2017.

[32] Gabriel Braun and Julien Narboux. From Tarski to Hilbert. In Tetsuo Ida and Jacques Fleuriot, editors, *Post-Proceedings of Automated Deduction in Geometry 2012*, volume 7993 of *LNCS*, pages 89–109, Edinburgh, United Kingdom, September 2012. Springer.

[33] Gabriel Braun and Julien Narboux. A synthetic proof of Pappus' theorem in Tarski's geometry. *Journal of Automated Reasoning*, 58(2):23, February 2017.

[34] Christophe Brun, Jean-François Dufourd, and Nicolas Magaud. Designing and proving correct a convex hull algorithm with hypermaps in Coq. *Computational Geometry*, 45(8):436–457, 2012.

[35] Christophe Brun, Jean-François Dufourd, and Nicolas Magaud. Formal Proof in Coq and Derivation of an Imperative Program to Compute Convex Hulls. In Tetsuo Ida and Jacques D. Fleuriot, editors, *Automated Deduction in Geometry*, number 7993 in Lecture Notes in Computer Science, pages 71–88. Springer Berlin Heidelberg, 2012. DOI: 10.1007/978-3-642-40672-0_6.

[36] Bruno Buchberger, Adrian Crăciun, Tudor Jebelean, Laura Kovács, Temur Kutsia, Koji Nakagawa, Florina Piroi, Nikolaj Popov, Judit Robu, Markus Rosenkranz, and others. Theorema: Towards Computer-Aided Mathematical Theory Exploration. *Journal of Applied Logic*, 4(4):470–504, 2006.

[37] Lukas Bulwahn. Intersecting Chords Theorem. *Archive of Formal Proofs*, October 2016.

[38] Lukas Bulwahn. Ptolemy's Theorem. *Archive of Formal Proofs*, August 2016.

[39] Ugo Cassina. Ancora sui fondamenti della geometria secondo Hilbert. I, II, III. *Istituto Lombardo di Scienze e Lettere, Rendiconti, Classe di Scienze Matematiche e Naturali*, 12(81):71–94, 1948.

[40] Amine Chaieb and Tobias Nipkow. Verifying and reflecting quantifier elimination for Presburger arithmetic. In G. Sutcliffe and A. Voronkov, editors, *Logic for Programming, Artificial Intelligence, and Reasoning (LPAR 2005)*, volume 3835 of *LNCS*, pages 367–380. Springer, 2005.

[41] Amine Chaieb and Makarius Wenzel. Context aware Calculation and Deduction — Ring Equalities via Gröbner Bases in Isabelle. In M. Kauers, M. Kerber, R. Miner, and W. Windsteiger, editors, *CALCULEMUS 2007*, volume 4573 of *Lecture Notes in Computer Science*, pages 27–39. Springer, 2007.

[42] Wenpai Chang, Yatsuka Nakamura, and Piotr Rudnicki. Inner products and angles of complex numbers. *Formalized Mathematics*, 11(3):275–280, 2003.

[43] Xiaoyu Chen and Dongming Wang. Formalization and Specification of Geometric Knowledge Objects. *Mathematics in Computer Science*, 7(4):439–454, December 2013.

[44] XueFeng Chen and DingKang Wang. The Projection of Quasi Variety and Its Application on Geometric Theorem Proving and Formula Deduction. In Franz Winkler, editor, *Automated Deduction in Geometry, 4th International Workshop, ADG 2002, Hagenberg Castle, Austria, September 4-6, 2002, Revised Papers*, volume 2930 of *Lecture Notes in Computer Science*, pages 21–30. Springer, 2004.

[45] Shang-Ching Chou. *Proving and discovering geometry theorems using Wu's method.* PhD thesis, The University of Texas, Austin, December 1985.

[46] Shang-Ching Chou and Xiao-Shan Gao. A class of geometry statements of constructive type and geometry theorem proving. In *Proceeding of CADE 92*. Academia Sinica, 1992.

[47] Shang-Ching Chou and Xiao-Shan Gao. Automated Generation of Readable Proofs with Geometric Invariants I. Multiple and Shortest Proof Generation. *J. Autom. Reasoning*, 17(3):325–347, 1996.

[48] Shang-Ching Chou and Xiao-Shan Gao. Automated Reasoning in Geometry. In John Alan Robinson and Andrei Voronkov, editors, *Handbook of Automated Reasoning*, pages 707–749. Elsevier and MIT Press, 2001.

[49] Shang-Ching Chou, Xiao-Shan Gao, and Jing-Zhong Zhang. Automated Geometry Theorem Proving by Vector Calculation. In Manuel Bronstein, editor, *Proceedings of the 1993 International Symposium on Symbolic and Algebraic Computation, ISSAC '93*, pages 284–291. ACM, 1993.

[50] Shang-Ching Chou, Xiao-Shan Gao, and Jing-Zhong Zhang. Automated production of traditional proofs for constructive geometry theorems. In Moshe Vardi, editor, *Proceedings of the Eighth Annual IEEE Symposium on Logic in Computer Science LICS*, pages 48–56. IEEE Computer Society Press, June 1993.

[51] Shang-Ching Chou, Xiao-Shan Gao, and Jing-Zhong Zhang. *Machine Proofs in Geometry.* World Scientific, Singapore, 1994.

[52] Shang-Ching Chou, Xiao-Shan Gao, and Jing-Zhong Zhang. Automated Production of Traditional Proofs in Solid Geometry. *Journal of Automated Reasoning*, 14:257–291, 1995.

[53] Shang-Ching Chou, Xiao-Shan Gao, and Jing-Zhong Zhang. Automated Generation of readable proofs with geometric invariants, Theorem Proving with Full Angle. *Journal of Automated Reasoning*, 17:325–347, 1996.

[54] Shang-ching Chou, Xiao-shan Gao, and Jing-zhong Zhang. A Deductive Database Approach to Automated Geometry Theorem Proving and Discovering. *Journal of Automated Reasoning*, 25:219–246, 2000.

[55] Chou Shang-Ching and Gao Xiao-Shan. A Survey of Geometric Reasoning Using Algebraic Methods. pages 97–119. Birkhäuser Boston, Boston, MA, 1996. DOI: 10.1007/978-1-4612-4088-4_5.

[56] H. Coelho and L. M. Pereira. GEOM: A Prolog Geometry Theorem Prover. Memórias 525, Laboratório Nacional de Engenharia Civil, Ministério de Habitação e Obras Públicas, Portugal, 1979.

[57] H. Coelho and L. M. Pereira. Automated reasoning in geometry theorem proving with Prolog. *Journal of Automated Reasoning*, 2(4):329–390, 1986.

[58] Roland Coghetto. Morley's Trisector Theorem. *Formalized Mathematics*, 23(2):75–79, 2015.

[59] Roland Coghetto. Altitude, Orthocenter of a Triangle and Triangulation. *Formalized Mathematics*, 24(1):27–36, 2016.

[60] Roland Coghetto. Circumcenter, circumcircle and centroid of a triangle. *Formalized Mathematics*, 24(1):17–26, 2016.

[61] Roland Coghetto. Pascal's Theorem in Real Projective Plane. 25:110, 2017.

[62] Roland Coghetto and Adam Grabowski. Tarski Geometry Axioms — Part II. *Formalized Mathematics*, 24:157–166, 2016.

[63] George E. Collins. Quantifier Elimination for Real Closed Fields by Cylindrical Algebraic Decomposition. volume 33 of *Lecture Notes In Computer Science*, pages 134–183. Springer-Verlag, 1975.

[64] George E. Collins. Quantifier Elimination for Real Closed Fields by Cylindrical Algebraic Decomposition: a synopsis. *SIGSAM Bull.*, 10(1):10–12, February 1976.

[65] Alain Connes. A new proof of Morley's theorem. *Publications Mathématiques de l'IHÉS*, 88:43–46, 1998.

[66] Thierry Coquand and Gérard Huet. The Calculus of Constructions. In *Information and Computation*, volume 76. 1988.

[67] Thierry Coquand and Christine Paulin-Mohring. Inductively defined types. In P. Martin-Löf and G. Mints, editors, *Proceedings of Colog'88*, volume 417 of *Lecture Notes in Computer Science*. Springer-Verlag, 1990.

[68] H. S. M. Coxeter. *Introduction to Geometry*. Wiley, 2nd edition, 1969.

[69] Li Dafa, Peifa Jia, and Xinxin Li. Simplifying von Plato's Axiomatization of Constructive Apartness Geometry. In *Annals of Pure and Applied Logic*, volume 102, pages 1–26. 2000.

[70] Christophe Dehlinger and Jean-François Dufourd. Formalizing generalized maps in Coq. *Theoretical Computer Science*, 323(1-3):351–397, 2004.

[71] Christophe Dehlinger, Jean-François Dufourd, and Pascal Schreck. Higher-Order Intuitionistic Formalization and Proofs in Hilbert's Elementary Geometry. In *Automated Deduction in Geometry*, volume 2061 of *Lecture Notes in Computer Science*, pages 306–324. Springer, 2001.

[72] David Delahaye. A Tactic Language for the System Coq. In *Proceedings of Logic for Programming and Automated Reasoning (LPAR), Reunion Island (France)*, volume 1955 of *Lecture Notes in Artificial Intelligence*, pages 85–95. Springer-Verlag, November 2000.

[73] David Delahaye and Micaela Mayero. Dealing with Algebraic Expressions over a Field in Coq using Maple. In *Journal of Symbolic Computation: special issue on the integration of automated reasoning and computer algebra systems*, volume 39, pages 569–592, 2005.

[74] René Descartes. *La géométrie*. Open Court, Chicago, 1925.

[75] Jean-François Dufourd. Définition et preuve d'un algorithme fonctionnel de segmentation d'image basé sur les hypercartes. In *Proceedings of JFLA 2007*, Aix-les-Bains, January 2007.

[76] Jean-François Dufourd. Design and formal proof of a new optimal image segmentation program with hypermaps. *Pattern Recognition*, 2007.

[77] Jean-François Dufourd. A hypermap framework for computer assisted proofs in surface subdivisions. In *Symposium on Applied Computing*, Seoul, March 2007.

[78] Jean-François Dufourd. Polyhedra Genus Theorem and Euler Formula: A Hypermap-formalized Intuitionistic Proof. *Theor. Comput. Sci.*, 403(2-3):133–159, August 2008.

[79] Jean-François Dufourd. An intuitionistic proof of a discrete form of the Jordan curve theorem formalized in Coq with combinatorial hypermaps. *Journal of Automated Reasoning*, 43(1):19–51, 2009.

[80] Jean-François Dufourd and Yves Bertot. Formal study of plane Delaunay triangulation. In Lawrence Paulson and Matt Kaufmann, editors, *Interactive Theorem Proving*, volume 6172, pages 211–226, Edinburgh, United Kingdom, July 2010. Springer.

[81] Jean Duprat. Proof assistant for plane geometry. In *Workshop on Lambda Calculus, Type Theory and Mathematical Logic*, volume 12 of *Schedea Informatica*, Cracovie, June 2003.

[82] Jean Duprat. Une axiomatique de la géométrie plane en Coq. In *JFLA (Journées Francophones des Langages Applicatifs)*, pages 123–136, Etretat, France, January 2008. INRIA.

[83] Jean Duprat. The Euclid's Plane: Formalization and Implementation in Coq. In *ADG 2010*, 2010.

[84] Jean Duprat. Fondements de géométrie euclidienne. 2010.

[85] Roy Dyckhoff and Sara Negri. Geometrization of first-order logic. *The Bulletin of Symbolic Logic*, 21:123–163, 2015.

[86] Jerzy Dydak. Axiomatization of geometry employing group actions. January 2015.

[87] Manuel Eberl. Basic Geometric Properties of Triangles. *Archive of Formal Proofs*, December 2015.

[88] Philip Ehrlich. From Completeness to Archimedean Completeness: An Essay in the Foundations of Euclidean Geometry. *A Symposium on David Hilbert*, 110:57–76, 1997.

[89] E. W. Elcock. Representation of knowledge in geometry machine. *Machine Intelligence*, 8:11–29, 1977.

[90] Alfredo Ferro and Giovanni Gallo. Automated theorem proving in elementary geometry. *Le Matematiche*, 43(1,2):195–224, 1990.

[91] Jacques D. Fleuriot. Nonstandard Geometric Proofs. In *Automated Deduction in Geometry, Third International Workshop, ADG 2000, Zurich, Switzerland, September 25-27, 2000, Revised Papers*, pages 246–267, 2000.

[92] Jacques D. Fleuriot. Theorem Proving in Infinitesimal Geometry. *Logic Journal of the IGPL*, 9(3):447–474, 2001.

[93] Jacques D. Fleuriot. Exploring the foundations of discrete analytical geometry in Isabelle/HOL. In *International Workshop on Automated Deduction in Geometry*, volume 6877 of *LNCS*, pages 34–50. Springer, 2010.

[94] Jacques D. Fleuriot and Lawrence C. Paulson. Proving Newton's Propositio Kepleriana Using Geometry and Nonstandard Analysis in Isabelle. In *Automated Deduction in Geometry, Second International Workshop, ADG'98, Beijing, China, August 1-3, 1998, Proceedings*, pages 47–66, 1998.

[95] Laurent Fuchs and Laurent Théry. A Formalization of Grassmann-Cayley Algebra in COQ and Its Application to Theorem Proving in Projective Geometry. In Pascal Schreck, Julien Narboux, and Jürgen Richter-Gebert, editors, *Automated Deduction in Geometry*, volume 6877 of *Lecture Notes in Computer Science*, pages 51–67. Springer Berlin Heidelberg, 2011.

[96] Herve Gallaire, Jack Minker, and Jean-Marie Nicolas. Logic and Databases: A Deductive Approach. *ACM Comput. Surv.*, 16(2):153–185, June 1984.

[97] Xiao-Shan Gao. Chapter 10 — Search methods revisited. In Xiao-Shan Gao and Dongming Wang, editors, *Mathematics Mechanization and Applications*, pages 253–271. Academic Press, London, 2000. DOI: 10.1016/B978-012734760-8/50011-9.

[98] Xiao-Shan Gao and Qiang Lin. MMP/Geometer — A Software Package for Automated Geometric Reasoning. In *Proceedings of Automated Deduction in Geometry (ADG02)*, volume 2930 of *Lecture Notes in Computer Science*, pages 44–66. Springer-Verlag, 2004.

[99] Herbert Gelernter. Realization of a geometry theorem proving machine. In *Proceedings of the International Conference Information Processing*, pages 273–282, Paris, 1959.

[100] Herbert Gelernter. Realisation of a Geometry-Proving Machine. In *Automation of Reasoning*. Springer-Verlag, Berlin, 1983.

[101] Herbert Gelernter, J. R. Hansen, and Donald Loveland. Empirical explorations of the geometry theorem machine. In *Papers presented at the May 3-5, 1960, western joint IRE-AIEE-ACM Computer Conference*, IRE-AIEE-ACM '60 (Western), pages 143–149, San Francisco, California, 1960. ACM.

[102] Jean-David Genevaux, Julien Narboux, and Pascal Schreck. Formalization of Wu's Simple Method in Coq. In Jean-Pierre Jouannaud and Zhong Shao, editors, *CPP 2011 First International Conference on Certified Programs and Proofs*, volume 7086 of *Lecture Notes in Computer Science*, pages 71–86, Kenting, Taiwan, December 2011. Springer-Verlag.

[103] Paul C. Gilmore. An Examination of the Geometry Theorem Machine. *Artif. Intell.*, 1(3):171–187, 1970.

[104] Eduardo N. Giovannini. Completitud y Continuidad En Fundamentos de la Geometría de Hilbert (Completeness and Continuity in Hilbert's Foundations of Geometry). *Theoria: Revista de Teoría, Historia y Fundamentos de la Ciencia*, 28(1):139–163, 2013.

[105] Ira Goldstein. Elementary Geometry Theorem Proving. AI Lab memo 280, MIT, April 1973.

[106] Georges Gonthier. *A Computer Checked Proof of the Four Colour Theorem.* 2004.

[107] Georges Gonthier. Formal Proof — The Four-Color Theorem. *Notices of the American Mathematical Society*, 55(11):1382–1393, 2008.

[108] Adam Grabowski. Tarski's geometry modelled in Mizar computerized proof assistant. In *Computer Science and Information Systems (FedCSIS), 2016 Federated Conference on*, pages 373–381. IEEE, 2016.

[109] James G. Greeno, Maria E. Magone, and Seth Chaiklin. Theory of constructions and set in problem solving. *Memory and Cognition*, 7(6):445–461, 1979.

[110] Benjamin Grégoire, Loïc Pottier, and Laurent Théry. Proof Certificates for Algebra and Their Application to Automatic Geometry Theorem Proving. In *Post-proceedings of Automated Deduction in Geometry (ADG 2008)*, volume 6301 of *Lecture Notes in Artificial Intelligence*, pages 42–59. Springer, 2011.

[111] Charly Gries, Pierre Boutry, and Julien Narboux. Somme des angles d'un triangle et unicité de la parallèle : une preuve d'équivalence formalisée en Coq. In *Les vingt-septièmes Journées Francophones des Langages Applicatifs (JFLA 2016)*, Actes des Vingt-septièmes Journées Francophones des Langages Applicatifs (JFLA 2016), page 15, Saint Malo, France, January 2016. Jade Algave and Julien Signoles.

[112] Frédérique Guilhot. Formalisation en Coq d'un cours de géométrie pour le lycée. In *Journées Francophones des Langages Applicatifs*. INRIA, January 2004.

[113] Haragauri Narayan Gupta. *Contributions to the Axiomatic Foundations of Geometry*. PhD thesis, University of California, Berkley, 1965.

[114] Thomas Hales. *Dense Sphere Packings: A Blueprint for Formal Proofs*. Cambridge University Press, New York, NY, USA, 2012.

[115] Thomas Hales, Mark Adams, Gertrud Bauer, Dang Tat Dat, John Harrison, Hoang Le Truong, Cezary Kaliszyk, Victor Magron, Sean Mclaughlin, Nguyen Tat Thang, Nguyen Quang Truong, Tobias Nipkow, Steven Obua, Joseph Pleso, Jason Rute, Alexey Solovyev, Ta Thi Hoai An, Tran Nam Trung, Trieu Thi Diep, Josef Urban, Vu Khac Ky, and Roland Zumkeller. A Formal Proof of the Kepler Conjecture. *Forum of Mathematics, Pi*, 5, 2017.

[116] Thomas C. Hales. Introduction to the Flyspeck Project. In *Mathematics, Algorithms, Proofs*, volume 05021 of *Dagstuhl Seminar Proceedings*. Internationales Begegnungs- und Forschungszentrum für Informatik (IBFI), Schloss Dagstuhl, Germany, 2006.

[117] Thomas C. Hales. The Jordan Curve Theorem, Formally and Informally. *The American Mathematical Monthly*, 114(10):882–894, 2007.

[118] Thomas C. Hales, John Harrison, Sean McLaughlin, Tobias Nipkow, Steven Obua, and Roland Zumkeller. A Revision of the Proof of the Kepler Conjecture. *Discrete & Computational Geometry*, 44(1):1–34, July 2010.

[119] John Harrison. HOL Light: A Tutorial Introduction. In Mandayam K. Srivas and Albert John Camilleri, editors, *Formal Methods in Computer-Aided Design*, volume 1166 of *Lecture Notes in Computer Science*. Springer, 1996.

[120] John Harrison. A HOL Theory of Euclidean space. In Joe Hurd and Tom Melham, editors, *Theorem Proving in Higher Order Logics, 18th International Conference, TPHOLs 2005*, volume 3603 of *Lecture Notes in Computer Science*, pages 114–129, Oxford, UK, August 2005. Springer-Verlag.

[121] John Harrison. The HOL Light Theory of Euclidean Space. *Journal of Automated Reasoning*, 50(2):173–190, 2013.

[122] Robin Hartshorne. *Geometry:Euclid and Beyond*. Undergraduate texts in mathematics. Springer, 2000.

[123] Olaf Helmer-Hirschberg. Axiomatischer Aufbau der Geometrie in formalisierter Darstellung. *Schriften des mathematische Seminars und des Instituts fur angewandte Mathematik der Universität Berlin*, 2:175–201, 1935.

[124] David Hilbert. *Grundlagen der Geometrie*. Leipzig, 1899.

[125] Tetsuo Ida. Interactive vs. automated proofs in computational origami. In *2012 14th International Symposium on Symbolic and Numeric Algorithms for Scientific Computing (SYNASC)*, pages 7–7, September 2012.

[126] Tetsuo Ida, Asem Kasem, Fadoua Ghourabi, and Hidekazu Takahashi. Morley's theorem revisited: Origami construction and automated proof. *Journal of Symbolic Computation*, 46(5):571–583, May 2011.

[127] Tetsuo Ida, Hidekazu Takahashi, and Mircea Marin. Computational Origami of a Morley's Triangle. In Michael Kohlhase, editor, *Mathematical Knowledge Management*, number 3863 in Lecture Notes in Computer Science, pages 267–282. Springer Berlin Heidelberg, July 2005. DOI: 10.1007/11618027_18.

[128] Predrag Janičić. GCLC - A Tool for Constructive Euclidean Geometry and More Than That. In Andrés Iglesias and Nobuki Takayama, editors, *Mathematical Software - ICMS 2006, Second International Congress on Mathematical Software, Castro Urdiales, Spain, September 1-3, 2006, Proceedings*, volume 4151 of *Lecture Notes in Computer Science*, pages 58–73. Springer, 2006.

[129] Predrag Janičić and Stevan Kordić. EUCLID — the Geometry Theorem Prover. *FILOMAT*, 9(3):723–732, 1995.

[130] Predrag Janičić, Julien Narboux, and Pedro Quaresma. The Area Method : A Recapitulation. *Journal of Automated Reasoning*, 48(4):489–532, 2012.

[131] Jianguo Jiang and Jingzhong Zhang. A review and prospect of readable machine proofs for geometry theorems. *Journal of Systems Science and Complexity*, 25(4):802–820, 2012.

[132] Cezary Kaliszyk and Tetsuo Ida. Proof Assistant Decision Procedures for Formalizing Origami. In James H. Davenport, William M. Farmer, Josef Urban, and Florian Rabe, editors, *Intelligent Computer Mathematics*, number 6824 in Lecture Notes in Computer Science, pages 45–57. Springer Berlin Heidelberg, July 2011. DOI: 10.1007/978-3-642-22673-1_4.

[133] A. B. Kempe. On the Geographical Problem of the Four Colours. *American Journal of Mathematics*, 2(3):193–200, 1879.

[134] Clark Kimberling. *Triangle Centers and Central Triangles.* 2001.

[135] Felix C. Klein. *A comparative review of recent researches in geometry.* PhD thesis, 1872.

[136] Kenneth R. Koedinger and John R. Anderson. Abstract Planning and Perceptual Chunks: Elements of Expertise in Geometry. *Cognitive Science*, 14(4):511–550, 1990.

[137] Artur Kornilowicz. Jordan curve theorem. *Formalized Mathematics*, 13(4):481–491, 2005.

[138] Eugeniusz Kusak. Desargues Theorem In Projective 3-Space. *Formalized Mathematics*, 2(1), 1991.

[139] Imre Lakatos, John Worrall, and Elie Zahar, editors. *Proofs and Refutations.* Cambridge University Press, 1976.

[140] Wojciech Leonczuk and Krzysztof Prazmowski. Projective Spaces. *Journal of Formalized Mathematics*, 1(4):767–776, 1990.

[141] A. Letac. Solutions (Morley's triangle). Problem N 490. *Sphinx: Revue Mensuelle des Questions Récréatives*, 9, 1939.

[142] Tielin Liang and Dongming Wang. Towards a Geometric-Object-Oriented Language. In *Automated Deduction in Geometry*, pages 130–155, 2004.

[143] Dongxi Liu, NealeL. Fulton, John Zic, and Martin de Groot. Verifying an Aircraft Proximity Characterization Method in Coq. In Lindsay Groves and Jing Sun, editors, *Formal Methods and Software Engineering*, volume 8144 of *Lecture Notes in Computer Science*, pages 86–101. Springer Berlin Heidelberg, 2013.

[144] Sha Ma, Zhiping Shi, Zhenzhou Shao, Yong Guan, Liming Li, and Yongdong Li. Higher-Order Logic Formalization of Conformal Geometric Algebra and its Application in Verifying a Robotic Manipulation Algorithm. *Advances in Applied Clifford Algebras*, 26(4):1305–1330, 2016.

[145] Nicolas Magaud, Agathe Chollet, and Laurent Fuchs. Formalizing a discrete model of the continuum in Coq from a discrete geometry perspective. *Annals of Mathematics and Artificial Intelligence*, 74(3-4):309–332, October 2014.

[146] Nicolas Magaud, Julien Narboux, and Pascal Schreck. Formalizing Desargues' Theorem in Coq using Ranks. In *SAC*, pages 1110–1115, 2009.

[147] Nicolas Magaud, Julien Narboux, and Pascal Schreck. Formalizing Projective Plane Geometry in Coq. In Thomas Sturm, editor, *Post-proceedings of Automated Deduction in Geometry (ADG) 2008*, volume 6301 of *LNAI*, pages 141–162, Shanghai, China, 2011. Springer.

[148] Nicolas Magaud, Julien Narboux, and Pascal Schreck. A Case Study in Formalizing Projective Geometry in Coq: Desargues Theorem. *Computational Geometry*, 45(8):406–424, 2012.

[149] Timothy James McKenzie Makarios. A Mechanical Verification of the Independence of Tarski's Euclidean Axiom. Master Thesis, Victoria University of Wellington, 2012.

[150] Filip Marić and Danijela Petrović. Formalizing complex plane geometry. *Annals of Mathematics and Artificial Intelligence*, 74(3-4):271–308, 2015.

[151] Filip Marić, Ivan Petrović, Danijela Petrović, and Predrag Janičić. Formalization and Implementation of Algebraic Methods in Geometry. In Pedro Quaresma and Ralph-Johan Back, editors, *Proceedings First Workshop on CTP Components for Educational Software, Wroclaw, Poland, 31th July 2011*, volume 79 of *Electronic Proceedings in Theoretical Computer Science*, pages 63–81. Open Publishing Association, 2012.

[152] Vesna Marinković. Proof Simplification in the Framework of Coherent Logic. *Computing and Informatics*, 34(2):337–366, 2015.

[153] Vesna Marinković, Predrag Janičić, and Pascal Schreck. Computer Theorem Proving for Verifiable Solving of Geometric Construction Problems. In Francisco Botana and Pedro Quaresma, editors, *Automated Deduction in Geometry - 10th International Workshop, ADG 2014, Coimbra, Portugal, July 9-11, 2014, Revised Selected Papers*, volume 9201 of *Lecture Notes in Computer Science*, pages 72–93. Springer, 2015.

[154] G. E. Martin. *The Foundations of Geometry and the Non-Euclidean Plane*. Undergraduate Texts in Mathematics. Springer, 1998.

[155] Noboru Matsuda and Kurt Vanlehn. GRAMY: A Geometry Theorem Prover Capable of Construction. *Journal of Automated Reasoning*, 32:3–33, 2004.

[156] William McCune. Solution of the Robbins Problem. *Journal of Automated Reasoning*, 19(3):263–276, 1997.

[157] Laura Meikle and Jacques D. Fleuriot. Formalizing Hilbert's Grundlagen in Isabelle/Isar. In *Theorem Proving in Higher Order Logics*, volume 2758 of *Lecture Notes in Computer Science*, pages 319–334. Springer, 2003.

[158] Laura Meikle and Jacques D. Fleuriot. Mechanical Theorem Proving in Computational Geometry. In Hoon Hong and Dongming Wang, editors, *Proceedings of Automated Deduction in Geometry 2004*, volume 3763 of *Lecture Notes in Computer Science*, pages 1–18. Springer-Verlag, November 2005.

[159] Richard S Millman and George D Parker. *Geometry, A Metric Approach with Models*. Springer Science & Business Media, 1991.

[160] Leonardo Mendonça de Moura and Nikolaj Bjørner. Z3: An Efficient SMT Solver. In C. R. Ramakrishnan and Jakob Rehof, editors, *Tools and Algorithms for the Construction and Analysis of Systems, 14th International Conference, TACAS 2008, Held as Part of the Joint European Conferences on Theory and Practice of Software, ETAPS 2008, Budapest, Hungary, March 29-April 6, 2008. Proceedings*, volume 4963 of *Lecture Notes in Computer Science*, pages 337–340. Springer, 2008.

[161] Julien Narboux. A Decision Procedure for Geometry in Coq. In Slind Konrad, Bunker Annett, and Gopalakrishnan Ganesh, editors, *Proceedings of TPHOLs'2004*, volume 3223 of *Lecture Notes in Computer Science*. Springer-Verlag, 2004.

[162] Julien Narboux. *Formalisation et automatisation du raisonnement géométrique en Coq*. PhD thesis, Université Paris Sud, September 2006.

[163] Julien Narboux. A Graphical User Interface for Formal Proofs in Geometry. *Journal of Automated Reasoning*, 39(2):161–180, 2007.

[164] Julien Narboux. Mechanical Theorem Proving in Tarski's Geometry. In Francisco Botana Eugenio Roanes Lozano, editor, *Post-Proceedings of Automated Deduction in Geometry 2006*, volume 4869 of *Lecture Notes in Computer Science*, pages 139–156, Pontevedra, Spain, 2008. Springer.

[165] Julien Narboux and David Braun. Towards A Certified Version of the Encyclopedia of Triangle Centers. In J. Rafael Sandra, Dongming Wang, and Jing Yang, editors, *Special Issue on Geometric Reasoning*, pages 1–17. Springer, 2016.

[166] A.J. Nevis. Plane geometry theorem proving using forward chaining. *Artificial Intelligence*, 6(1):1–23, 1975.

[167] Mladen Nikolić and Predrag Janičić. CDCL-Based Abstract State Transition System for Coherent Logic. In Jeuring J. et al., editor, *Intelligent Computer Mathematics - CICM 2012*, volume 7362 of *Lecture Notes in Computer Science*. Springer, 2012.

[168] Tobias Nipkow, Lawrence C. Paulson, and Markus Wenzel. *Isabelle HOL: a Proof Assistant for Higher-Order Logic*. Springer, 2005. \sc url: \tt http://www.cl.cam.ac.uk/research/hvg/Isabelle/dist/Isabelle/doc.

[169] Hans de Nivelle and Jia Meng. Geometric Resolution: A Proof Procedure Based on Finite Model Search. In *Automated Reasoning, Third International Joint Conference, IJCAR*, volume 4130 of *Lecture Notes in Computer Science*, pages 303–317. Springer, 2006.

[170] Victor Pambuccian. Groups and Plane Geometry. *Studia Logica*, 81(3):387–398, 2005.

[171] Victor Pambuccian. Axiomatizations of hyperbolic and absolute geometries. In *Non-Euclidean Geometries*, volume 581, pages 119–153. Springer, 2006.

[172] Victor Pambuccian. Axiomatizing geometric constructions. *Journal of Applied Logic*, 6(1):24–46, 2008.

[173] Victor Pambuccian. The axiomatics of ordered geometry: I. Ordered incidence spaces. *Expositiones Mathematicae*, 29(1):24–66, 2011.

[174] Moritz Pasch. *Vorlesungen über neuere Geometrie*. Teubner, Leipzig, 1882.

[175] Lawrence C. Paulson and Tobias Nipkow. Isabelle Tutorial and User's Manual. Technical Report 189, University of Cambridge, Computer Laboratory, January 1990.

[176] Giuseppe Peano. *Principii de Geometria*. Fratelli Bocca, Torino, 1889.

[177] Danijela Petrović and Filip Marić. Formalizing analytic geometries. In *Proceedings of Automated Deduction in Geometry*, September 2012.

[178] Tuan Minh Pham. Similar triangles and orientation in plane elementary geometry for Coq-based proofs. In *Proceedings of the 2010 ACM Symposium on Applied Computing*, pages 1268–1269. ACM, 2010.

[179] Tuan Minh Pham. *Formal Description of Geometrical Properties*. Theses, Univeristé Nice Sophia Antipolis, November 2011.

[180] Tuan Minh Pham and Yves Bertot. A Combination of a Dynamic Geometry Software With a Proof Assistant for Interactive Formal Proofs. *Electron. Notes Theor. Comput. Sci.*, 285:43–55, September 2012.

[181] Tuan Minh Pham, Yves Bertot, and Julien Narboux. A Coq-based Library for Interactive and Automated Theorem Proving in Plane Geometry. In *Proceedings of the 11th International Conference on Computational Science and Its Applications (ICCSA 2011)*, volume 6785 of *Lecture Notes in Computer Science*, pages 368–383. Springer-Verlag, 2011.

[182] David Pichardie and Yves Bertot. Formalizing Convex Hulls Algorithms. In *Proc. of 14th International Conference on Theorem Proving in Higher Order Logics (TPHOLs'01)*, volume 2152 of *Lecture Notes in Computer Science*, pages 346–361. Springer-Verlag, 2001.

[183] Jan von Plato. The axioms of constructive geometry. *Annals of Pure and Applied Logic*, 76:169–200, 1995.

[184] Jan von Plato. A constructive theory of ordered affine geometry. In *Indagationes Mathematicae*, volume 9, pages 549–562. 1998.

[185] Andrew Polonsky. *Proofs, Types and Lambda Calculus*. PhD thesis, University of Bergen, 2011.

[186] Loïc Pottier. Connecting Gröbner Bases Programs with Coq to do Proofs in Algebra, Geometry and Arithmetics. In G. Sutcliffe, P. Rudnicki, R. Schmidt, B. Konev, and S. Schulz, editors, *Knowledge Exchange: Automated Provers and Proof Assistants*, volume 418 of *CEUR Workshop Proceedings*, Doha, Qatar, 2008. CEUR-WS.org.

[187] François Puitg. *Preuves en modélisation géométrique par le calcul des constructions inductives*. PhD thesis, Université de Strasbourg, 1999.

[188] FranÇois Puitg and Jean FranÇois Dufourd. Formal specification and theorem proving breakthroughs in geometric modeling. In Jim Grundy and Malcolm Newey, editors, *Theorem Proving in Higher Order Logics: 11th International Conference, TPHOLs'98 Canberra, Australia September 27–October 1, 1998 Proceedings*, pages 401–422, Berlin, Heidelberg, 1998. Springer Berlin Heidelberg. DOI: 10.1007/BFb0055149.

[189] Art Quaife. Automated Development of Tarski's Geometry. *Journal of Automated Reasoning*, 5(1):97–118, 1989.

[190] Pedro Quaresma. Thousands of Geometric Problems for Geometric Theorem Provers (TGTP). In Pascal Schreck, Julien Narboux, and Jürgen Richter-Gebert, editors, *Automated Deduction in Geometry*, volume 6877 of *Lecture Notes in Computer Science*, pages 169–181. Springer, 2011.

[191] Pedro Quaresma. Towards an Intelligent and Dynamic Geometry Book. *Mathematics in Computer Science*, pages 1–11, 2017.

[192] Marco Riccardi. Heron's Formula and Ptolemy's Theorem. *Formalized Mathematics*, 16(1-4):97–101, 2008.

[193] William Richter. Formalizing Rigorous Hilbert Axiomatic Geometry Proofs in the Proof Assistant Hol Light.

[194] William Richter, Adam Grabowski, and Jesse Alama. Tarski Geometry Axioms. *Formalized Mathematics*, 22(2):167–176, 2014.

[195] Jürgen Richter-Gebert. Meditations on Ceva's Theorem. In Ellers E. Davis, C., editor, *The Coxeter Legacy: Reflections and Projections*, pages 227–254. American Mathematical Society, 2006.

[196] John Alan Robinson and Andrei Voronkov, editors. *Handbook of Automated Reasoning (Vol 2)*. Elsevier and MIT Press, 2001.

[197] Judit Robu. *Automated Geometric Theorem Proving*. PhD thesis, RISC, Johannes Kepler University, Linz, Austria, 2002.

[198] Ralph Romanos and Lawrence Paulson. *Proving the Impossibility of Trisecting an Angle and Doubling the Cube*. 2014.

[199] Karel Rössler. Géométrie abstraite mécanisée. *Publications de la Faculté des Sciences de l'Université Charles (Praha)*, (134):29, 1934.

[200] Wolfram Schwabhäuser. Uber dire Vollständigkeit der elementaren euklidischen Geometrie. *Zeitschrift für mathematische Logik und Grundlagen der Mathematik*, 2:137–165, 1956.

[201] Wolfram Schwabhäuser, Wanda Szmielew, and Alfred Tarski. *Metamathematische Methoden in der Geometrie*. Springer-Verlag, Berlin, 1983.

[202] Phil Scott. *Mechanising Hilbert's Foundations of Geometry in Isabelle*. Master Thesis, University of Edinburgh, 2008.

[203] Phil Scott and Jacques D. Fleuriot. An Investigation of Hilbert's Implicit Reasoning through Proof Discovery in Idle-Time. In *Automated Deduction in Geometry - 8th International Workshop, ADG 2010, Munich, Germany, July 22-24, 2010, Revised Selected Papers*, pages 182–200, 2010.

[204] Phil Scott and Jacques D. Fleuriot. A Combinator Language for Theorem Discovery. In *Intelligent Computer Mathematics - 11th International Conference, AISC 2012, 19th Symposium, Calculemus 2012, 5th International Workshop, DML 2012, 11th International Conference, MKM 2012, Systems and Projects, Held as Part of CICM 2012, Bremen, Germany, July 8-13, 2012. Proceedings*, pages 371–385, 2012.

[205] Phil Scott and Jacques D. Fleuriot. Compass-free Navigation of Mazes. In James H. Davenport and Fadoua Ghourabi, editors, *SCSS 2016. 7th International Symposium on Symbolic Computation in Software Science*, volume 39 of *EPiC Series in Computing*, pages 143–155. EasyChair, 2016.

[206] Wojciech Skaba. The Collinearity Structure. *Formalized Mathematics*, 1(4):3, 1990.

[207] Konrad Slind and Michael Norrish. A Brief Overview of HOL4. In Otmane Ait Mohamed, César Muñoz, and Sofiène Tahar, editors, *Theorem Proving in Higher Order Logics: 21st International Conference, TPHOLs 2008, Montreal, Canada, August 18-21, 2008. Proceedings*, pages 28–32. Springer Berlin Heidelberg, Berlin, Heidelberg, 2008. DOI: 10.1007/978-3-540-71067-7_6.

[208] Sana Stojanović, Julien Narboux, Marc Bezem, and Predrag Janičić. A Vernacular for Coherent Logic. In Stephen M. Watt, James H. Davenport, Alan P. Sexton, Petr Sojka, and Josef Urban, editors, *Intelligent Computer Mathematics*, volume 8543 of *Lecture Notes in Computer Science*, pages 388–403. Springer International Publishing, 2014.

[209] Sana Stojanović, Vesna Pavlović, and Predrag Janičić. A Coherent Logic Based Geometry Theorem Prover Capable of Producing Formal and Readable Proofs. In *Automated Deduction in Geometry*, volume 6877 of *Lecture Notes in Computer Science*, pages 201–220. Springer, 2011.

[210] Sana Stojanović Durdević, Julien Narboux, and Predrag Janičić. Automated Generation of Machine Verifiable and Readable Proofs: A Case Study of Tarski's Geometry. *Annals of Mathematics and Artificial Intelligence*, page 25, 2015.

[211] J. Strommer. Über die Kreisaxiome. *Periodica Mathematica Hungarica*, 4:3–16, 1973.

[212] Alfred Tarski. *A Decision Method for Elementary Algebra and Geometry*. University of California Press, 1951.

[213] Alfred Tarski. What is Elementary Geometry? In P. Suppes L. Henkin and A. Tarski, editors, *The axiomatic Method, with special reference to Geometry and Physics*, pages 16–29, Amsterdam, 1959. North-Holland.

[214] Alfred Tarski and Steven Givant. Tarski's System of Geometry. *The Bulletin of Symbolic Logic*, 5(2):175–214, June 1999.

[215] Andrzej Trybulec. Mizar. In Freek Wiedijk, editor, *The Seventeen Provers of the World*, volume 3600 of *Lecture Notes in Computer Science*. Springer, 2006.

[216] Josef Urban and Robert Veroff. Experiments with State-of-the-art Automated Provers on Problems in Tarskian Geometry. In Boris Konev, Stephan Schulz, and Laurent Simon, editors, *IWIL-2015. 11th International Workshop on the Implementation of Logics*, volume 40 of *EPiC Series in Computing*, pages 122–126. EasyChair, 2016.

[217] Dongming Wang. Elimination Procedures for Mechanical Theorem Proving in Geometry. *Ann. Math. Artif. Intell.*, 13(1-2):1–24, 1995.

[218] Dongming Wang. Geometry machines: From AI to SMC. In Jacques Calmet, John Campbell, and Jochen Pfalzgraf, editors, *Artificial Intelligence and Symbolic Mathematical Computation*, volume 1138 of *Lecture Notes in Computer Science*, pages 213–239. Springer Berlin / Heidelberg, 1996. DOI: 10.1007/3-540-61732-9_60.

[219] Dongming Wang, Xiaoyu Chen, Wenya An, Lei Jiang, and Dan Song. OpenGeo: An Open Geometric Knowledge Base. In Hoon Hong and Chee Yap, editors, *Mathematical Software – ICMS 2014*, volume 8592 of *Lecture Notes in Computer Science*, pages 240–245. Springer Berlin Heidelberg, 2014. DOI: 10.1007/978-3-662-44199-2_38.

[220] Ke Wang and Zhendong Su. Automated Geometry Theorem Proving for Human-readable Proofs. In *Proceedings of the 24th International Conference on Artificial Intelligence*, IJCAI'15, pages 1193–1199, Buenos Aires, Argentina, 2015. AAAI Press.

[221] Markus Wenzel. *Isabelle/Isar — A Versatile Environment for Human-Readable Formal Proof Documents*. PhD thesis, Institut für Informatik, Technische Universität München, 2002.

[222] Freek Wiedijk, editor. *The Seventeen Provers of the World*, volume 3600 of *Lecture Notes in Computer Science*. Springer, 2006.

[223] Sean Wilson and Jacques D. Fleuriot. Combining Dynamic Geometry, Automated Geometry Theorem Proving and Diagrammatic Proofs. In *ETAPS Satellite Workshop on User Interfaces for Theorem Provers (UITP)*, Edinburgh, 2005. Springer.

[224] Wen-Tsun Wu. On the Decision Problem and the Mechanization of Theorem-Proving in Elementary Geometry. In *Automated Theorem Proving: After 25 Years*, volume 29 of *Contemporary Mathematics*, pages 213–234. American Mathematical Society, 1984.

[225] Lu Yang, Xiao-Shan Gao, Shang-Ching Chou, and Jing-Zhong Zhang. Automated Production of Readable Proofs for Theorems in Non-Euclidian Geometries. In Dongming Wang, editor, *International Workshop on Automated Deduction in Geometry, Selected Papers*, volume 1360 of *Lecture Notes in Computer Science*, pages 171–188, Toulouse, 1996. Springer.

[226] Lu Yang, Xiao-Shan Gao, Shang-Ching Chou, and Jing. Zhong. Zhang. Automated Proving and Discovering of Theorems in Non-Euclidean Geometries. In *Proceedings of Automated Deduction in Geometry (ADG98)*, Lecture Notes in Artificial Intelligence, pages 171–188, Berlin, Heidelberg, 1998. Springer-Verlag.

[227] Zheng Ye, Shang-Ching Chou, and Xiao-Shan Gao. Visually Dynamic Presentation of Proofs in Plane Geometry. *Journal of Automated Reasoning*, 45(3):243–266, December 2009.

[228] Paul Yiu. *Introduction to the Geometry of the Triangle*. Department of Mathematics, Florida Atlantic University, 2001.

[229] Jing-Zhong Zhang, Shang-Ching Chou, and Xiao-Shan Gao. Automated Production of Traditional Proofs for Theorems in Euclidean Geometry. *Ann. Math. Artif. Intell.*, 13(1-2):109–138, 1995.

[230] Yu Zou and Jingzhong Zhang. Automated Generation of Readable Proofs for Constructive Geometry Statements with the Mass Point Method. In Pascal Schreck, Julien Narboux, and Jürgen Richter-Gebert, editors, *8th International Workshop, ADG 2010, Revised Selected Papers*, volume 6877 of *Lecture Notes In Computer Science*, pages 221–258, Munich, 2011. Springer.

# Chapter 3

# *Coordinate-Free Theorem Proving in Incidence Geometry*

**Jürgen Richter-Gebert**

*Faculty of Mathematics, Technical University of Munich, Germany*

**Hongbo Li**

*Academy of Mathematics and Systems Science, Chinese Academy of Sciences; University of Chinese Academy of Sciences, China*

## CONTENTS



### Introduction

The objects under consideration in this chapter are the classical objects of elementary geometry: points, lines, planes, circles, conics, etc. Many geometric theorems deal with the interdependence of properties that hold between such objects. A point may lie on a line or on a circle, a circle and line may touch tangentially, two lines may enclose a certain angle, two points may be at a certain distance, etc. A geometric theorem is typically stated in a way where certain such relations (the hypotheses) imply another such relation (typically under the presence of certain non-degeneracy assumptions). *Incidence geometry* exclusively deals with properties that do not refer to measurements and are of purely qualitative kind. Thus, the incidence of a point and a line is a prototypical elementary property that may occur in a theorem of incidence geometry, while the reference to a certain angle typically is not of this type.

In what follows standard techniques based on bracket algebra and Grassmann-Cayley algebra for the coordinate-free treatment of incidence statements and constructions will be explained. We shall focus on the case of two-dimensional projective incidence geometry; the extension to high dimensions is straightforward. See Section 2 of [30] for a general introduction to projective space. For an elaborate introduction to this topic we recommend [18, 25].

## 3.1 Incidence Geometry

### 3.1.1 Incidence Geometry in the Plane

#### Glossary

**Homogeneous coordinates:** For $p \in \mathbb{K}^n$ let $[v] := \{\lambda v \mid \lambda \in \mathbb{K} - \{0\}\}$. Vectors that are identified by scalar multiples are called homogeneous coordinates. We abbreviate $[(p_1, p_2, \dots, p_n)] =: (p_1 : p_2 : \cdots : p_n)$.

**Projective plane over $\mathbb{K}$:** For a field $\mathbb{K}$ let $\mathcal{P}_\mathbb{K}$ (points) and $\mathcal{L}_\mathbb{K}$ (lines) be two disjoint copies of the set $\{[p] \mid p \in \mathbb{K}^3 - \{(0,0,0)\}\} =: \mathbb{KP}^2$ of equivalence classes of non-zero three-dimensional vectors. An incidence relation $\mathcal{I}_\mathbb{K} \subset \mathcal{P}_\mathbb{K} \times \mathcal{L}_\mathbb{K}$ is defined by $[p]\mathcal{I}_\mathbb{K}[l] \iff \langle p, l \rangle = 0$, where the angular brackets denote the pairing between vector space $\mathbb{K}^3$ and its dual. The triple $(\mathcal{P}_\mathbb{K}, \mathcal{L}_\mathbb{K}, \mathcal{I}_\mathbb{K})$ is the projective plane over $\mathbb{K}$.

**Standard embedding:** For a point $(x,y) \in \mathbb{K}^2$ its homogenization is defined as $(x : y : 1)$. For a line $l \subset \mathbb{K}^2$ defined by the equation $ax + by + c = 0$ its homogeneous coordinates are defined by $(a : b : c)$. Incidences between points and lines in $\mathbb{K}^2$ translate to incidences in the projective plane of the standard embedding.

---

We consider the projective plane $(\mathcal{P}_\mathbb{K}, \mathcal{L}_\mathbb{K}, \mathcal{I}_\mathbb{K})$ over a field $\mathbb{K}$. Points in that plane are represented by their homogeneous coordinates $[p] \in \mathcal{P}_\mathbb{K}$. By this each point $p$ is represented by a non-zero vector $(x,y,z)$ such that two vectors represent the same point if and only if they differ by a non-zero scalar multiple $\lambda$. Equivalently the points are identified with the one-dimensional linear subspaces of $\mathbb{K}^3$. Since most operations and relations that we will consider are represented by linear operations, it is often sufficient to perform the calculations in terms of the representants. When we speak of the *point* $p$ with $p \in \mathbb{K}^3$ we mean the projective point represented by the equivalence class $[p]$. Sometimes it is necessary to refer to the *vector* $p$ itself.

In the case of $\mathbb{K} = \mathbb{R}$ it is instructive to consider the points as the intersection of the corresponding one-dimensional linear subspaces with the unit sphere $S^2 := \{(x,y,z) \mid x^2 + y^2 + z^2 = 1\}$. In this view each point in $\mathbb{RP}^2$ corresponds to an antipodal pair of points on the unit sphere. A line $[l] \in \mathcal{L}_K$ may be identified with the two-dimensional subspace of $\mathbb{K}^3$ normal to the vector $l$. Incidence between a point $(x : y : z)$ and a line $(a : b : c)$ corresponds to incidence between the corresponding subspaces and is algebraically resembled by the vanishing of the scalar product $ax + by + zc = 0$. In the case of $\mathbb{K} = \mathbb{R}$ a subspace representing a line may be intersected with the unit sphere and by this corresponds to a great circle on $S^2$.

In the standard embedding each point $(x,y) \in \mathbb{R}^2$ is represented by its *homogenization* $(x : y : 1)$. This covers almost all points of $\mathcal{P}_\mathbb{R}$ except for those of the form $(x : y : 0)$. They correspond to points that are infinitely far away and have no direct correspondence in $\mathbb{R}^2$. For each direction (equivalence class of parallels in $\mathbb{R}^2$) there is exactly one such point at infinity incident to all lines of a parallel class. All infinite points lie on a common line with parameter $(0:0:1)$ which is the *line at infinity*. Literally the same interpretation applies to all fields $\mathbb{K}$. See Section 2.2 of [30] for more on homogeneous coordinates and points at infinity.

---

**Glossary**

**Join:** For different points $[p],[q] \in \mathcal{P}_{\mathbb{K}}$ the unique line in $\mathcal{L}_{\mathbb{K}}$ incident to both. Can be calculated by $meet(p,q) = p \times q$, where the cross symbol denotes the vector product in vector algebra.

**Meet:** For different lines $[l],[m] \in \mathcal{L}_{\mathbb{K}}$ the unique point in $\mathcal{P}_{\mathbb{K}}$ incident to both. Can be calculated by $join(l,m) = l \times m$.

**Grassmann-Cayley algebra:** A non-associative algebra with two multilinear and graded-antisymmetric products: the wedge and vee products. For example, the operations *meet* and *join* are denoted by the wedge and vee products, respectively:

$$meet(p,q) = p \wedge q \quad \text{and} \quad join(l,m) = l \vee m.$$

They satisfy the following graded-antisymmetric properties: for $p$ of grade $d_p$ and $q$ of grade $d_q$, if $n$ is the dimension of the base vector space, then

$$p \wedge q = (-1)^{(n-d_p)(n-d_q)} q \wedge p; \quad p \vee q = (-1)^{d_p d_q} q \vee p.$$

As to the grade of an element, a scalar is of grade 0, a nonzero vector in the base vector space is of grade 1, and the wedge product of $r$ elements of grade $d_1, d_2, \ldots, d_r$ respectively, if nonzero, is of grade $d_1 + d_2 + \cdots + d_r$.

**Collinear:** Three points $[p],[q],[r] \in \mathcal{P}_{\mathbb{K}}$ are collinear if they are incident to the same line. This happens if and only if $\det(p,q,r) = 0$.

**Concurrent:** Three lines $[l],[m],[g] \in \mathcal{P}_{\mathbb{K}}$ are concurrent if they are incident to the same point. This happens if and only if $\det(l,m,g) = 0$.

**Bracket:** Shorthand for a determinant: $[p,q,r] := \det(p,q,r)$.

**Non-degeneracy condition:** A relation indicating that a certain dependency (typically an incidence or a higher algebraic dependency) does *not* occur.

**Incidence theorem:** A true statement of the form $(H_1, \ldots H_k, N_1, \ldots, N_r) \implies C$, where $H_i$ (the hypotheses) are statements of incidence, $N_i$ are non-degeneracy conditions and $C$ (the conclusion) is an incidence statement.

---

If one considers incidence relations from a structural point of view, the projective plane is way more natural than the usual plane $\mathbb{K}^2$. While in $\mathbb{K}^2$ it may happen that two affine lines do not have an intersection (in case they are parallel), in the projective plane over $\mathbb{K}$ two distinct lines will *always* have exactly one projective point in common: their *meet*. Likewise any pair of distinct points in $\mathbb{KP}^2$ have exactly one line incident to both of them: their *join*. (Remark: in this way the points and lines of $\mathbb{KP}^2$ satisfy the usual axioms of an abstract projective plane claiming that *join* and *meet* always exist and are unique; see Section 4 of [30] for an algebraic definition of the meet). The fact that $p$ is incident to $l$ if and only if the scalar product $\langle p,l \rangle$ vanishes directly translates into the fact that the join and meet operations can be algebraically expressed by the *vector product* in $\mathbb{K}^3$. See Section 4.2 of [30] for an explicit algebraic computation relating the join and vector products.

If the vector product $p \times q$ turns out to be the zero vector, then $p$ and $q$ must be linearly dependent and hence indicate an inadmissible (degenerate) operation of joining two identical points.

Collinearity of three points $p_1, p_2, p_3$ corresponds to the existence of a line that is simultaneously incident to all three points. In this case the points are linearly dependent. Hence collinearity can be expressed by the vanishing of the determinant $\det(p_1, p_2, p_3)$. Similarly, the fact that three lines $l_1, l_2, l_3$ meet at a common point corresponds to the vanishing of the determinant $\det(l_1, l_2, l_3)$. For better readability we will abbreviate a $3 \times 3$ determinant $\det(a, b, c)$ by the bracket notation $[a, b, c]$.

A minimal instance of a theorem of projective incidence geometry is Pappus' Theorem (Figure 3.1, left). In textbooks this theorem is often naively stated in the following way.

*If $a, b, c$ and $A, B, C$ are two triples of collinear points, then the points $Z = (a \vee B) \wedge (b \vee A)$, $X = (b \vee C) \wedge (c \vee B)$, $Y = (c \vee A) \wedge (a \vee C)$ are collinear as well.*

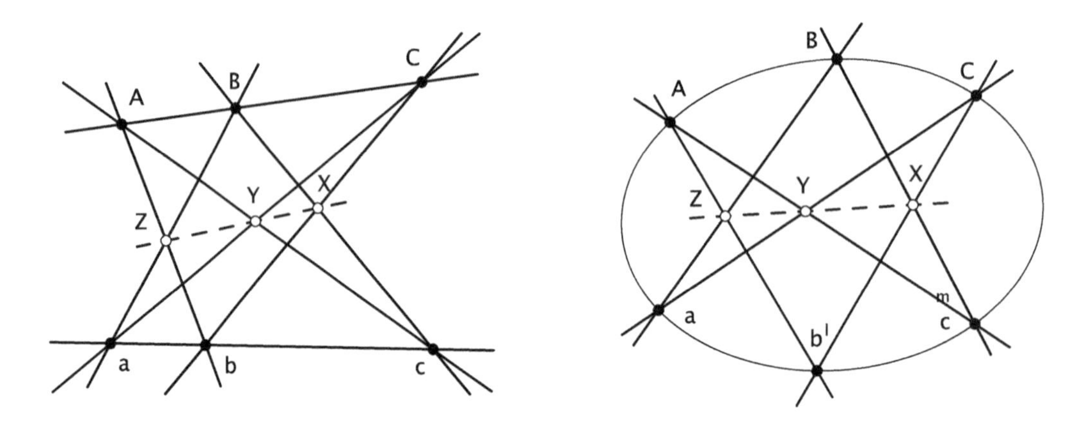


**Figure 3.1**
The theorems of Pappus and Pascal.

However, a bit of care is necessary here. It might happen that the initial six points are located in a way that some of the join and meet operators are degenerate. In this case the above formulation makes no sense. Thus, it is a wise choice to furthermore require that each join and meet operation is indeed non-degenerate. (This can be achieved by imposing that $a, b, c, A, B, C$ are mutually different.)

In principle the algebra of *meet* and *join* operations is very well suited to perform direct calculations in which three-dimensional vectors represent geometric points and lines. One crucial ingredient of such calculations comes from the fact that nested vector products can be reformulated as linear combinations of expressions involving geometric objects and determinants. The following very useful formula can be easily proved by expansion of the left and right sides:

$$(p \vee q) \wedge (r \vee s) = [p, q, r]\, s - [p, q, s]\, r. \tag{3.1}$$

A first approach to proving Pappus' Theorem is by transforming the join and meet operations into pure bracket expressions. We want to prove the collinearity of the points $X, Y, Z$ under the hypotheses of the theorem. We check whether the determinant $[X, Y, Z]$ vanishes. From the construction we obtain:

$$[X,Y,Z] = \big[[a,B,b]A - [a,B,A]b,\ [b,C,c]B - [b,C,B]c,\ [c,A,a]C - [c,A,C]a\big].$$

Expanding this formula we obtain eight summands:

$$
\begin{array}{rr}
+[a,B,b][b,C,c][c,A,a][A,B,C] & \underbrace{-[a,B,b][b,C,c][c,A,C][A,B,a]} \\
\overbrace{-[a,B,A][b,C,c][c,A,a][b,B,C]} & \underline{-[a,B,b][b,C,B][c,A,a][A,c,C]} \\
-[a,B,A][b,C,B][c,A,C][a,b,c] & \underline{+[a,B,b][b,C,B][c,A,C][A,c,a]} \\
\underbrace{+[a,B,A][b,C,c][c,A,C][b,B,a]} & \overbrace{+[a,B,A][b,C,B][c,A,a][b,c,C]}.
\end{array}
$$

Observe that the six terms cancel pairwise by summands of the same underbrace, or lowerbrace, or underline, and we are left with the expression:

$$+[a,B,b][b,C,c][c,A,a]\underline{[A,B,C]} - [a,B,A][b,C,B][c,A,C]\underline{[a,b,c]}.$$

This expression vanishes if the hypotheses of Pappus' Theorem are satisfied since in this case the two underlined determinants must be zero.

A closer look at this proof of Pappus' Theorem reveals two principal problems with this kind of coordinate-free calculation. First of all the calculation heavily depends on the fact that there is a nice cancellation pattern between the expanded bracket monomials. However, it might happen that bracket expressions vanish for less obvious reasons than simply being equal. We will deal with this phenomenon in depth in Section 3.2.1. The second issue that arises comes from the fact that this technique heavily relied on the expansion of the multiplication of sums of terms. This may easily lead to an exponential blowup of the calculations in the middle of the calculation (before cancellation occurs). Already for theorems slightly larger than Pappus' Theorem this may lead to practical intractability of such calculations even with the aid of computers.

### 3.1.2 Other Primitive Operations

One might think that building a system that exclusively uses points, lines, and incidences as primitive objects and relations leads to a relatively restricted expressiveness. Fortunately this is not the case. More elaborate constructions (or equivalently algebraic expressions) may be used to express more advanced geometric relations. To exemplify this let us consider the expression

$$+[a,B,b][b,C,c][c,A,a][A,B,C] - [a,B,A][b,C,B][c,A,C][a,b,c]$$

we obtained in our previous calculation. To get a more symmetric form of this formula we apply a cyclic shift to the variable names $a \to c \to b \to a$ and change it into

$$+[c,B,a][a,C,b][b,A,c][A,B,C] - [c,B,A][a,C,B][b,A,C][c,a,b].$$

Reordering the indices in the brackets yields the following very symmetric form:

$$+[a,b,c][a,B,C][A,b,C][A,B,c] - [A,B,C][A,b,c][a,B,c][a,b,C] \quad (*)$$

We have seen that this expression vanishes if both triples $(a,b,c)$ and $(A,B,C)$ are collinear. However this is not the only situation in which this expression vanishes. In fact, vanishing of this expression is equivalent to the six points $a,b,c,A,B,C$ being on a common conic.

To see this we first define a conic to be the solution of a quadratic equation in homogeneous coordinates. A point $(x,y,z)^T$ lies on a conic given by parameters $(\alpha,\beta,\gamma,\delta,\varepsilon,\zeta)$ if it satisfies the equation

$$\alpha x^2 + \beta y^2 + \gamma z^2 + \delta xy + \varepsilon xz + \zeta yz = 0.$$

In the standard affine embedding, conics correspond to ellipses, hyperbolas, parabolas, or may degenerate into pairs of lines (which may even coincide). For $p = (x,y,z)^T$ we define the quadratic operator

$$sq(p) := (x^2, y^2, z^2, xy, xz, yz).$$

We first observe that if for five points $p_1, \dots, p_5$ the vectors $sq(p_1), \dots, sq(p_5)$ are linearly independent then there is a unique conic passing through these five points, which is given by the solution of a system of linear equations. A sixth point lying on this conic must satisfy a quadratic equation. Now let us consider the above equation $(*)$. We first observe that each point occurs exactly twice in each expression. Hence if we fix five of the points (say $a,b,c,A,B$) the expression (if suitably non-degenerate) being zero gives a quadratic relation in the last point $C$. Here the non-degeneracy condition is that the formula does not automatically vanish for all possible choices of $C$. This can be achieved by making sure that no quadruple of the points $a,b,c,A,B$ are collinear (we omit the proof of this fact here). It turns out that the quadratic equation in $C$ describes the unique conic that passes through the other five points. For this it is sufficient to show that if $C$ is any of the points in the set $\{a,b,c,A,B\}$ then the expression $(*)$ will vanish. This is trivial for $C \in \{a,b,A,B\}$ since in this case one determinant in each summand becomes zero. If $C = c$ then it is easy to check that both summands are identical and hence their difference will vanish as well. We can rephrase the last result in the following way, that is essentially an algebraic formulation of Pascal's Theorem [25] (Figure 3.1, right):

*Six points $a,b,c,A,B,C$ are on a common conic if the expression $(*)$ vanishes. This is the case if $Z = (a \vee B) \wedge (b \vee A)$, $X = (b \vee C) \wedge (c \vee B)$, $Y = (c \vee A) \wedge (a \vee C)$ are collinear.*

### 3.1.3 Projective Invariance

---

**Glossary**

**Projective transformation:** A map $\tau\colon \mathcal{P}_{\mathbb{K}} \to \mathcal{P}_{\mathbb{K}}$ induced by the map $p \mapsto M \cdot p$ for a non-singular matrix $M$. The action of this map is well defined.

**Collineation:** A map $\tau\colon \mathcal{P}_{\mathbb{K}} \to \mathcal{P}_{\mathbb{K}}$ is a collineation if three points $a,b,c$ are collinear whenever so are $\tau(a), \tau(b), \tau(c)$ and vice versa. Over $\mathbb{R}$ every collineation is a projective transformation. In general a collinearity is a projective transformation composed with a field automorphism.

**Projective property:** Consider the space $(\mathcal{P}_{\mathbb{K}})^n$ of configurations of $n$ points. A property among these points is a subset $X \subseteq (\mathcal{P}_{\mathbb{K}})^n$ of this space with the interpretation that a point configuration has the property $X$ if it lies in this subset.

**Projectively invariant property:** A property $X$ of $n$ points with

$$([p_1], \dots, [p_n]) \in X \Leftrightarrow ([M \cdot p_1], \dots, [M \cdot p_n]) \in X,$$

for all projective transformations $M$.

**Determinant map:** Let $E = \{1, \dots, n\}$ be the index set of a list of vectors $P = (p_1, \dots, p_n)$ with underlying point configuration $([p_1], \dots, [p_n])$. The determinant map associates to every triple the corresponding determinant

$$\begin{array}{rcl} \Xi_P\colon E^3 & \to & \mathbb{K} \\ (i,j,k) & \mapsto & [p_i, p_j, p_k] \end{array}$$

**Proper configuration:** A configuration $P$ is proper if not all points are collinear. In this case the map $\Xi_P$ is not identically zero.

It turns out that projectively invariant properties are in close relation to the determinants of a point configuration.

**Theorem 3.1** *Up to projective equivalence a proper configuration* $([p_1],\dots,[p_n])$ *of* $P = (p_1,\dots,p_n)$ *is uniquely determined by the map* $\Xi_P$.

A proof of this theorem may be found in [25]. The basic idea of the proof is that after fixing a projective basis the positions of the points that satisfy the determinant values may be uniquely reconstructed. As a consequence every projectively invariant property may be decided entirely by the knowledge of the map $\Xi$. In view of this theorem it is not surprising that the fact that six points lie on a conic can be decided by the evaluation of a polynomial in the determinants among the vectors representing the points.

**Remark 3.1** In fact there are much stronger versions of the above theorem that also preserve the algebraic structure of such properties. In a stronger form (known as the First Fundamental Theorem of Invariant Theory, compare [12, 32]) it states that every projectively invariant property that may be encoded as the vanishing of a polynomial in the coordinate entries of the points may be rewritten as the vanishing of a polynomial of the determinants among vectors representing these points. See Section 3.1 of [30] for more on group actions and invariant polynomials.

## 3.2 Bracket Algebra: Straightening, Division, and Final Polynomials

### 3.2.1 Bracket Algebra and Straightening

**Glossary**

**Grassmann-Plücker relation in $\mathbb{K}^3$:** Let $a,b,c,d,e,f$ be six vectors in $\mathbb{K}^3$ then

$$[a,b,c][d,e,f]-[a,b,d][c,e,f]+[a,b,e][c,d,f]-[a,b,f][c,d,e]=0.$$

Similar expressions hold for all dimensions.

**Formal bracket:** A symbol $[\![ijk]\!]$ with $i,j,k \in E = \{1,2,\dots,n\}$ that will play the role of variables mimicking the role of formal determinants. The comma punctuation among the indices will be used only when an expression occurs as a vector variable in the bracket.

**Bracket polynomial:** A polynomial in the polynomial ring

$$\mathbf{R} := \mathbb{K}[\ [\![ijk]\!];(i,j,k)\in E^3\,].$$

**Alternating ideal:** An ideal mimicking the alternating determinant rules

$$\mathbf{Alt} := \Big\langle \big\{ [\![ijk]\!]+[\![jik]\!] \;\Big|\; i,j,k\in E \big\} \cup \big\{ [\![ijk]\!]+[\![ikj]\!] \;\Big|\; i,j,k\in E \big\} \Big\rangle.$$

In the quotient ring $\overline{\mathbf{R}} = \mathbf{R}/\mathbf{Alt}$ brackets that have the same index set are identified (with appropriate sign changes in accordance to the alternating determinant rules).

**Grassmann-Plücker ideal:** An ideal in $\overline{\mathbf{R}}$ mimicking the Grassmann-Plücker relations.

$$\mathbf{GP} := \Big\langle \big\{ [\![abc]\!][\![def]\!] - [\![abd]\!][\![cef]\!] + [\![abe]\!][\![cdf]\!] - [\![abf]\!][\![cde]\!] \;\big|\; a,b,c,d,e,f \in E \big\} \Big\rangle.$$

**Bracket ring (or algebra):** The quotient [35]

$$\mathbf{BR} := \mathbf{R}/\langle \mathbf{Alt} \cup \mathbf{GP} \rangle = \overline{\mathbf{R}}/\langle \mathbf{GP} \rangle.$$

---

In the light of Remark 3.1 it is a natural step to express properties of projective geometry entirely on the basis of determinants (brackets) among the vectors representing the points of a configuration. For this a natural question to ask is what kind of maps $\Xi \colon E^3 \to \mathbb{K}$ can be created as images from point configurations.

In fact the determinant values satisfy polynomial relations, the so-called Grassmann-Plücker relations that can be easily proved by expansion of the brackets. They impose natural restrictions on the map $\Xi_P$ that comes from a vector configuration $P$. If $\Xi$ comes from a configuration then it must satisfy all Grassmann-Plücker relations on subsets of points.

The Second Fundamental Theorem of Invariant Theory [22, 32] provides a close relation of the possible maps $\Xi$ that come from point configurations and Grassmann-Plücker relations.

**Theorem 3.2** *Let $\Xi : E^3 \to \mathbb{K}$ be an alternating map that does not vanish identically, i.e., for all indices $a,b,c \in E$, $\Xi(a,b,c) = -\Xi(b,a,c) = -\Xi(c,b,a) = -\Xi(a,c,b)$, and there exist $a,b,c \in E$ such that $\Xi(a,b,c) \neq 0$. If for all indices $a,b,c,d,e,f \in E$,*

$$\Xi(a,b,c)\Xi(d,e,f) - \Xi(a,b,d)\Xi(c,e,f) + \Xi(a,b,e)\Xi(c,d,f) - \Xi(a,b,f)\Xi(c,d,e) = 0$$

*then there is a vector configuration $P$ with $\Xi = \Xi_P$.*

The close relation of Grassmann-Plücker relations to possible maps $\Xi$ provides us with the possibility to describe a projective configuration entirely in terms of (formal) determinants. For this we fix an index set $E = \{1,2,\ldots,n\}$ and mimic the behavior of determinants by formal symbols.

Since determinants coming from a vector configuration are characterized by the fact that they are alternating and satisfy the Grassmann-Plücker relations, the bracket ring consists of bracket polynomials such that two polynomials are identified if they evaluate to the same value for every vector configuration. To make this statement more precise we consider the map $\Xi_P$ and reinterpret it with the formal brackets as arguments by setting $\Xi_P([\![ijk]\!]) := \Xi_P(i,j,k)$. We extend this map by the natural homomorphism to polynomials in $\mathbf{R}$ by setting $\Xi_P(x+y) = \Xi_P(x) + \Xi_P(y)$ and $\Xi_P(x \cdot y) = \Xi_P(x) \cdot \Xi_P(y)$. The bracket ring has the following crucial property:

**Theorem 3.3** *Let $X$ and $Y$ be polynomials in $\mathbf{R}$. The following statements are equivalent:*

*(i)* $\Xi_P(X) = \Xi_P(Y)$ *for all vector configurations $P$.*

*(ii)* $X = Y$ *in the bracket ring* **BR**.

For example, consider the two bracket polynomials

$$X = [\![abc]\!][\![def]\!] - [\![abd]\!][\![cef]\!] \quad \text{and}$$
$$Y = [\![abe]\!][\![cdf]\!] - [\![abf]\!][\![cde]\!].$$

They are equal to each other in **BR** since their sum is a Grassmann-Plücker relation. If the corresponding determinant polynomial $\Xi_P(X)$ vanishes for a configuration $P$ then so does $\Xi_P(Y)$. Both polynomials encode the same projectively invariant property (if they vanish), namely that the lines

$join(a,b), join(c,d), join(e,f)$ meet. See Section 3.2 of [30] for an explicit algebraic description of the relations on brackets.

As there are algebraic relations among the formal brackets taken as arguments, two drastically different polynomials in the same bracket ring may still be equal to each other. The *normal form* of a bracket polynomial $f$ refers to another bracket polynomial $N(f)$ such that any two bracket polynomials $f,g$ are equal if and only if $N(f)$ and $N(g)$ are identical. The normal form $N(f)$ of $f$ is said to be *straight*, and any algorithm changing $f$ into $N(f)$ for arbitrary input bracket polynomial $f$ is called a *straightening algorithm* [1, 2].

The term "straight" comes from the following unique property of bracket monomials in normal form. A bracket monomial

$$f = [\![a_1b_1c_1]\!][\![a_2b_2c_2]\!]\cdots[\![a_nb_nc_n]\!]$$

can be written in the following *tableau form* by piling up its bracket arguments:

$$f = \begin{bmatrix} a_1 & b_1 & c_1 \\ a_2 & b_2 & c_2 \\ \vdots & \vdots & \vdots \\ a_n & b_n & c_n \end{bmatrix}.$$

Given a total order among the vector variables (or indices of vector variables bearing the same name) $a_i, b_j, c_k$, then $f$ is in normal form if and only if in its tableau form, along each row, the order is increasing, and along each column, the order is non-decreasing.

For example, when $d_1 \prec d_2 \prec \cdots \prec d_5$, then $\begin{bmatrix} d_1 & d_3 & d_5 \\ d_2 & d_4 & d_5 \end{bmatrix}$ is straight, while $\begin{bmatrix} d_1 & d_4 & d_5 \\ d_2 & d_3 & d_5 \end{bmatrix}$ is not, because its second column is not "straight."

The classical *Young's straightening algorithm* [36, 37] is based on the so-called *van der Waerden relations*. They include not only the Grassmann-Plücker relations, but also the following:

$$\begin{array}{rl} [\![ab_1b_2]\!][\![b_3b_4c]\!] - [\![ab_1b_3]\!][\![b_2b_4c]\!] + [\![ab_1b_4]\!][\![b_2b_3c]\!] & \\ +[\![ab_2b_3]\!][\![b_1b_4c]\!] - [\![ab_2b_4]\!][\![b_1b_3c]\!] + [\![ab_3b_4]\!][\![b_1b_2c]\!] & = 0. \end{array} \tag{3.2}$$

By the Grassmann-Plücker relations, the two lines on the left side of the above equality are both equal to $[\![ab_1c]\!][\![b_2b_3b_4]\!]$, so the van der Waerden relations are in the ideal $\langle \mathbf{GP} \rangle$.

In [33], it is shown that all van der Waerden relations form a Gröbner basis of $\langle \mathbf{GP} \rangle$, and the classical Young's straightening algorithm is a top reduction procedure with respect to the Gröbner basis. When the order among bracket monomials is the row-deglex order (to be introduced below), then Young's straightening algorithm reduces the row-deglex order of the input bracket polynomial every time a van der Waerden relation is employed.

---

**Glossary**

**Straight bracket monomial:** A bracket monomial in tableau form where along each row from left to right, the order of vector variables is increasing, and along each column from top to bottom, the order of vector variables is non-decreasing. The procedure changing a bracket polynomial into its straight form (normal form) is called straightening.

**van der Waerden relation:** An identity in the bracket ring obtained from the fact that the antisymmetric tensor product of any four vector variables $b_1, b_2, b_3, b_4$ is zero. Given two other vector variables $a, c$, there are six different ways of dividing the 4-tuple of vector variables into two pairs, with the first pair allocated to $a$ and the second pair allocated to $c$, such that each 3-tuple form a bracket, and a bracket monomial of degree two is generated by each allocation. As the antisymmetric tensor product of the original 4-tuple is zero, so is the signed sum (3.2) of the degree-2 bracket monomials over all possible allocations.

**Row-deglex order:** A total order among such monic bracket monomials that within each bracket, the order of vector variables is increasing. The row-deglex order is the degree lexicographical order among the resulting sequences of vector variables obtained by removing all bracket symbols from the monic bracket monomials. For example, if $a_1 \prec a_2 \prec \ldots \prec a_6$, then $\left[\begin{array}{cc} a_1 & a_4 \\ a_2 & a_5 \\ a_3 & a_6 \end{array}\right] \prec \left[\begin{array}{cc} a_1 & a_5 \\ a_2 & a_3 \\ a_4 & a_6 \end{array}\right]$ in row-deglex order, because when scanned by row, $a_1a_4a_2a_5a_3a_6 \prec a_1a_5a_2a_3a_4a_6$ degree lexicographically.

**Admissible order:** An admissible order among straight monic bracket monomials is a total order that is preserved by multiplication and then straightening. For example, the row-deglex order is not an admissible order among straight monic bracket monomials.

---

### 3.2.2 Division

---

**Glossary**

**Negative column-deglex order:** A total order among monic bracket monomials where within each bracket, the order of vector variables is increasing. It is the negative of the degree lexicographical order among the resulting sequences of vector variables obtained by scanning the tableau forms of the monic bracket monomials columnwise from the first column to the last. For example, if $a_1 \prec a_2 \prec \ldots \prec a_6$, then $\left[\begin{array}{cc} a_1 & a_4 \\ a_2 & a_5 \\ a_3 & a_6 \end{array}\right] \succ \left[\begin{array}{cc} a_1 & a_5 \\ a_2 & a_3 \\ a_4 & a_6 \end{array}\right]$ in negative column-deglex order, because when scanned by column, $a_1a_2a_3a_4a_5a_6 \prec a_1a_2a_4a_5a_3a_6$ degree lexicographically.

**Columnwise straightening:** Given a monic bracket monomial $f$, its columnwise straightening is denoted by $f^{\downdownarrows}$, and is a monic bracket monomial obtained from the tableau form of $f$ by sorting the entries of each column of $f$ so that the entries become non-decreasing within each column. The operator "$\downdownarrows$" can be extended linearly to all bracket polynomials.

**Columnwise reduction:** Given two monic bracket monomials $f, g$, for any column of $g$, if all its entries belong to the same column of $f$, then $f$ is said to be columnwise reducible with respect to $g$. Let $h$ be the monic bracket monomial whose tableau form is obtained from that of $f$ by removing the entries of $g$ columnwise from $f$, called the columnwise quotient of $f$ with respect to $g$. Then $f = gh + r$ for some bracket polynomial $r$ called the columnwise remainder, and the procedure is called the columnwise reduction of $f$ by $g$.

**Homogeneous bracket polynomial:** A bracket polynomial that is homogeneous in its every vector variable.

---

For example, when $f = \left[\begin{array}{ccc} d_1 & d_4 & d_5 \\ d_2 & d_3 & d_5 \end{array}\right]$, then $f^{\downdownarrows} = \left[\begin{array}{ccc} d_1 & d_3 & d_5 \\ d_2 & d_4 & d_5 \end{array}\right]$; it is clearly straight. Given $g = \left[\begin{array}{ccc} d_2 & d_4 & d_5 \end{array}\right]$, then $f$ is columnwise reducible with respect to $g$, the columnwise quotient is $h = \left[\begin{array}{ccc} d_1 & d_3 & d_5 \end{array}\right]$, and the columnwise remainder is

$$r = f - gh = \left[\begin{array}{ccc} d_1 & d_4 & d_5 \\ d_2 & d_3 & d_5 \end{array}\right] - \left[\begin{array}{ccc} d_1 & d_3 & d_5 \\ d_2 & d_4 & d_5 \end{array}\right] = -\left[\begin{array}{ccc} d_1 & d_2 & d_5 \\ d_3 & d_4 & d_5 \end{array}\right].$$

The last step follows a three-summand Grassmann-Plücker relation.

**Theorem 3.4** *[8] Given a homogeneous bracket polynomial $f$, the leading term of its normal form $N(f)$ under the negative column-deglex order is the same with the leading term of $f^{\downarrow\downarrow}$. In particular when $f$ is a bracket monomial, then the leading term of $N(f)$ is $f^{\downarrow\downarrow}$, and the latter is always straight.*

**Theorem 3.5** *[19] The negative column-deglex order is an admissible order among straight homogeneous bracket polynomials: let $f,g$ be two straight homogeneous bracket polynomials and $f \prec g$ in the negative column-deglex order, then for any nonzero homogeneous bracket polynomial $h$, $N(fh) \prec N(gh)$.*

Given two homogeneous bracket polynomials $f,g$, the *division* of $f$ by $g$ is defined as the reduction of $f$ with respect to $\langle g, \mathbf{GP}\rangle$. Usually this can be done by first computing a Gröbner basis **GB** [31, 41] of the ideal and then using it to make top reduction to $f$. The number of elements of **GB** is usually greater than one, and the elements cannot be predicted from the explicit form of $g$, so the division by Gröbner basis is dramatically different from the canonical division between two polynomials by top reduction, where the divisor contains only one polynomial all the time.

Now that the negative column-deglex order "$\prec$" is an admissible order among straight homogeneous bracket polynomials, for two homogeneous bracket polynomials $f \succeq g$ in straight form, the columnwise reduction of the leading term $\mathbf{lt}(f)$ of $f$ by the leading term $\mathbf{lt}(g)$ of $g$ results in a columnwise quotient $h$ that is a bracket monomial, such that $\mathbf{lt}(f) = h \cdot \mathbf{lt}(g) + r$, where the columnwise remainder $r$ satisfies $N(r) \prec \mathbf{lt}(f)$. Then $f = h \cdot g + \tilde{r}$, where

$$\tilde{r} = r + (f - \mathbf{lt}(f)) - h \cdot (g - \mathbf{lt}(g))$$

satisfies $N(\tilde{r}) \prec \mathbf{lt}(f)$. This is called an *invariant top reduction* of $f$ by $g$. The invariant top reduction of $N(\tilde{r})$ by $g$ can be done similarly. By successive invariant top reductions with respect to $g$, we get $f = h' \cdot g + r'$, where $h', r'$ are bracket polynomials, $N(r') \prec f$, and $\mathbf{lt}(N(r'))$ is not columnwise reducible with respect to $\mathbf{lt}(g)$. This procedure is called an *invariant division* of $f$ by $g$, and $h', r'$ are called the *invariant quotient* and *invariant remainder*, respectively.

### 3.2.3 Final Polynomials

---

**Glossary**

**Non-degeneracy monomial:** A bracket monomial $N \in \overline{\mathbf{R}}$ such that $N = b_1 \cdot b_2 \cdots b_l$ with $b_d = [\![ijk]\!]$ and $(i,j,k) \notin \mathcal{T}$ for all $d \in \{1, \ldots, l\}$, where $\mathcal{T}$ is a finite set of collinear triples.

**$\mathcal{T}$-vanishing polynomial:** A bracket monomial $A \in \overline{\mathbf{R}}$ in the ideal

$$\left\langle \{[\![ijk]\!] \mid (i,j,k) \in \mathcal{T}\} \right\rangle.$$

**Final polynomial:** A polynomial $X \in \mathbf{GP}$ such that $X = A + N$ for a $\mathcal{T}$-vanishing polynomial $A$ and a non-degeneracy monomial $N$.

---

Being able to express projective geometry entirely on the level of determinants implies that there must be ways to prove geometric theorems without ever referring to concrete points and their specific coordinates. Usually such *coordinate-free* calculations tend to be considerably shorter and more instructive than calculations that refer to a concrete embedding and to a particular choice of a basis. The technique for this is called *final polynomials* and was introduced by Bokowski and Sturmfels [6] and independently Whiteley [40] in the context of non-realizability proofs for matroids and oriented matroids.

We will start by exemplifying the method of final polynomials with the example of a proof of Pappus' Theorem. For this we consider the following version of Pappus' Theorem which in its statement is slightly more symmetric than our original formulation. Our final polynomial will in particular be of the type of bi-quadratic final polynomials which are algorithmically easier to find [5, 24] and admit additional structural properties [26].

*Let* $1, 2, \ldots, 9$ *be the indices of nine points in the projective plane. Consider the nine triples of points* $(1,2,3)$, $(1,5,9)$, $(1,6,8)$, $(4,5,6)$, $(4,3,8)$, $(4,2,9)$, $(7,2,6)$, $(7,5,3)$ *and* $(7,8,9)$. *Under the non-degeneracy assumptions that all other triples are non-collinear, the collinearity of eight of the nine triples implies the collinearity of the last one.*

To give a proof that is entirely based on calculations with determinants consider the following polynomial $X$ in formal determinants. It is a linear combination of Grassmann-Plücker relations. Hence it must be zero in the bracket ring **BR**. Equivalently we can state that the evaluation $\Xi_P(X)$ (where we replace each formal bracket with the value of the corresponding determinant) must vanish for every point configuration $P$.

$$
\begin{aligned}
X \;=\; &+(\llbracket 714\rrbracket\llbracket 735\rrbracket - \llbracket 713\rrbracket\llbracket 745\rrbracket + \llbracket 715\rrbracket\llbracket 743\rrbracket)\cdot\llbracket 148\rrbracket\llbracket 127\rrbracket\llbracket 149\rrbracket\llbracket 467\rrbracket\\
&+(\llbracket 471\rrbracket\llbracket 438\rrbracket - \llbracket 473\rrbracket\llbracket 418\rrbracket + \llbracket 478\rrbracket\llbracket 413\rrbracket)\cdot\llbracket 157\rrbracket\llbracket 127\rrbracket\llbracket 149\rrbracket\llbracket 467\rrbracket\\
&+(\llbracket 147\rrbracket\llbracket 132\rrbracket - \llbracket 143\rrbracket\llbracket 172\rrbracket + \llbracket 142\rrbracket\llbracket 173\rrbracket)\cdot\llbracket 157\rrbracket\llbracket 478\rrbracket\llbracket 149\rrbracket\llbracket 467\rrbracket\\
&-(\llbracket 147\rrbracket\llbracket 195\rrbracket - \llbracket 149\rrbracket\llbracket 175\rrbracket + \llbracket 145\rrbracket\llbracket 179\rrbracket)\cdot\llbracket 478\rrbracket\llbracket 124\rrbracket\llbracket 137\rrbracket\llbracket 467\rrbracket\\
&-(\llbracket 714\rrbracket\llbracket 798\rrbracket - \llbracket 719\rrbracket\llbracket 748\rrbracket + \llbracket 718\rrbracket\llbracket 749\rrbracket)\cdot\llbracket 145\rrbracket\llbracket 124\rrbracket\llbracket 137\rrbracket\llbracket 467\rrbracket\\
&-(\llbracket 471\rrbracket\llbracket 492\rrbracket - \llbracket 479\rrbracket\llbracket 412\rrbracket + \llbracket 472\rrbracket\llbracket 419\rrbracket)\cdot\llbracket 145\rrbracket\llbracket 178\rrbracket\llbracket 137\rrbracket\llbracket 467\rrbracket\\
&+(\llbracket 471\rrbracket\llbracket 465\rrbracket - \llbracket 476\rrbracket\llbracket 415\rrbracket + \llbracket 475\rrbracket\llbracket 416\rrbracket)\cdot\llbracket 178\rrbracket\llbracket 247\rrbracket\llbracket 149\rrbracket\llbracket 137\rrbracket\\
&+(\llbracket 147\rrbracket\llbracket 168\rrbracket - \llbracket 146\rrbracket\llbracket 178\rrbracket + \llbracket 148\rrbracket\llbracket 176\rrbracket)\cdot\llbracket 457\rrbracket\llbracket 247\rrbracket\llbracket 149\rrbracket\llbracket 137\rrbracket\\
&+(\llbracket 714\rrbracket\llbracket 762\rrbracket - \llbracket 716\rrbracket\llbracket 742\rrbracket + \llbracket 712\rrbracket\llbracket 746\rrbracket)\cdot\llbracket 457\rrbracket\llbracket 148\rrbracket\llbracket 149\rrbracket\llbracket 137\rrbracket.
\end{aligned}
$$

Within the ring $\overline{\mathbf{R}}$ we can apply alternating determinant rules and calculations with polynomials to rewrite the above polynomial in the following way (after expanding the summands, reordering brackets and cancelling terms):

$$
\begin{aligned}
X = \;&+\llbracket 147\rrbracket\underline{\llbracket 735\rrbracket}\llbracket 148\rrbracket\llbracket 127\rrbracket\llbracket 149\rrbracket\llbracket 467\rrbracket + \llbracket 147\rrbracket\underline{\llbracket 438\rrbracket}\llbracket 157\rrbracket\llbracket 127\rrbracket\llbracket 149\rrbracket\llbracket 467\rrbracket\\
&+\llbracket 147\rrbracket\underline{\llbracket 132\rrbracket}\llbracket 157\rrbracket\llbracket 478\rrbracket\llbracket 149\rrbracket\llbracket 467\rrbracket - \llbracket 147\rrbracket\underline{\llbracket 195\rrbracket}\llbracket 478\rrbracket\llbracket 124\rrbracket\llbracket 137\rrbracket\llbracket 467\rrbracket\\
&-\llbracket 147\rrbracket\underline{\llbracket 798\rrbracket}\llbracket 145\rrbracket\llbracket 124\rrbracket\llbracket 137\rrbracket\llbracket 467\rrbracket - \llbracket 147\rrbracket\underline{\llbracket 492\rrbracket}\llbracket 145\rrbracket\llbracket 178\rrbracket\llbracket 137\rrbracket\llbracket 467\rrbracket\\
&+\llbracket 147\rrbracket\underline{\llbracket 465\rrbracket}\llbracket 178\rrbracket\llbracket 247\rrbracket\llbracket 149\rrbracket\llbracket 137\rrbracket + \llbracket 147\rrbracket\underline{\llbracket 168\rrbracket}\llbracket 457\rrbracket\llbracket 247\rrbracket\llbracket 149\rrbracket\llbracket 137\rrbracket\\
&+\llbracket 147\rrbracket\underline{\llbracket 762\rrbracket}\llbracket 457\rrbracket\llbracket 148\rrbracket\llbracket 149\rrbracket\llbracket 137\rrbracket.
\end{aligned}
$$

We know that also the above expression must evaluate to zero if we replace formal determinants by the values of real determinants coming from any vector configuration. The underlined brackets refer to the triples that play a role in our formulation of Pappus' Theorem. Now assume that in contradiction to Pappus' Theorem there is a point configuration $P$ in which only eight of the collinearities are satisfied but not the ninth. Our non-degeneracy assumptions imply that all other brackets are non-zero. This immediately leads to a contradiction to the above expression being zero, since in this case all summands except for one would vanish.

The decisive point is that this proving strategy works in general and provides us with a general tool for carrying out geometric proofs on the level of determinants. To see this consider the following setup. Assume that we want to prove the non-existence of a configuration $P$ which has a certain set of collinear triples $\mathcal{T}$ and all other triples are non-collinear (like in our proof an assumed configuration that satisfies only eight of the Pappus triples).

The existence of a final polynomial $X$ proves the non-existence of a configuration that has exclusively the collinearities indicated by $\mathcal{T}$. $X$ being in **GP** implies that this polynomial evaluates to zero for any given configuration. The property $X = A + N$ implies it to be non-zero if the triples in $\mathcal{T}$ are realized as collinear triples. Over algebraically closed fields the converse is also true.

**Theorem 3.6** *Over an algebraically closed field, if a configuration is not realizable then it must admit a final polynomial.*

The non-existence of a configuration in the presence of a final polynomial can also be explored in general fields. However the converse direction relies on Hilbert's Nullstellensatz. There are versions of final polynomials that apply to the real numbers field. Details may be found in [6, 32].

## 3.3 Cayley Expansion and Factorization

### 3.3.1 Cayley Expansion

**Glossary**

**Cayley bracket:** Any expression in Grassmann-Cayley algebra that contains only the meet and join operations is called a monic Cayley monomial. If in addition the expression is scalar-valued, then it is called a monic Cayley bracket. A Cayley bracket is the scaling of a monic Cayley bracket by a factor of $\mathbb{K}$.

**Cayley expansion:** By the First Fundamental Theorem of Invariant Theory, any Cayley bracket equals a homogeneous bracket polynomial. Changing a Cayley bracket into an equal bracket polynomial is called Cayley expansion.

**Binomial Cayley expansion:** It refers to changing a Cayley bracket into an equal homogeneous bracket binomial. Formula (3.1) provides the following *shuffle formula* of join-splitting binomial Cayley expansion:

$$\begin{aligned}(a_1 \vee a_2) \wedge (b_1 \vee b_2) \wedge (c_1 \vee c_2) &= [\![a_1 a_2 b_1]\!][\![b_2 c_1 c_2]\!] - [\![a_1 a_2 b_2]\!][\![b_1 c_1 c_2]\!] \\ &= -[\![a_1 b_1 b_2]\!][\![a_2 c_1 c_2]\!] + [\![a_2 b_1 b_2]\!][\![a_1 c_1 c_2]\!] \\ &= -[\![a_1 a_2 c_1]\!][\![b_1 b_2 c_2]\!] + [\![a_1 a_2 c_2]\!][\![b_1 b_2 c_1]\!].\end{aligned} \tag{3.3}$$

The three different binomial Cayley expansions are obtained by alternatively distributing the two vector variables of $b_1 \vee b_2$, or $a_1 \vee a_2$, or $c_1 \vee c_2$, to the other two joins, respectively. The equality of the three different results follows Grassmann-Plücker relations.

**Monomial Cayley expansion:** It refers to changing a Cayley bracket into another equal expression that is the multiplication of more than one Cayley bracket. For example in (3.3), when $a_1 = b_1$, then the first two expansion results are identically the following monomial:

$$(a_1 \vee a_2) \wedge (a_1 \vee b_2) \wedge (c_1 \vee c_2) = -[\![a_1 a_2 b_2]\!][\![a_1 c_1 c_2]\!]. \tag{3.4}$$

Changing a Cayley bracket into a bracket polynomial is a procedure of eliminating the meet operation, and algebraically this is a procedure of simplification, as the meet and join operations are not associative when both occur in the same Cayley monomial. Cayley expansion has a prominent feature that the expansion result is generally not unique, and in some cases, even for monomial Cayley expansion.

Although formula (3.3) gives three different expansion results, all of them are by distributing entries of a join operation to the other join operations. In duality, for some Cayley monomials one can distribute entries of a meet operation to other meet operations [3, 9]. For example, let

$$f = [\![(a_1 \vee a_2) \wedge (a_1' \vee a_2'),\ (a_1 \vee a_3) \wedge (a_1' \vee a_3'),\ (a_2 \vee a_3) \wedge (a_2' \vee a_3')]\!], \tag{3.5}$$

then by distributing the two entries $a_1 \vee a_2$ and $a_1' \vee a_2'$ of the first meet operation to the other two meet operations, we get

$$\begin{aligned} f &= -((a_1 \vee a_2) \wedge (a_1 \vee a_3) \wedge (a_1' \vee a_3')) \cdot ((a_1' \vee a_2') \wedge (a_2 \vee a_3) \wedge (a_2' \vee a_3')) \\ &\quad +((a_1 \vee a_2) \wedge (a_2 \vee a_3) \wedge (a_2' \vee a_3')) \cdot ((a_1' \vee a_2') \wedge (a_1 \vee a_3) \wedge (a_1' \vee a_3')) \\ &= [\![a_1a_2a_3]\!][\![a_1'a_2'a_3']\!] \cdot ([\![a_1a_1'a_3']\!][\![a_2a_3a_2']\!] - [\![a_1a_3a_1']\!][\![a_2a_2'a_3']\!]). \end{aligned} \tag{3.6}$$

Alternatively one can use the join-splitting shuffle formula to expand $(a_1 \vee a_2) \wedge (a_1' \vee a_2')$ first, as follows:

$$\begin{aligned} f &= -[\![a_1a_1'a_2']\!][\![a_2,\ a_1a_3 \vee a_1'a_3',\ a_2a_3 \vee a_2'a_3']\!] \\ &\quad +[\![a_2a_1'a_2']\!][\![a_1,\ a_1a_3 \vee a_1'a_3',\ a_2a_3 \vee a_2'a_3']\!] \\ &= [\![a_1a_1'a_2']\!][\![a_2a_2'a_3']\!][\![a_2,\ a_1a_3 \vee a_1'a_3',\ a_3]\!] \\ &\quad -[\![a_2a_1'a_2']\!][\![a_1a_1'a_3']\!][\![a_1,\ a_3,\ a_2a_3 \vee a_2'a_3']\!] \\ &= [\![a_1a_2a_3]\!](-[\![a_1a_1'a_2']\!][\![a_2a_2'a_3']\!][\![a_3a_1'a_3']\!] + [\![a_2a_1'a_2']\!][\![a_1a_1'a_3']\!][\![a_3a_2'a_3']\!]). \end{aligned} \tag{3.7}$$

Besides (3.3) and (3.4), there are other formulas for binomial Cayley expansion and monomial Cayley expansion. For example we have the following six formulas [16]:

$$\begin{aligned}
\text{i.}\quad & ((a_1 \vee a_2) \wedge (a_3 \vee a_4) \wedge (a_5 \vee a_6)) \cdot ((a_1 \vee a_3) \wedge (a_2 \vee a_4) \wedge (a_5 \vee a_6)) \\
&= [\![a_1a_2a_4]\!][\![a_1a_3a_4]\!][\![a_2a_5a_6]\!][\![a_3a_5a_6]\!] - [\![a_1a_2a_3]\!][\![a_2a_3a_4]\!][\![a_1a_5a_6]\!][\![a_4a_5a_6]\!]; \\[6pt]
\text{ii.}\quad & [\![a_1,\ (a_1' \vee a_2') \wedge (a_3' \vee a_4'),\ (a_1' \vee a_2') \wedge (a_3'' \vee a_4'')]\!] \\
&= [\![a_1a_1'a_2']\!]((a_1' \vee a_2') \wedge (a_3' \vee a_4') \wedge (a_3'' \vee a_4'')); \\[6pt]
\text{iii.}\quad & [\![(a_1 \vee a_2) \wedge (a_3 \vee a_4),\ (a_1 \vee a_2) \wedge (a_3' \vee a_4'),\ (a_1'' \vee a_2'') \wedge (a_3'' \vee a_4'')]\!] \\
&= -((a_1 \vee a_2) \wedge (a_3 \vee a_4) \wedge (a_3' \vee a_4')) \cdot ((a_1 \vee a_2) \wedge (a_1'' \vee a_2'') \wedge (a_3'' \vee a_4'')); \\[6pt]
\text{iv.}\quad & [\![(a_1 \vee a_2) \wedge (a_3 \vee a_4),\ (a_1 \vee a_2') \wedge (a_3' \vee a_4'),\ (a_2 \vee a_2') \wedge (a_3'' \vee a_4'')]\!] \\
&= -[\![a_1a_2a_2']\!]\ ([\![a_1a_3a_4]\!][\![a_2a_3''a_4'']\!][\![a_2'a_3'a_4']\!] - [\![a_1a_3'a_4']\!][\![a_2a_3a_4]\!][\![a_2'a_3''a_4'']\!]); \\[6pt]
\text{v.}\quad & [\![(a_1 \vee a_2) \wedge (a_3 \vee a_4),\ (a_1 \vee a_3) \wedge (a_2 \vee a_4),\ (a_1 \vee a_4) \wedge (a_3' \vee a_4')]\!] \\
&= [\![a_1a_2a_4]\!][\![a_1a_3a_4]\!]\ ([\![a_1a_2a_3]\!][\![a_4a_3'a_4']\!] + [\![a_1a_3'a_4']\!][\![a_2a_3a_4]\!]); \\[6pt]
\text{vi.}\quad & [\![(a_1 \vee a_2) \wedge (a_3 \vee a_4),\ (a_1 \vee a_2') \wedge (a_3 \vee a_4'),\ (a_2 \vee a_2') \wedge (a_4 \vee a_4')]\!] \\
&= -[\![a_1a_2a_2']\!][\![a_3a_4a_4']\!]((a_1 \vee a_3) \wedge (a_2 \vee a_4) \wedge (a_2' \vee a_4')).
\end{aligned} \tag{3.8}$$

As the result of Cayley expansion of a Cayley polynomial is generally not unique, the Cayley expansions leading to factored and shortest results are often critical for effective symbolic computing of incidence configurations. Their finding and classification are the main content of Cayley expansion theory [16].

How diverse can different Cayley expansion results of the same Cayley bracket be? For generic vector variables, while $(a_1 \vee a_2) \wedge (a_3 \vee a_4) \wedge (a_5 \vee a_6)$ has only three different Cayley expansion results in bracket polynomials,

- $((a_1 \vee a_2) \wedge (a_3 \vee a_4) \wedge (a_5 \vee a_6)) \cdot ((a'_1 \vee a'_2) \wedge (a'_3 \vee a'_4) \wedge (a'_5 \vee a'_6))$ has 45 different expansion results in bracket polynomials;
- $[\![a_1,\, (a'_1 \vee a'_2) \wedge (a'_3 \vee a'_4),\, (a''_1 \vee a''_2) \wedge (a''_3 \vee a''_4)]\!]$ has 46 different expansion results in bracket polynomials;
- $[\![(a_1 \vee a_2) \wedge (a_3 \vee a_4),\, (a'_1 \vee a'_2) \wedge (a'_3 \vee a'_4),\, (a''_1 \vee a''_2) \wedge (a''_3 \vee a''_4)]\!]$ has 16,847 different expansion results in bracket polynomials.

### 3.3.2 Cayley Factorization

---

**Glossary**

**Cayley factorization:** The inverse of Cayley expansion: changing a homogeneous bracket polynomial into an equal Cayley bracket.

**Rational Cayley factorization:** Changing a homogeneous bracket polynomial into an equal rational function where the numerator is a Cayley bracket, and the denominator is a bracket monomial.

---

If a homogeneous bracket polynomial can be converted into a Cayley bracket, then the incidence geometric interpretation of the bracket polynomial can be read directly from the expression of the Cayley bracket. Cayley factorization is a procedure of generating geometric interpretation for a projective invariant by incidence constructions. Unfortunately, for a general homogeneous bracket polynomial, its Cayley factorization usually does not exist. See Section 5 of [30] for the special case of multilinear Cayley factorization.

Besides geometric translation, Cayley factorization is also useful in Cayley expansion. Suppose that a Cayley bracket is changed into different factored forms of bracket polynomials. Then Cayley factorization may be used to unify different factors of the Cayley expansion results. This technique is particularly useful in robust symbolic computing of projective conic theorems, in that the defect of diverse Cayley expansion results may be significantly reduced or even eliminated.

For example, (3.6) and (3.7) are two different Cayley expansion results of the same input (3.5). Although they are both binomial, the two results are from different shuffle formulas. To unify the different factors of the results, equality (3.3) from right to left, and the fourth equality of (3.8) from right to left, can be used to make Cayley factorization, and the results are unified to

$$[\![a_1 a_2 a_3]\!]\, [\![a'_1 a'_2 a'_3]\!]\, (a_1 \vee a'_1) \wedge (a_2 \vee a'_2) \wedge (a_3 \vee a'_3). \tag{3.9}$$

Cayley factorization is a difficult task. Only the simplest case where a bracket polynomial is linear with respect to every vector variable of it is solved [37]. Even the following seemingly simple question remains open: is the following *Crapo's binomial*

$$[\![a_1 a'_2 a'_3]\!]\, [\![a_2 a'_3 a'_4]\!] \cdots [\![a_k a'_1 a'_2]\!] + (-1)^{k-1} [\![a_1 a'_1 a'_2]\!]\, [\![a_2 a'_2 a'_3]\!] \cdots [\![a_k a'_k a'_1]\!]$$

Cayley factorizable?

In [34], the problem of "rational Cayley factorizability" was investigated. The problem is as follows: Given a bracket polynomial that is not Cayley factorizable, is it Cayley factorizable after being multiplied with a suitable bracket monomial? An affirmative answer was given in [31]. The technique *rational Cayley factorization* is very important in projective geometric computing. Still there is no general algorithm for this factorization.

The following is a simple example of rational Cayley factorization. By equality (3.3) from right to left,

$$[\![a_1a_3a_4]\!][\![a_2a_5a_6]\!] - [\![a_2a_3a_4]\!][\![a_1a_5a_6]\!] = (a_1 \vee a_2) \vee (a_3 \vee a_4) \vee (a_5 \vee a_6)$$

is a Cayley factorization. If the minus sign on the left side is changed to a plus sign, the bracket binomial is no longer Cayley factorizable. Instead, it is rational Cayley factorizable, by the fifth equality of (3.8) from right to left:

$$\begin{aligned} & [\![a_1a_3a_4]\!][\![a_2a_5a_6]\!] + [\![a_2a_3a_4]\!][\![a_1a_5a_6]\!] \\ = \; & \frac{[\![(a_1 \vee a_2) \vee (a_5 \vee a_6),\ (a_1 \vee a_3) \vee (a_2 \vee a_4),\ (a_1 \vee a_4) \vee (a_2 \vee a_3)]\!]}{[\![a_1a_2a_3]\!][\![a_1a_2a_4]\!]} \\ = \; & \frac{[\![(a_1 \vee a_2) \vee (a_3 \vee a_4),\ (a_1 \vee a_5) \vee (a_2 \vee a_6),\ (a_1 \vee a_6) \vee (a_2 \vee a_5)]\!]}{[\![a_1a_2a_5]\!][\![a_1a_2a_6]\!]}. \end{aligned}$$

### 3.3.3 Cayley Expansion and Factorization in Geometric Theorem Proving

Cayley expansion and factorization have various applications in combinatorics [10, 20], computer vision [15, 17], robotics [38, 39], etc., and can be generalized to superalgebras [11, 13, 27, 28, 29].

Cayley expansion and factorization are particularly useful in proving incidence theorems having linear construction sequences [18]. In a linear construction sequence of a geometric configuration, any geometric entity is constructed by equations linear in the vector variable representing the geometric entity. A constrained geometric entity in a linear construction sequence is usually represented by a Cayley monomial. To prove a conclusion in the form of an equality in Cayley brackets, we only need to substitute the Cayley monomial representations of the geometric entities into the corresponding vector variables in the Cayley brackets in the reverse order of the linear construction sequence, and simplify the result by Cayley expansion and factorization. This method works very well, and for all projective incidence theorems tested so far, each theorem can be given a *binomial proof*, i.e., the maximal number of terms of the conclusion expression during elimination and simplification is two.

The benefit of the Cayley expansion and factorization method is not restricted to providing short proofs. If one constraint of equality type is removed from the original geometric configuration, the Cayley expansion and factorization approach usually provides quantitative description of the relationship between a measurement of the discrepancy of the conclusion from being true and a measurement of the discrepancy of the missing constraint from being satisfied. The following Desargues' Theorem [4] is a typical example.

*For two triangles $a_1a_2a_3$ and $a'_1a'_2a'_3$ in the plane, if the three lines $a_1a'_1, a_2a'_2, a_3a'_3$ concur, then the three points of intersection $a = a_1a_2 \cap a'_1a'_2$, $b = a_1a_3 \cap a'_1a'_3$, and $c = a_2a_3 \cap a'_2a'_3$ are collinear.*

There is only one equality constraint in the hypothesis, which is the concurrence of the three lines $a_1a'_1, a_2a'_2, a_3a'_3$. In Grassmann-Cayley algebra, the hypothesis is represented by

$$f := (a_1 \vee a'_1) \wedge (a_2 \vee a'_2) \wedge (a_3 \vee a'_3) = 0,$$

and the conclusion is represented by

$$g := [\![(a_1 \vee a_2) \wedge (a'_1 \vee a'_2),\ (a_1 \vee a_3) \wedge (a'_1 \vee a'_3),\ (a_2 \vee a_3) \wedge (a'_2 \vee a'_3)]\!] = 0.$$

We disclose the relationship between the hypothesis expression and the conclusion expression by Cayley expansion and factorization.

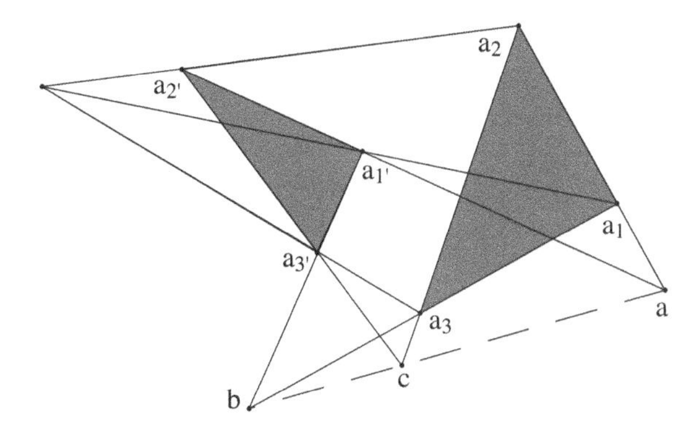


**Figure 3.2**
Desargues' Theorem.

By Cayley expansion (3.6) or (3.7), then Cayley factorization (3.9), we get the following identity, which is just the last equality of (3.8):

$$\begin{aligned} & [\![(a_1 \vee a_2) \wedge (a_1' \vee a_2'),\ (a_1 \vee a_3) \wedge (a_1' \vee a_3'),\ (a_2 \vee a_3) \wedge (a_2' \vee a_3')]\!] \\ = \ & [\![a_1 a_2 a_3]\!] [\![a_1' a_2' a_3']\!]\ (a_1 \vee a_1') \wedge (a_2 \vee a_2') \wedge (a_3 \vee a_3'). \end{aligned} \tag{3.10}$$

So Desargues' Theorem and its converse are direct corollaries of (3.10). The two geometric theorems when represented in this identity form can be directly used in symbolic manipulations as term rewriting rules. This is a much higher level of algebraization of geometric theorems.

When changed into polynomials of homogeneous coordinates, $f$ contains 48 terms, while $g$ contains as many as 1,290 terms. The effect of controlling input expression size and middle expression swell by Cayley expansion and factorization is obvious.

### 3.3.4 Rational Invariants and Antisymmetrization

A projective space has various subspaces, and each subspace has its own invariants. An invariant of a projective subspace is no longer an invariant of the whole space. For example, on a projective line, the bracket of two points is an invariant, but no longer an invariant of the projective plane, instead it becomes the covariant representing the line: the join operation of the two points.

Nevertheless, the ratio of two invariants of a projective subspace is always an invariant, called an *invariant ratio*. For example, the ratio of two brackets $[a_1, a_2]$ and $[a_3, a_4]$ for collinear points $a_1, a_2, a_3, a_4$ is an invariant: for any point $a_5$ not collinear with the four points,

$$\frac{[a_1, a_2]}{[a_3, a_4]} = \frac{[\![a_1 a_2 a_5]\!]}{[\![a_3 a_4 a_5]\!]}$$

is independent of the choice of *dummy vector variable* $a_5$. This is a ubiquitous property of all invariant ratios in projective geometry.

---

**Glossary**

**Invariant ratio:** The signed ratio of two collinear line segments. Invariant ratios are as fundamental as brackets in projective geometry. They are the direct heritage of low dimensional invariants.

**Rational monomial invariant:** A rational monomial function of brackets and invariant ratios.

**Completion of rational monomial invariant:** It refers to changing a rational monomial invariant into an equal rational bracket monomial.

**Antisymmetrization of rational monomial invariant:** A method of completing rational monomial invariants by constructing meet operations in the numerator and denominator of a rational monomial invariant, respectively, then making monomial Cayley expansions.

---

The completion of a rational monomial invariant is always possible by adding dummy vector variables to the components of invariant ratios, so the aim should be to find the simplest completion.

Since the numerator and denominator of an invariant ratio are both covariants, they can connect with other covariants by the meet operation. A monomial of invariant ratios is then changed into a monomial of ratios of meet operations. This transformation is called a partial *antisymmetrization* of the rational monomial invariant. After the antisymmetrization, by monomial Cayley expansions, a monomial of ratios of meet operations can be changed into a rational bracket monomial. This is the *antisymmetrization approach* [18] to the completion of rational monomial invariants. It is very useful in proving incidence theorems involving ratios. The following is an illustrative example.

*Let $a_1', a_2', a_3'$ be points collinear with sides $a_2a_3, a_1a_3, a_1a_2$ of triangle $a_1a_2a_3$, respectively.*

*(a) [Ceva's Theorem and its converse (Fig 3.3, left)] Lines $a_1a_1', a_2a_2', a_3a_3'$ concur if and only if*

$$\frac{a_1'a_2}{a_3a_1'}\frac{a_2'a_3}{a_1a_2'}\frac{a_3'a_1}{a_2a_3'} = 1. \tag{3.11}$$

*(b) [Menelaus' Theorem and its converse (Fig 3.3, right)] Points $a_1', a_2', a_3'$ are collinear if and only if*

$$\frac{a_1'a_2}{a_3a_1'}\frac{a_2'a_3}{a_1a_2'}\frac{a_3'a_1}{a_2a_3'} = -1. \tag{3.12}$$

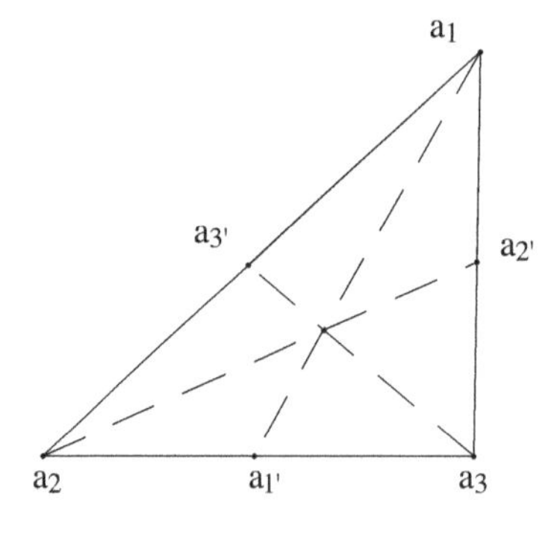

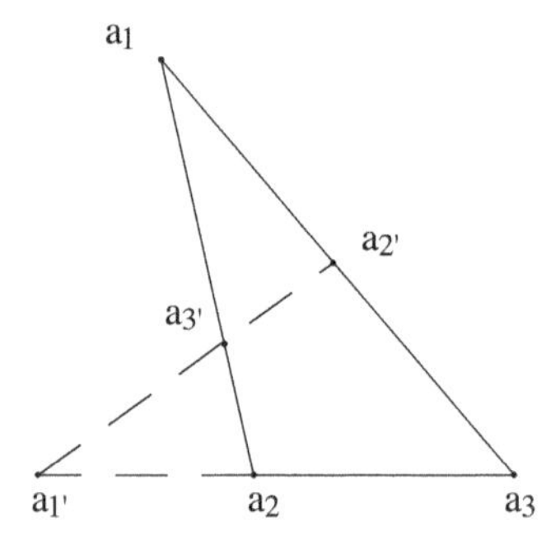


**Figure 3.3**
Ceva's Theorem (left) and Menelaus' Theorem (right).

Denote the left side of (3.11) by $p$. It has a natural antisymmetrization

$$p = \frac{(a_1' \vee a_2) \wedge (a_2' \vee a_3) \wedge (a_3' \vee a_1)}{(a_3 \vee a_1') \wedge (a_1 \vee a_2') \wedge (a_2 \vee a_3')}.$$

The following Cayley expansion of $p$, where the priorities in the expansion are described by the parentheses, leads to Ceva's Theorem:

$$\frac{((a_1' \vee a_2) \wedge (a_2' \vee a_3)) \wedge (a_3' \vee a_1)}{((a_2 \vee a_3') \wedge (a_3 \vee a_1')) \wedge (a_1 \vee a_2')} = \frac{[\![a_2a_1'a_2']\!][\![a_1a_3a_3']\!]}{[\![a_3a_1'a_3']\!][\![a_1a_2a_2']\!]}.$$

It changes (3.11) into

$$[\![a_2a_1'a_2']\!][\![a_1a_3a_3']\!]-[\![a_1a_2a_2']\!][\![a_3a_1'a_3']\!]=-(a_1\vee a_1')\vee(a_2\vee a_2')\vee(a_3\vee a_3')=0.$$

The following Cayley expansion of $p$ leads to Menelaus' Theorem:

$$\frac{((a_1'\vee a_2)\wedge(a_2'\vee a_3))\wedge(a_3'\vee a_1)}{((a_3\vee a_1')\wedge(a_1\vee a_2'))\wedge(a_2\vee a_3')}=-\frac{[\![a_2a_1'a_2']\!][\![a_1a_3a_3']\!]}{[\![a_1a_1'a_2']\!][\![a_2a_3a_3']\!]}.$$

It changes (3.12) into

$$\begin{aligned}[\![a_2a_1'a_2']\!][\![a_1a_3a_3']\!]-[\![a_1a_1'a_2']\!][\![a_2a_3a_3']\!] &= (a_1\vee a_2)\wedge(a_1'\vee a_2')\wedge(a_3\vee a_3')\\ &= -[\![a_1a_2a_3]\!][\![a_1'a_2'a_3']\!]\\ &= 0.\end{aligned}$$

## 3.4 Bracket Algebra for Euclidean Geometry

So far all our considerations apply to configurations and theorems in the projective plane. The only properties that might be encoded are those that are invariant under the group of projective transformations. One might be misled to think that the algebraic expressiveness and the proving techniques introduced in the previous sections cannot be applied to theorems that involve measurements and refer to distances and angles.

In a sense rather the opposite is the case. In a suitable setup all properties and relations of Euclidean (and even hyperbolic or elliptic) geometry can be expressed in terms of projective geometry. The working horse behind this is the so-called Cayley-Klein geometries that can be traced back to the works of Beltrami, Cayley, and Klein. A detailed account of them can be found in [14, 23, 25]. Here we will only roughly sketch how Euclidean properties can be expressed in projective terms.

The key to the embedding of Euclidean properties into a projective setup is the use of complex numbers in a very controlled way. We consider the real projective plane $\mathbb{RP}^2$ and introduce two extra points with *complex* homogeneous coordinates: $\mathtt{I}=(1,-i,0)^T$ and $\mathtt{J}=(1,i,0)^T$.

### 3.4.1 The Points I and J

**Glossary**

**I and J:** Two points with special homogeneous coordinates $\mathtt{I}=(1,-i,0)^T$ and $\mathtt{J}=(1,i,0)^T$ with respect to the standard embedding, where $i=\sqrt{-1}$. They are both on the line at infinity $l_\infty$ of the plane.

**Concyclicity:** In the standard embedding, all conics passing simultaneously through I and J are exactly the circles (including the limit situations of the union of a line and the line at infinity).

**Euclidean similarity:** A transformation that leaves absolute values of angles as well as ratios of lengths invariant. Euclidean similarities are those projective transformations that stabilize the pair $\{\mathtt{I},\mathtt{J}\}$.

**Cross ratio:** A rational monomial invariant for four points $A,B,C,D$ on a line $l$. If point $X$ is not incident to line $l$ then $cr(A,B;C,D):=\frac{[\![XAC]\!][\![XBD]\!]}{[\![XAD]\!][\![XBC]\!]}$. The result is independent of the choice of $X$.

**Laguerre's formula for angles:** The angle from line $l$ to line $m$ in homogeneous coordinates can be calculated by $\frac{1}{2i}\ln(\mathrm{cr}(L,M;\mathtt{I},\mathtt{J}))$ with $L = l \wedge l_\infty$ and $M = m \wedge l_\infty$.

---

If we consider the standard embedding $\mathbb{R}^2 \to \mathbb{RP}^2$ via $(x,y) \mapsto (x,y,1)$ the two points $\mathtt{I}$ and $\mathtt{J}$ are exactly those points that are obtained by intersecting an arbitrary circle with the line at infinity. To see this consider a circle $(x-m_x)^2 + (y-m_y)^2 = r^2$ with center $m$ and radius $r$. Written in homogeneous coordinates this circle is a conic given by the equation

$$x^2 + y^2 - 2m_x \cdot xz - 2m_y \cdot yz + (m_x^2 + m_y^2 - r^2) \cdot z^2 = 0.$$

Intersecting this conic with the line at infinity $l = (0,0,1)^T$ asks for those solutions of the conic equation that also satisfy $z = 0$. Thus, one has to solve $x^2 + y^2 = 0$. Up to scalar multiples the unique two solutions are $\mathtt{I} = (1,-i,0)^T$ and $\mathtt{J} = (1,i,0)^T$.

It is remarkable that the solution is independent of the particular choice of the initial circle. In other words: All circles pass through $\mathtt{I}$ and $\mathtt{J}$. Conversely it can be shown that any conic passing through the two points either is a proper circle (in the standard embedding), or decomposes into the union of a line $l$ and the line at infinity. The latter situation can be considered as the $\mathbb{RP}^2$ equivalent of a circle with infinite radius (compare [25]). In other words: *circles are the conics through* $\mathtt{I}$ *and* $\mathtt{J}$.

Transformations that map circles to circles in Euclidean geometry are those that leave ratios of lengths invariant. They turn out to be rotations, reflections, translations, scalings, and arbitrary compositions of those (summarized under the term *Euclidean similarity transformations*). In particular, these transformations preserve the absolute values of angles (which is an alternative characterization). With respect to the standard embedding of $\mathbb{R}^2$ into $\mathbb{RP}^2$ this group forms a subgroup of the projective transformations. Combining these considerations with the above characterization of circles we can say that Euclidean similarity transformations are exactly those transformations that leave the pair $\{\mathtt{I},\mathtt{J}\}$ invariant.

In fact this is a special case of a more general construction. The pair $\{\mathtt{I},\mathtt{J}\}$ can be considered as a special degenerate conic. The theory of Cayley-Klein geometries characterizes measurement of angles and distances by referring the calculation of cross ratios to general conics. This construction is remarkably rich, and within the theory of Cayley-Klein geometries Euclidean, hyperbolic, elliptic, and relativistic geometries occur as special cases [14, 23, 25].

The above characterization of Euclidean similarities implies that every property that is invariant under Euclidean similarity transformations can be expressed as a projective invariant in which $\mathtt{I}$ and $\mathtt{J}$ play a special role. Rephrased in a different way Euclidean similarity geometry is projective geometry with two points $\mathtt{I}$ and $\mathtt{J}$ distinguished. We exemplify this general principle by Laguerre's formula that allows us to calculate the angle between two lines.

**Theorem 3.7 (Laguerre's formula)** *Given two lines $l$ and $m$ in the Euclidean plane (in the standard embedding), the absolute value of the angle between them can be calculated by*

$$\left|\frac{1}{2i}\ln(cr(L,M;\mathtt{I},\mathtt{J}))\right|, \quad \text{with } L = l \wedge l_\infty, \quad M = l \wedge m_\infty.$$

### 3.4.2 Proving Euclidean Theorems

The considerations of the previous subsection show that it is possible to express properties of Euclidean geometry by performing projective calculations and expressing Euclidean properties with projective invariant measures in which $\mathtt{I}$ and $\mathtt{J}$ play a distinguished role. Thus the projective proving techniques introduced previously may also be applied to purely Euclidean situation. For this it is necessary to express every Euclidean relation by a projectively invariant property that involves I and J. We are going to exemplify this philosophy by working one concrete example.

We consider Miquel's Theorem about six circles and eight points (Figure 3.4):

*Let* $1,2,\ldots,8$ *be the indices of eight distinct points in the Euclidean plane. Consider the six quadruples of points* $(1,2,3,4)$, $(1,2,7,8)$, $(5,2,3,8)$, $(5,6,3,4)$, $(1,6,7,4)$, $(5,6,7,8)$. *If five of these quadruples are concyclic so is the last one.*

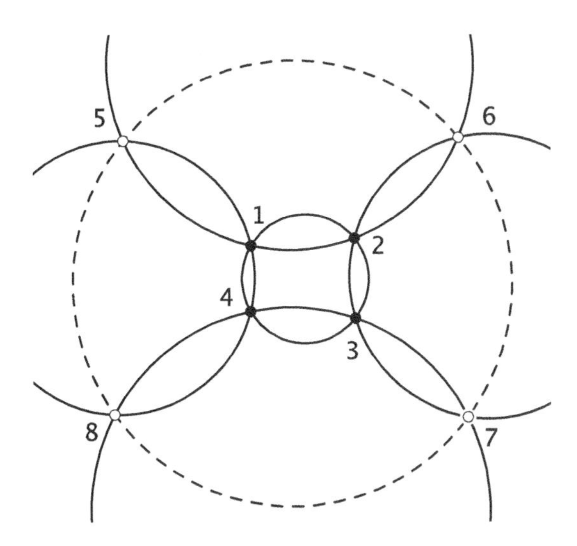


**Figure 3.4**
The Theorem of Miquel.

First we observe that we can characterize the concyclicity of four points $1,2,3,4$ by expressing the fact that $1,2,3,4,\mathtt{I},\mathtt{J}$ are on a common conic. Using our characterization of conics by bracket binomials we see that this can be algebraically expressed by

$$[\![12\mathtt{I}]\!][\![34\mathtt{I}]\!][\![14\mathtt{J}]\!][\![32\mathtt{J}]\!] = [\![12\mathtt{J}]\!][\![34\mathtt{J}]\!][\![14\mathtt{I}]\!][\![32\mathtt{I}]\!].$$

Observe that the following transformation property arises, that finally justifies this characterization. Assume that $(1,2,3,4)$ are concyclic and by this satisfy the above equation. Now consider a projective transformation $\tau$ that leaves the pair $\{\mathtt{I},\mathtt{J}\}$ invariant (i.e., a Euclidean similarity). Without loss of generality assume that $\tau(\mathtt{I}) = \mathtt{I}$ and $\tau(\mathtt{J}) = \mathtt{J}$. Since the transformation is projective and the above formula is stable under projective transformations, we get

$$[\![1'2'\mathtt{I}]\!][\![3'4'\mathtt{I}]\!][\![1'4'\mathtt{J}]\!][\![3'2'\mathtt{J}]\!] = [\![1'2'\mathtt{J}]\!][\![3'4'\mathtt{J}]\!][\![1'4'\mathtt{I}]\!][\![3'2'\mathtt{I}]\!],$$

with $p' = \tau(p)$ for $p \in \{1,2,3,4\}$. Thus as expected, the transformed points again satisfy our algebraic concyclicity characterization. The above-mentioned Theorem of Miquel is now an easy consequence.

Assume that the first five concyclicities are satisfied; this gives the following five equations:

$$[\![12\mathtt{I}]\!][\![34\mathtt{I}]\!][\![14\mathtt{J}]\!][\![32\mathtt{J}]\!] = [\![12\mathtt{J}]\!][\![34\mathtt{J}]\!][\![14\mathtt{I}]\!][\![32\mathtt{I}]\!],$$
$$[\![12\mathtt{J}]\!][\![\mathbf{65}\mathtt{J}]\!][\![15\mathtt{I}]\!][\![62\mathtt{I}]\!] = [\![12\mathtt{I}]\!][\![\mathbf{65}\mathtt{I}]\!][\![15\mathtt{J}]\!][\![62\mathtt{J}]\!],$$
$$[\![62\mathtt{J}]\!][\![37\mathtt{J}]\!][\![\mathbf{67}\mathtt{I}]\!][\![32\mathtt{I}]\!] = [\![62\mathtt{I}]\!][\![37\mathtt{I}]\!][\![\mathbf{67}\mathtt{J}]\!][\![32\mathtt{J}]\!],$$
$$[\![\mathbf{87}\mathtt{J}]\!][\![34\mathtt{J}]\!][\![84\mathtt{I}]\!][\![37\mathtt{I}]\!] = [\![\mathbf{87}\mathtt{I}]\!][\![34\mathtt{I}]\!][\![84\mathtt{J}]\!][\![37\mathtt{J}]\!],$$
$$[\![15\mathtt{J}]\!][\![84\mathtt{J}]\!][\![14\mathtt{I}]\!][\![\mathbf{85}\mathtt{I}]\!] = [\![15\mathtt{I}]\!][\![84\mathtt{I}]\!][\![14\mathtt{J}]\!][\![\mathbf{85}\mathtt{J}]\!].$$

The situation is similar to that of manipulating the conclusion part of a bi-quadratic final polynomial. Multiplying all the left sides and all the right sides and cancelling everything that occurs on both sides (we can do this since any line containing two distinct real points does contain neither I nor J), we are left with

$$[\![65\mathtt{J}]\!][\![87\mathtt{J}]\!][\![67\mathtt{I}]\!][\![85\mathtt{I}]\!] = [\![65\mathtt{I}]\!][\![87\mathtt{I}]\!][\![67\mathtt{J}]\!][\![85\mathtt{J}]\!],$$

an equation expressing the concyclicity of $5,6,7,8$.

For more elaborate discussion of Euclidean theorems from a projective viewpoint including many elaborated examples, we refer to [7, 25].

**Acknowledgment.** The first author is supported by the DFG Collaborative Research Center TRR 109, "Discretization in Geometry and Dynamics"; the second author is supported by the NSFC Project 11671388 and the CAS Frontier Key Project QYZDJ-SSW-SYS022.

## References

[1] Abhyankar, S.S. Invariant theory and enumerative combinatorics of young tableaux. In: Mundy, J.L. and Zisserman, A. *(eds.) Geometric Invariance in Computer Vision.* MIT Press, Cambridge, MA, pp. 45–76, 1992.

[2] Anick, D. & Rota, G.-C. Higher-order syzygies for the bracket algebra and for the ring of coordinates of the Grassmannian. *Proc. Nat. Acad. Sci.* **88**(18): 8087–8090, 1991.

[3] Barnabei, M., Brini, A., & Rota, G.-C. On the exterior calculus of invariant theory. *J. of Algebra* **96**: 120–160, 1985.

[4] Berger, M. *Geometry I, II.* Springer, 1987.

[5] Bokowski, J. & Richter-Gebert, J. On the finding of final polynomials. *Europ. J. Combinatorics* **11**: 21–34, 1990.

[6] Bokowski, J. & Sturmfels, B. Computational synthetic geometry. In: *Lecture Notes in Mathematics* **1355**, Springer-Verlag, Berlin, Heidelberg, 1989.

[7] Crapo, H. & Richter-Gebert, J. Automatic proving of geometric theorems. In: *Invariant Methods in Discrete and Computational Geometry*, White, N. (ed.), Kluwer Academic Publishers, Dodrecht, 107–139, 1995.

[8] Désarménien, J. An algorithm for the Rota Straightening Formula, *Discrete Mathematics* **30**: 51–68, 1980.

[9] Doubilet, P., Rota, G.-C., & Stein, J. On the foundations of combinatorial theory IX: Combinatorial Methods in Invariant Theory. *Stud. Appl. Math.* **57**: 185–216, 1974.

[10] Dress, A. & Wenzel, D. Grassmann-Plücker relations and matroids with Coefficients. *Advances in Mathematics* **86**: 68–110, 1991.

[11] Grosshans, F.D., Rota, G.-C., & Stein, J. *Invariant Theory and Superalgebras*, AMS, 1987.

[12] Gurevich, G.B. *Foundations of the Theory of Algebraic Invariants,* P. Noordhoff, Groningen, 1964.

[13] Huang, R.Q., Rota, G.-C., & Stein, J. Supersymmetric bracket algebra and invariant theory, *Acta Appl. Math.* **21**: 193–246, 1990.

[14] Kowol, G. *Projektive Geometrie und Cayley-Klein Geometrien der Ebene,* Birkhäuser, 2009.

[15] Li, H. & Sommer, G. Coordinate-free projective geometry for computer vision. In: Sommer, G. (*ed.*), *Geometric Computing with Clifford Algebras*, Springer, Heidelberg, pp. 415–454, 2001.

[16] Li, H. & Wu, Y. Automated short proof generation in projective geometry with Cayley and bracket algebras I, II. *J. Symb. Comput.* **36**(5): 717–762, 763–809, 2003.

[17] Li, H., Zhao, L., & Chen, Y. A symbolic approach to polyhedral scene analysis by parametric calotte propagation. *Robotica* **26**(4): 483-501, 2008.

[18] Li, H. *Invariant Algebras and Geometric Reasoning*. World Scientific, Singapore, 2008.

[19] Li, H., Shao, C., Huang, L., & Liu, Y. Reduction among bracket polynomials. In: *Proc. ISSAC'14*, ACM Press, pp. 304–311, 2014.

[20] Mainetti, M. & Yan, C.H. Arguesian Identities in Linear Lattices. *Advances in Mathematics* **144**: 50–93, 1999.

[21] McMillan, T. & White, N. The dotted straightening algorithm. *J. Symb. Comput.* **11**: 471–482, 1991.

[22] Olver, P.J. *Classical Invariant Theory*. Cambridge University Press, Cambridge, 1999.

[23] Onishchik, A.L. & Sulanke, R. *Projective and Cayley-Klein Geometries, Springer Monographs in Mathematics*, Springer, 2006.

[24] Richter-Gebert, J. Mechanical theorem proving in projective geometry. *Annals of Mathematics and Artificial Intelligence* **13**: 139–172, 1995.

[25] Richter-Gebert, J. *Perspectives on Projective Geometry,* Springer, 2011.

[26] Richter-Gebert, J. Meditations on Ceva's Theorem. In: *The Coxeter Legacy: Reflections and Projections*, Davis, C. & Ellers, E. (Eds.), American Mathematical Society, Fields Institute, 227–254, 2006.

[27] Rota, G.-C. & Stein, J. Applications of Cayley Algebras. Academia Nazionale dei Lincei atti dei Convegni Lincei **17**, Colloquio Internazionale sulle Teorie Combinatoire, Tomo **2**, Roma, 1976.

[28] Rota, G.-C. & Stein, J. Symbolic method in invariant theory, *Proc. Nat. Acad. Sci.* **83**: 844–847, 1986.

[29] Rota, G.-C. & Sturmfels, B. Introduction to invariant theory in superalgebras. In: Staton, D. (*ed.*), *Invariant Theory and Tableaux*, Springer, New York, pp. 1–35, 1990.

[30] Sidman, J. & Traves, W. Cayley factorization and special positions. In: Sitharam, M., St. John, A., and Sidman, J. (*ed.*), *Handbook of Geometric Constraints Systems: Principles*.

[31] Sturmfels, B. Computational algebraic geometry of projective configurations. *J. Symbolic Computation* **11**: 595–618, 1991.

[32] Sturmfels, B. *Algorithms in Invariant Theory,* Springer-Verlag, Wien, New York, 1993.

[33] Sturmfels, B. & White, N. Gröbner bases and invariant theory. *Adv. Math.* **76**(2): 245–259, 1989.

[34] Sturmfels, B. & Whiteley, W. On the synthetic factorization of homogeneous invariants. *J. Symbolic Computation* **11**: 439–454, 1991.

[35] White, N. The bracket ring of combinatorial geometry I. *Transactions of AMS* **202**: 79–95, 1975.

[36] White, N. Implementation of the straightening algorithm of classical invariant theory. In: Staton, D. (*ed.*), *Invariant Theory and Tableaux*, Springer, New York, pp. 36–45, 1990.

[37] White, N. Multilinear Cayley factorization. *J. Symb. Comput.* **11**: 421–438, 1991.

[38] White, N. Geometric applications of the Grassmann-Cayley algebra. In: Goodman, J.E. and O'Rourke, J. (*eds.*), *Handbook of Discrete and Computational Geometry*, CRC Press, Boca Raton, FL, 1997.

[39] White, N. Grassmann-Cayley algebra and robotics applications. In: Bayro-Corrochano, E. (*ed.*), *Handbook of Geometric Computing*, Springer, Heidelberg, pp. 629–656, 2005.

[40] Whiteley, W. Logic and invariant computation for analytic geometry. In: *Symbolic Computations in Geometry*, I.M.A. Preprint # 389, University of Minnesota, January 1988.

[41] Whiteley, W. Invariant computations for analytic projective geometry. *J. Symb. Comput.* **11**: 549–578, 1991.

# Chapter 4

# *Special Positions of Frameworks and the Grassmann-Cayley Algebra*

**Jessica Sidman and William Traves**

*Mount Holyoke College and US Naval Academy*

**CONTENTS**



## 4.1 Introduction: the Grassmann-Cayley Algebra and Frameworks

Given a framework determined by a system of geometric constraints, a fundamental question is to determine whether the framework is rigid or flexible. If the framework is a bar-and-joint or body-and-bar framework, the constraints are all given by fixing the distance between selected pairs of points with *bars*. The theorems of Laman [2], (actually discovered first by Pollaczek-Geiringer [5]),

Tay [12], and White-Whiteley [13] characterize rigidity combinatorially in various settings, but these combinatorial characterizations only determine the behavior of sufficiently generic realizations of a framework with given combinatorics.

The pure condition of a generically rigid framework, introduced by White and Whiteley in [13] and [14], is a polynomial in *brackets*. The goal of this chapter is to give an introduction to the Grassmann-Cayley algebra, which can be used to give a geometric interpretation of the vanishing of bracket polynomials in order to better understand when a generically rigid framework admits nontrival internal motions.

In order to motivate the development of the Grassmann-Cayley algebra we present the simplest example of a generically rigid structure in the plane, two rigid bodies connected by three bars, which we represent by a multigraph with a vertex for each body and an edge for each bar in Figure 4.1.

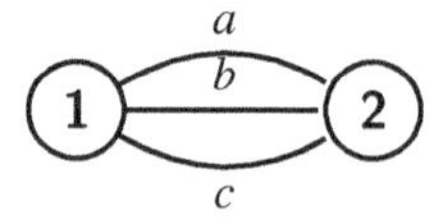


**Figure 4.1**
Vertices 1 and 2 correspond to rigid bodies in $\mathbb{P}^2$. Three bars with generic attachment points create a rigid framework.

Following [13], we embed the framework in the projective plane $\mathbb{P}^2$ and label each bar with a vector corresponding to the line in the direction of the bar. This vector is obtained by taking the cross product of the vectors corresponding to the endpoints of the bars. Equivalently, the bars are labeled by their Plücker coordinates in a projective plane that is dual to the plane in which the framework is embedded. (See Section 4.2 for a more detailed introduction to projective space, Section 4.2.5 for an introduction to Plücker coordinates, and Section 4.2.4 for a discussion of duality.) The resulting body-and-bar framework is generically infinitesimally rigid and has a nontrivial infinitesimal motion exactly when the three bars are parallel or meet at a point (see Figure 4.2).

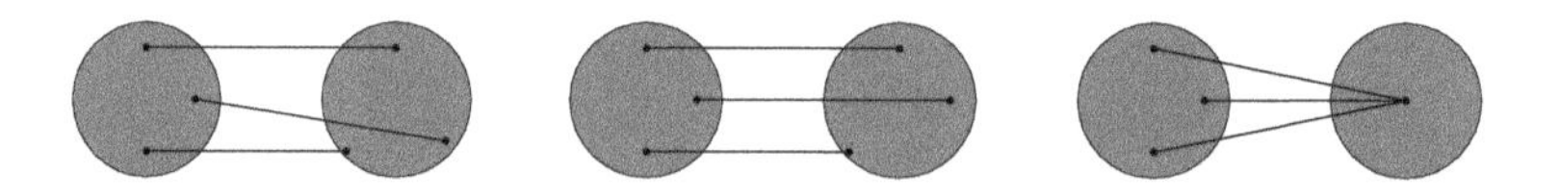

**Figure 4.2**
Three realizations of the graph from Figure 4.1. The first is generic. The middle figure depicts an embedding in which the bars are parallel, and the lines along the bars meet at a point on the line at infinity. In the third example the point of coincidence is in the finite plane.

In $\mathbb{P}^2$ lines are parallel exactly when they meet at a point on the line at infinity, hence the framework has an infinitesimal motion exactly when the lines along the bars are coincident. Three lines in $\mathbb{P}^2$ are coincident when their Plücker coordinates in the dual projective plane are collinear. This is precisely the condition that the $3 \times 3$ matrix $[abc]$ has determinant zero.

In Section 4.2 we give an introduction to projective space, the Grassmannian, and Plücker coordinates. In Section 4.3 we discuss the bracket ring as the ring of invariant polynomials on $\mathbb{P}^n$ and as a quotient ring with relations given by Plücker relations and Van der Waerden syzygies. We treat the Grassmann-Cayley algebra in Section 4.4 and Cayley factorization in Section 4.5, illustrating the theory with examples from the rigidity theory of body-and-bar frameworks. In the final section we include a new result, Theorem 4.1 due to the first author, Cai, St. John, and Theran, characterizing which body-and-bar frameworks have pure conditions which may be represented by a bracket monomial. We use the notation and conventions of Sturmfels [9] throughout, so that the reader who would like a fuller exposition may transition easily between this chapter and [9].

## 4.2 Projective Space

### 4.2.1 Motivation

The frameworks that we wish to consider live naturally in $\mathbb{R}^n$, a real vector space of dimension $n$, but it is often useful to work in the augmented $n$-dimensional projective space $\mathbb{P}^n$. To construct $\mathbb{P}^n$ we attach new points to $\mathbb{R}^n$ *at infinity*; these new points encode the ways that a sequence of points can diverge to infinity. The resulting projective space has many properties not found in our Euclidean space. For instance, Bézout's Theorem for the projective plane says that any two curves defined by polynomial equations of degrees $d$ and $e$ meet in $de$ points, suitably interpreted. Indeed, our points of intersection may lie at infinity and can even have complex coordinates; if the curves are tangent at some of the intersection points, we count these points with multiplicity (for details, see Schenk [7] or Smith et al. [8]). As a result of Bézout's Theorem, any pair of lines in the plane, each described by the vanishing of a polynomial of degree 1, must meet; when the lines are parallel, the point of intersection lies at infinity. Indeed, each point at infinity can be thought of as the point of intersection of a family of parallel lines. In the setting of body-and-bar frameworks this point of view unifies how we think about translations and rotations, as an infinitesimal translation perpendicular to these lines can then be thought of as an infinitesimal rotation about the point at infinity.

**Projective space:** *Projective space* $\mathbb{P}^n$ is obtained by adding a hyperplane of points at infinity to an $n$-dimensional vector space $V$. The projective space $\mathbb{P}^n$ may be viewed as parameterizing 1-dimensional subspaces of $V$.

**Homogeneous coordinates:** *Homogeneous coordinates* give a way to describe points in projective space. A polynomial $P$ on $V$ can be extended to all of $\mathbb{P}^n$ by *homogenizing* $P$ with respect to an additional variable. Though such polynomials are no longer functions on $\mathbb{P}^n$, their vanishing sets are well-defined.

**Duality:** There is an incidence-preserving *duality* between points in the projective plane $\mathbb{P}^2$ and lines in $\mathbb{P}^2$. Lines in $\mathbb{P}^2$ are elements of a *dual projective plane*, $\mathbb{P}^{2^*}$.

**Grassmannian:** The Grassmannian $G(d+1,n+1)$ is a variety parameterizing $(d+1)$-dimensional subspaces of an $(n+1)$-dimensional vector space. This is equivalent to $\mathbb{G}(d,n)$, parameterizing $d$-dimensional linear spaces of $\mathbb{P}^n$. Both varieties are manifolds of dimension $(d+1)(n-d)$.

**Plücker coordinates:** The *Plücker coordinates* give an embedding of $G(d+1,n+1)$ into a projective space. Each subspace is represented by a $d+1$ by $n+1$ matrix $X$ whose rows form a basis of the subspace. The Plücker coordinates are then just the determinants of the $(d+1)$ by $(d+1)$ submatrices of $X$. The Plücker coordinates are also called brackets and the determinant of the submatrix given by the columns $i_1, i_2, \ldots, i_{d+1}$ is denoted $[i_1 i_2 \ldots i_{d+1}]$. The vanishing sets of *homogeneous* polynomials in the brackets form well-defined subsets of the Grassmannian.

### 4.2.2 Homogeneous Coordinates and Points at Infinity

August Ferdinand Möbius introduced coordinates on projective space that allow us to discuss finite and infinite points using the same notation. For each nonzero point $(a_1,\dots,a_n) \in \mathbb{R}^n$ we add a (superfluous) coordinate $[1 : a_1 : \dots : a_n]$. More generally, points in projective space have the form $[a_0 : a_1 : \dots : a_n]$ (with not all coordinates $a_i$ equal to zero), but we identify any two points that differ by a nonzero scalar multiple,

$$[a_0 : a_1 : \dots : a_n] = [a_0\lambda : a_1\lambda : \dots : a_n\lambda] \text{ for } \lambda \neq 0.$$

The notation uses the colon rather than the comma to suggest that the ratio of coordinates is essential to the location of a point in $\mathbb{P}^n$, rather than the actual coordinates themselves, which can be simultaneously scaled by a common nonzero multiple $\lambda$. With this convention, points in $\mathbb{P}^n$ whose initial coordinate is nonzero (and hence can be scaled to 1) correspond to points in $\mathbb{R}^n$. For an introduction to the projective plane, $\mathbb{P}^2$ as a quotient of the sphere $S^2$, see Section 4.1.1 of Chapter 3.

Points in $\mathbb{P}^n$ whose initial coordinate is zero correspond to points at infinity. For example, let's look for the intersection of two parallel lines in the plane, e.g. the lines $y = 2x+3$ and $y = 2x+5$. These lines don't meet at all in $\mathbb{R}^2$; their intersection in $\mathbb{P}^2$ must lie among the points at infinity. To see which point, we take the limit of the points $[1 : x : 2x+3]$ for $x \in \mathbb{R}$ as $x$ goes to infinity:

$$\lim_{x\to\infty} [1 : x : 2x+3] = \lim_{x\to\infty} \left[\frac{1}{x} : 1 : 2+\frac{3}{x}\right] = [0 : 1 : 2].$$

A similar limit computation shows that the points on the second line also approach $[0 : 1 : 2]$ as $x$ approaches infinity. Indeed, the points on any line in $\mathbb{R}^2$ with slope equal to $m$ approach the point $[0 : 1 : m]$ as $x$ approaches infinity. Moreover, it does not matter whether $x$ approaches $\infty$ or $-\infty$, the points on the lines approach the same limit point. Though one might first imagine that we need to add a circle of limit points at infinity to $\mathbb{R}^2$ – one for each direction pointing from the origin – in fact, the antipodal points on this circle are identified in $\mathbb{P}^2$. Instead, we see that our homogeneous coordinates represent points of $\mathbb{R}^2$ together with an additional point for each line through the origin in $\mathbb{R}^2$. As a point of interest, we remark that if you were to thicken this curve of points at infinity to a band, the resulting surface would have just one edge (and one side) – the Möbius band. It is curious that this portion of the projective plane bears Möbius's name rather than the coordinates he introduced.

In the following sections we focus on low-dimensional examples such as $\mathbb{P}^2$ and $\mathbb{P}^3$. By specializing we are able to make our computations concrete. As well, our low-dimensional focus is well suited to applications involving physical body-and-bar frameworks.

### 4.2.3 Equations on Projective Space

In the Euclidean plane $\mathbb{R}^2$ with coordinates $(a_1, a_2)$ a polynomial equation such as $a_1^2 + 2a_2 - 3 = 0$ implicitly describes a curve but the solutions of this equation don't interact well with our new coordinates on $\mathbb{P}^2$; after all, if we multiply $[1 : a_1 : a_2]$ by an arbitrary nonzero scalar $\lambda$ we don't change the projective point but the resulting coordinates no longer satisfy the equation. We can deal with this issue by *homogenizing* the equation, multiplying each term by a power of $a_0$ so that each term has the same degree as the original polynomial. For example, the homogenization of $a_1^2 + 2a_2 - 3$ is $a_1^2 + 2a_0a_2 - 3a_0^2$. Scaling each coordinate by a nonzero constant $\lambda$ causes the evaluation of the homogeneous polynomial to be multiplied by a power of $\lambda$, so the homogeneous polynomial evaluates to zero independently of the representative chosen for the point in projective space. For this reason we use the vanishing of homogeneous polynomials to define objects in projective space and the points in $\mathbb{P}^n$ satisfying the homogenization of a polynomial condition on $\mathbb{R}^n$ can be thought of as the points on the extension of the original object to projective space. Indeed, the curves appearing in Bézout's Theorem must be projective plane curves defined by the vanishing of homogeneous polynomials.

### 4.2.4 Duality Between Lines and Points in $\mathbb{P}^2$

The vanishing of a linear polynomial describes a line in $\mathbb{P}^2$. Of particular interest is the curve defined by $a_0 = 0$; this consists of the points at infinity in $\mathbb{P}^2$, which form a line in $\mathbb{P}^2$ rather than the circle that one might naively picture from our discussion in Section 4.2.2. It is worth recomputing the intersection of the parallel lines $y = 2x+3$ and $y = 2x+5$ using the homogenized versions of the polynomials. Using the homogenizing variable $w$, the homogenized equations are $y-2x-3w=0$ and $y-2x-5w=0$ and their intersection consists of the nonzero points $[w:x:y] = [0:x:2x]$, any of which represent the intersection point $[0:1:2]$ in $\mathbb{P}^2$. This example illustrates yet another way of looking at points and lines in $\mathbb{P}^2$. Points in $\mathbb{P}^2$ correspond to 1-dimensional linear subspaces of $\mathbb{R}^3$ and lines in $\mathbb{P}^2$ correspond to 2-dimensional linear subspaces.

Lines in the projective plane have a peculiar property: if $a_0$, $a_1$, and $a_2$ are the coordinates on $\mathbb{P}^2$ then the line

$$c_0a_0 + c_1a_1 + c_2a_2 = 0 \tag{4.1}$$

is completely determined by the tuple of constants $(c_0, c_1, c_2)$. Moreover, the line doesn't change if we multiply its equation by a nonzero scalar, so the line corresponds uniquely to the projective point $[c_0 : c_1 : c_2]$. That is, the lines in $\mathbb{P}^2$ can be identified with points in another copy of $\mathbb{P}^2$, which we denote by $\mathbb{P}^{2^*}$. Recalling our earlier model of the projective plane in which points in $\mathbb{P}^2$ correspond to 1-dimensional linear subspaces and lines correspond to 2-dimensional linear subspaces and fixing the usual inner product on 3-space, we see that the dual of a projective line given by equation (4.1) is the point $[c_0 : c_1 : c_2] \in \mathbb{P}^{2^*}$ corresponding to the subspace orthogonal to the plane in 3-space determined by equation (4.1).

### 4.2.5 Grassmannians and Plücker Coordinates

The dual space $\mathbb{P}^{2^*}$ parameterizes lines in the projective plane. Describing lines in higher dimensional projective spaces requires a new perspective. Each point in $\mathbb{P}^n$ corresponds to a 1-dimensional subspace of $\mathbb{R}^{n+1}$. When we picked a nonzero $(n+1)$-tuple to represent our point in $\mathbb{P}^n$, we were really picking a basis for the underlying subspace. And when we identified all nonzero scalar multiples of this tuple we were really identifying all tuples that span the same subspace. Now each line in $\mathbb{P}^n$ can be thought of as a 2-dimensional subspace of $\mathbb{R}^{n+1}$. Given a basis of this space (equivalently, given two distinct points on the line), we can arrange the vectors in a $2\times(n+1)$ matrix $M$ of full rank. Multiplying $M$ on the left by any invertible $2\times 2$ matrix $A$ implements elementary row operations on the matrix $M$ and gives a new basis for the subspace corresponding to the line. Just as in our construction of projective space by identifying vectors that differ by a nonzero scalar multiple, we'd like to break the set of $2\times(n+1)$ matrices of full rank into equivalence classes, where two matrices are identified if their rows span the same subspace (equivalently, if there is an invertible $2\times 2$ matrix $A$ so that multiplication of one matrix on the left by $A$ gives the other matrix). Each equivalence class represents a 2-dimensional subspace in $\mathbb{R}^{n+1}$ and the set of these equivalence classes is called the Grassmannian $G(2,n+1)$. Alternatively, if we associate each equivalence class with a line in $\mathbb{P}^n$, the set of equivalence classes is then referred to as the Grassmannian $\mathbb{G}(1,n)$. We use the bold font $\mathbb{G}$ to indicate that we are thinking about objects projectively.

The coordinates on the Grassmannian $\mathbb{G}(1,n)$ are called Plücker coordinates. There is one coordinate for each of the $\binom{n+1}{2}$ $2\times 2$ submatrices of the $2\times(n+1)$ matrix $M$. Each coordinate takes a value given by the determinant of the corresponding $2\times 2$ submatrix. For example, there are six coordinates for lines in $\mathbb{P}^3$; each is denoted by a bracket of numbers representing the columns in a

$2 \times 2$ submatrix. For instance, the coordinates of the matrix

$$M = \begin{bmatrix} 1 & 2 & 1 & 2 \\ 0 & 1 & 1 & 3 \end{bmatrix} \quad \text{are} \quad \begin{matrix} [12] = 1 & [13] = 1 & [14] = 3 \\ [23] = 1 & [24] = 4 & [34] = 1. \end{matrix}$$

As with homogeneous coordinates on $\mathbb{P}^n$, it is not the actual values of the coordinates that matter, but rather their ratio relative to the other coordinates. Indeed, multiplying $M$ by an invertible $2 \times 2$ matrix $A$ changes the representative matrix for our line but does not change the line itself. At the same time, multiplying $M$ by $A$ multiplies each of the determinant coordinates by $det(A) \neq 0$. So the Plücker coordinates give an embedding of $\mathbb{G}(1,n)$ into the projective space $\mathbb{P}^N$, where $N = \binom{n+1}{2} - 1$.

We've explained how to get coordinates from the line; now we explain how to recover the line from the coordinates. Since the matrix $M$ has full rank, one of the coordinates $[ij]$ is nonzero. It follows that we can multiply $M$ by an appropriate invertible matrix $A$ so that the $(i,j)$ submatrix of $AM$ is the identity matrix. The newly scaled values of the coordinates $[ik]$ and $[kj]$ (for $k \neq i,j$) record the values of the remaining entries in $AM$, determining a basis for the line corresponding to the original matrix $M$. Though there are $\binom{n+1}{2}$ coordinates, the Grassmannian $\mathbb{G}(1,n)$ has much lower dimension. If we change basis so that the $(i,j)$ minor is the identity matrix, the remaining $2(n-1)$ coordinates can take arbitrary values. The space $\mathbb{G}(1,n)$ is a manifold of dimension $2(n-1)$. In particular, the collection $\mathbb{G}(1,3)$ of lines in 3-space is 4 dimensional. This seems intuitively reasonable: imagine two parallel planes in 3-space; most lines in 3-space are determined by where they intersect these two planes, and these two points of intersection are determined by their 4 coordinates.

To parameterize the set of linear spaces of $\mathbb{P}^n$ of dimension $d$ (corresponding to linear subspaces of $\mathbb{R}^{n+1}$ of dimension $d+1$) we consider full-rank $(d+1) \times (n+1)$ matrices modulo left multiplication by $(d+1) \times (d+1)$ invertible matrices. The homogeneous coordinates on $\mathbb{G}(d,n)$ or $G(d+1,n+1)$ are the determinants of the $(d+1) \times (d+1)$ submatrices. There are $\binom{n+1}{d+1}$ such submatrices so these determinants embed $\mathbb{G}(d,n)$ as a subset of $\mathbb{P}^N$, where $N = \binom{n+1}{d+1} - 1$. As in the construction of $\mathbb{G}(1,3)$, since one of the subdeterminants is nonzero we can left multiply by an invertible matrix $A$ to force that submatrix to be the identity. The remaining $(d+1)(n+1-(d+1)) = (d+1)(n-d)$ entries can take arbitrary values so the spaces $\mathbb{G}(d,n)$ and $G(d+1,n+1)$ have dimension $(d+1)(n-d)$.

Just as we require our polynomials to be homogeneous in order to define curves in the projective plane, the polynomials that define algebraic subsets of the Grassmannian $\mathbb{G}(d,n)$ must be homogeneous in the sense that each monomial contains the same number of Plücker coordinates.

### 4.2.6 More About Lines in 3-space

In this section we consider some bracket polynomials (polynomials in the Plücker coordinates) that describe the behavior of lines in 3-space.

**Example 1** *Suppose that we want to find an algebraic characterization of when two lines in $\mathbb{P}^3$ intersect. In other words, given $\ell_1 \in \mathbb{G}(1,3)$ and $\ell_2 \in \mathbb{G}(1,3)$, we would like to find a bracket polynomial that evaluates to zero when these two lines intersect. Thinking of each line as a 2-dimensional subspace in $\mathbb{R}^4$, we pick a basis $v_1$, $v_2$ for $\ell_1$ and $w_1$, $w_2$ for $\ell_2$. Using these bases we can construct the Plücker coordinates for each line. Let $[ij]_k$ stand for the $[ij]$ coordinate for line $\ell_k$ ($k = 1,2$, respectively). The two lines meet if and only if the span of their basis vectors is a 3-dimensional subspace of $\mathbb{R}^4$. This can be detected by seeing if the determinant of the $4 \times 4$ matrix with rows $v_1$, $v_2$, $w_1$ and $w_2$ vanishes. When we expand this determinant via Laplace expansion along the top two rows we obtain an expression for the determinant in terms of the Plücker coordinates for the lines:*

$$[12]_1[34]_2 - [13]_1[24]_2 + [14]_1[23]_2 + [23]_1[14]_2 - [24]_1[13]_2 + [34]_1[12]_2.$$

*This expression equals zero precisely when the two lines intersect or coincide.*

**Example 2** *Now we present a similar example where we will see that the equations derived using analogous geometric reasoning vanish on a set containing several components, only one of which corresponds to the geometry that we are trying to characterize. We'll consider three lines $\ell_1$, $\ell_2$, and $\ell_3$ and try to find equations in the brackets that guarantee that the three lines all meet at a single point. First we'll try to determine the dimension of the space parameterizing three lines that meet in a common point. Each of the three lines sits in $\mathbb{G}(1,3)$, a four-dimensional variety, so without any restrictions, the three lines are parameterized by 12 parameters. To describe a single line $\ell \subset \mathbb{P}^3$ that goes through a point P, it is enough to specify any other point Q on the line $\ell$. There is a 3-dimensional space of possible values for $Q \in \mathbb{P}^3$ but lots of choices for Q give the same line; Q and $Q'$ give rise to the same line $\ell$ precisely when Q, $Q'$ and P are collinear. Identifying points Q and $Q'$ that are collinear with P we obtain a 2-dimensional space parameterizing lines through P. One way to parameterize the lines $\ell_1$, $\ell_2$, and $\ell_3$ that meet at a point is to first choose the point P (3 dimensions of choice) and then choose each of the three lines going through P (2 dimensions of choice for each line). This leads to a parameter space of dimension $3+2+2+2=9$.*

*We might expect the set of lines that meet in a point to be cut out by $12-9=3$ bracket equations. Three such equations spring to mind! Just take the equations that indicate that $\ell_1$ meets $\ell_2$, $\ell_1$ meets $\ell_3$, and $\ell_2$ meets $\ell_3$.*

*These do cut out a 9-dimensional set of lines in $\mathbb{P}^3$ but it isn't quite the correct 9-dimensional set. Indeed, sitting inside this set is the subset $\Delta$ where all three lines are equal. That isn't what we wanted. More generally we don't want lines in the set $\Delta_{ij}$, where $\ell_i = \ell_j$ either. We could write down equations that characterize this situation and then require one of those equations to be nonzero. That would eliminate these unwanted sets but it also makes the description of the set by equations considerably more complicated. Moreover, the 3 equations that we used to cut out our 9-dimensional set of lines actually contain two 9-dimensional components. One of them corresponds to the set of lines that meet in a point and the other corresponds to the set of lines that are coplanar (such lines must always meet each other in projective space). We could add more bracket equations that vanish on the first component but not on the second. Adding enough equations intelligently will eventually remove the second component from our set but again the description of the space of 3 coincident lines becomes more complicated. This may be unavoidable but perhaps the three equations that we started with were just a bad choice and some other set of three equations cuts out the required set of lines. In algebraic geometry we say that a set of codimension-k (i.e., a set that has dimension k less than its ambient space) is a complete intersection if it is cut out by k equations. However, not all codimension-k sets are complete intersections. It can be very difficult to tell if a particular set is a complete intersection.*

## 4.3 The Bracket Algebra and Rings of Invariants

Many properties of geometric objects in $\mathbb{P}^d$ don't depend on the coordinate system that we choose for the ambient projective space. For instance, whether two lines in $\mathbb{P}^3$ intersect depends on the lines but not the coordinate system in which we represent the lines. As another example, whether the linearized distance constraints in a body-and-bar framework are satisfied should not depend on the coordinate system we use to describe the ambient space. Such conditions are said to be *invariant* under change of coordinates. In this section we'll lay out the mathematical framework used to describe invariant conditions, a topic that, perhaps surprisingly, links back up to Grassmannians and relations among the Plücker coordinates. For an alternative introduction in the plane, see Section 3.1.1 of Chapter 3.

**Automorphism Group:** The automorphism group $PGL_d$ of $\mathbb{P}^d$ consists of all invertible linear transformations on $\mathbb{C}^{d+1}$ modulo the multiples of the identity matrix. Equivalently, it consists of all linear transformations from $\mathbb{P}^d$ to itself.

**Coordinates and brackets:** Given $n$ points in $\mathbb{P}^d$ we form the $d+1$ by $n$ matrix $X$ of their coordinates. In contrast to the earlier construction of Grassmannians, here we think of each point as a column of $X$. Again we consider *brackets*, determinants of $d+1$ by $d+1$ submatrices of the matrix $X$ of coordinates. We label each point (and each column of $X$) by a symbol drawn from a set $\Lambda$. At various times we use numbers, letters, or subscripted letters to denote the elements of $\Lambda$, whichever is most convenient for the application at hand.

**Homogeneity:** A polynomial in the brackets is *homogeneous of degree d* if every term is the product of $d$ brackets. It is *multilinear* if every symbol in the polynomial appears in every monomial exactly once. More generally, it is *multihomogeneous* if each symbol in the polynomial appears in every monomial an equal number of times.

**Algebraic and invariant properties:** An *algebraic property* of the points is one that can be characterized by the vanishing of a system of well-defined polynomials in the coordinates given by $X$. Such a property is *invariant* if it holds independent of the coordinates chosen for the ambient space $\mathbb{P}^d$. The set of collections of points where a given set of multihomogeneous bracket polynomials evaluate to zero is an invariant set in the sense that membership in the set is an invariant property.

**Bracket algebra:** Every invariant set can be written as the vanishing set of a collection of multihomogeneous bracket polynomials. The ring of invariants is also referred to as the *bracket algebra*.

**Plücker relations:** The brackets are not independent. The relations on the brackets are generated by the alternating property: if $\sigma$ is a permutation of $1,\ldots,d+1$, $[i_1\cdots i_{d+1}] = \operatorname{sgn}(\sigma)[\sigma(i_1)\cdots\sigma(i_{d+1})]$ and the quadratic *Plücker relations*.

Each Plücker relation is determined by a choice $\alpha = (\alpha_1,\ldots,\alpha_d)$ of $d$ points and a second choice $\beta = (\beta_1,\ldots,\beta_{d+2})$ of $d+2$ points. The Plücker relation has the form

$$\sum_{t=1}^{d+2}(-1)^{t-1}[\alpha_1\ldots\alpha_d\beta_t][\beta_1\ldots\hat{\beta}_t\ldots\beta_{d+2}] = 0,$$

where the notation $\hat{\beta}_t$ means to omit the term $\beta_t$ from the bracket.

**Van der Waerden syzygies:** To determine whether two polynomials in the brackets are the same, we reduce them to a normal form using *Van der Waerden* syzygies. Given an integer $s$ with $1 \le s \le d+1$ and tuples $\alpha = (\alpha_1,\ldots,\alpha_{s-1})$, $\beta = (\beta_1,\ldots,\beta_{d+2})$, and $\gamma = (\gamma_1,\ldots,\gamma_{d+1-s})$ of points, we define $[[\alpha\dot{\beta}\gamma]]$ to be

$$\sum_{\substack{\tau=\{\tau_1<\ldots<\tau_s\}\\ \subset\{1,\ldots,d+2\}}} \operatorname{sgn}(\tau^*,\tau)\cdot[\alpha_1\ldots\alpha_{s-1}\beta_{\tau_1^*}\ldots\beta_{\tau_{d+2-s}^*}]\cdot[\beta_{\tau_1}\ldots\beta_{\tau_s}\gamma_1\ldots\gamma_{d+1-s}],$$

where $\tau^* = \{1,\ldots,d+2\}\setminus\tau$ is the complement of $\tau$ and $\operatorname{sgn}(\tau^*,\tau)$ is 1 if an even number of transpositions suffices to permute the $(d+2)$-tuple $(\tau^*,\tau)$ to $(1,\ldots,d+2)$ and $-1$ otherwise. The Van der Waerden syzygy is the relation $[[\alpha\dot{\beta}\gamma]] = 0$. The *dotted notation* helps identify the symbols split among the two brackets. The Plücker relations are the Van der Waerden syzygies with $s = d+1$.

**Tableau:** Monomials in the brackets are ordered *lexicographically*. Given a monomial in the brackets we write the brackets as rows in a *tableau* in lexicographic order. A tableau is *standard* if its columns are sorted and *nonstandard* otherwise.

**Normal form:** Polynomials in the brackets whose monomials are all standard are said to be *reduced to normal form.* Polynomials in the brackets can be iteratively reduced to normal form by adding a multiple of an appropriate Van der Waerden syzygy to the lexicographically largest nonstandard monomial.

### 4.3.1 Group Actions and Invariant Polynomials

Given a column vector $P$ consisting of $(d+1)$ entries representing the coordinates of a point in projective space we can change basis in $\mathbb{P}^d$, obtaining new coordinates $P'$ for the point, by multiplying on the left by an invertible $(d+1)\times(d+1)$ matrix $A$:

$$P' = AP.$$

Viewed in a slightly different way, we can imagine that the matrix $A$ acts on the projective space, moving the point with coordinates $P$ to the point with coordinates $P'$. However, not all invertible matrices $A$ actually move points: the matrices that are scalar multiples of the $(d+1)\times(d+1)$ identity matrix just scale the coordinates of the point $P$ and so don't move the point at all in projective space. In order to recognize these matrices that stabilize all points in $\mathbb{P}^d$, mathematicians speak of the automorphism group $PGL_d$ of $\mathbb{P}^d$; it consists of all linear transformations of $\mathbb{P}^d$ to itself. $PGL_d$ is the space obtained by taking the collection of invertible $(d+1)\times(d+1)$ matrices and identifying those that are scalar multiples of one another.

Now suppose that we are given $n$ points $P_1, \ldots, P_n$ in a projective space $\mathbb{P}^d$. We organize these points into a $(d+1)\times n$ matrix

$$X = (x_{ij}) \quad (0 \le i \le d;\ 1 \le j \le n),$$

where $x_{ij}$ represents the $i^{th}$ coordinate of point $P_j$; that is, $P_j = [x_{0j} : x_{1j} : \ldots : x_{dj}]^{\mathrm{T}}$. An algebraic property of the $n$ points is a property that can be expressed as the vanishing of a well-defined polynomial function $F(x_{01},\ldots,x_{dn})$ of the $x_{ij}$. The polynomial $F$ defines a well-defined function precisely if for each row $i$, the number of variables $x_{ij}$ (counted with multiplicity) from that row appearing in each of the monomials in $F$ is constant. The property is invariant if it does not depend on the choice of coordinates for $\mathbb{P}^d$; that is, the property holds for $P_1,\ldots,P_n$ if and only if for each invertible matrix $A$ the property holds for all $P'_1,\ldots,P'_n$, where $P'_j = AP_j$.

**Example 3** *Given 3 or more points $P_1$, $P_2$, $P_3$, … in $\mathbb{P}^2$, the polynomial function*

$$F = \det \begin{bmatrix} x_{01} & x_{02} & x_{03} \\ x_{11} & x_{12} & x_{13} \\ x_{21} & x_{22} & x_{23} \end{bmatrix}$$

*is an invariant polynomial that only depends on the first three points. If $A \in GL_3$ and $P'_j = AP_j$ then evaluating $F$ at the transformed points $P'_j$ equals the product of* $\det(A)$ *with the evaluation of $F$ at the original points $P_j$,*

$$F(P'_1,P'_2,P'_3,\ldots) = \det(A)F(P_1,P_2,P_3,\ldots),$$

*so the property defined by $F=0$ holds independent of the coordinates. We often write* $[123]$ *for the determinant defining this polynomial $F$. The invariant property can be stated succinctly: the polynomial* $[123]$ *vanishes precisely when the three points $P_1$, $P_2$, and $P_3$ are collinear. Stated in this way* *it is clear that the property doesn't depend on the coordinates on the ambient space. More generally, for points in* $\mathbb{P}^d$*, all determinants* $[j_0 j_1 \dots j_d]$ *of submatrices consisting of* $d+1$ *of the columns of* $X$ *are invariant. Products of the brackets are invariant and any multihomogeneous polynomial (each symbol* $j_k$ *appears the same number of times in every monomial) is invariant. For instance, the polynomial* $[123][245]+[125][234]$ *is invariant but the polynomial* $[123][245]+[124][345]$ *is not.*

In fact the First Fundamental Theorem of Invariant Theory states that the only invariant algebraic conditions on a collection of points in $\mathbb{P}^d$ are the multihomogeneous polynomials built out of the brackets $[j_0 j_1 \dots j_d]$. Even when an invariant algebraic condition is presented to us as a multihomogeneous polynomial it can be extremely difficult to determine a corresponding condition on the geometry of the points. This topic is treated in more detail in Section 4.5. One can also ask which maps from a finite set of $d+1$ elements to $\mathbb{C}$ (or any other field) arise from configurations of points, and this point of view is discussed in Section 3.1.3 of Chapter 3.

The ring of invariants $\mathbb{C}[x_{01},\dots,x_{dn}]^{PGL_d}$ consists of all sums and products of polynomials that are invariant under a linear change of coordinates. Every element of the ring of invariants can be expressed in terms of the brackets $[j_0 j_1 \dots j_d]$. The brackets themselves are not algebraically independent. The Second Fundamental Theorem of Invariant Theory describes the relations among the brackets, a topic treated in the next section. However, the main result is that all the relations come from identities that follow from the interpretation of the brackets as determinants. It follows that the ring of invariants $\mathbb{C}[x_{01},\dots,x_{dn}]^{PGL_d}$ can be identified with the coordinate ring of the Grassmannian $\mathbb{G}(d,n-1)$. We'll refer to this ring as the bracket algebra, in honor of its generators.

The two different interpretations of the bracket algebra – as the invariant ring $\mathbb{C}[x_{01},\dots,x_{dn}]^{PGL_d}$ of $n$ points in $\mathbb{P}^d$ and as the coordinate ring of the Grassmannian $\mathbb{G}(d,n-1)$ of $(d+1)$-dimensional linear spaces of $\mathbb{P}^{n-1}$ – arise from two different interpretations of the matrix $X$ of coordinates. When we view $X$ as a collection of columns, we get the interpretation in terms of points in $\mathbb{P}^d$ and when the view $X$ as a collection of rows we get the interpretation in terms of linear spaces in $\mathbb{P}^{n-1}$.

### 4.3.2 Relations Among the Brackets

The relations among the brackets are generated by special relations called Plücker relations. The Plücker relations all come from Cramer's rule, a well-known result about determinants. For instance, for 4 points in $\mathbb{P}^1$, the only Plücker relation is

$$[12][34]-[13][24]+[14][23]=0.$$

We explain how this relation arises, following the line of reasoning set out in Richter-Gebert [6]. Since the first three points are collinear we can express the first point as a linear combination of the second and third points; that is, there is a nontrivial solution to the matrix equation

$$\begin{bmatrix} x_{02} & x_{03} \\ x_{12} & x_{13} \end{bmatrix} \begin{bmatrix} a_2 \\ a_3 \end{bmatrix} = \begin{bmatrix} x_{01} \\ x_{11} \end{bmatrix}. \tag{4.2}$$

Using Cramer's rule we find the solutions to be

$$a_2 = \frac{[13]}{[23]} \text{ and } a_3 = \frac{[21]}{[23]}.$$

Substituting the solution into (4.2) gives a vector equation and multiplying by $[23]$ gives a linear dependence relation with brackets as coefficients

$$[23]\begin{bmatrix} x_{01} \\ x_{11} \end{bmatrix} - [13]\begin{bmatrix} x_{02} \\ x_{12} \end{bmatrix} - [21]\begin{bmatrix} x_{03} \\ x_{13} \end{bmatrix} = \begin{bmatrix} 0 \\ 0 \end{bmatrix}.$$

Writing $[23]\cdot 1-[13]\cdot 2-[21]\cdot 3$ for the left-hand side of this equation, we bracket with the last point and use the linearity of the determinant to compute the Plücker relation:

$$0=[[23]\cdot 1-[13]\cdot 2-[21]\cdot 3,4]=[14][23]-[13][24]+[12][34].$$

For higher dimensional projective spaces there are multiple Plücker relations. Given a set of points in $\mathbb{P}^d$ each Plücker relation is determined by a choice $\alpha=(\alpha_1,\ldots,\alpha_d)$ of $d$ points and a second choice $\beta=(\beta_1,\ldots,\beta_{d+2})$ of $d+2$ points. The Plücker relation has the form

$$\sum_{t=1}^{d+2}(-1)^{t-1}[\alpha_1\ldots\alpha_d\beta_t][\beta_1\ldots\hat{\beta}_t\ldots\beta_{d+2}]=0,$$

where the notation $\hat{\beta}_t$ means to omit the term $\beta_t$ from the bracket. The Plücker relations generate the *ideal* of all relations among the brackets in the sense that every relation is obtained by summing multiples of the Plücker relations by polynomials in the brackets. Some authors use the term Plücker relation more generally to refer to any quadratic relation among the brackets, but we won't follow this convention.

The same argument that establishes the Plücker relation on $\mathbb{P}^1$ verifies the relation in $\mathbb{P}^d$. Specifically, consider the following matrix equation that expresses a linear dependence relation among $d+2$ points $\beta_1,\ldots,\beta_{d+2}$ in $\mathbb{P}^d$,

$$\begin{bmatrix} x_{0\beta_2} & \cdots & x_{0\beta_{d+2}} \\ \vdots & & \vdots \\ x_{d\beta_2} & \cdots & x_{d\beta_{d+2}} \end{bmatrix}\begin{bmatrix} a_2 \\ \vdots \\ a_{d+2} \end{bmatrix}=\begin{bmatrix} x_{0\beta_1} \\ \vdots \\ x_{d\beta_1} \end{bmatrix}.$$

Use Cramer's rule to find the solution,

$$a_t=[\beta_2\ldots(\beta_1\hat{\beta}_t)\ldots\beta_{d+2}]/[\beta_2\ldots\beta_{d+2}]\ \ (2\le t\le d+2),$$

where the notation $(\beta_1\hat{\beta}_t)$ means replace the term $\beta_t$ in $\beta_2\ldots\beta_{d+2}$ with $\beta_1$. Since transposition of columns multiplies a determinant by $-1$ we can rewrite this as

$$a_t=(-1)^t[\beta_1\ldots\hat{\beta}_t\ldots\beta_{d+2}]/[\beta_2\ldots\beta_{d+2}].$$

It follows that

$$\sum_{t=2}^{d+2}(-1)^t[\beta_1\ldots\hat{\beta}_t\ldots\beta_{d+2}]\cdot\beta_t=[\beta_2\ldots\beta_{d+2}]\cdot\beta_1.$$

Moving all terms to the right-hand side, we obtain the relation

$$\sum_{t=1}^{d+2}(-1)^{t-1}[\beta_1\ldots\hat{\beta}_t\ldots\beta_{d+2}]\cdot\beta_t=\mathbf{0}.$$

Now we place the vector in the left-hand side as the right-most vector in the $(d+1)\times(d+1)$ matrix whose first $d$ columns come from the coordinates of the points $\alpha_1,\ldots,\alpha_d$ and take the determinant (which must evaluate to zero). Using the linearity of the determinant we can write the determinant as the Plücker relation

$$\sum_{t=1}^{d+2}(-1)^{t-1}[\beta_1\ldots\hat{\beta}_t\ldots\beta_{d+2}][\alpha_1\ldots\alpha_d\beta_t]=0.$$

The presence of multiple Plücker relations makes it difficult to determine if two expressions in the brackets are equal. One way to deal with such questions is to first reduce each expression $f$ to a normal form $N(f)$, a special representative for each collection of equivalent expressions. (The normal form of the bracket of 3 points is also discussed in Section 3.2.1 of Chapter 3.) Two expressions $f$ and $g$ are equivalent if and only if their normal forms are equal, $N(f) = N(g)$. The reduction to normal form can be viewed as an implementation of the normal form algorithm from Gröbner bases (see Sturmfels [9] or Sturmfels and White [10] for details). It can also be viewed as an implementation of the straightening algorithm due to Alfred Young [15]. We'll explain how to reduce (or straighten) an arbitrary expression in the brackets to its normal form. The method uses a more general relation that holds in the bracket algebra called a Van der Waerden syzygy. The word syzygy is used by algebraic geometers to refer to relations among polynomial expressions such as the Plücker relations.

A Van der Waerden syzygy on $\mathbb{P}^d$ is characterized in the following way. Let $s$ be an integer between 1 and $d$ and let $\alpha = (\alpha_1, \ldots, \alpha_{s-1})$ be a tuple of $s-1$ points in $\mathbb{P}^d$. Further let $\beta = (\beta_1, \ldots, \beta_{d+2})$ be a tuple of $d+2$ points in $\mathbb{P}^d$ and let $\gamma = (\gamma_1, \ldots, \gamma_{d+1-s})$ be a tuple of $d+1-s$ points in $\mathbb{P}^d$. Define $[[\alpha\dot{\beta}\gamma]]$ to be

$$\sum_{\substack{\tau = \{\tau_1 < \ldots < \tau_s\} \\ \subset \{1,\ldots,d+2\}}} \operatorname{sgn}(\tau^*, \tau) \cdot [\alpha_1 \ldots \alpha_{s-1} \beta_{\tau_1^*} \ldots \beta_{\tau_{d+2-s}^*}] \cdot [\beta_{\tau_1} \ldots \beta_{\tau_s} \gamma_1 \ldots \gamma_{d+1-s}],$$

where $\tau^* = \{1, \ldots, d+2\} \setminus \tau$ is the complement of $\tau$ and $\operatorname{sgn}(\tau^*, \tau)$ is 1 if an even number of transpositions suffices to permute the $(d+2)$-tuple $(\tau^*, \tau)$ to $(1, \ldots, d+2)$ and $-1$ otherwise. The Van der Waerden syzygy is the relation $[[\alpha\dot{\beta}\gamma]] = 0$. The notation is set up so that if $s = d$ and $\gamma = ()$, then $[[\alpha\dot{\beta}\gamma]]$ is a Plücker relation. Hongbo Li gives a nice elementary proof (using a clever averaging method combined with induction) that each Van der Waerden syzygy is in the ideal generated by the Plücker relations (see Li [3, Proposition 2.9]).

To describe the straightening algorithm we first need to define an ordering on bracket monomials in $\mathbb{P}^d$ and standard and nonstandard monomials. There is an ordering on brackets: $[\lambda_1 \ldots \lambda_{d+1}]$ is less than $[\omega_1 \ldots \omega_{d+1}]$ if, reading left-to-right, the first place they differ has smaller symbol in $\lambda$: that is, there is an index $t$ so that $\lambda_i = \omega_i$ for $1 \le i < t$ and $\lambda_t < \omega_t$. Given two monomials in the brackets we order them lexicographically. That is, we first compare their smallest brackets and if these are different then the monomial with the smaller bracket is smaller. If these brackets are the same we move on to compare their next smallest brackets and continue until a decision has been reached. This is precisely how we decide on the order of two words with the same length in a dictionary, which is why the ordering is called lexicographic. For example, $[123][146][235] < [123][156][234]$. Given a monomial in the brackets, e.g., $[145][234][156]$ we first arrange the brackets in lexicographic order $[145] < [156] < [234]$ and then write them as rows in a tableau $T$ (really, a matrix, though this is the common term in the mathematical literature), in increasing order:

$$T = \begin{bmatrix} 1 & 4 & 5 \\ 1 & 5 & 6 \\ 2 & 3 & 4 \end{bmatrix}.$$

The tableau $T$ is standard if its columns are all sorted, that is, reading down each column, the numbers never decrease; otherwise $T$ is nonstandard. In our example above, $T$ is nonstandard; the first column is sorted but the second is not and there is a violation in the second and third rows since $5 > 3$. If $T$ is nonstandard it can be straightened by adding an appropriate Van der Waerden polynomial. Just find the highest position with a violation in the left-most unsorted column. Suppose that the violation occurs in column $s$ and label the higher row involved in the violation as

$$[\alpha_1 \ldots \alpha_{s-1} \beta_1 \ldots \beta_{d+2-s}]$$

and the following row as

$$[\beta_{d+3-s} \cdots \beta_{d+2} \gamma_1 \cdots \gamma_{d+1-s}].$$

In our example, the rows are $[\alpha_1 \beta_1 \beta_2] = [156]$ and $[\beta_3 \beta_4 \gamma_1] = [234]$. To straighten the nonstandard monomial we subtract a multiple of the Van der Waerden polynomial $[[\alpha \dot{\beta} \gamma]]$. In our example, the Van der Waerden polynomial is $[[1\dot{5}\dot{6}\dot{2}\dot{3}4]]$, which equals

$$\begin{aligned}&[156][234]-[152][634]+[162][534]+[153][624]-[163][524]+[123][564]\\&=[156][234]+[125][346]-[126][345]-[135][246]+[136][245]+[123][456].\end{aligned}$$

Let's take a moment to pause and observe the utility of the dotted notation; the dotted terms in the notation $[[\alpha \dot{\beta} \gamma]]$ denote the $\beta$'s, the terms that are split in the expansion of the Van der Waerden syzygy. There are 6 terms in the syzygy since there are $\binom{4}{2} = 6$ ways to pick 2 of the 4 $\beta's$ to put in the first bracket.

To help straighten the monomial $[145][234][156]$ we solve for the nonstandard monomial $[234][156]$ in the Van der Waerden syzygy and substitute the resulting expression into the monomial; equivalently, in this instance, we subtract a multiple of the Van der Waerden syzygy (we need to multiply by the brackets not appearing in the violation) from the monomial. The result is

$$\begin{array}{llllll} & [145][234][156] & -[145][[1\dot{5}\dot{6}\dot{2}\dot{3}4]] & & & \\ = & -[145][125][346] & +[145][126][345] & +[145][135][246] & -[145][136][245] & -[145][123][456] \\ = & -[125][145][346] & +[126][145][345] & +[135][145][246] & -[136][145][245] & -[123][145][456]. \end{array}$$

The second and fourth monomials in the resulting polynomial are nonstandard. We straighten the largest nonstandard monomial. In this case we straighten $[136][145][245]$ by adding a multiple of the appropriate Van der Waerden syzygy; the reader can check that this is $[245][[13\dot{6}\dot{1}\dot{4}\dot{5}]]$. This has the effect of producing a new polynomial in which some of the monomials might be nonstandard but these new monomials can only be smaller than the one we've eliminated. Continuing in this way we eventually reach a polynomial with only standard monomials. This is the normal or straightened form of the original polynomial. The normal form for our example polynomial $[145][234][156]$ is $[123][145][456] - [124][145][356] + [134][145][256]$.

**Example 4** *Consider four points a, b, c, and d in* $\mathbb{P}^1$. *The second monomial in the bracket polynomial* $P(a,b,c,d) = [ac][bd] - [ad][bc]$ *is nonstandard with the violation occurring in the second column. To straighten P we use the Plücker relation*

$$[[a\dot{d}\dot{b}\dot{c}]] = [ad][bc] - [ab][dc] + [ac][db] = [ad][bc] + [ab][cd] - [ac][bd] = 0$$

*to express the nonstandard monomial* $[ad][bc]$ *in a different manner and substitute into P. We obtain a standard monomial* $[ab][cd]$ *equivalent to P.*

## 4.4 The Grassmann-Cayley Algebra

### 4.4.1 Motivation

Let $V$ be a vector space of dimension $d$ over a field $k$. The Grassmann-Cayley algebra is generated by elements that correspond to subspaces of $V$ and has two operations, the join and meet, that correspond to subspace sum and intersection under suitable conditions. We introduce the Grassmann-Cayley algebra with $d$ arbitrary here, and take an algebraic point of view. For a more geometric discussion in the plane, please see Chapter 3 of this handbook by Li and Richter-Gebert.

**Exterior product:** We denote the *exterior product* of vectors $v_1, \ldots, v_m \in V$ by $v_1 \vee \cdots \vee v_m$. (Typically, the exterior product is denoted by "$\wedge$", but we reserve that symbol for the meet operation in the Grassmann-Cayley algebra.) In the literature on the Grassmann-Cayley algebra the $\vee$ is frequently omitted, and we write $v_1 \cdots v_m$ for the exterior product.

**Extensor:** The exterior product of $m$ vectors is an *extensor of step m.*

**Join:** The *join* of extensors $\alpha$ and $\beta$ is $\alpha \vee \beta$. If $\alpha$ is an extensor of step $m$ and $\beta$ is an extensor of step $n$, and the vector space spanned by the factors of $\alpha$ and $\beta$ has dimension $m+n$, then their join has step $m+n$; otherwise their join is zero.

**Exterior algebra:** The *exterior algebra* of $V$ is the $k$-algebra generated by all of the extensors on $V$; that is, $V = \oplus V^{\vee i}$, where $V^{\vee i}$ is the span of the extensors of step $i$. The exterior algebra is the quotient of the tensor algebra obtained by setting $v \vee v = 0$ for all $v \in V$, hence the name "extensor" (standing for "exterior tensor").

**Meet:** The *meet* of extensors $v_1 \cdots v_m$ and $w_1 \cdots w_n$ is defined if $m+n \geq d$. In this, case $v_1 \cdots v_m \wedge w_1 \cdots w_n$ is defined to be the following extensor of step $m+n-d$:

$$\sum_{\sigma} \operatorname{sgn}(\sigma)[v_{\sigma(1)} \cdots v_{\sigma(d-n)} w_1 \cdots w_n] v_{\sigma(d-n+1)} \cdots v_{\sigma(m)}.$$

The sum is taken over all permutations $\sigma$ of $\{1,2,\ldots,m\}$ such that $\sigma(1) < \sigma(2) < \cdots < \sigma(d-n)$ and $\sigma(d-n+1) < \sigma(d-n+2) < \cdots < \sigma(m)$. Each such permutation $\sigma$ is called a shuffle. Note that when the step of the meet is zero, the resulting element is a bracket.

**Dotted notation:** As in Section 4.3.2, we use the *dotted* notation and write

$$[\dot{v}_1 \cdots \dot{v}_{d-n} w_1 \cdots w_n] \dot{v}_{d-n+1} \cdots \dot{v}_m$$

for

$$\sum_{\sigma} \operatorname{sgn}(\sigma)[v_{\sigma(1)} \cdots v_{\sigma(d-n)} w_1 \cdots w_n] v_{\sigma(d-n+1)} \cdots v_{\sigma(m)},$$

where the sum is over all shuffles $\sigma$.

**Grassmann-Cayley algebra:** The *Grassmann-Cayley algebra* of $V$ is the exterior algebra of $V$ together with the meet operation.

### 4.4.2 The cross product as a Join

We show that the join of vectors in $\mathbb{R}^3$ is the familiar cross product, suitably interpreted.

Let $V = \mathbb{R}^3$ with standard basis vectors $e_1, e_2$, and $e_3$. Note that $V^{\vee 2} = V \vee V$ has basis $e_2 \vee e_3, e_1 \vee e_3, e_1 \vee e_2$ and the join of $v = v_1e_1 + v_2e_2 + v_3e_3$ and $w = w_1e_1 + w_2e_2 + w_3e_3$ is

$$\begin{aligned} &(v_1e_1 + v_2e_2 + v_3e_3) \vee (w_1e_1 + w_2e_2 + w_3e_3) \\ &= (v_2w_3 - v_3w_2)e_2 \vee e_3 - (v_1w_3 - v_3w_1)e_1 \vee e_3 + (v_1w_2 - v_2w_1)e_1 \vee e_2. \end{aligned}$$

Note that the coefficients of $v \vee w$ are the coefficients of the cross product of $v$ and $w$. The set $\{e_i^*\}$ is a dual basis for $V^*$, and if we define a linear map given by $e_1^* \mapsto e_2 \vee e_3$, $e_2^* \mapsto e_1 \vee e_3$ and $e_3^* \mapsto e_1 \vee e_2$, we have an isomorphism from $V^*$ to $V^{\vee 2}$. Thus, $v \vee w$ gives a linear function on $V$ whose kernel is the plane spanned by $v$ and $w$. So, here we can see that the join of $v$ and $w$ corresponds to the subspace spanned by $v$ and $w$ in a natural way.

If we think of the points $[v_1 : v_2 : v_3]$ and $[w_1 : w_2 : w_3]$ in $\mathbb{P}^2$, then the same computation above shows that the extensor represents the line spanned by the points, and that the coordinates of the extensor in this basis give the coefficients of the equation of the line. Thus, we see that lines in $\mathbb{P}^2$ are parameterized by a dual $\mathbb{P}^2$ whose coordinates correspond to step 2 extensors.

### 4.4.3 Properties of the Exterior Product

The exterior product is associative, alternating (or equivalently, antisymmetric or anticommutative) and multilinear, and we can see that together these two properties imply that the exterior product of a set of vectors is nonzero if and only if they are linearly independent.

***Lemma 4.1***
*Let $V$ be a vector space over a field $k$.*

(a) *If $v \in V$, then $v \vee v = 0$.*

(b) *If $v = r_1 v_1 + \cdots r_m v_m$, where $r_i \in k$ and $v_i \in V$, then $v \vee v_1 \vee v_2 \vee \cdots \vee v_m = 0$.*

**Proof 4.1** For (1), note that the alternating property states that if we transpose the order of two factors in an exterior product then its sign changes. As a consequence, $v \vee v = 0$ for any $v \in V$, and any exterior product with two factors the same must be zero.

For (2), by linearity $v \vee v_1 \vee v_2 \vee \cdots \vee v_m = \sum r_i(v_i \vee v_1 \vee \cdots \vee v_m)$, and we can see that each summand has two factors that are the same.

From (2) of 4.1 we see that each extensor of step $m$ gives rise to a unique $m$-dimensional linear subspace of $V$, as the factors in the product are a basis for a vector space. In fact, given a vector space $V$, the exterior product of the vectors in a basis for $V$ is unique up to a scalar multiple.

**Example 5 (Extensors and the Grassmannian)** *Suppose that we work in $V = \mathbb{R}^4$ with standard basis $e_1, e_2, e_3, e_4$ and $U$ is the 2-dimensional subspace spanned by $e_1$ and $e_2$. Any other basis for $U$ has the form $ae_1 + be_2$, $ce_1 + de_2$ where $ad - bc \neq 0$. The extensor $(ae_1 + be_2) \vee (ce_1 + de_2)$ expands to*

$$(ac)e_1 \vee e_1 + (ad)e_1 \vee e_2 + (bc)e_2 \vee e_1 + (bd)e_2 \vee e_2 = (ad - bc)e_1 \vee e_2,$$

*and we see that $U$ is indeed represented by a unique extensor up to a scalar multiple.*

*In fact, we see that for any dimension 2 vector subspace $U$ in $\mathbb{R}^4$, the exterior product of a pair of basis vectors is an extensor in the span of the 6 extensors $e_1 \vee e_2, \ldots, e_3 \vee e_4$ obtained by joining pairs of the four standard basis vectors. Since this extensor representing $U$ is only unique up to scalar multiple, we see that in fact each rank 2 vector space of $\mathbb{R}^4$ maps to a point in $\mathbb{P}^5$. The extensors $e_i \vee e_j$ are the Plücker coordinates of $G(2,4)$, and the sole Plücker relation arises by looking at relations on the $2 \times 2$ minors of the $2 \times 4$ matrix whose rows are given by a basis for $U$. This is exactly how the Plücker relation on 4 points in $\mathbb{P}^1$ was derived in Section 4.3.2 and indeed the Grassmannian $G(2,4)$ is a degree 2 hypersurface in $\mathbb{P}^5$ defined by this same equation.*

This example illustrates an important general phenomenon first observed in Section 4.3.1. The bracket ring on $n$ symbols with brackets consisting of $d$ symbols is both the ring of invariants for $n$ points on $\mathbb{P}^{d-1}$ and the coordinate ring of the Grassmannian $G(d,n)$ parameterizing $d$-dimensional subspaces of $\mathbb{C}^n$ (or $d-1$-dimensional linear spaces in $\mathbb{P}^{n-1}$).

### 4.4.4 Brackets and the Grassmann-Cayley algebra

When we write an extensor of step $d$ in terms of the standard basis we can see how the alternating and multilinear properties force brackets to arise. For example, suppose that we have $d = 2$, $a = a_1e_1 + a_2e_2$, and $b = b_1e_1 + b_2e_2$. Then

$$\begin{aligned} a \vee b &= (a_1e_1 + a_2e_2) \vee (b_1e_1 + b_2e_2) \\ &= a_1e_1 \vee (b_1e_1 + b_2e_2) + a_2e_2 \vee (b_1e_1 + b_2e_2) \\ &= (a_1b_2 - a_2b_1)e_1 \vee e_2 \end{aligned}$$

and we see that we have the determinant of the matrix with $a$ and $b$ as column vectors. In other words, the coefficient of $e_1 \vee e_2$ is the bracket $[ab]$.

More generally we have

***Lemma 4.2***
*If $a_1, \ldots, a_d \in V$, then $a_1 \vee \cdots \vee a_d = [a_1 \cdots a_d] e_1 \vee \cdots \vee e_d$.*

**Proof 4.2** As in the example above, if we write the $a_i$ in terms of the basis vectors $e_i$, and use the alternating and multilinear properties to write the resulting extensor in terms of the exterior product of the standard basis vectors in order, the result follows.

### 4.4.5 The Join and Meet

The join and meet in the Grassmann-Cayley algebra were designed to capture the operations of subspace sum and intersection. Let $U$ and $W$ be two subspaces of a vector space $V$ with bases $u_1, \ldots, u_m$ and $w_1, \ldots, w_n$. Note that $u_1 \cdots u_m \vee w_1 \cdots w_n$ represents $U + W$ if and only if $U \cap W = 0$, otherwise the join is zero even though the subspace sum is not.

Since we have $m + n \geq d$ in the definition of the meet, the meet only captures the intersection of subspaces whose span is all of $V$. If $U + W$ is indeed all of $V$, then $u_1 \cdots u_m \wedge w_1 \cdots w_n$ is a sum of extensors of step $d - (m + n)$. In order to interpret the meet as the intersection of $U$ and $W$ we need to show that the meet of extensors is again an extensor.

**Example 6** *Let $V = \mathbb{R}^4$ with the standard basis. Let $u = e_1 \vee e_2 \vee e_3$ and $w = e_2 \vee e_3 \vee e_4$. Then*

$$u \wedge w = [e_1 e_2 e_3 e_4] e_2 \vee e_3 - [e_2 e_2 e_3 e_4] e_1 \vee e_3 + [e_3 e_2 e_3 e_4] e_1 \vee e_2 = [e_1 e_2 e_3 e_4] e_2 \vee e_3.$$

*We see that $u \wedge w$ is a scalar multiple of $e_2 \vee e_3$, and that $e_2$ and $e_3$ form a basis for the intersection of the subspaces corresponding to the extensors $u$ and $w$.*

***Lemma 4.3***
*Let $V$ be a vector space with vectors $u_1 \ldots, u_m$ forming a basis for a subspace $U$, and vectors $w_1 \ldots, w_n$ forming a basis for a subspace $W$. Let $u = u_1 \cdots u_m$ and $w = w_1 \cdots w_n$. If $u \wedge w \neq 0$, then $u \wedge w$ is indeed an extensor.*

**Proof 4.3** Let $C$ be a basis for $U \cap W$ and expand $C$ to a basis $B_U$ of $U$. We can also expand $C$ to a basis $B_W$ for $W$ so that $C \cup (B_U \backslash C) \cup (B_W \backslash C)$ is a basis for $V$. Let $\alpha$ denote the exterior product of the elements of $B_U$ and $\beta$ denote the exterior product of the elements in $B_W$. Then because of the alternating property, when we expand the sum in the definition of $\alpha \wedge \beta$ we find that either the entire sum is zero or the only nonzero term is supported on the exterior product of the elements of $C$. Since the exterior product of the elements of a basis for a vector space is unique up to a scalar multiple, we see that $\alpha \wedge \beta$ is a scalar multiple of $u \wedge w$, which must then be an extensor and hence corresponds to a subspace (the subspace $U \cap W$).

## 4.5 Cayley Factorization

### 4.5.1 Motivation

The vanishing of a simple Grassman-Cayley expression has a geometric interpretation. For example, in $\mathbb{P}^2$, the join $a \vee b \vee c$ is zero if and only if the three points $a, b$, and $c$ are collinear. If we pick a basis $\{e_i\}$ for the underlying vector space, then $a \vee b \vee c = [abc](e_1 \vee e_2 \vee e_3)$, and we can see that the vanishing of the join is equivalent to the vanishing of a bracket monomial. In rigidity theory bracket polynomials arise as the pure conditions of frameworks, and we want to understand when such a bracket polynomial vanishes in terms of the geometry of the points in the brackets. In this section we review an algorithm that implements Grassmann-Cayley factorization. It takes a multihomogeneous bracket polynomial $P$ as input and attempts to find a simple Grassmann-Cayley expression, which expands to $P$ if it exists. We follow the treatment in [9]. See Chapter 3.3 which contains a discussion of Cayley factorization and Cayley *expansion* in the plane.

**Bracket monomial:** A *bracket monomial* is a product of brackets.

**Simple Grassmann-Cayley expression:** A *simple Grassmann-Cayley expression* is an element of the Grassmann-Cayley algebra in which only meets and joins appear (i.e., sums are forbidden).

**Atomic extensors:** Given a multilinear bracket polynomial $P(a_1,\ldots,a_n)$, we partition the set $\{a_1,\ldots,a_n\}$ into equivalence classes which we call *atomic extensors*. Two points $a_i$ and $a_j$ are in the same atomic extensor if $P$ vanishes when $a_i$ and $a_j$ are equal.

### 4.5.2 The Pure Condition as a Bracket Monomial

The simplest generalization of the vanishing of a single bracket is the vanishing of a bracket monomial. Since the bracket ring is a quotient ring, if $P$ is a bracket polynomial there are (infinitely) many different coset representatives of $P$, each of which may be obtained by adding a element of the defining ideal of the quotient ring to $P$. In rigidity theory, bracket polynomials arise as the pure conditions of body-and-bar frameworks. Via the $k$-fan diagrams of White and Whitely [13], we can construct a bracket polynomial from a tie-down of a body-and-bar framework. The example below shows that a judicious choice of tie-down may greatly simplify the analysis.

**Example 7** *The multigraph below is an isostatic planar body-and-bar framework in the plane.*

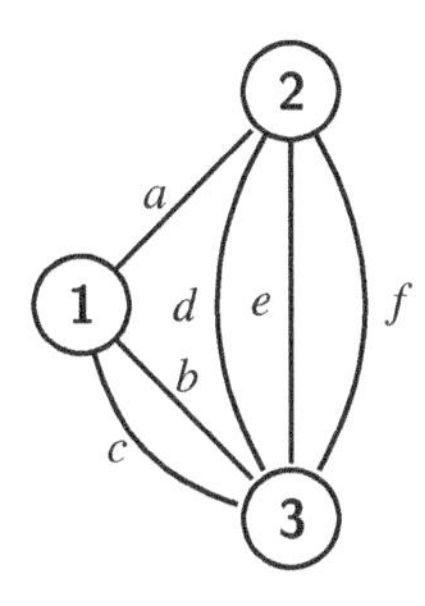


*If we tie down vertex 1 then there are three 3-fan diagrams, and the pure condition is represented by*

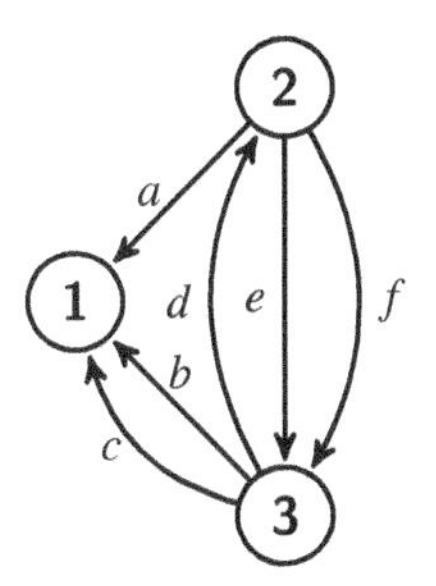

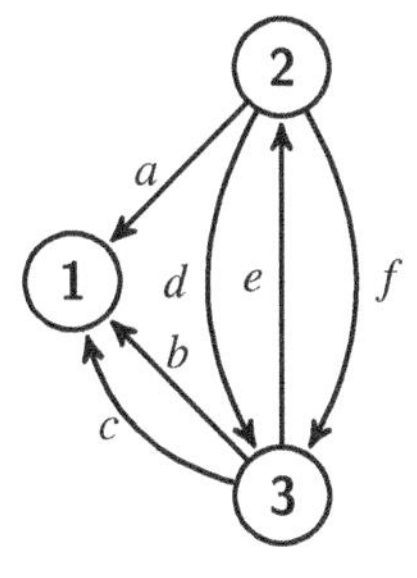

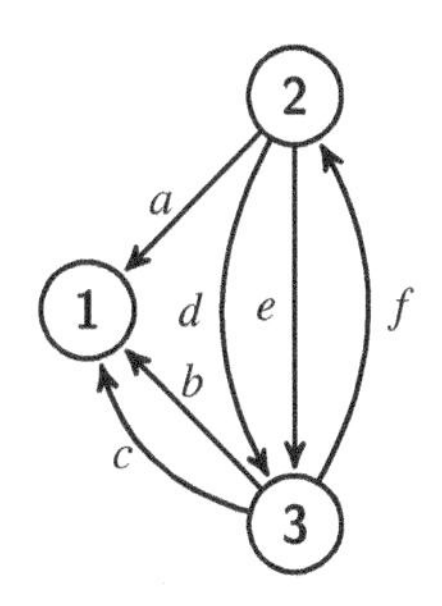

$$[aef][bcd] - [adf][bce] + [ade][bcf].$$

*If we tie down vertex 2 there is a unique 3-fan diagram and we obtain the representative*

$$[abc][def].$$

*Of course these two representations are equivalent; one can add* $[[bc\dot{a}\dot{d}\dot{e}\dot{f}]]$ *to the first representation to get the second.*

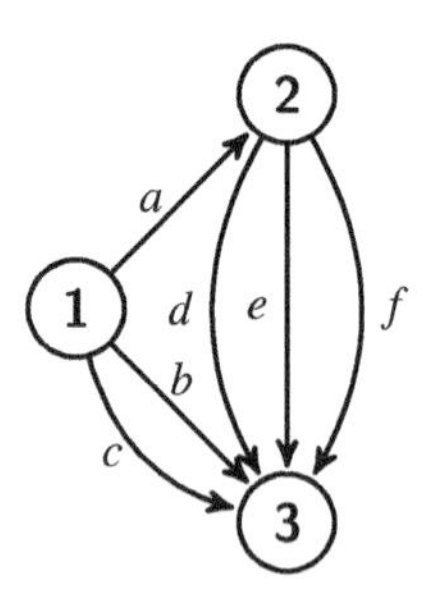


In fact, the first author, together with Ruimin Cai, Audrey St. John, and Louis Theran have characterized the body-and-bar frameworks whose pure condition can be represented by a bracket monomial. We digress briefly to give the proof of this result here, as we think it may be of interest to those interested in rigidity theory, and it is not a lengthy proof.

Recall that if we fix a positive integer $d$, a graph $G$ is a $(d,d)$-Henneberg graph if there exists a sequence of graphs, $G_0,\dots,G_m = G$ such that $G_0$ is a graph consisting of a single vertex, and the graph $G_i$ is obtained by adding a new vertex $v_i$ to the vertex set of $G_{i-1}$ together with $d$ edges from $v_i$ to some vertices in $G_{i-1}$ (where multi-edges are allowed).

We define a multigraph $G$ to be a $(d,d)$-recursively Henneberg graph if either $G$ is a $(d,d)$-Henneberg graph or if there exists a subgraph $G'$ such that $G'$ and $G/G'$ are $(d,d)$-recursively Henneberg graphs. The notion of a recursively Henneberg graph is related to the construction of quadratically solvable graphs due to Owen in Theorem 3.2 of [4].

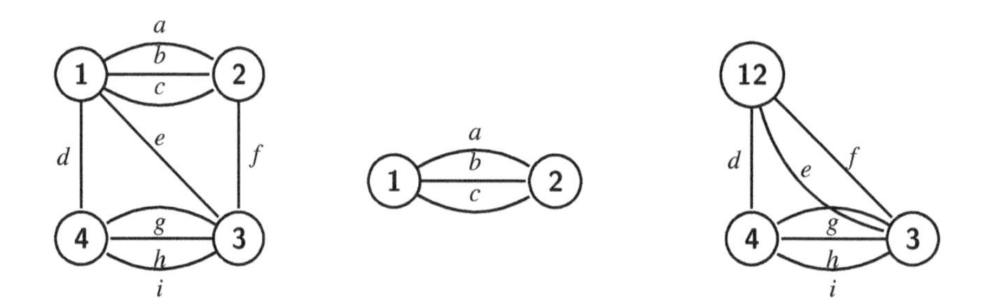


**Figure 4.3**
A recursively $(3,3)$-Henneberg graph $G$ is depicted on the left. The middle graph $G'$ is a Henneberg subgraph of $G$, and the graph on the right is $G/G'$.

**Theorem 4.1 (Cai, Sidman, St. John, Theran)** *If $G$ is a multigraph representing an isostatic body-and-bar framework labeled by $d$-vectors, then the pure condition of $G$ can be represented by a bracket monomial if and only if $G$ is a recursively $(d,d)$-Henneberg graph.*

**Proof 4.4** Suppose that the pure condition $P$ of $G$ has a representation as a bracket monomial, and proceed by induction on the number of vertices of $G$. If $P$ is a single bracket, then we are in the base case where $G$ is a graph with two vertices joined by $d$ edges, and $G$ is a recursively $(d,d)$-Henneberg graph. If $P$ is a product of more than one bracket, then $P$ factors, and by Proposition 4.4 in [13], $G$ has an isostatic subgraph $G'$. The pure condition of $G$ is the product of the pure conditions of $G'$ and $G/G'$, so these graphs both have pure conditions that are products of brackets. By induction, they are recursively $(d,d)$-Henneberg graphs, so $G$ is as well.

For the reverse direction, suppose that $G$ is a recursively Henneberg $(d,d)$-graph. Then it either consists of $d$ edges between two vertices (in which case the pure condition is a single bracket) or it has a proper subgraph $G'$ such that $G'$ and $G/G'$ are recursively Henneberg $(d,d)$-graphs. By induction, the pure conditions of $G'$ and $G/G'$ are bracket monomials, and the product of these bracket monomials is the pure condition of $G$.

### 4.5.3 White's Algorithm for Multilinear Grassmann-Cayley Factorization

The bracket polynomials that occur as the pure condition of a body-and-bar framework are homogeneous and multilinear. White's algorithm implements Grassmann-Cayley factorization for multilinear expressions in the bracket algebra. We follow the presentation in [9], using the same notation for consistency.

**White's algorithm:** Let $P$ be a bracket polynomial.

(a) Partition the set of points in $P$ into a set $Q$ of atomic extensors. If there do not exist atomic extensors whose step sizes sum to $d$, then the algorithm terminates.

(b) Choose an atomic extensor $A$ of step $d$ in $Q$, and straighten $P$ with the variables in $A$ first. This yields a new coset representative of $P = A \cdot P'$. Repeat this step with atomic extensors $Q' = Q \backslash \{A\}$ and bracket polynomial $P'$ while atomic extensors of step $d$ remain. If $P'$ is a scalar, the algorithm terminates with the Cayley factorization of the original polynomial.

(c) If there exist atomic extensors $E$ and $F$ of steps $j$ and $k$ respectively, where $j,k < d$ and $j+k > d$, order the variables so that the variables in $E$ come first, followed by the variables in $F$. If the result of straightening has $E$ first in each tableaux, followed by shuffles of $F$, followed by the rest of the variables, then $E \wedge F$ is part of the factorization. If we cannot find such $E$ and $F$, then the algorithm terminates and the polynomial cannot be factored.

(d) If $E \wedge F$ is a factor, then delete $E$ and $F$ from $Q$. Define a new symbol $G = E \wedge F$, add $G$ to $Q$, and recompute the atomic extensors. Replace $P$ by the bracket polynomial whose tableaux have one fewer row by deleting the top row and replacing occurrences of $F$ in the next row by dummy variables. Start over with step 2 now.

**Example 8** *We illustrate the first two steps in White's algorithm, finding atomic extensors and straightening, on the polynomial* $P(a,b,c,d) = [ac][bd] - [ad][bc]$. *Computing*

- $P(a,a,c,d) = [ac][ad] - [ad][ac] = 0,$
- $P(a,b,c,c) = [ac][bc] - [ac][bc] = 0,$
- $P(a,b,a,d) = [aa][bd] - [ad][ba] = [ad][ab].$

*we see that the atomic extensors of* $P$ *are* $ab$ *and* $cd$.

*We put* $a > b > c > d$ *and straighten using the syzygy* $S = [[a\dot{b}\dot{c}\dot{d}]] = [ab][cd] - [ac][bd] + [ad][bc]$. *Since* $P + S = [ab][cd]$, *we can see that the Cayley factorization is* $ab \wedge cd$, *and we are done.*

We illustrate the full algorithm on an isostatic body-and-bar framework with four bodies in the plane.

**Example 9 (Sidman, St. John, Theran)**

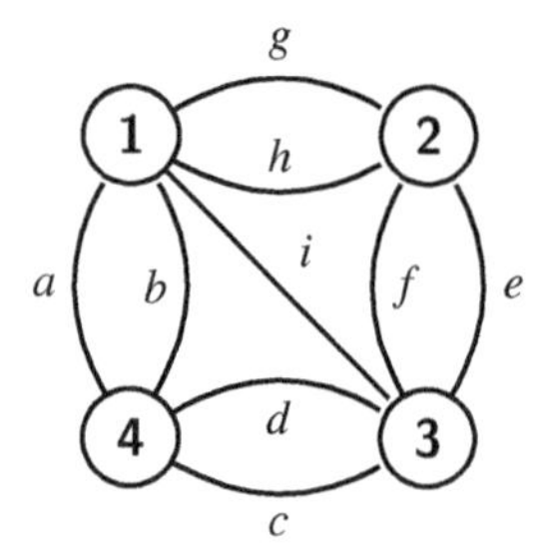


$$P = -[abc][dfi][egh] + [abd][cfi][egh] + [abc][dei][fgh] - [abd][cei][fgh].$$

*(a) Atomic extensors are* $Q = \{ab, cd, ef, gh, i\}$.

*(b) There are no atomic extensors of step* $d = 3$.

*(c) If we put* $a > b > c > d > e > f > g > h > i$, *then* $P$ *is already straightened with* $a > b > c > d$. *Indeed,* $P = -[ab\dot{c}][\dot{d}fi][egh] + [ab\dot{c}][\dot{d}ei][fgh]$.

*(d) Store* $ab \wedge cd$ *and continue with bracket polynomial* $P' = -[ufi][egh] + [uei][fgh]$.

*(e) The atomic extensors of* $P'$ *are* $Q' = \{ui, ef, gh\}$.

*(f) Still no atomic extensors of step* $d = 3$.

*(g) Straighten with the order* $u > i > e > f > g > h$. *We have* $[uif][egh] - [uie][fgh] = [ui\dot{f}][\dot{e}gh]$.

*(h) Store* $ui \wedge ef$ *and continue with* $[vgh]$. *Since this is a bracket of step* $d = 3$, *we are done.*

*(i) Unwinding all of this, we have* $(((ab \wedge cd) \vee i) \wedge ef) \vee gh$.

*(j) This can be interpreted as saying that the condition* $P$ *is equivalent to the requirement that three points must be collinear: the three points are given by intersecting lines* $e$ *and* $f$, *lines* $g$ *and* $h$, *and line* $i$ *with the line through the intersection of* $a$ *with* $b$ *and* $c$ *with* $d$.

**Example 10 (Sidman, St. John, Theran)** *We give an isostatic planar body-and-bar framework whose pure condition does not have a Grassmann-Cayley factorization. The pure condition of the graph below is* $[abf][cdi][egh] - [abg][cdi][efh] + [abi][cde][fgh]$. *We checked that it is not factorable using an implementation of White's algorithm in Macaulay 2 [1].*

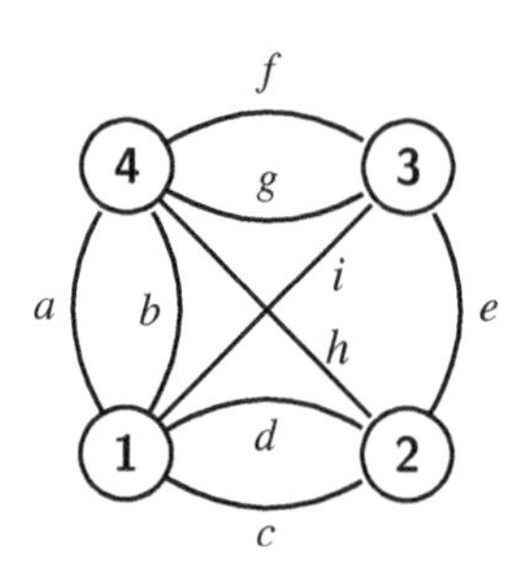

It remains an open problem to develop an algorithm that produces a Grassmann-Cayley factorization of an arbitrary bracket polynomial. One interesting result regarding the Grassmann-Cayley factorization of arbitrary bracket polynomials is due to Sturmfels and Whiteley.

**Theorem 4.2 (Sturmfels-Whiteley, [11])** *If $P$ is a multihomogeneous bracket polynomial there exists a bracket monomial $M$ so that $MP$ has a Grassmann-Cayley factorization.*

As Sturmfels notes in [9], the proof is constructive but the degree of the monomial $M$ may be very large, so a factorization algorithm based on this theorem would be impractical in many cases. If we are given $M$ and then factor $MP$, we see that if $M \neq 0$ then $P = 0$ is equivalent to geometry encoded in the Grassmann-Cayley factorization. That is, as long as the degeneracies encoded in the vanishing of $M$ do not arise, then the vanishing of the bracket polynomial is equivalent to the geometric interpretation of the Grassmann-Cayley factorization of $MP$.

**Acknowledgments.** Ruimin Cai worked on Theorem 4.1 as part of her undergraduate research at Mount Holyoke College, and the proof presented here is due to joint work with the first author, Audrey St. John, and Louis Theran. The examples of body-and-bar frameworks appearing in the text were produced jointly with Audrey St. John and Louis Theran. Code for Grassmann-Cayley factorization was implemented by the first author and Josephine Yu at the August 2010 Macaulay 2 workshop funded by NSF grant number 0964128 and NSA grant number H98230-10-1-0218.

## References

[1] Daniel R. Grayson and Michael E. Stillman. Macaulay2, a software system for research in algebraic geometry. Available at `http://www.math.uiuc.edu/Macaulay2/`.

[2] Gerard Laman. On graphs and rigidity of plane skeletal structures. *J. Engrg. Math.*, 4:331–340, 1970.

[3] Hongbo Li. *Invariant algebras and geometric reasoning*. World Scientific, 2008.

[4] John C. Owen and Steve C. Power. The non-solvability by radicals of generic 3-connected planar Laman graphs. *Trans. Amer. Math. Soc.*, 359(5):2269–2303, 2007.

[5] Hilda Pollaczek-Geiringer. Über die gliederung ebener fachwerk. *Zeitschrift fur Angewandte Mathematik und Mechanik (ZAMM)*, 7:58–72, 1927.

[6] Jürgen Richter-Gebert. *Perspectives on projective geometry*.

[7] Hal Schenck. *Computational algebraic geometry*, volume 58 of *London Mathematical Society Student Texts*. Cambridge University Press, Cambridge, 2003.

[8] Karen E. Smith, Lauri Kahanpää, Pekka Kekäläinen, and William Traves. *An invitation to algebraic geometry*. Universitext. Springer-Verlag, New York, 2000.

[9] Bernd Sturmfels. *Algorithms in invariant theory*. Texts and Monographs in Symbolic Computation. Springer, New York, second edition, 2008.

[10] Bernd Sturmfels and Neil White. Gröbner bases and invariant theory. *Adv. Math.*, 76(2):245–259, 1989.

[11] Bernd Sturmfels and Walter Whiteley. On the synthetic factorization of projectively invariant polynomials.

[12] Tiong-Seng Tay. On the generic rigidity of bar-frameworks. *Adv. in Appl. Math.*, 23(1):14–28, 1999.

[13] Neil White and Walter Whiteley. The algebraic geometry of motions of bar-and-body frameworks. *SIAM J. Algebraic Discrete Methods*, 8(1):1–32, 1987.

[14] Neil L. White and Walter Whiteley. The algebraic geometry of stresses in frameworks. *SIAM J. Algebraic Discrete Methods*, 4(4):481–511, 1983.

[15] Alfred Young. *The Collected Papers of Alfred Young (1873–1940).*

# Chapter 5

# *From Molecular Distance Geometry to Conformal Geometric Algebra*

**Timothy F. Havel**

*Energy Compression Inc., Boston, Massachusetts*

**Hongbo Li**

*Academy of Mathematics and Systems Science, Chinese Academy of Sciences; School of Mathematics, University of Chinese Academy of Sciences, China*

**CONTENTS**



## 5.1 Euclidean Distance Geometry

**Glossary**

**Semimetric space:** A set $\mathsf{A}$ with nonnegative symmetric function $\rho : \mathsf{A} \times \mathsf{A} \to \mathbb{R}_{\geq 0}$ that vanishes only on the diagonal: $\rho(\mathsf{a},\mathsf{b}) = 0 \Leftrightarrow \mathsf{a} = \mathsf{b}$ for all $\mathsf{a},\mathsf{b} \in \mathsf{A}$.

**Metric space:** The domain $\mathsf{A}$ of a semimetric $\rho$ that satisfies the triangle inequality: $\rho(\mathsf{a},\mathsf{b}) \leq \rho(\mathsf{a},\mathsf{c}) + \rho(\mathsf{b},\mathsf{c})$ for all $\mathsf{a},\mathsf{b},\mathsf{c} \in \mathsf{A}$.

**Metric vector space:** A real vector space $\mathcal{V}$ equipped with a quadratic form $\mathcal{Q}\colon \mathcal{V} \to \mathbb{R}$, or equivalently, a bilinear form $R\colon \mathcal{V} \times \mathcal{V} \to \mathbb{R}$, where $R(\mathbf{v},\mathbf{w}) = \frac{1}{2}\big(\mathcal{Q}(\mathbf{v}) + \mathcal{Q}(\mathbf{w}) - \mathcal{Q}(\mathbf{v}-\mathbf{w})\big)$ for all $\mathbf{v},\mathbf{w} \in \mathcal{V}$.

**Metric affine space:** A set $\mathsf{P}$ together with an antisymmetric *translation* operator $\overrightarrow{\cdot\cdot} : \mathsf{P} \times \mathsf{P} \to \mathcal{V}$ onto a metric vector space $(\mathcal{V},\mathcal{Q})$ which satisfies $\overrightarrow{\mathsf{pq}} = \mathbf{0} \Leftrightarrow \mathsf{p} = \mathsf{q}$ and $\overrightarrow{\mathsf{pq}} + \overrightarrow{\mathsf{qr}} = \overrightarrow{\mathsf{pr}}$ for all $\mathsf{p},\mathsf{q},\mathsf{r} \in \mathsf{P}$.

**Euclidean space:** A metric affine space $\mathbb{E} = (\mathsf{P},(\mathcal{V},\mathcal{Q}),\overrightarrow{\cdot\cdot})$ on which the quadratic form is positive-definite, meaning $\mathcal{Q}(\overrightarrow{\mathsf{pq}}) \geq 0$ and $\mathcal{Q}(\overrightarrow{\mathsf{pq}}) = 0 \Leftrightarrow \mathsf{p} = \mathsf{q}$ for all $\mathsf{p},\mathsf{q} \in \mathsf{P}$.

**Euclidean embedding:** An injective mapping $\varepsilon : \mathsf{A} \hookrightarrow \mathsf{P}$ of a semimetric space $(\mathsf{A}, \rho)$ into a Euclidean space $\mathbb{E}$ such that $\rho\big(\varepsilon^{-1}(\mathsf{p}), \varepsilon^{-1}(\mathsf{q})\big)^2 = \mathcal{Q}(\overrightarrow{\mathsf{pq}})$ for all $\mathsf{p}, \mathsf{q} \in \varepsilon(\mathsf{A})$; when such a mapping exists, we say that $\rho$ is *embeddable* in Euclidean space.

**$n$-dimensional Euclidean distance function:** The metric $\rho : \mathsf{A} \times \mathsf{A} \to \mathbb{R}_{\geq 0}$ of a metric space $(\mathsf{A}, \rho)$ which can be embedded in a Euclidean space $\mathbb{E}_n = (\mathsf{P}, (\mathcal{V}, \mathcal{Q}), \overrightarrow{\cdot\cdot})$ of dimension $n = \dim(\mathcal{V})$ but not $n-1$.

**$n$-dimensional Euclidean group:** Given a Euclidean space $\mathbb{E}_n$ of dimension $n$, a fixed origin $\mathsf{o} \in \mathsf{P}$, a translation $\mathbf{t} \in \mathcal{V}$ and a rotation $\mathbf{R} \in \mathcal{O}$, where $\mathcal{O}$ is the group of isometries of $(\mathcal{V}, \mathcal{Q})$, the $n$-dimensional Euclidean group $\mathcal{E}_n$ may be defined via its action $\omega$ on $\mathsf{P}$, namely $\mathsf{q} = \omega_{\mathbf{t},\mathbf{R}}(\mathsf{p}) \Leftrightarrow \overrightarrow{\mathsf{oq}} = \mathbf{t} + \mathbf{R}\overrightarrow{\mathsf{op}}$; composition in $\mathcal{E}_n$ is given by $\omega_{\mathbf{t}',\mathbf{R}'} \circ \omega_{\mathbf{t},\mathbf{R}} = \omega_{\mathbf{t}'+\mathbf{R}'\mathbf{t},\mathbf{R}'\mathbf{R}}$, which shows that this definition is origin-independent.

---

*Distance geometry* may be regarded as the study of "geometric" metric spaces [39]. These are metric spaces $(\mathsf{A}, \rho)$ with a transitive group action such that the metric is invariant with respect to the action induced on the unordered pairs of $\mathsf{A}$. Early investigators including Karl Menger, Johan Seidel, and Leonard Blumenthal demonstrated that the classical elliptic, Euclidean, and hyperbolic geometries are completely characterized by their metrics alone [4, 40, 54]. Implicit in their work is the fact that all the invariant geometric quantities and relations of these geometries can be expressed in a coordinate-free fashion using the interpoint distances as the primitive notion.

For $n$-dimensional Euclidean geometry, the group in question is the $n$-dimensional Euclidean group $\mathcal{E}_n$ of translations composed with proper and improper rotations. The $n = 2$ case dates back at least to Heron of Alexandria, who in AD 60 gave us what is now known as Heron's formula for the area $A$ of a triangle with vertices $\mathsf{a}, \mathsf{b}, \mathsf{c}$ in terms of the lengths of its sides $d_{\mathsf{a},\mathsf{b}}, d_{\mathsf{b},\mathsf{c}}, d_{\mathsf{a},\mathsf{c}}$. Letting $2s = d_{\mathsf{a},\mathsf{b}} + d_{\mathsf{b},\mathsf{c}} + d_{\mathsf{a},\mathsf{c}}$ be its perimeter, this is given by:

$$A^2 \;=\; s(s - d_{\mathsf{a},\mathsf{b}})(s - d_{\mathsf{a},\mathsf{c}})(s - d_{\mathsf{b},\mathsf{c}}) \tag{5.1}$$

This turns out to be the first in a sequence of determinants of matrices involving *squared* interpoint distances, known as *Cayley-Menger determinants*. These were first shown by Arthur Cayley to be proportional to the squared volume $D$ of the simplex spanned by its $N$ vertices $\mathsf{a}, \mathsf{b}, \mathsf{c}, \ldots, \mathsf{d}$:

$$-(-2)^{N-1}((N-1)!)^2 D(\mathsf{a},\mathsf{b},\mathsf{c},\ldots,\mathsf{d}) \;:=\; \begin{vmatrix} 0 & 1 & 1 & 1 & \cdots & 1 \\ 1 & 0 & d^2_{\mathsf{a},\mathsf{b}} & d^2_{\mathsf{a},\mathsf{c}} & \cdots & d^2_{\mathsf{a},\mathsf{d}} \\ 1 & d^2_{\mathsf{a},\mathsf{b}} & 0 & d^2_{\mathsf{b},\mathsf{c}} & \cdots & d^2_{\mathsf{b},\mathsf{d}} \\ 1 & d^2_{\mathsf{a},\mathsf{c}} & d^2_{\mathsf{b},\mathsf{c}} & 0 & \cdots & d^2_{\mathsf{c},\mathsf{d}} \\ \vdots & \vdots & \vdots & \vdots & \ddots & \vdots \\ 1 & d^2_{\mathsf{a},\mathsf{d}} & d^2_{\mathsf{b},\mathsf{d}} & d^2_{\mathsf{c},\mathsf{d}} & \cdots & 0 \end{vmatrix} \tag{5.2}$$

Note in particular that $D(\mathsf{a},\mathsf{b}) = d^2_{\mathsf{a},\mathsf{b}}$, an alternative notation which we will frequently use.

Given a semimetric $\rho : \mathsf{A} \times \mathsf{A} \to \mathbb{R}_{\geq 0}$, the Cayley-Menger determinant on a finite sequence of $N$ points $[\mathsf{a}, \mathsf{b}, \mathsf{c}, \ldots, \mathsf{d}] \in \mathsf{A}^N$ is obtained by setting $d_{\mathsf{x},\mathsf{y}} \equiv \rho(\mathsf{x}, \mathsf{y})$ for all subscripts $\mathsf{x}, \mathsf{y}$ in Equation (5.2). Menger's intrinsic characterization of the Euclidean distance function may now be stated as follows.

**Theorem 5.1** – *A semimetric $\rho : \mathsf{A} \times \mathsf{A} \to \mathbb{R}_{\geq 0}$ on a (not necessarily finite) set $\mathsf{A}$ is Euclidean of dimension $n$ if and only if there exists a sequence of $n+1$ distinct points $[\mathsf{a}, \mathsf{b}, \mathsf{c}, \ldots, \mathsf{d}] \in \mathsf{A}^{n+1}$ such that the Cayley-Menger determinants on those points satisfy*

$$D(\mathsf{a},\mathsf{b}), D(\mathsf{a},\mathsf{b},\mathsf{c}), \ldots, D(\mathsf{a},\mathsf{b},\mathsf{c},\ldots,\mathsf{d}) \;>\; 0, \tag{5.3}$$

*and for every pair of points* $\mathsf{f}, \mathsf{g} \in \mathsf{A}$ *we also have*

$$0 \;=\; D(\mathsf{a},\mathsf{b},\mathsf{c},\ldots,\mathsf{d},\mathsf{f}) \;=\; D(\mathsf{a},\mathsf{b},\mathsf{c},\ldots,\mathsf{d},\mathsf{f},\mathsf{g})\,. \tag{5.4}$$

Theorem 5.1 is actually a corollary of a stronger theorem, also due to Menger.

**Theorem 5.2** – *[Menger's Theorem] A semimetric* $\rho : \mathsf{A} \times \mathsf{A} \to \mathbb{R}_{\geq 0}$ *is Euclidean of dimension* $n$ *if and only if all Cayley-Menger determinants on* $M \leq n+1$ *points are nonnegative, at least one determinant on* $n+1$ *points is strictly positive, all determinants on* $n+2$ *points vanish, and a Cayley-Menger determinant on at least one set of* $n+3$ *points is nonnegative (in which case it is necessarily zero).*

Another way of stating this theorem is that if every subset of $n+2$ points can be isometrically embedded in an $n$- but not generally $(n-1)$-dimensional Euclidean space, then the semimetric is Euclidean of dimension $n$ unless $\mathsf{A}$ consists of exactly $n+3$ points and the Cayley-Menger determinant on those $n+3$ points is strictly negative. Such semimetrics are called *pseudo-Euclidean,* and can be obtained by a general construction known as isogonal conjugation, starting from an arbitrary full-dimensional $(n+2)$-point subset of $n$-dimensional Euclidean space [4].

Note that for metric spaces the restriction $\rho(\mathsf{x},\mathsf{y}) = 0 \Leftrightarrow \mathsf{x} = \mathsf{y}$ in the foregoing is trivial since an easy analysis of Equation (5.1) with $A^2 = D(\mathsf{x},\mathsf{y},\mathsf{z})$ shows that $0 = d_{\mathsf{x},\mathsf{y}}$ only if $d_{\mathsf{x},\mathsf{z}} = d_{\mathsf{y},\mathsf{z}}$ for all $\mathsf{z} \in \mathsf{A}$, so the points $\mathsf{x}$ and $y$ may be identified. Henceforth, by a slight abuse of notation, whenever a semimetric is Euclidean we will not distinguish between a set of points in its domain and the embedded image thereof. Indeed we shall generally restrict ourselves to Euclidean spaces, since what is of interest to us is the Euclidean interpretation of Cayley-Menger determinants and related homogeneous polynomials in the interpoint distances squared $D(\mathsf{x},\mathsf{y})$.

Another, at first glance rather different, intrinsic characterization of the Euclidean metric is the following theorem due to I. J. Schoenberg [53]:

**Theorem 5.3** – *[Schoenberg's Theorem] A semimetric* $\rho : \mathsf{A} \times \mathsf{A} \to \mathbb{R}_{\geq 0}$ *on a finite set* $\mathsf{A} = \{\mathsf{a}_1,\ldots,\mathsf{a}_N\}$ *is Euclidean dimension* $n < N$ *if and only if Schoenberg's quadratic form*

$$Q_D(\mathbf{x}) \;:=\; -\frac{1}{2}\sum_{k=1}^{N}\sum_{\ell=1}^{N} D(\mathsf{a}_k,\mathsf{a}_\ell)\,x_k x_\ell\;, \tag{5.5}$$

*with* $D(\mathsf{a},\mathsf{a}') = d^2_{\mathsf{a},\mathsf{a}'} := \rho(\mathsf{a},\mathsf{a}')^2$ *as before, is positive semidefinite of rank* $n$ *on the subspace of* $[x_1,\ldots,x_N]^\top =: \mathbf{x} \in \mathbb{R}^N$ *defined by* $\mathbf{1}^\top\mathbf{x} = \sum_{k=1}^{N} x_k = 0$*, where* $\mathbf{1}$ *is a column of* 1*'s.*

Now let $\mathbf{D}$ be the matrix of squared distances $\left[D(\mathsf{a}_j,\mathsf{a}_k)\right]_{j,k=1,\ldots,N}$, and $\tilde{\mathbf{D}}$ be $\mathbf{D}$ augmented by a 0-th row and column of 1's with a 0 on the diagonal, as in Cayley-Menger determinants. Then on the subspace $\mathbf{1}^\top\mathbf{x} = 0$ the augmented form

$$\tilde{Q}_D(\tilde{\mathbf{x}}) \;:=\; -\tfrac{1}{2}\,\tilde{\mathbf{x}}^\top\tilde{\mathbf{D}}\,\tilde{\mathbf{x}} \tag{5.6}$$

is numerically equal to Schoenberg's quadratic form $Q_D$ for all $\tilde{\mathbf{x}} = [x_0;\mathbf{x}] \in \mathbb{R}^{N+1}$ with $x_0 = 0$. Moreover, given any set of $n+1$ independent points in $n$-dimensional Euclidean space, say $\{\mathsf{p}_1,\ldots,\mathsf{p}_n,\mathsf{p}_{n+1}\}$, the barycentric coordinates $\mathbf{x} \in \mathbb{R}^{n+1}$ $(\mathbf{1}^\top\mathbf{x} = 1)$ of any point $\mathsf{x}$ versus this basis can be found by solving the linear system of equations:

$$\begin{bmatrix} 0 & 1 & 1 & \cdots & 1 \\ 1 & 0 & D(\mathsf{p}_1,\mathsf{p}_2) & \cdots & D(\mathsf{p}_1,\mathsf{p}_{n+1}) \\ 1 & D(\mathsf{p}_1,\mathsf{p}_2) & 0 & \cdots & D(\mathsf{p}_2,\mathsf{p}_{n+1}) \\ \vdots & \vdots & \vdots & \ddots & \vdots \\ 1 & D(\mathsf{p}_1,\mathsf{p}_{n+1}) & D(\mathsf{p}_1,\mathsf{p}_2) & \cdots & 0 \end{bmatrix} \begin{bmatrix} x_0 \\ x_1 \\ x_2 \\ \vdots \\ x_{n+1} \end{bmatrix} = \begin{bmatrix} 1 \\ D(\mathsf{x},\mathsf{p}_1) \\ D(\mathsf{x},\mathsf{p}_2) \\ \vdots \\ D(\mathsf{x},\mathsf{p}_{n+1}) \end{bmatrix} \tag{5.7}$$

The 0-th coordinate $x_0$ is equal to the determinant of the augmented matrix for the linear system obtained by dropping the 0-th column from the matrix above.

In the special case that $D(\mathsf{p}_k, \mathsf{p}_{n+1}) = 1$ for $k = 1, \ldots, n$ and $D(\mathsf{p}_j, \mathsf{p}_k) = 2$ for all $j$ and $k$ with $1 < j < k \leq n$, the basis constitutes an orthonormal frame with origin $\mathsf{p}_{n+1}$, and the barycentric coordinates $x_1, \ldots, x_n$ are just the usual Cartesian coordinates of $\mathsf{x}$ with respect to this frame. From these observations it may be seen that Schoenberg's form is nothing but the squared Euclidean lengths of the interpoint vectors obtained by taking the differences of their barycentric coordinates [8]. It is striking nonetheless that these facts are naturally expressed in a space of $n+2$ dimensions.

We will presently provide a more complete geometric interpretation of this $(n+2)$-dimensional space. Meanwhile, to further spur the reader's curiosity, we note that the rank of the matrix of squared distances $\mathbf{D}$ also has something to say about the geometry of the embedded points. For example, the four-point determinant expands to:

$$\begin{vmatrix} 0 & d_{\mathsf{p,q}}^2 & d_{\mathsf{p,r}}^2 & d_{\mathsf{p,s}}^2 \\ d_{\mathsf{p,q}}^2 & 0 & d_{\mathsf{q,r}}^2 & d_{\mathsf{q,s}}^2 \\ d_{\mathsf{p,r}}^2 & d_{\mathsf{q,r}}^2 & 0 & d_{\mathsf{r,s}}^2 \\ d_{\mathsf{p,s}}^2 & d_{\mathsf{q,s}}^2 & d_{\mathsf{r,s}}^2 & 0 \end{vmatrix} = \begin{matrix} -\left(d_{\mathsf{p,q}}d_{\mathsf{r,s}}+d_{\mathsf{p,r}}d_{\mathsf{q,s}}+d_{\mathsf{p,s}}d_{\mathsf{q,r}}\right)\left(-d_{\mathsf{p,q}}d_{\mathsf{r,s}}+d_{\mathsf{p,r}}d_{\mathsf{q,s}}+d_{\mathsf{p,s}}d_{\mathsf{q,r}}\right) \\ \left(d_{\mathsf{p,q}}d_{\mathsf{r,s}}-d_{\mathsf{p,r}}d_{\mathsf{q,s}}+d_{\mathsf{p,s}}d_{\mathsf{q,r}}\right)\left(d_{\mathsf{p,q}}d_{\mathsf{r,s}}+d_{\mathsf{p,r}}d_{\mathsf{q,s}}-d_{\mathsf{p,s}}d_{\mathsf{q,r}}\right) \end{matrix} \tag{5.8}$$

in analogy to Heron's formula (5.1). The non-negativity of the last three factors on the right is known as *Ptolemy's inequality*, which holds as an equality if and only if the four points lie on a circle or a line in a plane of dimension 2. More generally, the rank of an $N \times N$ matrix of squared Euclidean distances is $m < N$ if and only if the $N$ points lie on a $(m-2)$-dimensional hypersphere in Euclidean space (of possibly infinite radius, in which case it is a $(m-2)$-dimensional Euclidean subspace).

More recent results in Euclidean distance geometry, including the fact that the inverse of the bordered matrix appearing in Cayley-Menger determinants is a matrix of inner products of outer normals to the corresponding $n$-dimensional simplex bordered by the barycentric coordinates of its orthocenter, and the applications of this fact to the classification of such simplices, may be found in the late Miroslav Fiedler's book [14].

## 5.2 The Distance Geometry Theory of Molecular Conformation

**Glossary**

**Distance constraint:** Lower and upper bounds $\ell_{\mathsf{p,q}} \leq u_{\mathsf{p,q}}$ on the distance between a pair of points $\mathsf{p}, \mathsf{q}$; the constraint is *exact* when $\ell_{\mathsf{p,q}} = u_{\mathsf{p,q}}$.

**Chirality constraint:** A subset $\tilde{\chi}(\mathsf{p}, \mathsf{q}, \ldots, \mathsf{r})$ of the power set $2^{\{-1,0,+1\}}$ containing the possible signs of the oriented volume of the $n$-dimensional simplex spanned by the ordered $(n+1)$-tuple of points $\mathsf{p}, \mathsf{q}, \ldots, \mathsf{r}$.

**Conformation:** Any element of the $(nN - n(n+1)/2)$-dimensional manifold of orbits of a set of $N$ points in an $n$-dimensional Euclidean space, under the action of the Euclidean group $\mathcal{E}_n$, that satisfies the given distance and chirality constraints.

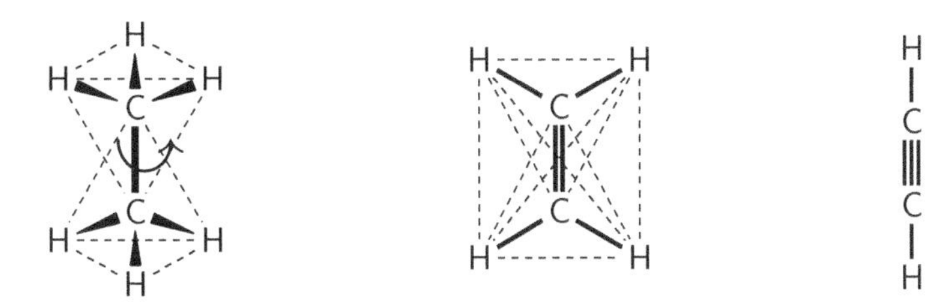


**Figure 5.1**
Standard chemical depictions of the three most common local structures in organic (carbon-based) chemistry, wherein carbon atoms are labeled C, hydrogen H, chemical bonds are drawn as solid lines and other fixed distances are drawn with light dashed lines: At a carbon-carbon single bond, as in the ethane molecule shown on the left, the tetrahedra of atoms around each carbon can rotate around the bond's axis; at a double bond, as in the ethylene molecule in the center, the atoms bonded to the carbons prefer to be coplanar with the carbons and one another; and at a triple bond, as in the acetylene molecule on the right, all atoms are collinear with the carbon bond.

**Conformation space:** The set of orbits, with boundary and singularities, consisting of all conformations as above; it may be obtained by letting the Euclidean group act on the real semialgebraic variety defined by the distance and chirality constraints relative to an arbitrary orthonormal coordinate frame.

**Triangle inequality limits:** The infimum and supremum of each individual distance over all $N$-point metric spaces consistent with the distance constraints.

**Euclidean distance limits:** The infimum and supremum of each individual distance over all sets of $N$ points in Euclidean space the distances of which satisfy the distance constraints.

**$n$-dimensional distance limits:** The infimum and supremum of each individual distance over all sets of $N$ points in $n$-dimensional Euclidean space ($n < N$) which satisfy the distance *and* chirality constraints.

---

Euclidean distance geometry is the natural foundation of a coordinate-free approach to the study of molecular conformation [8]. A molecule is a set of $N > 1$ atoms which are held together in three-dimensional Euclidean space by chemical bonds, despite being constantly jostled by random, thermally driven collisions with other molecules. Over time, the atoms of the molecule undergo a random, but highly biased, walk within a real but "fuzzy" variety, with boundaries and singularities, embedded within the $(3N-6)$-dimensional space of orbits (unless the conformation is collinear, in which case it's $3N-5$) under the action of the 3D (three-dimensional) Euclidean group of translations and rotations [11]. This fuzzy variety is often referred to as the "conformation space" of the molecule. In the case of large molecules such as the proteins and nucleic acids commonly encountered in biological organisms, the number of atoms may be in the thousands or even millions, and the dimensionality of conformation space by itself is daunting.

The usual approach to understanding this fuzzy variety is based on computational brute force. Starting from a given (perhaps experimentally determined) conformation, Newton's equations of motion are solved in an iterative process known as molecular dynamics, using an analytic (semi-empirical) or quantum mechanical (first principles) approximation to the internal energy of the molecule [52]. This produces a time-ordered sequence of thousands or even millions of conformations, each represented with essentially meaningless precision by a set of e.g., Cartesian coordinates for the atoms versus an arbitrary coordinate frame. In order to obtain an approximation to a possible physical trajectory of an single molecule, a small time step of order femtoseconds must be used.

As a result, successive conformations in such sequences differ only slightly from one another, and only in the (often unattainable, even with today's supercomputers) limit of simulations spanning milliseconds or more can one hope to have obtained a reasonably complete sample of the conformation space of large molecules. Although the trajectory itself can be viewed as a movie, the problem of making sense of the massive amounts of data in a trajectory is also highly nontrivial.

The preceding paragraph hardly does justice to an immense area of applied computational science, which recently earned the 2013 Nobel Prize in Chemistry [58]. Yet the fact remains that despite many heroic attempts to sample conformation space exhaustively and to render this high-dimensional set of chemically relevant conformations comprehensible, we have a long way to go. In contrast to analytic approaches such as molecular dynamics, the distance geometry approach to the study of molecular conformation is synthetic in nature: A "molecule" is simply defined by the totality of invariant geometric properties associated with its conformation space. While many different geometric properties could be (and are) used, considerable simplicity and generality is attained by focusing on the most elementary yet fundamental, namely the interatomic distances. For example, the distances $d_{\mathsf{a},\mathsf{b}}$ between bonded pairs of atoms $\{\mathsf{a},\mathsf{b}\}$ are generally the same to within a few percent in all possible conformations of the molecule, and so may be treated as fixed constants to a good approximation. Similarly, the angles between pairs of chemical bonds $\{\mathsf{a},\mathsf{b}\}$ and $\{\mathsf{b},\mathsf{c}\}$ with a common atom $\mathsf{b}$ are often also nearly constant, and given that the bond lengths themselves are constant may be determined by treating the "geminal" distance $d_{\mathsf{a},\mathsf{c}}$ as constant. More generally, all the distances in any essentially rigid portion of the molecule may be treated as fixed constants (see Figure 5.1).

In any molecule with a nontrivial conformation space there will also be distances that are not even approximately constant, but that does not mean they can assume any combination of values consistent with embedability in 3D Euclidean space. If nothing else, atoms repel one another at short distances even when there are no chemical bonds between them, and this effectively imposes lower bounds on all nonbonded and nongeminal pairs of atoms. These lower bounds are typically modeled as a sum of hard-sphere radii associated with the chemical types of the atoms. While the attractive forces between pairs of atoms become insignificant at large separations, the cumulative effects of many such interactions, possibly combined with the molecule's interactions with its surroundings, may also lead to nontrivial upper bounds on nonbonded, nongeminal distances in any of its possible conformations. The distance geometry approach to the study of molecular conformation is predicated on the empirical observation that one can do a reasonable job of describing the conformation space of just about any molecule by specifying lower and upper bounds on its interatomic distances (together with the chirality constraints introduced in the next paragraph). This implicit characterization of conformation space is much more compact, and also more comprehensible, than a molecular dynamics trajectory. It may in fact be used to summarize the results of such a simulation, simply by setting the bounds to the minimum and maximum values of the distances observed in the trajectory.

By themselves, however, neither the distances nor their bounds can distinguish a molecule from its mirror image, since the distances are invariant under reflections. A *chiral* molecule is one that cannot be superimposed on its mirror image by translation, rotation and intramolecular motions such as bond rotations, in which case its *chirality* specifies which mirror image it is. Chirality is most often encountered in molecules wherein four chemically distinct kinds of atoms are bonded to a common carbon atom, which hence lie on the vertices of a tetrahedron with the carbon atom in its interior. The chirality of this tetrahedron can be specified by the sign of its oriented volume $V$ relative to a given ordering of its vertex atoms $\mathsf{a},\mathsf{b},\mathsf{c},\mathsf{d}$, namely

$$\chi(\mathsf{a},\mathsf{b},\mathsf{c},\mathsf{d}) \ := \ sign\big(V(\mathsf{a},\mathsf{b},\mathsf{c},\mathsf{d})\big). \tag{5.9}$$

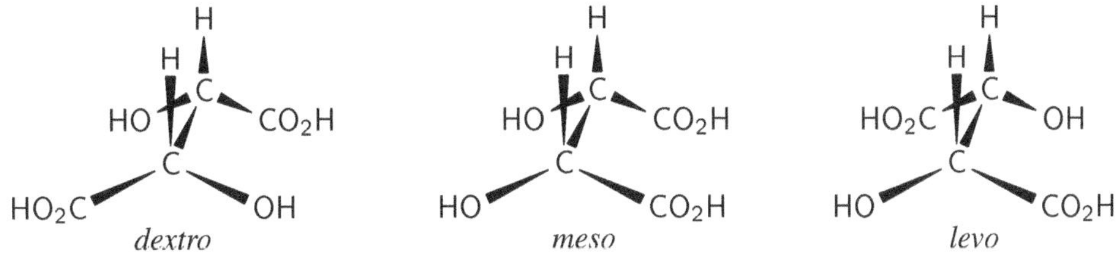

**Figure 5.2**
The *dextro* and *levo* forms of tartaric acid are *enantiomers*, or mirror images, of one another and hence have the same interatomic distances in at least one pair of their possible conformations, whereas the *meso* form has an internal plane of symmetry in at least one of its possible conformations and so is achiral.

The oriented volume of the tetrahedron, in turn, is given by the determinant

$$V(\mathsf{a},\mathsf{b},\mathsf{c},\mathsf{d}) := \frac{1}{3!}\begin{vmatrix} 1 & 1 & 1 & 1 \\ x_{\mathsf{a}} & x_{\mathsf{b}} & x_{\mathsf{c}} & x_{\mathsf{d}} \\ y_{\mathsf{a}} & y_{\mathsf{b}} & y_{\mathsf{c}} & y_{\mathsf{d}} \\ z_{\mathsf{a}} & z_{\mathsf{b}} & z_{\mathsf{c}} & z_{\mathsf{d}} \end{vmatrix}, \tag{5.10}$$

where $x,y,z$ are the Cartesian coordinates of the subscript atoms. Note this definition of chirality applies to any rigid quadruple of chemically distinct atoms in a molecule, not merely those bonded to a common carbon atom. It can be generalized to nonrigid quadruples whenever these have a common chirality in all possible conformations of the molecule. The chirality of the molecule as a whole is determined by the chirality of any one of its chiral quadruples, or *chiral centers* as they are known. Note however that if any conformation of the molecule has an internal plane of symmetry, then the molecule will be achiral even if it has chiral centers (see Figure 5.2).

As might be expected, the distances and chiralities are not wholly independent. Instead, the *relative* chirality of any pair of chiral quadruples is given by

$$\chi(\mathsf{a}_1,\mathsf{a}_2,\mathsf{a}_3,\mathsf{a}_4)\,\chi(\mathsf{b}_1,\mathsf{b}_2,\mathsf{b}_3,\mathsf{b}_4) = sign\big(D(\mathsf{a}_1,\mathsf{a}_2,\mathsf{a}_3,\mathsf{a}_4;\,\mathsf{b}_1,\mathsf{b}_2,\mathsf{b}_3,\mathsf{b}_4)\big)\,, \tag{5.11}$$

where $D(\mathsf{a}_1,\ldots,\mathsf{a}_M;\,\mathsf{b},\ldots,\mathsf{b}_M)$ is more generally a nonsymmetric $M\times M$ Cayley-Menger determinant of the form:

$$-(-2)^{M-1}\left((M-1)!\right)^2 D(\mathsf{a}_1,\ldots,\mathsf{a}_M;\,\mathsf{b}_1,\ldots,\mathsf{b}_M) := \begin{vmatrix} 0 & 1 & 1 & \cdots & 1 \\ 1 & d^2_{\mathsf{a}_1,\mathsf{b}_1} & d^2_{\mathsf{a}_1,\mathsf{b}_2} & \cdots & d^2_{\mathsf{a}_1,\mathsf{b}_M} \\ 1 & d^2_{\mathsf{a}_2,\mathsf{b}_1} & d^2_{\mathsf{a}_2,\mathsf{b}_2} & \cdots & d^2_{\mathsf{a}_2,\mathsf{b}_M} \\ \vdots & \vdots & \vdots & \ddots & \vdots \\ 1 & d^2_{\mathsf{a}_M,\mathsf{b}_1} & d^2_{\mathsf{a}_M,\mathsf{b}_2} & \cdots & d^2_{\mathsf{a}_M,\mathsf{b}_M} \end{vmatrix} \tag{5.12}$$

These determinants may be interpreted geometrically as the products of oriented volumes whenever the $2M$ points $\mathsf{a}_1,\ldots,\mathsf{a}_M,\mathsf{b}_1,\ldots,\mathsf{b}_M$ are embeddable in a Euclidean space of dimension $n = M-1$. Surprisingly, they contain none of the distances $D(\mathsf{a}_j,\mathsf{a}_k)$ or $D(\mathsf{b}_j,\mathsf{b}_k)$ for $j$ and $k$ with $1 \le j,k \le M$. The following determinantal identity, which may be derived from Jacobi's theorem on a minor of the adjugate, shows conversely that the consistency of the distances as a whole with this interpretation, together with the nonnegativity of all the symmetric Cayley-Menger determinants on $n+1$ or fewer

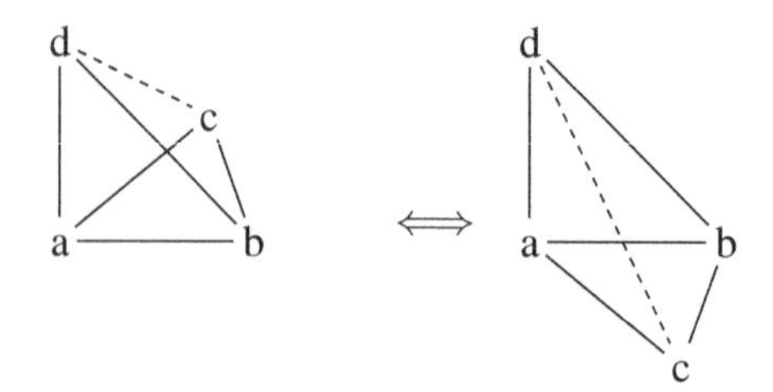


**Figure 5.3**
The two roots of the four-point Cayley-Menger determinant $D(\mathsf{a},\mathsf{b},\mathsf{c},\mathsf{d})$, viewed as a quadratic in $d^2_{\mathsf{c},\mathsf{d}}$, correspond to planar conformations in which the atoms $\mathsf{c},\mathsf{d}$ are on the same (left) or opposite (right) sides of the line of the bond connecting atoms $\mathsf{a},\mathsf{b}$. Chemists refer to these as the *cis* and *trans* conformations of the bond $\mathsf{a}-\mathsf{b}$.

points, implies that the semimetric $\rho(\mathsf{a},\mathsf{b}) := d_{\mathsf{a},\mathsf{b}}$ is $n$-dimensional Euclidean:

$$\begin{aligned}\tfrac{n^2}{(n-1)^2} D(\mathsf{a}_1,\dots,\mathsf{a}_{n-1})D(\mathsf{a}_1,\dots,\mathsf{a}_{n-1},\mathsf{b},\mathsf{c}) &= D(\mathsf{a}_1,\dots,\mathsf{a}_{n-1},\mathsf{b})D(\mathsf{a}_1,\dots,\mathsf{a}_{n-1},\mathsf{c}) \\ &\quad - D(\mathsf{a}_1,\dots,\mathsf{a}_{n-1},\mathsf{b};\, \mathsf{a}_1,\dots,\mathsf{a}_{n-1},\mathsf{c})^2 \qquad (5.13)\end{aligned}$$

Moreover the following "cross-swap" relation

$$D(\mathsf{E};\mathsf{F})D(\mathsf{G};\mathsf{H}) = D(\mathsf{E};\mathsf{G})D(\mathsf{F};\mathsf{H}) \qquad (5.14)$$

holds among any four $(n+1)$-tuples of points $\mathsf{E},\mathsf{F},\mathsf{G},\mathsf{H}$ in an $n$-dimensional Euclidean space.

We now consider the implications of these relations among the distances for bounds on the distances among the set of atoms $\mathsf{A}$ of a molecule. Given upper bounds $u_{\mathsf{a},\mathsf{b}}$, $u_{\mathsf{b},\mathsf{c}}$ on two of the distances among three atoms, the triangle inequality implies that the third distance can be no greater than their sum. This implicit bound $\bar{u}_{\mathsf{a},\mathsf{c}} := u_{\mathsf{a},\mathsf{b}} + u_{\mathsf{b},\mathsf{c}}$ is called the upper triangle inequality *limit* on the third distance. Any explicit bound $u_{\mathsf{a},\mathsf{c}} > \bar{u}_{\mathsf{a},\mathsf{c}}$ on the third distance is therefore redundant, in that we can set $u_{\mathsf{a},\mathsf{c}} := \bar{u}_{\mathsf{a},\mathsf{c}}$ without changing the set of conformations consistent with the bounds. Similarly, the inverse triangle inequality imposes lower limits on $d_{\mathsf{a},\mathsf{c}}$ of the form $\bar{\ell}_{\mathsf{a},\mathsf{c}} := \max(\ell_{\mathsf{a},\mathsf{b}} - u_{\mathsf{b},\mathsf{c}}, \ell_{\mathsf{b},\mathsf{c}} - u_{\mathsf{a},\mathsf{b}})$, and if $\ell_{\mathsf{a},\mathsf{c}} < \bar{\ell}_{\mathsf{a},\mathsf{c}}$ we may set $\ell_{\mathsf{a},\mathsf{c}} := \bar{\ell}_{\mathsf{a},\mathsf{c}}$ without changing the conformation space consistent with the bounds.

Iterating on these relations, we find that the "tightest" upper limits $\bar{u}$ among the atoms $\mathsf{A}$ are equal to the shortest paths in a graph with edge lengths equal to the given upper bounds $u$, whereas the lower limits are given by

$$\bar{\ell}_{\mathsf{a},\mathsf{b}} = \max_{\mathsf{c}\in\mathsf{A}} \left( \max(\ell_{\mathsf{a},\mathsf{c}} - \bar{u}_{\mathsf{b},\mathsf{c}},\ \ell_{\mathsf{b},\mathsf{c}} - \bar{u}_{\mathsf{a},\mathsf{c}}) \right), \qquad (5.15)$$

for all $\mathsf{a},\mathsf{b} \in \mathsf{A}$. Thus the lower and upper limits can be computed in polynomial time by finding the shortest paths in a graph on two disjoint copies $\mathsf{A}$, $\mathsf{A}'$ of the atoms. The lengths of the edges within each copy are equal to the given upper distance bounds, whereas the *directed* edge lengths connecting each pair of atoms $\mathsf{a} \in \mathsf{A}$, $\mathsf{b}' \in \mathsf{A}'$ are equal to the *negatives* of the corresponding lower bounds, $-\ell_{\mathsf{a},\mathsf{b}'}$ [8].

The polynomial-time computability of the triangle inequality limits is a benefit of the factorizability of the three-point Cayley-Menger determinants into linear factors (Equation 5.1). The non-negativity of the four-point Cayley-Menger determinants also impose limits on the distances for any given set of lower and upper bounds, but no polynomial-time or even closed (provably finite) algorithm for their computation is known. In the simplest case that all but one of the distances among four atoms are known, the sixth distance must lie between the square roots of the roots of the quadratic in the squared distance $d_{\mathrm{c,d}}^2$ obtained by expanding the four-point Cayley-Menger determinant along its last row and column:

$$288\,D(\mathsf{a,b,c,d}) = -2D(\mathsf{a,b})\,d_{\mathrm{c,d}}^4 + 32D(\mathsf{a,b,c;a,b,d})|_{d_{\mathrm{c,d}}^2=0}\,d_{\mathrm{c,d}}^2 + 288\,D(\mathsf{a,b,c,d})|_{d_{\mathrm{c,d}}^2=0} \quad (5.16)$$

These correspond to the two conformations shown in Figure 5.3.

The implications of this "tetrangle inequality" for the lower and upper limits on one of the distances among any four atoms, given lower and upper bounds on the other five distances, are also rather more complicated than the triangle inequality limits, and are illustrated in Figure 5.4. The extension of these results to include chirality constraints may be found in Ref. [51]. Similar results can be derived for the pentangle and, in principle, even higher-order inequalities, but are too complicated to be presented here. These higher-order inequalities, moreover, are valid in $n$D Euclidean space for arbitrary $n \leq N-1$, and hence do not have much to say about the 3D case that is of interest to chemists. In particular, they cannot derive the implications of the sphere packing constraints mentioned above, due to the fact that nonbonded, nongeminal pairs of atoms can be viewed as hard nonoverlapping spheres. In chemistry, the consequences of these ubiquitous lower distance bounds are called *excluded volume* effects, and in large molecules they can eliminate the vast majority of otherwise possible conformations. To some extent, they could be derived by taking account of the fact that the pentangle inequality holds as an equality in three dimensions. In this case, however, the active distance bounds depend on the relative chiralities of the five quadruples of atoms in each quintuple, further complicating the combinatorics of a search for what we call the three-dimensional Euclidean distance limits. Thus, despite its potential relative to an analytical approach, there are many open problems that need to be solved before a purely distance theoretic approach to molecular conformation can deal with complex large molecules. This has led to the hybrid approach introduced in the next section.

## 5.3 Inductive Geometric Reasoning by Random Sampling

**Glossary**

**Conformational ensemble:** A random sample from conformation space.

**Inductive geometric reasoning:** Imputing the common geometric features of a conformational ensemble to be necessary consequences of the distance and chirality constraints that define it.

**Metrization:** The computation of a random metric space from between the upper and lower triangle inequality limits.

**EMBED algorithm:** An approach (not a true "algorithm") to computing the Cartesian coordinates of a conformational ensemble based on diagonalization of an estimate of the metric matrix derived from a distance matrix obtained by metrization.

**Metric matrix:** The matrix of inner products of all pairs of vectors between an arbitrary origin, usually taken as the centroid of the atoms, and the atoms themselves.

**Error function:** A function of the Cartesian coordinates of the atoms that increases monotonically

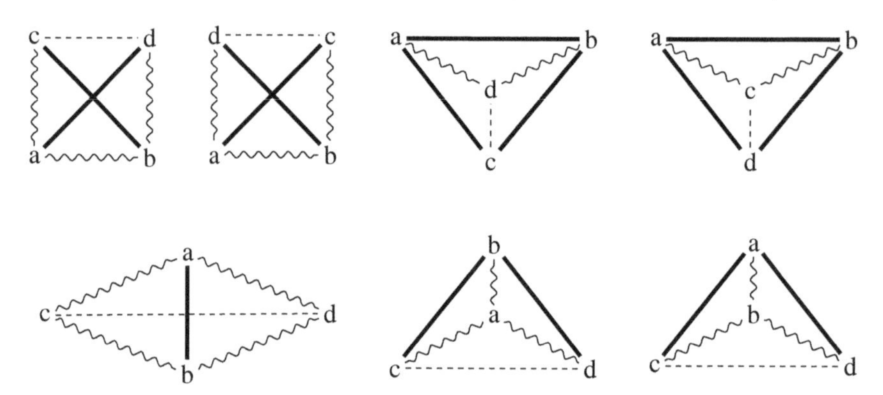


**Figure 5.4**
The combinations of bounds among four atoms $\mathsf{a},\mathsf{b},\mathsf{c},\mathsf{d}$ that can be active constraints in the conformations wherein the lower (top row) and upper (bottom row) tetrangle inequality limits on $d_{\mathsf{c},\mathsf{d}}$ are attained. Distances at their lower bounds are indicated with a heavy line, while distances at their upper bounds are indicated with wavey lines. If we interpret the lower bounds as the struts and the upper bounds as cables in a tensegrity framework, then the frameworks of the two configurations on the left and right sides of this figure are *globally rigid* [7].The combinations of bounds among four atoms $\mathsf{a},\mathsf{b},\mathsf{c},\mathsf{d}$ that can be active constraints in the conformations wherein the lower (top row) and upper (bottom row) tetrangle inequality limits on $d_{\mathsf{c},\mathsf{d}}$ are attained. Distances at their lower bounds are indicated with a heavy line, while distances at their upper bounds are indicated with wavey lines. If we interpret the lower bounds as the struts and the upper bounds as cables in a tensegrity framework, then the frameworks of the two configurations on the left and right sides of this figure are *globally rigid* [7].

with increasing violations of the distance constraints as well as the oriented volume constraints derived from the distance and chirality constraints.

---

In this section we introduce an inductive approach to geometric reasoning, which can be used to gain insight into the distance geometry descriptions of even very large and complex molecules such as proteins and nucleic acids. This is done by computing a random sample of conformation space, called a *conformational ensemble*, and then analyzing it to determine the common geometric features of all its members. Given an ensemble that is sufficiently large and random, those common geometric features will, with high probability, be necessary consequences of the distance and chirality constraints. The geometric features of interest need not themselves be either distances or oriented volumes, but can include aggregate properties such as the size and shape of the molecule, which atoms are on its surface versus buried under surface atoms, or even kinematic features such as correlations in atomic motions. Whereas one certainly could obtain a conformational ensemble from a molecular dynamics simulation of sufficient duration, the techniques we shall now describe can produce random ensembles of diverse conformations in far less time [8].

Let $\underline{\mathsf{A}} = [\mathsf{a},\mathsf{b},\ldots,\mathsf{z}]$ be a total ordering of the set of atoms $\mathsf{A}$, and let $\mathbf{L} = [\ell^2_{\mathsf{a},\mathsf{b}}]_{\mathsf{a},\mathsf{b}\in\underline{\mathsf{A}}}$, $\mathbf{U} = [u^2_{\mathsf{a},\mathsf{b}}]_{\mathsf{a},\mathsf{b}\in\underline{\mathsf{A}}}$ be symmetric, nonnegative and zero-diagonal matrices containing the lower and upper distance bounds squared ($\mathbf{L} \le \mathbf{U}$). Also let $\mathbf{C} = [\tilde{\chi}_{\mathsf{a},\mathsf{b},\mathsf{c},\mathsf{d}}]_{\mathsf{a},\mathsf{b},\mathsf{c},\mathsf{d}\in\underline{\mathsf{A}}}$ be the antisymmetric tensor of possible chiralities for its quadruples of atoms, which has its values in the power set $2^{\{-1,0,+1\}}$. Given this input, the *EMBED* algorithm proceeds through the following steps, which are described in greater detail in the ensuing text:

(a) Compute the triangle inequality limits $\bar{\ell}, \bar{u}$ from the interatomic distance bounds $\ell, u$, as described in the previous section; to the extent computationally practical, it is advantageous to also compute selected tetrangle or higher-order distance limits;

(b) Generate a random metric space within the distance limits, i.e., a symmetric zero-diagonal matrix $\mathbf{R} = [r_{a,b}]_{a,b \in \underline{A}}$ such that $\bar{\ell}_{a,b} \leq r_{a,b} \leq r_{a,c} + r_{b,c} \leq \bar{u}_{a,c} + \bar{u}_{b,c}$ for all $a, b, c \in A$;

(c) Let $\mathbf{D} := [r^2_{a,b}]_{a,b \in \underline{A}}$ be the corresponding matrix of squared metrics, and let $\mathbf{M} := -\frac{1}{2}\mathbf{PDP}$ be the projection on the subspace orthogonal to the vector $\mathbf{1}$ of 1's, i.e. $\mathbf{P} := \mathbf{I} - \mathbf{1}\mathbf{1}^\top / N$;

(d) Let $\mathbf{Q}\Lambda\mathbf{Q}^\top$ be an eigenvalue decomposition of $\mathbf{M}$, with the eigenvalues $\lambda_1 \geq \lambda_2 \geq \cdots \geq \lambda_N$ arranged in nonincreasing order;

(e) If the third eigenvalue $\lambda_3 < 0$, return to step 2; otherwise, let $[\mathbf{q}_1, \mathbf{q}_2, \ldots, \mathbf{q}_N]$ be the columns of the orthogonal matrix $\mathbf{Q}$;

(f) The vectors $\mathbf{x} = \sqrt{\lambda_1}\,\mathbf{q}_1$, $\mathbf{y} = \sqrt{\lambda_2}\,\mathbf{q}_2$, $\mathbf{z} = \sqrt{\lambda_3}\,\mathbf{q}_3$ can now be taken as center-of-mass (assuming all atoms have equal masses), principal-axis (ditto), three-dimensional Cartesian coordinates for the atoms, with the property that the squared distances computed from these coordinates approximate those in $\mathbf{D}$ and hence do not seriously violate the distance bounds, since $\mathbf{L} \leq \mathbf{D} \leq \mathbf{U}$ by construction;

(g) Compute the chirality error function $C(\mathbf{x}, \mathbf{y}, \mathbf{z})$ (see below for its definition) for the embedded coordinates and the mirror image thereof, and keep the mirror image that minimizes this function;

(h) Using your favorite global optimization algorithm, minimize the violations of the distance bounds and chirality constraints with respect to the Cartesian coordinates until the residual violations are acceptably small;

(i) Repeat steps 2 through 8 until you have as many conformations as desired for your ensemble.

Following this overview, we now explain the individual steps in greater detail.

The random metric space given by the matrix $\mathbf{R}$ in step 2 is obtained by setting the lower and upper bounds $\ell_{a,b}$, $u_{a,b}$ on the distance between an arbitrary pair of atoms equal to some random value $\bar{\ell}_{a,b} \leq r_{a,b} \leq \bar{u}_{a,b}$ between the corresponding limits, and then computing the new triangle inequality limits implied by the modified bounds $\ell_{a,b} := r_{a,b} =: u_{a,b}$. Iterating in this fashion over all pairs of atoms, one attains the desired metric space once all the lower and upper bounds have become equal. Adopting a term from topology, this process has been dubbed *metrization*. Given that the majority of triangle inequality limits are unchanged by equating the lower and upper bounds between one pair of atoms, the new distance limits can be obtained in far less time than would be needed to compute the limits from scratch in each iteration. Indeed, if one treats an arbitrary atom as the root of a shortest-path tree, and chooses the distances to each of the other, as-yet untreated atoms sequentially, the shortest-path tree can be updated in time linear in $N$ on each iteration. This enables the random metric spaces $\mathbf{R}$ to be computed in time proportional to at most $N^3$, just as with a single set of triangle inequality limits. Restricting the iteration to all distances from a single root atom to all other as-yet untreated atoms does however introduce correlations in the values of the metrics and hence a bias in the sampling, unless the order in which the atoms are taken as root atoms is chosen randomly on each pass over steps 2 through 8.

The projection in step 3 may be shown to transform the entries of the matrix of Schönberg's quadratic form $-\frac{1}{2}\mathbf{D} = -\frac{1}{2}[d^2_{a,b}]_{a,b \in \underline{A}} := -\frac{1}{2}[r^2_{a,b}]_{a,b \in \underline{A}}$ into a matrix $\mathbf{M} = [m_{a,b}]_{a,b \in \underline{A}}$ of inner products, called the *metric matrix*, given via the law of cosines as

$$m_{a,b} = \tfrac{1}{2}\left(d^2_{0,a} + d^2_{0,b} - d^2_{a,b}\right), \tag{5.17}$$

where the squared distances to the new point 0 are given by

$$d_{0,\mathsf{a}}^2 = \frac{1}{N}\sum_{\mathsf{f}\in\mathsf{A}} d_{\mathsf{a,f}}^2 - \frac{1}{2N^2}\sum_{\mathsf{f,g}\in\mathsf{A}} d_{\mathsf{f,g}}^2 \,. \tag{5.18}$$

In the three-dimensional Euclidean case, it was first shown by Lagrange that $d_{0,\mathsf{a}}^2$ is the squared distance between the point a and the centroid 0 of the points. This interpretation also holds for any set of points embedded in a metric affine space of any dimension, and hence for an arbitrary semimetric space (cf. Theorem 5.3).

In steps 4 through 6, it can be shown that the metric matrix $\mathbf{M}' = \mathbf{x}\mathbf{x}^\top + \mathbf{y}\mathbf{y}^\top + \mathbf{z}\mathbf{z}^\top$ minimizes the Frobenius norm $\|\mathbf{M} - \mathbf{M}''\|_F$ over all rank 3 positive semidefinite matrices $\mathbf{M}''$. This is why the matrix of three-dimensional squared Euclidean distances, obtained from the embedded coordinates using the Hadamard (or entrywise) matrix product "$\odot$" as $\mathbf{D}'$ =

$$(\mathbf{x}\mathbf{1}^\top - \mathbf{1}\mathbf{x}^\top)\odot(\mathbf{x}\mathbf{1}^\top - \mathbf{1}\mathbf{x}^\top) + (\mathbf{y}\mathbf{1}^\top - \mathbf{1}\mathbf{y}^\top)\odot(\mathbf{y}\mathbf{1}^\top - \mathbf{1}\mathbf{y}^\top) + (\mathbf{z}\mathbf{1}^\top - \mathbf{1}\mathbf{z}^\top)\odot(\mathbf{z}\mathbf{1}^\top - \mathbf{1}\mathbf{z}^\top), \tag{5.19}$$

is a good approximation to the non-Euclidean estimate $\mathbf{D} = [r_{\mathsf{a,b}}^2]_{\mathsf{a,b}\in\underline{\mathsf{A}}}$ thereof obtained from metrization. As a result, when the input distance bounds are reasonably "tight," the embedded atomic coordinates $\mathbf{x},\mathbf{y},\mathbf{z}$ will be close to those of a conformation that satisfies the distance constraints, and the relative chiralities of the chiral centers will also be largely correct.

Step 7 attempts to also get the absolute chiralities $\chi$ largely correct, essentially by choosing the mirror image of the embedded coordinates which minimizes the volume-weighted sum of the chirality constraint violations over all chiral centers. Then in step 8, a real-valued *error function*, which measures the violations of the distance and chirality constraints, is minimized to eliminate the residual constraint violations. In the case of the distance constraints, the function

$$G(\mathbf{x},\mathbf{y},\mathbf{z}) = \sum_{\mathsf{a,b}\in\mathsf{A}}\left(\max\left(0, \frac{\bar{\ell}_{\mathsf{a,b}}^2}{d_{\mathsf{a,b}}(\mathbf{x},\mathbf{y},\mathbf{z})^2} - 1\right)^2 + \max\left(0, \frac{d_{\mathsf{a,b}}(\mathbf{x},\mathbf{y},\mathbf{z})^2}{\bar{u}_{\mathsf{a,b}}^2} - 1\right)^2\right), \tag{5.20}$$

where $d_{\mathsf{a,b}}(\mathbf{x},\mathbf{y},\mathbf{z})^2 = (x_{\mathsf{a}} - x_{\mathsf{b}})^2 + (y_{\mathsf{a}} - y_{\mathsf{b}})^2 + (z_{\mathsf{a}} - z_{\mathsf{b}})^2$, has been found to be relatively free of local minima and to give chemically reasonable structures even when (as is generally the case) the residual distance constraint violations cannot be eliminated entirely. The total error function is then obtained by adding the chirality error function $C$ to $G$. Letting $V_{\mathsf{a,b,c,d}}(\mathbf{x},\mathbf{y},\mathbf{z})$ be the oriented volumes computed from the coordinates as in Equation (5.10), and $\bar{L}_{\mathsf{a,b,c,d}}$, $\bar{U}_{\mathsf{a,b,c,d}}$ be the minimum and maximum *unsigned* volumes consistent with the distance and chirality constraints among the corresponding quadruple, this is defined as

$$\begin{aligned}
C(\mathbf{x},\mathbf{y},\mathbf{z}) = {} & \sum_{\substack{\underline{\mathsf{a,b,c,d}}\in\underline{\mathsf{A}} \\ 0\notin\tilde{\chi}_{\mathsf{a,b,c,d}}}} \sum_{\chi\in\tilde{\chi}_{\mathsf{a,b,c,d}}} \Bigg( \min\left(0,\, \chi\cdot V_{\mathsf{a,b,c,d}}(\mathbf{x},\mathbf{y},\mathbf{z}) - \bar{L}_{\mathsf{a,b,c,d}}\right)^2 + \\
& \qquad\qquad \max\left(0,\, \chi\cdot V_{\mathsf{a,b,c,d}}(\mathbf{x},\mathbf{y},\mathbf{z}) - \bar{U}_{\mathsf{a,b,c,d}}\right)^2 \Bigg) \\
& + \sum_{\substack{\underline{\mathsf{a,b,c,d}}\in\underline{\mathsf{A}} \\ 0\in\tilde{\chi}_{\mathsf{a,b,c,d}}}} \sum_{\substack{\chi\in\tilde{\chi}_{\mathsf{a,b,c,d}} \\ \chi\neq 0}} \max\left(0,\, \chi\cdot V_{\mathsf{a,b,c,d}}(\mathbf{x},\mathbf{y},\mathbf{z}) - \bar{U}_{\mathsf{a,b,c,d}}\right)^2 \\
& + \sum_{\substack{\underline{\mathsf{a,b,c,d}}\in\underline{\mathsf{A}} \\ \tilde{\chi}_{\mathsf{a,b,c,d}}=\{0\}}} V_{\mathsf{a,b,c,d}}(\mathbf{x},\mathbf{y},\mathbf{z})^2 \,,
\end{aligned} \tag{5.21}$$

where $\underline{\mathsf{a,b,c,d}} \in \underline{\mathsf{A}}^4$ implies a sum over all quadruples $\mathsf{a,b,c,d} \in \mathsf{A}$ with $\mathsf{a} < \mathsf{b} < \mathsf{c} < \mathsf{d}$ with respect to its total ordering $\underline{\mathsf{A}}$. Note that in the last line of this equation, we are treating planarity as a "chirality" constraint, enforced by penalizing any oriented volume other than zero.

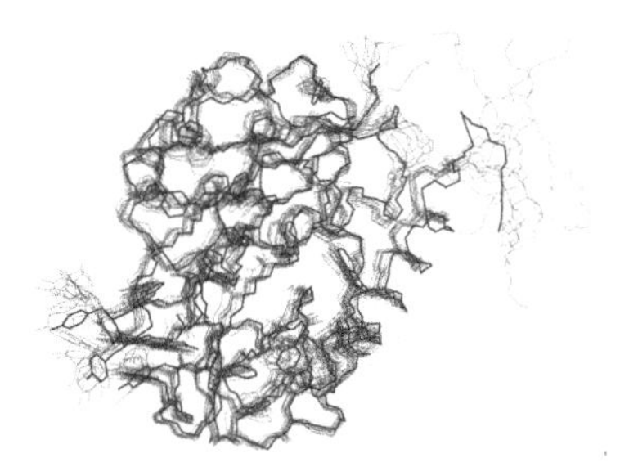

**Figure 5.5**
A conformational ensemble for the protein *E. coli* Flavodoxin, obtained by homology modeling using distance constraints derived from multiple sequence alignments with four homologous (evolutionarily related) Flavodoxins, superimposed upon the crystal structure of the protein which was subsequently determined (heavy lines). Since each conformation contains nearly 1000 atoms, only the main-chain atoms, aromatic side-chains, and flavin cofactor are shown for simplicity. Reproduced with permission from [19].

The EMBED algorithm is best known as a means of determining "the" conformation of proteins and other large biomolecules in aqueous solution from Nuclear Magnetic Resonance (NMR) spectroscopy [57, 59]. The most important geometric parameters that can be extracted from these spectra are extensive lists of distances between pairs of hydrogen atoms that are less than 0.5 nm in all conformations. While many initially doubted that such imprecise constraints could suffice to determine the relative positions of the polypeptide backbone atoms of the protein with a precision approaching that of crystallographic methods, another Nobel prize in chemistry proved them wrong [21, 62]. In the years since, other spectroscopic parameters have gained prominence which, like all Euclidean invariants, can be expressed as functions of the interatomic distances, but these functions are relatively complicated [48]. As a result, the propagation of experimental uncertainties makes it difficult to derive reasonably tight bounds on individual distances from them. Distance-based conformational sampling has however also been applied to many other problems in computational chemistry, including ring closure [23], drug design [9], and protein structure prediction from sequence alignments with homologous proteins of known structure [19]. Figure 1.5 dipicts a conformational ensemble for the protein Flavodoxin obtained from the latter application, known as homology modeling.

Early implementations of the EMBED algorithm minimized the error function by gradient descent methods such as conjugate gradients [17, 22]. Subsequently, considerably more robust convergence was obtained by dynamical simulated annealing, wherein the error function is treated as if it were the energy function in a molecular dynamics simulation [16, 18]. Starting from a high temperature ($T \propto 2(C+G)/3R$, where $C$ and $G$ are the distance and chirality error functions as above and $R$ is the gas constant), the temperature is gradually reduced to zero by scaling down the velocities at each point along the trajectory. While it may seem ironic to be using molecular dynamics in a method designed to improve upon it for conformational= sampling purposes, the error functions turn out to be much softer and to have far fewer local minima than physically realistic energy functions. Moreover, since the intent is not to produce a physically meaningful trajectory but merely to simulate the natural tendency of thermodynamic systems to fall into a low-energy state upon annealing, a time step of order milliseconds can be utilized, thereby covering a far larger region of conformation space in a given amount of real time. Although convergence to zero error is seldom attained in large problems, the residual violations of the distance and chirality constraints are often small enough to be of no chemical significance.

Over the last 25 years, a great many different global optimization algorithms have been applied to the minimization of such error functions [38, 43]. Some are able to solve large problems much faster than dynamical simulated annealing, even starting from random rather than embedded coordinates. In many cases, unfortunately, these algorithms have made chemically unrealistic assumptions about the problem to be solved, particularly regarding the precision of the distances. Indeed a great deal of this work has focused on the distance matrix completion problem, wherein all distances are either known exactly or are completely unknown, and the goal is to fill in the missing entries of the distance matrix so as to make it embeddable in a low, if not three-dimensional, Euclidean space. Even when reasonably general distance and chirality constraints are assumed, however, the optimization problem was solved well enough for most chemical intents and purposes years ago. In the hope of spurring future work on more important issues, we close this section by giving two open problems, the solutions of which would considerably broaden the scope of distance geometry's chemical applications.

**Problem I: Tight Outer Estimates of the 3D Euclidean Distance Limits.** Typically, if one takes the minimum and maximum values of the distances over a conformational ensemble as the bounds from which to compute a new conformational ensemble using the *EMBED* algorithm, one finds that the fit of the distances in the unoptimized conformations to the new, as well as the old, distance bounds is so good that very little, and certainly no demanding global, optimization is needed. This means that improved estimates of the 3D Euclidean distance limits would be an effective means of reducing the *EMBED* algorithm's dependence on global optimization. A computationally tractable means of estimating these limits would moreover help one to discover errors in the distance bounds; in the simple case of the triangle inequality, for example, contradictions of the form

$$u_{\mathsf{a,b}} + u_{\mathsf{b,c}} \;<\; l_{\mathsf{a,b}} \tag{5.22}$$

are always discovered when computing the triangle inequality limits. The ability to discover, and pinpoint, such contradictions using the higher order inequalities and equalities of distance geometry would be very useful to chemists, who must gather large quantities distance information from diverse sources in the course of studying the conformation of a given molecule. It also enables an approach to geometric reasoning via contradiction. One could, for example, attempt to prove that two atoms were necessarily in close proximity to one another by imposing a large lower bound on the distance between them, and then testing the combined constraints for contradictions.

The estimate of the distance limits $\bar{\mathbf{L}}$, $\bar{\mathbf{U}}$ obtained by taking the minimum and maximum values of the distances over a large conformational ensemble is what we call an *inner estimate*: The true 3D distance limits $\bar{\mathbf{L}}_{3\mathrm{D}}$, $\bar{\mathbf{U}}_{3\mathrm{D}}$ will always be looser than $\bar{\mathbf{L}}$, $\bar{\mathbf{U}}$, i.e. $\bar{\mathbf{L}}_{3\mathrm{D}} \leq \bar{\mathbf{L}}$ and $\bar{\mathbf{U}}_{3\mathrm{D}} \geq \bar{\mathbf{U}}$ (because it is impossible for a finite ensemble to completely sample the infinitude of possible combinations of values for the distances). In order to prove the existence of contradictions ($\bar{u} < \bar{\ell}$) with no false positives (erroneous contradictions), these inequalities need to point the other way, i.e. $\bar{\mathbf{L}}_{3\mathrm{D}} \geq \bar{\mathbf{L}}$ and $\bar{\mathbf{U}}_{3\mathrm{D}} \leq \bar{\mathbf{U}}$. Estimates of the 3D Euclidean distance limits with this property will be called *outer estimates*. In principle they can be obtained from any suitable (computationally tractible) relaxation of the system of inequalities and equalities which the distances and chiralities mutually satisfy in three dimensions [41]. In practice, such a relaxation must not be overly simplistic, or at least must provide a significant improvement over that obtained from the triangle inequality alone. Most importantly, it must provide information about the lower limits on the distances which result from the hard sphere packing constraints characteristic of atoms in 3D molecules, as described previously. The fact that packing and covering problems have long been studied in mathematics and yet still contain many open questions implies the problem of computing such tight outer estimates will be challenging. The payoff for such an *algorithmic* approach will nevertheless be comparably large, and could well impact mathematics as much as chemistry. Some preliminary work on the one-dimensional version of this problem has already been performed [13].

**Problem II: Computations in Rings of Symmetric Distance Polynomials.** Many experimental sources of information on molecules yield information on the distances among atoms but are ambiguous concerning which atoms those distances stem from. For example, X-ray diffraction provides a 3D map of the density of electrons in a molecule, which has peaks at the positions of its atoms. With large molecules or low-resolution data, fitting a conformation with all the correct bond lengths and angles to such an electron density map so that all the atoms are near the correct peaks can be difficult or impossible. It is much easier to find the peaks and compute all the distances between them, although these distances may contain significant errors from peak overlaps due to the limited resolution of the map. A representation of the distance matrix that did not assume a total ordering of the atoms, or equivalently, which was invariant to permutations of the underlying set, could help solve such problems. This would be done by fitting a conformation of the molecule, represented by a distance matrix versus a total ordering of its atoms, to the permutation-invariant representation so as to make the permutation-invariant representation calculated by symmetrization of the former match the latter to within the error bounds on its values. As usual in distance geometry, it will be easier to do this using the squared distances rather than the distances themselves.

In the simplest nontrivial case of three vertices (points, atoms), the polynomials in question are:

$$\begin{aligned} P_1(\mathsf{a,b,c}) &= d^2_{\mathsf{a,b}} + d^2_{\mathsf{b,c}} + d^2_{\mathsf{a,c}} \\ P_2(\mathsf{a,b,c}) &= d^2_{\mathsf{a,b}}d^2_{\mathsf{b,c}} + d^2_{\mathsf{a,b}}d^2_{\mathsf{a,c}} + d^2_{\mathsf{b,c}}d^2_{\mathsf{a,c}} \\ P_3(\mathsf{a,b,c}) &= d^2_{\mathsf{a,b}}d^2_{\mathsf{b,c}}d^2_{\mathsf{a,c}} \end{aligned} \tag{5.23}$$

These are the same as the elementary symmetric polynomials among three indeterminates, so there's nothing new here. It is worth noting, nonetheless, that $4P_1(\mathsf{a,b,c}) - P_0(\mathsf{a,b,c})^2 = 16D(\mathsf{a,b,c}) \geq 0$, so these polynomials inherit dependencies from those of Euclidean distances.

The four-point elementary symmetric distance polynomials are something new:

$$\begin{aligned} P_1(\mathsf{a,b,c,d}) &= d^2_{\mathsf{a,b}} + d^2_{\mathsf{b,c}} + d^2_{\mathsf{a,d}} + d^2_{\mathsf{b,c}} + d^2_{\mathsf{b,d}} + d^2_{\mathsf{c,d}} \\ P_2(\mathsf{a,b,c,d}) &= d^2_{\mathsf{a,b}}d^2_{\mathsf{c,d}} + d^2_{\mathsf{a,c}}d^2_{\mathsf{b,d}} + d^2_{\mathsf{a,d}}d^2_{\mathsf{b,c}} \\ P_3(\mathsf{a,b,c,d}) &= d^2_{\mathsf{a,b}}d^2_{\mathsf{a,c}} + d^2_{\mathsf{a,b}}d^2_{\mathsf{a,d}} + d^2_{\mathsf{a,c}}d^2_{\mathsf{a,d}} + d^2_{\mathsf{a,b}}d^2_{\mathsf{b,c}} + d^2_{\mathsf{a,b}}d^2_{\mathsf{b,d}} + d^2_{\mathsf{b,c}}d^2_{\mathsf{b,d}} + \\ &\quad d^2_{\mathsf{a,c}}d^2_{\mathsf{b,c}} + d^2_{\mathsf{a,c}}d^2_{\mathsf{c,d}} + d^2_{\mathsf{b,c}}d^2_{\mathsf{c,d}} + d^2_{\mathsf{a,d}}d^2_{\mathsf{b,d}} + d^2_{\mathsf{a,d}}d^2_{\mathsf{c,d}} + d^2_{\mathsf{b,d}}d^2_{\mathsf{c,d}} \\ P_4(\mathsf{a,b,c,d}) &= d^2_{\mathsf{a,b}}d^2_{\mathsf{a,c}}d^2_{\mathsf{a,d}} + d^2_{\mathsf{a,b}}d^2_{\mathsf{b,c}}d^2_{\mathsf{b,d}} + d^2_{\mathsf{a,c}}d^2_{\mathsf{b,c}}d^2_{\mathsf{c,d}} + d^2_{\mathsf{a,d}}d^2_{\mathsf{b,d}}d^2_{\mathsf{c,d}} \\ P_5(\mathsf{a,b,c,d}) &= (d^2_{\mathsf{a,b}} + d^2_{\mathsf{c,d}})(d^2_{\mathsf{a,c}}d^2_{\mathsf{b,d}} + d^2_{\mathsf{a,d}}d^2_{\mathsf{b,c}}) + (d^2_{\mathsf{a,c}} + d^2_{\mathsf{b,d}})(d^2_{\mathsf{a,b}}d^2_{\mathsf{c,d}} + d^2_{\mathsf{a,d}}d^2_{\mathsf{b,c}}) \\ &\quad + (d^2_{\mathsf{a,d}} + d^2_{\mathsf{b,c}})(d^2_{\mathsf{a,b}}d^2_{\mathsf{c,d}} + d^2_{\mathsf{a,c}}d^2_{\mathsf{b,d}}) \\ P_6(\mathsf{a,b,c,d}) &= d^2_{\mathsf{a,b}}d^2_{\mathsf{a,c}}d^2_{\mathsf{b,c}} + d^2_{\mathsf{a,b}}d^2_{\mathsf{b,d}}d^2_{\mathsf{b,d}} + d^2_{\mathsf{a,c}}d^2_{\mathsf{a,d}}d^2_{\mathsf{c,d}} + d^2_{\mathsf{a,d}}d^2_{\mathsf{b,c}}d^2_{\mathsf{c,d}} \\ P_7(\mathsf{a,b,c,d}) &= (d^2_{\mathsf{a,b}} + d^2_{\mathsf{a,c}} + d^2_{\mathsf{a,d}})d^2_{\mathsf{b,c}}d^2_{\mathsf{b,d}}d^2_{\mathsf{c,d}} + (d^2_{\mathsf{a,b}} + d^2_{\mathsf{b,c}} + d^2_{\mathsf{b,d}})d^2_{\mathsf{a,c}}d^2_{\mathsf{a,d}}d^2_{\mathsf{c,d}} + \\ &\quad (d^2_{\mathsf{a,c}} + d^2_{\mathsf{b,c}} + d^2_{\mathsf{c,d}})d^2_{\mathsf{a,b}}d^2_{\mathsf{a,d}}d^2_{\mathsf{b,c}} + (d^2_{\mathsf{a,d}} + d^2_{\mathsf{b,d}} + d^2_{\mathsf{c,d}})d^2_{\mathsf{a,b}}d^2_{\mathsf{a,c}}d^2_{\mathsf{b,c}} \\ P_8(\mathsf{a,b,c,d}) &= d^2_{\mathsf{a,c}}d^2_{\mathsf{a,d}}d^2_{\mathsf{b,c}}d^2_{\mathsf{b,d}} + d^2_{\mathsf{a,b}}d^2_{\mathsf{a,d}}d^2_{\mathsf{b,c}}d^2_{\mathsf{c,d}} + d^2_{\mathsf{a,b}}d^2_{\mathsf{a,c}}d^2_{\mathsf{b,d}}d^2_{\mathsf{c,d}} \\ P_9(\mathsf{a,b,c,d}) &= d^2_{\mathsf{a,c}}d^2_{\mathsf{a,d}}d^2_{\mathsf{b,c}}d^2_{\mathsf{b,d}}d^2_{\mathsf{c,d}} + d^2_{\mathsf{a,b}}d^2_{\mathsf{a,d}}d^2_{\mathsf{b,c}}d^2_{\mathsf{b,d}}d^2_{\mathsf{c,d}} + d^2_{\mathsf{a,b}}d^2_{\mathsf{a,c}}d^2_{\mathsf{b,c}}d^2_{\mathsf{b,d}}d^2_{\mathsf{c,d}} + \\ &\quad d^2_{\mathsf{a,b}}d^2_{\mathsf{a,c}}d^2_{\mathsf{a,d}}d^2_{\mathsf{b,d}}d^2_{\mathsf{c,d}} + d^2_{\mathsf{a,b}}d^2_{\mathsf{a,c}}d^2_{\mathsf{a,d}}d^2_{\mathsf{b,c}}d^2_{\mathsf{c,d}} + d^2_{\mathsf{a,b}}d^2_{\mathsf{a,c}}d^2_{\mathsf{a,d}}d^2_{\mathsf{b,c}}d^2_{\mathsf{b,d}} \\ P_{10}(\mathsf{a,b,c,d}) &= d^2_{\mathsf{a,b}}d^2_{\mathsf{a,c}}d^2_{\mathsf{a,d}}d^2_{\mathsf{b,c}}d^2_{\mathsf{b,d}}d^2_{\mathsf{c,d}} \end{aligned} \tag{5.24}$$

It may be observed that these polynomials are in a one-to-one correspondence with the loopless, undirected and unlabeled graphs on four vertices, save for the empty graph. In this correspondence, $P_9$ is the complementary graph of $P_1$ (paths of length 1), $P_8$ is the complement of $P_2$ (pairs of disjoint edges), $P_7$ the complement of $P_3$ (paths of length 2), $P_6$ the complement of $P_4$ (stars), and $P_5$ is self-complementary (paths of length 3). It is readily shown that

$$-P_1(\mathsf{a},\mathsf{b},\mathsf{c},\mathsf{d})P_2(\mathsf{a},\mathsf{b},\mathsf{c},\mathsf{d})+P_5(\mathsf{a},\mathsf{b},\mathsf{c},\mathsf{d})-2P_6(\mathsf{a},\mathsf{b},\mathsf{c},\mathsf{d}) = 288D(\mathsf{a},\mathsf{b},\mathsf{c},\mathsf{d}) \geq 0, \tag{5.25}$$

and these polynomials inherit other dependencies from those of Euclidean distances as well, e.g.,

$$\begin{aligned} & -P_1(\mathsf{a},\mathsf{b},\mathsf{c},\mathsf{d})^2+2P_2(\mathsf{a},\mathsf{b},\mathsf{c},\mathsf{d})+P_3(\mathsf{a},\mathsf{b},\mathsf{c},\mathsf{d}) \\ = {} & 4\big(D(\mathsf{b},\mathsf{c},\mathsf{d})+D(\mathsf{a},\mathsf{c},\mathsf{d})+D(\mathsf{a},\mathsf{b},\mathsf{d})+D(\mathsf{a},\mathsf{b},\mathsf{c})\big) \geq 0\,. \end{aligned} \tag{5.26}$$

Moreover, since we have 10 polynomials in only 6 indeterminates, it is clear they will satisfy 4 other algebraically independent equality relations as well. These appear to be rather too complicated to be found simply by inspection.

In order to work in the ring generated by these polynomials, a straightening (or normal form) algorithm is needed that allows one to determine if one expression can be transformed into another using the equality relations (syzygies). Ideally, this should be done in the ring of symmetric distance functions, obtained in the limit as the number of vertices $N \to \infty$. In addition to the fully symmetric case introduced above, it would also be of some chemical interest to be able to do the same for subgroups of the full symmetric group, particularly those which constitute a restriction of the symmetric group to one or more subsets of the vertices [10, 46]. Finally, there are several generalizations of purely mathematical interest. These include the analogous systems of polynomials for general symmetric matrices, which correspond to unlabeled graphs with loops, and nonsymmetric matrices, which correspond to unlabeled digraphs with loops. The boolean case also promises a novel approach to the graph isomorphism problem.

## 5.4 From Distances to Advanced Euclidean Invariants

### Glossary

**Null vector:** A nonzero vector of a metric vector space whose inner product with itself is zero.

**Positive/Negative vector:** A vector of a real metric vector space whose inner product with itself is positive/negative.

**Conformal point at infinity:** An extraneous point outside an $n$-dimensional Euclidean affine space obtained by pinching the hyperplane at infinity (tangent space) of the affine space to a single point, resulting in a compact topological space homeomorphic to an $n$-dimensional sphere.

**Conformal model:** A realization of $n$-dimensional Euclidean distance geometry in an $(n+2)$-dimensional Minkowski vector space, or more accurately, as an $n$-dimensional quadratic surface of the Minkowski space.

In a Euclidean vector space, for two vectors $\mathbf{a},\mathbf{b}$ representing two points relative to an arbitrary origin, their difference $\mathbf{a}-\mathbf{b}$ represents the displacement from $\mathbf{b}$ to $\mathbf{a}$ in the Euclidean affine space. The displacement has Euclidean distance $d_{\mathbf{ab}}$, so

$$d_{\mathbf{ab}}^2 = (\mathbf{a}-\mathbf{b})^2 = (\mathbf{a}-\mathbf{b})\cdot(\mathbf{a}-\mathbf{b}) = \mathbf{a}^2+\mathbf{b}^2-2\mathbf{a}\cdot\mathbf{b}. \tag{5.27}$$

What are these inner products geometrically: $\mathbf{a}^2$, $\mathbf{b}^2$, and $\mathbf{a}\cdot\mathbf{b}$? They always depend on the reference point (the origin of all vectors representing points), and are geometrically meaningless.

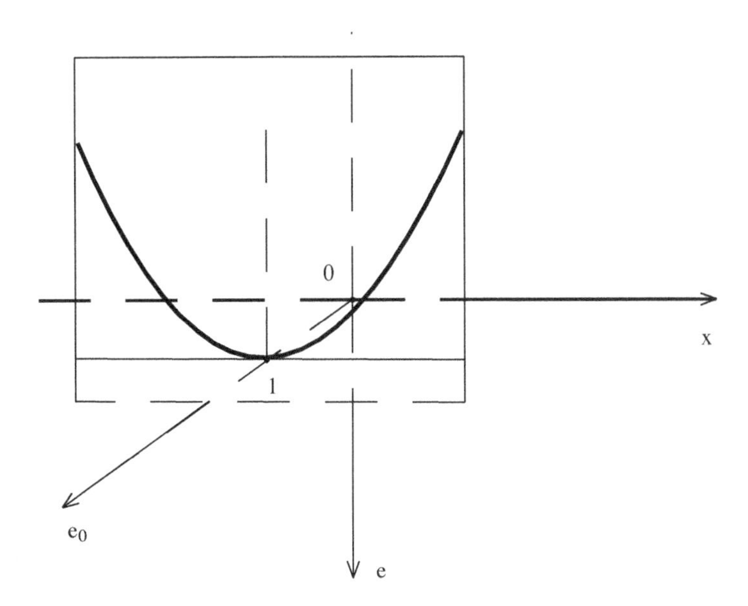


**Figure 5.6**
Illustration of the map **f** from a 1-space (the $x$-axis) of $\mathbb{R}^n$ to the space $\mathbb{R}^{n+1,1}$ spanned by $\mathbb{R}^n, \mathbf{e}, \mathbf{e}_0$. The $x$-axis is mapped to a parabola in an affine plane normal to the $\mathbf{e}_0$-axis.

If there were a vector space model of Euclidean geometry in which $\mathbf{a}\cdot\mathbf{b}$ had geometric meaning, what could it be? It would have to be some geometric relation between the two points. The only possibility is the distance, the unique invariant between the two points. Then $\mathbf{a}\cdot\mathbf{a}$ has to be zero, as it measures the distance between a point and itself. So "point" **a** should be represented by a *null vector*. From (5.27) we get

$$\mathbf{a}\cdot\mathbf{b} = -\frac{(\mathbf{a}-\mathbf{b})^2}{2} = -\frac{d^2_{\mathbf{ab}}}{2}. \tag{5.28}$$

Let $\mathbf{e}_1,\mathbf{e}_2,\ldots,\mathbf{e}_n,\mathbf{e}_0,\mathbf{e}$ be a basis of $\mathbb{R}^{n+1,1}$ with metric $\begin{pmatrix} 1 & & & & \\ & \ddots & & & \\ & & 1 & & \\ & & & 0 & -1 \\ & & & -1 & 0 \end{pmatrix}$. Then

$$\mathbf{f}:\ (x_1,x_2,\ldots,x_n)\in\mathbb{R}^n \mapsto \left(x_1,x_2,\ldots,x_n,1,-\frac{x_1^2+x_2^2+\cdots+x_n^2}{2}\right)\in\mathbb{R}^{n+1,1} \tag{5.29}$$

is an isometry from $n$-dimensional Euclidean space into $(n+2)$-dimensional Minkowski space, such that (5.28) is satisfied (Figure 5.6). The basis vector **e** denotes the *conformal point at infinity* of the Euclidean space, and $\mathbf{e}_0 = \mathbf{f}(0)$ denotes the origin of $\mathbb{R}^n$. Denote the orthogonal complement of the plane spanned by **e** and $\mathbf{e}_0$ in $\mathbb{R}^{n+1,1}$ by $\mathbb{R}^n$. Then

$$\mathbf{f}(\mathbf{x}) := \mathbf{x}+\mathbf{e}_0+\frac{\mathbf{x}^2}{2}\mathbf{e},\quad \forall\mathbf{x}\in\mathbb{R}^n. \tag{5.30}$$

The image space of isometry **f** is

$$\mathcal{N}_{\mathbf{e}} := \{\mathbf{y}\in\mathbb{R}^{n+1,1}\,|\,\mathbf{y}\cdot\mathbf{y}=0,\ \mathbf{y}\cdot\mathbf{e}=-1\}. \tag{5.31}$$

This unusual nonlinear realization of Euclidean distance geometry turns out to bring an amazing amount of benefits in both geometric modeling and symbolic manipulation [30]. It is a classical result [3] that any orthogonal transformation of $\mathbb{R}^{n+1,1}$ induces a conformal transformation in $\mathcal{N}_{\mathbf{e}} \cup \{\mathbf{e}\} = \mathbb{E}^n \cup \{\text{conformal point at infinity}\}$ via the conformal model, and conversely, any conformal transformation can be generated in this way. This justifies the name "conformal" of the model.

---

**Glossary**

**Minkowski representation:** A conformal object such as line, circle, plane, sphere is represented by an extensor of Minkowski signature in the Grassmann algebra over the conformal model.

**Dual Minkowski representation:** Hodge dual of the Minkowski representation.

**Affine representation:** A point "**a**" (null vector) in the conformal model has affine representation $\mathbf{e} \vee \mathbf{a}$, where "$\vee$" denotes the outer product in Grassmann algebra. Suitable for affine geometry.

**Dual affine representation:** Hodge dual of the affine representation.

**Homogeneous model:** Discard the constraint $\mathbf{a} \cdot \mathbf{e} = -1$ for null vector **a** representing an affine point. Two null vectors represent the same point (including the conformal point at infinity) if and only if they differ at most by scale [32].

---

The following are Minkowski representations of planes/lines/circles/spheres:

- Line **ab**: $\mathbf{e} \vee \mathbf{a} \vee \mathbf{b}$. Point **d** on line **C**: $\mathbf{d} \vee \mathbf{C} = 0$.
- Plane **abc**: $\mathbf{e} \vee \mathbf{a} \vee \mathbf{b} \vee \mathbf{c}$.
- Circle **abc** (circum-circle of the triangle): $\mathbf{a} \vee \mathbf{b} \vee \mathbf{c}$.
- Sphere **abcd** (circum-sphere of the tetrahedron): $\mathbf{a} \vee \mathbf{b} \vee \mathbf{c} \vee \mathbf{d}$.

The following are dual representations:

- Plane normal to vector $\mathbf{n} \in \mathbb{R}^3$ and passing through point **a**:
  positive vector $\mathbf{p} := \mathbf{n} + (\mathbf{a} \cdot \mathbf{n})\mathbf{e}$. Point **c** on plane **p**: $\mathbf{c} \cdot \mathbf{p} = 0$.
- Sphere with center $\emptyset$ and radius $R$: positive vector $\emptyset - R^2\mathbf{e}/2$.
- Line of intersection of two planes $\mathbf{p}_1, \mathbf{p}_2$: $\mathbf{p}_1 \vee \mathbf{p}_2$.
- Circle as intersection of two spheres, or a sphere and a plane $\mathbf{p}_1, \mathbf{p}_2$: $\mathbf{p}_1 \vee \mathbf{p}_2$.

The following are Euclidean interpretations of the inner products of positive vectors and null vectors in the conformal model [29]:

- For two points $\mathbf{c}_1$ and $\mathbf{c}_2$ (null vector representation),

$$\mathbf{c}_1 \cdot \mathbf{c}_2 = -\frac{d^2_{\mathbf{c}_1\mathbf{c}_2}}{2}. \tag{5.32}$$

- For a point **c** and a hyperplane $\mathbf{n} + \delta\mathbf{e}$ (positive vector representation),

$$\mathbf{c} \cdot (\mathbf{n} + \delta\mathbf{e}) = \mathbf{c} \cdot \mathbf{n} - \delta. \tag{5.33}$$

  The inner product is positive, zero, or negative, if and only if the vector from the hyperplane to the point is along **n**, zero, or along $-\mathbf{n}$, respectively. Its absolute value equals the distance between the point and the hyperplane.

- For a point $\mathbf{c}_1$ and a sphere $\mathbf{c}_2 - R^2\mathbf{e}/2$,

$$\mathbf{c}_1 \cdot (\mathbf{c}_2 - \frac{R^2}{2}\mathbf{e}) = \frac{R^2 - d^2_{\mathbf{c}_1\mathbf{c}_2}}{2}. \tag{5.34}$$

  This inner product, known as the *power of the point with respect to the sphere*, is positive, zero, or negative, if and only if the point is inside, on, or outside the sphere respectively. Let $d_{\max}$ and $d_{\min}$ be respectively the maximal distance and minimal distance between point $\mathbf{c}_1$ and the points on the sphere. Then the absolute value of (5.34) equals $d_{\max}d_{\min}/2$.

- For two hyperplanes $\mathbf{n}_1 + \delta_1\mathbf{e}$ and $\mathbf{n}_2 + \delta_2\mathbf{e}$,

$$(\mathbf{n}_1 + \delta_1\mathbf{e}) \cdot (\mathbf{n}_2 + \delta_2\mathbf{e}) = \mathbf{n}_1 \cdot \mathbf{n}_2. \tag{5.35}$$

- For a hyperplane $\mathbf{n} + \delta\mathbf{e}$ and a sphere $\mathbf{c} - R^2\mathbf{e}/2$,

$$(\mathbf{n} + \delta\mathbf{e}) \cdot (\mathbf{c} - \frac{R^2}{2}\mathbf{e}) = \mathbf{n} \cdot \mathbf{c} - \delta. \tag{5.36}$$

  It is positive, zero, or negative, if and only if the vector from the hyperplane to the center $\mathbf{c}$ of the sphere is along $\mathbf{n}$, zero, or along $-\mathbf{n}$ respectively. Its absolute value equals the distance between the center and the hyperplane.

- For two spheres $\mathbf{c}_1 - R_1^2\mathbf{e}/2$ and $\mathbf{c}_2 - R_2^2\mathbf{e}/2$,

$$(\mathbf{c}_1 - \frac{R_1^2}{2}\mathbf{e}) \cdot (\mathbf{c}_2 - \frac{R_2^2}{2}\mathbf{e}) = \frac{R_1^2 + R_2^2 - d^2_{\mathbf{c}_1\mathbf{c}_2}}{2}. \tag{5.37}$$

  It is zero if the two spheres are *perpendicular* to each other, *i.e.*, they intersect at 90°. More generally, when the two spheres intersect, (5.37) equals the cosine of the angle of intersection multiplied by $R_1R_2$.

Any negative vector $\mathbf{s} \in \mathbb{R}^{n+1,1}$ after rescaling, can be written in the standard form $\mathbf{s} = \mathbf{c} + (R^2/2)\,\mathbf{e}$, where $\mathbf{c}^2 = 0$. The Euclidean geometric meaning of the inner product of a negative vector with another vector in $\mathbb{R}^{n+1,1}$ is as follows [34]:

- Let $\mathbf{c}_1, \mathbf{c}_2$ be points represented by null vectors in $\mathcal{N}_\mathbf{e}$. Then

$$\mathbf{c}_1 \cdot (\mathbf{c}_2 + \frac{R^2}{2}\mathbf{e}) = -\frac{d^2_{\mathbf{c}_1\mathbf{c}_2} + R^2}{2}. \tag{5.38}$$

  Let $\mathbf{y}$ be any point on the sphere of center and radius $(\mathbf{c}_2, R)$ such that line $\mathbf{c}_2\mathbf{y} \perp$ line $\mathbf{c}_1\mathbf{c}_2$, then (5.38) equals $-d^2_{\mathbf{c}_1\mathbf{y}}/2$.

- Let $\mathbf{n}$ be a unit vector in $\mathbb{R}^n$, then

$$(\mathbf{n} + \delta\mathbf{e}) \cdot (\mathbf{c} + \frac{R^2}{2}\mathbf{e}) = \mathbf{c} \cdot \mathbf{n} - \delta. \tag{5.39}$$

  Let $\mathbf{y}$ be any point on sphere $(\mathbf{c}_2, R)$ such that line $\mathbf{c}_2\mathbf{y} \perp$ direction $\mathbf{n}$, then (5.39) equals the signed distance from hyperplane $\mathbf{n} + \delta\mathbf{e}$ to point $\mathbf{y}$.

- Let $\mathbf{c}_1, \mathbf{c}_2 \in \mathcal{N}_\mathbf{e}$, then

$$(\mathbf{c}_1 - \frac{R_1^2}{2}\mathbf{e}) \cdot (\mathbf{c}_2 + \frac{R_2^2}{2}\mathbf{e}) = \frac{R_1^2 - R_2^2 - d^2_{\mathbf{c}_1\mathbf{c}_2}}{2}. \tag{5.40}$$

  Let $\mathbf{y}$ be any point on sphere $(\mathbf{c}_2, R)$ such that line $\mathbf{c}_2\mathbf{y} \perp$ line $\mathbf{c}_1\mathbf{c}_2$, and let $d_{\max}, d_{\min}$ be respectively the maximal signed distance and the minimal signed distance from point $\mathbf{y}$ to the points on sphere $\mathbf{c}_1 - R_1^2\mathbf{e}/2$, then (5.40) equals $-d_{\max}d_{\min}/2$.

- Let $\mathbf{c}_1, \mathbf{c}_2 \in \mathcal{N}_\mathbf{e}$, then

$$(\mathbf{c}_1 + \frac{R_1^2}{2}\mathbf{e}) \cdot (\mathbf{c}_2 + \frac{R_2^2}{2}\mathbf{e}) \;=\; -\frac{R_1^2 + R_2^2 + d_{\mathbf{c}_1\mathbf{c}_2}^2}{2}. \tag{5.41}$$

Let $\mathbf{y}$ be any point on sphere $(\mathbf{c}_2, R_2)$ such that line $\mathbf{c}_2\mathbf{y} \perp$ line $\mathbf{c}_1\mathbf{c}_2$, and let $\mathbf{z}$ be any point on sphere $(\mathbf{c}_1, R_1)$ satisfying $\mathbf{c}_1\mathbf{z} \perp \mathbf{c}_1\mathbf{y}$. Then (5.41) equals $-d_{\mathbf{yz}}^2/2$.

While the conformal model is suitable for geometric interpretation, the homogeneous model is convenient for symbolic computation.

---

Glossary

**Second point of intersection:** For two circles/lines $\mathbf{ab}_1\mathbf{c}_1$ and $\mathbf{ab}_2\mathbf{c}_2$ in the plane, "point" $\mathbf{a}$ (may be either a finite point or the conformal point at infinity) is trivially a point of intersection; besides this "point," the other "point" of intersection is denoted by $\mathbf{b}_1\mathbf{c}_1 \cap_\mathbf{a} \mathbf{b}_2\mathbf{c}_2$. In particular, if $\mathbf{b}_1\mathbf{c}_1 \cap_\mathbf{a} \mathbf{b}_2\mathbf{c}_2 = \mathbf{a}$, then the two circles/lines are tangent to each other at "point" $\mathbf{a}$; in the case of two lines, they are parallel to each other (so that they are "tangent" to each other at the conformal point at infinity $\mathbf{a} = \mathbf{e}$).

**Reduced meet product:** For two extensors $\mathbf{a} \vee \mathbf{b}_1 \vee \mathbf{c}_1$ and $\mathbf{a} \vee \mathbf{b}_2 \vee \mathbf{c}_2$ in the Grassmann-Cayley algebra (also known as *Grassmann algebra*, *cf.* the previous chapter and the notations therein, or [60]) over the conformal model of the Euclidean plane,

$$(\mathbf{b}_1 \vee \mathbf{c}_1) \wedge_\mathbf{a} (\mathbf{b}_2 \vee \mathbf{c}_2) \;:=\; [\mathbf{ab}_1\mathbf{c}_1\mathbf{c}_2]\mathbf{b}_2 - [\mathbf{ab}_1\mathbf{c}_1\mathbf{b}_2]\mathbf{c}_2 \;=\; [\mathbf{ab}_1\mathbf{b}_2\mathbf{c}_2]\mathbf{c}_1 - [\mathbf{ac}_1\mathbf{b}_2\mathbf{c}_2]\mathbf{b}_1$$

is the reduced meet product [33] of the two extensors with respect to $\mathbf{a}$. It can be generalized to the conformal model of $n$-dimensional Euclidean space.

**Clifford algebra:** The Clifford algebra [6, 49] over a metric vector space $\mathcal{V}^n$ is the quotient of $\otimes(\mathcal{V}^n)$ modulo the two-sided ideal generated by elements of the form $\mathbf{a} \otimes \mathbf{a} - \mathbf{a} \cdot \mathbf{a}$ for all $\mathbf{a} \in \mathcal{V}^n$. The quotient of the tensor product is called the Clifford multiplication (or geometric product), denoted by the juxtaposition of elements. The Clifford multiplication is associative, multilinear, and satisfies $\mathbf{ab} = \mathbf{a} \cdot \mathbf{b} + \mathbf{a} \vee \mathbf{b}$ for all $\mathbf{a}, \mathbf{b} \in \mathcal{V}^n$.

**Monic Clifford monomial:** The Clifford multiplication of a sequence of vector variables of $\mathcal{V}^n$.

**Geometric algebra:** Clifford himself called the algebra now bearing his name "geometric algebra." This term is nowadays reserved specifically for the version of Clifford algebra [24, 25, 28] where the geometric product is defined on the Grassmann-Cayley algebra over the base metric vector space $\mathcal{V}^n$.

**Conformal geometric algebra:** It refers to the geometric algebra (Clifford algebra + Grassmann-Cayley algebra) over $\mathbb{R}^{n+1,1}$, together with the $n$-dimensional conformal interpretations of algebraic elements in this algebra [1, 2, 12, 15, 26, 27, 30, 50, 56]. The first application of Clifford algebra to the study of the conformal model dates back to 1892 in the work of E. Müller [44, 45].

---

In the Grassmann-Cayley algebra over the conformal model $\mathbb{R}^{3,1}$ of the Euclidean plane, the *meet product* of two circles/lines $\mathbf{ab}_1\mathbf{c}_1$ and $\mathbf{ab}_2\mathbf{c}_2$ in the plane is

$$(\mathbf{a} \vee \mathbf{b}_1 \vee \mathbf{c}_1) \wedge (\mathbf{a} \vee \mathbf{b}_2 \vee \mathbf{c}_2) \;=\; \mathbf{a} \vee \{(\mathbf{b}_1 \vee \mathbf{c}_1) \wedge_\mathbf{a} (\mathbf{b}_2 \vee \mathbf{c}_2)\}.$$

The resulting outer product has two ingredients: the trivial point of intersection $\mathbf{a}$ (null vector), and another vector $(\mathbf{b}_1 \vee \mathbf{c}_1) \wedge_\mathbf{a} (\mathbf{b}_2 \vee \mathbf{c}_2)$ that is generally not null. The second vector needs to be changed into a null vector in order to represent the point $\mathbf{b}_1\mathbf{c}_1 \cap_{\mathbf{a}} \mathbf{b}_2\mathbf{c}_2$ in the conformal model. This can be realized by reflecting null vector $\mathbf{a}$ with respect to invertible vector $(\mathbf{b}_1 \vee \mathbf{c}_1) \wedge_{\mathbf{a}} (\mathbf{b}_2 \vee \mathbf{c}_2)$.

In Clifford algebra, the reflection of a vector $\mathbf{y}$ with respect to an invertible vector $\mathbf{x}$ is defined by

$$\mathbf{y} \mapsto Ad_{\mathbf{x}}(\mathbf{y}) := -\mathbf{x}\mathbf{y}\mathbf{x}^{-1}.$$

In our mission of representing point $\mathbf{b}_1\mathbf{c}_1 \cap_{\mathbf{a}} \mathbf{b}_2\mathbf{c}_2$ in the conformal model, the following *homogeneous reflection* is more convenient to use, as it allows for $(\mathbf{b}_1 \vee \mathbf{c}_1) \wedge_{\mathbf{a}} (\mathbf{b}_2 \vee \mathbf{c}_2)$ to be non-invertible, *i.e.*, null:

$$N_{\mathbf{y}}(\mathbf{x}) := \frac{1}{2}\mathbf{x}\mathbf{y}\mathbf{x}.$$

Then $\mathbf{b}_1\mathbf{c}_1 \cap_{\mathbf{a}} \mathbf{b}_2\mathbf{c}_2$ is represented multiplicatively by $N_{\mathbf{a}}((\mathbf{b}_1 \vee \mathbf{c}_1) \wedge_{\mathbf{a}} (\mathbf{b}_2 \vee \mathbf{c}_2))$.

---

**Glossary**

**Basic algebraic invariant:** An algebraic $G$-invariant is a polynomial of coordinate variables, such that for any coordinate transformation in a group $G$, the polynomial is invariant [47]. In oriented Euclidean geometry, the basic algebraic invariants are squared distances and signed volumes of simplexes [63].

**Advanced algebraic invariant:** It refers to a scalar-valued monomial of an algebra such that if the multiplications in the monomial are expanded relative to a coordinate system, then the monomial becomes a polynomial of basic algebraic invariants [34].

**Clifford bracket algebra:** The Clifford bracket algebra of vector variables $\mathbf{a}_1, \ldots, \mathbf{a}_m$ in metric vector space $\mathcal{V}^n$ is the commutative ring generated by two kinds of *Clifford brackets* [33], the "hyper-determinant" and "hyper-inner product," by prolonging the bracket of $n$ vector variables and inner product of two vector variables respectively with the Clifford multiplication: for all $k, l \geq 0$,

$$\begin{array}{lll} \mathbf{a}_1 \cdot \mathbf{a}_2 = \langle \mathbf{a}_1\mathbf{a}_2 \rangle & \text{prolonged to} & \langle \mathbf{a}_1\mathbf{a}_2 \ldots \mathbf{a}_{2k} \rangle := \langle \mathbf{a}_1\mathbf{a}_2 \ldots \mathbf{a}_{2k} \rangle_0, \\ [\mathbf{a}_1 \ldots \mathbf{a}_n] = (\mathbf{a}_1 \vee \cdots \vee \mathbf{a}_n)^{\sim} & \text{prolonged to} & [\mathbf{a}_1\mathbf{a}_2 \ldots \mathbf{a}_{n+2l}] := \langle \mathbf{a}_1\mathbf{a}_2 \ldots \mathbf{a}_{n+2l} \rangle_n^{\sim}. \end{array}$$

Here "$\sim$" is the Hodge dual operator in Grassmann algebra, and $\langle\ \rangle_i$ is the $i$-grading operator in Grassmann algebra that extracts the $i$th-graded part of the argument [25].

**Null bracket algebra:** Clifford bracket algebra of null vector variables [34], i.e., $\mathbf{aa} = 0$ for any vector variable $\mathbf{a}$.

---

For points $\mathbf{a}_1, \cdots, \mathbf{a}_M \in \mathbb{R}^n$, let the dot symbol denote the contraction (also called the *inner product*, or *interior product*) of the Clifford algebra inherited from the contraction of the tensor algebra over $\mathbb{R}^{n+1,1}$. This may also be defined via the following Laplace expansion:

$$(\mathbf{e} \vee \mathbf{f}(\mathbf{a}_1) \vee \ldots \vee \mathbf{f}(\mathbf{a}_M)) \cdot (\mathbf{e} \vee \mathbf{f}(\mathbf{a}_1) \vee \ldots \vee \mathbf{f}(\mathbf{a}_M)) = \begin{vmatrix} 0 & -1 & -1 & \cdots & -1 \\ -1 & 0 & -\frac{1}{2}d^2_{\mathbf{a}_1\mathbf{a}_2} & \cdots & -\frac{1}{2}d^2_{\mathbf{a}_1\mathbf{a}_M} \\ -1 & -\frac{1}{2}d^2_{\mathbf{a}_2\mathbf{a}_1} & 0 & \cdots & -\frac{1}{2}d^2_{\mathbf{a}_2\mathbf{a}_M} \\ \vdots & \vdots & \vdots & \ddots & \vdots \\ -1 & -\frac{1}{2}d^2_{\mathbf{a}_M\mathbf{a}_1} & -\frac{1}{2}d^2_{\mathbf{a}_M\mathbf{a}_2} & \cdots & 0 \end{vmatrix} \tag{5.42}$$

The right side is the Cayley-Menger determinant (5.2) up to a factor of $((M-1)!)^2$. When the left side is replaced by $(\mathbf{e}\vee\mathbf{f}(\mathbf{a}_1)\vee\ldots\vee\mathbf{f}(\mathbf{a}_M))\cdot(\mathbf{e}\vee\mathbf{f}(\mathbf{b}_1)\vee\ldots\vee\mathbf{f}(\mathbf{b}_M))$, then its Laplace expansion is the general nonsymmetric Cayley-Menger determinant (5.12). Some applications of the conformal model in distance geometry can be found in [20].

When the conformal point at infinity in (5.42) is replaced by a point, we get

$$(\mathbf{f}(\mathbf{a}_1)\vee\cdots\vee\mathbf{f}(\mathbf{a}_M))\cdot(\mathbf{f}(\mathbf{a}_1)\vee\cdots\vee\mathbf{f}(\mathbf{a}_M)) \;=\; \det(\mathbf{f}(\mathbf{a}_i)\cdot\mathbf{f}(\mathbf{a}_j))_{M\times M} \;=\; \det(-d^2_{\mathbf{a}_i\mathbf{a}_j}/2)_{M\times M},$$

from which we get the following $n$-dimensional *Ptolemy's Theorem* [3]:

*Let* $\mathbf{a}_1,\cdots,\mathbf{a}_{n+2}$ *be points in* $\mathbb{E}^n$*, then they are on a hypersphere or hyperplane if and only if* $\det(d^2_{\mathbf{a}_i\mathbf{a}_j})_{(n+2)\times(n+2)}=0$.

Furthermore, if we replace some/all points on the left side of (5.42) with positive/negative vectors of $\mathbb{R}^{n+1,1}$, we get various extensions of the Cayley-Menger determinant to conformal objects other than points [42].

Cayley-Menger determinants and their various extensions are advanced algebraic invariants: each is a monomial of the inner-product Grassmann algebra over the conformal model, and each when expanded equals a polynomial in the squared distances, or in the inner products of non-null vectors representing geometric objects other than points. The above proof of $n$-dimensional Ptolemy's Theorem shows the effectiveness of using advanced algebraic invariants in geometric reasoning.

Still the Cayley-Menger determinant and all its generalizations are limited to monomials of length (maximal number of inclusive vector variables) at most $2(n+2)$, as the outer product of any $n+3$ vector variables in $\mathbb{R}^{n+1,1}$ is zero. They are still inefficient in more complex geometric reasoning, and advanced algebraic invariants of unlimited length are called for.

The two Clifford brackets extend the determinant of $n+2$ vector variables of $\mathbb{R}^{n+1,1}$ to a "hyper-determinant" of $n+2l$ vector variables for arbitrary $l\geq 1$, and extend the inner product of two vector variables to a "hyper-inner product" of $2k$ vector variables for arbitrary $k\geq 1$, respectively. The respective anti-symmetry and symmetry of the determinant and inner product functions are replaced by the following symmetries:

- Reversion:

$$\begin{aligned}\langle\mathbf{a}_1\mathbf{a}_2\cdots\mathbf{a}_{2k}\rangle &= \langle\mathbf{a}_{2k}\mathbf{a}_{2k-1}\cdots\mathbf{a}_1\rangle,\\ [\mathbf{a}_1\mathbf{a}_2\cdots\mathbf{a}_{n+2l}] &= (-1)^{\frac{n(n-1)}{2}}[\mathbf{a}_{n+2l}\mathbf{a}_{n+2l-1}\cdots\mathbf{a}_1].\end{aligned}$$

- Shift:

$$\begin{aligned}\langle\mathbf{a}_1\mathbf{a}_2\cdots\mathbf{a}_{2k}\rangle &= \langle\mathbf{a}_{2k}\mathbf{a}_1\mathbf{a}_2\cdots\mathbf{a}_{2k-1}\rangle,\\ [\mathbf{a}_1\mathbf{a}_2\cdots\mathbf{a}_{n+2l}] &= (-1)^{n-1}[\mathbf{a}_{n+2l}\mathbf{a}_1\cdots\mathbf{a}_{n+2l-1}].\end{aligned}$$

The null bracket algebra over $\mathbb{R}^{n+1,1}$ is related to the Clifford bracket algebra over $\mathbb{R}^n$ as follows: Let $\mathbf{u},\mathbf{v}$ and the $\mathbf{a}_i$ be null vectors in $\mathbb{R}^{n+1,1}$. Let each $\mathbf{a}_i$ represent a point in the orthogonal complement $\mathbb{R}^n$ of the Minkowski plane spanned by $\mathbf{u},\mathbf{v}$, and let $\overrightarrow{\mathbf{a}_i\mathbf{a}_j}$ denote the displacement vector from point $\mathbf{a}_i$ to point $\mathbf{a}_j$. Then

$$\begin{aligned}\langle\mathbf{a}_1\mathbf{a}_2\cdots\mathbf{a}_{2k}\rangle &= \frac{1}{2}\langle\overrightarrow{\mathbf{a}_1\mathbf{a}_2}\,\overrightarrow{\mathbf{a}_2\mathbf{a}_3}\cdots\overrightarrow{\mathbf{a}_{2k-1}\mathbf{a}_{2k}}\,\overrightarrow{\mathbf{a}_{2k}\mathbf{a}_1}\rangle,\\ [\mathbf{a}_1\mathbf{a}_2\cdots\mathbf{a}_{n+2l}] &= (-1)^n\frac{1}{2}[\overrightarrow{\mathbf{a}_1\mathbf{a}_2}\,\overrightarrow{\mathbf{a}_2\mathbf{a}_3}\cdots\overrightarrow{\mathbf{a}_{n+2l-1}\mathbf{a}_{n+2l}}\,\overrightarrow{\mathbf{a}_{n+2l}\mathbf{a}_1}].\end{aligned}$$

In 2D trigonometry, the Clifford brackets of the conformal model of Euclidean plane have the

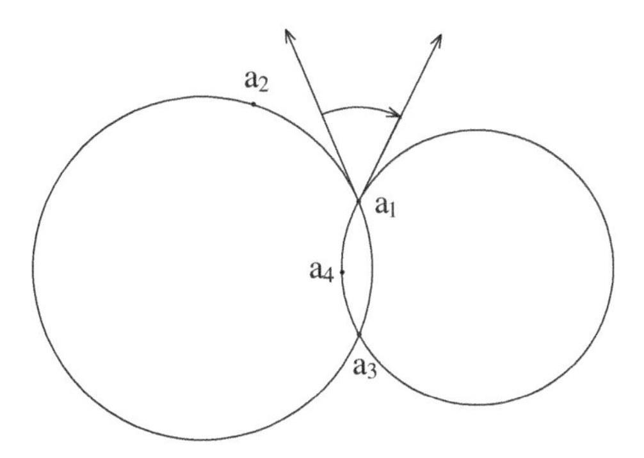


**Figure 5.7**
Illustration of the notation $\angle(\mathbf{a}_1\mathbf{a}_2\mathbf{a}_3,\mathbf{a}_1\mathbf{a}_3\mathbf{a}_4)$: circle $\mathbf{a}_1\mathbf{a}_2\mathbf{a}_3$ in the figure has anti-clockwise orientation (from $\mathbf{a}_1$ to $\mathbf{a}_2$ to $\mathbf{a}_3$), and the tangent direction of the circle at point $\mathbf{a}_1$ points from $\mathbf{a}_1$ to $\mathbf{a}_2$. Similarly, circle $\mathbf{a}_1\mathbf{a}_3\mathbf{a}_4$ in the figure has clockwise orientation, so the tangent direction of the circle at point $\mathbf{a}_1$ points from $\mathbf{a}_4$ to $\mathbf{a}_1$. The angle of rotation from the first tangent to the second one is $\angle(\mathbf{a}_1\mathbf{a}_2\mathbf{a}_3,\mathbf{a}_1\mathbf{a}_3\mathbf{a}_4)$.

following geometric interpretations [35]:

$$
\begin{aligned}
\langle \mathbf{a}_1\mathbf{a}_2\cdots\mathbf{a}_{2l+2}\rangle &= -\frac{d_{\mathbf{a}_1\mathbf{a}_2}d_{\mathbf{a}_2\mathbf{a}_3}\cdots d_{\mathbf{a}_{2l+1}\mathbf{a}_{2l+2}}d_{\mathbf{a}_{2l+2}\mathbf{a}_1}}{2}\cos(\angle(\mathbf{a}_1\mathbf{a}_2\mathbf{a}_3,\mathbf{a}_1\mathbf{a}_3\mathbf{a}_4)\\
&\quad +\angle(\mathbf{a}_1\mathbf{a}_4\mathbf{a}_5,\mathbf{a}_1\mathbf{a}_5\mathbf{a}_6)+\cdots+\angle(\mathbf{a}_1\mathbf{a}_{2l}\mathbf{a}_{2l+1},\mathbf{a}_1\mathbf{a}_{2l+1}\mathbf{a}_{2l+2}));\\
[\mathbf{a}_1\mathbf{a}_2\cdots\mathbf{a}_{2l+2}] &= -\frac{d_{\mathbf{a}_1\mathbf{a}_2}d_{\mathbf{a}_2\mathbf{a}_3}\cdots d_{\mathbf{a}_{2l+1}\mathbf{a}_{2l+2}}d_{\mathbf{a}_{2l+2}\mathbf{a}_1}}{2}\sin(\angle(\mathbf{a}_1\mathbf{a}_2\mathbf{a}_3,\mathbf{a}_1\mathbf{a}_3\mathbf{a}_4)\\
&\quad +\angle(\mathbf{a}_1\mathbf{a}_4\mathbf{a}_5,\mathbf{a}_1\mathbf{a}_5\mathbf{a}_6)+\cdots+\angle(\mathbf{a}_1\mathbf{a}_{2l}\mathbf{a}_{2l+1},\mathbf{a}_1\mathbf{a}_{2l+1}\mathbf{a}_{2l+2})),
\end{aligned}
$$

where $\angle(\mathbf{a}_1\mathbf{a}_2\mathbf{a}_3,\mathbf{a}_1\mathbf{a}_3\mathbf{a}_4)$ denotes the angle from the tangent direction of oriented circle $\mathbf{a}_1\mathbf{a}_2\mathbf{a}_3$ at point $\mathbf{a}_1$ to the tangent direction of oriented circle $\mathbf{a}_1\mathbf{a}_3\mathbf{a}_4$ at point $\mathbf{a}_1$, as shown in Figure 5.7. It should be remarked that the special role played by $\mathbf{a}_1$ in these formulas is not fundamental, in that any point can be made to serve that role by cyclic shift and reversion.

## 5.5 Geometric Reasoning in Euclidean Conformal Geometry

Glossary

**Monomial proof:** Throughout the process of verifying the conclusion equality of a theorem, the intermediate expression remains a monomial.

**Binomial proof:** The intermediate expression remains 2-termed during algebraic manipulations.

**Robust proof:** In theorem proving by an algebraic method, if there is more than one algebraic manipulation of the same quality available for the intermediate expression at one step (e.g., these algebraic manipulations all change the intermediate polynomial from $k$ terms to $l$ terms), then no matter which algebraic manipulation of the indicated quality is selected at the current step, it ultimately leads to an algebraic proof of the same size of the intermediate expression as

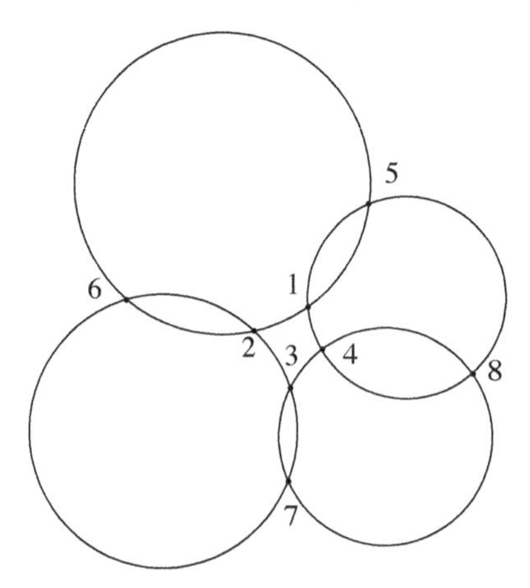


**Figure 5.8**
Miquel's 4-Circle Theorem, with the points of intersection labeled as in the text.

others (e.g., the intermediate polynomial is always within the same number of terms for each selection made).

**Quantitative extension of geometric theorem:** An algebraic description of a geometric theorem is usually the following: given a set of equality constraints $f_1 = 0, \ldots, f_k = 0$ and inequality constraints $g_1 \neq 0, \ldots, g_l \neq 0$ as the hypothesis, the conclusion $c = 0$ is true. If it is derived that when some equality constraints in the hypothesis are removed, say $f_1 = 0, \ldots, f_s = 0$ are missing where $s < k$, the conclusion expression and the missing constraints satisfy

$$tc = f_1 h_1 + \cdots + f_s h_s \qquad (5.43)$$

for some geometrically meaningful expressions $t, h_1, \ldots, h_k$, then an explicit form of (5.43) is called a *quantitative extension of the geometric theorem* [32], as it describes the quantitative relationship between the conclusion expression and the missing constraints, and provides a stronger and more general result than the original theorem.

---

Conformal geometric algebra and null bracket algebra together provide a powerful tool for geometric reasoning in Euclidean conformal geometry. Most theorems in classical Euclidean geometry involving lines, circles, planes, spheres are given robust monomial/binomial proofs. When one or more equality constraints are removed from the hypothesis of a theorem, usually a quantitative extension of the theorem can be found. We illustrate this by several examples.

**Example 1.** (*Miquel's 4-Circle Theorem* [61], Figure 5.8) Four circles intersect at eight points cyclically. If $\mathbf{1},\mathbf{2},\mathbf{3},\mathbf{4}$ are con-cyclic, so are $\mathbf{5},\mathbf{6},\mathbf{7},\mathbf{8}$.

There are five concyclicity constraints in the hypothesis, and the conclusion is the sixth concyclicity. For symmetry, we remove the concyclicity of $\mathbf{1},\mathbf{2},\mathbf{3},\mathbf{4}$, and see how the conclusion varies. The new geometric configuration is constructed as follows:

Free points: $\mathbf{1},\mathbf{2},\mathbf{3},\mathbf{4},\mathbf{5},\mathbf{7}$.

Intersections: $\mathbf{6} = \mathbf{15} \cap_{\mathbf{2}} \mathbf{37}$, $\mathbf{8} = \mathbf{15} \cap_{\mathbf{4}} \mathbf{37}$.

In conformal geometric algebra, their representations are, respectively,

$$\begin{aligned} \mathbf{6} &= 2^{-1}\{(\mathbf{1} \vee \mathbf{5}) \wedge_{\mathbf{2}} (\mathbf{3} \vee \mathbf{7})\}\mathbf{2}\{(\mathbf{1} \vee \mathbf{5}) \wedge_{\mathbf{2}} (\mathbf{3} \vee \mathbf{7})\}, \\ \mathbf{8} &= 2^{-1}\{(\mathbf{1} \vee \mathbf{5}) \wedge_{\mathbf{4}} (\mathbf{3} \vee \mathbf{7})\}\mathbf{4}\{(\mathbf{1} \vee \mathbf{5}) \wedge_{\mathbf{4}} (\mathbf{3} \vee \mathbf{7})\}. \end{aligned}$$

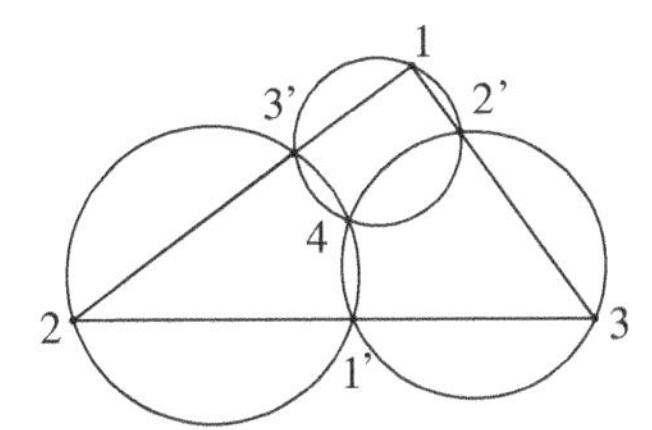


**Figure 5.9**
Miquel's 3-Circle Theorem, with the points of intersection labeled as in the text.

Conclusion expression: $[\mathbf{5678}]$. It equals zero if and only if $\mathbf{5,6,7,8}$ are con-cyclic.
We compute the expression by eliminating $\mathbf{6,8}$ and expanding the result. The following is a typical monomial proof/computing:

$$\begin{aligned}
[\mathbf{5678}] &\overset{\mathbf{6,8}}{=} 2^{-2}[\mathbf{5}\{(\mathbf{1}\vee\mathbf{5})\wedge_{\mathbf{2}}(\mathbf{3}\vee\mathbf{7})\}\mathbf{2}\{(\mathbf{1}\vee\mathbf{5})\wedge_{\mathbf{2}}(\mathbf{3}\vee\mathbf{7})\} \\
&\qquad \mathbf{7}\{(\mathbf{1}\vee\mathbf{5})\wedge_{\mathbf{4}}(\mathbf{3}\vee\mathbf{7})\}\mathbf{4}\{(\mathbf{1}\vee\mathbf{5})\wedge_{\mathbf{4}}(\mathbf{3}\vee\mathbf{7})\}] \\
&\overset{expand}{=} -2^{-2}[\mathbf{1257}][\mathbf{1457}][\mathbf{2357}][\mathbf{3457}][\mathbf{51237341}] \\
&\overset{monomial}{=} (\mathbf{1}\cdot\mathbf{5})(\mathbf{3}\cdot\mathbf{7})[\mathbf{1257}][\mathbf{1457}][\mathbf{2357}][\mathbf{3457}]\underline{[\mathbf{1234}]}.
\end{aligned} \tag{5.44}$$

In the second step, the Cayley expansion [31] is to eliminate the reduced meet product; in the third step, the monomial factorizations are by the identity $\mathbf{aba} = 2(\mathbf{a}\cdot\mathbf{b})\mathbf{a}$ for null vector $\mathbf{a}$, where $\mathbf{aba} = \mathbf{151}$ and $\mathbf{373}$ in the rightmost factor.

The removed concyclicity expression $[\mathbf{1234}]$ occurs naturally in the final result, so Miquel's 4-Circle Theorem is automatically discovered. In addition, the above computing together with similar computing of $(\mathbf{5}\cdot\mathbf{6})(\mathbf{7}\cdot\mathbf{8})$ yield the following *homogeneous identity* that is invariant under the rescaling of any vector variable, which is a *quantitative version* of the theorem and discloses how the conclusion depends on the concyclicity constraint of $\mathbf{1,2,3,4}$:

$$\frac{[\mathbf{5678}]}{(\mathbf{5}\cdot\mathbf{6})(\mathbf{7}\cdot\mathbf{8})} = \frac{[\mathbf{1234}]}{(\mathbf{1}\cdot\mathbf{2})(\mathbf{3}\cdot\mathbf{4})}\,\frac{[\mathbf{1257}][\mathbf{3457}]}{[\mathbf{1457}][\mathbf{2357}]}.$$

**Example 2.** (*Miquel's 3-Circle Theorem* [5], Figure 5.9) Let $\mathbf{1}',\mathbf{2}',\mathbf{3}'$ be points on lines $\mathbf{23},\mathbf{13},\mathbf{12}$ respectively. Then circles $\mathbf{12'3'}$, $\mathbf{1'23'}$, $\mathbf{1'2'3}$ meet at a common point $\mathbf{4}$.
Construct point $\mathbf{4}$ as the second intersection of circles $\mathbf{12'3'}$ and $\mathbf{1'23'}$. The hypothesis contains three collinearity constraints and two concyclicity constraints. The conclusion is the concyclicity of $\mathbf{1}',\mathbf{2}',\mathbf{3},\mathbf{4}$.

We remove all the collinearity constraints to see how the conclusion expression $[\mathbf{1'2'34}]$ depends on them. The following is a simple monomial computing:

$$\begin{aligned}
[\mathbf{1'2'34}] &= [\mathbf{1'42'3}] \\
&\overset{\mathbf{4}}{=} 2^{-1}[\mathbf{1'}\{(\mathbf{1}\vee\mathbf{2'})\wedge_{\mathbf{3'}}(\mathbf{1'}\vee\mathbf{2})\}\mathbf{3'}\{(\mathbf{1}\vee\mathbf{2'})\wedge_{\mathbf{3'}}(\mathbf{1'}\vee\mathbf{2})\}\mathbf{2'3}] \\
&\overset{expand}{=} 2^{-1}[\mathbf{11'2'3'}][\mathbf{21'2'3'}]\underline{[\mathbf{1'23'12'3}]}.
\end{aligned} \tag{5.45}$$

In the above computing, after rearranging the input $[\mathbf{1'2'34}]$ to $[\mathbf{1'42'3}]$ and substituting the expression of $\mathbf{4}$ into it, each meet product allows one and only one monomial Cayley expansion. For example, since the meet product $(\mathbf{1}\vee\mathbf{2'})\wedge_{\mathbf{3'}}(\mathbf{1'}\vee\mathbf{2})$ is a neighbor of null vector $\mathbf{1}'$, the Cayley expansion by separating $\mathbf{1}',\mathbf{2}$ automatically deletes the term of $\mathbf{1}'$, leaving only a multiple of $\mathbf{2}$ left in the expansion result. Finding a good neighbor makes good control of the size of Cayley expansion.

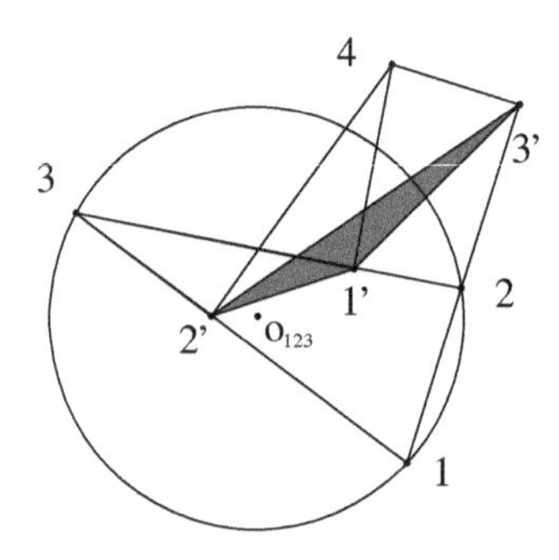


**Figure 5.10**
Generalized Simson's Theorem, with the points labeled as in the text and the triangle formed by the constructed points shaded.

If the three collinearities are restored, it is easily seen that generically $[\mathbf{11'2'3'}] \neq 0$ and $[\mathbf{21'2'3'}] \neq 0$. So it is factor $[\mathbf{1'23'12'3}] = 0$ that leads to the original conclusion under the two non-degeneracy conditions.

The removed collinearities can be reformulated as following: (i) three circles/lines $\mathbf{1'23}$, $\mathbf{12'3}$, $\mathbf{123'}$ concur; (ii) they concur at the conformal point at infinity (so that they are lines, not circles).

By a computing procedure similar to (5.45), we get that constraint (i) can be represented by $[\mathbf{12'31'23'}] = 0$ under some additional non-degeneracy conditions, and (ii) is redundant under these additional conditions. So the equivalence between the conclusion and the three collinearities is the following shift symmetry of Clifford bracket:

$$[\mathbf{1'23'12'3}] \;=\; -[\mathbf{12'31'23'}],$$

and Miquel's 3-Circle Theorem is a special case of this symmetry.

**Example 3.** (*Generalized Simson's Theorem* [3], Figure 5.10) In the plane, draw feet $\mathbf{1}', \mathbf{2}', \mathbf{3}'$ from point $\mathbf{4}$ to the three sides of triangle $\mathbf{123}$ respectively. Then

$$S_{\mathbf{1'2'3'}} \;=\; \frac{1}{4}\left(1 - \frac{d^2_{\mathbf{4}\phi_{\mathbf{123}}}}{R^2_{\mathbf{123}}}\right) S_{\mathbf{123}}, \tag{5.46}$$

where $S_{\mathbf{123}}, \phi_{\mathbf{123}}$, and $R_{\mathbf{123}}$ are the signed area, circum-center, and radius of the circum-circle of triangle $\mathbf{123}$ respectively. In particular, $\mathbf{1}, \mathbf{2}, \mathbf{3}, \mathbf{4}$ are con-cyclic if and only if $\mathbf{1}', \mathbf{2}', \mathbf{3}'$ are collinear, which is the original *Simson's Theorem*.
The construction of the geometric configuration is as follows:

Free points: $\mathbf{1}, \mathbf{2}, \mathbf{3}, \mathbf{4}$.

Feet: $\mathbf{1}' = \P_{\mathbf{4},\mathbf{23}}$ (projection of point $\mathbf{4}$ on line $\mathbf{23}$), $\mathbf{2}' = \P_{\mathbf{4},\mathbf{13}}$, $\mathbf{3}' = \P_{\mathbf{4},\mathbf{12}}$.
Their (reduced) representations in conformal geometric algebra are, respectively,

$$\begin{array}{rcll} \mathbf{1}' & = & (\mathbf{2} \vee \mathbf{3}) \wedge_{\mathbf{e}} (\mathbf{4} \vee \langle \mathbf{e23} \rangle_3^{\sim}) & \bmod \mathbf{e}, \\ \mathbf{2}' & = & (\mathbf{3} \vee \mathbf{1}) \wedge_{\mathbf{e}} (\mathbf{4} \vee \langle \mathbf{e31} \rangle_3^{\sim}) & \bmod \mathbf{e}, \\ \mathbf{3}' & = & (\mathbf{1} \vee \mathbf{2}) \wedge_{\mathbf{e}} (\mathbf{4} \vee \langle \mathbf{e12} \rangle_3^{\sim}) & \bmod \mathbf{e}, \end{array} \tag{5.47}$$

where "mod **e**" denotes that the two sides of the equality are equal up to an additional term $\lambda\mathbf{e}$ for some scalar $\lambda$. Equation (5.47) is geometrically readable: $\mathbf{e} \vee \mathbf{2} \vee \mathbf{3} = \langle \mathbf{e23} \rangle_3$ represents line $\mathbf{23}$, and the Hodge dual $\langle \mathbf{e23} \rangle_3^{\sim}$ represents the normal direction of the line, so the intersection of line $\mathbf{23}$ with the line passing through point $\mathbf{4}$ and the normal direction of line $\mathbf{23}$ is the foot $\mathbf{1}'$.

Conclusion expression: $[\mathbf{e1'2'3'}]$ (twice the signed area of triangle $\mathbf{1'2'3'}$).

We compute the conclusion expression by eliminating the three feet and making simplification, with the hope that $[\mathbf{e123}]$ and $[\mathbf{1234}]$ occur naturally as factors in the result, according to (5.46). The following is a typical binomial proof/computing:

$$
\begin{aligned}
[\mathbf{e1'2'3'}] &\overset{\mathbf{1',2',3'}}{=} [\mathbf{e}\{(\mathbf{2}\vee\mathbf{3})\wedge_{\mathbf{e}}(\mathbf{4}\vee\langle\mathbf{e23}\rangle_3^{\sim})\}\{(\mathbf{3}\vee\mathbf{1})\wedge_{\mathbf{e}}(\mathbf{4}\vee\langle\mathbf{e31}\rangle_3^{\sim})\}\ \{(\mathbf{1}\vee\mathbf{2})\wedge_{\mathbf{e}}(\mathbf{4}\vee\langle\mathbf{e12}\rangle_3^{\sim})\}]\\
&\overset{expand}{=} [\mathbf{e123}]([\mathbf{e34}\langle\mathbf{e31}\rangle_3^{\sim}][\mathbf{e124}][\mathbf{e}\langle\mathbf{e23}\rangle_3^{\sim}\mathbf{4}\langle\mathbf{e12}\rangle_3^{\sim}]-[\mathbf{e24}\langle\mathbf{e12}\rangle_3^{\sim}][\mathbf{e134}][\mathbf{e}\langle\mathbf{e23}\rangle_3^{\sim}\mathbf{4}\langle\mathbf{e31}\rangle_3^{\sim}])\\
&\overset{duality}{=} [\mathbf{e123}](\langle\mathbf{e34}\langle\mathbf{e31}\rangle_3\rangle[\mathbf{e124}][\mathbf{e}\langle\mathbf{e23}\rangle_3\mathbf{4}\langle\mathbf{e12}\rangle_3]-\langle\mathbf{e24}\langle\mathbf{e12}\rangle_3\rangle[\mathbf{e134}][\mathbf{e}\langle\mathbf{e23}\rangle_3\mathbf{4}\langle\mathbf{e31}\rangle_3])\\
&\overset{ungrading}{=} 2^{-3}[\mathbf{e123}](\langle\mathbf{e34e31}\rangle[\mathbf{e124}][\mathbf{e23e4e12}]-\langle\mathbf{e24e12}\rangle[\mathbf{e134}][\mathbf{e23e4e31}])\\
&\overset{monomial}{=} (\mathbf{e}\cdot\mathbf{2})(\mathbf{e}\cdot\mathbf{3})(\mathbf{e}\cdot\mathbf{4})[\mathbf{e123}]^2\ (\langle\mathbf{e134}\rangle[\mathbf{e124}]-\langle\mathbf{e124}\rangle[\mathbf{e134}])\\
&\overset{contract}{=} 2(\mathbf{e}\cdot\mathbf{1})(\mathbf{e}\cdot\mathbf{2})(\mathbf{e}\cdot\mathbf{3})(\mathbf{e}\cdot\mathbf{4})^2\underline{[\mathbf{e123}]^2[\mathbf{1234}]}.
\end{aligned}
$$

The second step is Cayley expansion to remove the reduced meet product. Notice that not all the meet products are eliminated at the same time, instead they are removed one by one. For example, if we expand the first meet product $(\mathbf{2}\vee\mathbf{3})\wedge_{\mathbf{e}}(\mathbf{4}\vee\langle\mathbf{e23}\rangle_3^{\sim})$ by separating $\mathbf{2},\mathbf{3}$, the result is a linear combination (binomial) of the two null vectors $\mathbf{2},\mathbf{3}$, and in successive Cayley expansions, any separation generating one of $\mathbf{2},\mathbf{3}$ leads to a monomial result (null vector advantage), so that the expansion result remains 2-termed.

The third step in the computing is to eliminate the Hodge dual operator; the fourth step is to remove the 3-grading operator; the fifth step is monomial factorizations; the last step is contraction to reduce the number of terms.

So factors $[\mathbf{e123}]^2[\mathbf{1234}]$ come up in the result automatically. The following quantitative result is also obtained by similar computing of $(\mathbf{e}\cdot\mathbf{1'})(\mathbf{e}\cdot\mathbf{2'})(\mathbf{e}\cdot\mathbf{3'})$:

$$
\frac{[\mathbf{e1'2'3'}]}{(\mathbf{e}\cdot\mathbf{1'})(\mathbf{e}\cdot\mathbf{2'})(\mathbf{e}\cdot\mathbf{3'})} = \frac{[\mathbf{1234}][\mathbf{e123}]^2}{4(\mathbf{e}\cdot\mathbf{1})(\mathbf{e}\cdot\mathbf{2})(\mathbf{e}\cdot\mathbf{3})(\mathbf{e}\cdot\mathbf{4})(\mathbf{1}\cdot\mathbf{2})(\mathbf{2}\cdot\mathbf{3})(\mathbf{1}\cdot\mathbf{3})}.
$$

In more geometrical terms, the above identity can be written as

$$
[\mathbf{e1'2'3'}] = \frac{[\mathbf{1234}]}{2R_{\mathbf{123}}^2},
$$

because $R_{\mathbf{123}}^2 = -2(\mathbf{1}\cdot\mathbf{2})(\mathbf{1}\cdot\mathbf{3})(\mathbf{2}\cdot\mathbf{3})[\mathbf{e123}]^{-2}$.

It is a common phenomenon that algebraic computing based on conformal geometric algebra representations and null bracket algebra manipulations are extremely short, and the results are geometrically meaningful. The simplicity may be partially attributed to the fact that many complicated syzygies among basic invariants are integrated into simple symmetries within advanced invariants.

The above examples have demonstrated the effectiveness of expansion and factorization in null bracket algebra. Besides these two symbolic manipulations, there are two other basic ones: normalization and division. The normalization converts a Clifford bracket polynomial into a unique normal form, and the division among Clifford bracket polynomials is drastically different from the coordinate polynomial case. Some recent advances can be found in [36, 37].

**Acknowledgment.** The second author is supported by the NSFC Project 11671388 and the CAS Frontier Key Project QYZDJ-SSW-SYS022.

## References

[1] E. Bayro-Corrochano and C. López-Franco. Omnidirectional vision: Unified model using conformal geometry. *ECCV'04* pages 536–548, 2004.

[2] E. Bayro-Corrochano, L. Reyes-Lozano, J. Zamora-Esquivel. Conformal geometric algebra for robotic vision. *J. Mathematical Imaging and Vision* **24**:55–81, 2006.

[3] M. Berger *Geometry I, II.* Springer Verlag, Berlin, Heidelberg, 1987.

[4] L. M. Blumenthal. *Theory and Applications of Distance Geometry.* Cambridge Univ. Press, Cambridge,1953. Reprinted by the Chelsea Publishing Co., 1970.

[5] S. C. Chou, X. S. Gao and J. Z. Zhang. *Machine Proofs in Geometry.* World Scientific, Singapore, 1994.

[6] W. K. Clifford. Application of Grassmann's extensive algebra. *Am. J. Math.* I:350–358, 1878.

[7] R. Connelly. Rigidity and energy. *Invent. Math.* 66:11–33, 1982.

[8] G. M. Crippen and T. F. Havel. *Distance Geometry and Molecular Conformation.* Research Studies Press, Taunton, UK, 1988.

[9] G. M. Crippen and S. A. Wildman. Quantitative structure-activity relationships (QSAR). In A. K. Ghose and V. N. Viswanadhan, editors, *Combinatorial Library Design and Evaluation,* pages 131–156. Marcel Dekker, New York, NY, 2001.

[10] G. M. Crippen. Cluster distance geometry of polypeptide chains. *J. Comput. Chem.*, 25:1305–1312, 2004.

[11] D. B. Dix. Polyspherical coordinate systems on orbit spaces with applications to biomolecular shape. *Acta Appl. Math.*, 90:247–306, 2006.

[12] L. Dorst, D. Fontijne and S. Mann. *Geometric Algebra for Computer Science.* Morgan Kaufmann Publishers, Elsevier Inc., 2007.

[13] A. W. M. Dress and T. F. Havel. Bound smoothing under chirality constraints. *SIAM J. Disc. Math.*, 4:535–549, 1990.

[14] M. Fiedler. *Matrices and Graphs in Geometry.* Cambridge University Press, 2011.

[15] D. Fontijne, L. Dorst. Modeling 3D Euclidean geometry–performance and elegance of five models of 3D Euclidean geometry in a ray tracing application. *Computer Graphics Appl.* 23:68-78, 2003.

[16] M. Fuhrmans, A. G. Milbradt, and C. Renner. Comparison of protocols for calculation of peptide structures from experimental NMR data. *J. Chem. Theory Comput.*, 2:201–208, 2006.

[17] T. F. Havel and K. Wüthrich. A distance geometry program for determining the structures of small proteins and other macromolecules from nuclear magnetic resonance measurements of $^1H - {}^1H$ proximities in solution. *Bull. Math. Biol.*, 46:673–698, 1984.

[18] T. F. Havel. An evaluation of computational strategies for use in the determination of protein structure from distance constraints obtained by nuclear magnetic resonance. In D. Nobel and T. L. Blundell, editors, volume 56 of *Progress in Biophysics and Molecular Biology*, pages 43–78. Permagon Press, Oxford, UK, 1991.

[19] T. F. Havel. Predicting the structure of the flavodoxin from *Erchericia coli* by homology modeling, distance geometry and molecular dynamics. *Molec. Simul.*, 10:175–210, 1993.

[20] T. F. Havel. Computational synthetic geometry with Clifford algebra. In: *Automated Deduction in Geometry*, D. Wang (ed.), Lect. Notes in Artif. Intellig. **1360**, pp. 102–114, Springer-Verlag, 1997.

[21] T. F. Havel. Metric matrix embedding in protein structure calculations, NMR spectra analysis, and relaxation theory. *Magn. Reson. Chem.*, 41:S37–S50, 2003.

[22] T. F. Havel, I. D. Kuntz, and G. M. Crippen. Theory and practice of distance geometry. *Bull. Math. Biol.*, 45:665–720, 1983.

[23] T. F. Havel and I. Najfeld. A new system of equations, based on geometric algebra, for ring closure in cyclic molecules. In J. Fleischer, J. Grabmeier, F. W. Hehl, and W. Küchlin, editors, *Computer Algebra in Science and Engineering*, pages 243–259. World Scientific, Singapore, 1995.

[24] D. Hestenes. *Space-Time Algebra*. Gordon and Breach, New York, 1966.

[25] D. Hestenes and G. Sobczyk. *Clifford Algebra to Geometric Calculus*, Kluwer Acad., Dordrecht NL, 1984.

[26] D. Hildenbrand. Geometric computing in computer graphics using conformal geometric algebra. *Computers and Graphics* 29:795–803, 2005.

[27] A. Lasenby and J. Lasenby. Surface evolution and representation using geometric algebra. In R. Cipolla and R. Martin, editors, *The Mathematics of Surfaces IX*, pages 144–168. Springer Verlag, London, 2000.

[28] A. Lasenby and C. Doran. *Geometric Algebra for Physicists*. Cambridge University Press, Cambridge, 2003.

[29] Li, H. Some applications of Clifford algebra to geometries. In X. S. Gao et al., editors, LNAI 1669, *Automated Deduction in Geometry*, pages 156–179. Springer Verlag, Heidelberg, 1998.

[30] H. Li, D. Hestenes and A. Rockwood. Generalized homogeneous coordinates for computational geometry. In G. Sommer, editor, *Geometric Computing with Clifford Algebras*, pages 27–60. Springer Verlag, Heidelberg, 2001.

[31] H. Li and Y. Wu. Automated short proof generation in projective geometry with Cayley and bracket algebras I. Incidence geometry. *J. of Symbolic Comput.* 36:717–762, 2003.

[32] H. Li. Symbolic computation in the homogeneous geometric model with Clifford algebra. In J. Gutierrez, editor, *Proc. ISSAC 2004*, pages 221–228. ACM Press, New York NY, 2004.

[33] H. Li. A recipe for symbolic geometric computing: Long geometric product, BREEFS and Clifford factorization. In C. W. Brown, editor, *Proc. ISSAC 2007*, pages 261–268. ACM Press, New York, NY, 2007.

[34] H. Li. *Invariant Algebras and Geometric Reasoning*. World Scientific, Singapore 2008.

[35] H. Li and L. Huang. Complex brackets, balanced complex differences, and applications in symbolic geometric computing. In *Proc. ISSAC 2008*, pages 181–188. ACM Press, New York, NY, 2008.

[36] H. Li. Normalization of polynomials in algebraic invariants of three-dimensional orthogonal geometry. arXiv: 1302.7194v1 [cs.SC], 2013.

[37] H. Li, C. Shao, L. Huang and Y. Liu. Reduction among bracket polynomials. In *Proc. ISSAC'14*, pages 304–311. ACM Press, New York NY, 2014.

[38] L. Liberti, C. Lavor, N. Maculan and A. Mucherino. Euclidean distance geometry and applications. *SIAM Rev.*, 56:3–69, 2014.

[39] K. Menger. Untersuchungen über allgemeine Metrik. *Math. Annal.*, 100:75–163, 1928.

[40] K. Menger. New foundation of Euclidean geometry. *Amer. J. Math.*, 53:721–745, 1931.

[41] D. S. Mitrinović, J. E. Pečarić and V. Volenec. *Recent Advances in Geometric Inequalities.* Springer Verlag (originally Kluwer Academic), Dordrecht, NL, 1989.

[42] B. Mourrain and N. Stolfi. Computational symbolic geometry. In: N. L. White, editor, *Invariant Methods in Discrete and Computational Geometry*, pages 107–139. D. Reidel, Dordrecht, NL, 1995.

[43] A. Mucherino, C. Lavor, L. Liberti and N. Maculan, editors. *Distance Geometry – Theory, Methods and Applications.* Springer Verlag, New York, NY, 2013.

[44] E. Müller. Die Kugelgeometrie nach den Principien der Grassmann'schen Ausdehnungslehre (Teil I). *Monatshefte Math. Phys.* **3**: 365–402, 1892.

[45] E. Müller. Die Kugelgeometrie nach den Principien der Grassmann'schen Ausdehnungslehre (Teil II). *Monatshefte Math. Phys.* **4**: 1–52, 1893.

[46] I. Najfeld and T. F. Havel. Embedding with a rigid substructure. *J. Math. Chem.*, 21:223–260, 1997.

[47] P. J. Olver. *Classical Invariant Theory*. Cambridge University Press, Cambridge, 1999.

[48] J. H. Prestegard, C. M. Bougault and A. I. Kishore. Residual dipolar couplings in structure determination of biomolecules. *Chem. Rev.*, 104:3519–3540, 2004.

[49] M. Riesz. *Clifford Numbers and Spinors*, 1958. From lecture notes made in 1957-8, edited by E. Bolinder and P. Lounesto. Kluwer Academic, Dordrecht NL, 1993.

[50] B. Rosenhahn and G. Sommer. Pose estimation in conformal geometric algebra I, II. *J. Mathematical Imaging and Vision* 22:27–48, 22:49–70, 2005.

[51] Aleix Rull, Josep M. Porta and Federico Thomas. Distance bound smoothing under orientation constraints. *IEEE Intnl. Conf. Robotics & Automation*, pages 1431–1436, 2015.

[52] Tamar Schlick. *Molecular Modeling and Simulation*, 2nd ed., Springer Media, New York, NY, 2010.

[53] I. J. Schoenberg. On certain metric spaces arising from euclidean spaces by a change of metric and their imbeddings in Hilbert space. *Annals Math.*, 38:787–793, 1937.

[54] J. J. Seidel. Distance-geometric development of two-dimensional euclidean, hyperbolic and spherical geometry. *Simon Stevin*, 29:32–50, 65–76, 1952.

[55] E. Snapper and R. J. Troyer. *Metric Affine Geometry*. Academic Press, 1971. Reprinted by Dover Publications, 1989.

[56] G. Sommer. A geometric algebra approach to some problems of robot vision. In J. Byrnes, editor, NATO Science Series 136, *Computational Non-commutative Algebra and Applications*, pages 309–338. Kluwer Acad., Dordrech,t NL, 2004.

[57] H. Takashima. High-resolution protein structure determination by NMR. *Ann. Rep. NMR Spect.*, 59:235–273, 2006.

[58] Walter Thiel and Genard Hummer. Nobel 2013 chemistry: Methods for computational chemistry. *Nature*, 504:96–97, 2013.

[59] G. Wagner, S. Hyberts, and T. F. Havel. NMR structure determination in solution: A critique and comparison with X-ray crystallography. *Ann. Rev. Biophys. Biomol. Struct.*, 21:167–198, 1992.

[60] N. White. Geometric applications of the Grassmann-Cayley algebra. In J. E. Goodman and J. O'Rourke, editors, *Handbook of Discrete and Computational Geometry*. CRC Press, Chapman & Hall, Boca Raton, FL, 1997.

[61] W. T. Wu. *Basic Principles of Mechanical Theorem Proving in Geometries I: Part of Elementary Geometries.* Science Press, Beijing, 1984; Springer Verlag, Vienna AT, 1994.

[62] K. Wüthrich. NMR structures of biological macromolecules. *Magn. Reson. Chem.*, 41:S89–S98, 2003.

[63] H. Weyl. *The Classical Groups*. Princeton Univ. Press, Princeton, NJ, 1939.

# Chapter 6

## *Tree-Decomposable and Underconstrained Geometric Constraint Problems*

**Ioannis Fudos**

*Department of Computer Science and Engineering, University of Ioannina, Ioannina, Greece*

**Christoph M. Hoffmann**

*Department of Computer Science, Purdue University, West Lafayette, IN*

**Robert Joan-Arinyo**

*Department of Computer Science, Universitat Politècnica de Catalunya, Barcelona, Catalonia*

### CONTENTS

## 6.1 Introduction, Concepts, and Scope

**Geometric constraint system (GCS):** Also known as a *geometric constraint problem*, is a pair $P = (U,F)$, where $U$ is a set of geometric objects (or geometric elements) on which we impose a set $F$ of geometric constraints.

**Constraint graph:** Given a GCS $P = (U,F)$, its *Constraint Graph* $G = (V,E)$ is a labeled undirected graph whose vertices $V$ are the geometric objects in $U$, each labeled with its degrees of freedom. An edge $e \in E$ represents a binary constraint between two geometric objects and is labeled by the number of equations formulating the geometric constraint.

**Deficit:** Deficit $\kappa$ of a constraint graph is defined as the sum of the labels of its vertices minus the sum of the labels of its edges. The deficit determines the degrees of freedom of the compound geometric object defined by the corresponding GCS.

**Quasi generically overconstrained:** A GCS problem $P$ is *quasi generically over-constrained* if there is an induced subgraph of the associated constraint graph, such that for the induced subgraph the following holds: $\sum l(v) - \sum l(e) < \kappa$ and none of the constraints fixes the geometric structure with respect to the global coordinate system, where $\kappa = 3$ in the plane and $\kappa = 6$ in 3-space.

**Quasi generically underconstrained:** A GCS problem $P$ in the Euclidean plane/space is quasi generically under-constrained if it is not overconstrained and $\sum l(v) - \sum l(e) > \kappa$.

**Instance solvability:** A GCS is *well-constrained* if the associated (algebraic) equation system has one or more discrete real solutions. It is *over-constrained* if it has no solution, and is *under-constrained* if it has a continuum of solutions.

A geometric constraint problem (we refer the reader to Chapter 1), also known as a geometric constraint system, consists of a finite set of geometric objects, such as points, lines, circles, planes, spheres, etc., and constraints upon them, such as incidence, distance, tangency, and so on. A solution of a geometric constraint problem $P$ is a coordinate assignment for each of the geometric objects of $P$ that places them in relation to each other such that all constraints of $P$ are satisfied. A problem $P$ may have a unique solution, it may have more than one solution, or it may have no solution.

In this chapter, we are concerned with geometric constraint solvers, i.e., with programs that find one or more solutions of a geometric constraint problem. If no solution exists, the solver is expected to announce that no solution has been found. Owing to the complexity, type or difficulty of a constraint problem, it is possible that the solver does not find a solution even though one may exist. Thus, there may be false negatives, but there should never be false positives. Intuitively, the ability to find solutions can be considered a measure of solver's competence.

We consider static constraint problems and their solvers. We do not consider dynamic constraint solvers, also known as dynamic geometry programs, in which specific geometric elements are moved, interactively or along prescribed trajectories, while continually maintaining all stipulated constraints. However, if we have a solver for static constraint problems that is sufficiently fast and competent, we can build a dynamic geometry program from it by solving the static problem for a sufficiently dense sampling of the trajectory of the moving element(s).

The work we survey has its roots in applications, especially in mechanical computer-aided design (MCAD). The constraint solvers used in MCAD took a quantum leap with the work by Owen [46]. Owen's algorithm solves a geometric constraint problem by an initial, graph-based structural analysis that extracts generic subproblems and determines how they would combine to form a complete solution. These subproblems are then handed to an algebraic solver that solves the specific instances of the generic subproblems and combines them. Owen's graph analysis is top down. A bottom-up analysis was proposed in [3]. Subsequent work expanded the knowledge of

(a) The structure and properties of the constraint graph, see Section 6.3;

(b) The geometric vocabulary, see Section 6.4;

(c) The understanding of spatial constraint systems, see Section 6.5.

We also look briefly at under-constrained problems, see Section 6.6.

Restricted to points and distances, the constraint graph analysis has deep roots in mathematics and combinatorics; see, e.g., [17, 38, 43] for some of these connections. Here, we limit the discussion to constraint problems that are motivated by MCAD applications, and that means that the geometric vocabulary has to be richer than points and distances between them. In the remainder of this section we introduce informally major concepts and methods for solving geometric constraint systems (GCS) with an application perspective in mind.

Note in particular that some definitions in this section may be *provisional*. Those definitions, although formally incorrect, are so given nevertheless because they facilitate understanding the material. They are clearly identified along with examples that show where they fall short — and what can be done about it.

### 6.1.1 Geometric Constraint Systems (GCS)

Fix the space in which to consider a geometric constraint system (GCS). We again refer the reader to Chapter 1. Typical examples are the Euclidean space in 2 or 3 dimensions. Each object to be placed by a GCS instance in that space has a specific number of degrees of freedom (dof), i.e., a specific number of independent coordinates. In Euclidean 2-space, points and lines each have 2 dof. In Euclidean 3-space, a point has 3 dof and a line has 4. A constraint on such objects corresponds to one or more equations expressing the constraints on the coordinates. So, requiring a distance $d$ between two points $A = (A_x, A_y)$ and $B = (B_x, B_y)$ in 2-space would be expressed by

$$(A_x - B_x)^2 + (A_y - B_y)^2 = d^2$$

Requiring that the two points are coincident would entail two equations:

$$\begin{aligned} A_x &= B_x \\ A_y &= B_y \end{aligned}$$

Accordingly, a GCS can be viewed simply as a system of equations: A solution of the GCS, if one exists, is a valuation of the variables, the coordinates, that satisfies all equations. Viewed in this

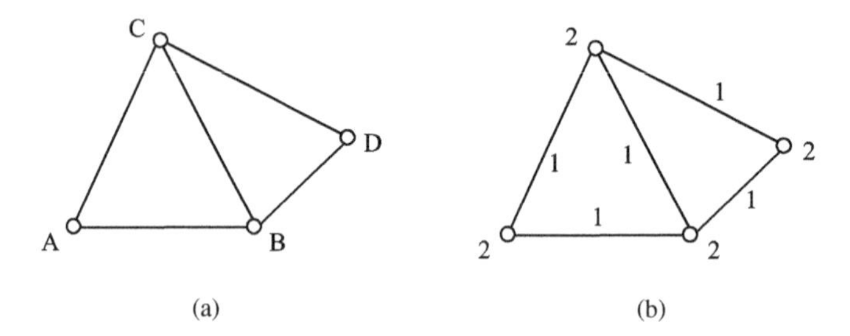


**Figure 6.1**
Constraint problem and associated constraint graph.

foundational way, solving a GCS boils down to formulating a system of equations in the coordinates of the geometric entities and solving the system by any means appropriate. The equations are almost always algebraic.

### 6.1.2 Constraint Graph, Deficit, and Generic Solvability

The approach of treating a GCS as an (unstructured) system of equations is inefficient and almost always unnecessary. Given a GCS, we analyze the structure of the corresponding equation system, and seek to identify smaller subsystems that can be solved independently and that admit especially efficient solution algorithms. Triangle decomposable systems, the main subject of this chapter, do this by analyzing a *constraint graph* that mirrors the equation structure. The constraint graph is recursively broken down into subgraphs that correspond to independently solvable subsystems. Once subsystems have been solved, their solutions are combined and expose, in the aggregate, additional subsystems that now can be solved separately. Given the GCS problem $P = (U, F)$, with the set of geometric objects $U$ and constraints $F$ upon them, we define the constraint graph $G(P)$ as follows.We again refer the reader to Chapter 1.

**Definition 6.1** Given the GCS $P = (U, F)$, its *constraint graph* $G = (V, E)$ is a labeled undirected graph whose vertices are the geometric objects in $U$, each labeled with its degrees of freedom. There is an edge $(u, v)$ in $E$ if there is a constraint between the geometric objects $u'$ and $v'$, corresponding to $u$ and $v$, respectively. The edge is labeled by the number of independent equations corresponding to the constraint between $u'$ and $v'$.

**Example 11** *Consider the constraint problem of Figure 6.1 that comprises four points in the plane, labeled A through D. A line between two points indicates a distance constraint on the points. Thus there are 5 distance constraints, shown left. The corresponding constraint graph is shown on the right.*

Since a GCS naturally corresponds to a system of equations, we expect that the number of independent equations should equal the number of variables. The number of independent variables equals the sum of degrees of freedom, i.e., the sum of vertex labels of the constraint graph. Moreover, the number of independent equations equals the sum of the edge labels, in most situations. Thus, we investigate ho the structure of the constraint graph reflects the structure of the equation system that corresponds to the GCS.

As stated, the GCS of Example 11 does not prescribe where on the plane to position and orient the points after solving the GCS. Thus, while this GCS is quasi well-constrained, the sum of vertex labels equals the sum of edge labels plus 3. Here, the *deficit* of 3 corresponds to 3 missing equations

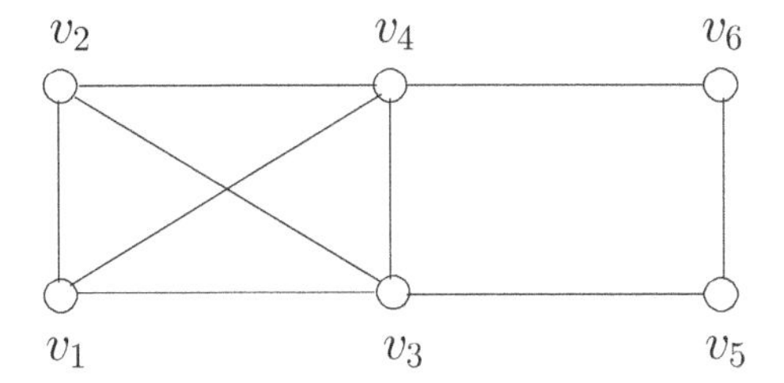


**Figure 6.2**
A GCS with an overconstrained and an underconstrained subgraph.

that fix where, in the plane, to place the solution. If necessary, the remaining degrees of freedom can be determined by placing the points with respect to a global coordinate system, for example by adding three equations that place $A$ at the origin and $B$ on the (positive) $x$-axis.

So, if the position and orientation of the solution is undetermined, we are assured that a constraint graph where $\sum l(v) - \sum l(e) < 3$ in the plane corresponds to an equation system in which at least one equation is not independent. Consequently, we conclude

**Definition 6.2**
A GCS problem $P$ is quasi generically over-constrained if there is an induced subgraph of the associated constraint graph, such that for the induced subgraph the following holds: $\sum l(v) - \sum l(e) < \kappa$ and none of the constraints fixes the geometric structure with respect to the global coordinate system, where $\kappa = 3$ in the plane and $\kappa = 6$ in 3-space.

Extrapolating this line of reasoning, we might be led to the conclusion that we can use this structural graph property to define

**Definition 6.3**
A GCS problem $P$ in the Euclidean plane/space is generically quasi under-constrained if it is not quasi overconstrained and $\sum l(v) - \sum l(e) > \kappa$.

**Definition 6.4 (Provisional)**
A GCS problem $P$ in the Euclidean plane/space is quasi generically well-constrained if: (i) it is not quasi over-constrained and (ii) the sum of vertex labels of the associated constraint graph equals the sum of the edge labels plus the deficit $\kappa$.

Note, however, for a graph to be well-constrained Definition 6.4 is not sufficient and will be refined in Section 6.3.7. The problem with Definition 6.4 is best illustrated by an example. The following example exhibits a constraint graph for which the sum of the labels of the vertices equals the sum of the labels of the edges plus $\kappa$ but it contains an overconstrained subgraph and therefore the GCS cannot be well-constrained.

**Example 12** *Consider the constraint graph of Figure 6.2. For specificity, let the graph vertices be points and the edges distances. The deficit is 3, so it seems that a problem with this constraint graph is quasi well-constrained. Now the subgraph induced by vertices $v_1, ..., v_4$ is overconstrained; drop any of the subgraph edges and you obtain a well-constrained subproblem with a deficit 3. On the other hand, the subgraph induced by the vertices $v_3 ... v_6$ has a deficit of 4 and is in fact underconstrained; vertices $v_5$ and $v_6$ cannot be placed. There is an extra constraint in the first subproblem that is not* available *for the second subproblem.*

Simple dof counting fails to identify all well-constrained graphs in 2D. Likewise, a GCS in 3D with deficit 6 that is not overconstrained does not always have a well-constrained graph. Chapter 1 partially resolves the abovementioned issues using a notion of minimal dof rigidity, which captures generic well constrained 2D distance constrained systems, as we will see in Section 6.3.7. Nevertheless, as pointed out in Chapter 1, in general, even this does not capture generically well constrained GCS.

Note that the definitions and properties explained assume in particular that the GCS solution does not have certain symmetries. For instance, consider placing concentrically two circles of given radii. The constraint graph has two vertices, labeled 2 each, and one edge labeled 2, because concentricity implies that the two centers coincide. Here the deficit $\sum l(v) - \sum l(e)$ is 2, less than $\kappa = 3$. The system appears to be quasi over-constrained, but the constraint system is well-constrained. The lower deficit reflects the rotational symmetry of the solution.

In the following, we exclude GCS that fix the structure with respect to the coordinate system, as well as GCS that exhibit symmetries that reduce $\kappa$.

### 6.1.3 Instance Solvability

Consider constructing a triangle $\triangle(A,B,C)$ with the three sides $a,b,c$ through each of the vertex pairs, in the Euclidean plane. The three points and three lines together have 12 degrees of freedom. Each vertex is incident to two lines, so there are six incidence constraints, each contributing one equation. We add three angle constraints, one for each line pair. The resulting constraint graph has a deficit of three and thus appears to be quasi generically well-constrained. However, the problem is not well-constrained: if the three angle constraints add up to $\pi$, then the GCS is actually under-constrained. If they add up to $\pi/4$, say, the GCS has no solution.

The problem arises from the interdependence of the three angle constraints. Thus generic solvability does not guarantee that problem instances are solvable. In particular, the constraint graph does not record specific dependencies among the equations. Such dependencies often arise from geometry theorems.

**Definition 6.5** A GCS is well-constrained if the associated (algebraic) equation system has one or more discrete real solutions. It is over-constrained if it has no solution, and is under-constrained if it has a continuum of solutions.

The definition of generic solvability is fair in the sense that dependencies among some of the constraints, and their equational form, can arise from geometry theorems. The example above is based on a simple theorem, but more complex theorems can arise and may be as hard to detect as solving the equations in the first place.

### 6.1.4 Root Identification and Valid Parameter Ranges

Consider Example 11. With the distance constraints as drawn, the GCS has multiple solutions, some shown in Figure 6.3. From the equational perspective none is distinguished. From an application perspective usually one is intended. Since the number of distinct solutions can grow exponentially with the number of constrained geometric objects, it is not reasonable to determine all solutions and let the application choose. This application-specific problem is the *root identification problem.* Some authors use the term *chirality*.

Solvability generalizes to the determination of *valid parameter ranges*: It is clear that the distances between the points $A,B,C$, and the distances between the points $B,C,D$ must satisfy the triangle inequality for there to be a (nondegenerate) solution. Considering the metric constraints (distance, angle) as coordinates of points in a configuration space, what is the manifold of points

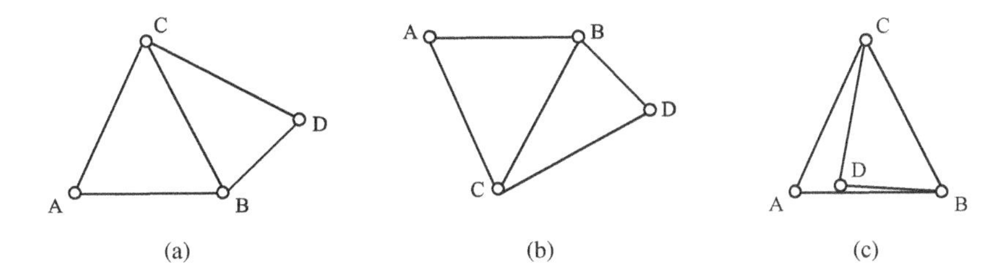


**Figure 6.3**
Different solutions of the constraint problem of Example 11.

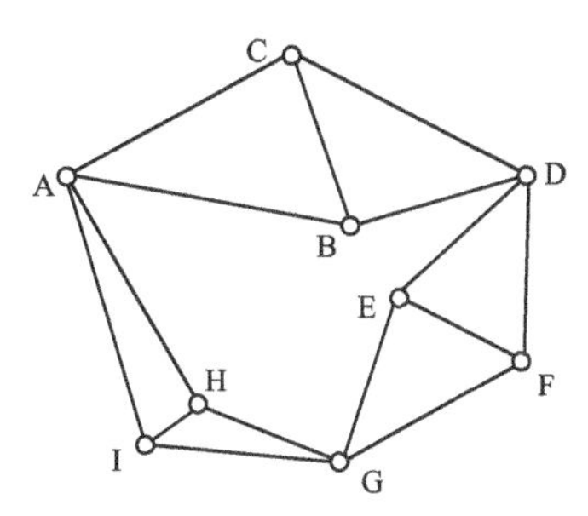


**Figure 6.4**
A variational constraint problem.

that are associated with solvable GCS instances? Is the manifold connected or are there solutions that cannot be reached from every starting configuration?

### 6.1.5 Variational and Serializable Constraint Problems

Some GCS have equation systems that are naturally triangular. That is, intuitively, the geometric objects can be ordered such that they can be placed one-by-one. Such GCS are *serializable or sequential constraint problems*. GCS that are not serializable have been called *variational constraint problems* in some of the application-oriented literature. The examples discussed so far are all sequential. The following example shows a variational problem.

**Example 13** *Place 9 points in the plane subject to the 15 distance constraints indicated by the lines, in Figure 6.4. The following three groups of 4 points each can be placed with respect to each other:* $(A,B,C,D)$, $(D,E,F,G)$, *and* $(G,H,I,A)$. *Each of these subproblems is sequential in nature. However, the overall placement problem of all points is not serializable; it is variational.*

Serializable constraint problems are of limited expressiveness. Due to their potential efficiency, however, they are used to great effect in some dynamic geometry packages such as Cinderella [6] and GeoGebra [15].

### 6.1.6 Triangle-Decomposing Solvers

The overall strategy of triangle-decomposing solvers is to construct the constraint graph and, analyzing the graph, to recursively isolate solvable subproblems. Then, the solved subproblems are recombined. The process is recursive.

For planar constraint systems, *triangle-decomposability* is understood to be a recursive property in which the base case consists of two vertices (2 dof each) and one edge between them eliminating one of the four dof's. Larger structures are then solved by isolating three solved structures that pairwise share a single geometric element and thus can be combined by placing the solved structures relative to each other. When this decomposition is done top-down, the solved structures are triconnected; [46]. When the decomposition is done bottom-up, triples of solved components are sought that pairwise share a geometric primitive; [3]. The act of combining such triples has been called a *cluster merge*.

For the problem of Example 13, illustrated in Figure 6.4, the graph is first decomposed into the three subproblems discussed. Next, each of these subproblems is further broken down. For the subproblem $A,B,C,D$, for example, this might be done by placing, with respect to each other, $A$ and $B$, $B$ and $C$, and $C$ and $A$. Those three subsystems are then combined into the triangle $A,B,C$. This merging step combines three subsystems that pairwise share a point, placing the three shared geometric elements in one geometric construct. The triangle can then be merged with the two subsystems $C$ and $D$, and $B$ and $D$. The result so obtained is the solved subsystem with vertices $A,B,C,D$. Similarly, we solve subproblems $D,E,F,G$ and $G,H,I,A$, each in two merge steps. Finally, the entire problem is solved by merging the three subsystem.

The recursion solving the problem of Figure 6.4 can be thought of either as a top-down decomposition, or as a bottom-up reconstruction.

Note that the top-level merge step places the shared objects, here the points $A$, $D$ and $G$. The distances between these points are not given but they can be obtained from the three subproblem solutions. We can think of this process as a plan that is formulated based on the constraint graph analysis. Two questions arise:

(a) Since the plan so formulated is not unique, do different decompositions arrive at different solutions? This question is settled by investigating the nature of the recursive decomposition and whether it satisfies a Church-Rosser property.

(b) What specific subsystems of equations must be solved and how? This question is approached by analyzing the basic subgraph configurations that can occur and are allowed by the geometric constraint solver, both when decomposing and when recombining.

Concerning the second question, in our example two key operations are needed: placing two points at a prescribed distance, and placing a third point at a required distance from two fixed points. There is also a third operation that places a rigid geometric configuration $X$ such that two points of $X$ match two given points, using translation and rotation. This operation is used when recombining the three subproblems.

Carrying out these operations entails solving equation systems of a fixed structure. Often, these equation systems are small and can be solved very efficiently. The simplest cases restrict the geometric repertoire to points, lines, and circles of fixed radius. Triangle decomposable systems then require at most solving univariate quadratic polynomials.

More complicated primitives can be added. They include circles of a-priori unknown radius, e.g., [37]; conic sections, e.g., [9]; and Bézier curves, e.g., [20]. Additions impact both the constraint graph analysis as well as the richness of algebraic equation systems that have to be solved. The impact on the graph analysis can be limited to adding more patterns to cluster merging. So extended, triangle decomposition becomes a more general analysis that can be called *tree decomposition*. The decomposition remains a tree, but the tree may have different numbers of subtrees associated with the growing repertoire of subgraph patterns. The number of possible patterns is infinite. Therefore, this approach becomes self-limiting. More than that, rigidity no longer has a simple characterization. The contributions to the equation solving algorithms is also considerable. Section 6.4.3 catalogs what is known about the equation systems required to add just circles of unknown radius.

Because of these rigors, more general techniques have to be considered. For the extended graph analysis, the DR-type algorithms solve the problem by dynamically identifying generically solvable subgraphs; [22, 23]. Likewise, the growth of irreducible algebraic equation systems increasingly motivates searching general equation solvers, including numerical solving techniques, such as Newton iteration, homotopy continuation, and other procedural techniques, for instance [50].

### 6.1.7 Scope and Organization

We begin with the graph analysis of triangle-decomposable constraint systems in 2D. These systems play an important role in linkage analysis and graph rigidity, [53, 54, 55]. But even in the Euclidean plane, applications of geometric constraint systems argue for more expressive systems. We can increase expressiveness by enlarging the repertoire of geometric objects, as well as by admitting more complex cluster merging operations.

In the plane, an extended repertoire includes foremost variable-radius circles; that is, circles whose radius is not given explicitly but must be inferred based on the constraints. More advanced objects, such as conics and certain parametric curves, have also been considered. Those additional geometries impact the subsystems of equations that must be solved.

More complex cluster merging operations affect both the constraint graph structures encountered as well as the equation systems that must be solved to position the geometric structures accordingly. We will discuss some examples.

After discussing constraint graphs and quasi generic over-, well-, and under-constrained graphs, we consider the equations that must be solved so as to obtain specific solutions. Here, we explain the structure of those systems as well as some tools to transform the equation systems into simpler ones. Questions of root identification, valid parameter ranges, and order type of solutions are to be discussed. Much is known about these topics in the Euclidean plane. For Euclidean 3-space much less is known. There the equation systems are in general much harder, and the number of cases that should be considered is larger. Even simple sequential problems can require daunting equation systems. A case in point is to find a line in 3-space that is at prescribed distances from four given points, discussed in Section 6.5.

## 6.2 Geometric Constraint Systems

**Solution of a GCS:** A *solution* of a geometric constraint system $(U,F)$ is an assignment of coordinates instantiating the elements of $U$ such that the constraints $F$ are all satisfied.

**Definition 6.6** A *geometric constraint system* (GCS) is a pair $(U,F)$ consisting of a finite set $U$ of geometric elements in an ambient space and a finite set $F$ of constraints upon them.

The ambient space typically is $n$-dimensional Euclidean space. The majority of applications require $n = 2$ or $n = 3$; e.g., [3, 31, 46]. Spherical geometry may also be considered, for instance in nautical applications.

The imposed constraints typically are binary relations. We do not consider higher-order constraints, such as "$C$ be the midpoint between points $A$ and $B$." Note, however, that such constraints can often be expressed by several binary constraints. This can be done in a variety of ways, with or without variable-radius circles.

**Definition 6.7** A *solution* of a geometric constraint system $(U,F)$ is an assignment of coordinates instantiating the elements of $U$ such that the constraints $F$ are all satisfied.

There may be several solutions [10]. Moreover, solutions may or may not be required to be in prescribed position and orientation, in a global coordinate system.

As defined, a GCS is a static problem in that solutions fix the geometric elements with respect to each other. The *dynamic geometry* problem asks to maintain constraints as some elements move with respect to each other. We consider only static constraint problems and their solvers.

By *geometric coverage* we understand the diversity of geometric elements admissible in $U$. Points, lines and circles of given radius are adequate for many applications in Euclidean 2-space [46]. For GCS in Euclidean 3-space, an analogous geometric coverage could be points, lines, planes, as well as spheres and cylinders of fixed radii. Here, the number of solutions even of simple GCS can be very large [13].

## 6.3 Constraint Graph

**Rigid body:** A *rigid body* is a set of geometric objects whose relative placement is fixed. A rigid body in 2D has in general 3 degrees of freedom while in 3D a rigid body has in general 6 degrees of freedom.

**Rigidity:** A GCS is *rigid* or has rigidity in 2D or 3D if the resulting set of geometric elements after placement with respect to the constraints is rigid. In graph theory a graph is rigid if the structure formed by replacing the edges by rigid rods and the vertices by flexible hinges is rigid.(links to infinitesimal rigidity and structural regidity).

**Serializable GCS:** A GCS is *serializable* if the geometric objects can be ordered in such a way that they can be placed sequentially one-by-one as function of preceding, already placed elements.

**Variational GCS:** A solvable GCS that is not serializable is *variational.*

**Triangle decomposable GCS:** A GCS is *triangle decomposable* if it can be solved by using as construction operations the placement of two or three geometric elements with pairwise constraints.

**Degrees of freedom:** Degrees of freedom (dof) are the number of independent, one-dimensional variables by which a geometric object can be instantiated and positioned. Geometric constraints consume one or more dof expressed by a number of independent equations.

**Compound (complex) geometric objects (elements):** A cluster of geometric objects constitutes a *compound geometric object (element).* The definition of a compound geometric object may also include a set of geometric constraints imposed on the geometric objects. Therefore a rigid body is a compound geometric object with deficit 3 in 2D or 6 in 3D.

**Quadratically solvable GCS:** A *quadratically solvable GCS* is a system of equations whose zeros can be obtained successively solving a sequence of quadratic equations where the coefficients in an equation may depend on the solution of preceding equations in the sequence.

**Laman graph and Laman's Theorem:** Laman's theorem and Laman graph provide a sufficient and necessary condition for characterizing a well constrained (rigid) GCS in 2D that consists of points and distances. A constraint graph that satisfies this condition is called *Laman graph.*

**Table 6.1**
Minimal GCS in the Euclidean plane; $d(...)$ denotes distance, $a(...)$ denotes angle.

| Geoms | Constraint | Notes |
|---|---|---|
| Points $A, B$ | $d(A,B)$ | Distance not zero |
| Point $A$, line $m$ | $d(m,A)$ | Zero distance allowed |
| Lines $m, n$ | $a(m,n)$ | Lines not parallel |

**Table 6.2**
Minimal GCS in Euclidean 3-space; $d(...)$ denotes distance, $a(...)$ denotes angle.

| Geoms | Constraint | Notes |
|---|---|---|
| Points $p_1, p_2, p_3$ | $d(p_i, p_k)$ | No zero distance |
| Point $p$, line $L$ | $d(p,L)$ | Distance not zero |
| Lines $L_1, L_2$ | $d(L_1,L_2), a(L_1,L_2)$ | Lines not parallel |
| Planes $E_1, E_2, E_3$ | $a(P_i, P_k)$ | No parallel planes |

The constraint graph of Definition 6.1 is an abstract representation of the equation system equivalent to the geometric constraint problem. The analysis of the graph yields a set of operations used to solve the equations. Ideally, those operations are simple, for instance univariate polynomials of degree 2 [3]. To start the graph analysis, we find *minimal constraint problems*. That is, constraint problems with a minimal number of geometric objects whose solution defines a local coordinate frame. Such problems depend on ambient space. Consider the following:

**Definition 6.8** Given a constraint problem in the Euclidean plane, consisting of two points $A$ and $B$ and a nonzero distance constraint between them. Such a problem is *minimal*. The associated constraint graph $G = (\{A,B\}, \{(A,B)\})$ is a *minimal constraint graph*.

This minimal constraint problem establishes a coordinate system of the Euclidean plane in which $A$ is the origin and the oriented line $\overrightarrow{AB}$ is the x-axis. This is not the only minimal constraint problem in the plane. Table 6.1 shows the minimal problems involving the basic geometric objects with 2 degrees of freedom. Note that fixed-radius circles can be used in lieu of points as long as the centers and points are not coincident. Two parallel lines, at prescribed distance in the plane, are not considered minimal because they do not establish a coordinate frame.

Table 6.2 shows the main cases for Euclidean 3-space. In the case of two lines that are skew, a third line $L_3$ is constructed that connects the two points of closest approach. Here, $L_1$ and $L_3$ define a plane that is oriented by $L_2$. If the lines $L_1$ and $L_2$ intersect, they lie on a common plane that is coordinatized by the two lines and is oriented by defining the third coordinate direction using a right-hand rule. If the two lines are parallel, they define a common plane but fail to coordinatize it.

### 6.3.1 Geometric Elements and Degrees of Freedom

A geometric element has a characteristic number of degrees of freedom that depends on the type of geometry and the dimensionality of ambient space.

**Definition 6.9** Degrees of freedom are the number of independent, one-dimensional variables by which a geometric object can be instantiated and positioned.

**Table 6.3**
Degrees of freedom for elementary objects and rigid objects.

| Geom | Geometric Meaning of dof | 2D | 3D |
|---|---|---|---|
| Point | Variables representing coordinates | 2 | 3 |
| Line | 2D: distance from origin and direction<br>3D: distance from origin, direction in 3D direction on the plane | 2 | 4 |
| Plane | Distance from origin, direction in 3D | | 3 |
| Circle, fixed-radius | Coordinates of center, orientation in 3D | 2 | 5 |
| Circle, variable-radius | Coordinates of center, radius, orientation in 3D | 3 | 6 |
| Rigid body | 2D: 2 displacements, 1 orientation<br>3D: 3 displacements, 3 orientation | 3 | 6 |
| Sphere, fixed- or variable-radius | 3 displacements or<br>3 displacements, radius | | 3/4 |
| Ellipse, variable axes | Center, axes lengths, axes orientation | 5 | 7 |
| Ellipsoid, variable axes | Center, axes lengths, axes orientation | | 9 |

Elementary objects are geometric objects that have a certain number of degrees of freedom and cannot be decomposed further into more elementary objects. Compound objects can be characterized by a GCS, and consist of one or several elementary objects placed in relation to each other according to a GCS solution instance. A complex object is rigid if its shape cannot change or, equivalently, if the relative position and orientation of the elementary objects that comprise the complex object cannot change.

The degrees of freedom for elementary geometric objects and for rigid objects in two and three dimensions are summarized in Table 6.3. Singularities of coordinatization can trigger robustness issues in GCS. Therefore, the representation of elementary geometric objects should be uniform, without singularities. For example, if we represent lines by the familiar $y = mx + b$ formula, lines parallel to the $y$-axis cannot be so represented. This problem can be avoided if we represent a line by its distance from the origin and the direction of the normal vector of the line.

### 6.3.2 Geometric Constraints

Geometric constraints consume one or more degrees of freedom and are expressed by an equal number of independent equations. A basic set of geometric constraints in the plane is summarized in Table 6.4. Constraints can be expressed directly. Alternatively, some can be reduced equivalently to other constraints. For example, suppose we stipulate that a circle of radius $r$ be tangent to a line $L$ in the plane. Then we can require equivalently that the center of the circle be at distance $r$ from $L$. For a comprehensive list of constraints involving planes, points and lines in 3D see for example [40, 57]. Finally we should note that distance 0 between two points takes out two degrees of freedom since this causes dimension reduction (from 1-dimensional to 0-dimensional). Similarly, requiring that two lines be parallel at distance $d$ in 3-space eliminates 3 dof, but if $d = 0$ the constraint eliminates 4 dof.

### 6.3.3 Compound Geometric Elements

Several geometric elements can be conceptually grouped into compound elements. A circular arc is an example. Compound geometric elements are convenient concepts for graphical user interfaces of GCS solvers. Internally, they are decomposed into geometric elements and implied constraints. We illustrate this dual view with the example of a circular arc shown in Figure 6.5 and explain why it has overall two intrinsic parameters.

Since the arc does not have a fixed radius, it lies on a variable-radius circle that has 3 dof. The two end points contribute an additional 4 dof. The two end points are incident to the circle, reducing the overall dof to 5. Position and orientation of the arc, in some coordinate system, requires 3 dof, leaving two intrinsic arc parameters. They can be interpreted as 1 dof for the radius of the arc, and 1 dof for the distance between the end points.

Note that this constraint problem has four solutions in general. Orienting the end points, and connecting them with an oriented line, or circle, the arc can be on one of two sides of the directed line. Moreover, the two end points divide the circle into a shorter and a longer arc, in general. Which one is chosen accounts for the other two solutions. Applications require selecting one of these solutions. Several conventions can be followed to determine a unique segment: preservation of the original configuration, preservation of the direction of the curve from the first endpoint towards the second endpoint, the shortest segment, and so on.

### 6.3.4 Serializable Graphs

When a GCS is serializable (Section 6.1.5) geometric objects can be ordered in such a way that they can be placed sequentially one-by-one as function of preceeding, already placed elements. This idea can be formalized as follows.

**Definition 6.10** Let $G = (V,E)$ be a GCS graph and $x_0, x_1, x_2 \ldots$ be elements in $V$. We say that $x_i$ depends on $x_k, x_r, \ldots$, written $x_i > x_k, x_r, \ldots$, if $x_i$ can be placed only after the $x_k, x_r, \ldots$ have been placed.

**Table 6.4**
Types of constraints and number of dof eliminated.

| Type | Constraint | 2D | 3D |
|---|---|---|---|
| point-point distance | One equation representing the distance between two points $p_1, p_2$ under a metric $\|\| \ \|\|$: $\|\|p_1 - p_2\|\| = d$ with $d > 0$. | 1 | 1 |
| Angle between lines and planes | Angle between two lines (can be represented by the angle between the normal vectors). **Exceptions in 3D:** Parallelism between lines eliminates 2 dof. Line-plane orthogonality in 3D eliminates 2 dof. | 1 | 1 |
| Point on point | For any metric $\|\|p_1 - p_2\|\| = 0$ is equivalent to $p_1 = p_2$. | 2 | 3 |
| Point-line distance | One equation expressing the point-line distance. | 1 | 1 |
| Line-line parallel distance | Parallelism and distance. | 2 | 3 |
| Plane-plane parallel distance | Parallelism and distance. | 2 | 3 |
| Point on line | Same as point-line distance In 3D an additional dof is canceled. | 1 | 2 |
| Line on line | Same as parallel distance between lines in 2D. In 3D an additional dof is canceled. | 2 | 4 |
| Point on plane | Same as point-plane distance. | - | 1 |
| Line on plane | Plane-line parallelism and zero distance. | - | 2 |
| Fixing elementary object | Fixing all or some of the dof of an elementary object. | dof | dof |

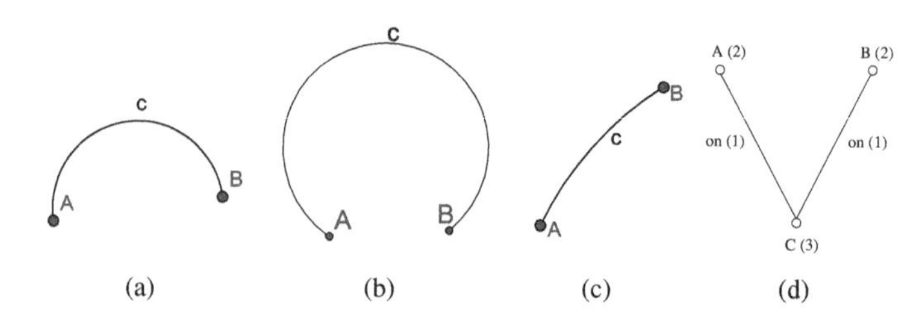


**Figure 6.5**
Building a complex object using one variable radius circle, two end points, and two incidence constraints.

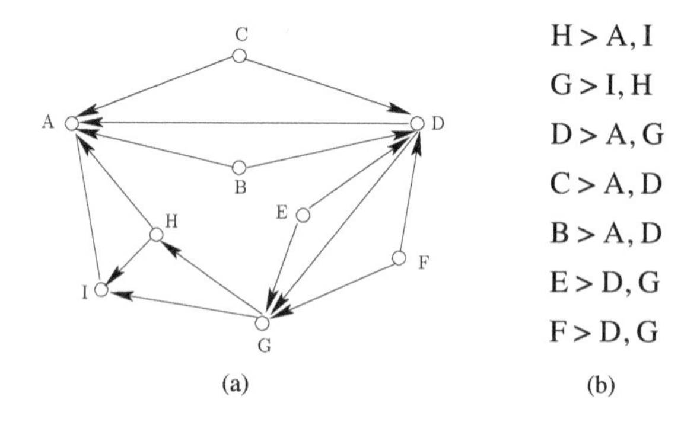


**Figure 6.6**
(a) DAG derived from a serializable graph, with $(A,I)$ as starting pair. (b) A construction sequence respecting the dependencies.

If the constraint graph is serializable, then the pair $(G,>)$ is a directed acyclic graph (DAG) and admits topological sorting [1]. See Example 13. More formally,

**Definition 6.11** Let $G(V,E)$ be the constraint graph associated with a GCS in Euclidean 2-space. Without loss of generality assume that $G$ is connected. We say that the GCS is serializable if $(G,>)$ describes a sequence of dependencies such that, under a suitable enumeration of $V$,

(a) There are elements $x_0$ and $x_1 \in V$ that induce a minimal constraint graph.

(b) Each subsequent element $x_i$, $2 \leq i \leq |V|$ depends on elements $x_j$ and $x_k$ where $j < i$ and $k < i$.

In general, the enumeration is not unique and depends on the pair $x_0, x_1 \in V$ that is placed first. However, as we will see later, different possible sequences derived from a given DAG are equivalent in the sense that they lead to the same final placement for all the objects in $V$ with respect to each other; see Section 6.3.7.

**Example 14** *Consider the graph in Figure 6.6(a). Edges have been directed to show dependencies of placement. Choosing $(A,I)$ as starting pair, a valid dependence relation is obtained. The list in Figure 6.6(b) gives a serial construction based on the graph $(G,>)$.*

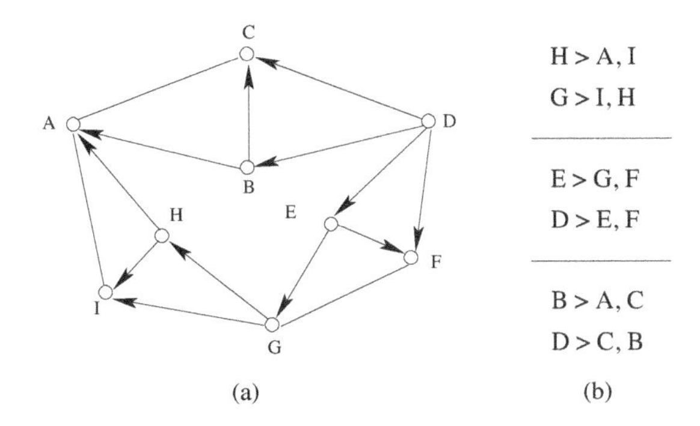


**Figure 6.7**
(a) DAG derived from the variational graph in Figure 6.4 with starting pairs $(A,I)$, $(G,F)$ and $(C,B)$.
(b) Three different subsequences of construction dependencies that can be identified in the DAG.

### 6.3.5 Variational Graphs

**Definition 6.12** A GCS which is not serializable is called *variational.*

When the GCS is serializable, starting with a minimal GCS and applying the dependence relationship to the constraint graph $G=(V,E)$ generates a sequence of dependencies that includes only a subset of elements in $V$. We call it a *subsequence* and the corresponding subgraph a *cluster.*

Assuming that the variational GCS is solvable, one repeatedly selects a minimal GCS and applies the dependence relationship, using graph edges not yet used, resulting in a collection of subsequences.

Intuitively, the situation described means that clusters corresponding to different sequences must be merged, usually applying translations and rotations defined by elements shared by subsequences. From an equational point of view, the existence of different subsequences reveals that there are several underlying equations that must be solved simultaneously.

**Example 15** *Figure 6.7a shows a variational DAG corresponding to the variational constraint problem in Figure 6.4. We choose three starting pairs $(A,I)$, $(G,F)$, and $(C,B)$. Each allows us to build a DAG from some of the graph vertices and edges. They are listed in Figure 6.7(b). Notice that each subsequence identifies a serializable subgraph.*

### 6.3.6 Triangle Decomposability

The strategy of triangle-decomposing solvers, as sketched in Section 6.1, is based on decomposing the constraint graph recursively. Decomposition splits a (sub)graph into three (sub)subgraphs that share one vertex pairwise. This is called a *triangle decomposition step.* More complex splitting configurations can be considered and called *tree decomposition.* Figure 6.8 shows a triangle and a more complex tree decomposition step.

From a practical point of view, in the case of GCS in Euclidean 2-space, triangle-decomposable constraint graphs suffice for solving many problems that arise in applications.

We define a *triangle decomposition step* as follows.

**Definition 6.13** Let $G=(V,E)$ be a graph. We say that three subgraphs of $G$, $G_i(V_i,E_i), 1 \le i \le 3$, define a triangle decomposition step of $G$ if

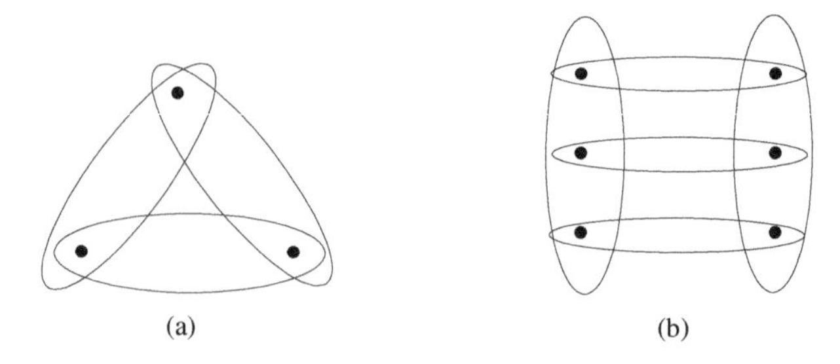


**Figure 6.8**
(a) Triangle decomposition step. (b) More complex tree decomposition step. Split subgraphs are shown as ovals, shared vertices as dots.

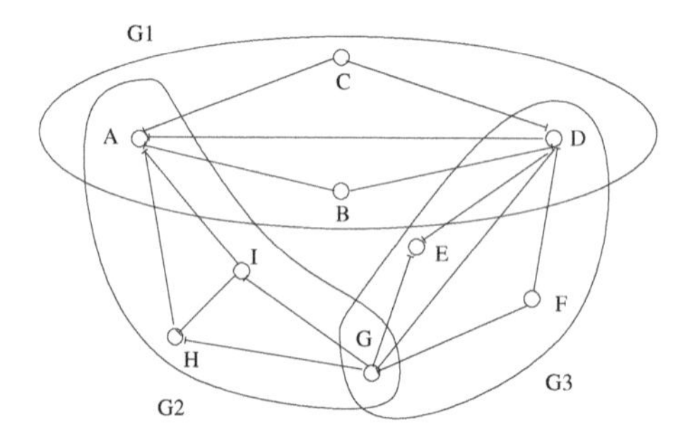


**Figure 6.9**
A triangle-decomposition step for the graph shown in Figure 6.7(a).

(a) $V_1 \cup V_2 \cup V_3 = V$, $E_1 \cup E_2 \cup E_3 = E$ and $E_i \cap E_j = \emptyset, i \neq j$, and

(b) There are three vertices, say $u, v, w \in V$, such that $V_1 \cap V_2 = \{u\}$, $V_2 \cap V_3 = \{v\}$ and $V_3 \cap V_1 = \{w\}$.

**Example 16** *Consider the graph $G$ in Figure 6.7(a). As shown in Figure 6.9, the subgraphs $G_1, G_2$ and $G_3$ define a triangle decomposition step of $G$. Vertices pairwise shared by the subgraphs are $A, D$ and $G$.*

**Definition 6.14** We say that a ternary tree $T$ is a *triangle decomposition* for the graph $G$ if

(a) The root of $T$ is the graph $G$.

(b) Each node of $T$ is a subgraph $G' \subset G$ which is either the root of a ternary tree generated by a triangle decomposition step of $G'$ or a leaf node with a minimal associated subgraph.

**Definition 6.15** A graph for which there is a triangle decomposition is called *triangle-decomposable.*

In general, a triangle decomposition of a graph is not unique. However, if the graph is triangle decomposable by one sequence of decomposition steps, then any legal sequence will decompose the graph [10].

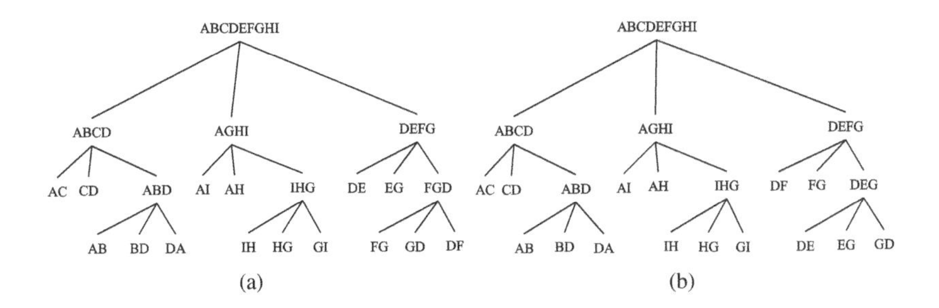


**Figure 6.10**
Two different triangle decompositions for the graph shown in Figure 6.7.

**Example 17** *Consider the graph and the triangle decomposition step shown in Figure 6.9. Now recursively apply decomposition steps to each of the subgraphs $G_1, G_2, G_3$ until reaching minimal subgraphs. Figure 6.10 shows two triangle decompositions for the graph considered that differ in the subtree rooted at node DEFG. Notice, however, that the set of terminal nodes is the same in both triangle decompositions.*

Assume that the three subgraphs $G_1, G_2, G_3$ are (graphs of) solvable GCS subproblems. Let $u$ and $v$ be the shared vertices of $G_2$ and there is no constraint between them. Then $u$ and $v$ constitute a virtual minimal GCS where the constraint relating the two elements is not given but can be deduced from the solution of $G_2$. A similar statement can be made about $G_1$ and $G_3$ and their shared vertices.

The triangle pattern is not the only decomposition construct [3]. Others include the pattern shown in Figure 6.8b. Intuitively, a decomposition pattern represents an equation system that must be solved simultaneously. Decomposition patterns are infinite in number; see also SECTION 2.2

### 6.3.7 Generic Solvability and the Church-Rosser Property

Recall Definitions 6.13, 6.15, and 6.4.

Triangle-decomposable graphs, of GCS in the Euclidean plane, are generically well-constrained, and can be solved either bottom-up [11] or top-down [36]. This assertion is based on the shared geometric elements, of each triangle decomposition step, having 2 dof and being in general position with respect to each other. Variable-radius circles have 3 dof and give rise to a special case illustrated in Figure 6.11.

Triangle decomposable graphs are also called *ternary-decomposable* on account of the topology of the decomposition tree. A (recursive) decomposition of a well-constrained triangle decomposable graph is not unique. However, it can be shown using the Church-Rosser property for reduction systems that if one triangle decomposition sequence fully reduces the graph, then all such decompositions must succeed [10]. The advantage of restricting to triangle-decomposable problems is that a fixed, finite repertoire of algebraic equation systems suffices to solve this class of problems.

Triangle decomposable graphs are a subset of the set of well-constrained constraint graphs of planar GCS. The entire set has been characterized by Laman in [38].

**Definition 6.16** Let $G = (V, E)$ be a connected, undirected graph whose vertices represent points in 2D and edges represent distances between points. $G$ is a well-constrained constraint graph of a GCS iff, the deficit of $G$ is 3 and, for every subset $U \subset V$, the induced subgraph $(U, F)$ has a deficit of no less than 3.

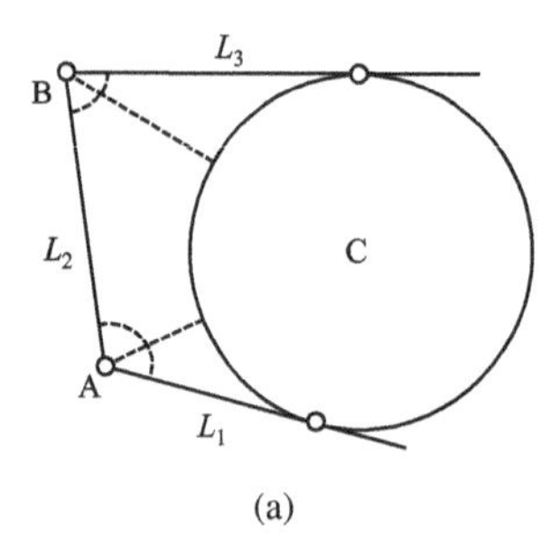


(a)

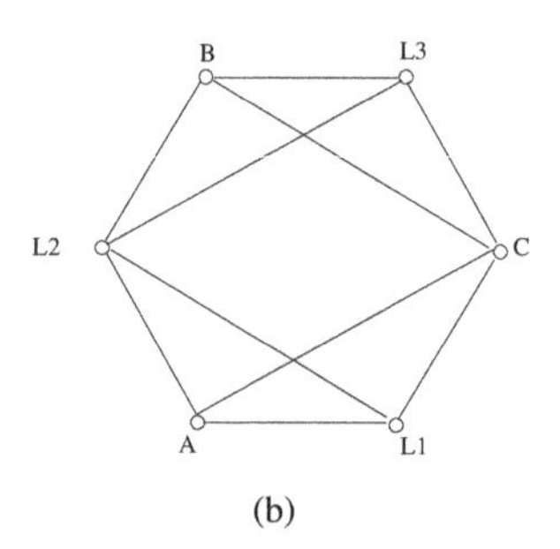


(b)

**Figure 6.11**
Well-constrained graph with three lines, two points, and a variable-radius circle. (a) Constraint problem. Dashed lines represent metric constraints. (b) Constraint graph. All vertices have 2 dof except C which has three. Solid lines represent incidence constraints.

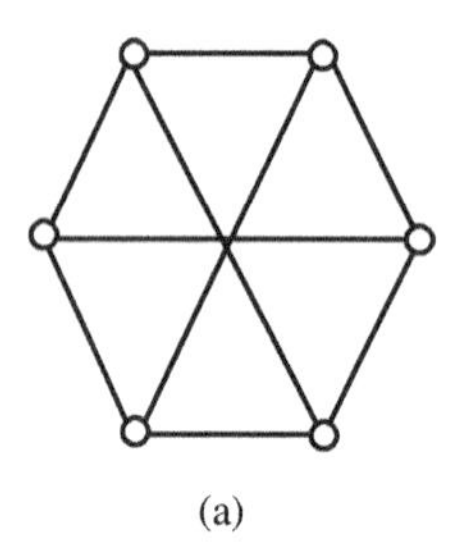

(a)

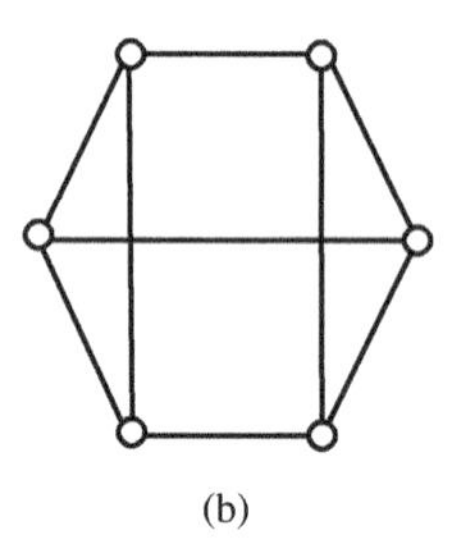

(b)

**Figure 6.12**
Well-constrained graphs that are not triangle-decomposable. (a) Graph $K_{3,3}$. (b) Desargue's graph. All vertices have 2 dof, all edges consume 1 dof.

Two examples of well-constrained graphs that are not triangle-decomposable are shown in Figure 6.12. In triangle decomposition, the irreducible constituents of the constraint graph are the minimal constraint graphs, Definition 6.8. For the general case, the set of irreducible constraint graphs has been characterized in [21] using a network flow approach. Conceptually, irreducible constraint graphs must be solved as a single equation system. Since the graphs can be arbitrarily complex, so can be the equation systems.

For the planar case, when we have only points and distances the set of triangle decomposable graphs coincides with the set of quadratically solvable graphs. However if we extend to lines and angles, there are quadratically solvable graphs which are not triangle decomposable. Consider the problem of finding a triangle from its three altitudes shown in Figure 6.13(a). The corresponding graph is shown in Figure 6.13(b) where the hexagon edges are point-on-line constraints and the diagonals are point-line distance constraints. The geometric problem is quadratically solvable, [32], but the graph, $K_{3,3}$, is not triangle decomposable.

Laman's theorem holds even if we extend the repertoire of geometries to any geometry having 2 degrees of freedom and the constraints to virtually any constraint of Table 6.4. However, if we extend the set of geometries to include for example variable radius circles, then the Laman condition is no longer sufficient.

**Example 18** *Consider the GCS of Figure 6.14. We have two rigid clusters* $C_1 = \{V_1, V_2, V_3, V_4\}$ *and* $C_2 = \{V_1, V_2', V_3', V_4'\}$*, where* $V_1$ *is a variable-radius circle,* $V_2, V_3, V_2', V_3'$ *are points, and* $V_4, V_4'$ *are lines. The constraints* $e_1, e_2, e_1', e_2'$ *are distances from the center of* $V_1$*,* $e_6, e_6'$ *are distances from the*

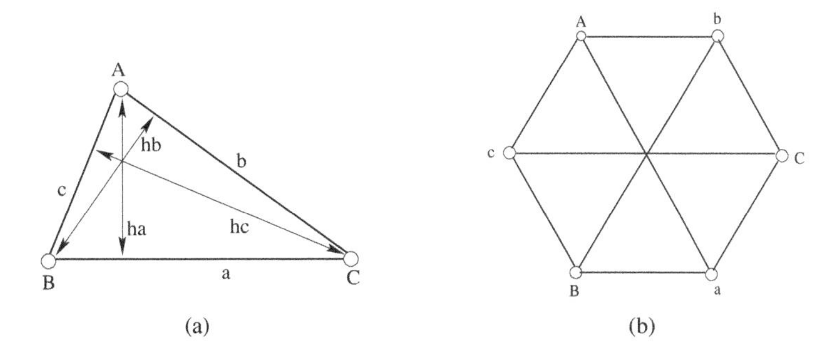


**Figure 6.13**
Deriving the geometry of a triangle given its three altitudes. (a) Formulating the three altitude problem. (b) The resulting constraint graph.

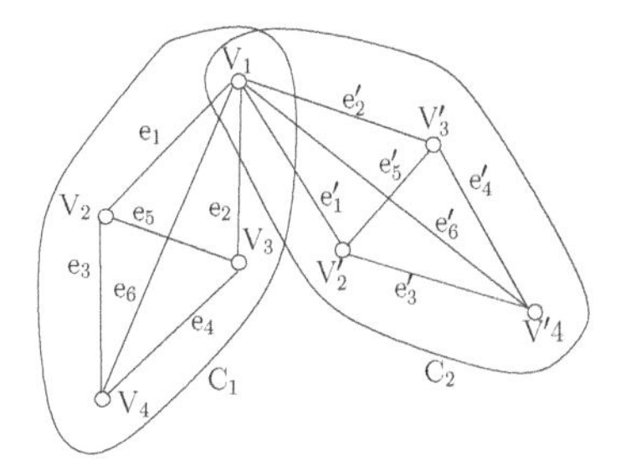


**Figure 6.14**
A Laman graph in 2D that is not rigid. $V_1$ is a variable radius circle. $V_2$,$V_3$,$V'_2$,$V'_3$ are points. $V_4$, $V'_4$ are lines. $e_1$,$e_2$,$e'_1$,$e'_2$ are points. $e_3$,$e_4$,$e'_3$,$e'_4$ are on constraints. $e_5$, $e'_5$ are distances. $e_6$, $e'_6$ are distances from cirlce ( circumference).

*circumference of $V_1$, $e_5, e'_5$ are distance constraints, and $e_3, e_4, e'_3, e'_4$ are incidence constraints. The two clusters share the variable-radius circle $V_1$.*

*The graph is clearly a Laman graph but is not rigid , since it is underconstrained — $C_1$ and $C_2$ can move independently around circle $V_1$. The problem is also overconstrained since the radius of $V_1$ can be derived independently from cluster $C_1$ and from cluster $C_2$.*

### 6.3.8 2D and 3D Graphs

Graph analysis for spatial constraint problems is not nearly as mature as the planar case. In the planar case, Definition 6.8 conceptualizes minimal GCS with which a bottom-up graph analysis would start and a top-down analysis would terminate. Table 6.1 lists the configurations. In 3-space, analogously, the corresponding minimal problem may consist of three points or three planes, with three constraints between them. However, the major configurations shown in Table 6.2 also include other cases, for example two lines.

In the plane, serially adding one additional point or line requires two constraints. In 3-space, adding a point or plane requires three. The inclusion of lines in 3-space leads to serious complications. For example, since lines have 4 dof, see Table 6.3, we have to be able to construct lines from given distances from 4 fixed points. Equivalently, we have to construct a common tangent to 4

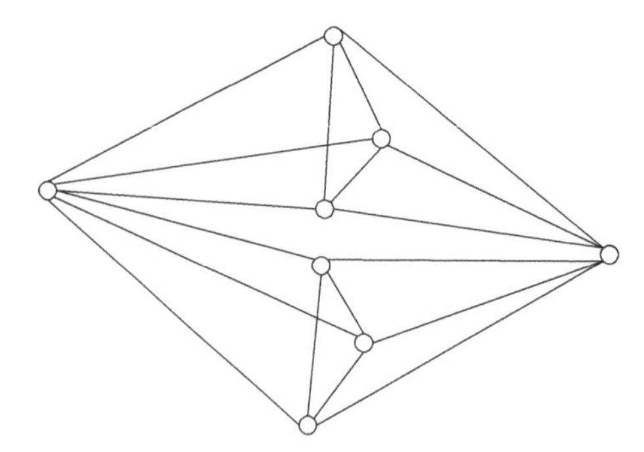

**Figure 6.15**
Two hexahedra sharing two points.

fixed spheres, a problem known to have up to 12 real solutions, [24]. Work in [63] has explored the optimization of the algebraic complexity of 3D subsystems.

In 3D, the Laman condition is not sufficient. Figure 6.15 illustrates two hexahedra sharing two vertices. If the length of the edges is given the GCS that arises is also known as the *double banana* problem. The graph is a Laman graph but the problem corresponds to two rigid bodies (each hexahedron is a rigid body) sharing two vertices and is thus non rigid in the sense that the two rigid bodies are free to rotate around the axis defined by the two shared points. The problem is also clearly overconstrained since the distance of the two shared geometries can be derived independently by each of the two rigid bodies.

Extending the theory in [56, 60], [16, 40] give a combinatorial condition that is necessary and sufficient for rigidity for extended types of 3D constraints including between point-point, point-line, plane etc. except for point-point coincidence. Note that this characterization does not capture the cases of Figures 6.14 and 6.15. The condition holds for both cases, but rigidity is not guaranteed since they involve point coincidences between rigid bodies directly or indirectly. In this approach, a multigraph $(V,(B,A))$ is formulated, where vertices $V$ represent rigid bodies and edges $(B,A)$ stand for primitive constraints that represent single equations between the two 6-vectors that describe the rigid body motion of the two vertices. Primitive constraints intuitively affect at most one degree of freedom. Each geometric constraint is translated to a number of primitive constraints (see Appendix C of [40]). A distinction is made between primitive angular and blind constraints: a primitive angular constraint may affect only a rotational degree of freedom. All other primitive constraints that may affect either a rotational or translational degree of freedom are called blind constraints.

Therefore edges are of two types: angular $(A)$ and blind $(B)$. Such a scheme is minimally rigid if and only if there is a subset $B'$ of the blind edges such that (i) $B-B'$ is an edge disjoint union of 3 spanning trees and (ii) $A \cup B'$ is an edge disjoint union of 3 spanning trees.

## 6.4 Solver

**Root Identification or Root Selection:** A well constrained GCS may have a large number of solutions. The problem of selecting the user intended solution is referred to as *root identification* or *root selection.*

After the constraint graph has been analyzed, the implied underlying equations are to be solved. We discuss now how to do that.

**Table 6.5**
Placing minimal GCS: $p$ represents points, $L$ represents lines, $d$ distance, $a$ angle.

| | |
|---|---|
| $d(p_1, p_2)$ | Set $p_1 = (0,0)$, $p_2 = (d,0)$ |
| $d(p,L)$ | Place $L$ on the $x$-axis, $p = (0,d)$ |
| $a(L_1, L_2)$ | Place $L_1$ on the $x$-axis, $L_2$ through the origin at angle $a$ |

### 6.4.1 2D Triangle-Decomposable Constraint Problems

We restrict to points and lines in the Euclidean plane. As discussed in Section 6.3.6, triangle-decomposable constraint systems in the plane require solvers that implement three operations:

(a) The two geometric elements of a minimal subgraph (Definition 6.8) are placed consistent with the constraint between them.

(b) A third geometric element is placed by two constraints on two geometric elements already placed.

(c) Given two geometric elements in fixed position, a rigid-body transformation is done that repositions the two elements elsewhere.

These operations are applied to the decomposition tree, progressing from the leaves of the tree to the root. The solving order is bottom-up regardless whether the decomposition tree was built top-down or bottom-up [3, 10, 46].

**Example 19** *Consider the constraint system of Example 11. We choose the subgraph induced by A and B as minimal and place A at the origin and B on the positive x-axis, at the stipulated distance from A, so executing Operation 1.*

*We place C by the two constraints on A and B, solving at most quadratic, univariate equations, executing Operation 2. The triangle $A,B,C$ is thereby constructed.*

*Assume that we have solved the triangle $B,C,D$ in like manner, separately and with B and D as vertices of the minimal subgraph. We can now assemble the two triangles by a rigid-body transformation that moves the triangle $B,C,D$ such that the points B and C are matched, using Operation 3.*

Note that we can extend the geometric vocabulary of points and lines, adding circles of given radius at no cost. A fixed-radius circle is replaced by its center. A point-on-circle constraint is replaced by a distance constraint between the point and the center, and a tangency constraint by a distance constraint between the tangent and the center.

#### Operation 1: Minimal GCS Placement

This operation chooses a default coordinate system. There are three pairs of geometric elements that occur in minimal GCS: (point, point), (point, line), and (line, line). The chosen placements are shown in Table 6.5. Other choices could be made.

#### Operation 2: Constructing one Element from two Constraints

Two elements $A$ and $B$ are given, a third element $C$ is to be placed by constraints upon them. There are six cases, with $p$ denoting a point and $L$ denoting a line.

$$\begin{array}{llllllll} (p,p) & \rightarrow & p & (p,L) & \rightarrow & p & (L,L) & \rightarrow & p \\ (p,p) & \rightarrow & L & (p,L) & \rightarrow & L & (L,L) & \rightarrow & L \end{array}$$

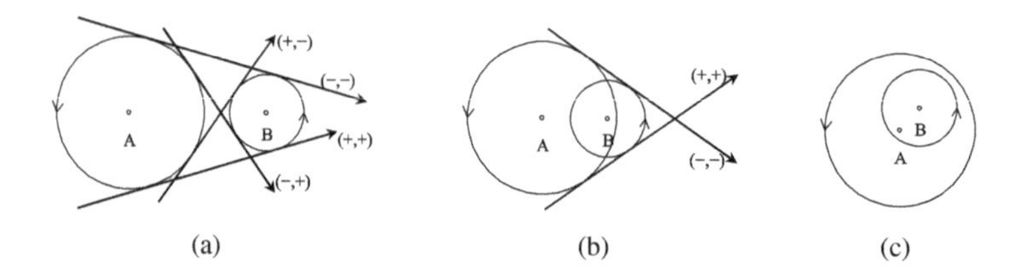


**Figure 6.16**
Constructing a line at specific distances from two points. Equivalently, finding common tangents of two circles with radius equal to the stipulated distances. The degenerate cases of 1 or 3 solutions are not shown.

The sixth case, $(L, L) \to L$ is underconstrained. See also [3]. We illustrate with the case $(p, p) \to L$.

**Example 20** *In the case* $(p, p) \to L$ *a line should be constructed by respective distance from two points. Consider the two circles with the given points as center and radius equal to the stipulated distance, as shown in Figure 6.16. Depending on the three distances, there will be 4, 2 or no real solution in general.*

*Order the points A and B, and orient the circles centered at those points counter-clockwise. Orient the line to be constructed so that the projection of A onto the line precedes the projection of B. Then we can distinguish the up to four tangents in a coordinate-independent way: observe whether the line orientation is consistent with the circle orientation (+), or whether the orientations are opposite (–). See also [3].*

*The degenerate cases where the two circles are tangent to each other yield three solutions or one. In theses cases there is one double solution that represents the coincidence of two solutions, with orientations (+,–) and (–,+), or with orientations (+,+) and (–,–).*

**Operation 3: Matching two Elements**

The operation requires that the geometric elements to be matched be congruent. The operation is a rigid-body transformation and is routine.

### 6.4.2 Root Identification and Order Type

We noted that constructing one element from two constraints can have multiple solutions. As a result, a well-constrained GCS has in general an exponential number of solution instances. We shall illustrate this with the simple construction $(p, p) \to p$.

**Example 21** *Consider placing n points, by* $2n - 3$ *distance constraints between them, and assume that the distance constraints are such that we can place the points by sequentially applying the construction* $(p, p) \to p$. *In general, each new point can be placed in two different locations: Let* $P_i$ *and* $P_j$ *be known points from which the new point Q is to be placed, at distance* $d_i$ *and* $d_j$, *respectively. Draw two circles, one about* $P_i$ *with radius* $d_i$, *the other about* $P_j$ *with radius* $d_j$ *as shown in Figure 6.17. The intersection of the two circles are the possible locations of Q. For n points, therefore, we could have up to* $2^{n-2}$ *solution instances.*

Note that not all construction paths derive real solutions. If, in Example 21, the distance between $P_i$ and $P_j$ is larger than the sum of $d_i$ and $d_j$, then there is no real solution for placing $Q$ and therefore any subsequent construction using this instance of $Q$ is not feasible. Therefore, one might argue that this pruning may result in polynomial algorithms. However, this is unlikely since the problem of

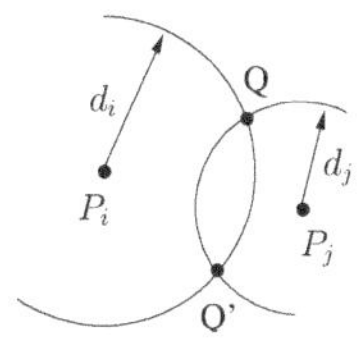


**Figure 6.17**
Placing one point $Q$ from two points $P_i, P_j$ already known.

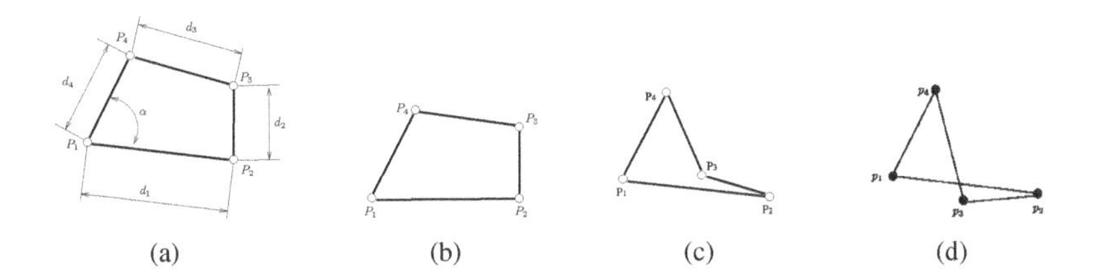


**Figure 6.18**
(a) GCS consisting of four points, four straight segments, four point-point distances and an angle. (b),(c) and (d) Three different solution instances to the GCS.

determining whether a well-constrained GCS has a real solution has been shown to be NP-complete [11].

In general, an application will require one specific solution, usually known as the *intended* solution. To identify it is not always a trivial undertaking. In [3] finding the intended solution is called the *Root Identification Problem.* Notice that, on a technical level, selecting the intended solution corresponds to selecting one among a number of different roots of a system of nonlinear algebraic equations.

A well-constrained GCS would not necessarily include enough information to identify which solution is the intended one. Consider the following example.

**Example 22** *The well-constrained GCS in Figure 6.18a consists of four points, four straight segments, four point-point distances and an angle. The solution includes four instances. Two correspond to the one shown in Figure 6.18b and to a symmetric arrangement of the same shape. Solution instances in Figures 6.18c and 6.18d are structurally different.*

Clearly, the GCS sketch in Figure 6.18a does not include any hint on which solution instance must be chosen to be displayed on the user's screen. Thus, additional information must be supplied to the solver. In [3], approaches applied to overcome this issue have been classified into five categories: Selectively moving geometric elements, adding extra constraints to narrow down the number of possible solution instances, placement heuristics, a dialogue with the constraint solver that identifies interactively the intended solution, and applying a design paradigm approach based on structuring the GCS hierarchically. Next we elaborate on each category.

### Moving Selected Geometry

In this approach, a solution is presented graphically to the user. The user selects, again graphically, certain geometric elements that are considered misplaced. The user then moves those elements where they should be placed in relation to other elements. This approach is used by the DCM solver [7, 46].

### Adding Extra Constraints

Adding a set of extra constraints to narrow down which is the intended solution instance is an intuitive and simple approach to solving the root identification problem. Extra constraints could capture domain knowledge from the application or could be just geometric — and actually over-specify the GCS. Unfortunately, both ideas result in NP-complete problems [3]. Nevertheless, extra constraints along with genetic algorithms have been applied to solve the root identification problem showing a promising potential. Authors in [61] argue that the approach is both effective and efficient in search spaces with up to $2^{100}$ solution instances. A different application of genetic algorithms to solve the root identification problem is described in [4]. The approach mixes a genetic algorithm with a chaos optimization method.

**Example 23** *Genetic algorithms described in [33, 41, 48], use extra geometric and topological constraints defined as logical predicates on oriented geometries. For example, assume that the polygon in Figure 6.18a is oriented counterclockwise. The solution shown in Figure 6.18c would be selected as the intended one by requiring the following two predicates to be fulfilled*

$$PointOnSide(P_3, \overrightarrow{P_1P_2}, left), \qquad Chirality(\overrightarrow{P_2P_3}, \overrightarrow{P_3P_4}, cw)$$

### Order-Based Heuristics

All solvers known to us derive information from the initial GCS sketch and use it to select a specific solution. This is reasonable, because one can expect that a user sketch is similar to what is intended. For instance, by observing on which side of an oriented line a specific point lies in the input sketch it is often appropriate to select solutions that preserve this sidedness. The solver described in [3] seeks to preserve the sidedness of the geometric elements in each construction step: The orientation of three points with respect to each other, of two lines and one point, and of one line and two points. The work described in [3] implements an additional heuristic for arc tangency which aims at preserving the type of tangency present in the sketch. See Figure 6.20. When the rules fail, the solver opens a dialogue to allow the user to amend the rules as the situation might require. These heuristics are also applied in the solver described in [34].

**Example 24** *Consider placing three points, $P_1$, $P_2$, and $P_3$, relative to each other. The points have been drawn in the initial sketch in the position shown in Figure 6.19a. The order defined by the points can be determined as follows. Determine where $P_3$ lies with respect to the line $\overrightarrow{P_1P_2}$. If $P_3$ is on the line, then determine whether it lies between $P_1$ and $P_2$, preceding $P_1$ or following $P_2$. The solver will preserve this orientation if possible. For this example, the solver will choose the point $P_3$ as shown in Figure 6.19b.*

### Dialog with the Solver

A useful paradigm for user-solver interaction has to be intuitive and must account for the fact that most application users will not be intimately knowledgeable about the technical working of the solver. So, we need a simple but effective communication paradigm by which the user can interact with the solver and direct it to a different solution, or even browse through a subset of solutions in case the one shown in the user's screen is not the intended one.

**Example 25** *SolBCN, a ruler-and-compass solver described in [34] offers a simple user-solver interaction tool. Figures 6.21a and b respectively show the GCS sketch input by the user and the solution instance selected by the heuristic rules implemented. Then, the user can trigger the solution selector by clicking on a button and a list with the set of construction steps where a quadratic*

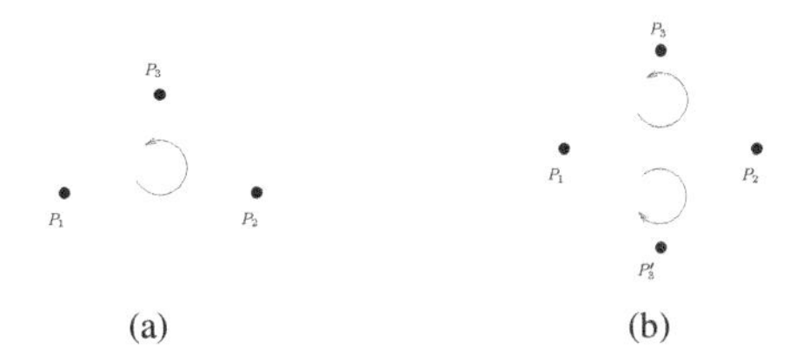


**Figure 6.19**
Placing three points, $P_1$, $P_2$, and $P_3$, relative to each other. (a) Points placed in the initial sketch and induced orientation. (b) $P_3$ and $P_3'$ are two possible placements for the third point. Preserving the orientation defined in the sketch leads to select $P_3$ as the intended placement.

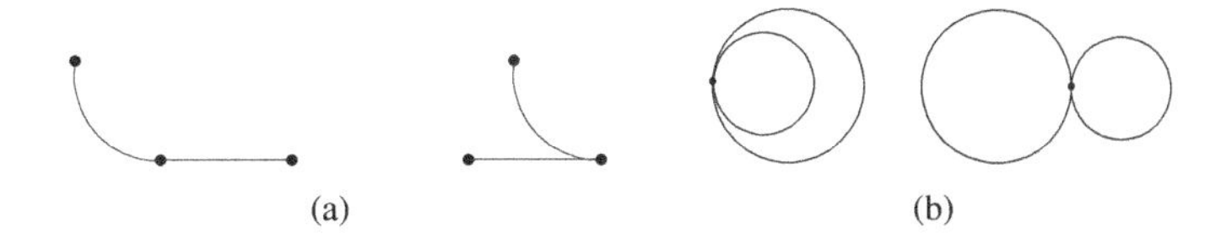


**Figure 6.20**
Tangency types. (a) Arc and segment tangency. (b) Circle-circle tangency.

*equation must be solved is displayed, Figure 6.21c. The user can then change the square root sign for each of these construction steps by either selecting it directly or navigating with the next/previous pair of buttons. Figure 6.21d shows a solution different from the first one so obtained.*

Navigating the GCS solution space using the approach illustrated in the Example 25 is simple. But it has obvious drawbacks. On the one hand, the number of items in the list selector grows exponentially with the number of quadratic construction steps in the GCS. On the other hand it is difficult to anticipate how choosing a root sign for a construction step will affect the solution selected by the next sign chosen by the user.

These problems are avoided by considering that, conceptually, all possible solution instances of a GCS can be arranged in a tree whose leaves are the different instances, and whose internal nodes correspond to stages in the placement of individual geometric elements. The different branches from a particular node are the different choices for placing the geometric element. Since the tree has depth proportional to the number of elements in the GCS, stepping from one solution instance to another is proportional to the tree depth only. Moreover, it is possible to define an incremental approach by allowing the user to select at each construction step which tree branch should be used. In solvers based on the DR-planning paradigm [28], this tree naturally is the construction plan generated by the solver.

**Example 26** *The Equation and Solution Manager [51] features a scalable method to navigate the solution space of GCS. The method incrementally assembles the desired solution to the GCS and avoids combinatorial explosion, by offering the user a visual walk-through of the solution instances to recursively constructed subsystems and by permitting the user to make gradual, adaptive solution choices. Figure 6.22 illustrates the approach.*

### Design Paradigm Approach

One of the difficulties in selecting the intended solution of a GCS stems from the fact that geometric elements in a problem sketch are not grouped into logical structures. Authors in [3] argue that hierarchically structuring the constraint problem would alleviate the complexity of solving the root

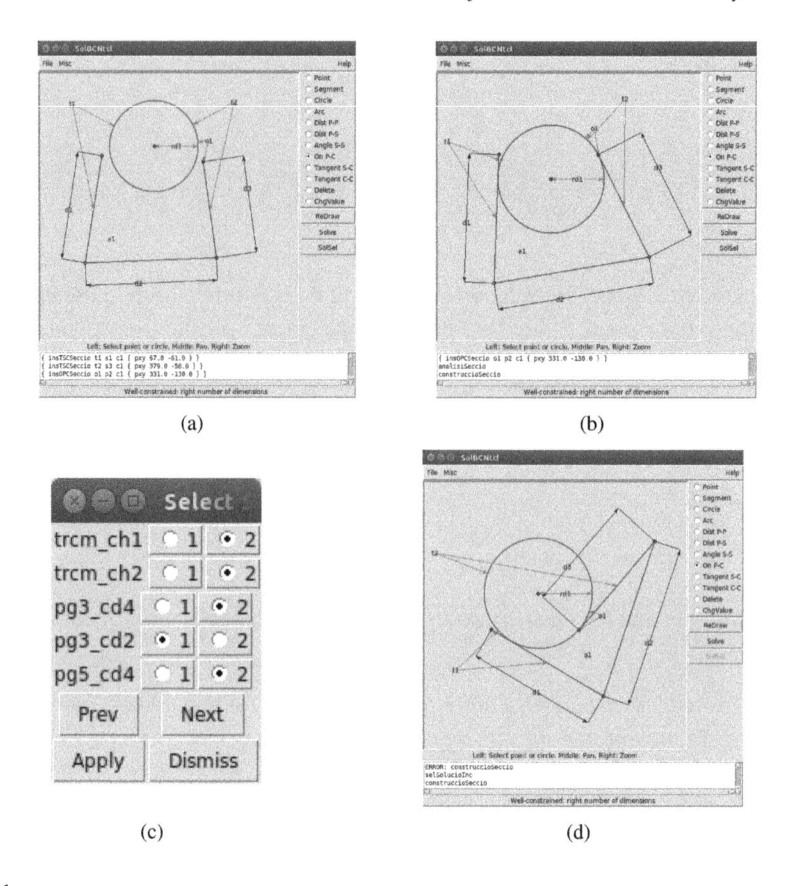


**Figure 6.21**
User-solver dialog offered by the ruler-and-compass solver described in [34]. (a) GCS sketch. (b) Solution instance selected by the heuristics implemented in the solver. (c) Solution instance selector. (d) Solution instance displayed after changing the square root signs of some construction steps.

identification problem, for example grouping geometric elements as design features. First a basic, dimension-driven sketch would be given. Then, subsequent dimension-driven steps would modify the basic sketch and add complexity. By doing so, the design intent would become more evident and some of the technical problems would be simplified.

**Example 27** *Consider solving the GCS in Figure 6.23a. The role of the arc is clearly to round the adjacent segments, and thus it is most likely that the solution shown in Figure 6.23b is the one the user meant rather than the one in Figure 6.23c, when changing the angles to* $30°$. *However, the solver would be unaware of the intended meaning of the arc. Instead, the user could sketch first the quadrilateral without the arc, and then add the arc to round a vertex. When changing some of the dimensional constraints, the role of the arc would remain that of a round, so preserving the user intent.*

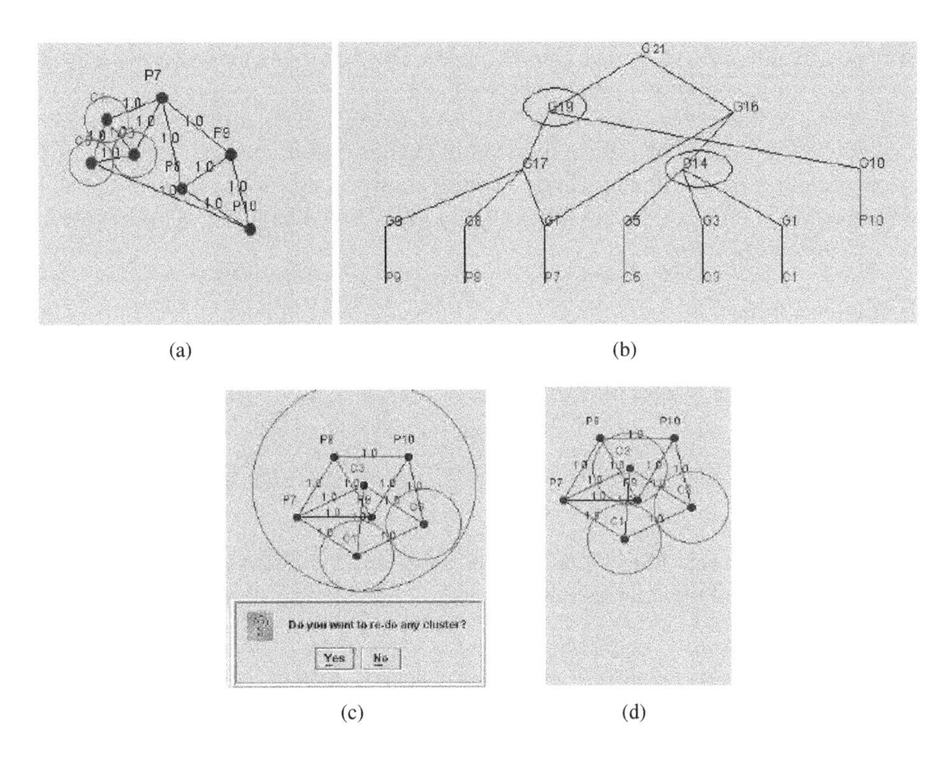


**Figure 6.22**
Incremental solution space navigation described in [51]. a) GCS problem including three circles. b) Construction plan graph for the GCS solution. c) GCS solution instance after choosing one of the possible solutions for each construction step. d) A different GCS solution instance after rebuilding the partial construction corresponding to the construction step labeled G14 in the tree.

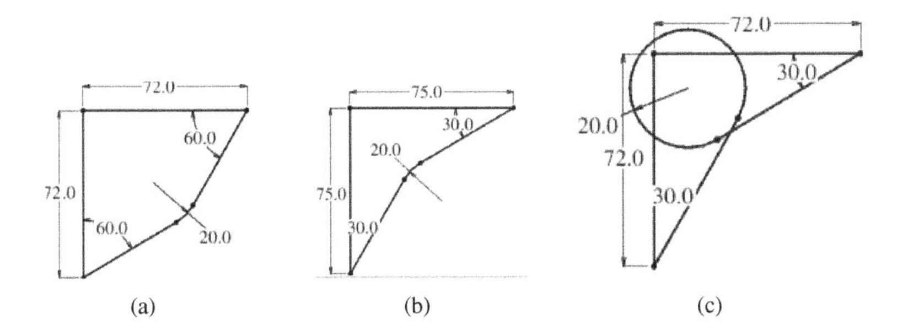


**Figure 6.23**
Solution selection by the design paradigm approach: Panel (a) shows the final GCS, panels (b) and (c) two different solution instances. If the arc is introduced as a rounding feature of a constrained quadrilateral, then selecting solution (b) over solution (c) is a logical choice.

### 6.4.3 Extended Geometric Vocabulary

So far we discussed 2D constraint solvers that use only points and lines, as well as circles of given radii. Now we will add other geometric element types. This has implications for both the equation solvers as well as for the constraint graph analysis.

The simplest addition allows GCS to include geometric elements of the new type, but only if these elements can be constructed sequentially, from explicit constraints on a set of already placed geometric elements. A more difficult addition would allow the use of elements of the added type in the same way as the core vocabulary of points and lines. Note that a new element type can have more than two degrees of freedom.

#### Variable-Radius Circles

Circles whose radii have not been given explicitly are arguably the most basic extension of the 2D core solver. Variable-radius circles have three degrees of freedom. We consider two ways in which they can arise:

(a) A variable-radius circle is to be constructed by a sequential step from three, already placed geometric entities.

(b) A variable-radius circle is to be determined in a step analogous to merging three clusters (Definition 6.13) and the circle acts as a cluster.

Note that a variable-radius circle cannot be a shared element in the sense of Definition 6.13. Shared elements have already been constructed; therefore a shared variable-radius circle has already become a fixed-radius circle.

In the following, we consider points to be circles of zero radius. Consequently, there are four ways in which a variable-radius circle can be constructed sequentially. Table 6.6 summarizes the equation systems.

It is advantageous to convert a line-distance constraint into two separate tangency constraints, so simplifying the equations that must be solved. For example, if the sought circle $C$ is to be at distance $d$ from line $L$, we work instead with two problems, one in which the circle $C$ is to be tangent to a parallel line $L_+$ of $L$. Here $L_+$ is at distance $d$ from $L$, on one side of $L$. In the other problem, $C$ is to be tangent to a parallel line $L_-$ also at distance $d$, but on the other side of $L$. Analogously, perimeter distance from a given circle reduces to tangency with a circle whose radius has been increased or reduced by said distance.

When deriving the algebraic equations, we work with *cyclographic maps*, orienting both lines and circles. Briefly, an oriented circle $C$ is mapped to a *normal cone* $\mu(C)$, with an axis parallel to the $z$-axis and a half angle of $\pi/4$. The cone intersects the $xy$-plane in $C$. Depending on the orientation, the apex of the cone is above or below the $xy$-plane. Considered as zero-radius circle, the point $P$ in the $xy$-plane maps to the normal cone $\mu(P)$ with apex in the $xy$-plane. Oriented lines $L$ in the $xy$-plane are mapped to planes through $L$ at an angle of $\pi/4$ with the $xy$-plane. This reformulates the constraint problem as a spatial intersection problem of planes and cones. See [5, 19, 25, 26, 47] for details and further reading.

The intersection of two normal cones is a conic, in affine space, plus a shared circle at infinity. The conic lies in a plane whose equation is readily obtained by subtracting the two cone equations: if $K_1 = 0$ and $K_2 = 0$ are the two normal cones, then $K_1 - K_2 = 0$ is the plane that contains the conic in which the two cones intersect.

**Example 28** *Consider the sequential construction problem of finding a circle that is tangent to three given circles. This is the classical Apollonius problem that has eight solutions in general.*

*We orient the circles and require that the sought circle be oriented consistently with the given circles at the points of tangency. After orienting the given circles, we can map the problem to the*

**Table 6.6**
Sequential construction of variable-radius circles. All constraints are tangent constraints. Equations formulated using cyclographic maps. For the cases *LCC* and *CCC* the linear equation(s) are from intersecting two cones.

| Given Elements | Equation System | Notes |
|---|---|---|
| *LLL* | (1,1) | Intersect two angle bisectors |
| *LLC* | (1,2) | Intersect two planes and a cone |
| *LCC* | (1,1,2) | Intersect two planes and a cone |
| *CCC* | (1,1,2) | Apollonius problem; Example 28 |

*intersection, in 3-space, of three normal cones, $C_1$, $C_2$, and $C_3$, each arising from an oriented circle. Intersect two cone pairs, say $C_1 \cap C_2$ and $C_1 \cap C_3$, obtaining two planes that, in turn, intersect in a line L in 3-space. Then intersect L with one of the cones, say $C_2$. Two points are obtained that, understood as the apex of a normal cone, map each to one (oriented) circle in the plane that is a solution; see [49]. Algebraically, the solution is obtained by solving linear equations plus one univariate quadratic equation.*

*There are 8 ways to orient the three circles, but they correspond pairwise, so only four such problems must be solved. If one or more circles are points, they must be considered oriented both ways. So, for each zero-radius circle, the number of solutions reduces by a factor of 2. The special cases of the Apollonius problem have been mapped out and solved in [49] using cyclographic maps.*

Now consider the determination of variable-radius circles in a cluster merge. Here, there are two constraints from each cluster to the variable-radius circle to be constructed, and the two clusters share a geometric element. The situation is analogous to the triangle merge characterized in Definition 6.13. The various cases and how to solve them have been studied and solved in [26, 25, 5]. Specifically, [26, 25] map out the cases in which the constraints are on the perimeter of the variable-radius circle; and [5] considers constraints on the center of the variable-radius circle as well.

Tables 6.7 and 6.8 summarize the results from those papers. The approach is conceptually as follows. Let $S_1 = \{E_0, E_1, E_2\}$ and $S_2 = \{E_0, E_3, E_4\}$ be the clusters constraining the variable-radius circle. The two clusters share element $E_0$, either a line denoted $L$, or a circle denoted $C$. The clusters can move relative to each other, translating along the shared line if $E_0 = L$, or rotating about the center of the shared circle if $E_0 = C$. Elements $E_1$ and $E_2$ belong to $S_1$ and constrain the sought circle. Likewise, $E_3$ and $E_4$ are the constraining elements of $S_2$. Proceed as follows:

(a) Fix the cluster $S_1$ that has the more difficult constraining elements $E_1$ and $E_2$; i.e., the cluster with the larger number of circles.

(b) Choose a convenient coordinate system: the shared line $E_0 = L$ as the $x$-axis, or the origin as the center of the shared circle if $E_0 = C$.

(c) Construct the cyclographic map of all constraining elements. The cones and planes of $S_2$ are parameterized by the distance $d$ between $S_1$ and $S_2$, or else by the angle $\theta$ between $S_1$ and $S_2$.

(d) Construct three planes from the constraining elements $E_1, \ldots, E_4$. They are either cyclographic maps of lines, or normal cone intersections. The elements of $S_2$ give rise to parameterized coefficients, by distance $d$ for translation along the $x$-axis, or by angle $\theta$ for rotation around the origin, of the moving cluster $S_2$. Intersect the planes, so obtaining a point with parameterized coordinates.

(e) Substitute the parameterized point into the equation of the element $E_1$ of the fixed cluster, so

**Table 6.7**
Cluster cases; all constraints on circle perimeter.

| Constraint type | Planes | Polynomial degree |
|---|---|---|
| $E(LL,LL)$ | $[L_2]$, $[L_3]^t$, $[L_4]^t$ | (1,2) |
| $E(CL,LL)$ | $[L_2]$, $[L_3]^t$, $[L_4]^t$ | (2,4) |
| $E(CL,CL)$ | $[L_2]$, $[L_4]^t$, $([C_1],[C_3]^t)$ | (4,4) |
| $E(CC,LL)$ | $([C_1],[C_2])$, $[L_3]^t$, $[L_4]^t$ | (2,4) |
| $E(CC,CL)$ | $([C_1],[C_2])$<br>$([C_1],[C_3]^t)$<br>$[L_4]$ | (4,4) |
| $E(CC,CC)$ | $([C_1],[C_2])$<br>$([C_1],[C_3]^t)$<br>$([C_3],[C_4])^t$ | (4,4) |

*Note*: $E(...)$ denotes whether clusters share a line $L$, the translational case, or share a circle $C$, rotational case. $[X]$ denotes the cyclographic map equation $\mu(X)$ of $X$; $[X]^t$ denotes the equation with coefficients parameterized by distance $t$ (translation case) or by angle $\theta$ (rotation case). $(X,Y)$ denotes the intersection plane equation of $X$ and $Y$. The parameterized point is substituted into the equation $[C_1]$, except for the first case where it is substituted into $[L_1]$. $(m,n)$ denotes the equation degrees, namely $m$ for the translation case $E=L$, and $n$ for the rotation case $E=C$.

obtaining a univariate polynomial that finds the intersection point(s) of the four cyclographic objects; a polynomial in $d$ or $\theta$.

(f) Solve the polynomial as described, each obtained by a particular configuration of orientations.

Some of the constraints can be on the center $c$ of the variable-radius circle, and [5] considers those cases. Note that there can be at most two constraints on the center of the variable-radius circle, for otherwise the relative position of $S_2$ to $S_1$ would be determined and the role of the variable-radius circle would be curtailed.

The problem is again solved in the same conceptual manner, but with a twist. When a constraint is placed on the center $c$ of the variable-radius circle, the constraint can be expressed by extending cyclographic maps with a map $\tau(X)$. Here, $\tau(L)$ is a vertical plane through the line $L$. Moreover, $\tau(C)$ is a cylinder through the circle $C$ with axis parallel to the $z$-axis. The results so obtain are summarized in Table 6.8.

A problem is denoted $E_0(E_1E_2,E_3E_4)$. $E_0$ is the shared element by the two clusters, a line $L$ or a circle $C$. $E_1$ and $E_2$ are the two elements of the fixed cluster constraining the variable-radius circle. $E_3$ and $E_4$ are the two elements of the moving cluster constraining the variable-radius circle. The numbers (m,n) are the equation degrees when $E_0 = L$ (m), and $E_0 = C$ (n). An element $E_k'$ constrains the center, an element $E$ the circumference, of the variable-radius circle.

## 6.5 Spatial Geometric Constraints

**Completable graph, completed graph and completion:** An under-constrained graph $G(V,E)$ is

**Table 6.8**
Cluster cases with constraints on the center of the variable-radius circle.

| One center constraint | | Two center constraints | |
|---|---|---|---|
| Problem | Degree | Problem | Degree |
| E(LL,LL') | (1,2) | E(LL,L'L') | (1,2) |
| | | E(LL',LL') | (1,2) |
| E(CL,LL') | (2,4) | E(CL,L'L') | (2,4) |
| E(CL',LL) | (2,4) | E(CL',LL') | (2,4) |
| E(C'L,LL) | (2,4) | E(C'L,LL') | (2,4) |
| | | E(C'L',LL) | (2,4) |
| E(CL,CL') | (4,4) | E(CL,C'L') | (4,8) |
| E(CL,C'L) | (4,16) | E(CL',CL') | (4,4) |
| | | E(C'L,CL') | (4,16) |
| | | E(C'L,C'L) | (4,4) |
| E(CC,LL') | (2,4) | E(CC,L'L') | (2,4) |
| E(CC',LL) | (4,4) | E(CC',LL') | (4,4) |
| | | E(C'C',LL) | (2,4) |
| E(CC,CL') | (4,4) | E(CC,C'L') | (4,8) |
| E(CC,C'L) | (4,32) | E(CC',CL') | (8,32) |
| E(CC',CL) | (8,32) | E(CC',C'L) | (8,32) |
| | | E(C'C',CL) | (4,8) |
| E(CC,CC') | (16,64) | E(CC,C'C') | (2,8) |
| | | E(CC',CC') | (16,64) |

*Note*: $(m,n)$ denotes the equation degree for $E = L$ and $E = C$, respectively. $L'$ denotes a constraint between a line and the center of the variable-radius circle; $C'$ denotes a constraint between a circle and the center of the variable-radius circle.

*completable* if there is a set of additional edges $E'$ such that the graph $G'(V, E \cup E')$ is well-constrained. We say that a minimal set $E'$ is a *completion* of $G$ and that $G'$ is a *completed* graph of $G$.

Compared to constraint problems in the plane, our knowledge of spatial constraint systems is relatively modest. The constraint graph analysis applies with some notable caveats. For example, Laman's characterization of rigidity does not apply in 3-space, not even when restricting to points only, and distances between them; see Section 6.3.8. Furthermore, the subsystems isolated by constraint graph decomposition can be complex, especially if lines are admitted to the geometric vocabulary. We illustrate the latter point with a few examples.

**Points and Planes**

Points and planes comprise the most elementary vocabulary in spatial constraint solving. Both have three degrees of freedom and are dual of each other. In analogy to the minimal constraint graph in 2D (Definition 6.8), a minimal constraint graph in 3D consists of three elements, points or planes, and three constraints, forming a triangle. The initial placement for the four combinations places the elements in canonical order, planes first, points second. Table 6.9 summarizes the method [8]. Note that the constraint between two planes is an angle, and the constraint between two points or a point

**Table 6.9**
Placement of three entities that are mutually constrained. $P$ denotes a plane $p$ a point.

| Canonical Placement of Three Points or Planes | |
|---|---|
| $P_1, P_2, P_3$ | $P_1$ placed as the plane $z = 0$<br>$P_2$ placed to intersect $P_1$ in the $y$-axis<br>$P_3$ placed to contain the origin |
| $P_1, P_2, p_3$ | $P_1$ placed as the plane $z = 0$<br>$P_2$ placed to intersect $P_1$ in the $y$-axis<br>$p_3$ is placed on the $xz$-plane |
| $P_1, p_2, p_3$ | $P_1$ placed as the plane $z = 0$<br>$p_2$ is placed on the positive $z$-axis<br>$p_3$ is placed on the $xz$-plane with $z \geq 0$ |
| $p_1, p_2, p_3$ | $p_1$ is placed at the origin<br>$p_2$ is placed on the positive $x$-axis<br>$p_3$ is placed on the $xz$-plane with $z \geq 0$ |

and a plane is a distance. The initial placement fails for the exceptional angles 0 and $\pi$, as well as for distance 0 between two points.

Sequential constructions of points and planes are straightforward. The locus of a third point $p$, at respective distances from two known points $p_1$ and $p_2$, is the intersection of two spheres centered at $p_1$ and $p_2$. It is a circle that is contained in a plane perpendicular to the line through $p_1$ ad $p_2$. As before, this fact can be used to simplify the algebra.

The simplest, nonsequential constraint system is the octahedron, consisting of 6 elements and 12 constraints [29, 30]. The name derives from the constraint graph that has the topology of the octahedron. There are 7 major configurations according to the number of planes. The configurations with 5 and with 6 planes are structurally underconstrained. Solutions of the octahedron constraint system have been proposed in [8, 29, 30, 44]. The number of distinct solutions is up to 16.

**Points, Lines, and Planes**

A line in 3-space has four degrees of freedom. Usually lines are represented with 6 coordinates, using Plücker coordinates, or with 8, using dual quaternions; e.g., [2]. Consequently, the implicit relationships between the coordinates have to be made explicit by additional equations when solving constraint systems with lines in 3-space.

The sequential construction of lines can be trivial, for instance determining a line by distance from two intersecting planes. But it can also be hard, for instance, when determining a line by distance from four given points in space. The latter problem, in geometric terms, asks for common tangents to four fixed spheres. This problem has up to 12 real solutions. The upper bound of 12 was shown in [31] using elementary algebra. The lower bound of 12 was established in [42] with an example. In contrast to the restricted problem of only points and planes, this sequential constraint problem involving lines is therefore much more demanding.

We can constrain two lines by distance from each other, by angle, and also by both distance and angle. Small constraint configurations that involve lines have been investigated [13, 14]. The papers show that there are two variational configurations of four lines. They are shown in Figure 6.24. The papers also show that the number of configurations with five geometric elements, including lines, is 17. Moreover, when six elements are considered, the number of distinct configurations grows to 683. Because of this daunting growth pattern, it is natural to seek alternatives. One such alternative has been proposed in [58, 59] where, instead of lines, only segments of lines are allowed. Consequently,

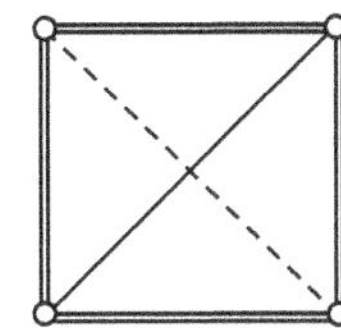
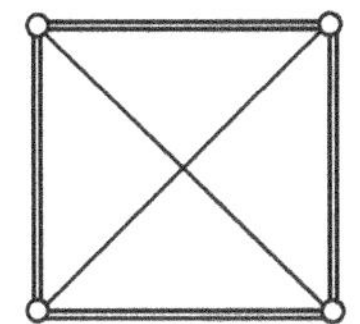

**Figure 6.24**
The two variational constraint problems with 4 lines. The 4 double lines represent both angle and distance, the solid diagonals distance, and the dashed diagonal angle constraints.

many constraints can be formulated as constraints on end points. Note that in many applications this is perfectly adequate.

## 6.6 Under-Constrained Geometric Constraint Problems

In general, existing geometric constraint solving techniques have been developed under the assumption that problems are well-constrained, that is, they adhere to Definition 6.4 given in Section 6.1. Put differently, the number of constraints and their placement on the geometric elements define a problem with a finite number of solution instances for nondegenerate configurations. However, there are a number of scenarios where the assumption of well-constrained does not apply. Examples are early stages of the design process when only a few parameters are fixed or in cooperative design systems where different activities in product design and manufacture examine different subsets of the information in the design model [27]. The problem then is under-constrained, that is, there are infinitely many solution instances for nondegenerate configurations.

**Example 29** *Consider the hook of a car trunk locker shown in Figure 6.25. Once distances $d_1$ and $d_2$ have placed the center of the exterior circle of the hook with respect to the hook's axis of rotation, the designer is mainly interested in finding a value for the angle $\alpha$ where the exterior circle is tangent to the small circle transitioning to the inner circle of the hook. The angle $\alpha$ has to be such that the hook smoothly rotates while closing and opening the hood. At this design stage, the stem shape of the hook is irrelevant.*

This, and other simple examples taken from computer-aided design, illustrate the need for efficient and reliable techniques to deal with under-constrained systems. The same need is found in other fields where geometric constraint solving plays a central role, such as kinematics, dynamic geometry, robotics, as well as molecular modeling applications.

Recent work on under-constrained GCS with one degree of freedom, has brought significant progress in understanding and formalizing generically under-constrained systems; [52, 54, 55]. The work focuses on GCS restricted to points and distances, also generically called *linkages*.

The goal of geometric constraint solving is to effectively determine realizations or embeddings of geometric objects in the ambient space in which the GCS problem is formulated. Thus, the current trend is that solving an under-constrained GCS should be understood as solving some well-constrained GCS derived from the given one.

There are two ways to transform an under-constrained GCS into a well-constrained one: adding to the GCS as many extra constraints as needed or removing from the GCS unconstrained geometric entities. Note that removing constrained entities makes little sense. Accordingly, the literature on

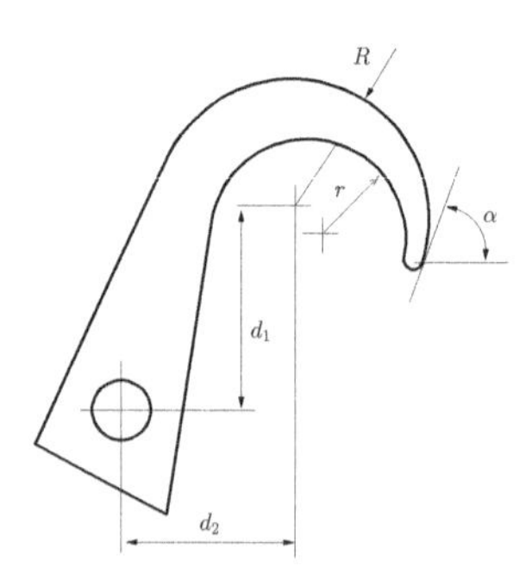


**Figure 6.25**
Hook of a car trunk locker.

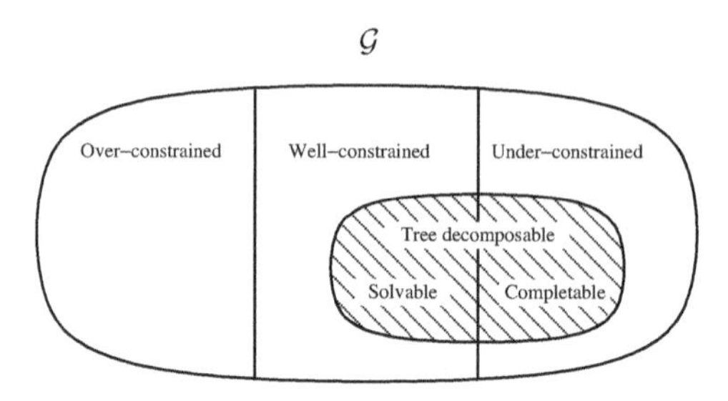


**Figure 6.26**
A partition of the geometric constraint graphs set, $\mathcal{G}$. The set of tree-decomposable graphs straddles over the sets of well- and over-constraint graphs.

under-constrained GCS advocates to transform an under-constrained GCS into a well-constrained one by adding new constraints. This technique was formally defined in [35] as follows.

**Definition 6.17** Let $G(V,E)$ be an under-constrained graph associated with a GCS problem. Let $E'$ be a set of additional edges each bounded by two distinct vertices in $V$ such that the graph $G'(V,E \cup E')$ is well-constrained. We say that $E'$ is a *completion* for $G$ and that $G'$ is the *completed* graph of $G$.

Let $\mathcal{G}$ denote the set of geometric constraint graphs. Definitions 6.4, 6.3 and, 6.2 given in Section 6.1 induce in $\mathcal{G}$ a partition as shown in Figure 6.26. The set of tree-decomposable graphs straddles over the sets of well- and over-constraint graphs. As described in Section 6.3.6, the set of well-constrained, tree-decomposable graphs are solvable by the tree decomposition approach.

Within the set of under-constrained graphs we can distinguish two families: Those which are not tree-decomposable and those which are. It is easy to see that there is no completion for a graph in the first family that could transform the graph into a tree-decomposable one. Considering graphs in the second family, Definition 6.4 fixes the number of extra constraints that must be added. However deciding which constraints should actually be added to the graph is not a straightforward matter because the resulting graph could be either over-constrained or well-constrained but not tree-decomposable.

**Example 30** *Figure 6.27a shows an under-constrained graph. To see that it is tree-decomposable just consider as the first decomposition step the subgraphs induced by the sets of vertices* $\{E,F,G\}$,

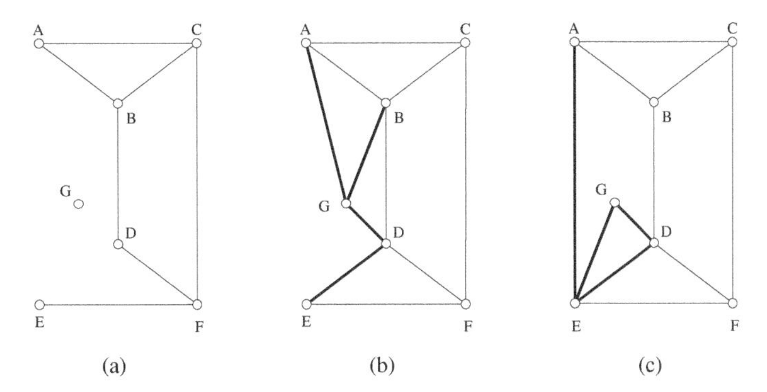


**Figure 6.27**
(a) An under-constrained, tree-decomposable graph $G$. (b) A tree-decomposable completion of $G$. (c) A non tree-decomposable completion of $G$.

*$\{A,G\}$ and $\{A,B,C,D,F\}$. Finding the steps needed to complete a tree-decomposition is routine. The completion $E = \{(A,G),(G,B),(G,D),(E,F)\}$ generates the well-constrained graph shown in Figure 6.27b. To see that the graph is tree-decomposable take as a first decomposition step the two minimal constraint graphs with edges $\{(E,D)\}$ and $\{(E,F)\}$ plus the subgraph induced by edges $\{(A,B),(A,C),(B,C),(B,D),(C,F),(D,F)\}$. However, completion $E = \{(A,E),(E,G),(E,D),(G,D)\}$ results in the graph depicted in Figure 6.27c which is no tree-decomposable.*

In what follows we restrict to the completion problem for triangle decomposable GCS.

**Definition 6.18** An under-constrained graph $G(V,E)$ is said to be *completable* if there is a set of edges $E'$ which is a completion for $G$.

Notice that completability does not require triangle-decomposability of the underconstrained graph, nor does it imply that a well-constrained, completed graph be tree-decomposable.

**Definition 6.19** Let $G(V,E)$ be a triangle-decomposable, under-constrained graph and let $E'$ be a completion for $G$. We say that $E'$ is a *triangle-decomposable completion*, (*td-completion* in short) for $G$ if $G'(V,E \cup E')$ is triangle-decomposable.

Reported techniques dealing with under-constrained GCS differ mainly in the way they figure out completions as well as whether they aim at figuring out td-completions or just completions. The work in [39] describes an algorithm where the constraint graph is captured as a bipartite connectivity graph whose nodes are either geometric entities or constraints. Each edge connects a constraint node with the constrained geometric node. In analogy to sequential solvers the graph edges are directed to indicate which constraints are used to fix (incident) geometric objects. The connectivity graph is analyzed according to the degrees of freedom of under-constrained geometric nodes. Each under-constrained geometric node is a candidate to support an additional edge to a new constraint, or if there is an edge that heads a propagation path to an existing constraint node that has an unused condition. When there are several candidates on which the new constraint can be established, the selection is left to the user. A similar approach to solve under-constrained GCS based on degrees of freedom analysis is described in [45].

Two different notions of td-completion were introduced in [35]. The first one is called *free completion* and is computed in three steps. First a triangle decomposition for the given graph is

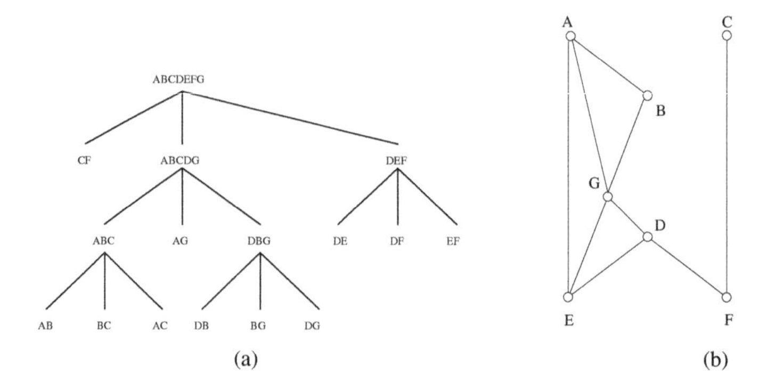


**Figure 6.28**
(a) A triangle-decomposition for the under-constrained graph $G(V,E)$ in Figure 6.27a. Pairs of vertices $\{A,G\},\{D,E\},\{B,G\}$ and $\{D,G\}$ in leaf nodes do not define graph edges. (b) A set of additional edges defined on $V(G)$. Edges $(A,E),(A,G),(B,G),(D,E),(D,G)$ and $(E,G)$ do not belong to $E$.

figured out. Then the set of under-constrained leaf nodes in the decomposition is identified. Notice that each node in this set stores a subgraph $G(V,E)$ where $|V|=2$ and $E$ is the empty set. Thus one edge is missing. Finally the completion is computed as the set of missing edges in the under-constrained leaf nodes of the triangle-decomposition.

**Example 31** *Figure 6.28a shows a triangle-decomposition for the under-constrained graph $G(V,E)$ in Figure 6.27a. We have $|V|=7$ and $|E|=7$. For 2D problems, the general property of a well-constrained graph described in Section 6.1 is the Laman theorem, [38], $2|V|-|E|=3$. Hence the number of additional edges required to complete $G$ to a well-constrained graph is $|E'|=2|V|-|E|-3=4$. The set of leaves in the triangle-decomposition corresponding to under-constrained minimal graphs includes exactly four elements: $\{A,G\},\{B,G\},\{D,E\}$ and $\{D,G\}$. Thus $E'=\{(A,G),(B,G),(D,E),(D,G)\}$ is a free completion for $G$. The completed well-constrained graph $G'(V,E\cup E')$ is shown in Figure 6.27b.*

The second td-completion is called *conditional completion*. The first and second steps are the same as in the free completion. However, in the third step, edges to complete under-constrained leaf nodes in the decomposition are drawn from an additional graph defined over a subset of vertices of the given graph. If the number of those edges that are not in the original graph is smaller than the number required by Definition 6.4, then the completed graph will remain under-constrained. However a free completion can eventually be applied to get a well-constrained completion.

**Example 32** *Consider again the under-constrained graph $G(V,E)$ in Figure 6.27a and its triangle-decomposition shown in Figure 6.28a. The set of under-constrained pairs of vertices in the triangle-decomposition is $\{(A,G),\ (B,G),\ (D,E),\ (D,G)\}$. Assume that the set of additional edges defined on $V(G)$ is*

$$\{(A,B),(A,E),(A,G),(B,G),(C,F),(D,F),(E,G),(E,D),(G,D)\}$$

*as shown in Figure 6.28b. Now, additional edges for the completion must be drawn from a set $E^*$ such that $E\cap E^*=\emptyset$. In the case at hand, $E^*=\{(A,E),(A,G),(B,G),(D,E),(D,G),(E,G)\}$. Thus, a completion for $G(V,E)$ is $E'=\{(A,G),(B,G),(D,E),(D,G)\}\subset E^*$. Figure 6.27b shows the completed graph $G'(V,E\cup E')$.*

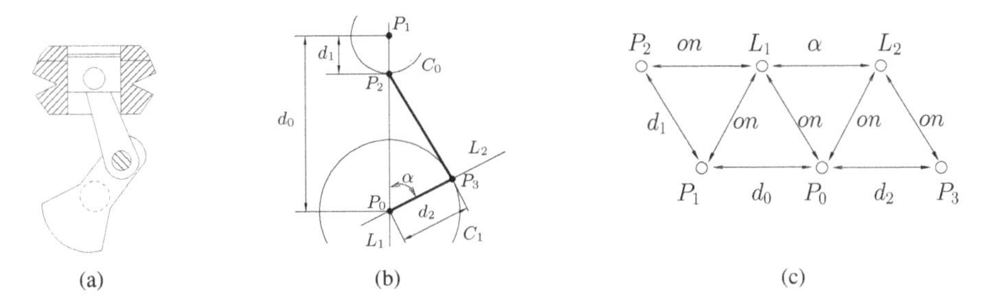


**Figure 6.29**
(a) A crankshaft and connecting rod in a reciprocating piston engine. (b) The crankshaft and connecting rod abstracted as a GCS. (c) Geometric constraint graph.

A technique to complete general under-constrained graphs is described in [35]. The approach is based on transforming the problem of computing a completion into a combinatorial optimization problem. Edges in the graph and in the additional set are assigned different weights. Then a greedy algorithm generates a well-constrained problem provided that there are enough edges in the additional set. A variant of this approach is reported in [62].

There is a class of 2D, triangle-decomposable, under-constrained GCS that occur in a number of fields like dynamic geometry or mechanical computer aided design. In these GCS the geometries are points, the constraints are usually point-point distances, and exactly one edge is missing in the associated constraint graph. Such GCS are known as *linkages*.

**Example 33** *Figure 6.29 a shows an illustration of a crankshaft and connecting rod in a reciprocating piston engine. The crankshaft and connecting rod can be abstracted as the GCS shown in Figure 6.29b. The GCS includes four points* $P_0, P_1, P_2, P_3$*, two lines* $L_1, L_2$*, three point-point distances,* $d_0, d_1, d_2$*, and one line-line angle,* $\alpha$*. Moreover, points* $P_0, P_1$ *and* $P_2$ *must be on the line* $L_1$*, and points* $P_0, P_3$ *must be on the line* $L_2$*. If, for example, values of either the distance* $d_0$ *or the angle* $\alpha$ *are freely assigned, then the GCS can be considered a linkage.*

Reachability is an important problem in fields such as dynamic geometry or conformational molecular geometry. It can be formalized as follows:

> Let Rs and Re be two realizations of a well defined geometric construction where Rs is called the starting instance and Re the ending instance. Are there continuous transformations that preserve the incidence relationships established in the geometric construction and transform Rs to Re?

The well-defined geometric construction in the reachability problem can be understood as a linkage, that is, a well-constrained GCS problem where values assigned to one specific constraint can freely change. In [18] the reachability problem is solved assuming that the underlying GCS is triangle-decomposable. The approach first computes the set of intervals of values that the free constraint can take for which the linkage is realizable. This set of intervals is known as the linkage *Cayley configuration space* [12]. When both Rs and Re are realizations with the free constraint taking values within one Cayley interval, the path is an interval arc. When Rs and Re realizations belong to different intervals, finding a path entails figuring out whether there are continuous transitions between consecutive intervals that permit the linkage to reach Re when starting at Rs. If more than one such path is found, one is chosen according to some predefined strategy. In [18] the minimal arc length path is the one chosen.

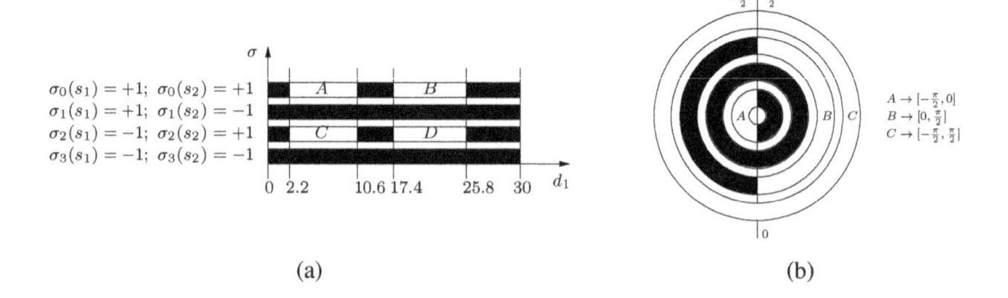


**Figure 6.30**
(a) Cayley configuration for distance constraint $d_1$. (b) Cayley configuration for angle constraint $\alpha$.

**Example 34** *The crankshaft and connecting rod GCS in Figure 6.29b is triangle-decomposable, ruler-and-compass solvable. A construction plan that places each geometric element with respect to each other is*

$$
\begin{array}{llll}
1.\ P_0 = origin() & & 5.\ P_2 = intLC(L_1, C_0, s_1) \\
2.\ P_1 = distPP(P_0, d_0) & & 6.\ L_2 = linePA(P_0, \alpha) \\
3.\ L_1 = line2P(P_0, P_1) & & 7.\ C_1 = circleCR(P_0, d_2) \\
4.\ C_0 = circleCR(P_1, d_1) & & 8.\ P_3 = intLC(L_2, C_1, s_2)
\end{array}
$$

*When a construction step has more than one solution, an orientation parameter is needed to select the desired solution instance. Parameters $s_1$ and $s_2$ in steps 5 and 8 respectively allow to select one of the two points in a line-circle intersection. Figure 6.30a shows the Cayley configuration space when the point-point distance constraint value $d_1$ changes. Intervals labeled A, B, C, and D define orientations and parameter values for which the GCS is realizable. Intervals A and C yield realizations consistent with the one depicted in Figure 6.29a. Intervals B and D yield realizations where point $P_1$ would be placed on the line $L_1$ opposite to $P_2$ with respect to $P_0$. The Cayley configuration space when the varying parameter is the angle $\alpha$ is shown in Figure 6.30b. Now orientations are represented along a radial axis and angle values for which the solution is realizable are represented as circular intervals.*

Linkages are extensively studied in [53, 54, 55]. The object of these works is to lay sound theoretical foundations for a reliable and efficient computation of Cayley configuration spaces for general tree decomposable linkages. New concepts like *size* and *computational complexity* are introduced and efficient algorithms are developed to answer a number of questions on linkages like effectively computing Cayley configuration spaces and solving the reachability problem. Methods so far applied to compute Cayley configuration spaces, like the one used in [18], suffer from potential combinatorial growth. The work in [54, 55] shows that for low Cayley complexity GCS problems, computing the configuration space is polynomial in the number of geometric elements of the problem.

# References


[1] A. Aho, J. Hopcroft, and J. Ullman. *The Design and Analysis of Computer Algorithms*. Addison-Wesley, 1974.

[2] W. Blaschke. *Kinematik und Quaternionen*. VEB Verlag der Wissenschaften, Berlin, Germany, 1960.

[3] W. Bouma, I. Fudos, C. Hoffmann, J. Cai, and R. Paige. Geometric constraint solver. *Computer-Aided Design*, 27(6):487–501, June 1995.

[4] C.H. Cao, W. Fu, and W. Li. The research of a new geometric constraint solver. In X.-T Yan, C.-Y Jiang, and N.P. Juster, editors, *Perspectives from Europe, and Asia on Engineering Design and Manufacturing*, A comparison of engineering design and manufacture in Europe and Asia, pages 205–214. Springer Scinence+Business Media, LLC., 2004.

[5] C.-S. Chiang and R. Joan-Arinyo. Revisiting variable radius circles in constructive geometric constraint solving. *Computer-Aided Geometric Design*, 221(4):371–399, 2004 2004.

[6] Cinderella. The interactive geometry software Cinderella, June 2007. http://cinderella.de/tiki-index.php.

[7] D-Cubed. *The Dimensional Constraint Manager*. Cambridge, England, 2003. 2D DCM Version 44.0, 3D DCM Version 28.0.

[8] C. Durand and C. Hoffmann. A systematic framework for solving geometric constraints analytically. *JSC*, 30:493–519, 2000.

[9] I. Fudos and C.M. Hoffmann. Constraint-based parametric conics for CAD. *Computer-Aided Design*, 28(2):91–100, 1996.

[10] I. Fudos and C.M. Hoffmann. Correctness proof of a geometric constraint solver. *International Journal of Computational Geometry and Applications*, 6(4):405–420, 1996.

[11] I. Fudos and C.M. Hoffmann. A graph-constructive approach to solving systems of geometric constraints. *ACM Transactions on Graphics*, 16(2):179–216, April 1997.

[12] H. Gao and M. Sitharam. Combinatorial classification of 2D underconstrained sytems. In *Proceedings of the Seventh Asian Symposium on Computer Mathematics (ASCM 2005)*, pages 118–127, September 6 2005.

[13] X.-S. Gao, C. M. Hoffmann, and W. Yang. Solving spatial basic geometric constraint configurations with locus intersection. *CAD*, 36:111–122, 2004.

[14] X.-S Gao, C.M. Hoffmann, and W.-Q Yang. Solving spatial basic geometric constraint configurations with locus intersection. In *Proceedings of Solid Modeling SM'02*, Saarbrucken, Germany, June, 17-21 2002. ACM Press.

[15] GeoGebra. http://www.geocebra.org/cms, July 2007.

[16] Kirk Haller, Audrey Lee-St.John, Meera Sitharam, Ileana Streinu, and Neil White. Body-and-cad geometric constraint systems. *Comput. Geom. Theory Appl.*, 45(8):385–405, October 2012.

[17] L. Henneberg. *Die graphische Statik der starren Körper*. Springer, 1908.

[18] M. Hidalgo and R. Joan-Arinyo. The reachability problem in constructive geometric constraint solving based dynamic geometry. *Journal of Automated Reasoning*, 44(7):709–720, 2013. DOI:10.1007/s10817-013-9280-y.

[19] C. M. Hoffmann. Computer vision, descriptive geometry and classical mechanics. In B. Falcidieno, I. Hermann, and C. Pienovi, editors, *Computer Graphics and Mathematics*, pages 229–224. Springer Verlag, Eurographics Series, 1992.

[20] C. M. Hoffmann and J. Peters. Geometric constraints for CAGD. In M. Dachlen, T. Lyche, and L. Schumaker, editors, *Mathematical Methods for Curves and Surfaces*, pages 237–254. Vanderbilt University Press, 1995.

[21] Christoph M. Hoffmann, Andrew Lomonosov, and Meera Sitharam. Finding solvable subsets of constraint graphs. In *Principles and Practice of Constraint Programming - CP97, Third International Conference, Linz, Austria, October 29 - November 1, 1997, Proceedings*, pages 463–477, 1997.

[22] Christoph M. Hoffmann, Andrew Lomonosov, and Meera Sitharam. Decomposition plans for geometric constraint problems, part ii: new algorithms. *Journal of Symbolic Computation*, 31(4):409–427, 2001.

[23] Christoph M. Hoffmann, Andrew Lomonosov, and Meera Sitharam. Decomposition plans for geometric constraint systems, part i: Performance measures for cad. *Journal of Symbolic Computation*, 31(4):367–408, 2001.

[24] Christoph M. Hoffmann and Bo Yuan. On spatial constraint solving approaches. In *Revised Papers from the Third International Workshop on Automated Deduction in Geometry*, ADG '00, pages 1–15, London, UK, 2001. Springer-Verlag.

[25] C.M. Hoffmann and C.-S. Chiang. Variable-radius circles of cluster merging in geometric constraints. Part II: Rotational clusters. *Computer-Aided Design*, 34:799–805, October 2002.

[26] C.M. Hoffmann and C.-X. Chiang. Variable-radius circles of cluster merging in geometric constraints. Part I: Translational clusters. *Computer-Aided Design*, 34:787–797, October 2002.

[27] C.M. Hoffmann and R. Joan-Arinyo. Distributed maintenance of multiple product views. *Computer-Aided Design*, 32(7):421–431, June 2000.

[28] C.M. Hoffmann, A. Lomonosov, and M. Sitharam. Decompostion Plans for Geometric Constraint Systems, Part I: Performance Measurements for CAD. *Journal of Symbolic Computation*, 31:367–408, 2001.

[29] C.M. Hoffmann and P.J. Vermeer. Geometric constraint solving in $R^2$ and $R^3$. In D.-Z. Du and F. Hwang, editors, *Computing in Euclidean Geometry*, pages 266–298. World Scientific Publishing, 1995.

[30] C.M. Hoffmann and P.J. Vermeer. A spatial constraint problem. In J.P. Merlet and B. Ravani, editors, *Computational Kinematics'95*, pages 83–92. Kluwer Academic Publ., 1995.

[31] C.M. Hoffmann and B. Yuan. On spatial constrint solving approaches. In J. Richter-Gebert and D. Wand, editors, *Proceedings of ADG'2000*, Zurich, Switzerland, 2000.

[32] R. Joan-Arinyo. Triangles, ruler and compass. Technical Report LSI-95-6-R, Department LiSI, Universitat Politècnica de Catalunya, 1995.

[33] R. Joan-Arinyo, M.V. Luzón, and A. Soto. Genetic algorithms for root multiselection in constructive geometric constraint solving. *Computer & Graphics*, 27(1):51–60, 2003.

[34] R. Joan-Arinyo and A. Soto. A ruler-and-compass geometric constraint solver. In M.J. Pratt, R.D. Sriram, and M.J. Wozny, editors, *Product Modeling for Computer Integrated Design and Manufacture*, pages 384–393. Chapman and Hall, London, 1997.

[35] R. Joan-Arinyo, A. Soto-Riera, S. Vila-Marta, and J. Vilaplana. Transforming an underconstrained geometric constraint problem into a wellconstrained one. In G. Elber and V.Shapiro, editors, *Eight Symposium on Solid Modeling and Applications*, pages 33–44, Seattle (WA), June 16-20 2003. ACM Press.

[36] R. Joan-Arinyo, A. Soto-Riera, S. Vila-Marta, and J. Vilaplana-Pasto. Revisiting decomposition analysis of geometric constraint graphs. In *Proceedings of the Seventh ACM Symposium on Solid Modeling and Applications*, SMA '02, pages 105–115, New York, NY, 2002. ACM.

[37] R.R. Kavasseri. Variable radius circle computations in geometric constraint solving. Master's thesis, Computyer Sciences. Purdue University, August 1966.

[38] G Laman. On graphs and the rigidity of plane skeletal structures. *J. Engineering Mathematics*, 4:331–340, 1970.

[39] R.S. Latham and A.E. Middleditch. Connectivity analysis: a tool for processing geometric constraints. *Computer-Aided Design*, 28(11):917–928, November 1996.

[40] Audrey Lee-St.John and Jessica Sidman. Combinatorics and the rigidity of CAD systems. *Computer-Aided Design*, 45(2):473 – 482, 2013. Solid and Physical Modeling 2012.

[41] M.V. Luzón, A. Soto, J.F. Gálvez, and R. Joan-Arinyo. Searching the solution space in constructive geometric constraint solving with genetic algorithms. *Applied Intelligence*, 22:109–124, 2005.

[42] I. Macdonald, J. Pach, and T. Theobald. Common tangents to four unit balls in R 3. *Discrete and Comp. Geometry*, 26:1–17, 2001.

[43] James Clerk Maxwell. On reciprocal figures and diagrams of forces. *Philosophical Magazine*, 27:250–261, 1864.

[44] D. Michelucci and S. Foufou. Using the Cayley-Menger determinants for geometric constraint solving. In *Solid Modeling and Applications*, pages 285–290. ACM, New York, 2004.

[45] A. Noort, M. Dohem, and W.F. Bronsvoort. Solving over and underconstrained geometric models. In *Geometric Constraint Solving*. Springer-Verlag, Berlin, Heidelberg, 1998.

[46] J.C. Owen. Algebraic solution for geometry from dimensional constraints. In R. Rossignac and J. Turner, editors, *Symposium on Solid Modeling Foundations and CAD/CAM Applications*, pages 397–407, Austin, TX, June 5–7 1991. ACM Press.

[47] H. Pottmann and M. Peternell. Applications of Laguerre geometry in cagd. *CAGD*, 15:165–188, 1998.

[48] E. Yeguas R. Joan-Arinyo, M.V. Luzón. Search space pruning to solve the root identification problem in geometric constraint solving,. *Computer-Aided Design and Applications*, 6(1):15–25, 2009.

[49] K. Ramanathan. Variable radius circle computations in geometric constraint solving. Master's thesis, Department of Computer Science. Purdue University, 1996.

[50] E. C. Sherbrooke and N. M. Patrikalakis. Computation of the solutions of nonlinear polynomial systems. *CAGD*, 10:379–405, 1993.

[51] M. Sitharam, A. Arbree, Y. Zhou, and N. Kohareswaran. Solution space navigation for geometric constraint systems. *ACM Transactions on Graphics*, 25(2):194–213, April 2006.

[52] M. Sitharam and H. Gao. Characterizing graphs with convex and connected Cayley configuration spaces. *Discrete & Computational Geometry*, 43(3):594–625, 2010.

[53] M. Sitharam and M. Wang. How the beast really moves: Cayley analysis of mechanisms realization spaces using CayMos. *Computer-Aided Design*, 46:205–210, January 2014.

[54] M. Sitharam, M. Wang, and H. Gao. Cayley configuration spaces of 1-dof tree-decomposable linkages, Part I: Structure and extreme points. arXiv:1112.6008v7[cs.CG], 26 Feb 2014.

[55] M. Sitharam, M. Wang, and H. Gao. Cayley configuration spaces of 1-dof tree-decomposable linkages, Part II: Combinatorial characterization of complexity. arXiv:1112.6009v4[cs.CG], 7 Nov 2012.

[56] Tiong-Seng Tay. Rigidity of multi-graphs. i. linking rigid bodies in n-space. *Journal of Combinatorial Theory, Series B*, 36(1):95 – 112, 1984.

[57] Gilles Trombettoni and Marta Wilczkowiak. Gpdof a fast algorithm to decompose under-constrained geometric constraint systems: Application to 3d modeling. *International Journal of Computational Geometry and Applications*, 16(05n06):479–511, 2006.

[58] H.A. van der Meiden. *Semantics of Families of Objects*. PhD thesis, Delft Technical University, 2008.

[59] H.A. van der Meiden and W.F. Bronsvoort. A constructive approach to calculate parameter ranges for systems of geometric constraints. *Computer-Aided Design*, 38(4):275–283, 2006.

[60] Neil White and Walter Whiteley. The algebraic geometry of motions of bar-and-body frameworks. *SIAM J. Algebraic Discrete Methods*, 8(1):1–32, January 1987.

[61] E. Yeguas, R. Joan-Arinyo, and M.V. Luzón. Modeling the performance of evolutionary algorithms on the root identification problem: A case study with PBIL and CHC algorithms. *Evolutionary Computation*, 19(1):107–135, 2011.

[62] G.-F. Zhang and X.-S Gao. Well-constrained completion and decomposition for under-constrained geometric consraint problems. *Int. Jour. of Computational Geometry & Applications*, 16(5-6):461–478, 2006.

[63] Y. Zhou. *Combinatorial decomposition, generic independence and algebraic complexity of geometric constraint systems. Applications in biology and engineering*. PhD thesis, University of Florida, 2006.

# Chapter 7

# *Geometric Constraint Decomposition: The General Case*

**Troy Baker**
*University of Florida*

**Meera Sitharam**
*University of Florida*

**Rahul Prabhu**
*University of Florida*

**CONTENTS**



## 7.1 Introduction

Solving or realizing geometric constraint systems requires finding real solutions to a large multivariate polynomial system (of equalities and inequalities representing the constraints); this requires double exponential time in the number of variables, even if the type or orientation of the solution is specified. Thus, to realize a geometric constraint system, it is crucial to perform recursive decomposition into locally rigid subsystems (which have finitely many solutions), and then apply the reverse process of recombining the subsystem solutions. The effectiveness of the decomposition is determined by the size of the largest (indecomposable) subsystem that must be solved. This size measures the complexity bottleneck. It is an NP-hard combinatorial optimization problem to find a DR-plan that optimizes this measure. This chapter briefly surveys a series of recursive decomposition algorithms and culminates in an algorithm to solve the optimal DR-planning problem in polynomial time for a broad class of minimally rigid or isostatic systems.

### 7.1.1 Terminology and Basic Concepts

Decomposition-recombination (DR-) planning has been studied since the mid 1980's. It was first precisely formulated in [8]. Several useful surveys have been compiled regarding DR-planning [11, 12, 17], to which the interested reader is referred.

DR-planning is a divide-and-conquer technique primarily used for determining the largest subsystem of polynomials that must be solved simultaneously, thereby reducing the overall time to realize the system. Informally, a DR-plan is a recursive decomposition of a rigid constraint graph into rigid subgraphs. These subgraphs can be solved independently of one another and then quickly recombined into the original graph.

The two main goals of a DR-plan [8] are that (1) the solutions of the small subsystems can be recombined into a solution for the entire system and (2) the subsystems should be geometrically meaningful. Consideration (1) is developed further in Section 7.2.2 and consideration (2) should be clear from the definitions (the subsystems are rigid graphs in the system). Different types of DR-plans further satisfy (2) by imposing more conditions on the subsystem, as discussed later in this section. The reader is referred to Chapter 1 for definitions of rigid graphs, independent graphs, dof rigid graphs, over/under-constrained graphs, etc.

**Definition 7.1 Decomposition-recombination (DR-) plan** A *DR-plan* of constraint graph $G$ is a forest where:

- Each node is a rigid subgraph of $G$.
- A root node is a maximal rigid subgraph.
- An internal node is the union of its children.
- A leaf node is a single constraint and the involved primitives.

See Figure 7.1 for two simple examples of DR-plans of a 2D bar-joint constraint graph.

Note that this definition permits the same rigid subgraph to appear in multiple nodes of the DR-plan. Not permitting such duplication would, in general, require the DR-plan to be defined as a directed acyclic graph instead of a forest. In this way, a DR-plan can also be thought of as a partial order on the nodes. The forest representation is used to simplify book keeping.

Observe also that the concept of a DR-plan is very general; it is usable for any type of constraint graphwith an underlying combinatorial (abstract) rigidity matroid and potentially broader classes of matroid-like structures with certain intersection properties. Furthermore, the DR-plan is combinatorial, concerned only with the underlying constraint graph. Therefore, a change in parameters has no impact on the DR-plan (see Section 7.4.2 for additional discussion.)

The *size* of a DR-plan is the total number of nodes in the forest. The *fan-in* of a node is the number of children, and the *fan-in* of a DR-plan is the maximum number of children over all nodes in the forest.

A DR-plan is *optimal* if it minimizes the fan-in over all possible DR-plans. This criterion for

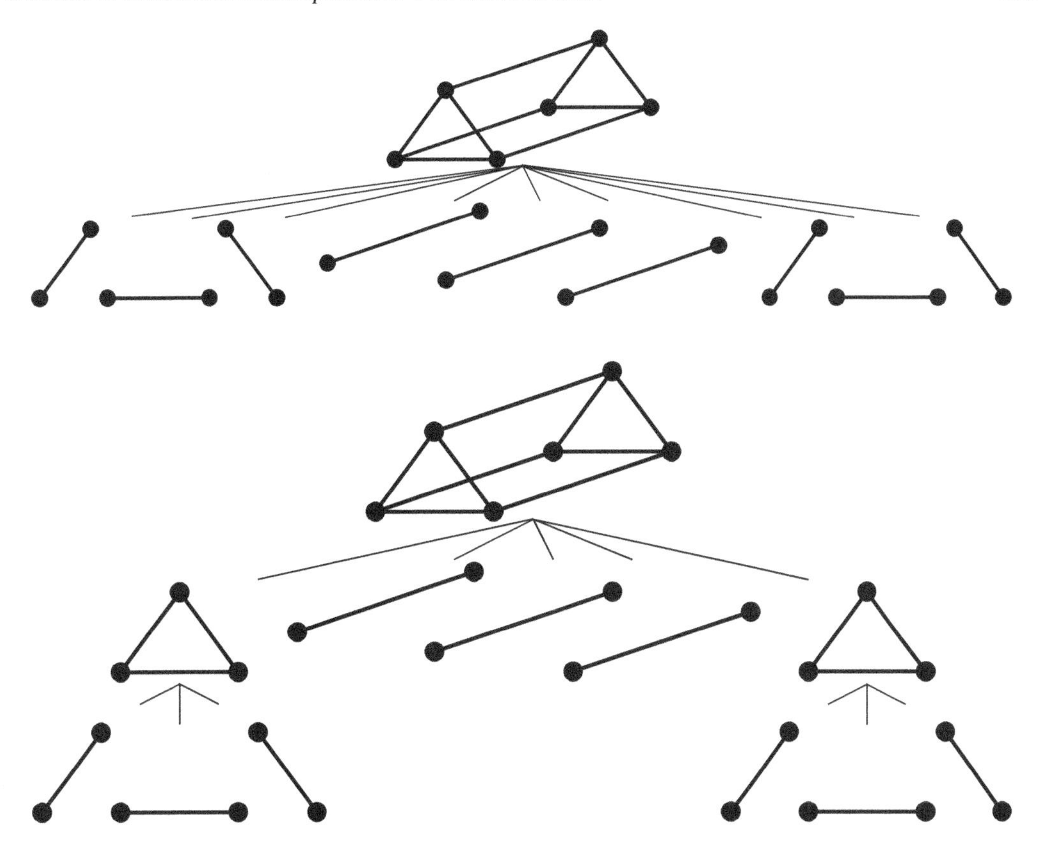

**Figure 7.1**
Two DR-plans of the same 2D bar-joint constraint graph (a $C_2 \times C_3$, or doublet.) The first is the worst possible DR-plan, decomposing into the individual edges of the graph. The second is much better, taking advantage of the rigid subgraphs (the two triangles.)

optimality is chosen because it measures how efficiently a GCS can be realized. The time is dominated by the largest subsystem to be recombined (the maximum fan-in of the DR-plan). If the optimal DR-plan has a fan-in of $|E|$, the number of constraints, the system is called *indecomposable*; the best that can be done is breaking the system down into its constituent constraints, and there is no improvement over solving the entire system simultaneously.

Finding an optimal DR-plan is NP-hard in general [12]. When dealing with a class for which optimal plans cannot be readily constructed, an algorithm can instead build DR-plans that satisfy other desirable properties. Several different meaningful properties have been studied. For example, see Section 7.1.5 for an algorithm that finds a DR-plan with *cluster minimality*, i.e. the union of no subset of children forms the parent. Section 7.1.6 discusses *proper vertex-maximality*, i.e., each child is a rigid vertex-maximal proper subgraph of the parent, which happens to be optimal if the input is independent.

In addition to optimality, a "good" DR-planner should adhere to the following principles [11]:

- Church-Rosser property (or Confluence) – If the DR-planner can follow multiple computational paths on the same input they all lead to the same outcome: either they all generate a valid DR-plan for the graph or they don't.
- Completeness – A DR-plan is found if one exists.
- Low time complexity.

- Correctness – The output is a valid DR-plan.

There are several natural use-cases of DR-plans, which include:

- Determining complexity of realization. The largest subsystem that must be solved simultaneously determines the time complexity.
- Guiding realization. The partial order of the tree establishes the partial-order in which to solve the system.
- Decomposing the rigidity matrix of the constraint graph, and thereby the self-stress and flex matrices. Notice also that the DR-planner is finding independent systems. Therefore, a DR-plan corresponds to a block decomposition of the rigidity matroid and matrix. As mentioned in Chapter 1, the left and right nullspaces of the rigidity matrix are the space of self-stresses and flexes, respectively. Therefore, this is also a decomposition of stresses and flexes.
- Completion of under-constrained graphs. See Section 7.3.
- Guided removal of constraints from over-constrained graphs. See Section 7.3.

The first and second items are the main motivation for DR-planning. The first was treated above and the second is discussed in Algorithm 1. This is a simple and naïve algorithm for using the DR-plan. More complex methods are discussed in Section 7.2. Next we survey different methods for DR-planning.

**Algorithm 1** Realization using a DR-plan

1: Recursively realize the children of the root. Base case: a child is a single constraint and is realized in constant time.
2: Recombine the children into a unified realization. This can be done by choosing a single child as a *home coordinate system* and then expressing the variables of the other children in terms of the chosen one. Then the polynomial system is solved.

### 7.1.2 Triangle-Decomposition

The basic algorithm (first appearing in [4] with proof of correctness in [6]) was originally worked for 2D point-line systems, but here it will be discussed in the context of 2D bar-joint. Triangle-, or tree-, decomposition is an older form of decomposition that only works on a specific class of graphs. In practice, this class arises often.

Algorithm 2 explains the bottom-up algorithm to construct the DR-plan. If Step 4 fails, then there is no triangle-decomposition. This algorithm will find a triangle-decomposition if one exists.

Conceptually, this can also be thought of in a top-down recursive manner. A triangle-decomposable system is one that can be separated into 3 subsystems that (a) pairwise intersect on a single vertex and (b) are each triangle-decomposable.

Realizing a graph with a triangle-decomposition has a simple, efficient algorithm. In fact, the class of graphs with a triangle-decomposition is exactly the class of graphs that are *quadratically radically solvable (QRS)*. Each node can be found with a simple ruler-and-compass construction. The constituent cluster has some distance between the participating vertices (those shared between other sibling clusters) which can be used in simply solving a simple triangle. A single cluster is chosen as a frame of reference and the other two are translated to line up with the vertex they share and then rotated to solve the triangle system.

**Algorithm 2** Triangle Decomposition

1: Every edge is registered as a *cluster*. ▷ Intuitively, a cluster will be a subgraph for which a realization is known. An edge is the base case.
2: Create a node for each cluster. ▷ These initial clusters will be the leaves of the DR-plan.
3: **while** there is more than one cluster **do**
4: Find three clusters, $C_i, C_j, C_k$, such that any two share a single vertex.
5: Add a new cluster $C_m$ that is a union of $C_i, C_j, C_k$.
6: Add a new node for $C_m$, and set it as the parent of $C_i, C_j, C_k$.
7: Remove $C_i, C_j, C_k$ from the list of clusters.
8: **end while**
▷ Ultimately, the graph will be reduced to a single cluster.

However, there are many rigid graphs that are not triangle-decomposable, including some simple ones such as $C2 \times C3$ and $K_{3,3}$. The idea of *shape recognition* is to look for specific shapes in the graph from a finite list of shapes, then recursively decompose the graph as discussed above. Triangle-decomposition is shape recognition with a triangle as the only shape in the list. I.e. the recursive definition would be to choose a shape from a list with $n$ edges and to find $n$ subsystems that (a) intersect on a single vertex in the same configuration as the shape and (b) are each decomposable by the same list of shapes. However, there is no finite list of shapes that can characterize rigidity in even two dimensions.

### 7.1.3 Dulmage-Mendelsohn Decomposition

Inspired by the algorithms for rigidity detection as discussed in Chapter 1, these algorithms use a form of maximum matching to generate the DR-plan. Such algorithms are referred to as *generalized maximum matching*, *generalized network flow*, or *Dulmage-Mendelsohn* decompositions.

Consider the *equation graph* $(Y \cup U, E)$ made by the system of equations of a GCS, where $|Y|$ is the number of equations, $|U|$ is the number of unknowns, and for all unknowns contained in an equation there is an edge between their associated vertices. This graph is naturally bipartite. However, this does not account for trivial motions. So often, when working with geometric constraint systems, the trivial motions should be factored out. This is done by pinning the system to the embedding space. This means that some of the variables are removed. In a bar-joint system, a single edge and the two associated vertices are removed, essentially fixing one vertex at the origin, and the other on some axis.

The system is well-constrained if and only if the equation graph satisfies the König-Hall relation: for any subset of equations $Y'$ of set $Y$, then $|\Gamma(Y')| \geq |Y'|$ where $\Gamma(Y')$ is the set of all neighbors of $Y'$, i.e. the set of unknowns appearing in $Y'$. This is equivalent to the existence of a perfect matching. A system is *irreducible* if every subsystem is under-constrained. An under-constrained system satisfies $|\Gamma(Y')| \gneqq |Y'|$

Ref. [1] introduced a new form of decomposition that works for any independent input. If the input $G$ is well-constrained, then it is bipartite and has a perfect matching, making the algorithm much simpler. The well-constrained variant is presented in Algorithm 3, their paper gives more details regarding the complete algorithm.

A directed graph is *strongly connected* if, for all $i$ and $j$ in the vertex set, there is a path from $i$ to $j$ and from $j$ to $i$. A *strongly connected component* of a graph is a maximal strongly connected subgraph. Finding all strongly connected components can be done in $O(|E|)$, via several well-known algorithms (such as Tarjan's.) Note that the strongly connected components are indecomposable (irreducible) subgraphs of $G$.

The resulting graph $R$ is a DAG. A DR-plan can be made from $R$ by Algorithm 4.

**Algorithm 3** Dulmage-Mendelsohn decomposition of well-constrained graphs

1: Find a maximum (perfect) matching $M$ of $G = (Y \cup U, E)$.
2: Build a directed graph $G'$ by replacing all edges $(s,t) \in M$ with the directed edges $(s,t)$ and $(t,s)$ and directing all other edges from $Y$ to $U$.
3: Compute the strongly connected components of $G'$.
4: Contract each strongly connected component into a single vertex to get $R$.

**Algorithm 4** DR-plan from DM decomposition

1: Run some topological sort to get a total ordering, with the largest element being some sink in the DAG.
2: Add a node whose children are the equations (constraints) in the largest element.
3: **for** each preceding element $a_{i-1}$ in the ordering **do**
4: Add a node whose children are the equations in $a_{i-1}$ and the node that was added for $a_i$.
5: **end for**

### 7.1.4 Assur Decomposition

Assur graphs (traditionally *Assur Groups*) were introduced by Leonid Assur in 1914 [2]. Originally a tool in mechanical engineering, the first paper to use the mathematics of rigidity theory was [15]. The terminology used in this section will be from the rigidity community and is mostly due to [15].

Since their introduction, they have become a mainstay in the design and analysis of *mechanisms* (1-dof linkages). Originally designed for 2D pinned linkages, the concepts have been generalized to all dimensions and symmetric settings [13].

The type of system considered in this section are bar-joint graphs with additional unary pin constraints and the restriction that no two vertices with a pin can share an edge. A *pin* is a unary constraint that fixes both degrees-of-freedom of a joint, giving it an absolute position relative to the embedding space. Joints without pin constraints are called *unpinned*, *free*, or *inner* vertices. The rigidity of a *pinned graph* (or *grounded graph* ) is different, as the system must not only be rigid but also have no trivial motions. Furthermore, a generically rigid pinned graph only requires that inner vertices be placed generically and pinned vertices be distinct. The rationale behind this is motivated by mechanical engineering, the trivial motions are uninteresting, only the internal motions of the mechanism itself are of concern. Also, while this section defines everything in terms of graphs, the theory is easily generalized to multigraphs.

A *d-Assur graph* is a minimally rigid (in $d$ dimensions) pinned graph whose proper subgraphs are all flexible. An *Assur decomposition* is a partial order of the Assur graphs in a minimally rigid pinned graph, which is unique [15].

Given input graph $G = (I \cup P, E)$, where $I$ is the set of inner vertices and $P$ is the set of pinned vertices; therefore, $E \subseteq (I \times I) \cup (I \times P)$. See Algorithm 5 for Assur decomposition.

**Algorithm 5** Assur Decomposition

1: Find a *d-directed orientation* of $G$ ▷ see definition below
2: Combine all pinned vertices into a single sink ▷ referred to as *grounding*
3: Find the strongly connected components ▷ see Section 7.1.3 for details

The first step is finding a *d-directed orientation* of $G$, which is an assignment of directions to the edges such that every inner vertex has out-degree $d$ and every pinned vertex has out-degree zero. Every minimally rigid $d$-dimensional pinned graph has a $d$-directed orientation. If $d = 2$, this can be found quickly via algorithms such as a slightly modified pebble game [10]. In higher dimensions, no combinatorial polynomial time algorithm is known.

While the $d$-directed orientations of a graph are not unique, they are unique modulo the orientation within a strongly connected component. Any two direction assignments on the same graph with the same out-degree at every vertex can be obtained from one another simply by the reversal of directed cycles. Thus, any $d$-directed orientation is equivalent to another modulo orientation of cycles. This means that the strongly connected components are the same and the decomposition is invariant to the choice of $d$-directed orientation on those graphs that admit them.

The strongly connected components are the $d$-Assur graphs of $G$. The directed edges between the components are a partial order on the Assur graphs to guide decomposition. A DR-plan can be made in a fashion similar to Algorithm 4. Simply perform a topological sort to get a total ordering, and at each level of the DR-plan recombine the growing graph with another component (Assur graph), namely the next one in the ordering.

When $d = 2$, a pinned graph admits a 2-directed orientation if and only if it is minimally rigid. Furthermore, the response of inner vertices to using a driver is not intuitive, as discussed above, unless $d = 2$. Nevertheless, Assur decomposition can be generalized to $d > 2$, as in [16]. In higher dimensions, a $d$-directed orientation is a necessary condition for $d$-dimensional rigidity of pinned graphs.

If, by the removal of a single single edge, all inner vertices of a $d$-Assur graph go into motion (i.e., there is some satisfying infinitesimal flex), the graph is called *strongly $d$-Assur*. In the case of 2-Assur graphs, the graph is always strongly 2-Assur; for dimensions greater than two, it is not always the case [16].

Originally used in mechanical engineering, Assur decomposition is well-suited to study the motion of under-constrained systems with 1 dof. A *non-edge* is an intuitive term that refers to a pair of vertices without an edge between them. In a system with 1 dof, setting the length of a non-edge that is independent of all edges gives a well-constrained graph from which a realization can be obtained. By continuously varying the length of this non-edge, the graph continuously changes realizations. Such a non-edge is called a *driver*, as it drives the motion of the mechanism. By performing Assur decomposition on minimally rigid graphs, the partial ordering imposed on the edges will describe the motion of the entire system given some edge were used as a driver.

Instead of using a rigid graph, one edge can be chosen to be a driver, to which some velocity is assigned. The position of all vertices can be easily calculated when changing the length of the driver. A driver only affects the positions of vertices in Assur graph components higher in the partial ordering. The velocities of the vertices in the component containing the driver can be calculated, and then used as driver velocities in the calculation of movement higher up the DR-plan tree.

For a better understanding of the configuration space of under-constrained systems (such as these 1 dof systems), see Chapter 10.

### 7.1.5 The Frontier Vertex Algorithm

Frontier is a DR-planner that operates via a bottom-up, network-flow based algorithm. Algorithm 6 describes Frontier. The problem of finding the minimum dof-rigid (see Chapter 1) subgraph is NP-hard [12] even for 2D. The algorithm *dense* can be used to find a *minimal*, non-trivial (i.e., more than a single edge) dof rigid subgraph (see Line 3 in Algorithm 6). See Chapter 1 for an explanation of dof-rigid subgraphs and [7] for an explanation of algorithm *dense*.

Some terminology used in the algorithm is defined here. Given some input $G$ and a subgraph $S = (I \cup F, E)$, those vertices $F$ which are adjacent to at least one vertex not in $I \cup F$ are called *frontier* vertices, and those vertices $I$ that are adjacent to only frontier and other vertices in $I$ are called *inner* vertices.

Frontier works for independent and over-constrained systems. For over-constrained systems, output is certainly not an optimal DR-plan, in that it does not guarantee optimal fan-in. For well-

**Algorithm 6** Frontier

1: $G_0 = G$.
2: **while** $G_i$ is not a single vertex **do**
3: Find $S_i = (I \cup F, E_{S_i})$, an arbitrary dense subgraph of $G_i$.
4: $G_{i+1} = G_i$.
5: Remove all edges in $E_{S_i}$ from $G_{i+1}$.
6: Remove all inner vertices $I$ from $G_{i+1}$.
7: Add a single vertex (the *core*) to $G_{i+1}$.
8: Add an edge from every frontier vertex $F$ to the core. The weight of the new edge is the sum of the weights of the original edges that got merged into the new edge, i.e. all those that went from this frontier vertex to some inner vertex. Call this set of edges $E_F$.
9: The core is given the weight $\sum_{v \in F} w(v) - \sum_{e \in E_F} w(e) - D$. ▷ i.e. the weight such that this new subgraph $(F \cup \text{core}, E_F)$ is dense.
10: **end while**

constrained systems, there is no guarantee of optimality, although whether or not the output is optimal is unknown. What Frontier does ensure is that the output DR-plan has cluster minimality. *Cluster minimality* ensures that there is no proper subset (of size greater than 1) of children whose union is dense.

### 7.1.6 Canonical Decomposition

Thus far, several DR-planning algorithms have been presented that create DR-plans in polynomial time for independent input graphs with an underlying sparsity matroid (see Chapter 1). However, the bottom-up nature of those algorithms made it difficult to provide formal guarantees about optimality; instead they made guarantees about other desirable properties, discussed previously. Here we discuss a top-down $O(|V|^3)$ algorithm that produces an optimal DR-plan for such graphs, written by the author of this survey [3].

In this section, we discuss only 2D bar-joint systems, although the paper provides extensions to body-bar, body-hyperpin, and pinned line-incidence systems. The unifying structure of these diverse systems is: a combinatorial abstract rigidity matroid, and for the algorithmic results an underlying sparsity matroid generalization.

We define a *canonical DR-plan* as:

**Definition 7.2 Canonical DR-plan** A DR-plan that satisfies the two additional properties:

(a) Children are rigid vertex-maximal proper subgraphs of the parents, and

(b) If all pairs of rigid vertex-maximal proper subgraphs intersect trivially then all of them are children, otherwise exactly two that intersect non-trivially are children.

A *trivial* subgraph is taken to mean a single vertex. See Figure 7.2 for an example of a canonical DR-plan. This definition gives gives the canonical DR-plan a surprisingly strong Church-Rosser property, stated in the following theorem.

**Theorem 7.1** *A canonical DR-plan exists for a graph G and any canonical DR-plan is an optimal DR-plan if G is independent.*

As a result of the preceding theorem, any canonical DR-plan we create will be optimal (of which there are potentially exponentially many.) Furthermore, if we construct a DR-plan where all nodes have fan-in less than or equal to the fan-in of some canonical DR-plan, we have made an optimal

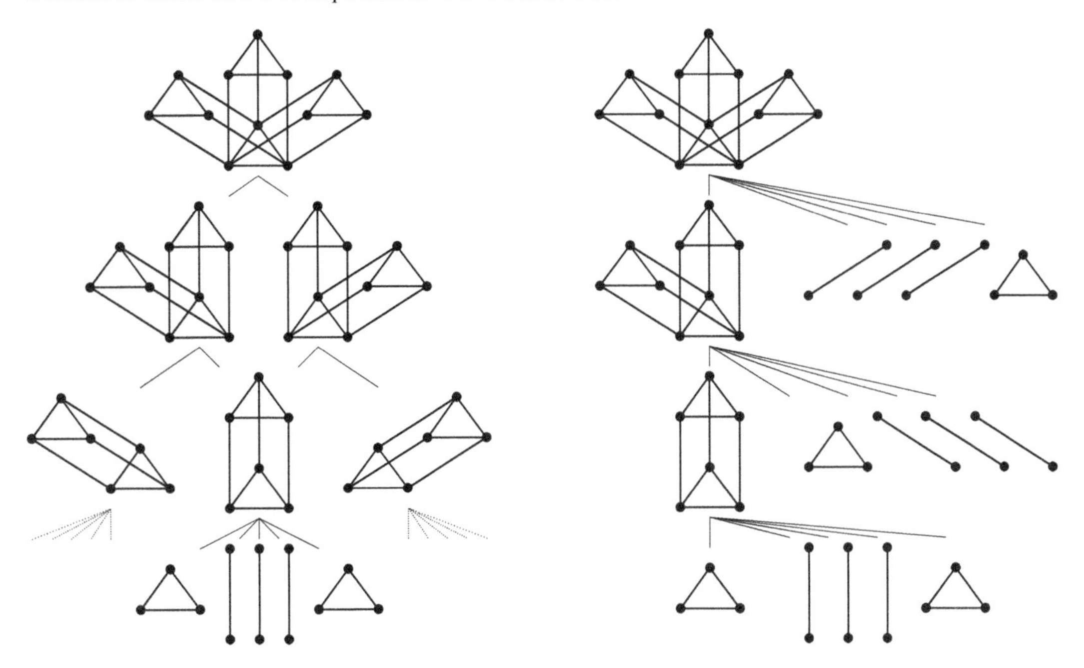

**Figure 7.2**
Two DR-plans of the same 2D bar-joint constraint graph (three $C_2 \times C_3$ graphs, or doublets, intersecting on a single triangle.) The first is a canonical DR-plan, the first and second levels demonstrating non-trivial intersections. The second is a pseudosequential DR-plan.

DR-plan. Such an alternative DR-plan (which we show to be optimal) is the *pseudosequential DR-plan*:

**Definition 7.3 Pseudosequential DR-plan** A DR-plan recursively defined as follows:

(a) Base case: When $G$ is an edge, the pseudosequential DR-plan for $G$ is $G$ itself.

(b) In case the pairwise intersections of the rigid vertex-maximal proper subgraphs $C_i$ of $G$ are all trivial, take the children of $G$ to be the roots of the pseudosequential DR-plans for $C_i$.

(c) In case there are two rigid vertex-maximal proper subgraphs $C_i$ and $C_j$ of $G$ with non-trivial intersection, take the children of $G$ to be the roots of pseudosequential DR-plans for $C_j \setminus C_i$ (called an *appendage*), and $C_i$ (called its *partner*).

See Figure 7.2 for an example of a pseudosequential DR-plan. Note the difference with canonical, when the intersection is nontrivial instead of the two proper vertex-maximal rigid subgraphs $C_i$ and $C_j$ both becoming children, the appendage and partner become children. This has the important affect of making every node in the DR-plan unique, leading to a much smaller size. Whereas canonical could be exponential in the size of the graph, pseudosequential will be linear while still being optimal. Additionally, with its much smaller size, finding a pseudosequential DR-plan has an efficient algorithm, as expressed by the following theorem.

**Theorem 7.2** *Computing a pseudosequential DR-plan for an independent graph $G = (V,E)$ has time complexity $O(|V|^3)$.*

The algorithm involves recursively "growing" the DR-plan tree one branch at a time. We call a

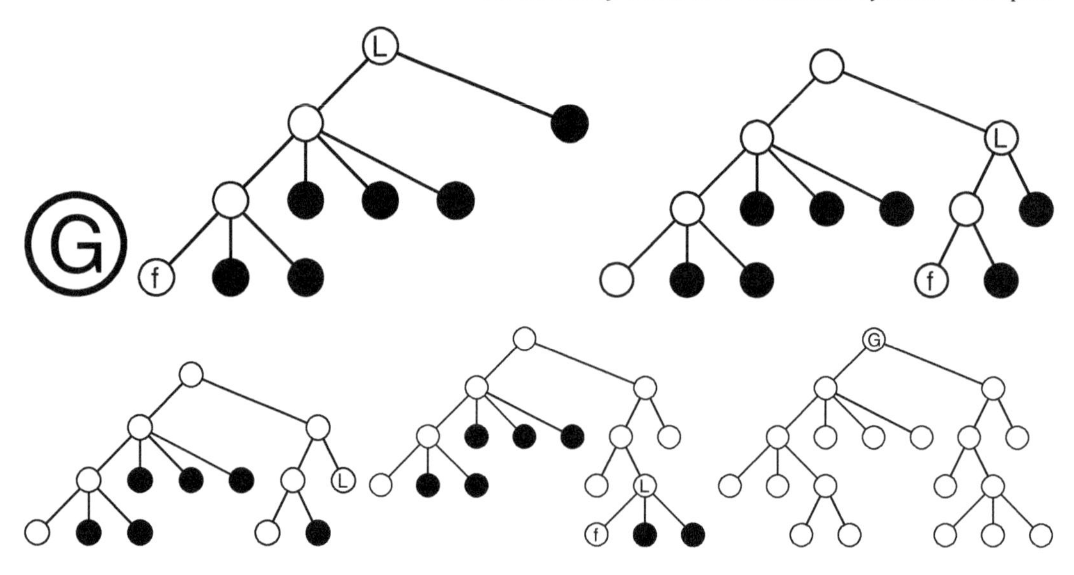


**Figure 7.3**
An illustration of the running of the algorithm to find a pseudosequential DR-plan. The DR-plan begins as just the input graph. A leaf is chosen (only can choose $G$) and the branch is computed down to arbitrary edge $f \in L$. Then another leaf is chosen, and its branch is computed, etc. In this way, the entire tree is built.

$branch(T,a,b)$ of tree $T$ the subtree that consists of the path from node $a$ to $b$ plus all children of the nodes on the path.

Given some graph $G$ and edge $e \in G$, there exists a pseudosequential DR-plan $P_G$ where the leaves of branch$(P_G, G, e)$ is exactly, ignoring some exceptions here, the rigid vertex-maximal proper subgraphs of $G \setminus e$. The branch $P_G$ can be built with Algorithm 7.

---

**Algorithm 7** Branch from rigid vertex-maximal proper subgraphs

1: Compute the rigid vertex-maximal proper subgraphs of $G \setminus e$, $\{L_i\}$.
2: **for** each $L \in \{L_i\}$ **do**
3: Choose an arbitrary edge $f \in L$ and compute the rigid vertex-maximal proper subgraphs of $G \setminus f$, $\{M_i\}$.
4: Compare the intersection of $L$ with each $M$ to get its position relative to the other leaves.
5: Compute nodes on the path from $G$ to $e$.
6: **end for**

---

Assuming a preprocessing step of finding all of the rigid vertex-maximal proper subgraphs of $G \setminus e$, for all $e$, then the above algorithm runs in time $O(|V|^2)$. The details of Step 4 are omitted, however, each loop $L$ is being used as a pivot; after running this on each leaf the position of each leaf relative to one another is determined. Step 5 is simple because the union of the nodes is the parent, and we have all but one node and we know how they intersect.

See Algorithm 8 for how to build the entire DR-plan $P_G$.

For an illustration of this algorithm, see Figure 7.3. Computing rigid vertex-maximal proper subgraphs is time $O(|V|^2)$, via the pebble game algorithm [10]; thus, preprocessing takes time $O(|V|^3)$.

**Algorithm 8** Pseudosequential DR-plan from branches

1: Preprocessing: Compute the rigid vertex-maximal proper subgraphs of $G \setminus e$, for all $e$.
2: Start with $G$ as the single node in the DR-plan.
3: Recursively compute a branch for each leaf in the DR-plan (that is not a single edge.)

Observe that the size of a pseudosequential DR-plan is $O(|V|)$; thus, the complexity of the last two steps will be $O(|V|^3)$ as well.

We conjecture that the, for independent input, Frontier (see Section 7.1.5) produces a pseudosequential DR-plan and is therefore optimal as well. However, the bottom-up method of computation makes analysis much more difficult. Both the complexity and the optimality in the case of well-constrained input is difficult to understand for Frontier.

## 7.2 Efficient Realization

Given a geometric constraint system, a realization can naïvely be found by directly finding the zeros of the corresponding polynomial system. If the independently solvable subsystems are found, then those can be solved separately and then recombined. This process applied recursively can be done with DR-planning. At this point, one may solve this system bottom-up; recombining constraints in the leaves into the parent node, working up to the root. The DR-plan tree is a partial order in which the solutions can be calculated.

While some subclasses of systems have efficient realization schemes, such as triangle-decomposable systems, this is not the case in general. Upon decomposition, there can be arbitrarily large systems that must be solved simultaneously. Even in an optimal DR-plan, the fan-in at any node can be as large as the number of constraints in the corresponding subgraph; such a system is called *indecomposable* . Therefore, it may be desirable to further reduce the algebraic complexity of the subsystems that must be solved.

### 7.2.1 Numerical Instability of Rigid Body Incidence and Seam Matroid

Consider bar-joint systems, where a *trivial* intersection is taken to mean a vertex in 2D, an edge in 3D, etc. When recombining rigid bodies whose intersection is non-trivial, and necessarily rigid, it can be done by simply rotating the reference frame of one system so that the overlapped primitives align. However, given a trivial intersection, there must be some complex system of rigid bodies. This system is its own polynomial system that can be solved.

Consider two rigid bodies in 3D that intersect on two vertices which represent six incidence constraints. The two bodies individually have six dof each, and the count indicates the system is well constrained . However, one of these incidence constraints is dependent as the two body system clearly has one degree of freedom. All six dependencies will result in numerical instability when the constraint system is being solved.

To address this, one incidence constraint can be dropped. As it turns out, there is an underlying matroid [18], the *seam matroid*, characterizing the dependencies of these constraints. This matroid admits an efficient greedy algorithm (the well-formed incidences algorithm) to find a maximal independent set of constraints.

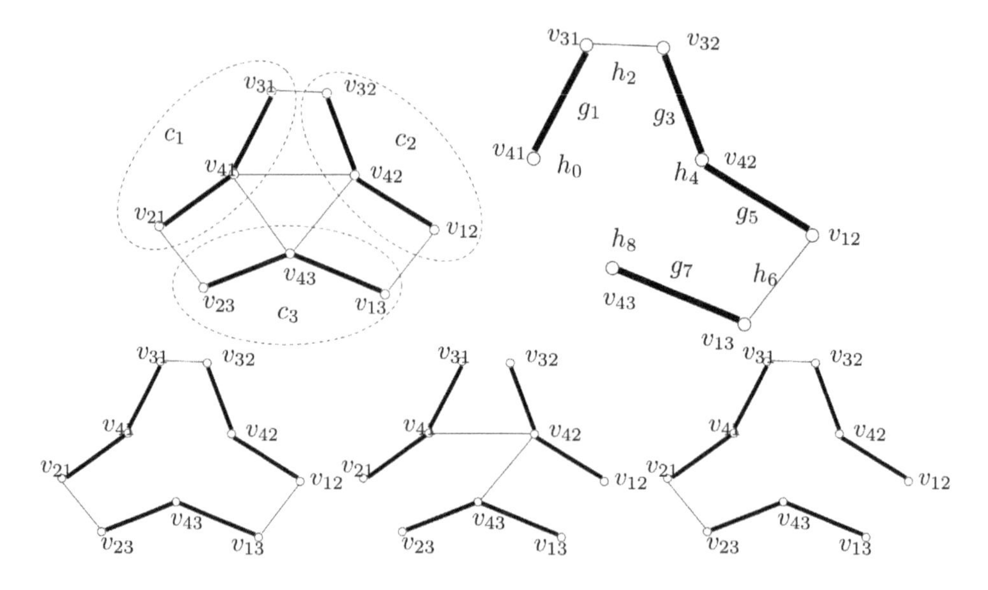


**Figure 7.4**
(Top,Left) shows the seam graph for a decomposition; this is the ground set of the seam matroid. (Top Right) shows a seam graph that is independent but not maximal. (Bottom, Left) shows a dependent seam graph. (Bottom, Middle) and (Bottom, Right) Show two maximally independent seam graphs.

### 7.2.2 Optimal Parameterization in Recombination

Obtaining a realization of the constraint system of an internal node in a DR-plan can be done by choosing a *home coordinate system*, that of some child node, and repositioning the rest of the children (by using the allowable rigid motions of the system) in the chosen coordinate system such that all incidences are satisfied.

To minimize the algebraic complexity of this recombination system, Sitharam et. al. in [20] introduce a combinatorial class of incidence tree parametrizations for a collection of rigid bodies. Minimizing the algebraic complexity over this class reduces this to a purely combinatorial optimization problem that is a special case of the set cover problem. They also quantify the exact improvement of algebraic complexity obtained by optimization.

Consider DR-plans of bar-joint systems where all pairwise intersections are trivial, e.g., a vertex in 2D, a vertex or edge in 3D, etc. Then the children of a node from an *overlap graph*, where vertices are the set of nodes and edges represent incidence between the child subgraphs. A spanning tree of this overlap graph, with a specified root, is an *incidence tree* [20]. Using an incidence tree, the system of polynomials describing the constraint system can be rewritten, leading to reduction in the complexity of teh resulting parameterized system.

The choice of origin and spanning tree can cause significant improvements in the complexity of the polynomial system. The total weight of the incidence tree corresponds to the total number of parameters in the new polynomial system. The depth of an incidence tree corresponds to the degree of the polynomial system. Ref. [20] shows how to simultaneously optimize over both of these properties. The overlap graph is used to choose and order the elimination of incidences, i.e., to select an incidence tree parametrization that minimizes the number of variables and the algebraic degree of the parametrized core system (minimizing algebraic complexity)

We first define some background terms and describe the algorithm using Figure 7.5

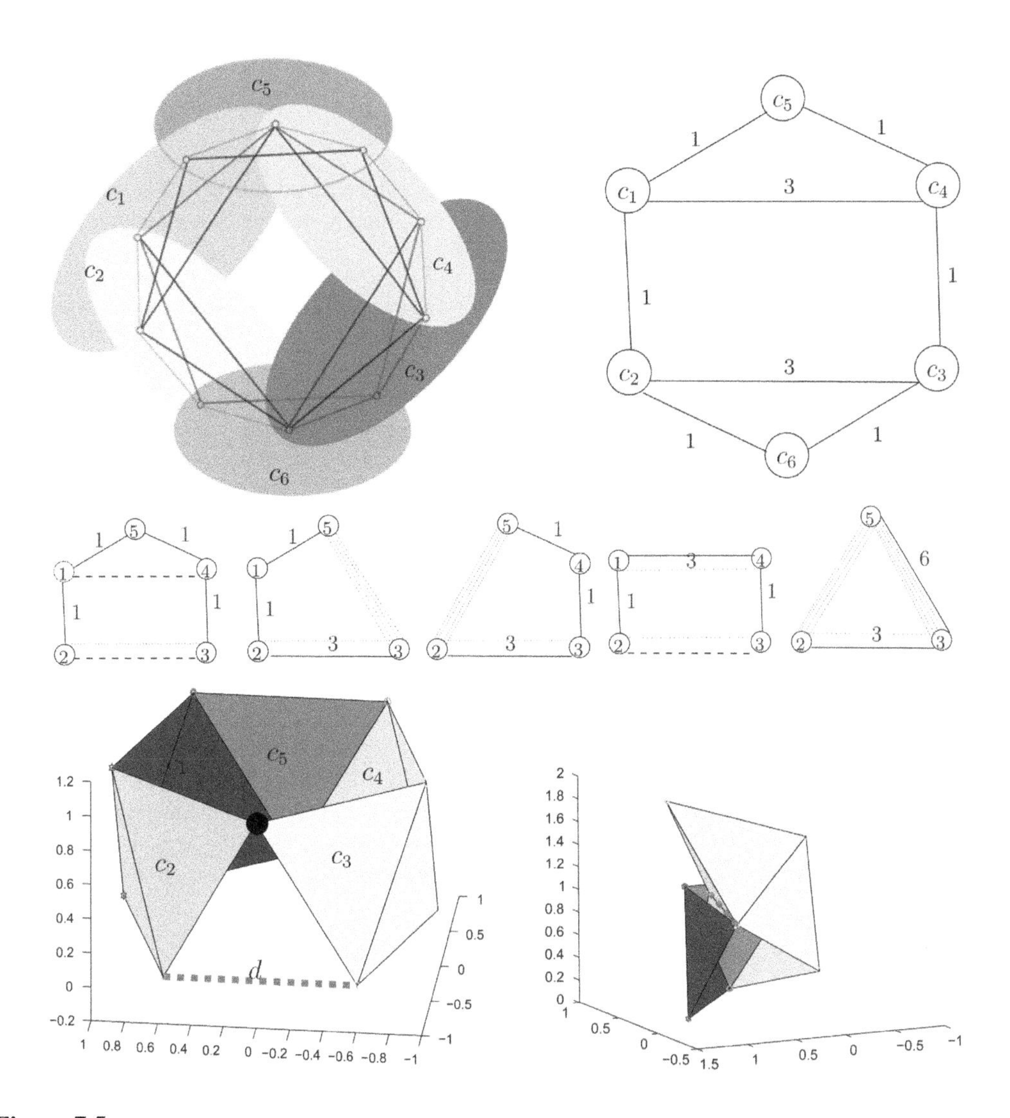


**Figure 7.5**
(Top, Left) Standard collections of rigid bodies and implied internal distances. (top, right) The corresponding weighted overlap graph (edges of weight 6 representing empty overlaps are omitted). (Middle) Five spanning trees of the overlap graphs of covering sets $S(c)$. The first and the fourth trees show implied non-tree distance constraints (shown in dotted lines) - they rigidify the remaining bodies in $c \setminus S(c)$. The label $i$ in a circle stands for $c_i$. (Bottom) Two of the eight possible solutions for the input collection of rigid bodies shown in (top, left). For each tetrahedral rigid body, $c_1$, $c_2$, $c_3$, $c_4$ , only two triangle facets are shown. The home rigid body $c_5$ is a triangle. The dashed line represents the parametrized distance constraint. The parametrized incidence constraint is marked as a black sphere. It forms a triangle with the distance constraint, i.e., together, they constitute the triangle rigid body $c_6$ (shown in top, left), that is not part of the current covering set (see Middle, Left). (Bottom, Left) This configuration can be recognized as representing two pentagons in the $z = 0$ and $z = 1$ planes respectively and rotated by $2pi/10$ with respect to one another. (Bottom, Right) This realization is self-penetrating.

**Collection of Rigid Bodies, Covering Set:** Let $c := c_1, \ldots, c_n$ be a set of rigid bodies and $X$ a set of shared (coordinate free) points in $\mathbb{R}^3$ that imply incidences, i.e., points that occur in two or more of the rigid bodies. The pair $(X, c)$ is a valid collection of rigid bodies (short: collection), if for $i \neq j$, $c_i$ contains at least one point in $X$ not in $c_j$, and $c_i \subsetneq X$ . We call $c' \subset c$ a covering set of $X$ if every point in $X$ lies in at least one rigid body in $c_0$.

**Proper-maximal:** A rigid body $c_i$ is proper-maximal in $X$ if there is no subset $u$ of $X$ with $c_i \subsetneq u \subsetneq X$ that represents a rigid body. That is, no such $u$ is rigidified by the incidences within $u$ and the fixed distances within the bodies outside $u$.

**Standard Collection of Rigid Bodies:** The collection $(X, c)$ is a standard collection of rigid bodies if the following hold.

(a) No pair $c_i \neq c_j$ intersects in more than two points.

(b) ($c$ is complete in $X$ ) All proper-maximal rigid bodies are in $c$.

(c) All $c_i$ are proper-maximal.

**Overlap Graph:** An *overlap graph* $\mathcal{G}(X,c)$ of a standard collection of rigid bodies $(X,c)$ is a weighted undirected graph whose vertices are the rigid bodies $c_j$ in $c$ and whose edges represent incidences between pairs of rigid bodies. If an edge between a pair $(c_i, c_j)$ represents $k$ incidences, the weight $w(k)$ assigned to the edge is the number of remaining dofs of $c_j$ after fixing $c_i$'s position and orientation and resolving the $k$ incidences between them.

Figure 7.5 shows an example where the *Optimized Incidence Tree Parametrization Algorithm* is used on an example standard collection of 6 rigid bodies. These rigid bodies can be viewed as the internal nodes in a DR-plan and the combined structure (bottom) as the parent of these nodes whose solution is being recombined from the solutions of its children. The figure in the top right shows its corresponding overlap graph. The figures in the middle show different spanning trees of the covering set. The figures at the bottom show the two realizations of the possible eight.

### 7.2.3 Reconciling Conflicting Combinatorial Preprocessors

The work in [21] reconciles the matroid of Section 7.2.1 with the work in this section to get a single matroid that can be used for an algorithm to solve both problems.

Given a collection of incident rigid bodies, a seam matroid is used to select a well-formed subset of incidences (maximal independent set), which ensures a correspondence between each solution of the original system and a solution to the numerically perturbed, well-formed system [19]. On the other hand, the overlap graph, defined earlier, is used to select incidence tree parameterization that minimizes the number of variables and hence algebraic complexity.

In [21], they show that while preprocessing, choosing parameters using seam graphs ignores algebraic complexity and choosing parameters, so as to reduce complexity does not guarantee well-formedness. They present a new algorithm, *Optimal Well-Formed Incidence Selection Algorithm*, that combines the key elements of the Optimal Incidence Tree Algorithm [20] with the Well-formed Incidences Algorithm [18] to guarantee both optimal algebraic complexity and well-formedness. This greedy and efficient algorithm is the result of a careful construction and proof of an independence-preserving map between the cycle matroid associated with the overlap graph and the seam matroid underlies well-formed systems of incidences.

## 7.3 Handling Under- and Over-Constrained Systems

Most of the content in Sections 7.1 and 7.2 considered independent systems (i.e., not over-constrained.) Even more so, the focus was predominantly on well-constrained systems; when dealing with under-constrained systems, the response was typically to operate only on the rigid components. However, there are several interesting concerns regarding under- and over-constrained systems.

For under-constrained systems the question of which constraints to add to make the system well constrained and what the constraint parameter values should be to guarantee a realization is discussed Chapter 10.

As discussed in Chapter 1, an over-constraint is a dependency in the rigidity matroid. An *n-over-constrained* system is a system with $n$ over-constraints. Detecting the minimal set of edges containing a 1-over-constrained system corresponds to finding a circuit in the rigidity matroid. As such, there is a unique minimal 1-over-constrained subsystem in a 1-over-constrained system. For an in-depth discussion on rigidity circuits, see [14].

Ref. [9] presents an algorithm for discovering this minimal subsystem in a 1-over-constrained system in the 2D bar-joint case. This algorithm uses the DR-plan for efficient detection, and moreover provides a method for quickly updating the DR-plan to reflect the dropped over-constraint.

The 1-over-constrained case arises naturally in practice; as a user incrementally adds constraints to a system single over-constraints can be identified as they occur. However, when considering a complete constraint graph, the $k$-over-constrained case can occur.

## 7.4 User Intervention in DR-Planning

In addition to mathematical properties that are important to satisfy, a useful DR-planner must also be able to handle interaction with a user. In practice, DR-planners are used in live environments where the designer is often tweaking and updating the geometric constraint system on which they are working. Often beginning with some under-constrained sketch of the final product and building up, they will frequently be adding and removing constraints. Additionally, the final product may be inconsistently over-constrained, requiring changes before the system can be solved.

Whenever intervention is required, while offering guidance to the user the constraint solver [9]:

- Should not require that the end-user be familiar with the mathematics involved behind the scenes.
- Should not artificially limit choices or include irrelevant choices.
- Should present intuitive and consistent choices that are unique in some well-defined manner.

One concern that is explicitly addressed in Section 7.1 and 7.2 is efficiency. As in all real-time applications, the user should wait as little as possible. Using optimal DR-plans, with quick recombination schemes and methods of solving, reduces the impact of redecomposing and solving the system and any subsystem when making changes.

### 7.4.1 Root Selection and Navigation

A rigid geometric constraint system may have more than one solution. In fact, it can be exponential in the number of geometric primitives. Consider a 2D bar-joint system, whose constraint graph is a 2-tree. At the placement of each vertex, there will be two positions that satisfy the distance constraint, resulting in $O(2^{|V|})$ realizations. This, of course, does not only happen with 2-trees. These non-congruent realizations are often called *orientations*. Choosing a specific realization may not only be of interest to the user, but may also be necessary for reasonable computation times.

In practice, often when designing systems with real-world application, self-intersection is a concern. Therefore, it may be natural to consider imposing that only non-intersection realizations are returned [5]. However, finding non-intersecting results for even a simple polygon is NP-hard. So this approach cannot be applied in general.

Another possible approach is to allow the user to offer additional over-constraints. This can significantly reduce the number of acceptable solutions [5]. While finding an optimal DR-plan is NP-hard in the presence of over-constraints, they can only improve the maximum fan-in since the original well-constrained systems DR-plan can be used by default. However, requiring additional work by the user may be undesirable.

Another option is to use heuristics to try to find the realization that most closely matches the user input sketch [5]. As the realization is being computed and the solver chooses orientations, make the choice that preserves the orientation in the sketch; if a point is on one side of an edge in the drawing, choose the solution that matches this. However, the orientation as it appears in the sketch may not have a real solution, in which case the algorithm must make an arbitrary choice.

A different approach is to instead involve the user in the recombination process. As the solver finds roots for the subsystems, the user may choose at this time. This choice then limits the space of real solutions as the subsystem is incorporated into its ancestors in the DR-plan. However, the choice the user made may not allow for any real solutions, so the algorithm must be able to identify the problematic subsystem.

### 7.4.2 Changing Constraint Parameters

A DR-plan is combinatorial, working with the underlying constraint graph. Therefore, a change in the constraint parameters of a geometric constraint system has no effect on its optimal DR-plan. Thus, the user is free to change these parameters without necessitating additional computation to modify the DR-plan. However, the realization will change, and in real-time applications it may be undesirable to have to re-solve the entire constraint system when changing a single constraint.

The structure of a DR-plan gives insight in how to eliminate repeated computation. The altered constraint appears in some leaf (or leaves) of the DR-plan. Only the ancestors of this leaf need be recalculated. So long as the subsystem in its own coordinate system is stored at each node in the DR-plan data structure, these results can be reused.

However, there is the underlying assumption that the system is generic, otherwise there is no guarantee that the nodes are truly rigid or that there is any realization at all. Therefore, this must be checked upon user input.

### 7.4.3 Dynamic Maintenance

*Dynamic maintenance* refers to the adding and removing of entire constraints. Unlike changing the values of parameters, this necessitates recomputation of the DR-plan. The challenges involved arise from the discussion in Section 7.3. Additional consideration comes from the assumption that a DR-plan already exists prior to the alterations, since this is happening in a live environment. Therefore, reusing portions of the DR-plan may be an option to save computation.

## 7.5 Conclusion

This survey discusses DR-planning and overviews several important DR-planning algorithms (Section 7.1.) Then, focusing on the primary use of DR-planning, i.e. realizing the graph, several different ways to further improve the time complexity were examined (Section 7.2.) Then, related questions regarding how to handle under- and over-constraints were addressed (Section 7.3.) Finally, the software design perspective was considered, regarding real-world concerns of user input and intervention (Section 7.4.)

The main goal of this survey was to motivate the importance of DR-planning. With its very general definition, DR-planning can be applied to any geometric constraint system. The types of primitives and constraints possible to use are varied, allowing for the modeling of a diverse cast of applied problems. Thus, work in the field has far-reaching impact.

Of particular interest to the author for future research is isolating the relevant combinatorial structure that allows for efficient DR-planning. What is the most minimal and still conceptually useful structure that unifies the diverse group of constraint systems for which DR-planning is efficient? Seemingly, the class must include those systems with an underlying abstract rigidity matroid, and further there must be an efficient algorithm for detecting rigidity. Is this so? Must there be an underlying sparsity matroid, as [3] requires? The author conjectures that even $(3,6)$-sparse graphs (which do not have and underlying sparsity matroid) may still admit a canonical DR-plan.

## References

[1] Samy Ait-Aoudia, Roland Jegou, and Dominique Michelucci. Reduction of constraint systems. *Compugraphics*, 25:83–92, December 1993.

[2] Leonid Assur. Issledovanie ploskih sterznevyh mehanizmov s nizsimi parami s tocki zreniya ih struktury i klassikacii. *Izdat. Akad. Nauk SSSR, Edited by I. I. Artobolevski*, 1952.

[3] T. Baker, M. Sitharam, M. Wang, and J. Willoughby. Optimal decomposition and recombination of isostatic geometric constraint systems for designing layered materials. *Computer-Aided Geometric Design*, 40:1–25, 2015.

[4] William Bouma, Ioannis Fudos, Christoph Hoffmann, Jiazhen Cai, and Robert Paige. Geometric constraint solver. *Computer Aided Design*, 27(6):487–501, 1995.

[5] Ioannis Fudos. Editable representations for 2D geometric design. Master's thesis, Purdue University, December 1993.

[6] Ioannis Fudos and Christoph M. Hoffmann. Correctness proof of a geometric constraint solver. *International Journal of Computational Geometry & Applications*, 06(04):405–420, 1996.

[7] Christoph M. Hoffman, Andrew Lomonosov, and Meera Sitharam. Decomposition plans for geometric constraint problems, part II: new algorithms. *Journal of Symbolic Computation*, 31(4):409–427, 2001.

[8] Christoph M. Hoffman, Andrew Lomonosov, and Meera Sitharam. Decomposition plans for geometric constraint systems, part I: Performance measures for CAD. *Journal of Symbolic Computation*, 31(4):367–408, 2001.

[9] Christoph M. Hoffmann, Meera Sitharam, and Bo Yuan. Making constraint solvers more usable: overconstraint problem. *Computer-Aided Design*, 36(4):377–399, 2004.

[10] Donald J. Jacobs and Bruce Hendrickson. An algorithm for two-dimensional rigidity percolation: The pebble game. *Journal of Computational Physics*, 137(2):346–365, 1997.

[11] Christophe Jermann, Gilles Trombettoni, Bertrand Neveu, and Pascal Mathis. Decomposition of geometric constraint systems: A survey. *International Journal of Computational Geometry & Applications*, 16(05n06):379–414, 2006.

[12] Andrew Lomonosov. *Graph and combinatorial analysis for geometric constraint graphs*. PhD thesis, University of Florida, 2004.

[13] Anthony Nixon, Bernd Schulze, Adnan Sljoka, and Walter Whiteley. Symmetry adapted Assur decompositions. *Symmetry*, 6(3):516, 2014.

[14] Brigitte Servatius. *Planar Rigidity*. PhD thesis, Syracuse University, 1987.

[15] Brigitte Servatius, Offer Shai, and Walter Whiteley. Combinatorial characterization of the Assur graphs from engineering. *European Journal of Combinatorics*, 31(4):1091–1104, 2010. Rigidity and Related Topics in Geometry.

[16] Offer Shai, Adnan Sljoka, and Walter Whiteley. Directed graphs, decompositions, and spatial linkages. *Discrete Applied Mathematics*, 161(18):3028–3047, 2013.

[17] Meera Sitharam. Combinatorial approaches to geometric constraint solving: Problems, progress, and directions. In Ravi Janardan, Michiel Smid, and Debasish Dutta, editors, *Geometric and Algorithmic Aspects of Computer-Aided Design and Manufacturing*, volume 67 of *DIMACS Series in Discrete Mathematics and Theoretical Computer Science*, chapter 5, pages 117–164. American Mathematical Society, 2005.

[18] Meera Sitharam. Well-formed systems of point incidences for resolving collections of rigid bodies. *International Journal of Computational Geometry & Applications*, 16(05n06):591–615, 2006.

[19] Meera Sitharam. Well-formed systems of point incidences for resolving collections of rigid bodies. *International Journal of Computational Geometry & Applications*, 16(05n06):591–615, 2006.

[20] Meera Sitharam, Jörg Peters, and Yong Zhou. Optimized parametrization of systems of incidences between rigid bodies. *J. Symb. Comput.*, 45(4):481–498, April 2010.

[21] Meera Sitharam, Yong Zhou, and Jörg Peters. Reconciling conflicting combinatorial preprocessors for geometric constraint systems. *International Journal of Computational Geometry & Applications*, 20(06):631–651, 2010.

# Part II

# Distance Geometry, Real Algebraic Geometry, and Configuration Spaces

# Chapter 8

# *Dimensional and Universal Rigidities of Bar Frameworks*

**A. Y. Alfakih**

*University of Windsor, Windsor, Ontario, Canada*

**CONTENTS**



## 8.1 Introduction

Suppose that we are given a configuration $p = (p^1, \ldots, p^n)$ of points in $r$-Euclidean space where a specified subset of the inter-point distances is to be preserved. A natural question to ask is whether there exists another configuration $q$ that satisfies these specified distances. To put it differently, we are interested in the problem of determining whether a given proper subset of the inter-point distances in a configuration $p$ uniquely determines $p$ up to a rigid motion. This problem can be conveniently described in terms of bar frameworks.

A bar framework $(G, p)$ (or simply a framework*) in dimension $r$ is a connected simple (no loops or multiple edges) graph $G$ whose vertices are points $p^1, \ldots, p^n$ in $r$-Euclidean space, and whose edges are line segments corresponding to the specified inter-point distances to be preserved. Two frameworks $(G, p)$ and $(G, q)$ are *equivalent* if the corresponding edges of $(G, p)$ and $(G, q)$ have the same (Euclidean) length; and they are *congruent* if all corresponding inter-point distances of $(G, p)$ and $(G, q)$ are the same. A given framework $(G, p)$ is *universally rigid* if every framework that is equivalent to $(G, p)$ is in fact congruent to it. Therefore, a proper subset of the inter-point distances in a configuration $p$ uniquely determines $p$, up to a rigid motion, if and only if the corresponding bar framework $(G, p)$ is universally rigid.

Representing a point configuration by a Gram matrix renders the universal rigidity problem amenable to semidefinite programming methodology [23]. In fact, many universal rigidity results are

*We are only interested in bar frameworks in this chapter.

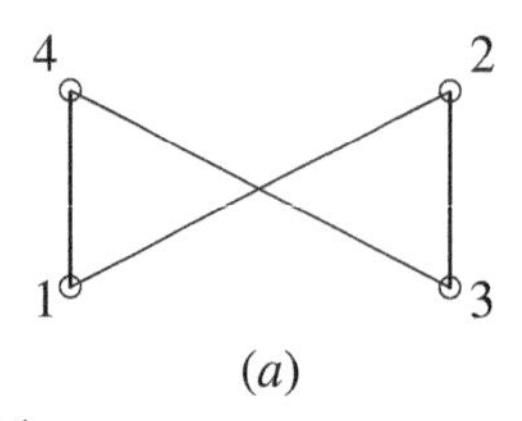


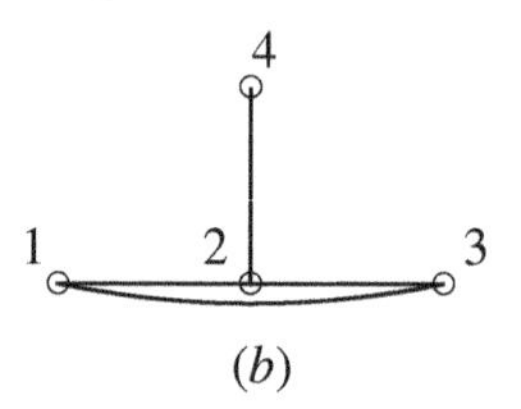


**Figure 8.1**
Two 2-dimensional bar frameworks in the plane. The edge $\{1,3\}$ in framework $(b)$ is drawn as an arc to make edges $\{1,2\}$ and $\{2,3\}$ visible. Framework $(a)$ is not dimensionally rigid and has no affine motions, while framework $(b)$ is dimensionally rigid and has an affine motion.

direct application of known facts in semidefinite programming. In addition to being a fundamental problem in distance geometry, the universal rigidity problem of bar frameworks has many important applications in multidimensional scaling in statistics, molecular conformations in chemistry and in ad hoc wireless sensor networks.

The notions of dimensional rigidity and affine motions turn out to be useful in the study of universal rigidity. Framework $(G,p)$ is $r$-dimensional if the points $p^1,\dots,p^n$ affinely span $r$-dimensional Euclidean space. An $r$-dimensional framework $(G,p)$, $r \leq (n-2)$, is *dimensionally rigid* if there is no $s$-dimensional framework $(G,q)$, equivalent to $(G,p)$, for any $s \geq r+1$. We say that $(G,p)$ has no *affine motions* if there is no framework $(G,q)$, equivalent but not congruent, to $(G,p)$ such that $q^i = Ap^i + b$ for all $i = 1,\dots,n$ for some matrix $A$ and vector $b$. Figure 8.1 depicts two 2-dimensional frameworks in the plane. Framework $(a)$ is not dimensionally rigid and has no affine motion. On the other hand, framework $(b)$ is dimensionally rigid and has an affine motion. Both frameworks are not universally rigid.

Universal rigidity is characterized in terms of dimensional rigidity and affine motions as follows.

**Theorem 8.1 (Alfakih [2])** *An $r$-dimensional bar framework $(G,p)$ on $n$ vertices, $n \leq (n-2)$, is universally rigid if and only if $(G,p)$ is both dimensionally rigid and has no affine motions.*

Thus, the universal rigidity problem can be addressed by separately addressing its two easier subproblems, namely dimensional rigidity and affine motions. For instance, sufficient conditions for dimensional rigidity and sufficient conditions for the absence of affine motions can be combined to establish sufficient conditions for universal rigidity.

## 8.2 Stress Matrices and Gale Matrices

The notion of a stress matrix is a useful tool in the study of dimensional and universal rigidities. An *equilibrium stress* (or simply a stress) of bar framework $(G,p)$ is a real-valued function $\omega$ on $E(G)$ such that:

$$\sum_{j:\{i,j\}\in E(G)} \omega_{ij}(p^i - p^j) = 0 \text{ for all } i = 1,\dots,n. \tag{8.1}$$

Let $\omega$ be a stress of $(G,p)$ and let $\overline{E}(G)$ denote the set of missing edges of $G$, i.e.,

$$\overline{E}(G) = \{\{i,j\} : i \neq j, \{i,j\} \notin E(G)\}. \tag{8.2}$$

Then the $n \times n$ matrix $\Omega$ where

$$\Omega_{ij} = \begin{cases} -\omega_{ij} & \text{if } \{i,j\} \in E(G) \\ 0 & \text{if } \{i,j\} \in \overline{E}(G) \\ \sum_{k:\{i,k\}\in E(G)} \omega_{ik} & \text{if } i = j \end{cases} \tag{8.3}$$

is called the *stress matrix* associated with $\omega$, or simply a *stress matrix* of $(G,p)$. Let $(G,p)$ be an $r$-dimensional bar framework on $n$ vertices in $r$-Euclidean space. The $n \times r$ matrix $P$ whose $i$th row is $(p^i)^T$, i.e.,

$$P = \begin{bmatrix} (p^1)^T \\ \vdots \\ (p^n)^T \end{bmatrix} \tag{8.4}$$

is called the *configuration matrix* of $(G,p)$. Observe that $P$ has full column rank since the points $p^1,\ldots,p^n$ affinely span the $r$-Euclidean space. Let $e$ denote the $n$-vector of all 1's, then it immediately follows from (8.3) that $e$ and the columns of $P$ are in the null space of $\Omega$. Consequently, a symmetric matrix $\Omega$ is a stress matrix of $(G,p)$ if and only if

$$\Omega P = 0,\ \Omega e = 0 \text{ and } \Omega_{ij} = 0 \text{ for all } \{i,j\} \in \overline{E}(G). \tag{8.5}$$

Hence, the rank of $\Omega$ is $\leq n-r-1$. In addition to being interesting in their own right, the following two theorems are frequently used in the proofs of subsequent results.

**Theorem 8.2 (Alfakih [4])** *Let $(G,p)$ be an r-dimensional bar framework on n vertices, $r \leq (n-2)$, and let $\Omega$ be a non-zero positive semidefinite stress matrix of $(G,p)$. Then $\Omega$ is a stress matrix for any bar framework $(G,q)$ that is equivalent to $(G,p)$.*

**Theorem 8.3 (Alfakih [4])** *Let $(G,p)$ be an r-dimensional bar framework on n vertices, $r \leq (n-2)$. Then $(G,p)$ admits a non-zero positive semidefinite stress matrix $\Omega$ if and only if there exists no $(n-1)$-dimensional bar framework $(G,q)$ that is equivalent to $(G,p)$.*

Note that if an $r$-dimensional framework $(G,p)$ on $n$ vertices, $r \leq (n-2)$, does not have a non-zero positive semidefinite stress matrix, then it follows from Theorem 8.3 that $(G,p)$ is not dimensionally rigid. As a result, Theorem 8.3 can be used to provide a necessary condition for dimensional rigidity and, consequently, universal rigidity.

Stress matrices are closely related to Gale matrices (or Gale transform) [16, 19]. Any $n \times (n-r-1)$ matrix $Z$ whose columns form a basis of the null space of $\begin{bmatrix} P^T \\ e^T \end{bmatrix}$ is called a *Gale matrix* of $(G,p)$. Sometimes it is more convenient to define Gale transform, as is usually done in the polytope literature [16, 19]. Let $(z^i)^T$ be the $i$th row of $Z$, then $z^i$ is called a *Gale transform* of $p^i$. The next theorem establishes the relationship between Gale matrices and stress matrices.

**Theorem 8.4 (Alfakih [3])** *Let $(G,p)$ be an r-dimensional bar framework on n vertices and let Z be a Gale matrix of $(G,p)$. Then $\Omega$ is a stress matrix of $(G,p)$ if and only if*

$$\Omega = Z\Psi Z^T \text{ and } \Omega_{ij} = 0 \text{ for all } \{i,j\} \in \overline{E}(G),$$

*where $\Psi$ is a symmetric matrix of order $n-r-1$.*

Theorem 8.4 shows that stress matrix $\Omega$ can be decomposed into a geometric part (Z), and a combinatorial one ($\Psi$).

Gale matrices have useful properties when the points $p^1,\ldots,p^n$ are in general position. We say that points $p^1,\ldots,p^n$ in $r$-Euclidean space are *in general position* if every $r+1$ of these points are affinely independent. For instance, points in the plane are in general position if no three of them are collinear.

***Lemma 8.1 [6]***

*Let $(G,p)$ be an r-dimensional bar framework and let $z^1,\ldots,z^n$ be Gale transforms of $p^1,\ldots,p^n$. Let J be a subset of $\{1,\ldots,n\}$ of cardinality $r+1$, and let the set of points $\{p^i : i \in J\}$ be affinely*

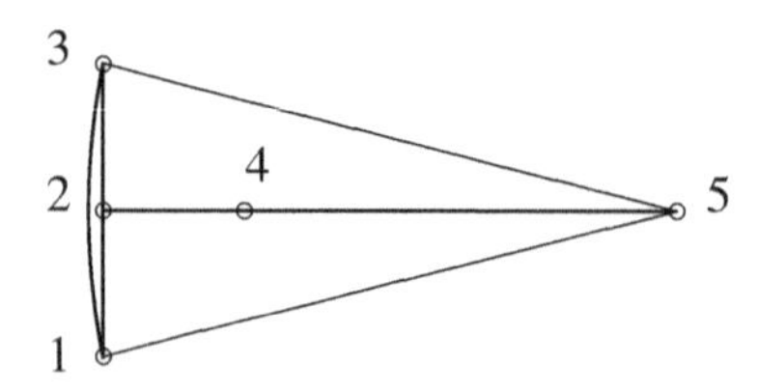


**Figure 8.2**
A dimensionally rigid 2-dimensional bar framework on 5 vertices which does not admit a positive semidefinite stress matrix of rank 2.

*independent. Then the set $\{z^i : i \in \bar{J}\}$ is linearly independent where $\bar{J} = \{1,\ldots,n\} \backslash J$, where $\backslash$ denotes the set theoretic difference.*

Hence, if $Z$ is a Gale matrix of a framework in general position, then every submatrix of $Z$ of order $n-r-1$ is nonsingular.

**Example 35** *Consider the 2-dimensional bar framework $(G,p)$ on 5 vertices depicted in Figure 8.2. The set of missing edges is given by $\overline{E}(G) = \{\ \{1,4\},\{3,4\},\{2,5\}\ \}$. Moreover, the configuration matrix and a Gale matrix of $(G,p)$ are given by:*

$$P = \begin{bmatrix} -1 & -1 \\ -1 & 0 \\ -1 & 1 \\ 0 & 0 \\ 3 & 0 \end{bmatrix} \text{ and } Z = \begin{bmatrix} 1 & 0 \\ -2 & 3 \\ 1 & 0 \\ 0 & -4 \\ 0 & 1 \end{bmatrix}.$$

*Notice that the points $p^1, p^3, p^5$ are affinely independent and the set $\{z^2 = \begin{bmatrix} -2 \\ 3 \end{bmatrix}, z^4 = \begin{bmatrix} 0 \\ -4 \end{bmatrix}\}$ is linearly independent. To find a stress matrix $\Omega$, we have to find a $2 \times 2$ symmetric matrix $\Psi$ such that:*

$$(z^1)^T \Psi z^4 = 0, (z^3)^T \Psi z^4 = 0, \text{ and } (z^2)^T \Psi z^5 = 0.$$

*Hence, $\Psi = \begin{bmatrix} 1 & 0 \\ 0 & 0 \end{bmatrix}$. Accordingly, $\Omega = Z\Psi Z^T = \begin{bmatrix} 1 \\ -2 \\ 1 \\ 0 \\ 0 \end{bmatrix} \begin{bmatrix} 1 & -2 & 1 & 0 & 0 \end{bmatrix}$. This should come as now surprise, since the only edges of $(G,p)$ with non-zero stresses are $\{1,2\},\{2,3\}$ and $\{1,3\}$.*

## 8.3 Dimensional Rigidity Results

This section presents sufficient and necessary conditions for dimensional rigidity. These conditions are expressed in terms of stress matrices or, equivalently, Gale transform.

**Theorem 8.5 (Alfakih [2])** *Let $(G,p)$ be an $r$-dimensional bar framework on $n$ vertices, $r \leq (n-2)$. If $(G,p)$ admits a positive semidefinite stress matrix $\Omega$ with rank $n-r-1$, then $(G,p)$ is dimensionally rigid.*

In fact, Theorem 8.5 in [2] is stated in terms of Gale transform as given next. Obviously, these two statements are equivalent.

**Theorem 8.6** *Let $(G,p)$ be an r-dimensional bar framework on n vertices, $r \leq (n-2)$. Let $z^i$ be a Gale transform of $p^i$ for $i = 1,\ldots,n$. If there exists a positive definite symmetric matrix $\Psi$ such that:*

$$(z^i)^T \Psi z^j = 0 \text{ for all } \{i,j\} \in \overline{E}(G),$$

*then $(G,p)$ is dimensionally rigid.*

Unfortunately, the sufficient condition in Theorem 8.5 is not necessary. Consider the bar framework depicted in Figure 8.2. Even though this framework is dimensionally rigid (in fact, it is also universally rigid), as shown in Example 35, this framework admits a positive semidefinite stress matrix of rank 1. But in this case, $n-r-1 = 2$.

A necessary and sufficient condition for dimensional rigidity was given by Connelly and Gortler [15] based on the Borwein–Wolkowicz facial reduction algorithm [10, 11]. In particular, they showed that an $r$-dimensional bar framework $(G,p)$ is dimensionally rigid if and only if $(G,p)$ admits a sequence of quasi-stress matrices such that: (i) the ranks of these matrices add up to $n-r-1$ and (ii) these matrices are positive semidefinite on a chain of nested subspaces.

A symmetric matrix $\Omega$ is called a *quasi-stress matrix* of a $(G,p)$ if and only if

$$P^T \Omega P = 0, \ \Omega e = 0 \text{ and } \Omega_{ij} = 0 \text{ for all } \{i,j\} \in \overline{E}(G), \tag{8.6}$$

where $P$ is the configuration matrix of $(G,p)$. Observe that every stress matrix is a quasi-stress matrix. On the other hand, if a quasi-stress matrix is positive semidefinite, then it is a stress matrix.

The following theorem is a refined version of Connelly-Gortler result. The positive semidefiniteness of a symmetric matrix $A$ is denoted by $A \succeq 0$, and the column space and the null space of $A$ are denoted by $\mathcal{R}(A)$, and $\mathcal{N}(A)$ respectively.

**Theorem 8.7 ([1])** *Let $(G,p)$ be an r-dimensional bar framework on n vertices in r-Euclidean space, $r \leq (n-2)$. Then $(G,p)$ is dimensionally rigid if and only if there exist nonzero quasi-stress matrices $\Omega^0, \Omega^1, \ldots, \Omega^k$, for some $k \leq n-r-2$, such that:*

*(a)* $\Omega^0 \succeq 0, \mathcal{U}_1^T \Omega^1 \mathcal{U}_1 \succeq 0, \ldots, \mathcal{U}_k^T \Omega^k \mathcal{U}_k \succeq 0,$

*(b)* $\text{rank } \Omega^0 + \text{rank } (\mathcal{U}_1^T \Omega^1 \mathcal{U}_1) + \cdots + \text{rank } (\mathcal{U}_k^T \Omega^k \mathcal{U}_k) = n-r-1,$

*(c)* $P^T \Omega^1 \rho_1 = 0, \ldots, P^T \Omega^k \rho_k = 0,$

*where $\rho_1, \mathcal{U}_1, \ldots, \mathcal{U}_k$ and $\xi_1, \ldots, \xi_k$ are full column rank matrices defined as follows.* $\mathcal{R}(\rho_1) = \mathcal{N}\left(\begin{bmatrix} \Omega^0 \\ P^T \\ e^T \end{bmatrix}\right)$. $\mathcal{R}(\xi_i) = \mathcal{N}(\rho_i^T \Omega^i \rho_i), \mathcal{U}_i = [P \ \rho_i]$ *for* $i = 1,\ldots,k$ *and* $\rho_{i+1} = \rho_i \xi_i$ *for all* $i = 1,\ldots,k-1$.

**Example 36** *Consider the bar framework $(G,p)$ depicted in Figure 8.2. Recall from Example 35 that a stress matrix $\Omega^0$ of $(G,p)$ is given by:*

$$\Omega^0 = \begin{bmatrix} 1 \\ -2 \\ 1 \\ 0 \\ 0 \end{bmatrix} \begin{bmatrix} 1 & -2 & 1 & 0 & 0 \end{bmatrix}.$$

*Thus,* $\rho_1 = [1\ 1\ 1\ -4\ 1]^T$ *and hence* $\mathcal{U}_1 = [P\ \rho_1] = \begin{bmatrix} -1 & -1 & 1 \\ -1 & 0 & 1 \\ -1 & 1 & 1 \\ 0 & 0 & -4 \\ 3 & 0 & 1 \end{bmatrix}$.

*Let*

$$\Omega^1 = \begin{bmatrix} 3 & -6 & 0 & 0 & 3 \\ -6 & 30 & 0 & -24 & 0 \\ 0 & 0 & -3 & 0 & 3 \\ 0 & -24 & 0 & 32 & -8 \\ 3 & 0 & 3 & -8 & 2 \end{bmatrix}. \text{ Hence, } \mathcal{U}_1^T \Omega^1 \mathcal{U}_1 = \begin{bmatrix} 0 & 0 & 0 \\ 0 & 0 & 0 \\ 0 & 0 & 800 \end{bmatrix}.$$

*Therefore,* $P^T\Omega^1 P = 0$, $\Omega^1 e = 0$ *and hence* $\Omega^1$ *is a quasi-stress matrix. Furthermore,* $P^T\Omega^1\rho_1 = 0$, $\mathcal{U}_1^T\Omega^1\mathcal{U}_1 \succeq 0$ *and rank* $\Omega^0$ + *rank* $(\mathcal{U}_1^T\Omega^1\mathcal{U}_1)$ = 2. *Therefore, framework* $(G, p)$ *is dimensionally rigid.*

## 8.4 Universal Rigidity

This section begins with a brief summary of affine motion results. These results are then combined with dimensional rigidity results in the previous section to obtain sufficient conditions for universal rigidity. The section ends with several universal rigidity results for some special classes of graphs, namely $r$-lateral graphs, chordal graphs and complete bipartite graphs.

### 8.4.1 Affine Motions

Affine motions of a bar framework $(G, p)$ can be characterized in terms of $p^1, \ldots, p^n$ and the edges of $G$.

***Lemma 8.2 [14]***

*Let* $(G, p)$ *be an r-dimensional bar framework on n vertices in r-Euclidean space,* $r \leq (n-2)$. *Then* $(G, p)$ *has an affine motion if and only if there exists a nonzero symmetric* $r \times r$ *matrix* $\Phi$ *such that:*

$$(p^i - p^j)^T \Phi (p^i - p^j) = 0 \text{ for all } \{i, j\} \in E(G),$$

*i.e.,* $(G, p)$ *has an affine motion if and only if its edges directions lie on a conic at infinity.*

Equivalently, affine motions can be characterized in terms of Gale matrix and the missing edges of $G$. Let $E^{ij}$ be the $n \times n$ symmetric matric with 1's in the $(i, j)$th and $(j, i)$th entries and 0's elsewhere.

***Lemma 8.3 [3]***

*Let* $(G, p)$ *be an r-dimensional bar framework on n vertices in r-Euclidean space,* $r \leq (n-2)$. *Let Z be a Gale matrix of* $(G, p)$. *Then* $(G, p)$ *has an affine motion if and only if there exists a nonzero* $y = (y_{ij})$ *such that:*

$$\sum_{\{i,j\} \in \overline{E}(G)} y_{ij} E^{ij} Z = e\xi^T,$$

*for some* $\xi$ *in* $(n - r - 1)$*-Euclidean space.*

**Example 37** *Consider framework (b) depicted in Figure 8.1. In this case* $p^1 - p^3 = 2(p^1 - p^2) = 2(p^2 - p^3) = \begin{bmatrix} -2 \\ 0 \end{bmatrix}$, *while* $p^2 - p^4 = \begin{bmatrix} 0 \\ -1 \end{bmatrix}$. *Let* $\Phi = \begin{bmatrix} 0 & 1 \\ 1 & 0 \end{bmatrix}$. *Then* $(p^i - p^j)^T \Phi (p^i - p^j) = 0$ *for all* $\{i,j\} \in E(G)$. *Hence, it follows from Lemma 8.2 that this framework has an affine motion.*

*On the other hand, the set of missing edges is given by* $\overline{E}(G) = \{ \{1,4\}, \{3,4\} \}$ *and a Gale matrix is* $Z = [1 \;\; -2 \;\; 1 \;\; 0]^T$. *Thus*

$$(y_{14}E^{14} + y_{34}E^{34})Z = \begin{bmatrix} 0 & 0 & 0 & y_{14} \\ 0 & 0 & 0 & 0 \\ 0 & 0 & 0 & y_{34} \\ y_{14} & 0 & y_{34} & 0 \end{bmatrix} \begin{bmatrix} 1 \\ -2 \\ 1 \\ 0 \end{bmatrix} = \begin{bmatrix} 0 \\ 0 \\ 0 \\ y_{14} + y_{34} \end{bmatrix} = \xi_1 e$$

*has the solution* $\xi_1 = 0$ *and* $y_{14} = -y_{34} = 1$. *Hence, again it follows from Lemma 8.3 that this framework has an affine motion.*

We say that $(G,p)$ is *generic* if the coordinates of $p^1, \ldots, p^n$ are algebraically independent over the integers, i.e., if these coordinates do not satisfy any nonzero polynomial with integer coefficients. The following lemmas present sufficient conditions for the absence of affine motions.

***Lemma 8.4 [14]***
*Let* $(G,p)$ *be an r-dimensional bar framework on n vertices in r-Euclidean space,* $r \leq (n-2)$. *Assume that: (i)* $(G,p)$ *is generic and (ii) each vertex of G has degree* $\geq r$. *Then* $(G,p)$ *has no affine motion.*

An alternative proof of Lemma 8.4 is given in [21]. Next, the generic assumption is replaced with the assumption of points in general position.

***Lemma 8.5 [8]***
*Let* $(G,p)$ *be an r-dimensional bar framework on n vertices in r-Euclidean space,* $r \leq (n-2)$. *Assume that: (i)* $(G,p)$ *admits a stress matrix* $\Omega$ *with rank* $n-r-1$ *and (ii)* $(G,p)$ *is in general position. Then* $(G,p)$ *has no affine motion.*

The assumptions in Lemma 8.5 are weakened in the following lemma.

***Lemma 8.6 [6]***
*Let* $(G,p)$ *be an r-dimensional bar framework on n vertices in r-Euclidean space,* $r \leq (n-2)$. *Assume that: (i)* $(G,p)$ *admits a stress matrix* $\Omega$ *with rank* $n-r-1$ *and (ii) for each vertex i, the set* $\{p^i\} \cup \{p^j : \{i,j\} \in E(G)\}$ *affinely span r-Euclidean space. Then* $(G,p)$ *has no affine motion.*

The following lemma is an immediate consequence of Lemma 8.6.

***Lemma 8.7 [6]***
*Let* $(G,p)$ *be an r-dimensional bar framework on n vertices in r-Euclidean space,* $r \leq (n-2)$. *Assume that: (i)* $(G,p)$ *admits a stress matrix* $\Omega$ *with rank* $n-r-1$ *and (ii) for each vertex i, the points* $\{p^i\} \cup \{p^j : \{i,j\} \in E(G)\}$ *are in general position. Then* $(G,p)$ *has no affine motion.*

### 8.4.2 Universal Rigidity Basic Results

The following theorem is an immediate consequence of Theorems 8.1 and 8.5.

**Theorem 8.8 (Connelly [12, 13], Alfakih [2])** *Let* $(G,p)$ *be an r-dimensional bar framework on n vertices,* $r \leq (n-2)$. *If the following two conditions hold:*

(a) $(G,p)$ *admits a positive semidefinite stress matrix* $\Omega$ *with rank* $n-r-1$,

(b) $(G,p)$ *has no affine motion,*

*then* $(G,p)$ *is universally rigid.*

Universal rigidity of bar frameworks has a complete characterization under the generic assumption.

**Theorem 8.9 (Alfakih [3], Connelly [14], Gortler and Thurston [18])** *Let* $(G,p)$ *be an r-dimensional generic bar framework on n vertices in r-Euclidean space,* $r \leq (n-2)$. *Then* $(G,p)$ *is universally rigid if and only if it admits a positive semidefinite stress matrix* $\Omega$ *with rank* $n-r-1$,

The sufficiency part of Theorem 8.9 was proved independently in [12] and [3] while the necessity part was conjectured in [3] and proved in [18].

If the generic assumption is replaced by the weaker assumption of points in general position, then only sufficient conditions are known. These conditions, which are not necessary, follow from Theorem 8.8 and Lemmas 8.5, 8.6, and 8.7.

**Theorem 8.10 (Alfakih and Ye [8])** *Let* $(G,p)$ *be an r-dimensional bar framework on n vertices in r-Euclidean space,* $r \leq (n-2)$. *If the following two conditions hold:*

(a) $(G,p)$ *admits a positive semidefinite stress matrix* $\Omega$ *with rank* $n-r-1$,

(b) $(G,p)$ *is in general position.*

*Then* $(G,p)$ *is universally rigid.*

**Theorem 8.11 (Alfakih and Nguyen [6])** *Let* $(G,p)$ *be an r-dimensional bar framework on n vertices in r-Euclidean space,* $r \leq (n-2)$. *If the following two conditions hold:*

(a) $(G,p)$ *admits a positive semidefinite stress matrix* $\Omega$ *with rank* $n-r-1$,

(b) *For each vertex i, the set* $\{p^i\} \cup \{p^j : \{i,j\} \in E(G)\}$ *affinely span r-Euclidean space.*

*Then* $(G,p)$ *is universally rigid.*

**Theorem 8.12 (Alfakih and Nguyen [6])** *Let* $(G,p)$ *be an r-dimensional bar framework on n vertices in r-Euclidean space,* $r \leq (n-2)$. *If the following two conditions hold:*

(a) $(G,p)$ *admits a positive semidefinite stress matrix* $\Omega$ *with rank* $n-r-1$,

(b) *For each vertex i, the points* $\{p^i\} \cup \{p^j : \{i,j\} \in E(G)\}$ *are in general position.*

*Then* $(G,p)$ *is universally rigid.*

### 8.4.3 Universal Rigidity for Special Graphs

We focus in this section on universal rigidity results for $r$-lateration graphs, chordal graphs and complete bipartite graphs. A graph $G$ on $n$ vertices is called an $r$-lateration graph if there exists a permutation $\pi$ of the vertices of $G$ such that: (i) the first $r$ vertices, $\pi(1),\ldots,\pi(r)$ induce a clique in $G$ and (ii) each remaining vertex $\pi(j)$, for $j = r+1,\ldots,n$, is adjacent to exactly $r$ vertices in $\{\pi(1),\ldots,\pi(j-1)\}$. It is known [24] that an $r$-dimensional bar framework $(G,p)$ in general position in $r$-Euclidean space, where $G$ is an $(r+1)$-lateration graph is universally rigid. It turns out that more can be said about such frameworks.

**Theorem 8.13 (Alfakih et al [7])** *Let $(G,p)$ be an $r$-dimensional bar framework on $n$ vertices in general position in $r$-Euclidean space, $r \leq (n-2)$, where $G$ is an $(r+1)$-lateration graph. Then $(G,p)$ admits a positive semidefinite stress matrix $\Omega$ of rank $n-r-1$.*

The proof of Theorem 8.13 is constructive. The desired stress matrix is obtained via an algorithm which starts with the matrix $ZZ^T$, where $Z$ is a Gale matrix, and iteratively zeros out the entries corresponding to the missing edges of $G$.

A graph $G$ is chordal if it has no induced cycle of length $\geq 4$. Chordal graphs have many well-known characterizations [9, 17]. An incomplete graph $G$ is $k$-vertex connected if $G$ has at least $k+2$ vertices and the deletion of any $k-1$ vertices leaves $G$ connected.

**Theorem 8.14 (Alfakih [5])** *Let $(G,p)$ be an $r$-dimensional bar framework on $n$ vertices in general position in $r$-Euclidean space, $r \leq (n-2)$, where $G$ is a chordal graph. Then the following statements are equivalent:*

*(a) $(G,p)$ is universally rigid,*

*(b) $G$ is $(r+1)$-vertex connected,*

*(c) $(G,p)$ admits a positive semidefinite stress matrix $\Omega$ of rank $n-r-1$.*

The proof of Theorem 8.14 is based on the characterization of chordal graphs in terms of perfect elimination order. Let $G^1=(V_1,E_1)$ and $G^2=(V_2,E_2)$ be two graphs such that $V_1 \cap V_2$ induces a clique in both $G^1$ and $G^2$. The *clique sum* of $G^1$ and $G^2$ is the graph $G=(V_1 \cup V_2, E_1 \cup E_2)$. The next theorem easily follows from the clique-sum characterization of chordal graphs, namely a graph is chordal if and only if it is the clique sum of cliques.

**Theorem 8.15 (Ratmanski [22])** *Let $(G^1,p^1)$ and $(G^2,p^2)$ be, respectively, two $r_1$ and $r_2$-dimensional universally rigid bar frameworks in general position in $r$-Euclidean space. Further, let $(G,p)$ be the bar framework such that $V(G)=V(G^1) \cup V(G^2)$ and $E(G)=E(G^1) \cup E(G^2)$. Then $(G,p)$ is universally rigid if and only if $V(G^1) \cap V(G^2)$ has at least $r+1$ vertices. Here, $V(G)$ denotes the vertex set of a graph.*

The complete bipartite graph $K_{m,n}$ is the graph whose nodes can be partitioned into two subsets $V_1$ and $V_2$ with $m$ and $n$, nodes, respectively such that: (i) every node in $V_1$ is adjacent to every node in $V_2$ and (ii) no two nodes in $V_1$ or in $V_2$ are adjacent. Thus $K_{m,n}$ has exactly $mn$ edges. The following theorem is a characterization of 1-dimensional bar frameworks on the line.

**Theorem 8.16 (Jordán and Nguyen [20])** *Let $(K_{m,n}, p \cup q)$ be a 1-dimensional complete bipartite bar framework on at least three vertices. If*

*(a) $p^1 < \cdots < p^m < q^1 < \cdots < q^n$, or*

*(b) $p^1 < \cdots < p^k < q^1 < \cdots < q^n < p^{k+1} < \cdots < p^m$.*

*Then $(K_{m,n}, p \cup q)$ is not universally rigid. Otherwise, if Conditions 1 and 2 do not hold, then $(K_{m,n}, p \cup q)$ is universally rigid.*

## 8.5 Glossary

**Bar Framework:** A bar framework $(G,p)$ in dimension $r$ is a connected simple graph $G$ whose vertices are points $p^1,\ldots,p^n$ in $r$-dimensional Euclidean space and whose edges are line segment between pairs of these points.

**Equivalent Frameworks:** Bar frameworks $(G,p)$ and $(G,q)$ are equivalent if $||p^i - p^j|| = ||q^i - q^j||$ for all edges $\{i,j\}$ of graph $G$.

**Congruent Frameworks:** Bar frameworks $(G,p)$ and $(G,q)$ are congruent if $||p^i - p^j|| = ||q^i - q^j||$ for all $i,j = 1,\dots,n$.

**Universal Rigidity:** Bar framework $(G,p)$ is universally rigid if every framework $(G,q)$ that is equivalent to $(G,p)$ is in fact congruent to it.

**Dimensional Rigidity:** An $r$-dimensional framework $(G,p)$ is dimensionally rigid if there is no $s$-dimensional framework $(G,q)$, equivalent to $(G,p)$, for any $s \geq r+1$.

**Affine Motion:** Bar framework $(G,p)$ has no affine motions if there is no framework $(G,q)$, equivalent but not congruent, to $(G,p)$ such that $q^i = Ap^i + b$ for all $i = 1,\dots,n$ for some matrix $A$ and vector $b$.

**Configuration matrix:** The configuration matrix $P$ of an $r$-dimensional bar framework in dimension $r$ is the $n \times r$ matrix whose its $i$th row is $(p^i)^T$.

**Stress Matrix:** A symmetric matrix $\Omega$ is a stress matrix of a bar framework $(G,p)$ if and only if $\Omega P = 0$, $\Omega e = 0$ and $\Omega_{ij} = 0$ for each missing edge $\{i,j\}$ of $G$, where $P$ is the configuration matrix of $(G,p)$ and $e$ is the vector of all 1's.

**Quasi-Stress Matrix:** A symmetric matrix $\Omega$ is a quasi-stress matrix of a bar framework $(G,p)$ if and only if $P^T \Omega P = 0$, $\Omega e = 0$ and $\Omega_{ij} = 0$ for each missing edge $\{i,j\}$ of $G$.

**Gale matrix:** A Gale matrix of an $r$-dimensional bar framework is any matrix $Z$ whose columns form a basis of the null space of $\begin{bmatrix} P^T \\ e^T \end{bmatrix}$.

**Points in General Position:** points $p^1,\dots,p^n$ in $r$-Euclidean space are in general position if every $r+1$ of these points are affinely independent.

**Generic Points:** points $p^1,\dots,p^n$ are generic if their coordinates are algebraically independent over the integers.

**$r$-lateration Graph:** A graph $G$ is an $r$-lateration graph if there exists a permutation $\pi$ of the vertices of $G$ such that: (i) the first $r$ vertices, $\pi(1),\dots,\pi(r)$ induce a clique in $G$ and (ii) each remaining vertex $\pi(j)$ is adjacent to exactly $r$ vertices in $\{\pi(1),\dots,\pi(j-1)\}$.

**Chordal Graph:** A graph $G$ is chordal if each cycle of $G$ of length $\geq 4$ has a chord.

**$k$-vertex Connected Graphs:** An graph $G$ is $k$-vertex connected iff $G$ is the complete graph on $k+1$ vertices or $G$ has at least $k+2$ vertices and the deletion of any $k-1$ vertices leaves $G$ connected

**Complete Bipartite Graph:** A complete bipartite graph is a graph whose nodes are partitioned into two subsets $V_1$ and $V_2$ such that: (i) every node in $V_1$ is adjacent to every node in $V_2$ and (ii) no two nodes in $V_1$ or in $V_2$ are adjacent.

**Acknowledgment:** Research partially supported by the Natural Sciences and Engineering Research Council of Canada.

## References

[1] A. Y. Alfakih. On Farkas lemma and dimensional rigidity of bar frameworks. arXiv/1405.2301.

[2] A. Y. Alfakih. On dimensional rigidity of bar-and-joint frameworks. *Discrete Appl. Math.*, 155:1244–1253, 2007.

[3] A. Y. Alfakih. On the universal rigidity of generic bar frameworks. *Contrib. Disc. Math.*, 5:7–17, 2010.

[4] A. Y. Alfakih. On bar frameworks, stress matrices and semidefinite programming. *Math. Program. Ser. B*, 129:113–128, 2011.

[5] A. Y. Alfakih. On stress matrices of chordal bar frameworks in general positions, 2012. arXiv/1205.3990.

[6] A. Y. Alfakih and V.-H. Nyugen. On affine motions and universal rigidity of tensegrity frameworks. *Linear Algebra Appl.*, 439:3134–3147, 2013.

[7] A. Y. Alfakih, N. Taheri, and Y. Ye. On stress matrices of $(d+1)$-lateration frameworks in general position. *Math. Program.*, 137:1–17, 2013.

[8] A. Y. Alfakih and Y. Ye. On affine motions and bar frameworks in general positions. *Linear Algebra Appl.*, 438:31–36, 2013.

[9] J. R. S. Blair and B. Peyton. An introduction to chordal graphs and clique trees. In J. A. George, J. R. Gilbert, and J. W-H. Liu, editors, *Graph Theory and Sparse Matrix Computation, IMA Vol. Math. Appl.,56*, pages 1–29. Springer, New York, 1993.

[10] J. M. Borwein and H. Wolkowicz. Facial reduction for a non-convex programming problem. *J. Austral. Math. Soc., Ser. A*, 30:369–380, 1981.

[11] J. M. Borwein and H. Wolkowicz. Regularizing the abstract convex program. *J. Math. Anal. Appl.*, 83:495–530, 1981.

[12] R. Connelly. Rigidity and energy. *Invent. Math*, 66:11–33, 1982.

[13] R. Connelly. Tensegrity structures: Why are they stable? In M. F. Thorpe and P. M. Duxbury, editors, *Rigidity theory and applications*, pages 47–54. Kluwer Academic/Plenum Publishers, 1999.

[14] R. Connelly. Generic global rigidity. *Discrete Comput. Geom.*, 33:549–563, 2005.

[15] R. Connelly and S. J. Gortler. Iterative universal rigidity. Technical report, arXiv:1401.7029v1, 2014.

[16] D. Gale. Neighboring vertices on a convex polyhedron. In *Linear inequalities and related system*, pages 255–263. Princeton University Press, 1956.

[17] M. C. Golumbic. *Algorithmic graph theory and perfect graphs.* Ann. Disc. Math 57, Elsevier, 2004.

[18] S. J. Gortler and D. P. Thurston. Characterizing the universal rigidity of generic frameworks. *Discrete Comput. Geom.*, 51:1017–1036, 2014.

[19] B. Grünbaum. *Convex polytopes.* John Wiley & Sons, 1967.

[20] T. Jordán and V-H. Nguyen. On universally rigid frameworks on the line. Technical report, Egerváry Research Group, 2012.

[21] M. Laurent and A. Varvitsiotis. Positive semidefinite matrix completion, universal rigidity and the strong Arnold property. *Linear Algebra Appl.*, 452:292–317, 2014.

[22] K. Ratmanski. Universally rigid framework attachments. Technical report, arXiv: 1011-4094v2, 2010.

[23] H. Wolkowicz, R. Saigal, and L. Vandenberghe, editors. *Handbook of Semidefinite Programming. Theory, Algorithms and Applications.* Kluwer Academic Publishers, Boston MA, 2000.

[24] Z. Zhu, A. M-C So, and Y. Ye. Universal rigidity: Towards accurate and efficient localization of wireless networks, 2010. Proc. IEEE INFOCOM.

# Chapter 9

# *Computations of Metric/Cut Polyhedra and Their Relatives*

**Mathieu Dutour Sikirić**

*Bošković Institute, Zagreb*

**Michel-Marie Deza**

*École Normale Supérieure, Paris*

**Elena I. Deza**

*Moscow State Pedagogical University*

**CONTENTS**



## 9.1 Introduction

Here we consider cones and polytopes, the points of which are various specifications, analogs and generalisations of *finite semimetrics*, i.e., for some finite set $X$ (usually, $X = \{1,\dots,n\}$), the symmetric functions $d = ((d_{ij})) : X^2 \to \mathbb{R}_{\geq 0}$ with $d_{ii} = 0$ and *triangle inequalities* $d_{ik} + d_{kj} - d_{ij} \geq 0$ for $i, j, k \in X$. We also will call a finite semimetric *n-points semimetric*, or *distance matrix*.

A semimetric is called *metric* if $d_{ij} > 0$ for $i \neq j$, and $(X,d)$ is called *metric space*. Metrics are ubiquitous in mathematics; the polyhedra considered here are important in analysis, measure theory, geometry of numbers, combinatorics, optimization, etc. Given two metric spaces $(X,d)$ and $(X',d')$, we say that $(X,d)$ is *isometrically embeddable into* (or is a *metric subspace of*) $(X',d')$ if there exists a mapping $\phi : X \to X'$ such that $d(x,y) = d(\phi(x),\phi(y))$ for all $x, y \in X$.

Given $1 \leq p \leq \infty$ and integer $m \geq 1$, the $l_p^m$*-space* is the metric space $(\mathbb{R}^m, d_p(x,y) = \|x-y\|_p)$, where the $l_p$*-norm* is

$$\|x\|_p = \left( \sum_{1 \le i \le m} x_i^p \right)^{\frac{1}{p}} \quad \text{for finite p and,} \quad \|x\|_\infty = \max_{1 \le i \le m} x_i.$$

A metric $d$ is called *$l_p$-metric* if $(X,d)$ is a metric subspace of the $l_p^m$-space, for some $X \subset \mathbb{R}^m$. A metric $d$ on $X$ is (cf. [13]) $l_p$-metric if and only if its restriction on any finite subset of $X$ is $l_p$-metric. The *unit ball* $B_p^m = \{x \in \mathbb{R}^m : \|x\|_p \le 1\}$ is $m$-hyperoctahedron, $m$-hypercube $[-1,1]^m$, $m$-sphere, $m$-spheroid, if $p = 1$, $\infty$, 2, otherwise. For any $p > 1$, it holds $B_1^m \subseteq B_p^m \subseteq B_\infty^m$, but $l_p^2$-spaces are isometric.

The most prominent is $l_2$-metric, called also *Euclidean* (or *Pythagorean, as-the-crow-flies, bee-line*) *distance*. The squared $n$-points $l_2$-semimetrics form the convex cone $EDM_n$ of *Euclidean Distance Matrices* $((d_2^2))_n$, introduced in Chapter 1 and discussed further in Chapters 8 and 10.

Every $l_2$-metric is a $l_p$-metric for any $p \ge 1$, while the $n$-points $l_\infty$-semimetrics form a convex cone, which coincides with the cone (called *metric cone* and denoted $MET_n$) of *all* $n$-points semimetrics. Indeed, any $n$-points semimetric $d$ is realized in the $l_\infty^{n-1}$-space by $n$ points $(d_{1i}, \ldots, d_{n-1\,i})$ for $i = 1, \ldots, n$.

The $2^{nd}$ by prominence is $l_1$-metric, also known as *taxicab, rectilinear, city block* or *Manhattan distance*. Every $l_p$-metric with $1 < p \le 2$ is ([51]) a $l_1$-metric. The $n$-points $l_1$-semimetrics form a convex cone, called *cut cone* and denoted $CUT_n$. In fact, it is [1, 4] generated by $2^{n-1} - 1$ *cut semimetrics* $\delta(S) = \delta(\overline{S}) = ((d_{ij}))$, defined, for any subset $S \subset \{1, \ldots, n\}$, by $d_{ij} = 1$ if $|\{i,j\} \cap S| = 1$ and $d_{ij} = 0$, otherwise.

Polyhedral (unlike to $EDM_n$) cones $CUT_n$ and (defined by $3\binom{n}{3}$ triangle inequalities) $MET_n$ are the central prototypes for polyhedra considered here. We will also consider corresponding polytopes, say, $CUTP_n$ and $METP_n$, since they have much more symmetries and so are easier to describe.

The cut cone and polytope also admit [3, 5] a characterization in terms of measure spaces: $d \in CUT_n$ (or $d \in CUTP_n$) if and only if there exist a measure (respectively, a probability) space $(\Omega, \mathcal{A}, \mu)$ and $n$ events $A_1, \ldots, A_n \in \mathcal{A}$ such that $d(i,j) = \mu(A_i \Delta A_j)$ for all $1 \le i, j \le n$.

Determining membership in $CUT_n$ is [6] NP-complete problem. Also, the connection of dual $CUT_n$ and $MET_n$ to multi-commodity flows was given in [6].

Under a linear mapping $d \to c(d)$ with $c_{ii} = d_{1n}$ for $1 \le i < n$ and $c_{ij} = \frac{1}{2}(d_{in} + d_{jn} - d_{ij})$ for $1 \le i < j < n$, we get the *correlation cone* $c(CUT_n)$ and *boolean quadric polytope* $c(CUTP_n)$, which are important in mathematical physics, quantum logic and optimization, say, in the *Boole problem* and *unconstrained quadratic* $\{0,1\}$*-programming problem*.

Connection of $CUT_n$ to geometry of numbers is given by the *hypermetrics*, i.e., distance matrices $d = ((d_{ij}))_n$ with $\sum_{1 \le i,j \le n} b_i b_j d_{ij} \le 0$ for any $(b_1, \ldots, b_n) \in \mathbb{Z}^n$. Clearly, permutations of $b = (1,1,-1,0,\ldots,0)$ give that $d$ is an $n$-points semimetric. Any $l_1$-metric is [54] a hypermetric. The hypermetrics are [2] exactly squared $l_2$-metrics of the *constellations* in Euclidean lattices, i.e., vertex sets of Delaunay polytopes.

So, $l_1$-metrics are $l_2^2$-metrics, while generalizing $l_2$-metrics. The study of $l_1$-metrics meets ongoing (esp., in Riemannian geometry, real analysis, approximation theory, statistics) process: to extend many prominent $l_2$-defined notions and results on more general $l_1$-setting.

The main connection of $CUT_n$ to combinatorics was given in [3]: the rational-valued $l_1$-metrics $d$ are exactly those, for which $\lambda d$ isometrically embeds into the path-metric of an $N$-cube $K_2^N$ for some integers $N, \lambda \ge 1$.

In this chapter, we collect known computations of cut and metric polyhedra and their relatives (hypermetric, quasi-metric, and $m$-hemimetric polyhedra) and generalizations of such polyhedra on general (instead of $K_n$) connected finite graphs. Such polyhedra over graphs are related to *maximum-cut problem*, which asks to find a cut of maximum weight in a graph. It is a prominent problem in combinatorial optimization with many applications.

In Chapter 8 a related viewpoint is considered for metric rigidities. There the distances are considered as faces of the cone of positive semidefinite forms. Our viewpoint is different since in the case of hypermetrics we consider positive definite matrices, not just positive semidefinite matrices. Also the algebraic structure is different since the lattice structures makes the relevant cones polyhedral.

*Quasi-metrics*, failing the condition $d_{ij} = d_{ji}$ for metrics, are encountered, for example, as quasi-distances on one-way routes, mountain walks or rivers with quick flow. Quasi-metrics and *oriented cuts* are used in semantics of computations and in computational geometry.

*Partial metrics*, where a weight $d_{ii} \geq 0$ is attributed to each point, are another generalization of metrics, which are equivalent to special quasi-metrics $((d_{ij} - d_{ii}))$. *Hemimetrics/supermetrics* are a generalization of metrics where instead of axiomatizing length, we axiomatize area and volume. For each of those generalizations, there is an analog of triangle inequality and corresponding cones.

The polytopal viewpoint, which is pursued within this chapter, can be useful for rigidity questions. Whenever a metric (or hypermetric, or any generalization considered here) in a cone $C$ is incident to an inequality defining $C$, the metric has less degree of freedom left, i.e., in other words, it is more rigid. This rigidity viewpoint is used in geometry of numbers for Delaunay polytopes: the extreme rays of the hypermetric cone correspond to rigid Delaunay polytopes. This viewpoint could be useful for other metric cones.

## 9.2 Metric and Cut Cones and Polytopes

**Metric cone $MET_n$:** the set of all *semimetrics on n points*, i.e., the symmetric functions $d : \{1,\dots,n\}^2 \to \mathbb{R}$ satisfying all $d_{ii} = 0$ and all *triangle inequalities*

$$d_{ik} \leq d_{ij} + d_{jk}.$$

**Metric polytope $METP_n$:** the set of all $d \in MET_n$ satisfying, in addition, all *perimeter inequalities*

$$d_{ik} + d_{ij} + d_{jk} \leq 2.$$

**Cut semimetric $\delta_S$ for a set $S \subseteq \{1,\dots,n\}$:** a vector defined as

$$\delta_S(i,j) = \begin{cases} 1 & \text{if } |S \cap \{i,j\}| = 1, \\ 0 & \text{otherwise.} \end{cases}$$

Clearly, $\delta_{\overline{S}} = \delta_S$, and $\delta_S$ is the adjacency matrix of a *cut subgraph* of $K_n$.

**Cut cone $CUT_n$:** the positive span of all $2^{n-1} - 1$ non-zero cut semimetrics.

**Cut polytope $CUTP_n$:** the convex hull of all $2^{n-1}$ cut semimetrics.

***m*-multicut $\delta_{S_1,\dots,S_m}$ for a partition $\cup_{i=1}^m S_i = \{1,\dots,n\}$:** a vector defined as

$$\delta_{S_1,\dots,S_m}(j_1,\dots,j_m) = \begin{cases} 1 & \text{if no two } j_i \text{ belong to the same } S_a, \\ 0 & \text{otherwise.} \end{cases}$$

The 2-multicuts are exactly cuts; all other multicuts on $n$ points are interior points of $CUT_n$, because of evident formula $2\delta_{S_1,\dots,S_m} = \sum_{i=1}^m \delta_{S_i,\overline{S_i}}$.

**Ridge graph of the cone $C$ or polytope $P$:** the skeleton of the dual cone $C^*$ or dual polytope $P^*$, respectively.

**Table 9.1**
The number of extreme rays and facets of cones $C = CUT_n, HYP_n, MET_n$ and the number of vertices and facets of polytopes $P = CUTP_n, HYPP_n, METP_n$

| $C/P$ | $n=3$ | $n=4$ | $n=5$ | $n=6$ | $n=7$ | $n=8$ |
|---|---|---|---|---|---|---|
| $CUT_n, e$ | 3(1) | 7(2) | 15(2) | 31(3) | 63(3) | 127(4) |
| $CUT_n, f$ | 3(1) | 12(1) | 40(2) | 210(4) | 38,780(36) | 49,604,520(2,169) |
| $HYP_n, e$ | 3(1) | 7(2) | 15(2) | 31(3) | 37,170(29) | 242,695,427(9,003) |
| $HYP_n, f$ | 3(1) | 12(1) | 40(2) | 210(4) | 3,773(14) | 298,592(86) |
| $MET_n, e$ | 3(1) | 7(2) | 25(3) | 296(7) | 55,226(46) | 119,269,588(3,918) |
| $MET_n, f$ | 3(1) | 12(1) | 30(1) | 60(1) | 105(1) | 168(1) |
| $CUTP_n, v$ | 4(1) | 8(1) | 16(1) | 32(1) | 64(1) | 128(1) |
| $CUTP_n, f$ | 4(1) | 16(1) | 56(2) | 368(3) | 116,764(11) | 217,093,472(147) |
| $HYPP_n, v$ | 4(1) | 8(1) | 16(1) | 32(1) | 13,152(6) | 1,388,383,872(581) |
| $HYPP_n, f$ | 4(1) | 16(1) | 56(2) | 68(3) | 10,396(7) | 1,374,560(22) |
| $METP_n, v$ | 4(1) | 8(1) | 32(2) | 554(3) | 275,840(13) | 1,550,825,600(533) |
| $METP_n, f$ | 4(1) | 16(1) | 40(1) | 80(1) | 140(1) | 224(1) |

The full symmetry group of the cones $CUT_n$ and $MET_n$ is the symmetric group $Sym(n)$ for $n \neq 4$ and $Sym(4) \times Sym(3)$ for $n = 4$. The polytopes $CUTP_n, METP_n$, are invariant, besides permutations, under the following *switching* operation:

$$U_S(d) = \begin{cases} \{1,\dots,n\}^2 & \to \quad \mathbb{R} \\ (i,j) & \mapsto \quad \begin{cases} 1 - d_{ij} & \text{if } |S \cap \{i,j\}| = 1, \\ d_{ij} & \text{otherwise.} \end{cases} \end{cases}$$

Together, those define a group of order $2^{n-1} \times n!$. For $n \neq 4$, this is the full symmetry group. For $n = 4$, the full group is $\text{Aut}(K_{4,4})$ and so its order is $2^3 \times 144$.

It holds $CUT_n \subseteq MET_n$ and $CUTP_n \subseteq METP_n$ with equality only for $n = 3, 4$.

Table 9.1 gives known results of the number $f$ of facets and extreme rays or vertices ($e$ or $v$) for small cut, metric and (considered in next section) hypermetric cones/polytopes. The numbers of orbits, given there in parentheses, is under $Sym(n)$ for cones and under $Sym(n)$ and $2^{n-1}$ switchings for polytopes.

The enumeration of orbits of facets of $CUT_n$ for $n \leq 7$ was done in [8, 43, 54] for $n = 5$, 6, and 7, respectively. For $CUT_8$ and $CUTP_8$, sets of facets were found in [15]; completeness of these sets was shown in [20]. The enumeration of orbits of extreme rays of $MET_n$ for $n \leq 8$ was done in [17, 18, 44].

## 9.3 Hypermetric Cone and Hypermetric Polytope

**Hypermetric cone** $HYP_n$**:** the set of all *hypermetrics on n points*, i.e., the symmetric functions $d : \{1,\dots,n\}^2 \to \mathbb{R}$ satisfying all $d_{ii} = 0$ and the inequalities

$$H(b,d) = \sum_{1 \leq i < j \leq n} b_i b_j d_{ij} \leq 0 \text{ for all } b \in \mathbb{Z}^n, \sum_i b_i = 1.$$

One obtains $MET_n$ using only $b$ of the form $(1,1,-1,0^{n-3})$.

**Table 9.2**
All types of Delaunay simplices $S$ in $\mathbb{R}^7$, their volume, the order of automorphism group and the number of facets of the cone $Bar_S$

| $S_i$ | Vol($S_i$) | $\lvert\text{Aut}(S_i)\rvert$ | No. of facets (orbits) |
|---|---|---|---|
| $S_1$ | 1 | 40,320 | 298,592(86) |
| $S_2$ | 2 | 40,320 | 5,768(9) |
| $S_3$ | 2 | 1,440 | 6,590(62) |
| $S_4$ | 3 | 540 | 966(9) |
| $S_5$ | 3 | 1,152 | 728(9) |
| $S_6$ | 3 | 240 | 640(39) |
| $S_7$ | 4 | 1,440 | 28(3) |
| $S_8$ | 4 | 240 | 153(11) |
| $S_9$ | 4 | 144 | 131(10) |
| $S_{10}$ | 5 | 72 | 28(6) |
| $S_{11}$ | 5 | 48 | 28(8) |

$(2k+1)$**-gonal inequality:** an inequality $H(b,d) \leq 0$ with $n = 2k+1$ and $\{0, \pm 1\}$-valued $b$. The case of general $b$ can be seen as such inequality on a multiset of $\sum_{i=1}^{n} |b_i|$ points with different points occuring $|b_1|, \ldots, |b_n|$ times, respectively.

**Hypermetric polytope** $HYPP_n$**:** the polytope defined by all *hypermetric inequalities*

$$\sum_{1 \leq i < j \leq n} b_i b_j d(i,j) \leq s(s+1),$$

where $b = (b_1, \ldots, b_n) \in \mathbb{Z}^n$ with $\sum_{i=1}^{n} b_i = 2s+1$ and $s \in \mathbb{Z}$.

One obtains $METP_n$ using only $b$ of the form $(1,1,-1,0^{n-3})$ and $(1,1,1,0^{n-3})$.

There is an infinity of inequalities, defining the hypermetric cone $HYP_n$; so, it is not obvious that this cone is polyhedral. The proof of this [30, 31, 32] was achieved through the connection with geometry of numbers that we now explain.

Given a quadratic form $q$, one can define the induced Delaunay tessellation with point set $\mathbb{Z}^n$ ([41, 40, 55, 52]). It is well known that there are only a finite number of such tessellations, up to the action of the group $\text{GL}_n(\mathbb{Z})$.

For a generic quadratic form, the tessellation is formed by simplices only; but, importantly, when it is not formed of simplices, this induces linear conditions on the coefficients of the quadratic form. There are a finite number of simplices, up to $\text{GL}_n(\mathbb{Z})$ action, and they have been classified in [38] for $n = 7$, extending previous classification for $n \leq 6$ [10, 49, 50]. Table 9.2 gives key information on all 11 types of simplices in $\mathbb{R}^7$.

Given a simplex $S$, denote by $Bar_S$ the set of quadratic forms, for which $S$ is contained in a Delaunay polytope of the Delaunay tessellation. It is a polyhedral cone, called a *Baranovskii cone* in [52].

For a quadratic form inside $Bar_S$, the Delaunay tessellation contains $S$ as a simplex. For a quadratic form on a facet of $Bar_S$, the simplex $S$ is a part of a *repartitioning polytope*, i.e., a Delaunay polytope with only $n+2$ vertices.

If the simplex $S$ has volume 1, then it is equivalent to the simplex formed by the vertices $v_1 = 0$, $v_2 = e_1, \ldots, v_{n+1} = e_n$. The quadratic form $q$ is described uniquely by the distance function $d(i,j) = q(v_i - v_j)$ on the vertices $v_i$. For a given positive definite quadratic form $q$, denote by $c(q)$ the center of the sphere, circumscribing $S$, and by $r(q)$ the radius of this sphere. Since $S$ is of volume 1, a given point $v \in \mathbb{Z}^n$ can be uniquely expressed in barycentric coordinates as $v = \sum_i b_i v_i$ with $1 = \sum_i b_i$ and $b_i \in \mathbb{Z}$. In [32], the following formula is proved:

$$H(b,d) = \|v - c(q)\|^2 - (r(q))^2.$$

So, the distance function $d$ corresponds to a quadratic form $q$, for which $S$ is a part of the Delaunay tessellation, if and only if it belongs to $HYP_{n+1}$.

With this connection, it is possible to prove the polyhedrality of the cone $HYP_n$. In general, we have $CUT_n \subseteq HYP_n$ with equality only for $3 \leq n \leq 6$ [32]. The first proper $HYP_n$, $HYP_7$, was described in [24]; $HYP_8$ was computed in [27].

The hypermetric polytope $HYPP_n$ is defined [27], by analogy with the metric and cut polytopes, as a polytope invariant under the switching operations $U_S$. It turns out, that it is polyhedral with the link being established with the centrally symmetric Delaunay polytopes of dimension $n$, see [27] for details. One can check easily whether a distance matrix $d$ belongs to $HYP_n$ or $HYPP_n$; this has been used in [27] to determine the facets and vertices of $HYPP_7$ and $HYPP_8$. See in Table 9.3 all 22 orbits of facets of $HYPP_8$; the simplicial ones there are marked by $*$.

The skeletons of $HYPP_n$, $HYP_n$ contain a clique consisting of all cuts and all nonzero cuts, respectively. We expect that any vertex is adjacent to a cut vertex (it holds for $n \leq 8$); if true, it will imply that each of the above skeletons has diameter 3. The ridge graphs of $METP_n$, $MET_n$ with $n \geq 4$ have diameter 2 [16]. We expect that any facet of $HYPP_n$, $HYP_n$ is adjacent to a triangle/perimeter facet (it holds for $n \leq 7$); if true, it will imply that the ridge graphs of $HYPP_n, HYP_n$ have diameter 4.

The hypermetric polytope $HYPP_8$ has 581 orbits of vertices, which are in details:

- 1 orbit of $2^7$ cuts, corresponding to the Delaunay polytope $[0,1]$;
- 24 orbits, corresponding to the Delaunay polytopes $2_{21}$ and $3_{21}$;
- 556 orbits, corresponding the Delaunay polytope $ER_7$.

## 9.4 Cut and Metric Polytopes of Graphs

**Cut polytope of the graph** $CUTP(G)$**:** the convex hull of all cut vectors of a given graph $G = (V,E)$, or, equivalently, the projections of $CUTP_{|V|}$ on the subspace $\mathbb{R}^{|E|}$ indexed by the edge-set $E$. *Cut cone* $CUT(G)$ is such projection of $CUT_{|V|}$.

**Metric polytope** $METP(G)$ **of the graph** $G$**:** the projection of $METP_{|V|}$ on $\mathbb{R}^{|E|}$ indexed by the edges of $G$. *Metric cone* $MET(G)$ is such projection of $MET_{|V|}$.

**Minor of graph** $G$**:** a graph, which can be obtained from $G$ by deleting edges and vertices and contracting edges.

Clearly, it holds $CUT(G) \subseteq MET(G)$ and $CUT(G) \subseteq MET(G)$.

**Theorem 9.1** *$CUT(G) = MET(G)$ or, equivalently, $CUTP(G) = METP(G)$ if and only if $G$ does not have any $K_5$-minor.*

It was proved in [53] for cones and in [9] for polytopes.

**Remark 9.1** By *Wagner's theorem* [56], a finite graph is planar if and only if it has no minors $K_5$ and $K_{3,3}$.

**Table 9.3**
Orbits of facets of the hypermetric polytope $HYPP_8$

| $F_i$ | Representative | $\frac{\|F_i\|}{32}$ | No. of classes | Inc.([0,1], $\{2_{21}, 3_{21}\}$, $ER_7$) |
|---|---|---|---|---|
| $F_1$ | $(0,0,0,0,0,1,1,1)$ | 7 | 2 | (96, 1598784, 80836608) |
| $F_2$ | $(0,0,0,1,1,1,1,1)$ | 28 | 3 | (80, 383040, 14300640) |
| $F_3$ | $(0,1,1,1,1,1,1,1)$ | 16 | 4 | (70, 131712, 3975552) |
| $F_4$ | $(0,0,1,1,1,1,1,2)$ | 168 | 6 | (60, 32160, 590960) |
| $F_5$ | $(0,1,1,1,1,1,2,2)$ | 336 | 9 | (52, 9600, 122160) |
| $F_6$ | $(1,1,1,1,1,1,1,2)$ | 32 | 8 | (56, 19656, 370272) |
| $F_7$ | $(0,1,1,1,1,1,1,3)$ | 112 | 7 | (42, 840, 1120) |
| $F_8$ | $(1,1,1,1,1,2,2,2)$ | 224 | 12 | (46, 3528, 39906) |
| $F_9$ | $(0,1,1,1,1,2,2,3)$ | 1,680 | 15 | (40, 656, 2686) |
| $F_{10}$ | $(1,1,1,1,1,1,2,3)$ | 224 | 14 | (42, 1323, 6489) |
| $F_{11}$ | $(1,1,1,1,1,1,1,4)$ | 32 | 8 | (28, 0, 0)* |
| $F_{12}$ | $(1,1,1,1,1,2,3,3)$ | 672 | 18 | (36, 252, 464) |
| $F_{13}$ | $(1,1,1,1,2,2,2,3)$ | 1,120 | 20 | (38, 585, 3210) |
| $F_{14}$ | $(1,1,1,1,1,2,2,4)$ | 672 | 18 | (32, 66, 36) |
| $F_{15}$ | $(1,1,1,2,2,2,3,3)$ | 2,240 | 24 | (33, 120, 302) |
| $F_{16}$ | $(1,1,1,1,2,2,3,4)$ | 3,360 | 30 | (31, 62, 82) |
| $F_{17}$ | $(1,1,1,1,2,2,2,5)$ | 1,120 | 20 | (25, 3, 0)* |
| $F_{18}$ | $(1,1,1,1,1,3,3,4)$ | 672 | 18 | (27, 0, 1)* |
| $F_{19}$ | $(1,1,1,2,2,3,3,4)$ | 6,720 | 36 | (28, 22, 22) |
| $F_{20}$ | $(1,1,1,1,2,3,3,5)$ | 3,360 | 30 | (25, 2, 1)* |
| $F_{21}$ | $(1,1,2,2,2,3,3,5)$ | 6,720 | 36 | (24, 3, 1)* |
| $F_{22}$ | $(1,1,1,2,2,3,4,5)$ | 13,440 | 48 | (24, 3, 1)* |

For a connected graph $G=(V,E)$, the polytope $CUTP(G)$ has $2^{|V|-1}$ vertices and dimension $|E|$. The switching $U_S$ still act on $CUTP(G)$ and the automorphisms of $G$ as well. Together, this defines a group $ARes(G)$ of automorphisms of $CUTP(G)$, and $|ARes(G)| = 2^{|V|-1}|\mathrm{Aut}(G)|$. The full group $\mathrm{Aut}(CUTP(G))$ might be larger than that. In [20] the dual description of $CUTP(G)$ was computed for several graphs; see data on some of them in Tables 9.4 and 9.5. In the $3^{rd}$ column there, $A(G)$ denotes $2^{1-|V|}|\mathrm{Aut}(CUTP(G))|$; it is $|\mathrm{Aut}(G)|$ for all, but $K_4$, graphs there.

For a cycle $C$ in $G$ and an odd-sized set $F$ of edges in $C$, the *cycle inequality* is

$$f_{C,F} = \sum_{e \in C-F} x_e - \sum_{e \in F} x_e \leq |F| - 1.$$

In fact, all facets of $METP(G)$ are defined [9] by all chordless cycle inequalities $f_{C,F}$ and all inequalities $0 \leq x_e \leq 1$ for edges $e$ not belonging to a triangle in $G$.

Besides above inequalities of $METP(G)$, some valid inequalities of hypermetric type show up on $CUTP(G)$. Below we give a way [27] to get such inequalities.

***Proposition 9.1***

*Let us take a valid inequality on $CUTP_n$ of the form*

$$f(x) = \sum_{1 \leq i < j \leq n} a_{ij} x_{ij} \leq A.$$

*Given a graph $G$ with the vertex-set $v_1, \ldots, v_n$, suppose that any two vertices $v_i$, $v_j$ are joined by a such path $P_{ij}$, that:*

**Table 9.4**
The number of facets of $CUTP(G) = METP(G)$ for some $K_5$-minor-free graphs $G$ (2 non-planar graphs and the skeletons of Platonic and some semiregular polyhedra)

| $G = (V,E)$ | $\|V\|,\|E\|$ | $A(G)$ | No. of facets (orbits) | Cycles |
|---|---|---|---|---|
| Wagner graph $M_8$ | 8,12 | 16 | 184(4) | 2,2,4,5 |
| $K_{3,3} = M_6$ | 6,9 | 72 | 90(2) | 2,4 |
| Dodecahedron | 20,30 | 120 | 167,164(8) | 2,5,9,10,10,11,12,12 |
| Icosahedron | 12,30 | 120 | 1,552(4) | 3,5,6,6 |
| Cube $K_2^2$ | 8,12 | 48 | 200(3) | 2,4,6 |
| Octahedron $K_{2,2,2}$ | 6,12 | 48 | 56(2) | 3,4 |
| Tetrahedron $K_4$ | 4,6 | 144 | 12(1) | 3 |
| Prism$_7$ | 14,21 | 28 | 7,394(6) | 2,2,4,7,9,9 |
| APrism$_6$ | 12,24 | 24 | 2,032(5) | 3,6,7,7,8 |
| Cuboctahedron | 12,24 | 48 | 1,360(5) | 3,4,6,6,8 |
| Tr. Tetrahedron | 12,18 | 24 | 540(4) | 2,3,6,8 |

- *the edge-sets of all paths $P_{ij}$ are disjoint;*
- *if $a_{ij} > 0$, then $P_{ij}$ is reduced to an edge.*

*Then, the following inequality is valid on $CUTP(G)$:*

$$H_{f,G}(x) = \sum_{1 \le i < j \le n} a_{ij} \left( \sum_{e \in P_{ij}} x_e \right) \le A.$$

**Remark 9.2** When applied to the triangle inequalities of $CUTP(K_n)$ and taking switchings, the above proposition gives us $METP(G)$. So, it is temping to define the hypermetric polytope $HYPP(G)$ for general graph $G$ by the switchings of the extensions of all hypermetric inequalities obtained from above proposition. What is not clear is when $CUTP(G) = HYPP(G)$ and whether there is a nice characterization of such hypermetrics. The above discussion is applied also to cones.

Any $K_n$-subgraph of $G$ will satisfy the hypothesis and the facets of $CUTP_n$ will give facets of $CUTP(G)$. Proposition 9.1 gives valid inequality induced by a class of graphs homeomorphic $K_n$. (A graph $H$ is *homeomorphic* to a subgraph of $G$ if $H$ can be mapped to $G$ so that the edges of $H$ are mapped to disjoint paths in $G$.) A graph homeomorphic to $K_n$ is a special case of a $K_n$-minor.

**Remark 9.3** The proof of Theorem 9.1 [53] does not give hypermetric inequalities, or their generalizations, in a straightforward way; it appears nonconstructive.

**Remark 9.4** In quantum information theory, *general Bell inequalities*, involving joint probabilities of two probabilistic events, are exactly valid inequalities of the *correlation polytope $CORP(G)$* (called also *Boolean quadric polytope*) of a graph, say, $G$. In particular, $CORP(K_{n,m})$ is seen there as the set of possible results of a series of Bell experiments with a *separable* (nonentangled) quantum state shared by two distant parties, where one party has $n$ choices of possible two-valued measurements and the other party has $m$ choices. A valid inequality of $CORP(K_{n,m})$ is called a *Bell inequality* and if facet inducing, a *tight Bell inequality*. This polytope is linearly isomorphic (via the *covari-*

**Table 9.5**
The number of facets of $CUTP(G)$ for some graphs $G$ with $K_5$-minor

| $G = (V,E)$ | $\lvert V\rvert, \lvert E\rvert$ | $A(G)$ | Number of facets (orbits) | Cycles |
|---|---|---|---|---|
| Heawood graph | 14, 21 | 336 | 5, 361, 194(9) | 2, 6, 8 |
| Petersen graph | 10, 15 | 120 | 3, 614(4) | 2, 5, 6 |
| Pyr(APrism$_4$) | 9, 24 | 16 | 389, 104(17) | 3, 3, 3, 4, 5 |
| Möbius ladder $M_{14}$ | 14, 21 | 28 | 369, 506(9) | 2, 2, 4, 8, 10 |
| Tr.Octahedron on $\mathbb{P}^2$ | 12, 18 | 48 | 62, 140(7) | 2, 2, 4, 6, 6 |
| $K_{5,5}$ | 10, 25 | $2(5!)^2$ | 16, 482, 678, 610(1, 282) | 2, 4 |
| $K_{4,7}$ | 11, 28 | 4!7! | 271, 596, 584(15) | 2, 4 |
| $K_{4,6}$ | 10, 24 | 4!6! | 23, 179, 008(12) | 2, 4 |
| $K_{4,5}$ | 9, 20 | 4!5! | 983, 560(8) | 2, 4 |
| $K_{3,3,3}$ | 9, 27 | $(3!)^4$ | 624, 406, 788(2, 015) | 3, 4 |
| $K_{1,4,4}$ | 9, 24 | $2(4!)^2$ | 36, 391, 264(175) | 3, 4 |
| $K_{1,3,5}$ | 9, 23 | 3!5! | 71, 340(7) | 3, 4 |
| $K_{1,3,4}$ | 8, 19 | 3!4! | 12, 480(6) | 3, 4 |
| $K_{1,1,3,3}$ | 8, 21 | $4(3!)^2$ | 432, 552(50) | 3, 3, 4 |
| $K_{1,1,2,m}, m > 2$ | $m$+4, 4$m$+5 | $4m!$ | $8 + 20m + 8\binom{m}{2}$ $(16m - 5(7)$ | 3, 3, 3, 4 |
| $K_{m+4} - K_m$, $m$>1 | $m$+4, 4$m$+6 | $4!m!$ | $8(8m^2 - 3m + 2)(4)$ | 3, 3 |
| $K_8 - K_3$ | 8, 25 | 360 | 2, 685, 152(82) | 3, 3 |
| $K_7 - K_2$ | 7, 20 | 240 | 31, 400(17) | 3, 3 |

*ance map*) to $CUTP(K_{1,n,m})$ [32, Section 5.2]. Similarly, $CUTP(K_{1,n,m,l})$ represents *three-party* Bell inequalities.

The symmetry group of $CORP(K_{n,n})$ and $CUTP(K_{1,n,m})$ has order $2^{1+n+m}n!m!$. Table 9.5 gives the number of facets of $CORP(K_{n,m})$ with $(n,m) = (4,4), (3,5), (3,4)$. The cases $(n,m) = (2,2), (3,3)$ were settled in [42] and [48], respectively.

In contrast to the Bell's inequalities, which probe entanglement between spatially-separated systems, the *Leggett-Garg inequalities* test the correlations of a single system measured at different times. The polytope, defined by those inequalities for $n$ observables, is equivalent ([7]) to the cut polytope $CUTP_n$.

**Remark 9.5** *Diversity cone $DIV_n$* is the set of all *diversities on n points*, i.e., [14] the functions $f : \{A : A \subseteq \{1,\dots,n\}\} \to \mathbb{R}$ satisfying all $f(A) \geq 0$ with equality if $|A| \leq 1$ and all

$$f(A \cup B) + f(B \cup C) \geq f(A \cup C) \text{ if } B \neq \emptyset.$$

The *induced diversity metric $d_{ij}$* is $f(\{i,j\})$.

*Cut diversity cone $CDIV_n$* is the positive span of all *cut diversities* $\delta(A)$, where $A \subseteq \{1,\dots,n\}$, which are defined, for any $S \subseteq \{1,\dots,n\}$, by

$$\delta_S(A) = \begin{cases} 1 & \text{if } A \cap S \neq \emptyset \text{ and } A \setminus S \neq \emptyset, \\ 0 & \text{otherwise.} \end{cases}$$

$CDIV_n$ is [14] the set of all diversities from $DIV_n$, which are isometrically embeddable into an $l_1$-

*diversity*, i.e., one defined on $\mathbb{R}^m$ with $m \leq \binom{n}{\lfloor \frac{n}{2} \rfloor}$ by

$$f_{m1}(A) = \sum_{i=1}^{m} \max_{a,b \in A} \{|a_i - b_i|\}.$$

These two cones are natural "hypergraph" extensions of $MET(G)$ and $CUT(G)$.

## 9.5 Quasi-Semimetric Polyhedra

**Quasi-semimetric cone $QMET_n$:** the set of all *quasi-semimetrics on n points*, i.e., the functions $q: \{1,\ldots,n\}^2 \to \mathbb{R}$ satisfying all $q_{ii} = 0$, all $q_{ij} \geq 0$ and all *oriented triangle inequalities*

$$q_{ik} \leq q_{ij} + q_{jk}.$$

**Quasi-semimetric polytope $QMETP_n$:** the set of all $q \in QMET_n$, satisfying all

$$q_{ki} + q_{ij} + q_{jk} \leq 2.$$

**Oriented $m$-multicut $\delta'_{S_1,\ldots,S_q}$ for an ordered partition $\cup_{i=1}^{m} S_i = \{1,\ldots,n\}$:** a vector (actually, a quasi-semimetric) defined as

$$\delta'_{S_1,\ldots,S_m}(i,j) = \begin{cases} 1 & \text{if } i \in S_a, j \in S_b \text{ and } a < b, \\ 0 & \text{otherwise.} \end{cases}$$

The number of all oriented multicuts on $n$ points is the *Fubini number* (called also *ordered Bell number*) $p'(n)$ of all ordered partitions of $n$.

**Oriented multicut cone $OMCUT_n$:** the positive span of all $p'(n) - 1$ nonzero oriented multicuts on $n$ points. By $\delta_{S_1,\ldots,S_m} = \delta'_{S_1,\ldots,S_m} + \delta'_{S_m,\ldots,S_1}$, we have

$$\{q + q^T : q \in OMCUT_n\} = CUT_n.$$

**Oriented cut cone $OCUT_n$:** the positive span of all $2^n - 2$ nonzero *oriented cuts*, i.e., 2-multicuts $\delta'_{S,\overline{S}}$, on $n$ points.

**Weightable quasi-semimetric cone $WQMET_n$:** the set of all *weightable quasi-semimetrics* on $n$ points, i.e., quasi-semimetrics $q \in QMET_n$, for which exists a (weight) function $w = (w_i)$ : $\{1,\ldots,n\} \to \mathbb{R}_{\geq 0}$ satisfying all

$$q_{ij} + w_i = q_{ji} + w_j.$$

**Weightable quasi-semimetric polytope $WQMETP_n$:** the set $WQMET_n \cap QMETP_n$.

$QMET_n$ and $OMCUT_n$ are full-dimensional cones in $R^{n(n-1)}$, while $WQMET_n$, $OCUT_n$ are of dimension $\binom{n+1}{2} - 1$ and $PMET_n$ is of dimension $\binom{n+1}{2}$.

Besides strict inclusions $WQMET_n \subset QMET_n$ and $OCUT_n \subset OMCUT_n$, we have, with equalities only for $n = 3$, the following inclusions

$$OMCUT_n \subseteq QMET_n \text{ and } OCUT_n \subseteq WQMET_n.$$

**Remark 9.6** Quasi-semimetrics were studied in [21, 22, 26, 33]. A quasi-semimetric $q$ is weightable if and only if it has *relaxed symmetry*, i.e., for distinct $i, j, k$ it holds

$$q_{ij} + q_{jk} + q_{ki} = q_{ik} + q_{kj} + q_{ji}.$$

Also, $WQMET_n$ consists of all $((d_{ij} + d_{io} - d_{jo}))$, where $((d_{ij}))$ is as a semimetric on $\{0, 1, \ldots, n\}$, but none of the inequalities $d_{io} + d_{jo} - d_{ij} \geq 0$ is required.

**Example 38** *Consider random walks on a connected graph $G = (V, E)$, where at each step walk moves with uniform probability from current vertex to a neighboring one. The* hitting time quasi-metric *$H(u, v)$ is defined as the expected number of steps (edges) for a walk on $G$ beginning at vertex $u$ to reach $v$ for the first time; put $H(u, u) = 0$. This quasi-metric is weightable. The* commuting time metric *is*

$$C(u, v) = H(u, v) + H(v, u) = 2|E|Q(u, v),$$

*where $Q(u, v)$ is called the* effective resistance metric.

It is shown in [29] that $OCUT_n$ and $WQMET_n$ on $\{1, \ldots, n\}$ are projections of $CUT_{n+1}$ and $MET_{n+1}$ (defined on $\{0, 1, \ldots, n\}$) on the subspace orthogonal to $\delta_{\{0\}}$. Moreover, the *quasi-hypermetric cone* $WQHYP_n$ is defined as such projection of $HYP_{n+1}$, and it holds $OCUT_n \subseteq WQHYP_n \subseteq WQMET_n$ with equalities only for $n = 3$ and $OCUT_n = WQHYP_n$ only for $3 \leq n \leq 5$. See data on $WQHYP_5$ in Table 9.6.

**Remark 9.7** $OCUT_n$ is the set of all $n$-vertex *$l_1$-quasi-semimetrics*, i.e., quasi-metrically embeddable ones into a quasi-metric space $(\mathbb{R}^m, \|x - y\|)$ with *oriented $l_1$-norm*

$$\|z = (z_1, \ldots, z_m)\| = \sum_{i=1}^{m} \max(z_i, 0).$$

On a measure space, such $\|x - y\|$ is $\mu(B \setminus A)$, where the sets $B, A$ correspond to $x, y$.

The cones $QMET_n$ and $OMCUT_n$ have $Sym(n)$ as a symmetry group. Another symmetry, called *reversal*, exists also: associate to each ray $q$ the ray $q^T$ defined by $q^T_{ij} = q_{ji}$, i.e., in matrix terms, the reversal corresponds to the transposition of matrices. This yields that $Z_2 \times Sym(n)$ is a symmetry group of the cones $QMET_n$ and $OMCUT_n$. We expect that this is their full symmetry group. It is so ([29]) for $OCUT_n$.

For $QMETP_n$, one can define the following analog of the switching operation of $METP_n$: given $S \subset \{1, \ldots, n\}$, call *oriented switching* the operation

$$U_S(q) = \begin{cases} \{1, \ldots, n\}^2 & \rightarrow \quad \mathbb{R} \\ (i, j) & \mapsto \quad \begin{cases} 1 - q_{ji} & \text{if } |S \cap \{i, j\}| = 1, \\ q_{ij} & \text{otherwise.} \end{cases} \end{cases}$$

This, together with reversal and $Sym(n)$, gives a group of order $2^n n!$, expected to be the full symmetry group of $QMETP_n$ and $WQMETP_n$ (checked for $n \leq 9$).

**Oriented cut polytope** $OCUTP_n$**:** the convex hull of all the oriented cuts and their images under oriented switchings.

**Oriented multicut polytope** $OMCUTP_n$**:** the convex hull of all the oriented multicuts and their images under oriented switchings.

**Table 9.6**
The number of extreme rays and facets in some quasi-metric cones for $3 \leq n \leq 6$

| Cone | Dimension | No. of ext. rays (orbits) | No. of facets (orbits) | Diameters |
|---|---|---|---|---|
| $OCUT_3$=$WQMET_3$ | 5 | 6(2) | 9(2) | 1; 2 |
| $OCUT_4$=$QHYP_4$ | 9 | 14(3) | 30(3) | 1; 2 |
| $OCUT_5$ | 14 | 30(4) | 130(6) | 1; 3 |
| $OCUT_6$ | 20 | 62(5) | 16,460(61) | 1; 3 |
| $QHYP_5$ | 14 | 70(6) | 90(4) | 2; 2 |
| $WQMET_4$ | 9 | 20(4) | 24(2) | 2; 2 |
| $WQMET_5$ | 14 | 190(11) | 50(2) | 2; 2 |
| $WQMET_6$ | 20 | 18,502(77) | 90(2) | 3; 2 |
| $OMCUT_3$=$QMET_3$ | 6 | 12(2) | 12(2) | 2; 2 |
| $OMCUT_4$ | 12 | 74(5) | 72(4) | 2; 2 |
| $OMCUT_5$ | 20 | 540(9) | 35,320(194) | 2; 3 |
| $QMET_4$ | 12 | 164(10) | 36(2) | 3; 2 |
| $QMET_5$ | 20 | 43,590(229) | 80(2) | 3; 2 |

In [28] the oriented quasi-semimetric cone and polytope on an undirected graph $G$ is defined and studied. A description by inequalities is given.

We expect, that $OCUTP_n$ have exactly $2^{2n-2}$ vertices, which is much higher than $2^n - 2$, the number of extreme rays of $OCUT_n$. Both polytopes have the same symmetry group as $QMETP_n$. In Tables 9.6 and 9.7, are given computation for small quasi-metric cones and polytopes, respectively; the orbits are under $Sym(n)$ for cones and under $Sym(n) \times Z_2$ for polytopes.

It is conjectured that the diameters of $OCUT_n$ and $OMCUT_n$ are 1 and 2, respectively. Furthermore, if $f \geq 0$ defines a facet of $OMCUT_n$, then the zero-extension of $f$ to $OMCUT_{n+1}$ is still a facet, just as in the $CUT_n$ case.

The extreme rays of $QMET_n$ have been studied in [19, 26]. There it was proved that they are not symmetric and have at least $n-1$ zeros, implying that no one is the directed path distance of an oriented graph. The oriented multicuts define extreme rays of $QMET_n$. Also, the vertex-splitting of an extreme ray is still an extreme ray.

A list of non-adjacencies between facets of $QMET_n$ is given in [26] and it is conjectured there that in all other cases the facets are adjacent, implying that the diameter of $QMET_n^*$ is 2. The diameter of $OMET_n$ is 3 for $n = 4, 5$.

## 9.6 Partial Metrics

**Partial semimetric cone** $PMET_n$**:** the set of all *partial semimetrics* on $n$ points, i.e., the symmetric functions $p : \{1, \ldots, n\}^2 \to \mathbb{R}$, satisfying all $0 \leq p_{ii} \leq p_{ij}$ and all *sharp triangle inequalities*

$$p_{ik} \leq p_{ij} + p_{jk} - p_{jj}.$$

**Partial semimetric convex body** $PMETP_n$**:** the set of all $d \in PMET_n$ satisfying, in addition, all $p_{ij} \leq 1 + p_{ii}$ and the *perimeter inequalities*

$$p_{ij} + p_{jk} + p_{ki} \leq 2 + p_{ii} + p_{jj} + p_{kk}.$$

**Table 9.7**
The number of vertices and facets in some quasi-metric polytopes for $3 \le n \le 6$

| Polytope | Dimension | No. of vertices (orbits) | No. of facets (orbits) |
|---|---|---|---|
| $OCUTP_3$=$WQMETP_3$ | 5 | 16(2) | 16(2) |
| $OCUTP_4$=$WQMETP_4$ | 9 | 64(3) | 40(2) |
| $OCUTP_5$ | 14 | 256(3) | 1,056(5) |
| $OCUTP_6$ | 20 | 1,024(4) | 1,625,068(97) |
| $WQMETP_5$ | 14 | 2,656(8) | 80(2) |
| $WQMETP_6$ | 20 | 1,933,760(120) | 140(2) |
| $OMCUTP_3$=$QMETP_3$ | 6 | 22(3) | 20(2) |
| $OMCUTP_4$ | 12 | 136(5) | 1,160(9) |
| $QMETP_4$ | 12 | 544(8) | 56(2) |
| $QMETP_5$ | 20 | 1,155,136(392) | 120(2) |

**Remark 9.8** Partial semimetrics were introduced in [47]; they are used for treatment of partially defined/computed objects in semantics of computation.

Clearly, $((p_{ij}))$ is a partial semimetric if and only if $((p_{ij} - p_{ii}))$ is a weighable quasi-semimetric.

The symmetry group of $PMET_n$ should be $Sym(n)$ (checked for $n \le 9$) but $PMETP_n$ has additional symmetries: for $S \subset \{1,\dots,n\}$ we define the *switching* $U_S$:

$$U_S(p) = \left\{ \begin{array}{ccl} \{1,\dots,n\}^2 & \to & \mathbb{R} \\ (i,j) & \mapsto & \left\{ \begin{array}{ll} 1 + p_{ii} + p_{jj} - p_{ji} & \text{if } |S \cap \{i,j\}| = 1, \\ p_{ij} & \text{otherwise.} \end{array} \right. \end{array} \right.$$

We expect that, together with $Sym(n)$, this defines the full symmetry group of $PMETP_n$ (checked for $n \le 9$).

**Remark 9.9** In $PMETP_n$, the entries $p_{ij}$ are controlled by the $p_{ii}$, but the $p_{ii}$ entries are not bounded; so, $PMETP_n$ is not a polytope. One may get a polytope by adding the inequalities $\sum_i p_{ii} \le 1$; then the vertices of the obtained polytope are the vertices of $METP_n$ together with the extreme rays of $PMETP_n$.

In the Table 9.8, the orbits of $PMET_n$ are under $Sym(n)$, while for $PMETP_n$, the orbits are under the group (of order $n!2^{n-1}$) generated by switchings and permutations. But both have 3 orbits of facets and, maybe, ridge graphs of diameter 2.

## 9.7 Supermetric and Hemimetric Cones

**$(m,s)$-supermetric cone $SMET_n^{m,s}$:** the set of all $(m,s)$-*supermetrics on n points*, i.e., for a number $s > 0$ and an integer $m \ge 1$, the functions $h : \{1,\dots,n\}^{m+1} \to \mathbb{R}_{\ge 0}$, such that

- $h_{i_1,\dots,i_{m+1}} = 0$ if $i_a = i_b$ for $a \ne b$;
- $h_{i_1,\dots,i_{m+1}} = h_{i_{\sigma(1)},\dots,i_{\sigma(m+1)}}$ for all $\sigma \in Sym(m+1)$ (total symmetry);

**Table 9.8**
The number of vertices and facets in partial metric cone $PMET_n$ and body $PMETP_n$ for $3 \leq n \leq 6$

| Cone/body | Dimension | No. of ext. rays (orbits) | No. of facets (orbits) | Diameters |
|---|---|---|---|---|
| $PMET_3$ | 6 | 13(5) | 12(3) | 3; 2 |
| $PMET_4$ | 10 | 62(11) | 28(3) | 3; 2 |
| $PMET_5$ | 15 | 1,696(44) | 55(3) | 3; 2 |
| $PMET_6$ | 21 | 337,092(734) | 96(3) | 3; 2 |
| $PMETP_3$ | 6 | 17(4) | 19(3) | 2; 2 |
| $PMETP_4$ | 12 | 97(6) | 44(3) | 2; 2 |
| $PMETP_5$ | 20 | 7953(24) | 85(3) | 3; 2 |
| $PMETP_6$ | 12 | 5090337(427) | 146(3) | ?; 2 |

- For all distinct $i_1, \dots, i_{m+2}$, it holds the $(s;m)$-*simplex inequality*

$$s \times h_{i_2,\dots,i_{m+2}} \leq \left\{h_{i_1,i_3,\dots,i_{m+2}} + \cdots + h_{i_1,\dots,i_{m+1}}\right\}.$$

**Binary $(m,s)$-supermetric cone $SCUT_n^{m,s}$:** the positive span of all $\{0,1\}$-valued extreme rays of $SMET_n^{m,s}$.

**$m$-hemimetric cone $HMET_n^m$:** the set of all *$m$-hemimetrics* on $n$ points, i.e., $(m,1)$-supermetrics.

**Multicut hemimetric cone $HCUT_n^m$:** the positive span of all $(m+1)$-multicuts.

In [28] a different definition of hemimetric cone is defined. Instead of $(1;m)$-simplex inequality a different inequality using manifold is defined and it allows to define a version of hemimetric cone on a simplicial complex.

Above notions has been studied in [19, 23, 34, 35].

The $m$-multicuts determine extreme rays of the cone $HMET_n^m$. An adjacency relation between $m$-multicuts is conjectured in [19, 23] and the dual-description of many such cones are computed there; see Table 9.6. The orbits in it are under $Sym(n)$.

Given an $m$-hemimetric $h$ on $X$, denote by $s(X,h)$ the maximal $s$, such that it is a $(m,s)$-supermetric. Easy to see that $s(X,h) \leq m+1$, if $d$ is not identically zero.

**Example 39** *For a set $X \subset \mathbb{R}^d$ and $x_1, \dots, x_{m+1} \in X$, denote by $\mu_m(x_1, \dots, x_{m+1})$ the $m$-dimensional volume of the convex hull of $\{x_1, \dots, x_{m+1}\}$. Clearly, $\mu_m$ is an $m$-hemimetric on $X$. Let $s_m(X) := s(X,\mu)$ and let $\alpha_n$, $\beta_n$, $\gamma_n$ denote the vertex-set of regular $n$-simplex, $n$-hyperoctahedron and $n$-cube, respectively. In [25], it was shown:*

- $s_m(\alpha_n) = m+1$ *for all* $n \geq 2$, $m \geq 1$.
- $s_2(\beta_n) = 1+\sqrt{3}$ *for all $n$ and* $s_m(\beta_n) = 3$ *for all $m \geq 3$ and $n$.*
- $s_2(\gamma_n) = \frac{1+2\sqrt{n-2}}{\sqrt{2n-3}}$ *for $n \leq 8$, while for any fixed $m$,* $\lim_{n\to\infty} s_m(\gamma_n) = 1$.

**Remark 9.10** We can define the *$m$-hemimetric polytope* by adding to the definition of $HMET_n^m$ the following *$m$-perimeter inequality*:

$$h(x_2, \dots, x_{m+2}) + h(x_1, x_3, \dots, x_{m+2}) + \cdots + h(x_1, \dots, x_{m+1}) \leq 2.$$

However, there is no equivalent of the switching in that case and so, the studies have been limited to cones so far.

**Table 9.9**
The number of extreme rays and facets of small supermetric cones

| Cone | Dim. | No. of ext. rays (orbits) | No. of facets (orbits) | Diameters |
|---|---|---|---|---|
| $SMET^{m,s}_{m+2}$, $1 \le s \le m-1$ | m+2 | $\binom{m+2}{s+1}(1)$ | 2m+4(2) | min(s+1,m-s+1); 2 but 2; 3 if m=2,s=1 |
| $SMET^{m,m}_{m+2}$ | m+2 | m+2(1) | m+2(1) | 1; 1 |
| $HCUT^2_5$ | 10 | 25(2) | 120(4) | 2; 3 |
| $HMET^2_5$ | 10 | 37(3) | 30(2) | 2; 2 |
| $HCUT^3_6$ | 15 | 65(2) | 4,065(16) | 2; 3 |
| $HMET^3_6$ | 15 | 287(5) | 45(2) | 3; 2 |
| $HCUT^4_7$ | 21 | 140(2) | 474,390(153) | 2; 3 |
| $HMET^4_7$ | 21 | 3,692(8) | 63(2) | 3; 2 |
| $HMET^5_8$ | 28 | 55,898(13) | 84(2) | 3; 2 |
| $HMET^6_9$ | 36 | 864,174(20) | 108(2) | ?; 2 |
| $HCUT^2_6$ | 20 | 90(3) | 2,095,154(3,086) | 2; ? |
| $HMET^2_6$ | 20 | 12,492(41) | 80(2) | 3; 2 |
| $SMET^{2,2}_5$ | 10 | 132(6) | 20(1) | 2; 1 |
| $SCUT^{2,2}_5$ | 10 | 20(2) | 220(6) | 1; 3 |
| $SMET^{3,3/2}_6$ | 15 | 331,989(596) | 45(2) | 6; 2 |
| $SMET^{3,2}_6$ | 15 | 12,670(40) | 45(2) | 4; 2 |
| $SMET^{3,5/2}_6$ | 15 | 85,504(201) | 45(2) | 6; 2 |
| $SMET^{3,3}_6$ | 15 | 1,138(12) | 30(1) | 3; 1 |
| $SCUT^{3,3}_6$ | 15 | 21(2) | 150(3) | 1; 3 |
| $SMET^{4,2}_7$ | 21 | 2,561,166(661) | 63(2) | ?; 2 |
| $SMET^{4,3}_7$ | 21 | 838,729(274) | 63(2) | ?; 2 |
| $SMET^{4,4}_7$ | 21 | 39,406(37) | 42(1) | 3; 1 |
| $SCUT^{4,4}_7$ | 21 | 112(2) | 148,554(114) | 1; 4 |
| $SMET^{5,5}_8$ | 28 | 775,807(92) | 56(1) | ?; 1 |

**Remark 9.11** Given a set $X$, the following cone on it was introduced in [46]: *regular multidistance cone* $MDIS_X$ is the set of all functions $D : \cup_{m>1} X^m \to \mathbb{R}_{\geq 0}$, such that for all $x_1, \dots, x_m, y \in X$ it holds

- $D(x_1, \dots, x_m) = 0$ if $x_1 = \cdots = x_m$;
- $D(x_1, \dots, x_m) = D(x_{\sigma(1)}, \dots, x_{\sigma(m)})$ for all $\sigma \in Sym(m))$ (total symmetry);
- $D(x_1, \dots, x_m) \leq \sum_{i=1}^{m} D(x_i, y)$ (*multidistance m-star inequality*);
- $D(x_1, \dots, x_m) \leq D(x_1, \dots, x_m, y)$ (*regularity*).

Clearly, the restriction of $D$ on $X^2$ is a semimetric.

## 9.8 Software Computations

The polyhedral cones and polytopes, that show up in the study of finite semimetrics, have a complexity that grows very fast with the number of points. For example, it does not appear hopeful that one could compute the facets of, say, $CUTP_{20}$ in any reasonable time. On the other hand, for too small number of points, no useful information is available. Therefore, when searching for new kind of objects, it is advantageous to search for what the limit of the computational power offers.

One striking feature of the considered polytopes, such as $CUTP_n$, is their fairly large symmetry group. The methodology of choice for this computation is the Adjacency Decomposition Method, which proceeds when one has a facet to compute the adjacent ones and add it to the list if not equivalent to a known one. The methodology was first introduced in [15] and applied in [23, 26] to quasi-semimetric and hemimetric cones. Then it was applied in [39, 40, 41] to several polytopes coming from geometry of numbers; see [12] for an overview of this kind of techniques. One should keep in mind that yes, symmetry helps the computational tasks but it is not a panacea: some cases are simply too hard to treat. Similarly to using a faster programming language (like C++) or having a parallel program, it helps but does not radically change the nature of the problem treated.

However, for other tasks the picture is not so bleak. One is computing automorphism groups of polytopes and cones, while knowing only their vertices (or facets). Cones defined by 1000 vertices do not pose any problems; see [11] for details on the procedure used. Another area which is not so much affected by combinatorial explosion is computing the adjacency between vertices (or facets). There are several techniques for doing it, that rely on linear programming and are able to scale up to dimension, say, 100

The software is programmed in the GAP programming language [45] and is available on [36]. A faster C++ parallel program has been written [37].

# References


[1] Patrice Assouad. Plongements isométriques dans $L^1$: aspect analytique. In *Initiation Seminar on Analysis: G. Choquet-M. Rogalski-J. Saint-Raymond, 19th Year: 1979/1980*, volume 41 of *Publ. Math. Univ. Pierre et Marie Curie*, pages Exp. No. 14, 23. Univ. Paris VI, Paris, 1980.

[2] Patrice Assouad. Sur les inégalités valides dans $L^1$. *European J. Combin.*, 5(2):99–112, 1984.

[3] Patrice Assouad and Michel-Marie Deza. Espaces métriques plongeables dans un hypercube: aspects combinatoires. *Ann. Discrete Math.*, 8:197–210, 1980. Combinatorics 79 (Proc. Colloq., Univ. Montréal, Montreal, Que., 1979), Part I.

[4] Patrice Assouad and Michel-Marie Deza. Metric subspaces of $l_1$. *Pub. Math. Orsay*, 82-03, 1982.

[5] David Avis. *Some Polyhedral Cones Related to Metric Spaces*. Ph.D. thesis, Stanford University, 1977.

[6] David Avis and Michel-Marie Deza. The cut cone, $L^1$ embeddability, complexity, and multicommodity flows. *Networks*, 21(6):595–617, 1991.

[7] David Avis, Patrick Hayden, and Mark M. Wilde. Leggett-Garg inequalities and the geometry of the cut polytope. *Phys. Rev. A (3)*, 82(3):030102, 4, 2010.

[8] David Avis and Mutt. All the facets of the six-point Hamming cone. *European J. Combin.*, 10(4):309–312, 1989.

[9] Francisco Barahona and Ali R. Mahjoub. On the cut polytope. *Math. Programming*, 36(2):157–173, 1986.

[10] Evgenii P. Baranovskiĭ. The conditions for a simplex of 6-dimensional lattice to be $l$-simplex (in russian). *Ivan. Univ.*, 2(3):18–24, 1999.

[11] David Bremner, Mathieu Dutour Sikirić, Dmitrii V. Pasechnik, Thomas Rehn, and Achill Schürmann. Computing symmetry groups of polyhedra. *LMS J. Comput. Math.*, 17(1):565–581, 2014.

[12] David Bremner, Mathieu Dutour Sikirić, and Achill Schürmann. Polyhedral representation conversion up to symmetries. In *Polyhedral computation*, volume 48 of *CRM Proc. Lecture Notes*, pages 45–71. Amer. Math. Soc., Providence, RI, 2009.

[13] Jean Bretagnolle, Didier Dacunha-Castelle, and Jean-Louis Krivine. Lois stables et espaces $L^p$. *Ann. Inst. H. Poincaré Sect. B (N.S.)*, 2:231–259, 1965/1966.

[14] David Bryant and Paul F. Tupper. Diversities and the geometry of hypergraphs. *Discrete Math. Theor. Comput. Sci.*, 16(2):1–20, 2014.

[15] Thomas Christof and Gerhard Reinelt. Decomposition and parallelization techniques for enumerating the facets of combinatorial polytopes. *Internat. J. Comput. Geom. Appl.*, 11(4):423–437, 2001.

[16] Antoine Deza and Michel-Marie Deza. The ridge graph of the metric polytope and some relatives. In *Polytopes: abstract, convex and computational (Scarborough, ON, 1993)*, volume 440 of *NATO Adv. Sci. Inst. Ser. C Math. Phys. Sci.*, pages 359–372. Kluwer Acad. Publ., Dordrecht, 1994.

[17] Antoine Deza, Michel-Marie Deza, and Komei Fukuda. On skeletons, diameters and volumes of metric polyhedra. In *Combinatorics and computer science (Brest, 1995)*, volume 1120 of *Lecture Notes in Comput. Sci.*, pages 112–128. Springer, Berlin, 1996.

[18] Antoine Deza, Komei Fukuda, Tomohiko Mizutani, and Cong Vo. On the face lattice of the metric polytope. In *Discrete and computational geometry*, volume 2866 of *Lecture Notes in Comput. Sci.*, pages 118–128. Springer, Berlin, 2003.

[19] Elena Deza, Michel Deza, and Mathieu Dutour Sikirić. *Generalizations of finite metrics and cuts*. World Scientific Publishing Co. Pte. Ltd., Hackensack, NJ, 2016.

[20] Michel Deza and Mathieu Dutour Sikirić. Enumeration of the facets of cut polytopes over some highly symmetric graphs. *Int. Trans. Oper. Res.*, 23(5):853–860, 2016.

[21] Michel-Marie Deza and Elena I. Deza. Cones of partial metrics. *Contrib. Discrete Math.*, 6(1):26–47, 2011.

[22] Michel-Marie Deza, Elena I. Deza, and Janoš Vidali. Cones of weighted and partial metrics. In *Proceedings of the International Conference on Algebra 2010*, pages 177–197. World Sci. Publ., Hackensack, NJ, 2012.

[23] Michel-Marie Deza and Mathieu Dutour. Cones of metrics, hemi-metrics and super-metrics. *Ann. Eur. Acad. Sci.*, 1:141–162, 2003.

[24] Michel-Marie Deza and Mathieu Dutour. The hypermetric cone on seven vertices. *Experiment. Math.*, 12(4):433–440, 2003.

[25] Michel-Marie Deza, Mathieu Dutour, and Hiroshi Maehara. On volume-measure as hemimetrics. *Ryukyu Math. J.*, 17:1–9, 2004.

[26] Michel-Marie Deza, Mathieu Dutour, and Elena I. Panteleeva. Small cones of oriented semimetrics. In *Forum for Interdisciplinary Mathematics Proceedings on Statistics, Combinatorics & Related Areas (Bombay, 2000)*, volume 22, pages 199–225, 2002.

[27] Michel-Marie Deza and Mathieu Dutour Sikirić. The hypermetric cone and polytope on eight vertices and some generalizations. preprint at `arxiv:1503.04554`, March 2013.

[28] Michel-Marie Deza and Mathieu Dutour Sikirić. Generalized cut and metric polytopes of graphs and simplicial complexes. preprint at `arxiv:1706.02516`, March 2017.

[29] Michel-Marie Deza, Vyacheslav P. Grishukhin, and Elena I. Deza. Cones of weighted quasimetrics, weighted quasi-hypermetrics and of oriented cuts. In *Mathematics of Distances and Applications, ITHEA, Sofia*, pages 31–53. 2012.

[30] Michel-Marie Deza, Vyacheslav P. Grishukhin, and Monique Laurent. Extreme hypermetrics and $L$-polytopes. In *Sets, graphs and numbers (Budapest, 1991)*, volume 60 of *Colloq. Math. Soc. János Bolyai*, pages 157–209. North-Holland, Amsterdam, 1992.

[31] Michel-Marie Deza, Vyacheslav P. Grishukhin, and Monique Laurent. Hypermetrics in geometry of numbers. In *Combinatorial optimization (New Brunswick, NJ, 1992–1993)*, volume 20 of *DIMACS Ser. Discrete Math. Theoret. Comput. Sci.*, pages 1–109. Amer. Math. Soc., Providence, RI, 1995.

[32] Michel-Marie Deza and Monique Laurent. *Geometry of cuts and metrics*, volume 15 of *Algorithms and Combinatorics*. Springer, Heidelberg, 2010. First softcover printing of the 1997 original [MR1460488].

[33] Michel-Marie Deza and Elena I. Panteleeva. Quasi-semi-metrics, oriented multi-cuts and related polyhedra. *European J. Combin.*, 21(6):777–795, 2000. Discrete metric spaces (Marseille, 1998).

[34] Michel-Marie Deza and Ivo G. Rosenberg. $n$-semimetrics. *European J. Combin.*, 21(6):797–806, 2000. Discrete metric spaces (Marseille, 1998).

[35] Michel-Marie Deza and Ivo G. Rosenberg. Small cones of $m$-hemimetrics. *Discrete Math.*, 291(1-3):81–97, 2005.

[36] Mathieu Dutour Sikirić. Polyhedral. `http://mathieudutour.altervista.org/Polyhedral/`.

[37] Mathieu Dutour Sikirić. Polyhedral cpp. `https://github.com/MathieuDutSik/DualDescriptionADM`.

[38] Mathieu Dutour Sikirić. The seven dimensional perfect Delaunay polytopes and Delaunay simplices. *Canad. J. Math.*, 69(5):1143–1168, 2017.

[39] Mathieu Dutour Sikirić, Achill Schürmann, and Frank Vallentin. Classification of eight-dimensional perfect forms. *Electron. Res. Announc. Amer. Math. Soc.*, 13:21–32 (electronic), 2007.

[40] Mathieu Dutour Sikirić, Achill Schürmann, and Frank Vallentin. Complexity and algorithms for computing Voronoi cells of lattices. *Math. Comp.*, 78(267):1713–1731, 2009.

[41] Mathieu Dutour Sikirić, Achill Schürmann, and Frank Vallentin. The contact polytope of the Leech lattice. *Discrete Comput. Geom.*, 44(4):904–911, 2010.

[42] Arthur Fine. Hidden variables, joint probability, and the Bell inequalities. *Phys. Rev. Lett.*, 48(5):291–295, 1982.

[43] Vyacheslav P. Grishukhin. All facets of the cut cone $\mathbf{C}_n$ for $n = 7$ are known. *European J. Combin.*, 11(2):115–117, 1990.

[44] Vyacheslav P. Grishukhin. Computing extreme rays of the metric cone for seven points. *European J. Combin.*, 13(3):153–165, 1992.

[45] The GAP group. *GAP — Groups, Algorithms, and Permutations, Version 4.4.6.*

[46] Martin Javier and Major Gaspar. Regular multidistances. In *XV Congresso Espagnol sombre Technologias y Logica Fuzzy, Huelva*, pages 297–301. 2010.

[47] Steve G. Matthews. Partial metric topology. In *Papers on general topology and applications (Flushing, NY, 1992)*, volume 728 of *Ann. New York Acad. Sci.*, pages 183–197. New York Acad. Sci., New York, 1994.

[48] Itamar Pitowsky and Karl Svozil. Optimal tests of quantum nonlocality. *Phys. Rev. A (3)*, 64(1):014102, 4, 2001.

[49] Sergey S. Ryshkov and Evgenii P. Baranovskiĭ. Repartitioning complexes in $n$-dimensional lattices (with full description for $n \leq 6$). In *Voronoi impact on modern science*, pages 115–124. Institute of Mathematics, Kyiv, 1998.

[50] Sergey S. Ryškov and Evgenii P. Baranovskiĭ. $C$-types of $n$-dimensional lattices and 5-dimensional primitive parallelohedra (with application to the theory of coverings). *Proc. Steklov Inst. Math.*, (4):140, 1978. Cover to cover translation of Trudy Mat. Inst. Steklov **137** (1976), Translated by R. M. Erdahl.

[51] Isaac J. Schoenberg. Metric spaces and positive definite functions. *Trans. Amer. Math. Soc.*, 44:522–536, 1938.

[52] Achill Schürmann. *Computational geometry of positive definite quadratic forms*, volume 48 of *University Lecture Series*. American Mathematical Society, Providence, RI, 2009. Polyhedral reduction theories, algorithms, and applications.

[53] Paul D. Seymour. Matroids and multicommodity flows. *European J. Combin.*, 2(3):257–290, 1981.

[54] Mikhail E. Tylkin (=M. Deza). On Hamming geometry of unitary cubes. *Soviet Physics. Dokl.*, 5:940–943, 1960.

[55] George Voronoi. Nouvelles applications des paramètres continus à la théorie des formes quadratiques. Deuxième Mémoire. Recherches sur les parallélloèdres primitifs. *J. Reine Angew. Math*, 134(1):198–287, 1908.

[56] Klaus Wagner. über eine Eigenschaft der ebene Komplexe. *Math. Annal.*, 114:570–590, 1937.

# Chapter 10

# *Cayley Configuration Spaces*

**Meera Sitharam**
*University of Florida*

**Menghan Wang**
*University of Florida*

**Joel Willoughby**
*University of Florida*

**Rahul Prabhu**
*University of Florida*

**CONTENTS**



## 10.1 Introduction

A *Euclidean Distance Constraint System (EDCS)* or *linkage* $(G,\delta)$ is a graph $G=(V,E)$ together with a distance vector $\delta$ that assigns a distance $\delta_e$ or a distance interval $[\delta_e^l,\delta_e^r]$ to each edge $e\in E$. A $d$-dimensional realization of the EDCS is the assignment $p$ of points in $\mathbb{R}^d$ to the vertices in $V$ such that the distance equality (resp. inequality) constraints are satisfied: $\delta_{(u,v)}=||p(u)-p(v)||$, respectively, $\delta_{(u,v)}^l\leq||p(u)-p(v)||\leq\delta_{(u,v)}^r$.

Describing, exploring, and sampling the realization space of a linkage is a difficult problem that arises in many classical areas of mathematics and computer science and has a wide variety of applications in computer aided design for mechanical engineering, robotics, and molecular modeling. Especially for under-constrained (or independent and not rigid) linkage whose realizations have one or more degrees of freedom of motion, progress on this problem has been very limited. Existing methods for sampling linkage realization spaces often use Cartesian representations, factoring out the Euclidean group by arbitrarily "pinning" or "grounding" some of the points' coordinate values. Even when the methods use "internal" representation parameters such as Cayley parameters (nonedges) or angles between unconstrained objects, the choice of these parameters is ad hoc. We characterize the graphs of linkages for which parameters can be judiciously chosen so that the parametrized configuration space is convex and therefore easier to explore.

We define a representation of the realization space of a linkage in $\mathbb{R}^d$ as (i) a choice of parameter set, specifically a choice of a set $F$ of nonedges of $G$ ($F \subseteq V \times V \setminus E$) called *Cayley parameters*, and (ii) a set $\Phi^d_F(G,\delta)$, called the *Cayley configuration space* (a term first coined in [31]), consisting of realizable distance vector $\delta_F$ for $F$. I.e., the augmented linkage $(G \cup F, \delta_E, \delta_F)$ has at least one $d$-dimensional realization. In the presence of inequalities, the Cayley configuration space is denoted $\Phi^d_F(G,[\delta^l,\delta^r])$ and the augmented linkage is: $(G \cup F,[\delta^l_E,\delta^r_E],\delta_F)$. Here $G \cup F$ refers to a graph $(V, E \cup F)$. In other words, our representations are in Cayley parameters or nonedge distances: the set $\Phi^d_F(G,\delta)$ is the projection onto the Cayley parameters in $F$, of the Cayley-Menger semi-algebraic set with fixed $(G,\delta)$ [10, 12]. Provided $G \cup F$ is minimally rigid, for every Cayley configuration in $\Phi^d_F(G,\delta)$, generically there exists at least one and at most finitely many Cartesian realizations, namely, the realizations of $(G \cup F, \delta, \delta_F)$.

We survey results on the study of exact, efficient representations of realization spaces of linkages. These results completely characterize linkages that have connected, convex Cayley configuration spaces with efficiently computable bounds based on precise and formal measures of efficiency. These results do not rely on the genericity of the linkages. Results presented here employ (a) convexity results and positive semi-definiteness of Euclidean distance matrices (we refer the reader to Chapter 1 for a discussion on Euclidean distance matrices), with (b) forbidden minor characterizations and algorithms related to $d$-flattenability of graphs.

### 10.1.1 Euclidean Distance Cone

The structure of the cone $\Phi_{n,l_p}$, its strata and faces are well-studied. For $l_2$ this is called the *Euclidean distance matrix (EDM)* cone (we refer the reader to Chapter 1 for a discussion on EDM) [14, 16, 37], which is a simple, relationship to the cone of positive semidefinite matrices, a fact first observed by Schoenberg [30]. Consequently, understanding its structure is important in semidefinite programming relaxations and the so-called sums of squares method with numerous applications [6, 20, 28]. Connections between combinatorial rigidity and the structure of the EDM have been investigated extensively by Alfakih [1, 2]. The reader is additionally referred to [15] for a comprehensive survey of key results about the EDM cone, including observations about the face structure and dimensional strata of the EDM cone.

**Organization:** The rest of the chapter is organized as follows. Section 10.2 introduces the various background definitions and terms. Section 10.3 lists the related chapters. Section 10.4 gives the characterization of graphs in 2D that have convex Cayley configuration spaces. Section 10.5 discusses the connection between inherent Cayley configuration space and flattenability, which extends beyond the Euclidean norm. The section further discusses genericity and independence in the context of flattenability. Section 10.6 discusses how a linkage can be modified for efficient realization. Section 10.7 discusses the Cayley configuration spaces of a special class of linkages, called 1-Dof tree-decomposable linkages, in 2D. Section 10.8 gives the conclusion.

## 10.2 Glossary

**Cayley Configuration Space:** The *d-dimensional Cayley configuration space* of a linkage $(G,\delta)$, over some set of nonedges, $F$, under norm $l_p$ is denoted $\Phi^d_{F,l_p}(G,\delta)$ and is defined as the set of vectors $\delta_F$ of lengths attained by the nonedges $F$ over all the realizations of the linkage $(G,\delta)$ in $\mathbb{R}^d$. This same space is also sometimes referred to as the *Cayley configuration space of a realization or framework* $(G,p)$ whose edge lengths are $\delta$.

**Graphs with convex $d$-dimensional Cayley configuration space:** Graph $G$ has a convex $d$-dimensional Cayley configuration space under norm $l_p$ if there exists a set of nonedges $F$ such that for all edge length vectors $\delta$, $\Phi^d_{F,l_p}(G,\delta)$, is convex.

**Graphs with inherent convex $d$-dimensional Cayley configuration space:** Graph $G$ has an inherent convex $d$-dimensional Cayley configuration space under norm $l_p$ if for every partition of the edges of $G$ into $H$ and $F$, and every edge length vector $\delta_H$, $\Phi^d_{F,l_p}(H,\delta)$, is convex.

**Graph Minor:** A graph $G$ has a graph $K$ as minor if $G$ can be reduced to $K$ via vertex/edge deletions and edge contractions (coalescing or identifying the two vertices of an edge).

**Minor Closed:** If a property of $G$ remains consistent under the operation of taking minors, that property is *minor-closed*. A useful result due to [29] is that if a property is minor-closed, then there is a finite set of *forbidden minors* (see next definition)

**Finite Forbidden Minor Characterization:** A class $C$ of graphs has a finite forbidden minor characterization, if there exists a fixed, finite set $M$ of minors, such that $G \in C$ if and only if $G$ doesn't contain any $K \in M$ as a minor.

**$k$-sum:** A $k$-sum is a way of combining two graphs by gluing them together at a $K_k$ or a complete graph on $k$ vertices.

**(Complete) $k$-Trees:** A graph is a (complete) $k$-tree if it is formed by $k$-sums of $K_{k+1}$s. A partial $k$-tree is a proper subgraph of a complete $k$-tree. These are often useful in forbidden minor characterization.

## 10.3 Related Chapters

**Chapter 8 Dimensional and Universal Rigidity of Bar Frameworks:** We use notions of rigidity, genericity, and independence defined in terms of the so-called rigidity matrix.

**Chapter 24 Rigidity with Polyhedral Norms:** Many of the results in this chapter extend readily to different norms with appropriate definitions of rigidity and realization developed by Kitson.

**Chapter 9 Computations of Metric/Cut Polyhedra and Their Relatives:** Many of the ideas related especially to $d$-flattenability rely heavily on the structure of the metric cone and its sections and stratum.

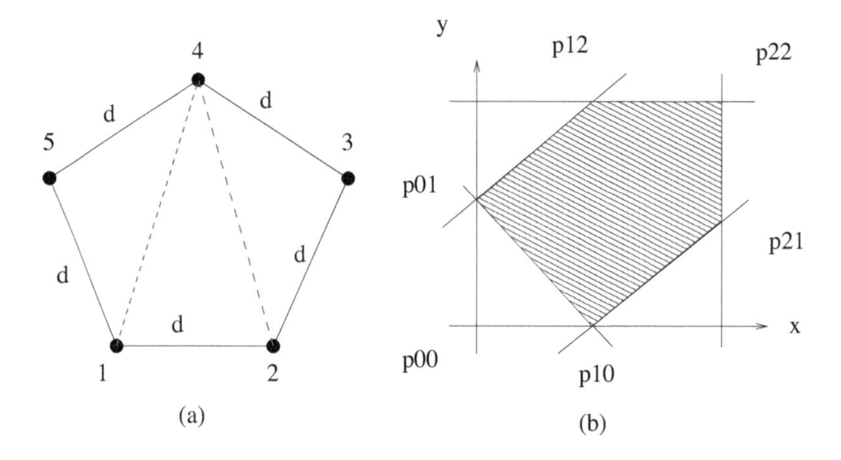


**Figure 10.1**
An example of a simple Cayley configuration space on 2 nonedges. On the left is a pentagon graph $G$. We choose to look at the Cayley configuration space $\Phi^2_F(G,\delta)$ where $F = \{(v_1,v_4),(v_2,v_4)\}$ and $\delta = (1,1,1,1,1)$. The result is the shaded region to the right where the $x$-axis are the edge lengths for $(v_1,v_4)$ and the $y$-axis edge lengths for $(v_2,v_4)$.

## 10.4 Characterizing 2D Graphs with Convex Cayley Configuration Spaces

For a graph, the choice of Cayley parameter nonedges is important, as this determines whether the resulting Cayley configuration space is convex or not. Figure 10.1 shows a choice of Cayley parameters for a graph which leads to a convex Cayley configuration space while Figure 10.2 shows an example where the choice of a different Cayley parameter for the same graph leads to a disconnected Cayley configuration space.

In this section, we present results from [31] that characterize 2D graphs that have convex Cayley configuration spaces. Specifically, we are interested in characterizing graphs $G$ for which there is a set $F$ of nonedges such that for all distance assignments $\delta_E$, the Cayley configuration space $\Phi^2_{f,}(G,\delta)$ is convex. We first define some background terms.

**Minimal 2-sum component:** Recursively decompose a graph using 2-sums. The resulting indecomposible graphs in the decomposition are minimal 2-sum components.

Theorem 10.1 characterizes 2D graphs in which, for a single nonedge chosen as a Cayley parameter, the Cayley configuration space is a single interval.

**Theorem 10.1** *[31] Given a graph $G = (V,E)$ and a nonedge $f$, the Cayley configuration space $\Phi^2_{f,}(G,\delta)$ is a single interval for all $\delta$ if and only if all the minimal 2-sum components of $G \cup f$ that contain $f$ are partial 2-trees.*

In 3-dimensions (and higher), however, the above theorem does not hold, as a simple counterexample shows (see Figure 10.3). In this example the specified nonedge must be contracted to obtain a forbidden minor from $G \cup f$. Thus, to fully characterize the property, the following conjecture was raised for graphs $G$ and nonedges $f$ such that $G \cup f$ is not $d$-flattenable.

**Conjecture 1** *An nonedge $f$ of a graph $G$ has a single interval of attainable values for all edge*

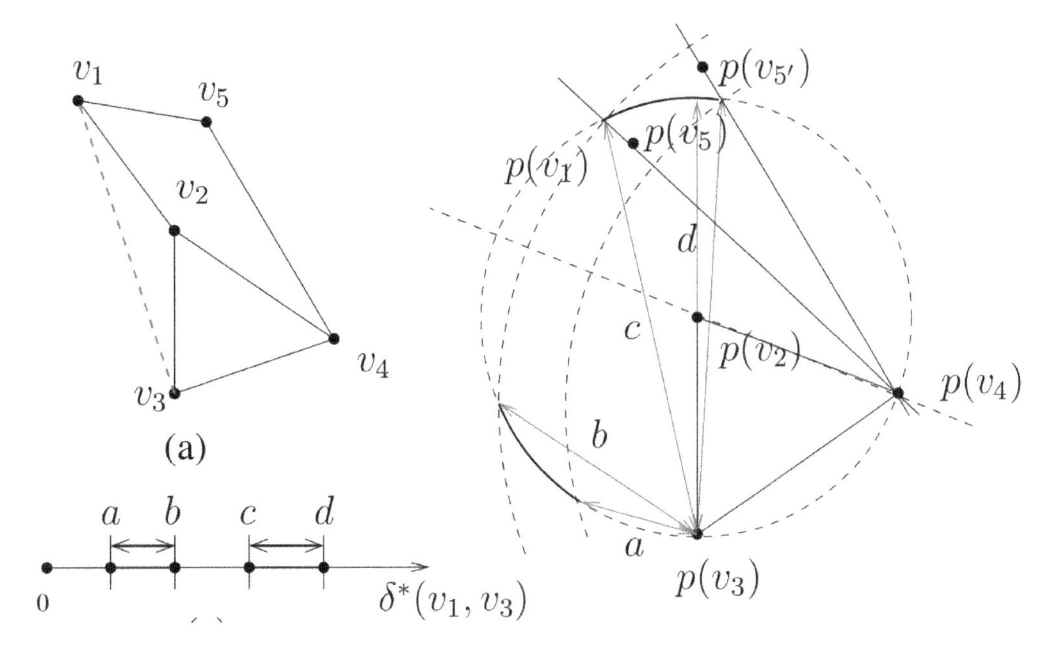


**Figure 10.2**
When parameters for the EDCS in (left) are chosen, to be the dashed edge, we get a disconnected Cayley configuration space shown in (right). The realization $p(v_1)$ can lie in either of the two solid arc segments of the circle, yielding a disconnected Cayley configuration space.

*length vectors of $G \setminus f$ in d-dimensions if and only if for any minimal 2-sum components of G containing f that are not d- flattenable, f must always (i) be removed or (ii) duplicated or (iii) contracted in order to obtain a forbidden minor of d-flattenability from G.*

Building on Theorem 10.1, Theorem 10.2 gives an exact characterization of the class of graphs whose corresponding linkages admit a 2D convex Cayley configuration space for a set of nonedges $F$.

**Theorem 10.2** *For a graph $G = (V,E)$, the following four statements are equivalent:*

1. *There exists a nonempty set of nonedges F such that for all $\delta$, $\Phi^2_{F,}(G, \delta_E)$ is connected;*
2. *There exists a nonempty set of nonedges F such that for all $\delta$, $\Phi^2_{F,}(G, \delta_E)$ is convex;*
3. *There exists a nonempty set of nonedges F such that for all $\delta$, $\Phi^2_{F,}(G, \delta_E)$ is a polytope linear faces.*
4. *G has a 2-sum component that is a flexible partial 2-tree.*

For the class of graphs described in Theorem 10.2, Theorem 10.3 gives an exact combinatorial characterization of the choices of Cayley parameters that ensure a convex Cayley configuration space.

**Theorem 10.3** *Given a graph $G = (V,E)$ and nonempty set of nonedges F, the 2D Cayley configuration space $\Phi^2_{F,}(G, \delta)$ is a linear polytope, connected or convex for all $\delta$ if and only if all the minimal 2-sum components of $G \cup F$ containing any subset of F are partial 2-trees.*

All of the above theorems hold with edge length intervals instead of edge lengths.

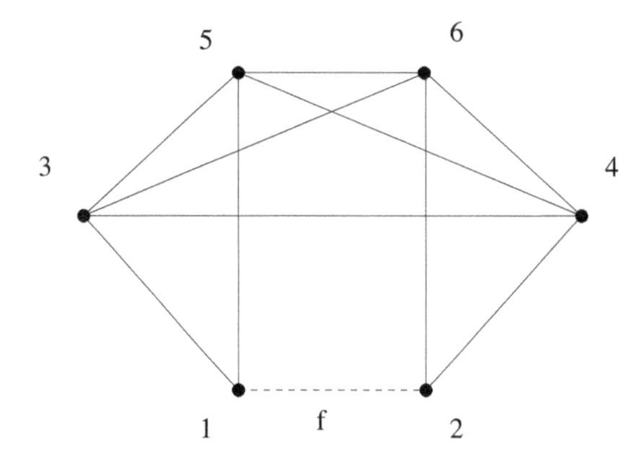


**Figure 10.3**
$G$'s 3D Cayley configuration space on nonedge $f$ is one interval, although $G \cup \{f\}$ has a $K_5$ minor.

**Theorem 10.4** *Theorems 10.1–10.3 hold with $\Phi_F^2(G,(\delta^l,\delta^r))$ replacing $\Phi_F^2(G,\delta)$.*

Convex Cayley configuration spaces can be extended to other norms. Many of these key results follow in the next sections. In particular, Section 10.5 gives the extension to general $l_p$ norms and its connection to flattenability.

## 10.5 Extension to other Norms, Higher Dimensions, and Flattenability

In this section we show the connection between flattenability and Cayley configuration spaces [36]. We start by defining some background terms.

**$d$-Flattenability:** A graph $G$ is *$d$-flattenable* if for every realization $r$ of $G$ under the $l_p$ norm, the linkage $(G,\delta)$ where $\delta_{vw} := ||r(v)-r(w)||_p$ also has a realization in the $d$-dimensional $l_p$-normed space. This definition does not imply that there is a continuous path of realizations starting from a realization of $(G,\delta)$ in some higher dimension to the realization in $d$-dimensions, nor does it refer to realizations of intrinsic dimension $d$ in some higher dimensional space.

The concept of $d$-flattenability was first introduced in a series of papers [8, 9] for the Euclidean or $l_2$ norm. However they called it "$d$-realizability," which can be confused with the realizability of a given linkage in $d$-dimensions. This is one of reasons we introduced the term: *flattenability.*

The term flattening has also been used by Matousek [21] in the context of non-isometric embeddings (with low distortion via Johnson-Lindenstrauss lemma in $l_2$ [24], impossibility of low distortion in $l_1$ [11], etc). We admit arbitrary distortions of nonedge lengths, but force edge lengths to remain undistorted.

Immediately by definition, $d$-flattenability is a minor-closed property under any norm. A full characterization for 3-flattenable graphs was given for the Euclidean or $l_2$ norm by [8, 9]. A summary of the known forbidden minor results for the $l_2$ and $l_1$ norms is presented in Table 10.1. The proofs for the $l_2$ 3-dimensional cases mostly involved exhaustion of all other possibilities. Later in this section, tools are given (see Theorem 10.5) which lead to more direct proofs of such properties. A close connection between Cayley configuration spaces and flattenability was shown in [36]:

**Table 10.1**
Summary of known forbidden minor characterizations for $l_1$ and $l_2$ norms.

| Norm | Dimension | Forbidden Minor(s) | Characterization | References |
|---|---|---|---|---|
| $l_2$ | 1 | $K_3$ | Forests | [9] |
| | 2 | $K_4$ | Partial 2-trees | [9] |
| | 3 | $K_5$, Octahedron | Partial 3-trees and others | [9] |
| | $n > 3$ | $K_{n+1}$ + others | Partial $n$-trees and others | [5] |
| $l_1$ | 1 | $K_3$ | Forests | [36] |
| | 2 | $W_4 + 2K_4 - 1$ | Partial 2-trees and others | [17] |

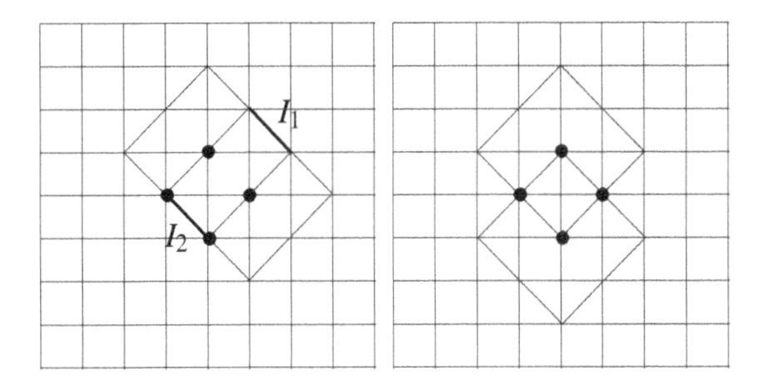


**Figure 10.4**
On the left is an illustration of an equidistant $K_4$ with the possible intervals of Case 1. On the right is the same for Case 2

**Theorem 10.5** *For $l_p$ norms, G is d-flattenable if and only if G has a d-*dimensional, inherent convex Cayley *configuration space. As a direct corollary, it follows that both properties are minor-closed for general $l_p$ norms.*

This "only if" direction of this statement was shown in [31] for the $l_2$ norm. The argument only required the fact that the cone of squared distance vectors is convex. Hence, we can use the same proof if we can show $\Phi_{n,l_p}$ is convex. The proof of the "if" direction requires that the cone is the convex hull of $l_p^p$ distance vectors in any dimension $d$.

The following result provides a nice link between $d$-flattenability and convex Cayley configuration spaces.

**Corollary 10.6** *Having a d-dimensional convex Cayley configuration space on all subgraphs is a minor-closed property.*

Another immediate result is that $d$-flattenability and convex Cayley configuration spaces have the same forbidden minor characterizations for given $d$ under the same $l_p$-norm. This gives us a nice tool when trying to find forbidden minors for other $l_p$ norms:

**Observation 1** *If for some assignment of distances $\delta$ to edges of G leads to a non-convex $\Phi^d_{F,l_p}(G,\delta)$, then G is not d-flattenable.*

**Example Usage:** This is an example to show how Theorem 10.5 can be used. We will use it to show a specific graph is not 2-flattenable under the $l_1$ (or Manhattan) norm. This example comes from [36].

**Theorem 10.7** *The so called "banana" graph or $K_5$ minus one edge is not 2-flattenable under the $l_1$ norm.*

**Proof 10.1** We will invoke Observation 1 to show this. Consider a distance vector for the banana with unit distances for all except one edge, $f$. This has a realization in 3-dimensions as $K_5$ is 3-flattenable for the $l_1$ norm (see [5]). Then, we have an equidistant $K_4$ as a subgraph. The only realization for such a $K_4$ in 2-dimensions is to have all 4 points arranged as the vertices of the unit ball centered at the origin. The two remaining unit edges then connect a new vertex to two of these points. Here we have 2 cases: the two vertices border the same quadrant or they lie across one of the axes from each other.

Case 1: Without loss of generality, we assume the two vertices are the upper right of the $K_4$. In Figure 10.4 (left), it can be seen that the new vertex can lie anywhere in $I_1$ or $I_2$. If it lies in $I_1$, the remaining edge of the banana can take lengths in the range $[0,1]$. If it lies in $I_2$, the only length it can be is two.

Case 2: Without loss of generality, assume the 2 vertices are the top-most and bottom-most. Again from Figure 10.4 (right), the new vertex only has 2 positions it can be in, each leading to a length of 1 for the remaining edge.

Hence, $\Phi^2_{F,l_1}(G\setminus F, \delta^{G\setminus F}) = [0,1] \cup \{2\}$, where $G$ is the banana and $F = \{f\}$. This is not convex and thus by Theorem 10.5, the banana is not 2-flattenable.

### 10.5.1 Computing Bounds of Convex Cayley Configuration Spaces in 3D for Partial 3-Trees

For the case of $l_2$, partial 3-trees are 3-flattenable and by Theorem 10.5 have inherent convex Cayley configuration spaces. Cayley configurations are bound by tetrahedral inequalities. However, it is possible to achieve tighter bounds, as shown in [13], where they give algorithms to find efficient bounds on Cayley parameters for partial 2-trees in 3D (a subclass of partial 3-trees), and show that bounds can be computed in quadratic time in the size of the graph. In cases where there are multiple Cayley parameters, they additionally show that the order in which Cayley parameters are fixed have an effect on the efficiency of the range computation.

To prove Theorem 10.5 and gain some insight into the connection, it is necessary to have some information on the so-called distance cone.

### 10.5.2 Some Background on the Distance Cone

**The $l_p^p$ Distance Cone:** Many of the proofs given for the results in this section rely heavily on the structure of the cone $\Phi_{n,l_p}$ consisting of vectors $\delta_q$ of pairwise $l_p^p$-distances of $n$-point configurations $q$. (A proof that this set is a cone can be found in [5], which also applies to infinite dimensional settings).

**Strata of the Cone :** The *$d$-dimensional stratum* of this cone consists of pairwise distance vectors of $d$-dimensional point configurations and is denoted $\Phi^d_{n,l_p}$.

**Projections of the Cone:** The projection or shadow of this cone (resp. stratum) on a subset of coordinates i.e., pairs corresponding to the edges of a graph $G$ is denoted $\Phi_{G,l_p}$ (resp. $\Phi^d_{G,l_p}$). This projection is the set of realizable edge-length vectors $\delta$ of linkages $(G,\delta)$ in $l_p^p$ (resp. in $d$-dimensions) (see Figure 10.5).

**Flattening Dimension:** The *$l_p$-flattening dimension* of a graph $G$ (resp. class $C$ of graphs) is the minimum dimension $d$ for which $G$ (resp. all graphs in $C$) is flattenable in $l_p$.

Notice that $\Phi_{n,l_p}$ is the same as $\Phi_{K_n,l_p}$, where $K_n$ is the complete graph on $n$ vertices. Let $n_p$ be the flattening dimension of $K_n$. It is not hard to show [15] that in fact $n_p \leq \mathbb{R}^{\binom{n}{2}}$. For the Euclidean or

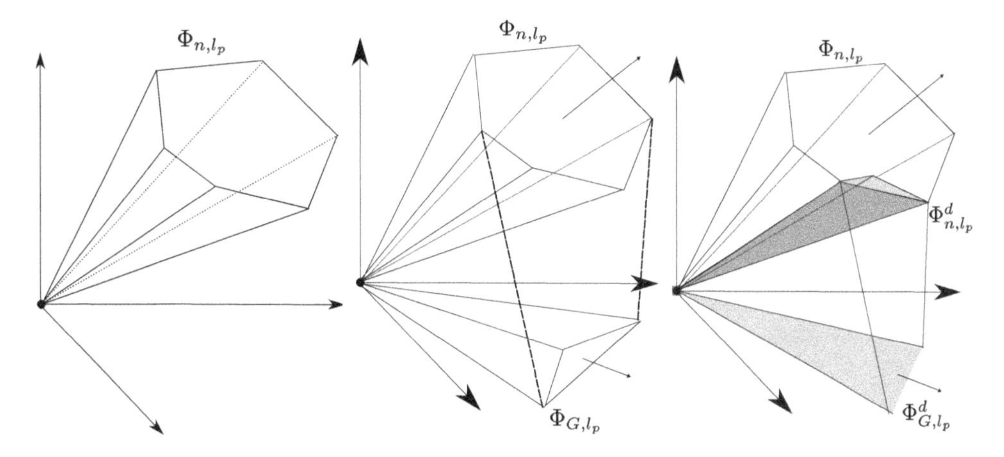


**Figure 10.5**
On the left we have the cone of realizable distance vectors under $l_p$. *It is shown here as a polytope, but in general that is not the case; these are not rigorous figures – their purpose is intuitive visualization.* The cone lives in $\binom{n}{2}$-dimensional space where each dimension is a pairwise distance among $n$ points. In the middle is a projection of the cone onto the edges of some graph. This yields a lower dimensional object (unless $G$ is complete). On the right, a $d$-dimensional stratum is highlighted and lines show the projection onto coordinates representing edges of a graph. In general this stratum is not just a single face. Note that this projection is equal to the projection as the whole cone (middle) iff $G$ is $d$-flattenable.

$l_2$ case, a further result of Barvinok [7] shows that the flattening dimension of any graph $G = (V,E)$ (although he did not use this terminology), is at most $O(\sqrt{|E|})$. Notice additionally that the $d$-dimensional Cayley configuration space on a set of non-edges $F$ of $G$ $\Phi^d_{F,l_p}(G,\delta)$ is the coordinate shadow of the $(G,\delta)$-fiber of $\Phi^d_{G\cup F,l_p}$, i.e., all linkages $(G\cup F,\delta_{G\cup F})$ that have $\delta$ assigned to the edges of $G$, on the coordinate set $F$ (see Figure 10.6).

Now we outline the proof of Theorem 10.5. The full proof can be found in [36]. We will be using the convexity properties of the cone and its strata described above to achieve the result.

***Proposition 10.1***
*$\Phi_{n,l_p}$ for general $l_p$ is contained in the convex hull of the $l_p^p$ distance vectors of the 1-dimensional n-point configurations in $\mathbb{R}$.*

To see this, take some realization of a complete linkage $(K_n,\delta)$, which from [5], we know to be finite dimensional. Then we can simply treat each dimension of the realization as a 1-dimensional point configuration in $\mathbb{R}$. We simply build the distance vector $\delta$ as a convex sum of vectors only defined by one dimension of the realization. It is well known that $\Phi_{n,l_p}$ is convex. This along with 10.1 leads to the following:

***Proposition 10.2***
*$\Phi_{n,l_p}$, $1\le p\le\infty$ is the convex hull of the $l_p^p$ distance vectors of the 1-dimensional, n-point configuration vectors in $\mathbb{R}$.*

Now we can move on to proving Theorem 10.5. Suppose $G$ is $d$-flattenable under some $l_p$-norm. Because $G$ is $d$-flattenable, $\Phi^d_{G,l_p}$ is convex. Given a subgraph $G'$ of $G$, if we break $G$ into $H$ and

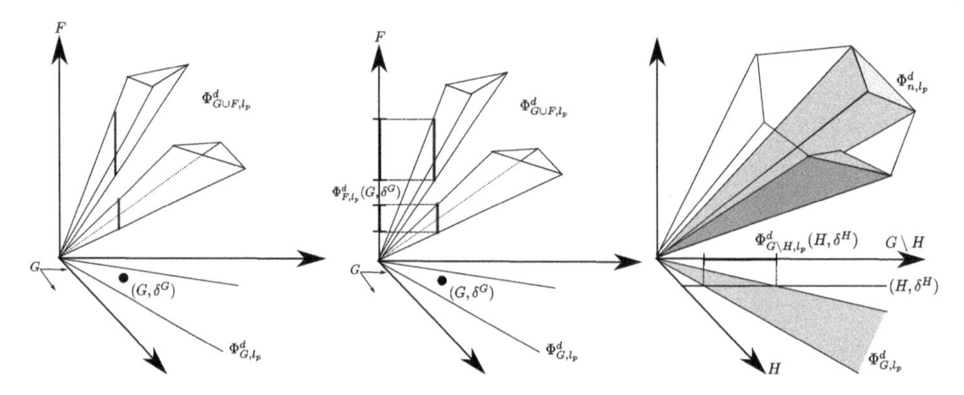


**Figure 10.6**
This is an example of $\Phi^d_{G\cup F,l_p}$ that is not convex. The linkage $(G,\delta)$ and its fiber in $\Phi^d_{G\cup F,l_p}$ are shown on the left. Note that the fiber is not convex. In the middle, this fiber is then projected onto the remaining edges of $G\cup F$ to form $\Phi^d_{F,l_p}(G,\delta)$. Note that it is not convex either. On the right, $\Phi^d_{n,l_p}$ is projected onto the edges of some $d$-flattenable $G$ (note that this is the same as projection of $\Phi_{n,l_p}$). The inherent Cayley configuration space corresponding to some subgraph $G\setminus H$ of $G$ is then shown projected onto the edges of $G\setminus H$. This projection is convex.

$G'$ and fix the values of $E$ corresponding to a linkage $(H,\delta_H)$, we are taking a section of $\Phi^d_{H\cup G',l_p}$, which is again convex. Then any inherent Cayley configuration of $G$ is convex.

For the other direction, we know the Cayley configuration space on the empty set is convex. This is just $\Phi^d_{G,l_p}$ and because it is convex, it is its own convex hull. From Proposition 10.2, we see that $\Phi_{G,l_p}$ is exactly the convex hull of its $d$-dimensional stratum, $\Phi^d_{G,l_p}$. Thus, $G$ is $d$-flattenable.

### 10.5.3 Genericity and Independence in the Context of Flattenability

**Generic $d$-Flattenability:** Given an $l_p$ framework $(G,r)$, with $n$ vertices, in arbitrary dimension, consider its pairwise length vector, $\delta_r$, in the cone $\Phi_{n,l_p}$. A framework $(G,r)$ of $n$ vertices is *generic* with respect to $d$-flattenability if the following hold:

(i) There is an open neighborhood $\Omega$ of $\delta_r$ in the (interior of the) cone $\Phi_{n,l_p}$, (recalling that $n_p$ is the flattening dimension of the complete graph $K_n$, $\Omega$ corresponds to an open neighborhood of $n_p$-dimensional point-configurations of $r$); and

(ii) $(G,r)$ is $d$-flattenable if and only if all the frameworks in $\Omega$ are.

**Generic Convex Cayley Configuration:** Let $\delta_r$ be as in the above definition. A framework $(G,r)$ of $n$ vertices in $d$-dimensions is *generic* with respect to the property of convexity of $\Phi^d_{F,l_p}(G,\delta)$ if

(i) There is an open neighborhood $\Omega$ of $\delta_r$ in the stratum $\Phi^d_{n,l_p}$ (this corresponds to an open neighborhood of $d$-dimensional point-configurations of $r$); and

(ii) $(G,r)$ has convex Cayley configuration space over $F$ if and only if all the frameworks in $\Omega$ do.

**Generic Property:** A *property* of frameworks is said to be *generic* if the *existence* of a generic framework with the property implies that the property holds for *all* generic frameworks.

We refer to Chapter 1 for the definitions of rigidity and independence. For frameworks in polyhedral norms (including the $l_p$ norms), Kitson [25] has defined properties such as *well-positioned, regular*, analogous to the above, which have been used to show (infinitesimal) rigidity to be a generic property of frameworks. Intuitively, a well-positioned $d$-dimensional framework under the $l_p$ norm is one in whose $d$-dimensional neighborhood in $l_p$-normed space the pairwise distances between points can be expressed in polynomial form. We refer the reader to Kitson's paper for a precise definition [25].

In this section we show relationships between $d$-flattenability (and thus existence of convex inherent Cayley configuration spaces) and combinatorial rigidity concepts via the cone $\Phi_{n,l_p}$. Note that the definition of $d$-flattenability of a graph $G$ in $l_p$ requires every $l_p$ framework of the graph $G$ – in an arbitrary dimension – to be $d$-flattenable. The next theorem weakens this requirement.

**Theorem 10.8** *Every generic framework of G is d-flattenable if and only if G is d-flattenable.*

The "if" direction follows immediately from the definition of $d$-flattenability. For the "only if" direction, notice that a nongeneric, (bounded) framework $(G,r)$ is a limit of a sequence $Q$ of generic, bounded frameworks $\{(G,r_i)\}_i$, with a corresponding sequence of pairwise distance vectors in $\Phi_{G,l_p}$, i.e., a sequence $Q'$ of bounded linkages of $G$. Because each $(G,r_i)$ is $d$-flattenable, each linkage in $Q'$ must be realizable as some generic bounded framework $(G,r_i')$ in $d$-dimensions The projection of the limit framework $(G,r)$ of the sequence $Q$ is the limit linkage of the projected sequence $Q'$ of linkages with bounded edge lengths.

Although $d$-flattenability is equivalent to the presence of an inherent convex Cayley configuration space for $G$, we now move beyond inherent convex Cayley configuration spaces to Cayley configuration spaces over specified nonedges $F$. These could be convex even if $G$ itself is not $d$-flattenable (simple examples can be found for $d = 2,3$ for $l_2$ in [31]). A complete characterization of such $G,F$ is shown in [31], in the case of $l_2$ norm for $d = 2$, conjectured for $d = 3$, and completely open for $d > 3$.

An analogous theorem to Theorem 10.8 can be proven for the property of a $d$-dimensional framework $(G,r)$ having a convex Cayley configuration space over specified non-edge set $F$.

**Theorem 10.9** *Every generic d-dimensional framework $(G,r)$ has a convex Cayley configuration space over F if and only if for all $\delta$, the linkage $(G,\delta)$ has a d-dimensional, convex Cayley configuration space over F.*

The proof of this theorem follows very closely the proof of Theorem 10.8.

Neither of the properties discussed above is a generic property of frameworks even for $l_2$.

**Theorem 10.10** *d-flattenability and convexity of Cayley configuration spaces over specific non-edges F are not generic properties of frameworks $(G,r)$.*

See Figure 10.7 for the $d$-flattenability argument. For convexity of Cayley configuration spaces: there are minimal, so-called Henneberg-I graphs [33] $G$, constructed on a *base* or initial edge $f$ with the following property: for some 2-dimensional frameworks (and neighborhoods) $(G,r)$ with edge length vector $\delta$, the 1-dimensional Cayley configuration space $\Phi^2_f(G \setminus f, \delta_{G\setminus f})$ (i.e, the attainable lengths for $f$) is a single interval, while for other such frameworks (and neighborhoods) it is two intervals. Please see Appendix in [33].

Next, we consider the implication of the *existence* of a generic $d$-flattenable framework. Specifically, we prove two theorems connecting the $d$-flattenability with independence in the rigidity matroid:

The "if" direction of this next theorem is a restatement of Proposition 2 in Asimow and Roth [3].

**Theorem 10.11** *For general $l_p$ norms, there exists a generic d-flattenable framework of G if and only if G is independent in the d-dimensional generic rigidity matroid.*

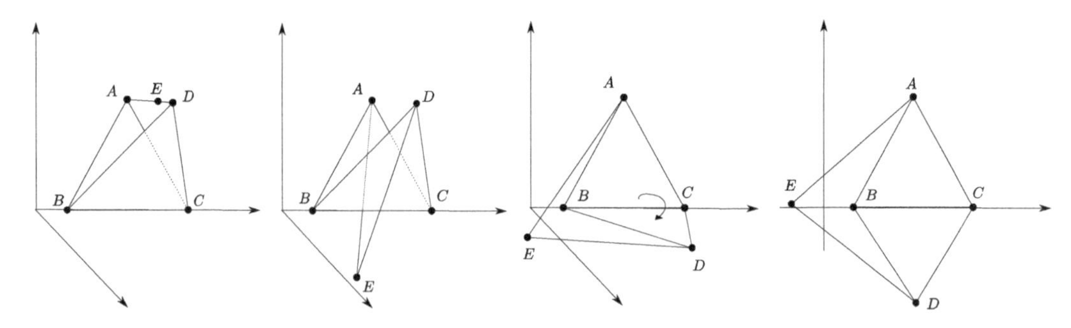


**Figure 10.7**
Two linkages of the same graph. In the first figure (left), we have edge lengths for $(a,e)$ and $(d,e)$ that do not allow $G$ to be flattened. The second graph is realized in 3-dimensions, but by "unfolding it" as shown, we can flatten it into 2-dimensions. It is easy to see that there is a full dimensional neighborhood of each linkage such that flattenability is maintained

Note that existence of a generic $d$-flattenable framework $(G,r)$ is equivalent to the statement that the pairwise distance vector $\delta_r$ has an open neighborhood $\Omega_r$ in the relative interior of the stratum $\Phi^d_{n,l_p}$. The fact that $\Omega_r$ is in the relative interior means it has full dimension.

Now observe that the generic rigidity matrix of $G$ is the Jacobian of the distance map from the $d$-dimensional point-configuration $s$ to the edge-length vector $\delta^G_s$ at the point $s$. For $l_2$ and integral $p > 1$, this map is clearly specified by polynomials. Because $\Omega_r$ has dimension equal to the number of edges in $G$, these polynomials are algebraically independent. Hence, their Jacobian has rank equal to the number of edges in $G$, meaning the rows of the generic rigidity matrix – that correspond to the edges of $G$ – are independent.

**Corollary 10.12** *For general $l_p$ norms, a graph $G$ is $d$-flattenable only if $G$ is independent in the $d$-dimensional rigidity matroid.*

The following theorem and corollary utilize the dimension of the projection of the $d$-dimensional stratum on the edges of $G$ from the above proof. Note that in the above proof, if $G$ is an $n$-vertex graph, the neighborhood $\Omega_r$ has dimension equal to the flattening dimension of $K_n$; $\Omega_s$ has dimension equal to that of the stratum $\Phi^d_{n,l_p}$, and $\Omega_G$ has dimension equal to the number of edges of $G$ (see Figure 10.9).

**Theorem 10.13** *For general $l_p$ norms, a graph $G$ is*

*(a) Independent in the generic $d$-dimensional rigidity matroid (i.e., the rigidity matrix of a well-positioned and regular framework has independent rows), if and only if coordinate projection of the stratum $\Phi^d_{n,l_p}$ onto $G$ has dimension equal to the number of edges of $G$;*

*(b) Maximal independent (minimally rigid) if and only if projection of the stratum $\Phi^d_{n,l_p}$ onto $G$ is maximal (i.e., projection preserves dimension) and is equal to the number of edges of $G$;*

*(c) Rigid in $d$-dimensions if and only if projection of the stratum $\Phi^d_{n,l_p}$ onto $G$ preserves its dimension;*

*(d) Not independent and not rigid in the generic $d$-dimensional rigidity matroid if and only if the projection of $\Phi^d_{n,l_p}$ onto $G$ is strictly smaller than the minimum of: the dimension of the stratum and the number of edges in $G$.*

**Corollary 10.14** *For $l_p$ norms, the rank of a graph $G$ in the $d$-dimensional rigidity matroid is equal to the dimension of the projection $\Phi^d_{G,l_p}$ on $G$ of the $d$-dimensional stratum $\Phi^d_{n,l_p}$.*

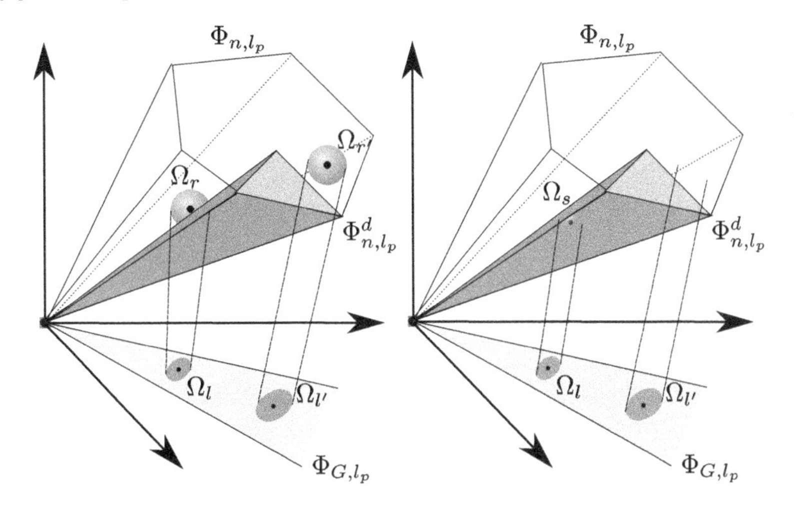


**Figure 10.8**
On the left we have two neighborhoods $\Omega_r$ and $\Omega_{r'}$ of two distance vectors $\delta_r$ and $\delta_{r'}$ in the cone. We then project $\Omega_r$ and $\Omega_{r'}$ onto the edges of $G$ to obtain $\Omega_l$ and $\Omega_{l'}$, which are essentially the neighborhoods of $(G, \delta_r^G)$ and $(G, \delta_{r'}^G)$. On the right, we then take the fiber of $\Omega_l$ and $\Omega_{l'}$ on $\Phi^d_{n,l_p}$. The fiber of $\Omega_l$ is completely contained in the stratum while that of $\Omega_{l'}$ misses (does not intersect) the stratum.

## 10.6 Efficient Realization through Optimal Cayley Modification

Realizing a *geometric constraint systems (GCS)* requires finding real solutions to a large multivariate polynomial system (of equalities and inequalities representing the constraints); this requires double exponential time in the number of variables, even if the type or orientation of the solution is specified. Thus, geometric constraint systems are typically recursively decomposed into locally rigid sub-systems and their solutions recombined to get the solution to the entire system. After decomposition, the complexity of realizing the system typically gets dominated by the largest indecomposible subsystem.

The *Optimal Cayley Modification (OCM)* [4] problem asks if the system to be realized can be modified, by dropping some constraints and adding certain others, so as to make the system easily realizable. This typically requires the modified system to have a convex Cayley configuration space. To find realizations of the original GCS, we find generate the Cayley configuration space of the modified GCS and search through it for realizations where the dropped edges have attained their original length. An important thing to note here is that in general, the modified graphs that yield "nice" Cayley spaces are also easy to realize. More formally:

**Definition 10.1 Optimal Cayley Modification (OCM) Problem** Given a geometric constraint system $G = (V, E)$ and two constants $k$ and $s$, find some set of at most $k$ constraints $E_{drop}$ and some set of constraints $E_{add}$ such that $H = (V, (E \setminus E_{drop}) \cup E_{add})$ is well-constrained and there exists some DR-plan with maximum fan-in of $s$.

Let $G' = (V, E \setminus E_{drop})$, with parameters $\delta'$, and $F = E_{add}$. If $\Phi_F(G', \delta')$ is restricted to convex Cayley configuration spaces, then the space can be searched efficiently. If $H$ is further restricted to be a class of efficiently realizable graphs, such as triangle decomposable graphs, then the Euclidean coordinates of each point in the configuration space can quickly be determined. In this way, the

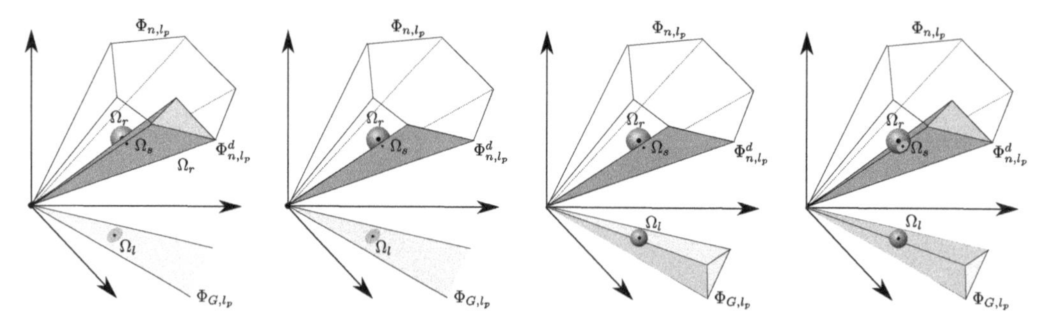


**Figure 10.9**
These are visualizations of when frameworks are isostatic and independent. In all of these cases $dim(\Omega_r) \geq \max\{dim(\Omega_s, dim(\Omega_l))\}$. We only show 2 and 3 dimensions here, but in general the dimensions will be much higher. See Figure 10.8 for explanation of what each is. In the following, when we use equality or inequality, we are referring to dimension. On the left, $\Omega_s = \Omega_l < \Phi^d_{n,l_p}$ meaning $\delta_r$ is independent but not isostatic. Middle left: $\Omega_s = \Omega_l = \Phi^d_{n,l_p}$, so $\delta_r$ is maximal independent or isostatic. Middle right: $\Omega_s = \Phi^d_{n,l_p} < \Omega_l$ meaning $\delta_r$ is rigid but not independent. Right: $\Omega_s < \Omega_l$ and $\Omega_s < \Phi^d_{n,l_p}$ meaning $\delta_r$ is neither independent nor rigid.

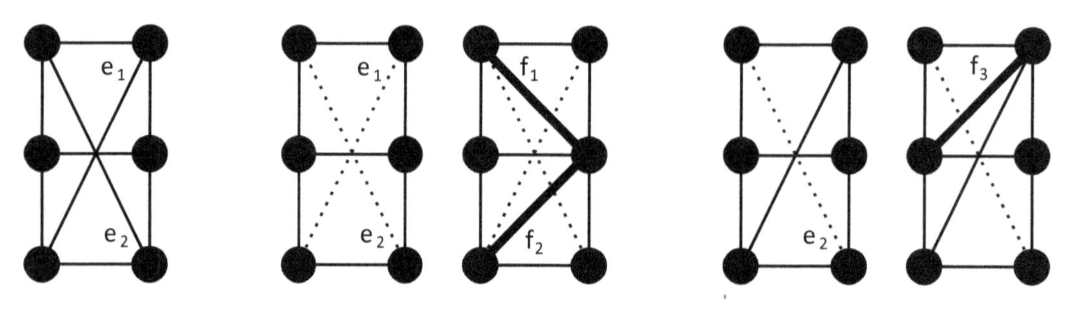


**Figure 10.10**
The first graph is a $K_{3,3}$. The second removes edges $e_1$ and $e_2$, making a partial 2-tree. The third adds in edges $f_1$ and $f_2$, making a 2-tree. The fourth removes just edge $e_2$. The fifth adds edge $f_3$, creating a low Cayley-complexity graph.

space can be efficiently searched for realizations that additionally satisfy the dropped constraints. This solution space must contain all solutions to the original constraint system $G$.

Such convex Cayley configuration spaces exist for partial 2-trees parametrized by edges that make a 2-tree; because 2-trees are triangle-decomposable, these have fast realization schemes as well. Alternatively, and perhaps more difficult to find, graphs whose added edges make low Cayley complexity graphs (see Section 10.7). Consider Figure 10.10 for an example of each of these OCM solutions.

## 10.7 Cayley Configuration Spaces of 1-Dof Tree-Decomposable Linkages in 2D

**Degree-of-Freedom (Dof) of a Linkage:** The degree-of-freedom of a linkage is the minimum number of bars that need to be added to the linkage to make it rigid.

**Tree-Decomposable Graphs:** A graph $G$ is tree-decomposable if it is a single edge, or it can be divided into three tree-decomposable subgraphs $G_1$, $G_2$ and $G_3$, such that $G = G_1 \cup G_2 \cup G_3$, $G_1 \cap G_2 = (\{v_3\}, \emptyset)$, $G_2 \cap G_3 = (\{v_2\}, \emptyset)$ and $G_1 \cap G_3 = (\{v_1\}, \emptyset)$, with $v_1$, $v_2$ and $v_3$ being distinct vertices

**1-Dof Tree-Decomposable Graphs:** A graph $G$ is 1-dof tree-decomposable if there exists a nonedge $f = (u, v)$ of $G$ such that $G \cup f$ is tree-decomposable.

**Base Nonedge of a 1-Dof Tree-Decomposable Graph:** A nonedge $f = (u, v)$ of a 1-dof tree-decomposable graph $G$ is a base nonedge of $G$ if $G \cup f$ is tree-decomposable.

**1-Dof Tree-Decomposable Linkages:** A linkage $(G, \delta)$ is 1-dof tree-decomposable if $G$ is 1-dof tree-decomposable.

**Quadratically-Radically Solvable (QRS) Values:** A real number is a QRS value if it is the solution a triangularized quadratic system with coefficients in $\mathbb{Q}$ (i.e., it belongs to an extension field over $\mathbb{Q}$ obtained by nested square-roots).

**Quadratically-Radically Solvable (QRS) graph:** A graph $G$ is QRS if for any linkage $(G, \delta)$ where $l$ is in $\mathbb{Q}$, the coordinate values of a realization of $(G, \delta)$ are QRS values.

**Generic 1-Dof Tree-Decomposable Linkages:** A 1-dof tree-decomposable linkage $(G, \delta)$ is generic if no bar length is zero, all bars have distinct lengths and at most one pair of adjacent bars can be collinear in any realization.

**Cluster:** A cluster $T$ of a 1-dof tree-decomposable graph $G$ is a maximal tree-decomposable subgraph of $G$.

**Graph Construction of a 1-Dof Tree-Decomposable Graph:** Any 1-dof tree-decomposable graph $G$ can be constructed iteratively as follows, starting from a given base nonedge $f$. At the $k^{th}$ construction step, two new clusters $T_1$ and $T_2$ sharing a single vertex $v_k$ are appended to the previously constructed graph $G_f(k-1)$, such that $T_1$ and $T_2$ each has exactly one shared vertex, $u_k$ and $w_k$, respectively, with $G_f(k-1)$, where $u_k \neq w_k$. Such a construction step is denoted as $v_k \triangleleft (u_k, w_k)$.

**Realization Type:** A construction step $v \triangleleft (u, w)$ of a 1-dof tree-decomposable graph $G$ can be associated with a local orientation for the corresponding realization step of a linkage $(G, \delta)$, which takes a value in $\{+1, -1, 0\}$ and represents the sign of the determinant $\Delta = \begin{vmatrix} p_w - p_u \\ p_v - p_u \end{vmatrix}$. A realization type is a set of local orientations for all construction steps of $G$. An oriented Cayley configuration space of a linkage $(G, \delta)$ is the Cayley configuration space of $(G, \delta)$ restricted to configurations respecting a given realization type.

**Base Pair of Vertices of a Construction Step:** The base pair of vertices of a construction step $v_k \triangleleft (u_k, w_k)$ is the pair $(u_k, w_k)$.

**Extreme Graphs and Extreme Edges:** The $k^{th}$ *extreme graph* for a base nonedge $f$ of a 1-dof tree-decomposable graph $G$, denoted $\hat{G}_f(k)$, is the graph obtained by adding a new edge $e_k = (u_k, w_k)$ in $G_f(k-1)$, where $(u_k, w_k)$ is the base pair of vertices of the $k^{th}$ construction step of $G$ from $f$. The edge $(u_k, w_k)$ is called the extreme edge of the extreme graph.

**Extreme Linkages:** For a 1-dof tree-decomposable linkage $(G, \delta)$, the $k^{th}$ extreme linkages are $(\hat{G}_f(k), l^{\min})$ and $(\hat{G}_f(k), l^{\max})$, where min and max represent the two possible extreme extensions of $l$ for the extreme edge $e_k = (u, w)$, obtained from triangle inequalities.

**Low Cayley Complexity:** A 1-dof tree-decomposable graph $G$ is said to have low Cayley (endpoint) on a base nonedge $f$ if all extreme graphs of $G$ for $f$ are tree-decomposable.

**Four-Cycle:** A four-cycle of clusters consists of four clusters $T_1, T_2, T_3, T_4$ such that each consecutive pair has exactly one shared vertex. In other words, we have $T_1 \cap T_2 = \{v_1\}$, $T_2 \cap T_3 = \{v_2\}$, $T_3 \cap T_4 = \{v_3\}$, $T_4 \cap T_1 = \{v_4\}$, where $v_1$, $v_2$, $v_3$, $v_4$ are distinct vertices.

**Last Level Vertex:** Given a 1-dof tree-decomposable graph $G$, a vertex $v$ is a last level vertex if (a) there are exactly two clusters $T_1$ and $T_2$ sharing $v$, and (b) each of $T_1$ and $T_2$ has exactly one shared vertex with the rest of the graph.

**1-Path Graph:** A 1-dof tree-decomposable graph $G$ has a 1-path construction from base nonedge $f = (v_0, v_{0'})$ if there is only one last level vertex, other than possibly $v_0$ and $v_0'$. A 1-dof tree-decomposable graph is 1-path if there exists a base nonedge permitting a 1-path construction.

**Minimal Complete Cayley Vector:** Given a general 1-dof tree-decomposable linkage $(G, \delta)$ with low Cayley complexity on a base nonedge $f = (v_0, v_0')$, a minimal *complete Cayley vector* $F$ is a list of $O(|V(G)|)$ nonedges of $G$, whose addition to $G$ as edges makes $G$ globally rigid (see the paper [34] for the detailed definition).

**Minimal Complete Cayley Distance Vector:** The minimal complete Cayley distance vector of a configuration of a 1-dof tree-decomposable linkage is the vector of distances for the nonedges in the complete Cayley vector.

The 1-degree-of-freedom (1-dof) tree-decomposable linkages [18, 23, 26] in $\mathbb{R}^2$, which generically has one degree-of-freedom, are the simplest natural generalization of partial 2-tree linkages that have nonconvex Cayley configuration spaces, obtained by measuring the attainable distances of a base nonedge. Since the underlying graphs are obtained by dropping an edge from the minimally rigid and QRS tree-decomposable graphs, there exists a simple linear time ruler and compass realization to convert from a Cayley configuration to a corresponding Cartesian configuration, given a specified realization type. For planar graphs, QRS and tree-decomposability have been shown [22, 27] to be equivalent, and the equivalence has been strongly conjectured for all graphs.

The following theorem, formalized in the paper [19], describes the structure of Cayley configuration spaces of generic 1-dof tree-decomposable linkages by associating a linkage configuration with each endpoint of an (oriented) Cayley configuration space. The proof of theorem follows from elementary algebraic geometry and can be considered folklore.

**Theorem 10.15 (Structure of Cayley configuration space [19])** *For a generic 1-dof tree-decomposable linkage $(G, \delta)$ with base nonedge $f = (v_0, v_0')$, the following hold:*

*(a) The (oriented) Cayley configuration space over $f$ is a set of disjoint closed real intervals or empty.*

*(b) Any interval endpoint in the (oriented) Cayley configuration space corresponds to the length of $f$ in a configuration of an extreme linkage.*

*(c) For any vertex $v$, $p_v$ is a continuous function of $l_f$ on each closed interval of the oriented Cayley configuration space. Consequently, for any nonedge $(u, w)$, $l(u, w)$ is a continuous function of $l_f$ on each closed interval of the oriented Cayley configuration space.*

Theorem 10.15 gives a straightforward algorithm [19, 34] which computes the (oriented) Cayley configuration space for a generic 1-dof tree-decomposable linkage $(G, \delta)$ by realizing all the extreme linkages. However, the algorithm could take time exponential in both the size of the linkage and the final Cayley configuration space.

The following theorem shows that low Cayley (endpoint) complexity is a robust measure of the complexity of Cayley configuration spaces of 1-dof tree-decomposable graphs, as it does not depend on the choice of base nonedge. That is, if we can characterize low Cayley complexity for a 1-dof tree-decomposable $G$ for a convenient base nonedge $f$, it translates to a characterization of low Cayley complexity of $G$ (for all possible base nonedges).

**Theorem 10.16 (equivalence of base nonedges [19, 35])** *A 1-dof tree-decomposable graph G either has low Cayley complexity on all possible base nonedges, or on none of them.*

The proof of Theorem 10.16 is via a minimal counterexample.

The following theorem characterizes 1-dof tree-decomposable graphs with low Cayley complexity by looking at the "base four-cycle" of each construction step.

**Theorem 10.17 (Four-cycle Theorem [35])** *Given a 1-dof tree-decomposable graph G on a base nonedge f with six or more clusters, G has low Cayley complexity, if and only if every construction step $v_k \lhd (u_k, w_k)$ from f has its base pair of vertices taken from an adjacent pair of clusters in a four-cycle of clusters. Note that a graph with less that six clusters trivially has low Cayley complexity.*

One direction of Theorem 10.17 directly follows from the structure of extreme graphs, and the other direction is proved by induction on the number of construction steps.

Theorem 10.17 yields an algorithm to verify whether a given 1-dof tree-decomposable graph $G$ has low Cayley complexity on base nonedge $f$, with $O(|V|^2)$ time complexity [35]. This is more efficient than the algorithm that follows the definition of low Cayley complexity, i.e., checking if all extreme graphs are tree-decomposable, which takes $O(|V|^3)$ time (checking $O(|V|)$ extreme graphs, each taking $O(|V|^2)$ time using existing algorithm [18]).

To go beyond the algorithmic characterization given by Theorem 10.17, it is desirable to obtain a finite forbidden minor characterization of low Cayley complexity. Unfortunately, such a characterization is shown to be impossible for general 1-dof tree-decomposable graphs [19, 35]. Nevertheless, for a natural subclass of 1-dof tree-decomposable graphs that are both 1-path and triangle-free, low Cayley (algebraic) complexity is shown to be equivalent to planarity.

**Theorem 10.18 (Low Cayley complexity and planarity [35])** *A 1-path, triangle-free, 1-dof tree-decomposable graph G has low Cayley complexity if and only if G is planar.*

The proof of Theorem 10.18 is by minimal counterexample and observations on the recursive structure of 1-path, triangle-free graphs.

The following theorem shows that for generic 1-dof tree-decomposable linkages with low Cayley complexity, a path of continuous motion between two given configurations can be efficiently obtained from the complete description of the Cayley configuration space.

**Theorem 10.19 (Continuous motion path Theorem [32, 34])** *For a generic linkage $(G, \delta)$ where G has low Cayley complexity on f,*

*(a) There exist* at most two *paths of continuous motion between any two given configurations, and the time complexity of finding such a path (provided one exists) is linear in the number of interval endpoints of oriented Cayley configuration spaces that the path contains.*

*(b) There is an algorithm that generates the entire set of connected components of the configuration space, where generating each connected component takes time linear in the number of interval endpoints of oriented Cayley configuration spaces that connected component contains.*

The proof of Theorem 10.19 is by showing certain local orientations of the configuration, under the genericity condition, can only be changed via at interval endpoints of oriented Cayley configuration spaces.

For a 1-dof tree-decomposable linkage with low Cayley complexity, assume that the clusters are globally rigid, and clusters sharing only two vertices with the rest of the graph be reduced into edges, there is a bijective correspondence between the Cartesian configuration space and a minimal complete Cayley distance vector. This yields a canonical representation and a meaningful visualization of the configuration space and continuous motion.

**Theorem 10.20 (bijectivity of representation [32, 34])** *(1) For a generic 1-path, 1-dof tree-decomposable linkage with low Cayley complexity, there exists a bijective correspondence between the set of Cartesian configurations and points on a curve in* $\mathbb{R}^2$*, whose points are the minimum complete Cayley distance vectors.*
*(2) For a generic 1-dof tree-decomposable linkage with low Cayley complexity, there exists a bijective correspondence between the set of Cartesian configurations and points on a curve in* $n$*-dimension, whose points are the minimal complete Cayley distance vectors, where* $n$ *is the number of last level vertices of the underlying graph.*

The complete Cayley distance vector also provides a meaningful way to define a canonical distance between any two configurations of 1-dof tree-decomposable linkages, as well as distances between different connected components of the configuration space [32, 34].

## 10.8 Conclusion

We characterized graphs whose linkages have convex Cayley configuration spaces, connected the property to flattenability for arbitrary $l_p$ norms and dimensions, applied the results towards efficient realization of linkages, studied genericity and independence in the context of $d$-flattenability, and characterized a larger class of graphs but with low complexity Cayley configurations.

## References

[1] Abdo Y. Alfakih. Graph rigidity via euclidean distance matrices. *Linear Algebra and its Applications*, 310(13):149–165, 2000.

[2] A.Y. Alfakih. On bar frameworks, stress matrices and semidefinite programming. *Mathematical Programming*, 129(1):113–128, 2011.

[3] L. Asimow and B. Roth. The rigidity of graphs. *Trans. Amer. Math. Soc.*, 245, 1978.

[4] T. Baker, M. Sitharam, M. Wang, and J. Willoughby. Optimal decomposition and recombination of isostatic geometric constraint systems for designing layered materials. *Computer Aided Geometric Design*, 40:1–25, 2015.

[5] Keith Ball. Isometric embedding in lp-spaces. *European Journal of Combinatorics*, 11(4):305–311, 1990.

[6] Boaz Barak, Prasad Raghavendra, and David Steurer. Rounding semidefinite programming hierarchies via global correlation. *CoRR*, abs/1104.4680, 2011.

[7] A.I. Barvinok. Problems of distance geometry and convex properties of quadratic maps. *Discrete & Computational Geometry*, 13(1):189–202, 1995.

[8] Maria Belk. Realizability of graphs in three dimensions. *Discrete Comput. Geom.*, 37(2):139–162, February 2007.

[9] Maria Belk and Robert Connelly. Realizability of graphs. *Discrete Comput. Geom.*, 37(2):125–137, February 2007.

[10] L.M. Blumenthal. *Theory and applications of distance geometry*. Chelsea Pub. Co., 1970.

[11] Bo Brinkman and Moses Charikar. On the impossibility of dimension reduction in l1. *J. ACM*, 52(5):766–788, September 2005.

[12] Arthur Cayley. On a theorem in the geometry of position. *Cambridge Mathematical Journal*, 2:267–271, 1841.

[13] Ugandhar Reddy Chittamuru. Efficient bounds for 3d cayley configuration space of partial 2-trees. Master's thesis, University of Florida, 2010.

[14] Jon Dattorro. *Convex Optimization & Euclidean Distance Geometry*. Meboo Publishing USA, 2011.

[15] Michel Marie Deza and Monique Laurent. *Geometry of Cuts and Metrics*. Algorithms and Combinatorics. Springer, Dordrecht, 2010.

[16] Dmitriy Drusvyatskiy, Gbor Pataki, and Henry Wolkowicz. Coordinate shadows of semidefinite and euclidean distance matrices. *SIAM Journal on Optimization*, 25(2):1160–1178, 2015.

[17] Samuel Fiorini, Tony Huynh, Gwenal Joret, and Antonios Varvitsiotis. The excluded minors for isometric realizability in the plane. *SIAM Journal on Discrete Mathematics*, 31(1):438–453, 2017.

[18] Ioannis Fudos and Christoph M. Hoffmann. A graph-constructive approach to solving systems of geometric constraints. *ACM Trans. Graph.*, 16(2):179–216, April 1997.

[19] Heping Gao and Meera Sitharam. Characterizing 1-dof Henneberg-I graphs with efficient configuration spaces. In *Proceedings of the 2009 ACM symposium on Applied Computing*, SAC '09, pages 1122–1126, New York, NY, 2009. ACM.

[20] Karin Gatermann and Pablo A. Parrilo. Symmetry groups, semidefinite programs, and sums of squares. *Journal of Pure and Applied Algebra*, 192(13):95–128, 2004.

[21] Piotr Indyk and Jiri Matousek. Low-distortion embeddings of finite metric spaces. In *in Handbook of Discrete and Computational Geometry*, pages 177–196. CRC Press, 2004.

[22] Bill Jackson and JC Owen. Radically solvable graphs. *arXiv preprint arXiv:1207.1580*, 2012.

[23] R. Joan-Arinyo, A. Soto-Riera, S. Vila-Marta, and J. Vilaplana-Pastó. Revisiting decomposition analysis of geometric constraint graphs. *Computer-Aided Design*, 36(2):123–140, 2004. Solid Modeling and Applications.

[24] William Johnson and Joram Lindenstrauss. Extensions of Lipschitz mappings into a Hilbert space. In *Conference in modern analysis and probability (New Haven, Conn., 1982)*, volume 26 of *Contemporary Mathematics*, pages 189–206. American Mathematical Society, 1984.

[25] D. Kitson. Finite and infinitesimal rigidity with polyhedral norms, 2014.

[26] J. C. Owen. Algebraic solution for geometry from dimensional constraints. In *Proceedings of the First ACM Symposium on Solid Modeling Foundations and CAD/CAM Applications*, SMA '91, pages 397–407, New York, NY, USA, 1991. ACM.

[27] J.C. Owen and S.C. Power. The non-solvability by radicals of generic 3-connected planar laman graphs. *Transactions of the American Mathematical Society*, 359(5):2269–2304, 2007.

[28] Pablo A. Parrilo and Bernd Sturmfels. Minimizing polynomial functions. In *Proceedings of the Dimacs Workshop on Algorithmic and Quantitative Aspects of Real Algebraic Geometry in Mathematics and Computer Science*, pages 83–100. American Mathematical Society, 2003.

[29] Neil Robertson and P.D. Seymour. Graph minors. xx. wagner's conjecture. *Journal of Combinatorial Theory, Series B*, 92(2):325 – 357, 2004. Special Issue Dedicated to Professor W.T. Tutte.

[30] I. J. Schoenberg. Remarks to maurice frchets article sur la dfinition axiomatique dune classe despaces distancis vectoriellement applicable sur lespace de hilbert. annals of mathematics 36(3), 1935.

[31] Meera Sitharam and Heping Gao. Characterizing graphs with convex and connected cayley configuration spaces. *Discrete & Computational Geometry*, 43(3):594–625, 2010.

[32] Meera Sitharam and Menghan Wang. How the beast really moves: Cayley analysis of mechanism realization spaces using caymos. *Computer-Aided Design*, 46(0):205 – 210, 2014. 2013 SIAM Conference on Geometric and Physical Modeling.

[33] Meera Sitharam, Menghan Wang, and Heping Gao. Cayley configuration spaces of 1-dof tree-decomposable linkages, part II: combinatorial characterization of complexity. *CoRR*, abs/1112.6009, 2011.

[34] Meera Sitharam, Menghan Wang, and Heping Gao. Cayley configuration spaces of 2d mechanisms Part I: extreme points, continuous motion paths and minimal representations. Submitted to *Discrete and Computational Geometry*, 2013.

[35] Meera Sitharam, Menghan Wang, and Heping Gao. Cayley configuration spaces of 2d mechanisms Part II: combinatorial characterization. Submitted to *Discrete and Computational Geometry*, 2013.

[36] Meera Sitharam and Joel Willoughby. On flattenability of graphs. In *Automated Deduction in Geometry - 10th International Workshop, ADG 2014, Coimbra, Portugal, July 9-11, 2014, Revised Selected Papers*, pages 129–148, 2014.

[37] Pablo Tarazaga. Faces of the cone of euclidean distance matrices: Characterizations, structure and induced geometry. *Linear Algebra and its Applications*, 408(0):1–13, 2005.

# Chapter 11

## *Constraint Varieties in Mechanism Science*

**Hans-Peter Schröcker, Martin Pfurner, and Josef Schadlbauer**
*University of Innsbruck*

**CONTENTS**



## 11.1 Introduction

In this chapter we look at constraint varieties of linkages from the viewpoint of mechanism science. There, not only abstract properties of linkages like flexibility (finite or infinitesimal) or configuration space topology are studied but also quantities that pertain to a concrete realization and to applications are of relevance. Examples include transmission ratios, joint forces, stiffness, collisions, or size. Moreover, additional problems like the construction of a linkage to accomplish a certain task ("linkage synthesis") or the computation and avoidance of "singular" configurations appear. Another specialty is the relevance of linkages with several degrees of freedom and the presence of different joint types. Linkages with as many degrees of freedom as the underlying motion group (three in case of SE(2) or SO(3) and six in case of SE(3)) are quite common. Redundant manipulators with even more degrees of freedom are an active research topic.

Mechanism science is a rewarding field for many branches of applied mathematics. Here, we mainly focus on algebraic and geometric aspects via Study parameters and dual quaternions. Progress in the field of computational algebraic geometry made this approach popular over recent years but, of course, traditional techniques from differential geometry, vector calculus or numerical mathematics remain indispensable as well.

## Glossary

**Fixed frame/base:** Space or coordinate frame attached to one link that is considered to be immobile.

**Joint:** Abstract representation of a mechanical device, attached to two or more links and constraining their relative positions. For example revolute or prismatic joint.

**Link:** Rigid body, abstractly representing part of a linkage.

**Linkgraph:** Graph representing linkage topology. The vertices are the links, the edges are the attached joints.

**Moving frame/end effector:** Space or coordinate attached to the link whose motion is of primary interest.

**Parallel linkage:** Linkage whose linkgraph has (several) cycles. Also called "parallel manipulator" or "parallel mechanism."

**Serial linkage:** Linkage whose linkgraph is linear. Also called "serial manipulator" or "serial mechanism."

### 11.1.1 Linkages and Joints

A *linkage* or *mechanism* is a set of rigid bodies, called *links* whose relative position is constrained by *joints* which are said to "connect" links. In order to fully specify a linkage, one must prescribe which links are connected by joints and also the type of the joint and its position with respect to the link. The abstract topology can be captured in the linkgraph whose vertices correspond to links. Two vertices are connected by an edge precisely if their links are connected by a joint. Mechanisms whose link graph have cycles are called *parallel* while mechanisms with linear link graph are called *serial.*

The predominant joint type in engineering applications is the revolute joint but other joint types are considered as well. The most common schemes classify joints by continuous subgroups of SE(3). Besides revolute joints, this gives rise to prismatic, helical, cylindrical, spherical and planar joints (Figure 11.1).

Note that some of these joints can be modelled as a concatenation of others. We may, for example, replace a cylindrical joint by a revolute joint and a prismatic joint provided the revolute axis is parallel to the translation direction. A spherical joint is equivalent to three revolute joints with intersecting axes. Other "compound joints" are possible as well. A common example is the *universal* or *Cardan joint* which is the composition of rotations about two perpendicular axes. In this text, we are mostly concerned with revolute and prismatic joints. Helical joints are not easily treated within an algebraic framework because they do not correspond to an algebraic subgroup of SE(3) (but see [3] for a remarkable exception).

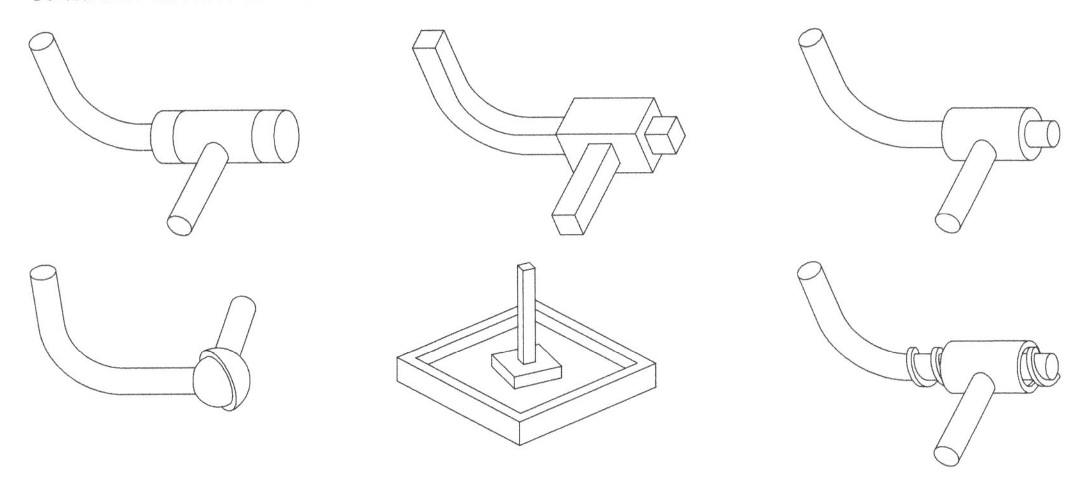

**Figure 11.1**
Revolute, prismatic, and cylindrical joint (top row); spherical, planar, and helical joint (bottom row).

### 11.1.2 Base and End-Effector Frame

When studying linkages from a purely mathematical viewpoint, it is good and common practice to consider isometric realizations of the linkage as equivalent. From the viewpoint of mechanism science, this is not always desirable. Industrial specifications of mechanisms or robots require coordinate frames attached to each link. Moreover, two links often play a particular role and deserve special treatment. The *end-effector,* is assumed to fulfill a certain task which is often described in terms of coordinates with respect to a *moving frame,* that is, a coordinate system attached to the end-effector link. Moreover, a mechanical linkage is usually attached to an immobile object via one of its links, the *base.* Its position is again described in some coordinate frame, often called *base frame, fixed frame,* or *world coordinate system.* The spaces attached to the moving and fixed frames, respectively, are called the *moving space* and *fixed space.*

## 11.2 Mechanisms and Algebraic Varieties

### Glossary

**Point model:** A usually projective or affine space or a subset thereof, whose elements (points) are in bijection with the elements of a motion group. For example, projective three-space $\mathbb{P}^3$ and the space of orthogonal $3 \times 3$ matrices with positive determinant are both point models for SO(3).

**Constraint equations:** Equations describing a subset of SE(3) in a suitable point model that can be reached by a linkage. Algebraic constraint equations give rise to constraint varieties.

**Dual quaternions:** Algebra obtained from quaternions via scalar extension from real to dual numbers. Can be viewed as Study parameters if they satisfy the Study condition.

**Four-bar linkage:** Linkage, usually with revolute joints, whose linkgraph is a cycle of length four.

**Homogeneous transformation matrix:** A $4 \times 4$ matrix representing a projective transformation of $\mathbb{P}^3$. Usually used for representing rigid body displacements in mechanism science.

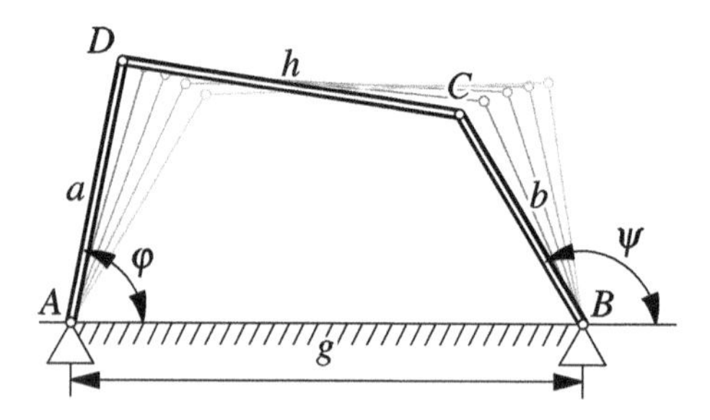


**Figure 11.2**
A planar four-bar linkage.

**Study parameters:** Homogeneous coordinates in $\mathbb{P}^7$ representing points on the Study quadric. May be identified with certain dual quaternions.

**Study quadric:** Hyperquadric in $\mathbb{P}^7$ providing a point model of SE(3) via Study's kinematic map.

### 11.2.1 Geometric Constraint Equations

A mathematical description of a linkage and its possible configurations can be accomplished by translating the joint constraints into *constraint equations.* We are particularly interested in constraining position and orientation of the moving end-effector link. Consider the planar four-bar linkage shown in Figure 11.2. It will serve as our model example in this section. Four links are cyclically connected by revolute joints at points $A$, $B$, $C$, and $D$. The link of $A$ and $B$ is assumed to be fixed while the link of $C$ and $D$ is the moving end-effector.

The possible positions of the end-effector are completely described by two "point on circle" constraints: The point $D$ lies on a circle of radius $a$ around $A$ and the point $C$ lies on a circle of radius $b$ around $B$. Thus, we may write

$$D = A + a(\cos\varphi, \sin\varphi)^\mathsf{T}, \quad C = B + b(\cos\psi, \sin\psi)^\mathsf{T},$$

where $\varphi$ and $\psi$ denote the angles the link $A$-$B$ forms with $A$-$D$ and $B$-$C$, respectively. The kinematics of the four-bar linkage follow from the equation $\|C - D\|^2 = h^2$ which relates the angles $\varphi$ and $\psi$. Note that this equation is trigonometric but can easily be converted to an algebraic form. One method is to introduce new variables $C_1 := \cos\varphi$, $S_1 := \sin\varphi$, $C_2 := \cos\psi$, $S_2 := \sin\psi$ and equations $S_1^2 + C_1^2 = S_2^2 + C_2^2 = 1$. Another possibility is the tangent half-angle substitution $u := \tan\frac{\varphi}{2}$, $v := \tan\frac{\psi}{2}$ whence $\cos\varphi = (1-u^2)/(1+u^2)$, $\sin\varphi = 2u/(1+u^2)$ and similar for cosine and sine of $\psi$.

We will present an alternative and more elaborate algebraic formulation later (see Equation (11.11)). Here it suffices to say that circle and sphere constraints, their spatial counterparts, are ubiquitous in mechanism science. Several examples will be given in Section 11.4. However, numerous other constraints like "point in plane" or "line through point" are conceivable as well.

### 11.2.2 Study Parameters

Now we introduce a mathematical tool to handle geometric constraints imposed by kinematic joints. The idea is that any geometric constraint restricts the relative positions of the attached links. Thus, it can be described by a subset of a suitable point model of SE(3). We primarily focus on geometric constraints that impose *algebraic* constraints on the motion group. Luckily, these comprise numerous constraints of practical relevance. The point model we propose is obtained via *Study's kinematic map,* – a bijection between SE(3) and the points of a hyperquadric $\mathcal{S} \subset \mathbb{P}^7$, the *Study quadric*, minus the points of a projective three-space $E$, the *exceptional generator.*

Recall that any rigid body displacement maps a point with Cartesian coordinate vector $x$ (with respect to the moving coordinate system) to a point with Cartesian coordinates $x'$ in the fixed system according to $x' = A \cdot x + a$. Here the matrix $A$ is orthogonal, of dimension $3 \times 3$, and has positive determinant. It is responsible for the orientation component of the displacement and the vector $a \in \mathbb{R}^3$ gives its translation. Often, the rotation matrix $A$ and translation vector $a$ are merged into a four-by-four matrix according to

$$\begin{bmatrix} x' \\ 1 \end{bmatrix} = \begin{bmatrix} A & a \\ 0 & 1 \end{bmatrix} \cdot \begin{bmatrix} x \\ 1 \end{bmatrix}.$$

In this way, it may also act on homogeneous coordinate vectors and one speaks of a *homogeneous transformation matrix.* Composition of rigid body displacements corresponds to matrix multiplication.

It can be shown that there exists a vector $p = (p_0, p_1, p_2, p_3)^\intercal \in \mathbb{R}^4$ such that

$$A = \frac{1}{\Delta} \begin{pmatrix} p_0^2 + p_1^2 - p_2^2 - p_3^2 & -2p_0p_3 + 2p_1p_2 & 2p_0p_2 + 2p_1p_3 \\ 2p_0p_3 + 2p_1p_2 & p_0^2 - p_1^2 + p_2^2 - p_3^2 & -2p_0p_1 + 2p_2p_3 \\ -2p_0p_2 + 2p_1p_3 & 2p_0p_1 + 2p_2p_3 & p_0^2 - p_1^2 - p_2^2 + p_3^2 \end{pmatrix}, \tag{11.1}$$

where $\Delta = p_0^2 + p_1^2 + p_2^2 + p_3^2$. Clearly, replacing $p$ by any non-zero real multiple gives the same matrix. Conversely, any non-zero vector $p \in \mathbb{R}^4$ yields, via (11.1), an orthogonal matrix with determinant one. Thus, we have established a *bijection between* SO(3) *and the points of real projective three-space* $\mathbb{P}^3$. The homogeneous coordinates $(p_0, p_1, p_2, p_3)^\intercal$ are called the rotation's *Euler parameters.*

Now we also include translations. For any translation vector $a = (a_1, a_2, a_3)^\intercal$, there exists a unique vector $q = (q_0, q_1, q_2, q_3)^\intercal \in \mathbb{R}^4$ such that

$$\begin{aligned} p_0q_0 + p_1q_1 + p_2q_2 + p_3q_3 &= 0 \\ 2(p_1q_0 - p_0q_1 + p_3q_2 - p_2q_3) &= a_1 \\ 2(p_2q_0 - p_3q_1 - p_0q_2 + p_1q_3) &= a_2 \\ 2(p_3q_0 + p_2q_1 - p_1q_2 - p_0q_3) &= a_3. \end{aligned} \tag{11.2}$$

Indeed, (11.2) is a linear system for computing $q$ and its determinant $-8(p_0^2 + p_1^2 + p_2^2 + p_3^2)^2$ is different from zero. Equations (11.1) and (11.2) together define a bijection

$$\begin{aligned} \varkappa\colon \mathcal{S} \setminus E &\to \mathrm{SE}(3) \\ (p_0, p_1, p_2, p_3, q_0, q_1, q_2, q_3) &\mapsto (A, a) \end{aligned} \tag{11.3}$$

from the *Study quadric* $\mathcal{S} \subset \mathbb{P}^7$: $p_0q_0 + p_1q_1 + p_2q_2 + p_3q_3 = 0$ minus the exceptional generator $E \subset \mathcal{S}$: $p_0 = p_1 = p_2 = p_3 = 0$ to SE(3). It is called *Study's kinematic map* [46]. The defining equation of $\mathcal{S}$ (first equation in (11.2)) is called the *Study condition.* The coefficients $p_0, \ldots, p_3, q_0, \ldots, q_3$ are called *Study parameters.* Note that they depend on the coordinate systems in the fixed and moving frame.

Study's construction may seem rather weird at first sight. However, it is well-motivated by two important properties. The point model for SE(3) requires a *minimal number of parameters* (eight homogeneous coordinates) while providing a *bilinear composition law* (see Section 11.2.3). Study showed that these two properties determine the point model up to isomorphism [46]. Other common point models violate the first or the second condition. For example the composition of homogeneous transformation matrices is bilinear but each matrix requires twelve parameters, even if we do not count the constant last row. In computer graphics and gaming industry, Euler angles ("roll," "pitch," and "yaw") plus translation vectors are popular. They require only six parameters but have a non-linear composition.

Let us proceed with some useful formulas for computations in Study parameters: A procedure for obtaining the vector $p=(p_0,p_1,p_2,p_3)^\intercal$ from the matrix $A=(a_{ij})_{i,j=1,\ldots,3}$ goes back to Study. The ratio of the coefficients of $p$ is given by any of the four proportions

$$\begin{aligned} p_0:p_1:p_2:p_3 &= 1+a_{11}+a_{22}+a_{33} : a_{32}-a_{23} : a_{13}-a_{31} : a_{21}-a_{12} \\ &= a_{32}-a_{23} : 1+a_{11}-a_{22}-a_{33} : a_{12}+a_{21} : a_{31}+a_{13} \\ &= a_{13}-a_{31} : a_{12}+a_{21} : 1-a_{11}+a_{22}-a_{33} : a_{23}+a_{32} \\ &= a_{21}-a_{12} : a_{31}+a_{13} : a_{23}+a_{32} : 1-a_{11}-a_{22}+a_{33}. \end{aligned} \tag{11.4}$$

In general, all four proportions of (11.4) give the same result. It is, however, possible that up to three proportions yield $0:0:0:0$ and become invalid. A related procedure paying more attention to numeric data is described in [34]. Having computed $p$, we may use the equations in (11.2) to find $q$ for given $a$.

It is useful to know the Study parameters of some fundamental rigid body displacements. The rotation about the unit vector $(v_1,v_2,v_3)^\intercal$ (line through the origin) with rotation angle $2\varphi$ corresponds to

$$p_0=\cos\varphi, \quad p_1=v_1\sin\varphi, \quad p_2=v_2\sin\varphi, \quad p_3=v_3\sin\varphi \tag{11.5}$$

and $q=(0,0,0,0)^\intercal$. The translation by the vector $(t_1,t_2,t_2)^\intercal$ corresponds to $p=(1,0,0,0)^\intercal$ and $q=-\frac{1}{2}(0,t_1,t_2,t_3)^\intercal$. Any displacement of SE(3) is a suitable and intuitive composition of rotations about lines through the origin and translations. Hence, we need the composition formula for two displacements in Study parameters. This we do by relating Study's kinematic map to quaternions and dual quaternions.

### 11.2.3 Dual Quaternions

In the Euclidean vector space $\mathbb{R}^4$ we pick an orthonormal basis $(1,\mathbf{i},\mathbf{j},\mathbf{k})$ and we embed $\mathbb{R}^3$ into $\mathbb{R}^4$ by identifying it with the subspace spanned by $\mathbf{i}$, $\mathbf{j}$ and $\mathbf{k}$ via $(x_1,x_2,x_3)^\intercal \in \mathbb{R}^3 \hookrightarrow (0,x_1,x_2,x_3)^\intercal \in \mathbb{R}^4$. Moreover, we embed $\mathbb{R}$ into $\mathbb{R}^4$ via $x\in\mathbb{R}\hookrightarrow(x,0,0,0)^\intercal\in\mathbb{R}^4$. The ambiguous use of the symbol "1" will be justified as soon as we have extend the multiplication of real numbers to $\mathbb{R}^4$. More precisely, we define an associative and bilinear multiplication that also extends the familiar multiplication of real numbers and vectors by the rules

$$\mathbf{i}^2=\mathbf{j}^2=\mathbf{k}^2=\mathbf{ijk}=-1. \tag{11.6}$$

From (11.6) we may derive a multiplication table and multiplication formula:

| | 1 | **i** | **j** | **k** |
|---|---|---|---|---|
| 1 | 1 | **i** | **j** | **k** |
| **i** | **i** | −1 | **k** | −**j** |
| **j** | **j** | −**k** | −1 | **i** |
| **k** | **k** | **j** | −**i** | −1 |

$$\begin{aligned} ab = {} & a_0b_0-a_1b_1-a_2b_2-a_3b_3 \\ & +(a_0b_1+a_1b_0+a_2b_3-a_3b_2)\mathbf{i} \\ & +(a_0b_2-a_1b_3+a_2b_0+a_3b_1)\mathbf{j} \\ & +(a_0b_3+a_1b_2-a_2b_1+a_3b_0)\mathbf{k}, \end{aligned} \tag{11.7}$$

where $a=a_0+a_1\mathbf{i}+a_2\mathbf{j}+a_3\mathbf{k}$ and $b=b_0+b_1\mathbf{i}+b_2\mathbf{j}+b_3\mathbf{k}$. The elements of the thus defined associative algebra are called *quaternions* and, in remembrance of their discoverer W. R. Hamilton, the algebra itself is denoted by $\mathbb{H}$. It contains two important and familiar sub-algebras: For $a$, $b\in\mathbb{R}$, quaternion multiplication reduces to the multiplication of real numbers. For $a=a_0+a_1\mathbf{i}$, $b=b_0+b_1\mathbf{i}$, quaternion multiplication reduces to the multiplication of complex numbers. This is also true if the quaternion unit $\mathbf{i}$ is replaced by $\mathbf{j}$ or $\mathbf{k}$.

Every quaternion can be written as $h=h_0+h_1\mathbf{i}+h_2\mathbf{j}+h_3\mathbf{k}$ where $h_0$, $h_1$, $h_2$, $h_3$ are real numbers.

The quaternion $\overline{h} := h_0 - h_1\mathbf{i} - h_2\mathbf{j} - h_3\mathbf{k}$ is called the *conjugate quaternion* of $h$, the *norm of h* is defined as $\|h\| := \sqrt{N(h)}$ where $N(h) := h\overline{h}$. This definition is possible because $N(h) = h_0^2 + h_1^2 + h_2^2 + h_3^2$ is a non-negative real number. For any two quaternions $a$ and $b$ we have $N(ab) = N(a)N(b)$, $\|ab\| = \|a\|\,\|b\|$, and $\overline{(ab)} = \overline{b}\,\overline{a}$. If $a \neq 0$ there exists a unique quaternion, denoted by $a^{-1}$, such that $aa^{-1} = 1$. It is given by $a^{-1} = \overline{a}/N(a)$ and also satisfies $a^{-1}a = 1$.

One of the main applications of quaternions in mechanism science is the modelling of SO(3). The action of a quaternion $p = p_0 + p_1\mathbf{i} + p_2\mathbf{j} + p_3\mathbf{k} \in \mathbb{H}$ on the vector $x = x_1\mathbf{i} + x_2\mathbf{j} + x_3\mathbf{k} \in \mathbb{R}^3 \subset \mathbb{H}$ is defined as

$$x \mapsto y = y_1\mathbf{i} + y_2\mathbf{j} + y_3\mathbf{k} := px\overline{p}/N^2(p). \tag{11.8}$$

The right-hand side is linear in $x_1$, $x_2$, and $x_3$. Hence, there exists a matrix $A$ such that $(y_1, y_2, y_3)^\intercal = A \cdot (x_1, x_2, x_3)^\intercal$. A short computation confirms that $A$ is the matrix given in (11.1). Combing with (11.5), this implies:

- The action (11.8) describes a rotation about the vector $(p_1, p_2, p_3)$ whose rotation angle $2\varphi$ satisfies $\cos\varphi = p_0/\|p\|$.
- The composition of rotations in Study parameters corresponds to *multiplication of quaternions.* Indeed, we have $(rp)x\overline{(rp)} = r(px\overline{p})\overline{r}$ for any quaternions $p$, $r \in \mathbb{H}$ and any vector $x \in \mathbb{R}^3$.

We thus see that Study parameters for displacements from SO(3) and quaternions are essentially the same. This provides us with a convenient algebraic interpretation for Study parameters.

The quaternion model exhibits quite some advantages over other orthogonal matrices. Apart from the already mentioned smaller number of parameters, the composition of rotations in Study parameters is more efficient than the multiplication of orthogonal matrices. Moreover, quaternions provide a very intuitive representation of SO(3). Their vector part already gives the oriented direction of the rotation axis. The cosine of half the rotation angle can be read off after normalization. Finally "orthogonal" matrices in applications usually have floating point entries and therefore require periodic re-orthogonalization. Similarly, quaternions require normalization because one usually wants to avoid the division in (11.8). However, the latter is a simple task while a geometrically meaningful orthogonalization of near-orthogonal matrices is not obvious.

Now we extend the standard quaternion algebra by allowing *dual numbers* as coefficients. This extension of scalars will provide us with the desired point model for SE(3). The resulting algebra $\mathbb{DH}$ of dual quaternions shares many properties with the quaternions $\mathbb{H}$ but some essential details are different. Most notably, non-invertible elements do exist.

The ring of *dual numbers* is defined as $\mathbb{D} := \mathbb{R}[\varepsilon]/\langle\varepsilon^2\rangle$. For every $d \in \mathbb{D}$ there exist unique $a$, $b \in \mathbb{R}$ (*primal* and *dual part*) such that $d = a + b \cdot \varepsilon$. The product of the two dual numbers $d_1 = a_1 + b_1\varepsilon$ and $d_2 = a_2 + b_2\varepsilon$ equals $d_1 d_2 = (a_1 + b_1\varepsilon)(a_2 + b_2\varepsilon) = a_1 a_2 + \varepsilon(a_1 b_2 + b_1 a_2)$. The inverse to $d = a + b\varepsilon$ exists if and only if $a \neq 0$. In this case it is $d^{-1} = a^{-1} - \varepsilon b a^{-2}$. The algebra of dual quaternions is defined as the scalar extension $\mathbb{DH} := \mathbb{D} \otimes_{\mathbb{R}} \mathbb{H}$. For every $h \in \mathbb{DH}$ there exist unique $p$, $q \in \mathbb{H}$ (*primal* and *dual part*) such that $h = p + \varepsilon q$. The product of two dual quaternions $h_1 = p_1 + \varepsilon q_1$ and $h_2 = p_2 + \varepsilon q_2$ equals

$$h_1 h_2 = (p_1 + \varepsilon q_1)(p_2 + \varepsilon q_2) = p_1 p_2 + \varepsilon(p_1 q_2 + q_1 p_2).$$

Note that addition and multiplication of dual quaternions commute with the projection on the primal part. The inverse to $h = p + \varepsilon q$ exists if and only if $p \neq 0$, whence we have $h^{-1} = p^{-1} - \varepsilon p^{-1} q p^{-1}$.

The dual quaternion conjugate to $h = p + \varepsilon q$ is $\overline{h} := \overline{p} + \varepsilon\overline{q}$. The dual quaternion norm is $\|h\| := \sqrt{N(h)}$, where $N(h) := h\overline{h}$. This definition of the dual quaternion norm requires extension of the analytic square root function to dual numbers because of $N(h) = h\overline{h} = (p + \varepsilon q)(\overline{p} + \varepsilon\overline{q}) = p\overline{p} + \varepsilon(p\overline{q} + q\overline{p}) \in \mathbb{D}$. Fortunately, this is not necessary in kinematics because for dual quaternions satisfying the Study condition we have $N(h) \in \mathbb{R}$.

The motivation for the scalar extension from quaternions to dual quaternions is the modelling of SE(3) and not just SO(3). To this end, we identify Euclidean three space $\mathbb{R}^3$ (with fixed Cartesian coordinates) with the affine three-space $1+\langle \varepsilon\mathbf{i}, \varepsilon\mathbf{j}, \varepsilon\mathbf{k}\rangle$ via the inclusion map $(x_1, x_2, x_3) \hookrightarrow 1+\varepsilon(x_1\mathbf{i}+x_2\mathbf{j}+x_3\mathbf{k})$. The action of the dual quaternion $h = p+\varepsilon q$ on $x = 1+\varepsilon(x_1\mathbf{i}+x_2\mathbf{j}+x_3\mathbf{k})$ is defined as

$$x \mapsto h_\varepsilon x \overline{h}, \tag{11.9}$$

where $h_\varepsilon := p-\varepsilon q$. With $x = 1+\varepsilon(x_1\mathbf{i}+x_2\mathbf{j}+x_3\mathbf{k})$, the right-hand side of (11.9) is linear in $x_1$, $x_2$, $x_3$ and it is elementary to verify that it may be written as $A \cdot (x_1, x_2, x_3)^\intercal + a$, where $A$ is given by (11.1) and $a = (a_1, a_2, a_3)^\intercal$ by (11.2). Hence, the *Study parameters for rigid body displacements may be interpreted as dual quaternion coefficients.* In particular, we obtain the composition law for Study parameters which again is given by dual quaternion multiplication due to

$$(hk)_\varepsilon x \overline{(hk)} = h_\varepsilon (k_\varepsilon x \overline{k}) \overline{h}$$

for all $h$, $k \in \mathbb{DH}$. Moreover, the restriction of (11.9) to $\mathbb{H}$ gives the action of quaternions on vectors via (11.8), as expected.

We may also use the composition of displacements in Study parameters via dual quaternion multiplication for coordinate changes in the moving and fixed frame, respectively. Suppose $y = h_\varepsilon x \overline{h}$ are the coordinates (in the fixed frame) of the image of a point $x$ (in the moving frame) and select new coordinates $y'$, and $x'$ in the fixed and moving frame, respectively. Then there exist dual quaternions $f$, $m$ that satisfy the Study condition and $y' = f_\varepsilon y \overline{f}$, $x = m_\varepsilon x' \overline{m}$. The same displacement in new coordinates then reads $y' = h'_\varepsilon x' \overline{h'}$ where $h' = fhm$ because of

$$y' = f_\varepsilon y \overline{f} = f_\varepsilon (h_\varepsilon x \overline{h}) \overline{f} = f_\varepsilon (h_\varepsilon (m_\varepsilon x' \overline{m}) \overline{h}) \overline{f} = (fhm)_\varepsilon x' \overline{(fhm)}.$$

We see that *coordinate changes in the fixed or moving coordinate frame are done by dual quaternion left or right multiplication, respectively.* These are projective transformations of $\mathbb{P}^7$

$$\varphi\colon \mathbb{P}^7 \to \mathbb{P}^7,\ [h] \mapsto [fh]; \quad \mu\colon \mathbb{P}^7 \to \mathbb{P}^7, [h] \mapsto [hm] \tag{11.10}$$

that leave invariant the Study quadric $\mathcal{S}$ and the exceptional generator $E$ by construction. A complete system of invariants of the transformation group generated by (11.10) has only recently been described in [39]. A more detailed introduction to Study parameters and dual quaternions may be found in [23] or [43].

### 11.2.4 Analyzing Mechanisms via Algebraic Varieties

In the previous section we have constructed a point model of SE(3) with rich algebraic and geometric structure (the algebra $\mathbb{DH}$ of dual quaternions and the Study quadric $\mathcal{S} \subset \mathbb{P}^7$, respectively). This model can be used to systematically analyze the kinematics of linkages based on their *constraint varieties.* A constraint variety is an algebraic variety $V$ in the Study quadric $\mathcal{S} \subset \mathbb{P}^7$ such that all rigid body displacements in $V$ satisfy certain kinematic conditions that are characteristic of the given linkage. This makes numerous aspects of computational kinematics accessible to the powerful framework of computational algebraic geometry. Note that the minimal number of parameters in Study's point model is advantageous in this context. Fundamental concepts of algebraic geometry like "dimension," "singularity," or "prime decomposition" have a concrete kinematic meaning and are important in theoretical and applied mechanism science.

In this section we mostly confine ourselves to a simple example, the planar four-bar linkage (Figure 11.2). More comprehensive studies of constraint varieties can be found in [44, 45] and also later in this chapter. We already recognized the importance of the "point on circle" constraint for the kinematic description of the planar four-bar. Hence, we start by looking at the constraint variety of a linkage with a single revolute joint. Since we are only considering planar kinematics,

**Figure 11.3**
Constraint varieties of four-bar linkages: General four-bar with two assembly modes; rational constraint variety; twisted cubic and straight line; two conics (from left to right).

we can restrict to the sub-algebra of $\mathbb{DH}$ which is generated by 1, $\mathbf{k}$, $\varepsilon\mathbf{i}$, and $\varepsilon\mathbf{j}$ (rotations about the third coordinate axis and translations in direction of first and second coordinate axis). Write $h = p_0 + p_3\mathbf{k} + \varepsilon(q_1\mathbf{i} + q_2\mathbf{j})$ and consider the action (11.9) on a point $x = 1 + \varepsilon(x_1\mathbf{i} + x_2\mathbf{j})$ in the moving plane. By appropriately choosing coordinate origins in the moving and the fixed frame, respectively, we may assume $x_1 = x_2 = 0$ and consider the condition that the image of $x$ lies on the circle with center $(0,0)$ and radius $r$. This gives the *prototype* of a circle constraint variety

$$V\colon (p_0^2 + p_3^2)r^2 - 4(q_1^2 + q_2)^2 = 0. \tag{11.11}$$

The general form, without referring to special coordinates, can be obtained in a similar fashion or by subjecting the variety (11.11) to the projective transformations (11.10).

We see that $V$ is a regular, ruled quadric that contains the conjugate complex points $(p_0, p_3, q_1, q_3)^\intercal = (0,0,1,\pm i)^\intercal$ – a property which is, in fact, invariant with respect to the transformations (11.10) and hence of kinematic relevance, see for example the direct kinematics problem of the planar RPR-manipulator in Section 11.4.1.

The constraint variety of a planar four-bar linkage is the intersection of two circle constraint varieties (Figure 11.3). This allows the immediate derivation of a number of relevant properties of four-bar linkages. We give a detailed discussion at an intuitive level. To begin with, the constraint variety of a planar four-bar linkage is an algebraic curve of dimension one and degree four. In general it is of genus one and may hence have two disconnected real components. These correspond to what is known as "assembly modes" in kinematics (leftmost image in Figure 11.3, left row in Figure 11.4).

The constraint variety of a planar four-bar linkage can be rational if it has precisely one singularity (second image from the left in Figure 11.3 and Figure 11.4, middle column). This is both a singularity in algebraic geometry and also in the kinematic sense. It corresponds to a folded position of the four-bar linkage where its behaviour is undefined. If one of the two arms is actuated, it is possible that the other arm follows in one or the other direction. For automatically controlled linkages, this is a serious problem.

Also reducible constraint varieties are possible. Two general quadrics can intersect in a twisted cubic and a straight line, two conic sections, a pair of lines and a conic section, or four straight lines. Here, only the first three cases can occur (two images on the right of Figure 11.3). They correspond to a deltoid linkage and a parallelogram linkage, respectively. The former has purely rotational motion mode which corresponds to a straight line in the constraint variety and a rational component corresponding to a twisted cubic. The latter has a translational mode along a circle and

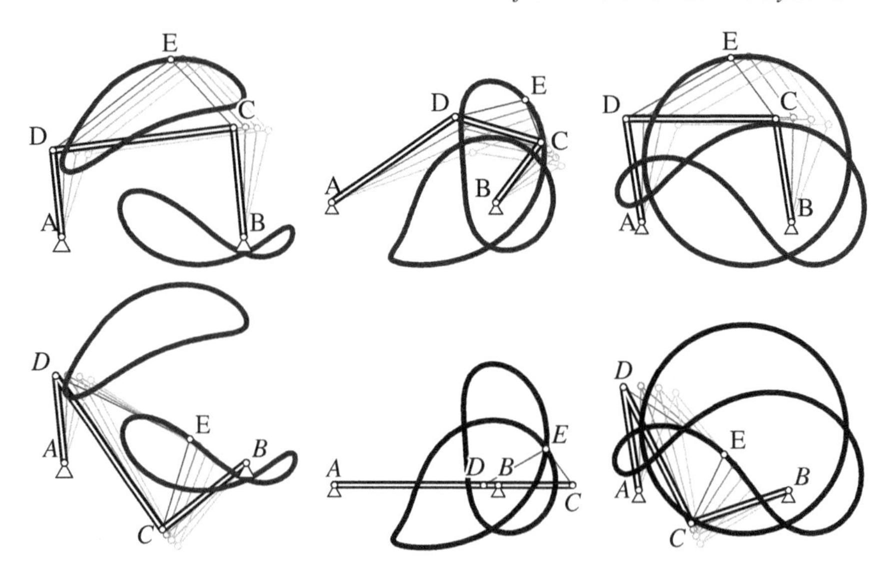


**Figure 11.4**
Four-bar linkages: Two assembly modes (left column), rational four-bar (middle column), reducible four bar (right column).

a second mode which is called "anti-parallelogram" mode (Figure 11.4, right column). Finding irreducible components of an algebraic variety amounts to a prime decomposition of their ideal. This is nowadays a standard tool of computational kinematics, see for example [42, 47].

## 11.3 Serial Manipulators

A serial mechanism is characterized by having a linear linkgraph. The number $n$ of joints equals the dimension of its constraint variety which is also called the mechanism's *degree of freedom*. Traditionally, one assumes $n \leq 6$ because this is already enough for producing a full-dimensional subset of SE(3). But recently redundant serial chains ($n > 6$) have gained attention of researchers and industry because one can use the extra degree of freedom for imposing additional requirements such as better joint or end-effector velocity or collision avoidance.

### Glossary

**Bennett linkage:** The only nontrivial spatial four-bar linkage with one degree of freedom.

**Closed $n$R linkage:** Linkage with only revolute joints whose linkgraph is a cycle of length $n$.

**Denavit-Hartenberg Parameters:** Geometric parameters providing a specification of a serial linkage independent of its configuration.

**Direct kinematics:** Task of finding the end-effector pose for given joint parameters (rotation angles, translation distances).

**Goldberg linkage:** The only nontrivial spatial five-bar linkage with one degree of freedom. Obtained by merging two Bennett linkages.

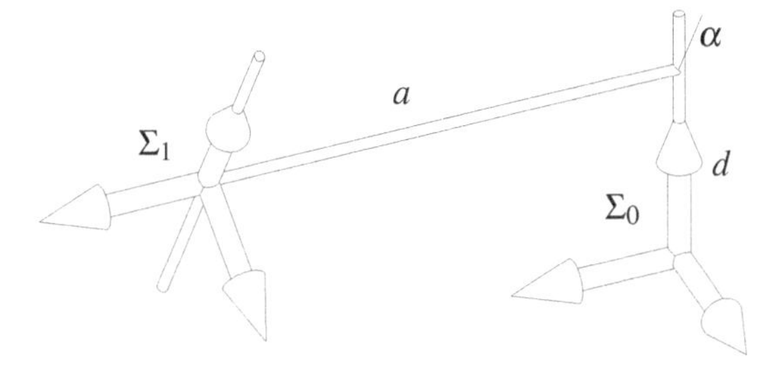


**Figure 11.5**
Relative position of two lines in space.

**Inverse kinematic:** Task of finding the joint parameters (rotation angles, translation distances) for given end-effector pose.

**Linkage synthesis:** Interpolation or approximation of motions by means of linkages. Construction of linkages to fulfill certain tasks.

**Redundant manipulator:** Linkage whose degrees of freedom exceed the dimension of the underlying motion group (six in case of SE(3)).

**Singular position:** "Problematic" position of a linkage that should be avoided. Depending on linkage type and description, several different definitions are used.

In a serial manipulator each link connects exactly two joints. If the joints are viewed as lines (the rotation axis for revolute joints, some line in translation direction for prismatic joints), the link may be attached to their common normal. They allow a convenient description of the manipulator's geometry that is independent of its current configuration. The relative transformation $G$ between two coordinate frames attached to the joints (see Figure 11.5) is then the composition of a rotation with angle $\alpha$ about the first coordinate axis and a translation by the vector $(a,0,d)^\intercal$. Here, $a$ is the length of the common normal, $d$ is the offset between the origin of the frame attached to the first line and the foot point of the common normals of the lines, and $\alpha$ denotes the twist angle of the two lines. The transformation $G$ is commonly written in terms of a homogeneous transformation matrix but may as well be given in terms of dual quaternions.

The Denavit Hartenberg (DH) parameters of a linkage consist of all common normal lengths, offsets, and twist angles of consecutive frames in a serial chain. They provide a geometrically invariant way for specifying a serial manipulator, regardless of its configuration and can also be used for a convenient description of the motion of the end-effector of a serial chain with revolute and prismatic joints or combinations of these.

After attaching a coordinate frame to each of the $n$ joints of the serial manipulator (Figure 11.6), it is possible to describe the motion of the end-effector of the serial manipulator with respect to the base frame as the product

$$E = M_1 G_1 M_2 G_2 \cdots M_{n-1} G_{n-1} M_n G_n. \tag{11.12}$$

Here, $G_i$ is the transformation between consecutive links $\Sigma_i$ and $\Sigma_{i+1}$ as illustrated by Figure 11.5, and $M_i$ is a rotation about the first coordinate axis in the local frame (the manipulator's axis) whose rotation angle $u_i$ is a motion parameter. In the matrix model, we have for example

$$G_i = \begin{pmatrix} 1 & 0 & 0 & 0 \\ a_i & 1 & 0 & 0 \\ 0 & 0 & \cos\alpha_i & -\sin\alpha_i \\ d_i & 0 & \sin\alpha_i & \cos\alpha_i \end{pmatrix}, \quad M_i = \begin{pmatrix} 1 & 0 & 0 & 0 \\ 0 & \cos u_i & -\sin u_i & 0 \\ 0 & \sin u_i & \cos u_i & 0 \\ 0 & 0 & 0 & 1 \end{pmatrix}. \tag{11.13}$$

An extension to include prismatic joints is quite straightforward but not unique.

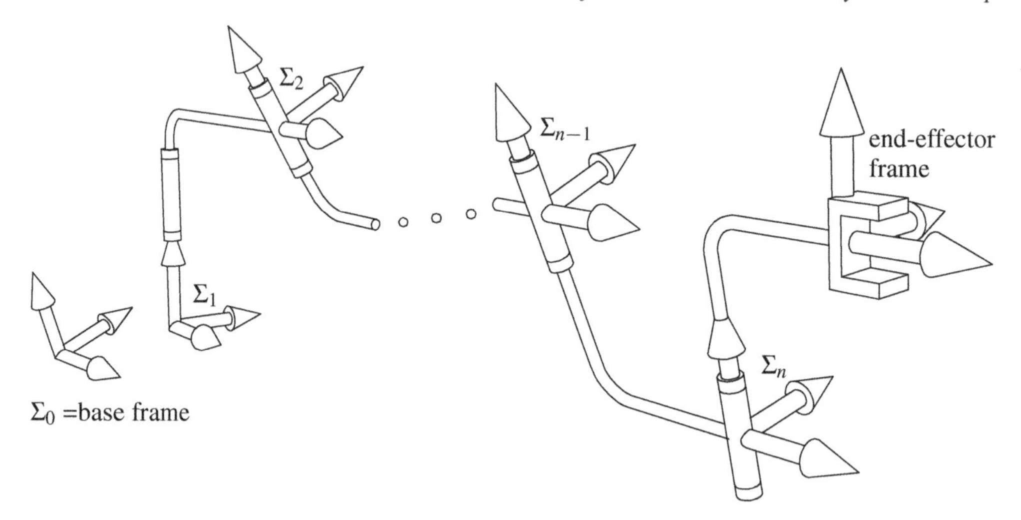


**Figure 11.6**
Coordinate frames attached to a general $n$R-mechanism.

### 11.3.1 Direct and Inverse Kinematics

In the *direct kinematics* problem the joint parameters $u_1, \ldots, u_n$ are known and the pose (position and orientation) of the end-effector is to be determined. Conversely in the *inverse kinematics* problem, the end-effector pose is known while one is looking for joint parameters that bring the mechanism's end-effector in exactly that given pose. In both cases the Denavit-Hartenberg parameters (normal distances $a_i$, offsets $d_i$, twist angles $\alpha_i$) are given. In calibration problems, these parameters are to be determined from information on joint parameters and end-effector pose. In the case of serial manipulators the direct kinematics is easily solved via (11.12) and has a unique solution.

In contrast, the inverse kinematics problem of serial manipulators is often a difficult and highly nonlinear problem. The end-effector position is known and the joint parameters are to be determined. If the manipulator has $n \leq 5$ degrees of freedom, this problem only has a solution if the end-effector position lies in the constraint variety of the corresponding $n$R chain. In the case of $n = 6$ only "practical" constraints like link lengths, joint limits, or self-collisions come into play. The inverse kinematics of the 6R manipulators once gained much attention and was known as the "Mount Everest" problem of kinematics. Nowadays it is considered as solved (see [36]). One possible solution algorithm makes use of constraint varieties [22]. The fundamental idea is to divide the mechanism into two chains of three revolute joints. These two chains generate (parametric) constraint varieties of open 3R chains that can be efficiently computed and intersected. This intersection yields exactly the configurations of the sub-chains that allow a re-assembling at the middle link, thus providing a configuration of the complete serial chain. There exist up to 16 solutions all of which can be real. Recently, hundreds of manipulators with 16 real inverse kinematics solutions were found [6]. Some of them have all real solutions on a whole path in their workspace. The possibility to select among 16 possible paths with different properties concerning collisions, required joint limits, or energy consumption is beneficial for the path planning.

### 11.3.2 Synthesis of Open and Closed Serial Chains

In this part we discuss the construction of serial chains. In mechanism science, this is called *synthesis*. We may also view it as an interpolation problem on the Study quadric by certain constraint varieties. Problems of this type have a long tradition in kinematics [33]. In the case of serial chains, synthesis means finding the Denavit-Hartenberg parameters of a manipulator for a prescribed task,

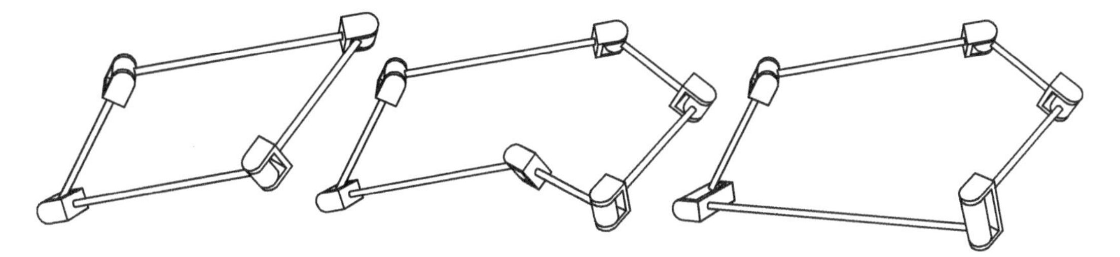

**Figure 11.7**
A Bennett mechanism (left), two Bennett mechanisms sharing one link and two joints (middle), and the corresponding Goldberg 5*R* mechanism (right).

for example the capability to reach a number of prescribed end-effector poses. One may be interested in approximate or exact solutions. In the second case, the prescribed task yield an algebraic system of equations which has to be solved. In case of approximate synthesis the task may lead to an over-constrained system of equations. Its approximate solution requires a concept for the "distance" between rigid body displacements. This is quite problematic as incompatible units (angles and distances) are involved. In this text we focus on exact and algebraic synthesis methods.

There exist several methods for the exact synthesis of 2R chains [1, 2, 8, 33]. For serial chains with three revolute joints, only numerical solutions are known. Homotopy continuation is used in [27, 28, 29] while [30] proposed interval analysis. The most recent solution [15] uses numerical algebraic geometry via the software package Bertini [4]. There it was shown that the exact synthesis of 3R chains with five prescribed end-effector poses can have at most 456 solutions over the complex numbers. The number of possible real solutions is still unknown. Virtually nothing seems to be known about the synthesis of arbitrary serial chains with four or five joints.

A task which is not yet been extensively explored is the synthesis of *closed* serial chains. One searches for mechanisms that are movable even if the end-effector is restricted to one fixed pose and in addition fulfill a certain prescribed task. Closed serial chains with seven or more joints are movable, whereas such chains with three or less joints are rigid. Therefore, the cases of four, five, or six joints are of particular interest. Their mobility requires a special geometric design. Flexible examples with only four revolute joints are planar and spherical four-bar linkages (all axes are parallel or concurrent, respectively), the Bennett linkage [5] and degenerate cases with coinciding axes. The Bennett linkage (see Figure 11.7 left) is a mobile spatial four-bar linkage. The geometric condition for flexibility is that the common normals between any two consecutive axes form a closed spatial parallelogram, that is, consecutive sides intersect and the distances between opposite point pairs are equal. An algebraic synthesis algorithm for four-bar linkages with revolute and prismatic joints using constraint varieties was recently presented in [37].

Essentially the only nontrivial flexible example for $n = 5$ is known as Goldberg's linkage. We may think of it as the concatenation of two Bennett linkages with one common link and two common joints such that the one Bennett linkage "drives" the other. Removing one common joint and merging certain links one obtains a 5R linkage with one degree of freedom. This is shown in Figure 11.7. The necessity of this construction was first proved in [25]. A simpler proof using bond theory is in [17].

The cases $n = 4$ and $n = 5$ are more or less fully understood. This is not true for $n = 6$. In fact, a full classification of all mobile closed 6R linkages is one of the big open problems in the field. A precise problem formulation can be found in [32], the current state of research is summarized in [31]. Recent progress was made using *bond theory* and *bond diagrams* [17, 32]. Only little is known about the synthesis of closed-loop linkages with six joints. The synthesis of certain closed six-bar linkages to four given poses is the topic of [18]. It is based on a factorization theory of motion polynomials (parametric equations of rational curves on the Study quadric) [16] where the factorization into linear factors corresponds to a decomposition of the parametric curve/motion into rotations. In general, there exist several factorizations, each of them giving rise to a serial $n$R chain. Combining two such chains it is possible to construct a closed chain with $2n$ revolute joints. The synthesis method of [18] can be specialized to the synthesis of closed five-bar linkages [40]. Still, there is a lot of room for future research. One of the reasons for this is the missing classification of movable closed 6R linkages.

### 11.3.3 Singularities and Path Planning

In industrial applications, it is necessary to carefully plan the desired path of a mechanical device. It is not only necessary to ensure a desired trajectory but also to limit velocities and accelerations [14]. In addition, it is necessary to avoid configurations of the mechanism where it is in a *singular position*. In case of serial manipulators such configurations can be defined in different ways. One possibility is coincidence of two of the inverse kinematics solutions. Alternatively, the Plücker coordinates of the joint axes satisfy a linear relation [24] or [38, Section 3.1]. Singularities are very important in practice for several reasons. The manipulator looses degrees of freedom in such configurations. Moreover, control of the mechanism in the vicinity of singularities is very difficult because numerical calculations become unstable. Therefore, singularity avoiding algorithms for path planning are an important topic.

### 11.3.4 Open Problems

- The synthesis of serial manipulators with three, four, or five joints to a prescribed number of task poses. Only numerical solutions for three revolute joints are known.
- The synthesis of closed serial chains which is related to yet another open issue, the complete classification of this linkage type.
- It is believed that algebraic methods could also be beneficial for singularity avoiding path planning.
- The factorization theory for motion polynomials still is not fully understood. Some non-generic cases are topic of current research and details may be rather tricky (see for example [13]). A factorization theory for multivariate motion polynomials would be highly interesting for applications in mechanism science but is probably rather difficult.

## 11.4 Parallel Manipulators

A *parallel manipulator* consists of a fixed base and a moving end-effector (platform), connected by multiple chains of links and joints called *limbs*. In other words, its linkgraph contains multiple loops. In many technical applications the limbs have the same structure. One of the most important parallel manipulators is the Stewart-Gough platform (Figure 11.9) used for example in flight simulators. The Stewart-Gough platform has six degrees of freedom and is hence called a "full degree of freedom parallel manipulator". A second example is the Delta-Robot, which is used for high-speed pick and place tasks [35]. The Delta-Robot has three degrees of freedom and is hence called a "lower degree of freedom parallel manipulator." Also planar parallel manipulators are interesting, as the following sections will show.

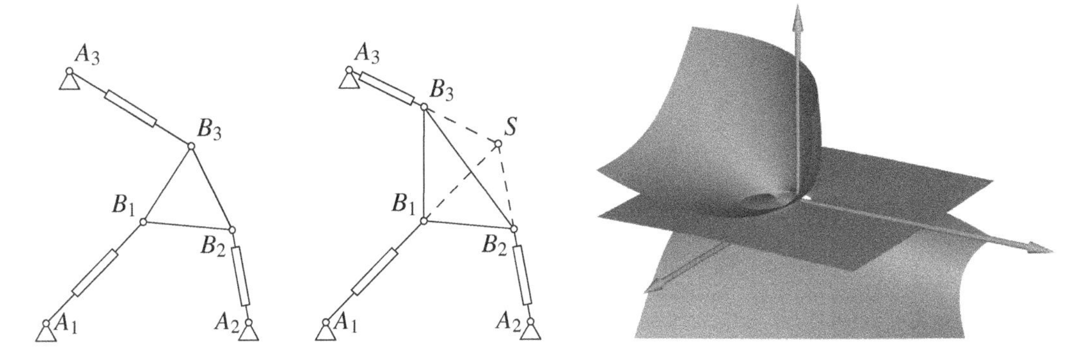


**Figure 11.8**
Planar 3-RPR parallel manipulator, singular configuration, and singularity surface.

## Glossary

**3-RPR manipulator:** Parallel manipulator with three limbs, each consisting of a revolute, a prismatic, and another revolute joint.

**Stewart-Gough platform:** Parallel manipulator with six limbs, each consisting of two passive (nonactuated) spherical joints connected by an active (actuated) prismatic joint.

**Self-motion:** Motion of a generically rigid parallel manipulator. Only possible in special singular configurations.

### 11.4.1 Direct and Inverse Kinematics

In contrast to serial manipulators, solving the direct kinematics problem of a parallel manipulator takes more effort than the inverse kinematic problem. In case of the direct kinematic problem, the input parameters, in this case some of the joint parameters, are given, and the position of the platform has to be determined. This will be discussed for two examples in more detail.

For the planar case, consider the 3-RPR parallel manipulator. This manipulator consists of a fixed base, given by the triangle $A_1A_2A_3$ as shown in Figure 11.8, left, and a moving platform given by the triangle $B_1B_2B_3$. Corresponding vertices are connected by RPR limbs, this means revolute joints are located in $A_i$ and $B_i$. To move the platform, the distance between $A_i$ and $B_i$ is changed by an actuated prismatic joint and the variable distance is denoted by $r_i$. The direct kinematics problem is: Determine the position of the platform for given $r_1$, $r_2$ and $r_3$. The vertex $B_1$ has to be located on a circle with center $A_1$ and radius $r_1$. Similar constraints hold for $B_2$ and $B_3$. So the reformulated geometric task is: Place the triangle $B_1B_2B_3$ on three circles centered at $A_1, A_2$ and $A_3$ with respective radii $r_1$, $r_2$ and $r_3$. To solve this task using Study's kinematic mapping, we have to intersect three circle constraint varieties with prototype equation (11.11). The intersection of three quadrics in projective three-space $\mathbb{P}^3$ gives, in general, eight points. But in Section 11.2 we saw that the invalid solutions $(0,0,1,\pm \mathrm{i})^\mathsf{T}$ are always among them. Hence, the direct kinematics problem has only six solutions.

A more general mathematical approach uses computational algebraic geometry for intersecting the constraint varieties of each limb. Often the number of possible solutions is of interest. In this case it is useful to eliminate all unknowns until a univariate polynomial is achieved. This maybe done by computing step wise resultants or a Gröbner basis [10]. In case of the 3-RPR manipulator this univariate polynomial has, indeed, degree six. In general the degree of the univariate polynomial is

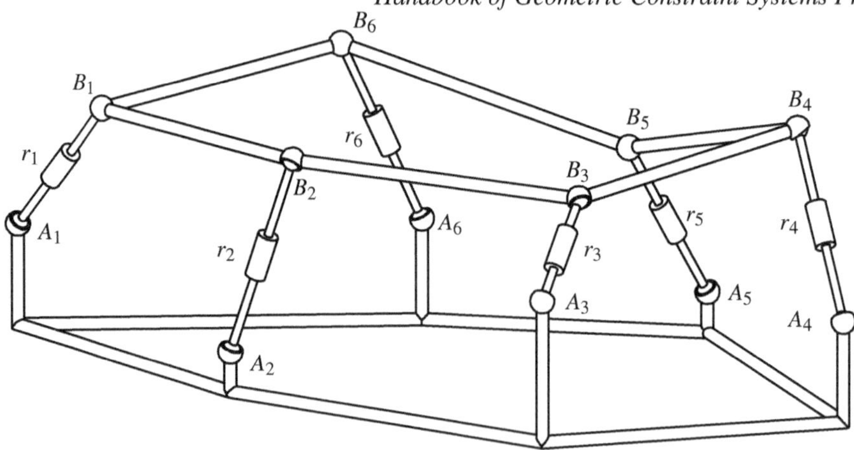


**Figure 11.9**
Stewart-Gough platform

only an upper bound, because not all solutions have to be real, but for the 3-RPR parallel manipulator there are examples with six real solutions as shown in [20].

A spatial example of a parallel manipulator is the Stewart-Gough platform shown in Figure 11.9. This parallel manipulator consists of six anchor points $A_i$, $i = 1,\dots,6$ in the fixed base and six points $B_i$, $i = 1,\dots,6$ in the moving platform. Any two points $A_i$ and $B_i$ are connected by an SPS-limb, i.e. a spherical joint in $A_i$ followed by a prismatic joint and again a spherical joint in $B_i$. The actuated joints are the prismatic joints and the variable lengths are denoted by $r_i$. In this case there are six sphere constraint equations $\|A_i - B_i\|^2 - r_i^2 = 0$, $i = 1,\dots,6$. The constraint equations in this case are again quadratic, but this time with eight homogeneous unknowns $p_0,\dots,p_3$ and $q_0,\dots,q_3$. A univariate polynomial for this problem was computed for the first time in [19]. It is of degree 40, so for given lengths $r_i$ there exist up to 40 positions of the platform. All of them can be real [12].

More recently, lower degree of freedom parallel manipulators draw attention in mechanism science. The reason is that for some tasks not the full six degrees of freedoms are required. One example is the already mentioned Delta-Robot which has three degrees of freedom and is specially designed for fast "pick and place" tasks.

### 11.4.2 Singularities and Self-Motions

Some advantages of parallel manipulators over serial manipulators are the high stiffness and the high accuracy. Because of this, a parallel manipulator is capable of taking high loads. But in some positions of the platform this stiffness is lost, causing some unwanted behaviour during the motion of the platform. In rigidity theory, one speaks of infinitesimally flexible configurations. In mechanism science, these positions are called "singular." Depending on the structure of the manipulator, computing singular positions can be a challenging task. In general, if the constraint equations of the parallel manipulator are known, they can be computed as the variety to the system's Jacobian. This gives rise to a singularity surface in kinematic image space (Figure 11.8, right-hand side).

Sometimes the singularity conditions have a geometrical meaning, like in case of the planar 3-RPR manipulator. Here, singular positions are characterised by concurrency of the three legs (Figure 11.8, middle). The spatial Stewart-Gough platform is in a singular position, iff the six lines joining corresponding vertices $A_i$, $B_i$ belong to a linear line complex [38, Section 3.1]. Some special singular configurations can result in *self-motions* of the parallel manipulator. These motions appear, when the input parameters are fixed, but the moving platform is able to perform a motion with at least one degree of freedom. For the planar 3-RPR manipulator these are described in [7], for the Stewart-Gough platform see [26].

### 11.4.3 Open Problems

- For lower degree of freedom parallel manipulators the path planing task is still an open problem. This is due to the fact that the workspace is a subset of SE(3) and the description via constraint equations is an implicit representation. In this case it has to be ensured that the path, a curve in the intersection of the constraint varieties, lies completely in the workspace. Avoiding singular positions is also necessary. Therefore, additional side conditions have to be included. In some special cases, like the 3-RPS considered in [41], a parameterization of the complete workspace is possible, which makes the task of path planing much easier and relates mechanism science to parameterization issues of algebraic varieties.
- Recently a new class of parallel manipulators, cable driven manipulators, became very popular. For cable driven manipulators, the rigid limbs of a parallel manipulator are replaced by cables that control the moving platform. A physical constraint of cables is, that they are only able to contribute to the motion if they are under tension. This fact leads to some extra constraints in each cable, giving rise to semi-algebraic equations. The planar case of the so-called "hanging box" problem was investigated in [9]. where a bound of six solutions for the direct kinematics problem is given.
- An interesting problem of parallel manipulators are "non-singular assembly mode changes". This means the continuous changing between two direct kinematics solutions without passing through a singular configuration [21]. In this way, it is possible to permute direct kinematics solutions. The thus defined permutation group should capture a lot of information on reconfigurability of the robot. Apparently, this has not yet been considered in mechanism but similar ideas appear in other classes of geometric problems [11].

## References

[1] Kassim Abdul-Sater, Franz Irlinger, and Tim C. Lueth. Two-configuration synthesis of origami-guided planar, spherical and spatial revolute-revolute chains. *ASME Journal of Mechanisms and Robotics*, 5(3):031005, 2013.

[2] Kassim Abdul-Sater, Manuel M. Winkler, Franz Irlinger, and Tim C. Lueth. Three-position synthesis of origami-evolved, spherically constrained spatial revolute-revolute chains. *ASME Journal of Mechanisms and Robotics*, 8(1):011012, 2015.

[3] Hamid Ahmadinezhad, Zijia Li, and Josef Schicho. An algebraic study of linkages with helical joints. *J. Pure Appl. Algebra*, 219(6):2245–2259, 2015.

[4] Daniel J. Bates, Jonathan D. Hauenstein, Andrew J. Sommese, and Charles W. Wampler. Bertini: Software for numerical algebraic geometry. Available at *bertini.nd.edu*.

[5] Geoffrey T. Bennett. A new mechanism. *Engineering*, 76:777–778, 1903.

[6] Mathias Brandstötter. *Adaptable Serial Manipulators in Modular Design*. PhD thesis, UMIT - University for Health Sciences, Medical Informaticvs and Technology, Hall, Austria, 2016.

[7] Sébastien Briot, Ilian Bonev, Damien Chablat, Philippe Wenger, and Vigen Arakelian. Self-motions of general 3-RPR parallel robots. *International Journal of Robotics Research*, 27(7):855–866, 2008.

[8] Katrin Brunnthaler, Hans-Peter Schröcker, and Manfred L. Husty. A new method for the synthesis of Bennett mechanisms. In *Proceedings of CK 2005, International Workshop on Computational Kinematics*, Cassino, 2005.

[9] Marco Carricato and Jean-Pierre Merlet. Stability analysis of underconstrained cable-driven parallel robots. *IEEE Transactions on Robotics*, 29(1):288–296, 2013.

[10] David Cox, John Little, and Donal O'Shea. *Ideals, Varieties, and Algorithms*. Springer, 2007.

[11] Abraham Martin del Campo Sanchez. *Galois groups of Schubert problems*. PhD thesis, Texas A&M University, 2012.

[12] Peter Dietmaier. The stewart-gough platform of general geometry can have 40 real postures. In Jadran Lenarčič and Manfred L. Husty, editors, *The 6th conference on Advances in Robot Kinematics 1998, Stobl*, pages 7–16, 1998.

[13] Matteo Gallet, Christoph Koutschan, Zijia Li, Georg Regensburger, Josef Schicho, and Nelly Villamizar. Planar linkages following a prescribed motion. *Mathematics of Computation*, 86(303):473–506, 2016.

[14] Francisco Geu Flores and Andrés Kecskeméthy. Time-optimal path planning along specified trajectories. In Hubert Gattringer and Johannes Gerstmayr, editors, *Multibody System Dynamics, Robotics and Control*, pages 1–16. Springer, 2013.

[15] Jonathan D. Hauenstein, Charles W. Wampler, and Martin Pfurner. Synthesis of three-revolute spatial chains for body guidance. *Mechanism and Machine Theory*, 2016.

[16] Gábor Hegedüs, Josef Schicho, and Hans-Peter Schröcker. Factorization of rational curves in the Study quadric. *Mechanism and Machine Theory*, 69:142152, 2013.

[17] Gábor Hegedüs, Josef Schicho, and Hans-Peter Schröcker. The theory of bonds: A new method for the analysis of linkages. *Mechanism and Machine Theory*, 70:407424, 2013.

[18] Gábor Hegedüs, Josef Schicho, and Hans-Peter Schröcker. Four-pose synthesis of angle-symmetric 6R linkages. *ASME Journal of Mechanisms and Robotics*, 7(4):041006, 2015.

[19] Manfred L. Husty. An Algorithm for Solving the Direct Kinematics of General Stewart-Gough Platforms. *Mechanism and Machine Theory*, 31(4):365–380, 1996.

[20] Manfred L. Husty. Non-singular assembly mode change in 3-RPR-parallel manipulators. In Andrés Kecskeméthy and Andreas Müller, editors, *Computational Kinematics*. Springer, 2009.

[21] Manfred L. Husty and Clément Gosselin. On the singularity surface of planar 3-RPR parallel mechanisms. *Mechanics Based Design of Structures and Machines*, 36(4):411–425, 2008.

[22] Manfred L. Husty, Martin Pfurner, and Hans-Peter Schröcker. Algebraic methods in mechanism analysis and synthesis. *Robotica*, 25(6):661–675, 2007.

[23] Manfred L. Husty and Hans-Peter Schröcker. Kinematics and algebraic geometry. In J. Michael McCarthy, editor, *21st Century Kinematics. The 2012 NSF Workshop*, pages 85–123. Springer, London, 2012.

[24] Adolf Karger. Singularity analysis of serial robot-manipulators. *Journal of Mechanical Design*, 118(4):520–525, 1996.

[25] Adolf Karger. Classification of 5R closed kinematic chains with self mobility. *Mechanism and Machine Theory*, 33(1-2):213222, 1998.

[26] Adolf Karger and Manfred L. Husty. Classification of all self-motions of the original Stewart-Gough platform. *Computer Aided Design*, 30(3):205–215, 1998.

[27] Eric Lee and Constantinos Mavroidis. Solving the geometric design problem of spatial 3R robot manipulators using polynomial homotopy continuation. *Journal of Mechanical Design*, 124:652–661, 2002.

[28] Eric Lee and Constantinos Mavroidis. An algebraic elimination based algorithm for solving the geometric design problem of spatial 3R manipulators. In *Proc. ASME Design Eng. Tech. Conf., Mech. & Robotics Conf.* ASME, 2004.

[29] Eric Lee and Constantinos Mavroidis. Geometric design of 3R robot manipulators for reaching four end-effector spatial poses. *The International Journal of Robotics Research*, 23(3):247 – 254, 2004.

[30] Eric Lee, Constantinos Mavroidis, and Jean-Pierre Merlet. Five precision point synthesis of spatial RRR manipulators using interval analysis. *Journal of Mechanical Design*, 126:842–849, 2004.

[31] Zijia Li. *Closed linkages with six revolute joints.* Phd thesis, Johannes Kepler University, Linz, 2015.

[32] Zijia Li, Josef Schicho, and Hans-Peter Schröcker. A survey on the theory of bonds. *IMA J. Math. Control Inform.*, 2016.

[33] J. Michael McCarthy. *Geometric Design of Linkages*, volume 11 of *Interdisciplinary Applied Mathematics*. Springer, 2000.

[34] Johan Ernest Mebius. Derivation of the Euler-Rodrigues formula for three-dimensional rotations from the general formula for four-dimensional rotations.

[35] Jean-Pierre Merlet. *Parallel Robots*. Kluwer Academic Publishers, 2000.

[36] Martin Pfurner. *Analysis of spatial serial manipulators using kinematic mapping*. PhD thesis, University of Innsbruck, Innsbruck, Austria, 2008.

[37] Martin Pfurner, Thomas Stigger, and Manfred L. Husty. Overconstrained single loop four link mechanisms with revolute and prismatic joints. In Philippe Wenger and Paulo Flores, editors, *New Trends in Mechanism and Machine Science, Theory and Industrial Applications*, pages 71–79, Nantes, France, 2016.

[38] Helmut Pottmann and Johannes Wallner. *Computational Line Geometry*. Springer, 2001.

[39] Tudor-Dan Rad, Daniel F. Scharler, and Hans-Peter Schröcker. The kinematic image of RR, PR, and RP dyads. Submitted for publication, 2016.

[40] Tudor-Dan Rad and Hans-Peter Schröcker. The kinematic image of 2R dyads and exact synthesis of 5R linkages. In *Proceedings of the IMA Conference on Mathematics of Robotics*, 2015.

[41] Josef Schadlbauer, Latifah Nurahmi, Manfred L. Husty, Stéphane Caro, and Philippe Wenger. Operation modes of lower-mobility parallel manipulators. In *Second Conference on Interdisciplinary Applications in Kinematics*, pages 3–10, 2013.

[42] Josef Schadlbauer, Dominic R. Walter, and Mandfred L. Husty. The 3-RPS parallel manipulator from an algebraic viewpoint. *Mechanism and Machine Theory*, 75:161176, 2014.

[43] Jon M. Selig. *Geometric Fundamentals of Robotics*. Monographs in Computer Science. Springer, 2 edition, 2005.

[44] Jon M. Selig. On the geometry of point-plane constraints on rigid-body displacements. *Acta Appl. Math.*, 116(2):133155, 2011.

[45] Jon M. Selig. Some rigid-body constraint varieties generated by linkages. In Jadran Lenarčič and Manfred L. Husty, editors, *Latest Advances in Robot Kinematics*, pages 293–300. Springer, 2012.

[46] Eduard Study. Von den Bewegungen und Umlegungen. *Math. Ann.*, 39:441–566, 1891.

[47] Dominic R. Walter, Manfred L. Husty, and Martin Pfurner. A complete kinematic analysis of the SNU 3-UPU parallel manipulator. In Daniel J. Bates, Gian M. Besana, Sandra Di Rocco, and Charles W. Wampler, editors, *Contemporary Mathematics*, volume 496, pages 331–346. American Mathematical Society, 2009.

# Chapter 12

# *Real Algebraic Geometry for Geometric Constraints*

**Frank Sottile**

*Texas A&M University, College Station, Texas*

**CONTENTS**



Real algebraic geometry adapts the methods and ideas from (complex) algebraic geometry to study the real solutions to systems of polynomial equations and polynomial inequalities. As it is the real solutions to such systems modeling geometric constraints that are physically meaningful, real algebraic geometry is a core mathematical input for geometric constraint systems.

## 12.1 Introduction

Algebraic geometry is fundamentally the study of sets, called varieties, which arise as the common zeroes of a collection of polynomials. These include familiar objects in analytic geometry, such as conics, plane curves, and quadratic surfaces. Combining intuitive geometric ideas with precise algebraic methods, algebraic geometry is equipped with many powerful tools and ideas. These may be brought to bear on problems from geometric constraint systems because many natural constraints, particularly prescribed incidences, may be formulated in terms of polynomial equations.

Consider a four-bar mechanism; a quadrilateral in the plane with prescribed side lengths $a$, $b$, $c$, and $d$, which may rotate freely at its vertices and where one edge is fixed as in Figure 12.2. The points $x$ and $y$ are fixed at a distance $a$ apart, the point $p$ is constrained to lie on the circle centered

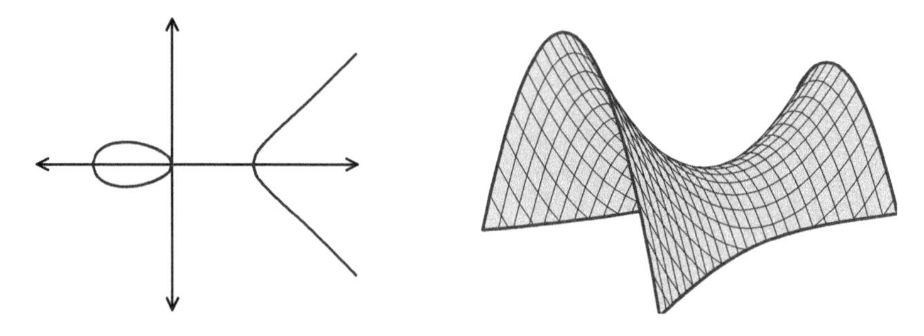

**Figure 12.1**
A cubic plane curve and a quadratic surface (a hyperbolic paraboloid).

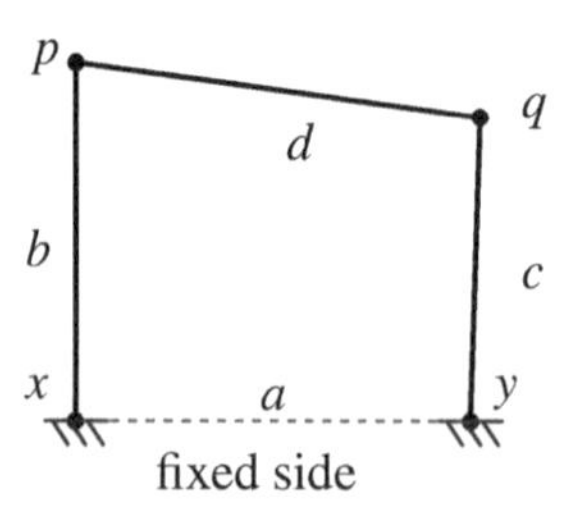


**Figure 12.2**
A four-bar mechanism.

at $x$ with radius $b$, the point $q$ lies on the circle centered at $y$ with radius $c$, and we additionally require that $p$ and $q$ are a distance $d$ apart. Squaring the distance constraints gives a system of three quadratic equations whose solutions are all positions of this four-bar mechanism.

Algebraic geometry works best over the complex numbers, because the geometry of a complex variety is controlled by its defining equations. (For instance, the Fundamental Theorem of Algebra states that a univariate polynomial of degree $n$ always has $n$ complex roots, counted with multiplicity.) Geometric constraint systems are manifestly real (as in real-number) objects. For this reason, the subfield of real algebraic geometry, which is concerned with the real solutions to systems of equations, is most relevant for geometric constraint systems. Working over the real numbers may give quite different answers than working over the complex numbers.

This chapter will develop some parts of real algebraic geometry that are useful for geometric constraint systems. Its main point of view is that one should first understand the geometry of corresponding complex variety, which we call the algebraic relaxation of the original problem. Once this is understood, we then ask the harder question about the subset of real solutions.

For example, $x^2+y^2=1$ and $x^2+y^2=-1$ define isomorphic curves in the complex plane—send $(x,y)\mapsto(\sqrt{-1}x,\sqrt{-1}y)$—which are quite different in the real plane. Indeed, $x^2+y^2=1$ is the unit circle in $\mathbb{R}^2$ and $x^2+y^2=-1$ is the empty set. Replacing $\pm1$ by 0 gives the pair of complex conjugate lines

$$x^2+y^2 \;=\; (x+\sqrt{-1}y)(x-\sqrt{-1}y) \;=\; 0,$$

whose only real point is the origin $(0,0)$. The reason for this radically different behavior amongst these three quadratic plane curves is that only the circle has a smooth real point—by Theorem 12.2, when a real algebraic variety has a smooth real point, the salient features of the underlying complex variety are captured by its real points.

## 12.2 Ideals and Varieties

**$\mathbb{C}[x_1,\dots,x_d]$:** The ring of all polynomials with complex coefficients in the variables $x_1,\dots,x_d$.

**Variety:** A subset $\mathcal{V}(S) \subset \mathbb{C}^d$ where all polynomials in a set $S \subset \mathbb{C}[x_1,\dots,x_d]$ vanish.

**Ideal:** Set of polynomials in $\mathbb{C}[x_1,\dots,x_d]$ closed under addition and multiplication by other polynomials.

**$\mathcal{I}(X)$:** For a subset $X$ of $\mathbb{C}^d$, this is the set of all polynomials that vanish identically on $X$.

**Hilbert's Nullstellensatz:** A fundamental theorem in algebraic geometry; the variety of a set $S$ of polynomials is empty if and only if the set $S$ generates the trivial idea $\langle 1\rangle = \mathbb{C}[x_1,\dots,x_d]$.

The best accessible introduction to algebraic geometry is the classic book of Cox, Little, and O'Shea [7]. Many thousands find this an indispensable reference. We assume a passing knowledge of some aspects of the algebra of polynomials, or at least an open mind. We work over the complex numbers, $\mathbb{C}$, for now. A collection $S \subset \mathbb{C}[x_1,\dots,x_d]$ of polynomials in $d$ variables defines a variety,

$$\mathcal{V}(S) := \{x \in \mathbb{C}^d \mid f(x) = 0 \text{ for all } f \in S\}.$$

We may add to $S$ any of its polynomial consequences, $g_1 f_1 + \cdots + g_s f_s$ where $g_i \in \mathbb{C}[x_1,\dots,x_d]$ and $f_i \in S$, without changing $\mathcal{V}(S)$. This set of polynomial consequences is the ideal generated by $S$, and so it is no loss to assume that $S$ is an ideal. Hilbert's Basis Theorem states that any ideal in $\mathbb{C}[x_1,\dots,x_d]$ is finitely generated, so it is also no loss to assume that $S$ is finite. We pass between these extremes when necessary.

Dually, given a variety $X \subset \mathbb{C}^d$ (or any subset), let $\mathcal{I}(X)$ be the set of polynomials which vanish on $X$. Any polynomial consequence of polynomials that vanish on $X$ also vanishes on $X$. Thus $\mathcal{I}(X)$ is an ideal in the polynomial ring $\mathbb{C}[x_1,\dots,x_d]$. Let $\mathbb{C}[X]$ be the set of functions on $X$ that are restrictions of polynomials in $\mathbb{C}[x_1,\dots,x_d]$. Restriction is a surjective ring homomorphism $\mathbb{C}[x_1,\dots,x_d] \twoheadrightarrow \mathbb{C}[X]$ whose kernel is the ideal $\mathcal{I}(X)$ of $X$, so that $\mathbb{C}[X] = \mathbb{C}[x_1,\dots,x_d]/\mathcal{I}(X)$. Call $\mathbb{C}[X]$ the coordinate ring of $X$.

To see this connection between algebra and geometry, consider the two plane curves $\mathcal{V}(y-x^2)$ and $\mathcal{V}(y^2-x^3)$ of Figure 12.3. One is the familiar parabola, which is smooth, and the other the

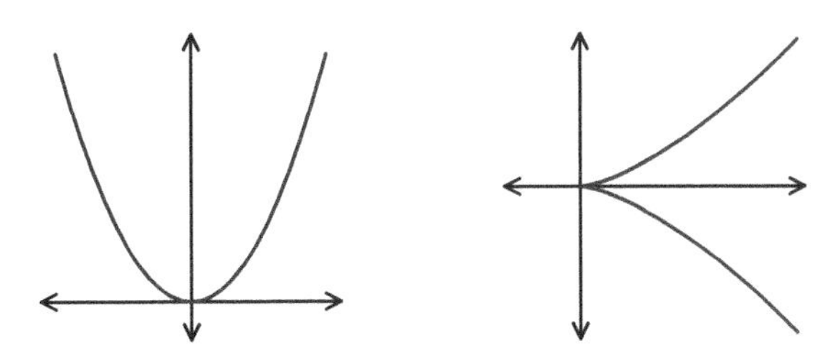

**Figure 12.3**
Plane curves $\mathcal{V}(y-x^2)$ and $\mathcal{V}(y^2-x^3)$.

semicubical parabola or cuspidal cubic, which is singular (see Section 12.4.2) at the origin. Their coordinate rings are $\mathbb{C}[x,y]/\langle y-x^2\rangle$ and $\mathbb{C}[x,y]/\langle y^2-x^3\rangle$, respectively. The first is isomorphic to $\mathbb{C}[t]$, which is the coordinate ring of the line $\mathbb{C}$, while the second is not—it is isomorphic to $\mathbb{C}[t^2,t^3]$.

The isomorphisms come from the parameterizations $t \mapsto (t, t^2)$ and $t \mapsto (t^2, t^3)$. This illustrates another way to obtain a variety—as the image of a polynomial map.

Thus begins the connection between geometric objects (varieties) and algebraic objects (ideals). Although they are different objects, varieties and ideals carry the same information. This is expressed succinctly and abstractly by stating that there is an equivalence of categories, which is a consequence of Hilbert's Nullstellensatz, whose finer points we sidestep. For the user, this equivalence means that we may apply ideas and tools either from algebra or from geometry to better understand the sets of solutions to polynomial equations.

## 12.3 ... and Algorithms

Because the objects of algebraic geometry have finiteness properties (finite-dimensional, finitely generated), they may be faithfully represented and manipulated on a computer. There are two main paradigms: symbolic methods based on Gröbner bases and numerical methods based on homotopy continuation. The first operates on the algebraic side of the subject and the second on its geometric side.

**Gröbner basis:** An algorithmically optimal generating set of an ideal.

**Buchberger Algorithm:** An algorithm to compute a Gröbner basis for an ideal from a generating set.

**Numerical algebraic geometry:** Computational paradigm that applies tools from numerical analysis to study algebraic varieties.

A consequence of the Nullstellensatz is that we may recover any information about a variety $X$ from its ideal $\mathcal{I}(X)$. By Hilbert's Basis Theorem, $\mathcal{I}(X)$ is finitely generated, so we may represent it on a computer by a list of polynomials. We emphasize computer because expressions for multivariate polynomials may be too large for direct human manipulation or comprehension. Many algorithms to study a variety $X$ through its ideal begin with a preprocessing: a given list of generators $(f_1, \ldots, f_m)$ for $\mathcal{I}(X)$ is replaced by another list $(g_1, \ldots, g_s)$ of generators, called a Gröbner basis for $\mathcal{I}(X)$, with optimal algorithmic properties. Buchberger's Algorithm is a common method to compute a Gröbner basis.

Many algorithms to extract information from a Gröbner basis have reasonably low complexity. These include algorithms that use a Gröbner basis to decide if a given polynomial vanishes on a variety $X$ or to determine the dimension or degree (see Section 12.4.1 and Section 12.4.2) of $X$. Consequently, a Gröbner basis for $\mathcal{I}(X)$ transparently encodes much information about $X$. We expect, and it is true, that computing a Gröbner basis may have high complexity (double exponential in $d$ in the worst case), and some computations do not terminate in a reasonable amount of time. Nevertheless, symbolic methods based on Gröbner bases easily compute examples of moderate size, as the worst cases appear to be rare.

Several well-maintained computer algebra packages have optimized algorithms to compute Gröbner bases, extensive libraries of implemented algorithms using Gröbner bases, and excellent documentation. Two in particular—Macaulay2 [11, 12] and Singular [8, 9]—are freely available with dedicated communities of users and developers. Commercial software, such as Magma, Maple, and Mathematica, also compute Gröbner bases and implement some algorithms based on Gröbner bases. Many find SageMath [10], an open-source software connecting different software systems together, also to be useful.

The other computational paradigm—numerical algebraic geometry—uses methods from numerical analysis to manipulate varieties on a computer [18]. Numerical homotopy continuation is used to solve systems of polynomial equations, and Newton's method may be used to refine the solutions. These methods were originally developed as a tool for mechanism design in kinematics [15], which is closely related to geometric constraint systems.

In numerical algebraic geometry, a variety $X$ of dimension $n$ (see Section 12.4.2) in $\mathbb{C}^d$ is represented on a computer by a witness set, which is a triple $(W,S,L)$, where $L$ is a general affine plane in $\mathbb{C}^d$ of dimension $d-n$, $S$ is a list of polynomials defining $X$, and $W$ consists of numerical approximations to the points of $X \cap L$ (the number of which is the degree of $X$, see Section 12.4.2). Following the points of $W$ as $L$ varies using homotopy continuation samples points of $X$, and may be used to test for membership in $X$. Other algorithms, including computing intersections and the image of a variety under a polynomial map, are based on witness sets.

Two stand-alone packages—PHCPack [22] and Bertini [2, 3]—implement the core algorithms of numerical algebraic geometry, as does the Macaulay2 package NAG4M2 [14]. Both PHCPack and Bertini may be accessed from Macaulay2, Singular, or SageMath.

Each computational paradigm, symbolic and numerical, has its advantages. Symbolic computations are exact and there are many implemented algorithms. The inexact numerical computations give refinable approximations, yielding a family of well-behaved relaxations to exact computation. Also, numerical algorithms are easily parallelized and in some cases the results may be certified to be correct [13, 17].

## 12.4 Structure of Algebraic Varieties

Varieties and their images under polynomial maps have well-understood properties that may be exploited to understand objects modeled by varieties. We discuss some of these fundamental and structural properties.

**Zariski topology:** Topology on $\mathbb{C}^n$ and on varieties whose closed sets are varieties.

**Subvariety:** A variety that is a subset of another.

**Irreducible variety:** A variety that is not the union of two proper subvarieties.

**Generic:** A property that holds on a dense Zariski open set.

**General:** A point of a variety where a generic property holds.

**Smooth:** A point of a variety where it is a manifold.

**Singular:** A point of a variety where it is not a manifold.

**Dimension of $X$:** The dimension of the smooth (manifold) points of a variety $X$.

**Degree of $X$:** The number of points in the intersection of a variety $X$ with a general affine subspace of dimension $n - \dim X$.

**Locally closed:** A set that is open in its closure.

**Constructible:** A set that is a finite union of locally closed sets.

### 12.4.1 Zariski Topology

Algebraic varieties in $\mathbb{C}^d$ are closed subsets in the usual (classical) topology because polynomial functions are continuous. Varieties possess a second, much coarser topology—the Zariski topology—whose value is that it provides the most natural language for expressing many properties of varieties. The Zariski topology is determined by its closed sets, which are simply the algebraic varieties, and therefore its open sets are complements of varieties.

Closure in the Zariski topology is easily expressed: The Zariski closure $\overline{U}$ of a set $U \subset \mathbb{C}^d$ is $\mathcal{V}(\mathcal{I}(U))$, the set of points in $\mathbb{C}^d$ where every polynomial that vanishes identically on $U$ also vanishes. A non-empty Zariski open subset $U$ of $\mathbb{C}^d$ is dense in $\mathbb{C}^d$ in the classical topology, and a classical open subset of $\mathbb{C}^d$ (e.g. a ball) is dense in the Zariski topology. The Zariski topology of a variety $X$ in $\mathbb{C}^d$ is induced from that of $\mathbb{C}^d$.

We use the Zariski topology to express the analog of unique factorization of integers for varieties. A variety $X$ is irreducible if cannot be written as a union of proper subvarieties. That is, if $X = Y \cup Z$ with $Y, Z$ subvarieties of $X$, then either $X = Y$ or $X = Z$. A variety $X$ has an irredundant decomposition into irreducible subvarieties, $X = X_1 \cup \cdots \cup X_m$, which is unique in that each $X_i$ is an irreducible subvariety of $X$ and if $i \neq j$, then $X_i \not\subset X_j$. We call the subvarieties $X_1, \ldots, X_m$ the (irreducible) components of $X$.

This decomposition for a hypersurface is equivalent to the factorization of its defining polynomial into irreducible polynomials. For the curve on the left in Figure 12.4 we have,

$$x^3 + x^2y - xy - y^2 \;=\; (x^2 - y)(x + y)\,,$$

showing that its components are the parabola $y = x^2$ and the line $y = -x$. Both $\mathcal{V}(x^3+x^2y-xy-y^2)$

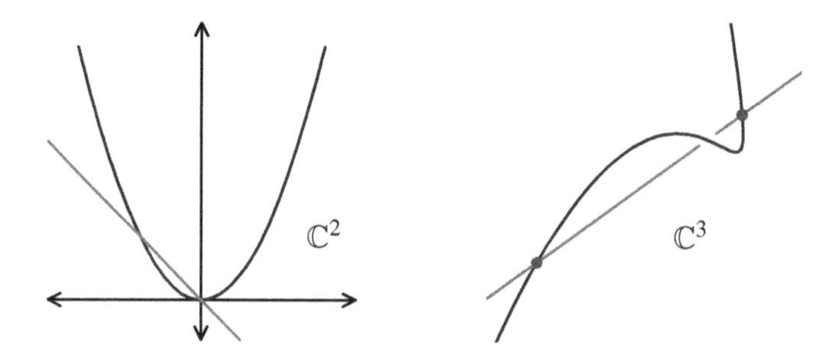


**Figure 12.4**
$\mathcal{V}(x^3+x^2y-xy-y^2)$ and $\mathcal{V}(z-xy, xz-y^2-x^2+y)$.

and $\mathcal{V}(z-xy, xz-y^2-x^2+y)$ are curves with two components, as we see in Figure 12.4.

Zariski open sets are quite large. Any nonempty Zariski open subset $U$ of an irreducible variety $X$ is Zariski dense in $X$. Indeed, $X = \overline{U} \cup (X \smallsetminus U)$, the union of two closed subsets. Since $X \neq X \smallsetminus U$, we have $\overline{U} = X$. In fact, $U$ is dense in the classical topology, and any subset of $X$ that is dense in the classical topology is Zariski dense in $X$.

A property of an irreducible variety $X$ is generic if the set of points where that property holds contains a Zariski open subset of $X$. Generic properties of $X$ hold almost everywhere on $X$ in a very strong sense, as the points of $X$ where they do not hold lie in a proper subvariety of $X$. A point of a variety where a generic property holds is general (with respect to that property).

### 12.4.2 Smooth and Singular Points

Algebraic varieties are not necessarily manifolds, as may be seen in Figures 12.3 and 12.4. However, the set of points where a variety fails to be a manifold is a proper subvariety. To see this, suppose that $X \subset \mathbb{C}^d$ is a variety whose ideal $\mathcal{I}(X)$ has generators $f_1, \ldots, f_s$. At each point $x$ of $X$, the Jacobian matrix $J = (\partial f_i / \partial x_j)_{i=1,\ldots,s}^{j=1,\ldots,d}$ has rank between 0 and $d$. The set $X_i$ of points of $X$ where the rank of $J$ is at most $i$ is a subvariety which is defined by the vanishing of all $(i+1) \times (i+1)$ minors of $J$. If $i$ is the smallest index such that $X_i = X$, so that $X_{i-1} \subsetneq X$, then at every point of $X_{\mathrm{sm}} := X \smallsetminus X_{i-1}$ the Jacobian has rank $i$. Differential geometry informs us that $X_{\mathrm{sm}}$ is a complex manifold of dimension $d-i$.

When $X$ is irreducible, $X_{\mathrm{sm}}$ is the set of smooth points of $X$ and $X_{\mathrm{sing}} := X \smallsetminus X_{\mathrm{sm}}$ is the singular locus of $X$. A point being smooth is a generic property of $X$. The dimension $\dim X$ of an irreducible variety $X$ is the dimension of $X_{\mathrm{sm}}$. When $X$ is reducible, its dimension is the maximum dimension of an irreducible component. The singular locus of a variety $X$ always has smaller dimension than $X$.

For algebraic varieties, dimension has the following properties. If $X$ and $Y$ are subvarieties of $\mathbb{C}^d$ of dimensions $m$ and $n$, respectively, then either $X \cap Y$ is empty or every irreducible component of $X \cap Y$ has at least the expected dimension $m+n-d$. For a general translate $Y'$ of $Y$, $\dim(X \cap Y') = m+n-d$. More precisely, there is a Zariski open subset $U$ of the group $\mathbb{C}^d \rtimes GL(d,\mathbb{C})$ of affine transformations of $\mathbb{C}^d$ such that if $g \in U$ then $X \cap gY$ has dimension $m+n-d$ and is as smooth as possible in that its singular locus is a subset of the union of $X_{\mathrm{sing}} \cap gY$ with $X \cap gY_{\mathrm{sing}}$.

Similarly, Bertini's Theorem states that there is a Zariski open subset $U$ of the set of polynomials of a fixed degree such that for $f \in U$, $X \cap \mathcal{V}(f)$ has dimension $\dim X - 1$ and is as smooth as possible. A consequence of all this is that if $L$ is a general affine linear subspace of dimension $d - \dim X$, then $X \cap L$ is a finite set of points contained in $X_{\mathrm{sm}}$. The number of points is the maximal number of isolated points in any intersection of $X$ with an affine plane of this dimension and is called the degree of $X$. These facts underlie the notion of witness set in numerical algebraic geometry from Section 12.3.

### 12.4.3 Maps

We often have a map $\varphi\colon \mathbb{C}^d \to \mathbb{C}^n$ given by polynomials, and we want to understand the image of a variety $X \subset \mathbb{C}^d$ under this map. Algebraic geometry provides a structure theory for the images of polynomial maps. We begin with an example. Consider the hyperbolic paraboloid $\mathcal{V}(y - xz)$ in $\mathbb{C}^3$ and its projection to the $xy$-plane, which is a polynomial map. This image is the union of all lines through the origin, except for the $y$-axis, $\mathcal{V}(x)$. Figure 12.5 shows both the hyperbolic paraboloid and a schematic of its image in the $xy$-plane. This image is $(\mathbb{C}^2 \smallsetminus \mathcal{V}(x)) \cup \{(0,0)\}$, the union of a Zariski open subset of $\mathbb{C}^2$ and the variety $\{(0,0)\} = \mathcal{V}(x,y)$.

A set is locally closed if it is open in its closure. In the Zariski topology, locally closed sets are Zariski open subsets of some variety. A set is constructible if it is a finite union of locally closed sets. What we saw with the hyperbolic paraboloid is the general case.

**Theorem 12.1** *The image of a constructible set under a polynomial map is constructible.*

Suppose that $X \subset \mathbb{C}^d$ and $\varphi\colon \mathbb{C}^d \to \mathbb{C}^n$ is a polynomial map. Then the closure $\overline{\varphi(X)}$ of the image of $X$ under $\varphi$ is a variety. When $X$ is irreducible, then so is $\overline{\varphi(X)}$. (The inverse image of a decomposition $\overline{\varphi(X)} = Y \cup Z$ under $\varphi$ is a decomposition of $X$.) Theorem 12.1 then implies that $\varphi(X)$ contains a nonempty Zariski open and therefore a Zariski dense subset of $\overline{\varphi(X)}$. Applying this to each irreducible component of a general variety $X \subset \mathbb{C}^d$ implies that each irreducible component of $\overline{\varphi(X)}$ has a dense open subset contained in the image $\varphi(X)$.

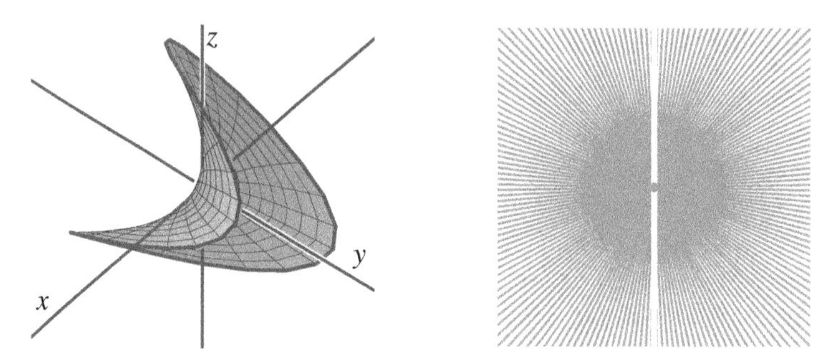


**Figure 12.5**
The hyperbolic paraboloid and its image in the plane.

## 12.5 Real Algebraic Geometry

Real algebraic geometry predates its complex cousin, having its roots in Cartesian analytic geometry in $\mathbb{R}^2$. Its importance for applications is evident, and applications have driven some of its theoretical development. A comprehensive treatment of the subject is given in the classic treatise of Bochnak, Coste, and Roy [4]. Real algebraic geometry has long enjoyed links to computer science through fundamental questions of complexity. There are also many specialized algorithms for treating real algebraic sets. The equally classic book by Basu, Pollack, and Roy [1] covers this landscape of complexity and algorithms.

**Real algebraic geometry:** Study of real solutions to systems of polynomial equations.

**Real algebraic variety:** A variety defined by real polynomials; its subset of real points.

**$X(\mathbb{R})$:** The real points of a variety $X$ defined by real polynomials.

**Semi-algebraic set:** A set defined by a system of polynomial equations and inequalities.

**Tarski-Seidenberg Theorem:** Quantifier elimination for semi-algebraic sets; the image of a semi-algebraic set under a polynomial map is a semi-algebraic set.

**Cylindrical algebraic decomposition:** Algorithm to decompose a semi-algebraic set into a cell complex of semi-algebraic cells adapted to quantifier elimination.

**Positivestellensatz:** Real-algebraic version of Nullstellensatz.

### 12.5.1 Algebraic Relaxation

A complex variety $X \subset \mathbb{C}^d$ defined by real polynomials has a subset $X(\mathbb{R}) := X \cap \mathbb{R}^d$ of real points. Both $X$ and (more commonly) $X(\mathbb{R})$ are referred to as real algebraic varieties. In the Introduction, we claimed that it is fruitful to study a real algebraic variety $X(\mathbb{R})$ by first understanding the complex variety $X$, and then asking about $X(\mathbb{R})$. We consider studying the complex variety $X$ to be an algebraic relaxation of the problem of studying the real variety. The fundamental reason this approach is often successful is the following result.

**Theorem 12.2** *Let $X \subset \mathbb{C}^d$ be an irreducible variety defined by real polynomials. If $X$ has a smooth real point, then $X(\mathbb{R})$ is Zariski dense in $X$.*

To paraphrase, suppose that $X \subset \mathbb{C}^d$ is an irreducible variety defined by real polynomials. If $X$ has a smooth real point, then all algebraic and geometric information about $X$ is already contained in $X(\mathbb{R})$, and vice-versa.

The reader may have noted that we used pictures of the real algebraic variety $X(\mathbb{R})$ to illustrate properties of the complex variety $X$ in most of our figures. Theorem 12.2 justifies this sleight of hand.

A proof of Theorem 12.2 begins by noting that when $X$ has a smooth real point, then the set of smooth real points $X_{\mathrm{sm}}(\mathbb{R})$ forms a real manifold of dimension $\dim X$. Consequently, the derivatives at a point of $X_{\mathrm{sm}}(\mathbb{R})$ of a polynomial $f$ restricted to $X$ are determined by the restriction of $f$ to $X_{\mathrm{sm}}(\mathbb{R})$, which implies that if a polynomial vanishes on $X_{\mathrm{sm}}(\mathbb{R})$, then it vanishes on $X$.

The two cones $\mathcal{V}(x^2+y^2-z^2)$ and $\mathcal{V}(x^2+y^2+z^2)$ serve to illustrate the hypotheses of Theorem 12.2. In $\mathbb{C}^3$, these cones are isomorphic to each other under the substitution $z \mapsto \sqrt{-1}z$. In $\mathbb{R}^3$, the first is the familiar double cone, with real smooth points the complement of the origin, while the other is the single isolated (and hence singular) point $\{(0,0,0)\}$. We display the double cone on the left in Figure 12.6. On the right is the Whitney umbrella. This is the Zariski closure of the image

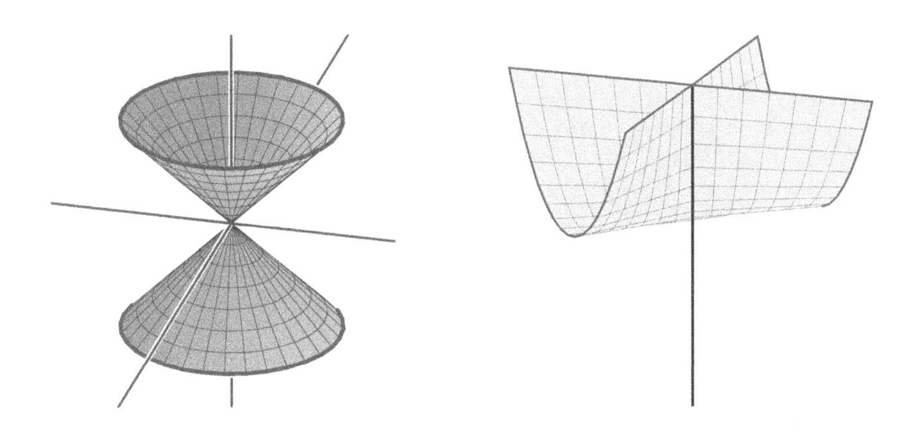

**Figure 12.6**
Double cone and Whitney umbrella in $\mathbb{R}^3$.

of $\mathbb{R}^2$ under the map $(u,v) \mapsto (uv, v, u^2)$, and is defined by the polynomial $x^2 - y^2z$. The image of $\mathbb{R}^2$ is the canopy of the umbrella. Its handle is the image of the imaginary part of the $u$-axis of $\mathbb{C}^2$, the points $(\mathbb{R}\sqrt{-1},0)$. The Whitney umbrella is singular along the $z$-axis, which is evident as the canopy has self-intersection along the positive $z$-axis. This singularity along the negative $z$-axis is implied by its having local dimension 1: were it smooth, it would have local dimension 2.

Theorem 12.2 also leads to the following cautionary example. The cubic $y^2-x^3+x$ is irreducible and its set of complex zeroes is a torus (with one point removed). Its set of real zeroes has two path-connected components. Each is Zariski-dense in the complex cubic. Thus the property $x \leq 0$ which holds on the oval is not a generic property, even though it holds on a Zariski dense subset, which is neither Zariski open or closed.

### 12.5.2 Semi-Algebraic Sets

The image of $\mathbb{R}^2$ in the Whitney umbrella is only its canopy, and not the handle. More interestingly, the image under projection to the $xy$-plane of the sphere $\mathcal{V}(x^2+y^2+z^2-1)$ of radius 1 and center $(0,0,0)$ is the unit disc $\{(x,y) \in \mathbb{R}^2 \mid 1-x^2-y^2 \geq 0\}$. Similarly, by the quadratic formula, the

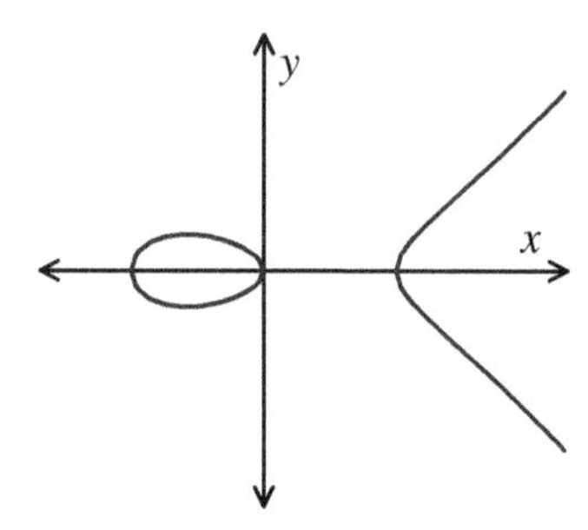


**Figure 12.7**
Reprise: cubic plane curve.

polynomial $x^2+bx+c$ in $x$ has a real root if and only if $b^2-4c \geq 0$. Thus, if we project the surface $\mathcal{V}(x^2+bx+c)$ to the $bc$-plane, its image is $\{(b,c) \in \mathbb{R}^2 \mid b^2-4c \geq 0\}$. We illustrate these examples in Figure 12.8. They show that the image of an irreducible real variety under a polynomial map need

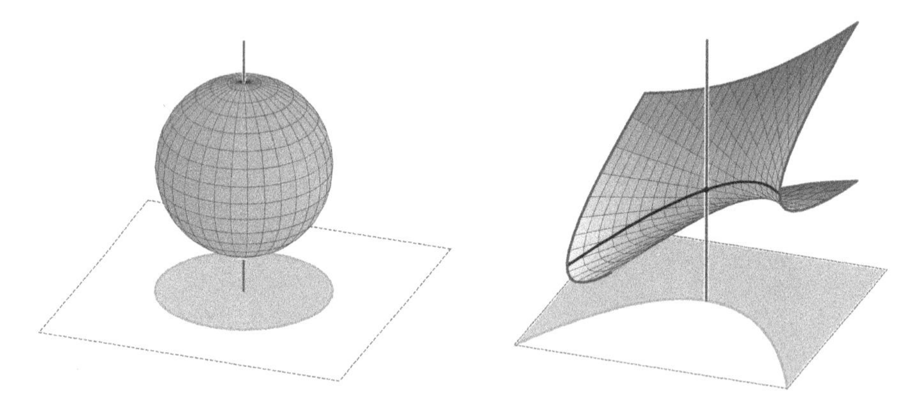

**Figure 12.8**
Projection of the sphere and the quadratic formula.

not be dense in the image variety, even though it will be dense in the Zariski topology. We describe the image of a real variety by enlarging our notion of a real algebraic set.

A subset $V$ of $\mathbb{R}^d$ is a semi-algebraic set if it is the union of sets defined by systems of polynomial equations and polynomial inequalities. Technically, a set $V$ is semi-algebraic if it is given by a formula in disjunctive normal form, whose elementary formulas are of the form $f(x)=0$ or $f(x)>0$, where $f$ is a polynomial with real coefficients. This is equivalent to $V$ being given by a formula that involves only the logical operations 'and' and 'or' and elementary formulas $f(x)=0$ and $f(x)>0$. Tarski showed that the image of a real variety under a polynomial map is a semi-algebraic set [20, 21], and Seidenberg gave a more algebraic proof [16].

**Theorem 12.3 (Tarski-Seidenberg)** *The image of a semi-algebraic set under a polynomial map is a semi-algebraic set.*

The astute reader will note that our definition of a semi-algebraic set was in terms of propositional logic, and should not be surprised that Tarski was a great logician. The Tarski-Seidenberg Theorem is known in logic as quantifier elimination: its main step is a coordinate projection, which is equivalent to eliminating existential quantifiers.

**Example 40** *We give a simple application from rigidity theory. Let $G$ be a graph with $n$ vertices $V$ and $m$ edges. An embedding of $G$ into $\mathbb{R}^d$ is simply a map $\rho\colon V \to \mathbb{R}^d$, and thus the space of embeddings is identified with $\mathbb{R}^{nd}$. The squared length of each edge of $G$ in an embedding $\rho$ defines a map $f\colon \mathbb{R}^{nd} \to \mathbb{R}^m$ with image some set $M$. By the Tarski-Seidenberg Theorem, $M$ is a semi-algebraic set and so it contains an open subset of the real points of its Zariski closure, $\overline{M}$. By Sard's Theorem, $M$ contains a smooth point of its Zariski closure, and thus $M$ has an open (and dense in the classical topology) set of smooth points. These are images of embeddings where the Jacobian of $f$ (which is the rigidity matrix) has maximal rank (among all embeddings).* ◇

**Remark 12.1** Semi-algebraic sets are needed to describe more general frameworks involving cables and struts. In an embedding, the length of an edge corresponding to a cable is bounded above by the length of that cable, and the length of an edge corresponding to a strut is bounded below by the length of that strut. In either case, inequalities are necessary to describe possible configurations.

The Tarski-Seidenberg Theorem is a structure theorem for images of real algebraic varieties under polynomial maps. Much later, this existential result was refined by Collins, who gave an effective version of quantifier elimination for semi-algebraic sets, called cylindrical algebraic decomposition [6]. This uses successive coordinate projections to build a description of a semi-algebraic set as a cell complex whose cells are semi-algebraic sets. While implemented in software [5], it suffers more than many algorithms in this subject from the curse of complexity and is most effective in low ($d \lesssim 3$) dimensions. There are however several software implementations of cylindrical algebraic decomposition. In the worst case, the complexity of a cylindrical algebraic decomposition is doubly exponential in $d$, and this is achieved for general real varieties. A focus of [1] and subsequent work is on stable algorithms with better performance to compute different representations of a semi-algebraic set.

### 12.5.3 Certificates

We close with the Positivestellensatz of Stengele [19], which states that a semi-algebraic set is empty if and only if there is a certificate of its emptiness having a particular form. A polynomial $\sigma$ is a sum of squares if it may be written as a sum of squares of polynomials with real coefficients. Such a polynomial takes only nonnegative values on $\mathbb{R}^d$. We may use semidefinite programming to determine if a polynomial is a sum of squares.

**Theorem 12.4 (Positivestellensatz)** *Suppose that $f_1,\dots,f_r$, $g_1,\dots,g_s$, and $h$ are real polynomials. Then the semi-algebraic set*

$$\{x \in \mathbb{R}^d \mid f_i(x) = 0,\ i = 1,\dots,r \text{ and } g_j(x) \geq 0,\ k = 1,\dots,s \text{ and } h(x) \neq 0\} \tag{12.1}$$

*is empty if and only if there exist polynomials $k_1,\dots,k_r$, sums of squares $\sigma_0,\dots,\sigma_s$, and a positive integer $n$ such that*

$$0 = f_1k_1 + \cdots + f_rk_r + \sigma_0 + g_1\sigma_1 + \cdots + g_s\sigma_s + h^{2n}. \tag{12.2}$$

**Remark 12.2** To see that (12.2) is a sufficient condition for emptiness, suppose that $x$ lies in the set (12.1), and then evaluate the expression (12.2) at $x$. The terms involving $f_i$ vanish, those involving $g_j$ are nonnegative, and $h(x)^{2n} > 0$, which is a contradiction. If $h$ does not appear in a description (12.1), then we take $h = 1$ in (12.2).

# References

[1] Saugata Basu, Richard Pollack, and Marie-Françoise Roy. *Algorithms in real algebraic geometry*, volume 10 of *Algorithms and Computation in Mathematics*. Springer-Verlag, Berlin, second edition, 2006.

[2] Daniel J. Bates, Jonathan D. Hauenstein, Andrew J. Sommese, and Charles W. Wampler. Bertini: Software for numerical algebraic geometry. `www.nd.edu/~sommese/bertini`.

[3] Daniel J. Bates, Jonathan D. Hauenstein, Andrew J. Sommese, and Charles W. Wampler. *Numerically solving polynomial systems with Bertini*, volume 25 of *Software, Environments, and Tools*. Society for Industrial and Applied Mathematics (SIAM), Philadelphia, PA, 2013.

[4] Jacek Bochnak, Michel Coste, and Marie-Françoise Roy. *Real algebraic geometry*, volume 36 of *Ergebnisse der Mathematik und ihrer Grenzgebiete (3)*. Springer-Verlag, Berlin, 1998.

[5] Christopher W Brown. Qepcad b: a program for computing with semi-algebraic sets using cads. *ACM SIGSAM Bulletin*, 37(4):97–108, 2003.

[6] George E. Collins. Quantifier elimination for real closed fields by cylindrical algebraic decomposition. In *Automata theory and formal languages (Second GI Conf., Kaiserslautern, 1975)*, pages 134–183. Lecture Notes in Comput. Sci., Vol. 33. Springer, Berlin, 1975.

[7] David Cox, John Little, and Donal O'Shea. *Ideals, varieties, and algorithms*. Undergraduate Texts in Mathematics. Springer, New York, third edition, 2007.

[8] Wolfram Decker, Gert-Martin Greuel, Gerhard Pfister, and Hans Schönemann. SINGULAR 4-0-2 — A computer algebra system for polynomial computations. `www.singular.uni-kl.de`, 2015.

[9] Wolfram Decker and Christoph Lossen. *Computing in algebraic geometry*, volume 16 of *Algorithms and Computation in Mathematics*. Springer-Verlag, Berlin, 2006.

[10] The Sage Developers. *Sage Mathematics Software*, 2015. `www.sagemath.org`.

[11] David Eisenbud, Daniel R. Grayson, Michael Stillman, and Bernd Sturmfels, editors. *Computations in algebraic geometry with Macaulay 2*, volume 8 of *Algorithms and Computation in Mathematics*. Springer-Verlag, Berlin, 2002.

[12] Daniel R. Grayson and Michael E. Stillman. Macaulay2, a software system for research in algebraic geometry. `www.math.uiuc.edu/Macaulay2/`.

[13] Jonathan D. Hauenstein and Frank Sottile. Algorithm 921: alphaCertified: certifying solutions to polynomial systems. *ACM Trans. Math. Software*, 38(4):Art. ID 28, 20, 2012.

[14] Robert Krone and Anton Leykin. NAG4M2: Numerical algebraic geometry for Macaulay 2. `people.math.gatech.edu/~aleykin3/NAG4M2`.

[15] Alexander Morgan. *Solving polynomial systems using continuation for engineering and scientific problems*. Prentice Hall Inc., Englewood Cliffs, NJ, 1987.

[16] Abraham Seidenberg. A new decision method for elementary algebra. *Ann. of Math. (2)*, 60:365–374, 1954.

[17] Stephen Smale. Newton's method estimates from data at one point. In *The merging of disciplines: new directions in pure, applied, and computational mathematics (Laramie, Wyo., 1985)*, pages 185–196. Springer, New York, 1986.

[18] Andrew J. Sommese and Charles W. Wampler, II. *The numerical solution of systems of polynomials.* World Scientific Publishing Co. Pte. Ltd., Hackensack, NJ, 2005.

[19] Gilbert Stengle. A nullstellensatz and a positivstellensatz in semialgebraic geometry. *Math. Ann.*, 207:87–97, 1974.

[20] Alfred Tarski. *A Decision Method for Elementary Algebra and Geometry.* RAND Corporation, Santa Monica, CA., 1948.

[21] Alfred Tarski. A decision method for elementary algebra and geometry. In *Quantifier elimination and cylindrical algebraic decomposition (Linz, 1993)*, Texts Monogr. Symbol. Comput., pages 24–84. Springer, Vienna, 1998.

[22] Jan Verschelde. Algorithm 795: PHCpack: general-purpose solver for polynomial systems by homotopy. *ACM Trans. Math. Software*, 25(2):251–276, 1999.

# Part III

# Geometric Rigidty

# Chapter 13

# *Polyhedra in 3-Space*

**Brigitte Servatius**
*WPI*

**CONTENTS**



## 13.1 Euler's Conjecture

We give a brief history of rigidity results on polyhedra and point to some open problems. In 1766, Euler conjectured: "A closed spatial figure allows no changes, as long as it is not ripped apart," see [12]. He ends the paragraph describing the problem with the sentence:

> Interim patet hemispherii figuram certe esse mutabilem; cujusmodi autem mutationes recipere possit, problema videtur difficillimum

indicating that Euler's approach to the problem of finding motion in the sphere, considering how to match the motions independently obtainable on the upper and lower hemispheres, anticipated the approaches of Cauchy, Bricard, Connelly, Steffen, and others.

## 13.2 Cauchy's Theorem

A first step toward proving the Euler Conjecture and the first major result published in rigidity theory was a theorem of Cauchy in the early 19'th century:

**Theorem 13.1 (A. L. Cauchy, 1813 [5])** *If there is an isometry between the surfaces of two strictly convex polyhedra which is an isometry on each of the faces, then the two polyhedra are congruent.*

By a strictly convex polyhedron, we mean a polyhedron which has the property that, for each vertex, there exists a plane which intersects the polyhedron at that vertex and no other point of the polyhedron. Cauchy's proof involved a mixture of topological arguments and arguments from elementary geometry. Unfortunately, there were several minor errors in the proof; the first complete proof was published by Steinitz and Rademacher in 1934 [21]. A simpler complete proof of a slight generalization of Cauchy's Theorem was given by Alexandrov in 1950 [1].

It is easy to see that the convexity condition cannot be dropped from the hypothesis of Cauchy's Theorem. Consider the pair of polyhedra constructed as follows. Take a cube and a pyramid with base congruent to a face of the cube and with height less than the height of the cube. Remove the base from the pyramid and remove one face from the cube. Then join these two surfaces along their boundaries, once with the apex of the pyramid pointing out and once with it pointing in. There is an obvious homeomorphism between these two surfaces which restricts to an isometry on each face; but these two polyhedra are not congruent. See Figure 13.1.

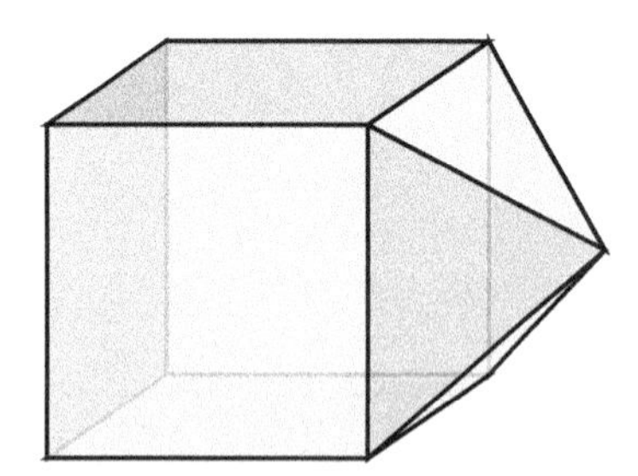

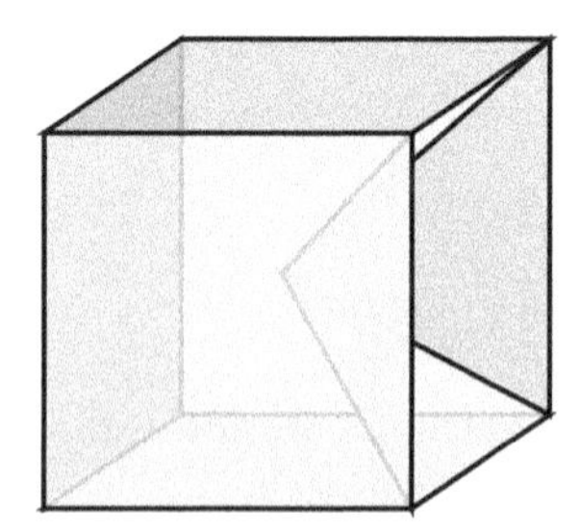

**Figure 13.1**
Two isomorphic but non-congruent polyhedra with corresponding faces congruent.

The boundary of the 3-dimensional polyhedron of Cauchy's theorem is the set of 0, 1, and 2-dimensional faces, the 2-*skeleton*. The following corollary is an immediate consequence of Cauchy's Theorem.

**Corollary 13.2** *The* 2*-skeleton of a strictly convex polyhedron in* 3*-space is rigid.*

## 13.3 Co-Dimension 2 Results – Bricard Octahedra

With the Euler Conjecture in mind, it is natural to ask if "strictly convex" may be dropped from the hypothesis of the corollary.

In 1897 Bricard published a complete study of the rigidity properties of the octahedron. The main result of that study can be stated as follows.

**Theorem 13.3 (R. Bricard, 1897, [4])** *The* 2*-skeleton of any polyhedral embedding of the octahedron in* 3*-space is rigid. However, the* 1*-skeleton of the octahedron has a non-rigid embedding in* 3*-space.*

At first glance the two parts of this theorem seem to be in contradiction to one another: since the faces of the octahedron are all triangular, deleting them could not change a rigid plate and hinge framework into a non-rigid rod and joint framework, see Chapter 13. It would then seem to follow from the first part of the theorem that any embedding of the 1-skeleton of the octahedron would also be rigid, however, not every embedding of the 1-skeleton extends to an embedding of the 2-skeleton. Non-rigid embeddings of the 1-skeleton of the octahedron are easy to describe. Note first that any quadrilateral framework in $\mathbb{R}^3$ whose opposite sides are parallel must have a 180° axis of symmetry. Consider a non-regular pyramid with, say, a square base and assume that the perpendicular projection of the apex onto the base does not lie any line of symmetry of the square; see Figure 13.2a, b. This last condition is just to ensure that the added bars do not intersect any of the

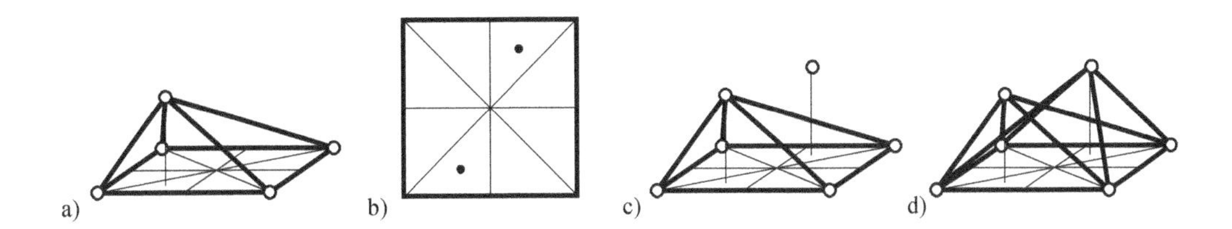


**Figure 13.2**
How to embed an octahedral framework in $\mathbb{R}^3$ such that it has a motion.

original bars in their interior. Now erect a second pyramid on the same base which is isometric to the first under the 180° rotation about the line perpendicular to the base at its center; see Figure 13.2c, d. Note that both apices are on the same side of their common base. The reason that this octahedral 1-skeleton is non-rigid is that the 180° rotation of the base framework will always interchange the pyramids, and the motion of the pyramid extends to the larger framework. In other words both pyramids produce the same family of distortions of the rectangular base as they flex.

To understand the relationship between the two parts of Bricard's Theorem, consider the 1-skeleton of the octahedron embedded as just described. Since the faces of the octahedron are triangles, adding these faces to this 1-skeleton can be done without destroying the planarity of those faces and without limiting any motions of the structure. We thus have a copy of the 2-skeleton of the octahedron in 3-space which is not rigid. However, it is clearly not embedded: it has self-intersections which always involve interior points of some triangles.

The next new result on the Euler Conjecture did not occur until many years later when Alexandrov gave his generalization of Cauchy's Theorem.

**Theorem 13.4 (A. D. Alexandrov, 1950 [1])** *If vertices are inserted in the edges of a strictly convex polyhedron and the faces are triangulated, then the* 1*-skeleton of the resulting polyhedron is infinitesimally rigid.*

This result has the following corollary:

**Corollary 13.5** *If a convex polyhedron in* 3*-space has the property that the collection of faces containing a given vertex do not all lie in the same plane, then the* 2*-skeleton of that polyhedron is infinitesimally rigid. See Figure 13.3.*

In 1978, Asimow and Roth showed that the triangulation condition in the hypothesis of Alexandrov's Theorem is necessary.

**Theorem 13.6 (L. Asimow and B. Roth, 1978 [3])** *The* 1*-skeleton of a strictly convex polyhedron embedded in* 3*-space which has at least one non-triangular face is not rigid.*

Asimow and Roth's proof of this theorem has two parts. The first combinatorial part uses Euler's formula for the sphere to show that the number of edges in such a 1-skeleton is less than $3v-6$, the number of edges in a triangulation of a sphere with with $v$ vertices. Then, by dimension arguments

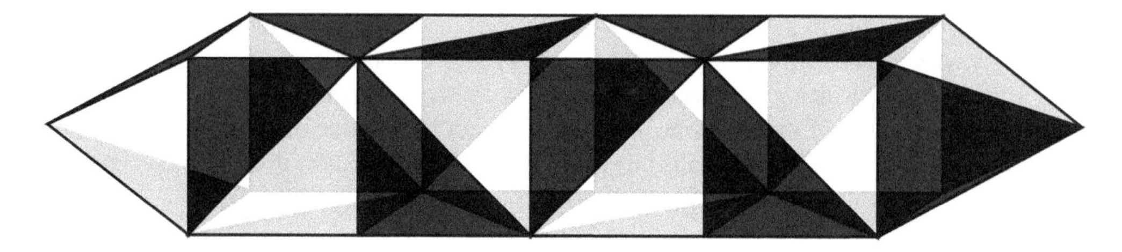

**Figure 13.3**
This convex but not strictly convex polyhedron has an infinitesimally rigid 2-skeleton by the Corollary to Alexandrov's Theorem. Since all its faces are triangles, the 1-skeleton is also infinitesimally rigid.

they conclude that there are not enough constraints to make this framework rigid unless it represents a singular point on the algebraic variety $C$ derived from the edge length constraints, see Chapter 18. In the second, geometric part of their proof, Asimow and Roth show that a strictly convex framework cannot correspond to a singular point.

To see that strict convexity is essential to their result, consider a tetrahedron with equilateral faces. Insert, centered in one face, a smaller equilateral triangle with sides parallel to the sides of the face. Now, join by edges the corresponding vertices; see Figure 13.4.

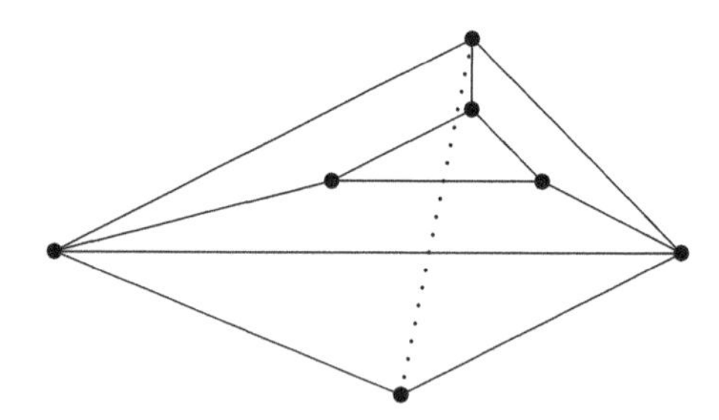

**Figure 13.4**
This is a convex non-triangulated polyhedron which is rigid in $\mathbb{R}^3$, but not infinitesimally rigid as a panel and hinge structure. The polyhedron has three non-triangular faces, yet the 1-skeleton is easily seen to be rigid (second order) with the small triangle held in place by tension.

## 13.4 Polyhedral Surfaces

Stoker [22] extends Cauchy's rigidity theorem to some more general 3-dimensional polyhedra, including special types of non-convex ones and polyhedra with boundary. Cauchy's original proof is refined, together with some lemmas of Alexandrov to obtain the results. Stoker views the underlying problem as that of determining when two polyhedra are *isogonal*, that is, when corresponding faces are parallel. Additional conditions are then imposed to ensure congruence.

**Conjecture 2 (J. J. Stoker)** *Two combinatorially equivalent polyhedra with equal corresponding dihedral angles are isogonal.*

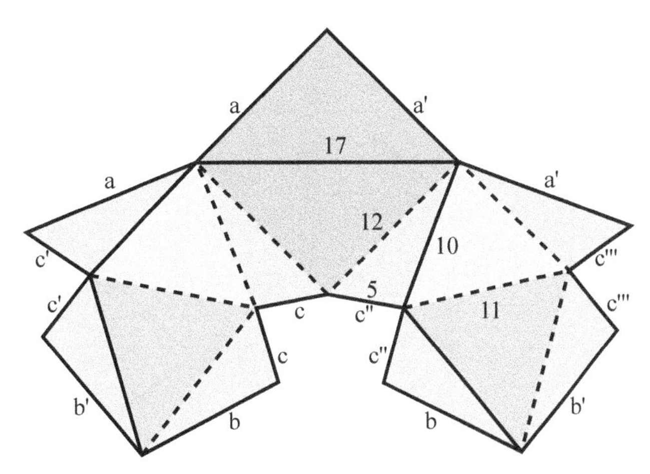


**Figure 13.5**
A net for Steffen's construction of a flexible sphere with only 9 vertices, 14 triangles in 4 congruence classes, and 21 edges. The valley folds are dashed.

Karcher [18] gives an affirmative answer to Stoker's problem in some special cases, in particular for simple 3-polytopes.

The Euler Conjecture was finally settled in 1977, but just prior to that a result was published which greatly extended the set of configurations for which the Euler Conjecture was known to be valid.

**Theorem 13.7 (H. Gluck, 1975 [15])** *Every closed simply connected polyhedral surface embedded in 3-space is generically rigid.*

Gluck's Theorem tells us that the Euler Conjecture is almost always true for closed simply connected polyhedral surfaces. Just two years later, a counterexample to the Euler Conjecture was found by R. Connelly [7]. The counterexample is based on Bricard's flexible octahedron and, of course, it is non-convex and non-generic. Topologically Connelly's surface is a sphere, thus showing that Gluck's Theorem is best possible. Connelly's flexible sphere, described in [8], has 11 vertices, 18 triangular faces and 27 edges. Klaus Steffen gave a smaller example of a "Connelly sphere" with only 9 vertices, and it is very often built according to Steffen's original instructions communicated to Connelly in a handwritten note in 1977. A net for Steffen's example is given in Figure 13.5 which may be somewhat challenging to assemble.

Connelly showed that the volume enclosed by his flexible spheres remains constant under the motion. The conjecture, by Connelly and Sullivan, that every orientable closed polyhedral surface flexes with constant volume [9] is known as the *bellows conjecture*, since it implies that a mathematical bellows is impossible to construct.

The motion of Steffen's polyhedron is not so easy to observe, he suggests to cut a hole in one of the triangles to watch the small motion that stops when self intersection occurs. Euler sensed the difficulty of finding a truly flexible sphere and Connelly's spheres are ingenious examples. One may, however, easily create families of infinitesimally flexible polyhedra. For example, starting with a regular tetrahedron of Figure 13.6, which in the illustration is placed in a unit cube for visual reference, then subdivide the top and bottom edges and displace by one unit as shown. The middle framework has an infinitesimal motion which may be extended to a new point, placed symmetrically and connected to all the vertices of one of the four quadrilaterals. If the point is placed at $(p,p,t)$, and has infinitesimal motion $(p',p',t')$, then the parameters must satisfy

$$t' = 6 - t, \quad ss' = (t-2)^2, \quad (s-1)(s'-1) = (t+1)(t-3)$$

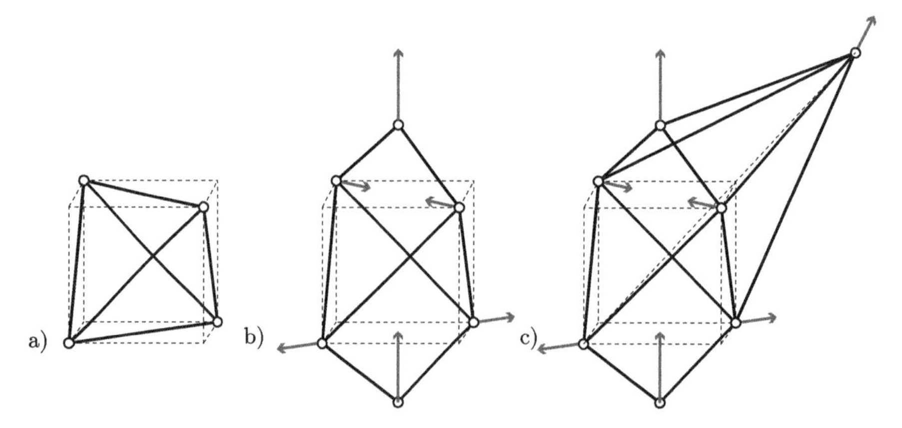


**Figure 13.6**
(a) Tetrahedron with vertices $(1,1,-1)$, $(1,-1,1)$, $(-1,1,-1)$, $(1,1,-1)$; (b) adding $(0,0,2)$ and $(0,0,-2)$. A symmetric infinitesimal motion with the top and bottom velocities both $(0,0,2)$ and the middle having zero vertical component; (c) a horn placed at $(p,p,p)$ with $p=(\frac{1+\sqrt{5}}{2})^2$.

If one places a point symmetrically for each of the other four quadrilaterals, then this yields a triangulated sphere, see Figure 13.7, which has dihedral symmetry and is infinitesimally flexible. In [24]

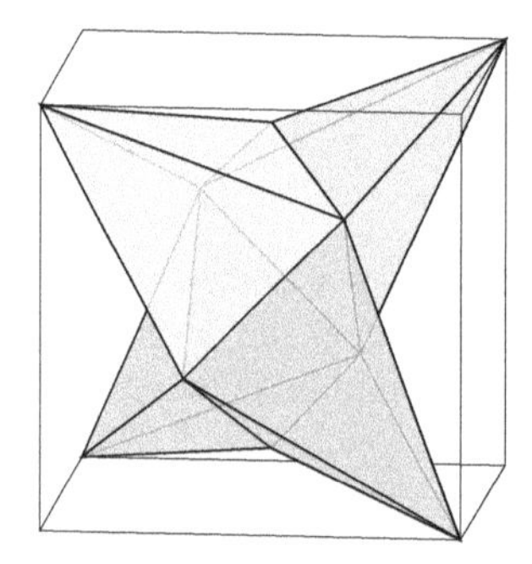

**Figure 13.7**
A four-horn. If one pinches the bottom horns infinitesimally together, the top two are infinitesimally pulled apart.

Wunderlich and Schwabe construct a family of polyhedra, called four-horns, which are combinatorially isomorphic to the one in Figure 13.7. Four-horns have congruent faces and are "almost movable" in the sense that each four-horn has three non-congruent embeddings, see Figure 13.8, and the "snapping" of the model from one form to the other requires just a small deformation of the edges, for some examples less than 1%. Two of the embeddings are perfectly flat, while the third embedding encloses positive volume. Cycling through the fake motion involves connecting four true conformations, since the non-flat conformation will be transformed into its mirror image, provided we allow the panels to pass through one another in the flat position. Like so many examples considered in this section, four-horns are constructed just as Euler suggested in the Latin sentence of the introduction, by matching a real motion of the top two horns with those of the bottom. A physical model appears more flexible than Steffen's polyhedron and feels like a genuine counterexample to the bellows conjecture.

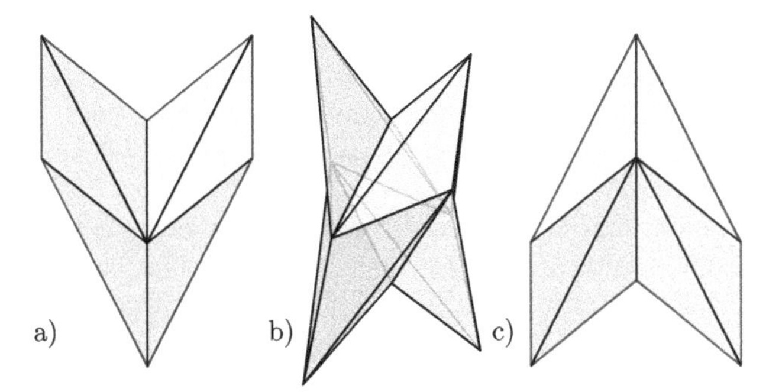


**Figure 13.8**
The three forms of the Wunderlich-Schwabe four-horn.

A similar phenomenon of apparent model flexibility is observed in the Jessen Icosahedron [17], see Figure 13.9. The Jessen icosahedron starts with an irregular icosahedron drawn in the unit cube.

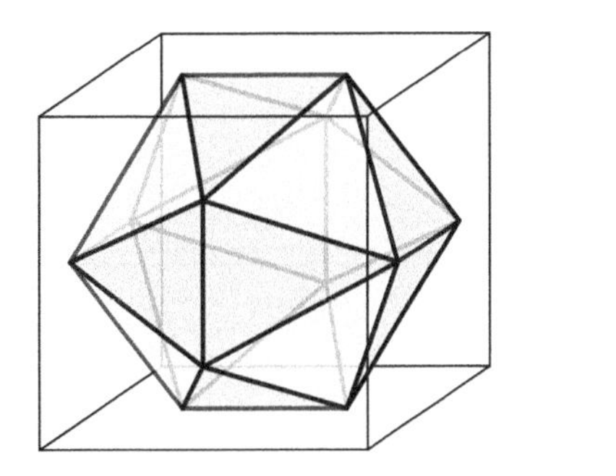

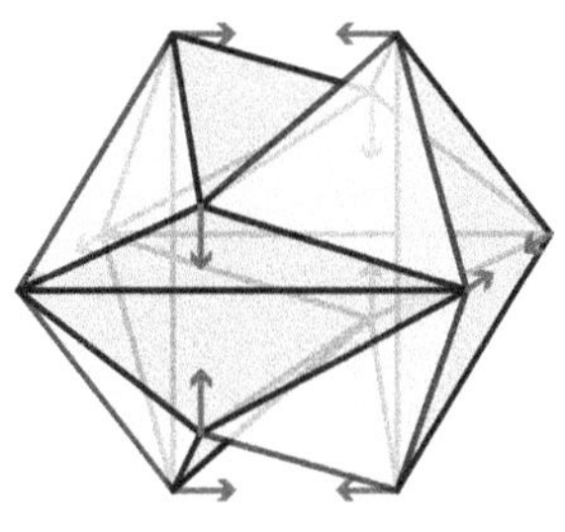

**Figure 13.9**
The Jessen icosahedron.

It may be taken to have vertices $(1,c,0)$ and all 12 permutations thereof by the rotational group of one of the tetrahedra embedded in the unit cube, just as for the regular icosahedron, except that instead of taking $c$ to be the golden ration, Jessen takes $c = .5$. The resulting irregular icosahedron is vertex but not face transitive, and has two congruence classes of triangles, equilateral at the corners of the cube, and six pairs of isosceles triangles, one at each cube face. To form the Jessen icosahedron, the isosceles pairs are removed and replaced with the pair of isosceles triangles formed by adding the other diagonal to each quadrilateral boundary. These six "beaks" have a right angled valley fold, and indeed all angles of the non-convex icosahedron are right angles, and there is an infinitesimal motion with velocities along the deleted edges, simultaneously closing all the beaks. It is interesting to note that this motion, infinitesimally closing the breaks and screwing-in the equilateral triangles, half clockwise and half counter-clockwise, obviously decreases infinitesimally the enclosed volume, so the Jessen icosahedron, unlike the four-horn, is an infinitesimal bellows. The Jessen icosahedron takes a special place in the family of Douday Shaddocks [11], which Douady calls shaddocks with six beaks. Combinatorially, members of the shaddock family are isomorphic to icosahedra. Just as the Wunderlich-Schwabe model is rigid and only infinitesimally flexible, all members of the Shaddock family are rigid, and only the Jessen Icosahedron is infinitesimally flexible. In [16] it is shown that a mere .01% change in edge length in the Jessen Icosahedron produces a change in the dihedral angles of almost 10%, which is a good explanation of the observed flexibility:

The change in the faces goes unnoticed, while the change in the angles is apparent in manipulating the model. As a second explanation, the beaks may be replaced by broken beaks, thereby changing the combinatorial properties through the introduction of new crease-lines, but these new crease-lines are arbitrarily close to the original edges. The new object is truly flexible, and again optically indistinguishable from the original.

By putting three 2-horns together Wunderlich and Schwabe obtain a 6-horn, which they show to be second order flexible. However, $2n$-horns for $n > 3$ are shown to be rigid.

The bellows conjecture was proven by Sabitov [19] for spherical polyhedra in 1995 and for the general case of triangulated oriented surfaces by Connelly, Sabitov, and Walz [6] in 1997. A short proof of the Bellows Theorem is given in [20], together with a good overview of related research following the original proof and many open problems.

It is well known that two embedded polyhedra with the same volume and the same Dehn invariants are scissors congruent [10, 23]. The strong bellows conjecture [2] states that if there is a continuous flex from one, possibly singular, polyhedron $P$ to another, possibly singular, polyhedron $Q$, then $P$ and $Q$ have the same Dehn invariants.

Gaĭfullin, in [14], proves the bellows conjecture for all odd dimensional Lobachevsky spaces. In [13] he finds, for all $n \geq 2$, embedded flexible cross polytopes in the unit sphere $\mathbb{S}^n$ with non-constant volume and formulates the modified bellows conjecture stating that some vertices of a flexible polyhedron in $\mathbb{S}^n$ can be replaced with their antipodes so that the generalized volume of the resulting flexible polyhedron will remain constant during the flex.

## Glossary

***n*-skeleton:** Then $n$ skeleton of a cell complex $n$ is the subcomplex consisting of all cells of dimension $n$ or less.

**Bricard octahedra:** One of the collection of bar and joint frameworks whose graph is that of the octahedron, and which flexes in $\mathbb{R}^3$.

**Connelly sphere:** A triangulated sphere, embedded in $\mathbb{R}^3$ which is flexible. It is necessarily non-convex.

**Dehn invariant:** An algebraic measurement assigned to a three-dimensional polyhedron in $\mathbb{R}^3$ which is defined using the dihedral angles, see [10]. It is invariant under the operation of dissecting the polyhedron into a finite number of pieces and reassembling them to form a new polyhedron, which is therefore *scissors congruent* to the first.

**isogonal:** Relation between two combinatorially isomorphic polyherda whose corresponding faces are parallel.

**Jessen icosahedron:** A construction of an infinitesimally flexible sphere whose 1-skeleton is the graph of the icosahedron.

**polyhedral surface:** A 2-dimensional cell complex homeomorphic to a sphere and embedded in $\mathbb{R}^3$ such that the faces are plane polygons.

**4-horn:** One of a collection of infinitesimally flexible polyhedral surfaces due to Wunderlich.

**scissors congruent:** Two polyhedra scissors-congruent if the first can be cut into finitely many polyhedral pieces that can be reassembled to yield the second. The Dehn invariant is invariant under scissors congruence.

**strictly convex polyhedron:** A polyhedron each of whose vertices is the sole point of the polyhedron which intersects some plane.

# References


[1] A. D. Alexandrov. *Convex polyhedra*. Springer Monographs in Mathematics. Springer-Verlag, Berlin, 2005. Translated from the 1950 Russian edition by N. S. Dairbekov, S. S. Kutateladze and A. B. Sossinsky, With comments and bibliography by V. A. Zalgaller and appendices by L. A. Shor and Yu. A. Volkov.

[2] Victor Alexandrov and Robert Connelly. Flexible suspensions with a hexagonal equator. *Illinois J. Math.*, 55(1):127–155 (2012), 2011.

[3] L. Asimow and B. Roth. The rigidity of graphs. *Trans. Amer. Math. Soc.*, 245:279–289, 1978.

[4] Raoul Bricard. Mémoire sur la théorie de l'octahèdre articulé. *J. math. pures et appliquées*, 3:113–150, 1897.

[5] A.L. Cauchy. *Recherches sur les (polygones et les) polyèdres: .... Mémoire I.* 1813.

[6] R. Connelly, I. Sabitov, and A. Walz. The bellows conjecture. *Beiträge Algebra Geom.*, 38(1):1–10, 1997.

[7] Robert Connelly. A counterexample to the rigidity conjecture for polyhedra. *Inst. Hautes Études Sci. Publ. Math.*, (47):333–338, 1977.

[8] Robert Connelly. A flexible sphere. *Math. Intelligencer*, 1(3):130–131, 1978/79.

[9] Robert Connelly. Conjectures and open questions in rigidity. In *Proceedings of the International Congress of Mathematicians (Helsinki, 1978)*, pages 407–414. Acad. Sci. Fennica, Helsinki, 1980.

[10] M. Dehn. Ueber den Rauminhalt. *Math. Ann.*, 55(3):465–478, 1901.

[11] A. Douady. Le shaddock à six becs. *Bulletin A.P.M.E.P.*, 281:699–701, 1971.

[12] L. Euler and Academiae Scientiarum Petropolitanae. *Leonhardi Euleri Opera postuma mathematica et physica: anno MDCCCXLIV detecta.* Number v. 2 in Leonhardi Euleri Opera postuma mathematica et physica: anno MDCCCXLIV detecta. Eggers, 1862.

[13] Alexander A. Gaifullin. Embedded flexible spherical cross-polytopes with nonconstant volumes. *Proc. Steklov Inst. Math.*, 288(1):56–80, 2015.

[14] A. A. Gaĭ fullin. The analytic continuation of volume and the bellows conjecture in Lobachevskiĭ spaces. *Mat. Sb.*, 206(11):61–112, 2015.

[15] Herman Gluck. Almost all simply connected closed surfaces are rigid. *Lecture Notes in Math.*, 438:225–239, 1975.

[16] V. Gorkavyy and D. Kalinin. On model flexibility of the Jessen orthogonal icosahedron. *Beitr. Algebra Geom.*, 57(3):607–622, 2016.

[17] Bø rge Jessen. Orthogonal icosahedra. *Nordisk Mat. Tidskr*, 15:90–96, 1967.

[18] Hermann Karcher. Remarks on polyhedra with given dihedral angles. *Comm. Pure Appl. Math.*, 21:169–174, 1968.

[19] I. Kh. Sabitov. On the problem of the invariance of the volume of a deformable polyhedron. *Uspekhi Mat. Nauk*, 50(2(302)):223–224, 1995.

[20] I. Kh. Sabitov. Algebraic methods for the solution of polyhedra. *Uspekhi Mat. Nauk*, 66(3(399)):3–66, 2011.

[21] Ernst Steinitz and Hans Rademacher. *Vorlesungen über die Theorie der Polyeder unter Einschluss der Elemente der Topologie*. Springer-Verlag, Berlin-New York, 1976. Reprint der 1934 Auflage, Grundlehren der Mathematischen Wissenschaften, No. 41.

[22] J. J. Stoker. Geometrical problems concerning polyhedra in the large. *Comm. Pure Appl. Math.*, 21:119–168, 1968.

[23] J.-P. Sydler. Conditions nécessaires et suffisantes pour l'équivalence des polyèdres de l'espace euclidien à trois dimensions. *Comment. Math. Helv.*, 40:43–80, 1965.

[24] W. Wunderlich and C. Schwabe. Eine Familie von geschlossenen gleichflächigen Polyedern, die fast beweglich sind. *Elem. Math.*, 41(4):88–98, 1986.

# Chapter 14

# *Tensegrity*

**Robert Connelly**

*Department of Mathematics, Cornell University, Ithaca, NY*

**Anthony Nixon**

*Department of Mathematics and Statistics, Lancaster University, U.K.*

**CONTENTS**



## 14.1 Introduction

Tensegrities were invented and implemented by Kenneth Snelson in 1947 and other artists who were intrigued by the almost magical way the sticks could be suspended in midair. They were called tensegrities by R. Buckminster Fuller [31, 32] because of their "tensional integrity" and he popularised their use.

As well as their mathematical interest and artistic beauty, tensegrities arise in a number of application areas. In engineering (see, among others, [7, 36]), tensegrity structures provide efficient solutions for applications in deployable structures [51, 59], mechanisms [8, 43], multi agent systems [45], interesting examples of form finding [42, 66], algorithms for synthesis and analysis [27, 52] and smart sensors [53] as well as being intriguingly light relative to their stability. In biology, tensegrity structures are employed as models underlying the behavior of entities such as the cytoskeleton [38, 39]. Let us also mention Skelton's Type 1 (when no pair of bars is adjacent) and Type 2 tensegrities (where bar adjacencies are permitted) [65]. Skelton's interest in tensegrities arose from engineering and control theory. We focus on the mathematics of tensegrities in this chapter. A popular attempt at a logical explanation of Fuller's idea's for tensegrities was given by Edmondson [29].

Mathematically, a fundamental problem in geometry is to determine when selected distance constraints, on a finite number of points, fix these points up to congruence, at least for small perturbations. In the language of rigidity theory, we consider a realization of a graph as a finite set of points where certain pairs (corresponding to the edges) represent one of three types of constraint: cables are constrained not to get further apart; struts not to get closer together; and bars are constrained to stay the same distance apart.

We are then interested first, in whether there is a continuous motion of the points satisfying these constraints which is not simply the restriction of a rigid motion of the ambient Euclidean space. This is the rigidity question. Stronger is the global rigidity question which asks if the realization is unique up to rigid motions. In this chapter we will review the theory of tensegrity frameworks and point out connections with other chapters.

## 14.2 Tensegrity Frameworks

**Tensegrity:** A *tensegrity framework* $(G,p)$ in $\mathbb{R}^d$ is a graph $G=(V,E)$ where $E$ has been partitioned into 3 sets $E_-, E_0$ and $E_+$, together with a map $p: V \to \mathbb{R}^d$ realising the vertices as points in $\mathbb{R}^d$. The members in $E_-$ are known as *cables*, the members in $E_0$ are *bars*, and the members in $E_+$ are *struts*. If $E = E_0$ then we have a standard bar-joint framework. Cables cannot increase in length, bars have fixed length and struts cannot decrease in length.

**Dominates:** A tensegrity framework $(G,p)$ *dominates* the tensegrity framework $(G,q)$, written $(G,p) \geq (G,q)$, if

$$\begin{aligned} \|p_i - p_j\| &\geq \|q_i - q_j\| \text{ when } ij \in E_-, \\ \|p_i - p_j\| &= \|q_i - q_j\| \text{ when } ij \in E_0, \\ \|p_i - p_j\| &\leq \|q_i - q_j\| \text{ when } ij \in E_+. \end{aligned}$$

**Locally rigid:** A tensegrity framework $(G,p)$ is *locally rigid* (or continuously rigid, or often just rigid) if there is an $\varepsilon > 0$ such that whenever $(G,p)$ dominates $(G,q)$ and $\|p-q\| < \varepsilon$ then $p$ is congruent to $q$.

**Infinitesimal flex:** An *infinitesimal flex* of a tensegrity $(G,p)$ is an assignment $p': V \to \mathbb{R}^d$ such that for each edge $ij \in E$ we have

$$\begin{aligned} (p_i - p_j)\cdot(p'_i - p'_j) &\leq 0 \text{ for cables } ij \in E_-, \\ (p_i - p_j)\cdot(p'_i - p'_j) &= 0 \text{ for bars } ij \in E_0, \\ (p_i - p_j)\cdot(p'_i - p'_j) &\geq 0 \text{ for struts } ij \in E_+. \end{aligned}$$

**Infinitesimally rigid:** A tensegrity framework $(G,p)$ is *infinitesimally rigid* if every infinitesimal flex is an infinitesimal isometry of $\mathbb{R}^d$.

**Stress:** A *stress* $\omega = (\ldots, \omega_{ij}, \ldots) \in \mathbb{R}^{|E|}$ on a tensegrity framework $(G,p)$ is an assignment of scalars to the edges.

**Equilibrium stress:** A stress is an *equilibrium stress* if

$$\sum_j \omega_{ij}(p_j - p_i) = 0 \text{ for all } i \in V,$$

where the sum is taken over all $j \in V - i$, and if $ij \notin E$ then $\omega_{ij} = 0$.

**Proper stress:** An equilibrium stress is a *proper stress* if $\omega_{ij} \geq 0$ for all $ij \in E_-$ and $\omega_{ij} \leq 0$ for all $ij \in E_+$ (there is no condition for the stresses on the bars).

**Strict proper stress:** An equilibrium stress is a *strict proper stress* if $\omega_{ij} > 0$ for all $ij \in E_-$ and $\omega_{ij} < 0$ for all $ij \in E_+$ (there is no condition for the stresses on the bars).

**Underlying bar-joint framework:** For a tensegrity framework $(G,p)$, let $(\bar{G},p)$ denote the *underlying bar-joint framework* that arises by replacing all members of $(G,p)$ by bars.

**Equilibrium load:** An *equilibrium load* on a tensegrity framework $(G,p)$ is an assignment $F : V \to \mathbb{R}^d$ such that for each trivial infinitesimal flex $p' \in \mathbb{R}^{d|V|}$ we have $F \cdot p' = 0$.

**Resolution:** A *resolution* of an equilibrium load $F$ by a tensegrity framework $(G,p)$ is an equilibrium stress $\omega$ such that $\sum_j \omega_{ij}(p_j - p_i) + F_i = 0$.

**Statically rigid:** A tensegrity framework $(G,p)$ is *statically rigid* if every equilibrium load is resolvable.

**Prestress stable:** A tensegrity framework $(G,p)$ in $\mathbb{R}^d$ is *prestress stable* if there is a proper equilibrium stress $\omega$ such that for every non-trivial infinitesimal flex $p'$ of $(G,p)$ we have $\sum_{i<j} \omega_{ij}(p'_i - p'_j)^2 > 0$.

**Second-order flex:** A *second-order flex* $(p', p'')$ for a tensegrity framework $(G,p)$ is a solution to the following constraints:
(a) for a bar $ij$: $(p_i - p_j) \cdot (p'_i - p'_j) = 0$ and $||p'_i - p'_j||^2 + (p_i - p_j) \cdot (p''_i - p''_j) = 0$;
(b) for a cable $ij$: $(p_i - p_j)(p'_i - p'_j) = 0$ and $||p'_i - p_j||^2 + (p_i - p_j)(p''_i - p'_j) \leq 0$ or $(p_i - p_j)(p'_i - p'_j) < 0$; and
(c) for a strut $ij$: $(p_i - p_j)(p'_i - p'_j) = 0$ and $||p'_i - p_j||^2 + (p_i - p_j)(p''_i - p'_j) \geq 0$ or $(p_i - p_j)(p'_i - p'_j) > 0$.

**Second-order rigid:** A tensegrity framework $(G,p)$ is *second-order rigid* if all second-order infinitesimal flexes $(p', p'')$ have $p'$ as a trivial infinitesimal flex.

Tensegrity frameworks arise from bar-joint frameworks by replacing some of the fixed distance constraints by inequalities. When the distance is forced not to increase we call the edges cables and when the distance is forced not to decrease we call the edges struts.

The basic definitions of rigidity and infinitesimal rigidity can be easily extended to the setting of tensegrities. In Figure 14.1 we give an example of a rigid tensegrity framework and a different realization of the same tensegrity graph which is not rigid. This example makes it immediately clear that the rigidity or infinitesimal rigidity of the underlying bar-joint framework is not sufficient for a tensegrity framework to be rigid or infinitesimally rigid (see also Figure 14.3). We also note the following partial equivalence of rigidity and infinitesimal rigidity.

**Theorem 14.1 (Connelly and Whiteley, 1996 [26])** *Let $(G,p)$ be a tensegrity framework where: the vertices of $G$ are realised as a strictly convex polygon; the bars form a Hamilton cycle on the boundary of this polygon; and there are no struts. Then $(G,p)$ is rigid if and only if it is infinitesimally rigid.*

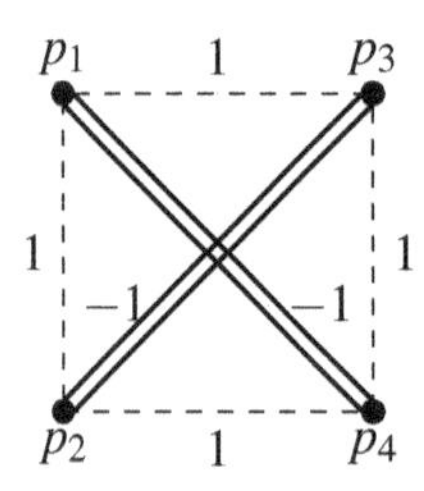


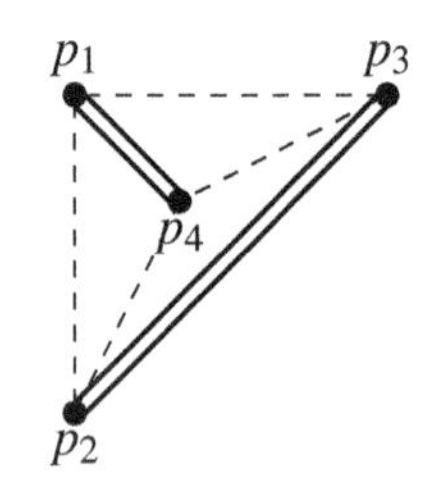


**Figure 14.1**
A simple tensegrity framework in $\mathbb{R}^2$ with a proper equilibrium stress indicated. Cables are represented by dashed lines and struts by doubled lines. This tensegrity is rigid. A second realization of the same tensegrity graph, again in $\mathbb{R}^2$, which is flexible.

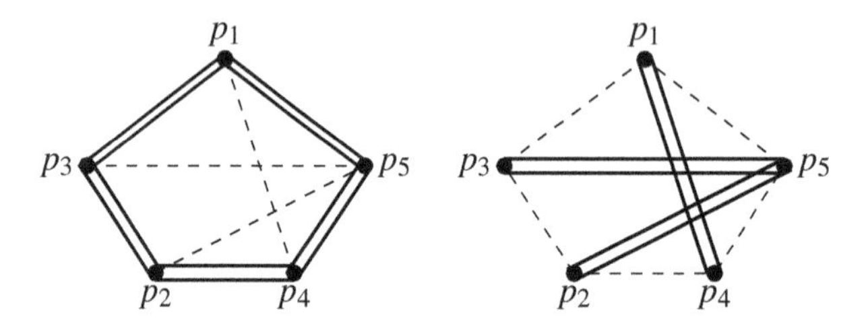


**Figure 14.2**
Two infinitesimally rigid tensegrities (in the plane). The tensegrity on the right comes from the one on the left by interchanging the cables and struts.

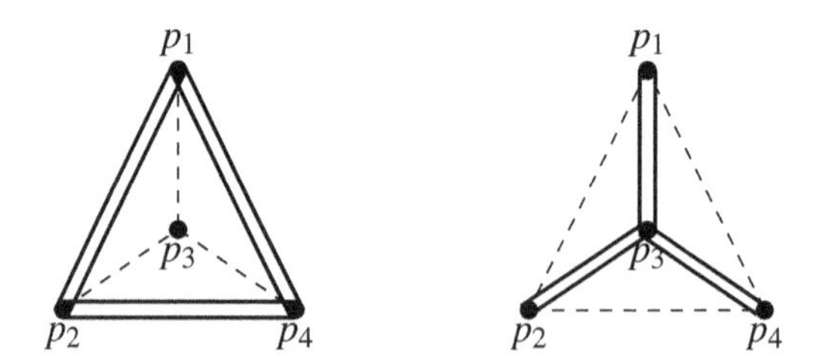


**Figure 14.3**
The tensegrity on the left is rigid but infinitesimally flexible (in $\mathbb{R}^3$). On the right the tensegrity with cables and struts interchanged is flexible and infinitesimally flexible (again in $\mathbb{R}^3$).

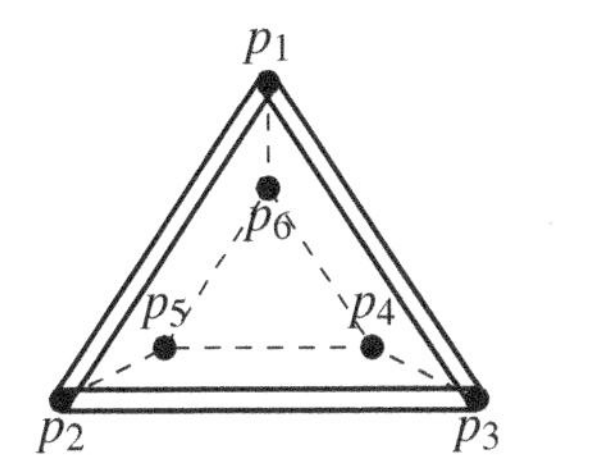


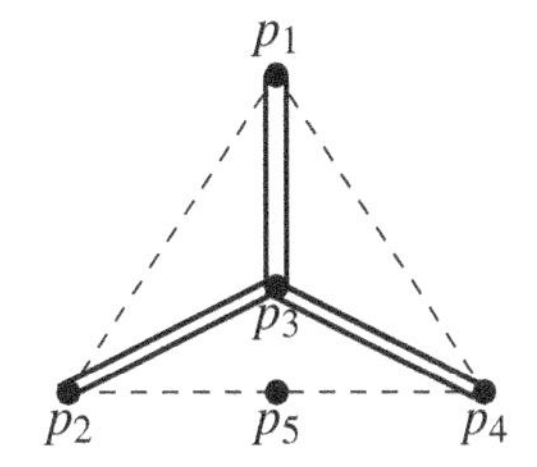


**Figure 14.4**
Examples of pre-stress stable tensegrity frameworks (in the plane) which are not infinitesimally rigid.

Interchanging cables and struts preserves infinitesimal rigidity, as we illustrate in Figure 14.2. Note however that rigidity is not preserved by this process, see Figure 14.3.

Equilibrium stresses (see Chapter 16), with their physical interpretation as forces on the edges, can also easily be extended to tensegrities (typically with positive stresses on cables and negative stresses on struts), in fact they were originially introduced to the rigidity literature in the context of tensegrities [12]. The extra condition is simply that for an equilibrium stress to be proper, the stress on each cable must be positive and the stress on each strut must be negative. The fundamental importance of equilibrium stresses to the rigidity of tensegrities is indicated in the following basic result.

**Theorem 14.2 (Connelly, 1982 [12])** *Let $(G, p)$ be a tensegrity framework with at least one cable or strut. Suppose $(G, p)$ is rigid, then there is a non-zero proper equilibrium stress.*

Connelly and Whiteley [26] proved that infinitesimal rigidity implies prestress stability which in turn implies second-order rigidity which then implies rigidity for any tensegrity framework. Examples show that none of these implications are reversible. Figure 14.4 shows examples of pre-stress stable tensegrity frameworks which are not infinitesimally rigid. In Figure 14.4a there is a rotational motion on $p_4, p_5, p_6$ and in Figure 14.4b there is the obvious motion on $p_5$. In both cases there is a strict proper equilibrium stress such that these non-trivial motion components only occur on vertices adjacent to members with positive stresses. The pre-stress stability follows from this.

It can be convenient to translate infinitesimal rigidity of a tensegrity framework into a statement about the underlying bar-joint framework as in the following theorem of Roth and Whiteley.

**Theorem 14.3 (Roth and Whiteley, 1981 [49])** *Let $(G, p)$ be a tensegrity framework with at least one cable or strut. Then $(G, p)$ is infinitesimally rigid if and only if the underlying bar framework $(\bar{G}, p)$ is infinitesimally rigid and there is a strict proper stress on $(G, p)$.*

A simple consequence of this theorem is that an infinitesimally rigid tensegrity framework $(G, p)$ is either a bar-joint framework or has at least one cable or strut and satisfies $|E| > d|V| - \binom{d+1}{2}$. (Contrast with the Maxwell counts, see Chapter 18.) A similar fact was observed by Roth and Whiteley [49]: if $(G, p)$ is an infinitesimally rigid tensegrity framework in $\mathbb{R}^d$ (with at least one cable or strut) and $(G', p)$ is obtained from $(G, p)$ by deleting a single cable or strut and then replacing all remaining cables and struts with bars, then $(G', p)$ is infinitesimally rigid in $\mathbb{R}^d$.

We can use equilibrium stresses to prove rigidity in the absence of infinitesimal rigidity. See also [21] for more information.

The following is an equivalent dual statement of second-order rigidity. The idea is that the framework has a demon who anticipates, for each possible non-trivial infinitesimal flex, a proper blocking equilibrium stress that prevents the extension to a second order flex. This is opposed to prestress stability, where one stress works for all non-trivial infinitesimal flexes.

**Theorem 14.4 (Connelly and Whiteley, 1996 [26])** *Let $(G, p)$ be a tensegrity framework in $\mathbb{R}^d$. Suppose that for all nontrivial infinitesimal motions $p'$ of $(G, p)$, there exists a proper equilibrium stress $\omega$ such that*

$$\sum_{ij} \omega_{ij}(p'_i - p'_j) \cdot (p'_i - p'_j) > 0.$$

*Then $(G, p)$ is second order rigid in $\mathbb{R}^d$.*

### 14.2.1 Combinatorics of Tensegrities

It is well known that (infinitesimal) rigidity of tensegrity frameworks is not a generic property, unlike bar-joint frameworks (see Chapter 18). In particular, as Figure 14.1 illustrates, there are graphs and assignments of cables, struts and bars to the edges such that there are open sets of realizations as a tensegrity framework which are rigid but also open sets of realizations such that the tensegrity framework is not rigid [60]. It is an open problem to characterize those graphs which have infinitesimally rigid realizations as tensegrity frameworks. See also [47].

Define a graph $G$ to be *strongly rigid* if every generic realization of $G$ in $\mathbb{R}^d$ as a tensegrity framework is infinitesimally rigid. Strongly rigid examples include any tensegrity which contains a spanning generic infinitesimally rigid bar-joint framework. See Figure 14.1 for an example which is not strongly rigid. It was shown in [40] that testing whether a graph is strongly rigid is NP-hard even in $\mathbb{R}^1$. Recski and Shai [48] characterised infinitesimal rigidity for 1-dimensional tensegrities and showed for such frameworks, infinitesimal rigidity can be decided in polynomial time.

In [41] it was shown that the edges of a graph can be assigned as cables and struts such that there is some realization in $\mathbb{R}^d$ as a tensegrity framework which is infinitesimally rigid if and only if the graph is redundantly rigid in $\mathbb{R}^d$ as a generic bar-joint framework (see Chapter 16 for information about redundantly rigid graphs).

### 14.2.2 Geometric Interpretations

Dual to the infinitesimal theory for tensegrities is the theory of statics for tensegrity frameworks which was developed by Whiteley [63] including results for placing tensegrities within faces of convex polyhedra without breaking static rigidity.

Whiteley considered one story buildings with cables as "roof" edges [61] and also explored links between tensegrities and scene analysis [62].

The projective invariance of infinitesimal rigidity for bar-joint frameworks extends to tensegrity frameworks in the following way. Suppose a projective transformation sends a cable to infinity, then we must replace it by a (finite) strut, and similarly a strut sent to infinity by the transformation must be replaced by a (finite) cable [63].

While taking a projective viewpoint, we mention very recent results of Eftekhari [30] developing a tensegrity theory for spherical frameworks and point-hyperplane frameworks (the higher dimensional analogue of point-line frameworks); see Chapters 22 and 17, Section 17.2.4.

Projective interpretations also lead to the problem of polarity for weaving lines [64]. Here tensegrity appears unexpectedly. The lines correspond to points, the intersection of the lines correspond to the edge constraints with a choice needed for the designation of whether the constraints are cables or struts.

### 14.2.3 Packings

Packings of spherical disks in a polyhedral container can be regarded as tensegrities with all struts connecting centers of touching disks to each other and the boundary of the container. With this viewpoint tensegrities have been used to prove results about packings, see, for example, [15, 19]. The local maximal density of the configuration of disks is determined by the rigidity and infinitesimal rigidity of the underlying tensegrity framework. Examples of this are [4, 5, 33] and [20, 25] for the periodic case (see Chapter 25 for various results on periodic rigidity for bar-joint frameworks).

A jammed packing thought of as a tensegrity can be detected, even for large sizes, using linear programming [28] to verify rigidity.

The Kneser-Poulsen conjecture states that a re-arrangement of a configuration of spherical balls in $\mathbb{R}^d$ in which the distance between every pair of centres (of the balls) does not decrease has the property that the volume of the union of the balls does not decrease. This was proved in the case when $d = 2$ by Bezdek and Connelly [6]. (See also [35].) One difficulty in the proof is that there are configurations of centers that have another expansive arrangement in $\mathbb{R}^d$ but moving between these arrangements requires using a path in $\mathbb{R}^{2d}$. In [6] it was shown that you can "leapfrog" one arrangement to the other expansively using a path in only 2 extra dimensions, thus proving the conjecture in dimension 2. In [3], examples were analysed using second order rigidity and in particular the "demon characterization." This is a condition for a first order infinitesimal motion to extend to a second order motion. In particular, suppose we have a first order motion $p'$ but the tensegrity framework is not second order rigid. If there exists an equilibrium stress $\omega$ such that $(p')^T(\Omega \otimes I_d)p' \neq 0$ (where $\Omega$ is the stress matrix defined in the following section and the operator $\otimes$ denotes the tensor product) then a brief calculation gives a contradiction. However it may be that there are different equilibrium stresses for different possible motions.

## 14.3 Types of Rigidity

**Energy function:** The *energy function* associated to a stress $\omega$ for a tensegrity framework $(G, p)$ is defined as

$$E_\omega(p) = \sum_{1 \leq i < j \leq n} \omega_{ij} \|p_i - p_j\|^2,$$

where $n = |V|$.

**Stress matrix:** The *stress matrix* $\Omega$ is the $n \times n$ symmetric matrix with off-diagonal entries $-\omega_{ij}$ and diagonal entries $\sum_j \omega_{ij}$.

**Globally rigid:** A tensegrity framework $(G, p)$ in $\mathbb{R}^d$ is *globally rigid* if every tensegrity framework $(G, q)$ in $\mathbb{R}^d$, with the same edge labeling, dominated by $(G, p)$ is congruent to $(G, p)$.

**Universally rigid:** A tensegrity framework $(G, p)$ in $\mathbb{R}^d$ is *universally rigid* if every tensegrity framework $(G, q)$ in $\mathbb{R}^D$ for any $D \geq d$, with the same edge labeling, dominated by $(G, p)$ is congruent to $(G, p)$.

**Super stable:** A tensegrity framework $(G, p)$ is *super stable* if it has a proper equilibrium stress $\omega$ such that $\Omega$ is positive semi-definite, rank $\Omega = n - d - 1$ and the underlying bar-joint framework $(\bar{G}, p)$ is rigid.

**Dimensionally rigid:** A tensegrity $(G, p)$ in $\mathbb{R}^d$ is *dimensionally rigid* if any other framework $(G, q)$ in $\mathbb{R}^D$, for any $D$ satisfying the edge constraints of $(G, p)$ has an affine span of dimension at most $d$.

**Conic at infinity:** A finite set of (non-zero) vectors in $\mathbb{R}^d$ lie on a *conic at infinity* if when regarded as points in projective $d-1$-dimensional space, they lie on a conic.

### 14.3.1 Global Rigidity and Stress Matrices

We now turn to global rigidity for tensegrities. Computationally, as with bar-joint frameworks, testing global rigidity for tensegrities is NP-hard [50], essentially equivalent to the subset-sum problem.

Note that when $(G,p)$ dominates $(G,q)$ and $\omega$ is a proper equilibrium stress for $(G,p)$ then $E_\omega(p) \geq E_\omega(q)$ and when $\omega$ is strict and $E_\omega(p) = E_\omega(q)$ then the $\|p_i - p_j\| = \|q_i - q_j\|$ for all $ij \in E$.

Consider now the energy function $E_\omega$ introduced in [12]. At a critical point we have $\sum_j \omega_{ij}(p_j - p_i) = 0$. This leads us to equilibrium stresses and the stress matrix. Specifically, by regarding $E_\omega$ as a quadratic form and considering the associated (symmetric) matrix we get the "big" stress matrix $\Omega \otimes I_d$. The fact that the stress matrix $\Omega$ is $n \times n$ thus arises from regarding the realization $p$ (in the tensegrity $(G,p)$ in $\mathbb{R}^d$) as a column vector. Notice further that with

$$P = \begin{pmatrix} p_1 & p_2 & \cdots & p_n \\ 1 & 1 & \cdots & 1 \end{pmatrix}$$

the critical point condition can be interpreted as the condition $P\Omega = 0$. See also Chapter 16.

For example, recall the tensegrity in Figure 14.1. With the equilibrium stress as indicated we have the stress matrix

$$\begin{pmatrix} 1 & -1 & -1 & 1 \\ -1 & 1 & 1 & -1 \\ -1 & 1 & 1 & -1 \\ 1 & -1 & -1 & 1 \end{pmatrix},$$

which has rank $1 = 4 - 3 = n - (d+1)$.

The following theorem extends a result of Connelly showing that equivalent generic bar-joint frameworks have the same equilibrium stresses, see Chapter 16, but it does use the additional assumption of positive semidefiniteness.

**Theorem 14.5 (Alfakih and Nguyen, 2013 [1])** *Let $(G,p)$ be a given tensegrity framework and let $\Omega$ be a proper positive semidefinite stress matrix of $(G,p)$. Then $\Omega$ is a proper stress matrix for all tensegrity frameworks $(G,q)$ dominated by $(G,p)$.*

Hendrickson's necessary conditions for global rigidity [37] (see Chapter 16) also apply to tensegrities. That is, if $(G,p)$ is a generic globally rigid tensegrity framework in $\mathbb{R}^d$ then $G$ is $(d+1)$-connected and $(G,p)$ is redundantly rigid in $\mathbb{R}^d$. We also have the following sufficient condition.

**Theorem 14.6 (Connelly, 2005 [14])** *Let $(G,p)$ be a generic tensegrity framework with a proper equilibrium stress $\omega$ and stress matrix $\Omega$ of rank $n-d-1$. Then $(G,p)$ is globally rigid in $\mathbb{R}^d$.*

### 14.3.2 Universal and Dimensional Rigidity

Next we discuss universal rigidity. We have the following fundamental result showing how to use stress matrices to confirm universal rigidity. This follows from the results of [12], see also [21].

**Theorem 14.7 (Connelly, 2013 [17])** *Let $(G,p)$ be a tensegrity framework whose affine span of $p$ is all of $\mathbb{R}^d$, with a proper equilibrium stress $\omega$ and stress matrix $\Omega$. Suppose further that*

*(a) $\Omega$ is positive semi-definite;*

*(b) the rank of $\Omega$ is $n-d-1$; and*

*(c) The member directions of $(G,p)$ with a non-zero stress, and bars, do not lie on a conic at infinity.*

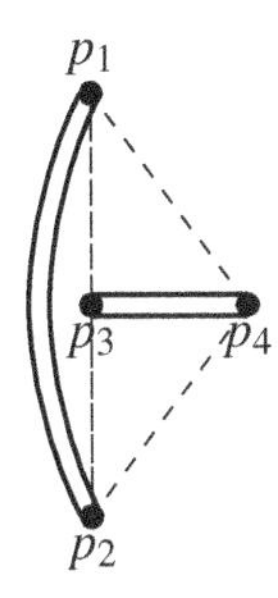


**Figure 14.5**
A universally rigid tensegrity in the plane.

*Then $(G, p)$ is universally rigid.*

Note that the third condition can be replaced by there being no non-trivial affine image of $(G, p)$. Consider the tensegrity in Figure 14.5. This tensegrity framework has a maximum rank positive semi-definite stress with non-zero stresses only on the collinear triangle. Therefore the stress directions of the members with a non-zero stress do lie on a conic at infinity. However this tensegrity is universally rigid.

Alfakih and Nguyen [1] used dimensional rigidity to help in understanding universal rigidity of tensegrities. In particular they proved that the assumptions of the above theorem without the conic condition give sufficient conditions for dimensional rigidity.

**Theorem 14.8 (Alfakih and Nguyen, 2013 [1])** *Let $(G, p)$ be a tensegrity framework whose affine span of $p$ is all of $\mathbb{R}^d$, with a proper equilibrium stress $\omega$ and stress matrix $\Omega$. Suppose further that*

*(a) $\Omega$ is positive semi-definite and*

*(b) the rank of $\Omega$ is $n-d-1$.*

*Then $(G, p)$ is dimensionally rigid.*

**Theorem 14.9 (Connelly, 2009 [16])** *Suppose a tensegrity framework $(G, p)$ in $\mathbb{R}^d$ has a proper equilibrium stress such that the underlying bar-joint framework $(\bar{G}, p)$ is super stable and infinitesimally rigid. Then $(G, q)$ is globally rigid for any generic $q$.*

We also have the following counterpoint to the characterisation of global rigidity for bar-joint frameworks in the plane, see Chapter 21.

**Theorem 14.10 (Connelly, 2009 [16])** *If $G$ is 3-connected and generically redundantly rigid as a bar-joint framework in $\mathbb{R}^2$, then there is a realization $(G, p)$ in $\mathbb{R}^2$ as a tensegrity framework which is super stable.*

In the next theorem we will need some definitions. An iterated affine set $\mathcal{C} = \mathcal{A}_0 \supset \mathcal{A}_1 \supset \mathcal{A}_2 \supset \dots \mathcal{A}_k$ is a sequence of affine sets. For each $\mathcal{A}_i$ we take a basis matrix $B_i$ and define a restricted stress matrix $\Omega_i^* = B_{i-1}\Omega_i B_{i-1}^T$. In particular each $\Omega_i^*$ is positive semi-definite.

**Theorem 14.11 (Connelly and Gortler, 2015 [22])** *Suppose $C = \mathcal{A}_0 \supset \mathcal{A}_1 \supset \mathcal{A}_2 \supset \dots \mathcal{A}_k$ is an iterated affine set for a tensegrity $(G, p)$ with $n$ vertices in $\mathbb{R}^d$, with an associated iterated proper positive semidefinite stress described by positive semidefinite restricted stress matrices $\Omega_i^*$. Let $r_i$*

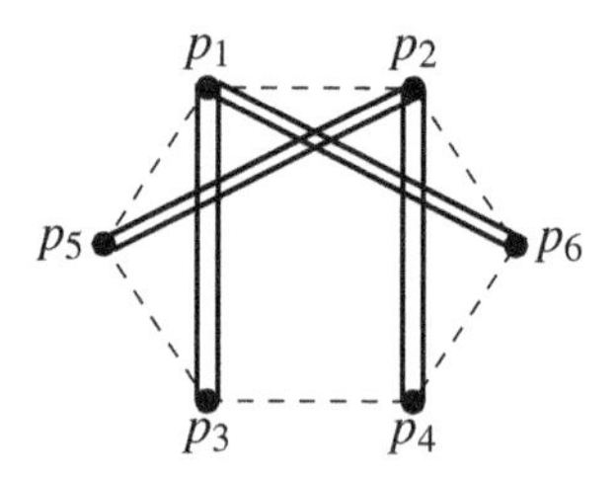


**Figure 14.6**
An example of a super stable tensegrity.

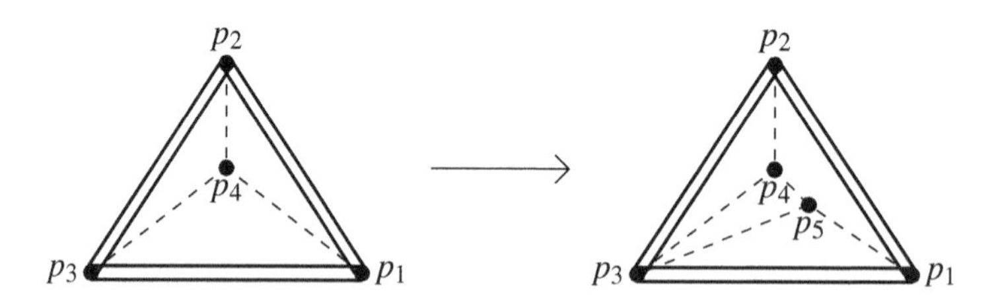


**Figure 14.7**
A 1-extension on the edge $wx$ adding new vertex $u$.

*be the rank of* $\Omega_i^*$. *If* $\sum_{i=1}^{k} r_i = n-d-1$ *and the member directions with non-zero stress directions and bars do not lie on a conic at infinity, then* $(G,p)$ *is universally rigid.*

*Conversely if* $(G,p)$ *is universally rigid in* $\mathbb{R}^d$, *then there is an iterated affine set with an associated iterated positive semidefinite stress determined by proper stresses, the dimension of* $\mathcal{A}_k$ *is* $(d+1)\binom{d+1}{2}$, *and the members with non-zero stress directions and bars do not lie on a conic at infinity.*

The theorem gives a certificate for universal rigidity.

Figure 14.6 gives an example of a super stable tensegrity. A second example comes from choosing the vertices of a cube with cables for each edge of the cube and 4 struts connecting antipodal vertices.

### 14.3.3 Operations on Tensegrities

We finish this section by commenting on methods for building larger tensegrities with certain rigidity properties from smaller ones. First let us discuss the 1-extension operation (see Figure 14.7 and Chapter 19, Section 19.2.1).

A simple three-node tensegrity is when $p_i, p_0$ and $p_j$ are ordered on a line so that $ij$ is a strut and both $i0$ and $0j$ are cables. This tensegrity framework has an equilibrium stress where $\omega_{ij} < 0, \omega_{i0} > 0$ and $\omega_{j0} > 0$. The stress matrix for this tensegrity is easily seen to be positive semidefinite with one positive eigenvalue and two zero eigenvalues. So in the 1-extension operation, before the perturbation, one can choose the insertion of $p_0$ and the two new edges as the addition of a small stress matrix to the original stress matrix in such a way that the cable stress say on the edge $ij$ cancels with the strut stress on the three node tensegrity. The effect is that the new eigenvalue will be positive, while the others will only be perturbed slightly. The same argument works when $ij$ is a strut and $p_0$ is on the line through $p_i$ and $p_j$, but not in the interval between them.

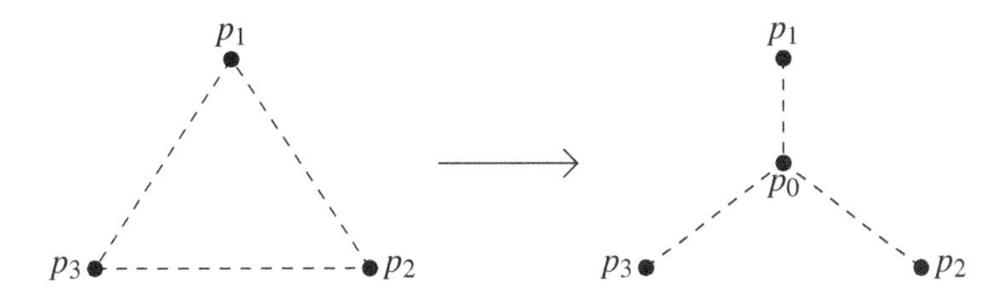


**Figure 14.8**
A delta-$Y$ operation adding a new point $p_0$.

Second we comment on vertex splitting (again see Chapter 19, Section 19.2.1). Recall that a vertex splitting operation on a vertex $v$, deletes $v$ and the edges $2,3\ldots,t$ incident to $v$ and adds two vertices $0,1$ along with edges $20,21,30,31,01$, $i0$ for $3 < i \leq k$ and $i1$ for $k < i \leq t$. The vertex split is *non-trivial* for any $t,k$ with $k > 2$ and $t > k$.

**Theorem 14.12 (Connelly, 2009 [16])** *Let $(G,p)$ be a tensegrity framework which is super stable in $\mathbb{R}^d$. Let $G'$ be formed from $G$ by a non-trivial vertex splitting operation which splits $v$ into two vertices $v_0, v_1$ and let $q$ be the realization of $G$ in which $q(x) = p(x)$ for all $x \in V - v$ and $q(v_0) = q(v_1) = p(v)$. Suppose that $(G' - v_0v_1, q)$ is infinitesimally rigid in $\mathbb{R}^d$. Then $(G', \hat{p})$ is super stable in $\mathbb{R}^d$ for any generic $\hat{p}$.*

The theorem also holds if we replace super stable by globally rigid since the proof keeps track of the rank of the stress matrix independently of the stress matrix being positive semi-definite.

It is also worth pointing out that the technique of coning (see Chapter 17, Section 17.2.2) extends to tensegrities by simply choosing all edges to the cone vertex as bars.

Next, let us discuss methods for combining globally rigid tensegrity frameworks. For rigidity we note that the 2-sum operation (see Chapter 19) preserves generic infinitesimal rigidity for tensegrities, see [41]. To move to global rigidity, it is easy to see that adding two positive semi-definite matrices results in a positive semi-definite matrix. Moreover it follows from the definition that adding two generic globally rigid tensegrity frameworks in $\mathbb{R}^d$ with at least $d+1$ vertices in common preserves generic global rigidity. In [9] it is shown that in this process, for bar-joint frameworks, a single common edge can also be removed. However, extending this to tensegrities is an open problem.

Finally, recall that the delta-Y operation on a graph $G$ removes a triangle of edges $12,23,31$ and adds a new vertex 0 and 3 new edges $01,02,03$, see Figure 14.8. Applying this operation to tensegrities is discussed in [17].

## 14.4 Examples and Applications

**Tensegrity polygon:** A *tensegrity polygon* is a planar tensegrity framework in which the cables form a convex polygon containing all the vertices and the set of bars is empty.

**Abstract tensegrity polygon:** An *abstract tensegrity polygon* is a graph $G = (V,E)$ where $E$ is partitioned into two sets $E_-$ and $E_+$ and the edges in $E_-$ form a Hamilton cycle of $G$.

**Convex realization:** A *convex realization* of an abstract tensegrity polygon is a planar realization in which the Hamilton cycle of cables forms a convex polygon.

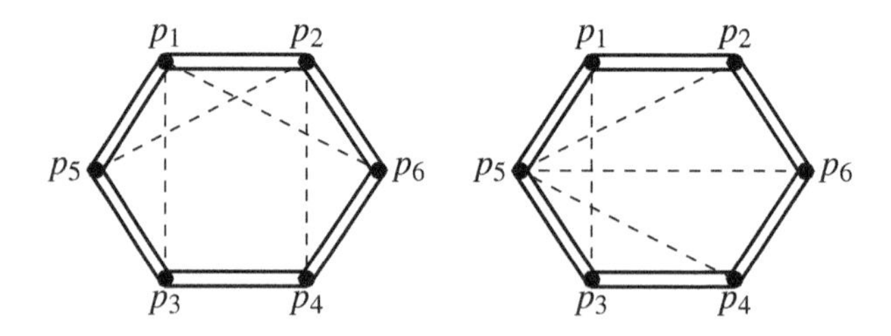


**Figure 14.9**
A Cauchy polygon and a Grunbaum polygon.

**Strong:** An abstract tensegrity polygon $G$ is *strong* if every convex realization $(G,p)$ has a non-trivial proper stress.

**Robust:** An abstract tensegrity polygon $G$ is *robust* if every convex realization $(G,p)$ has a strict proper stress.

**Stable:** An abstract tensegrity polygon $G$ is *stable* if every convex realization $(G,p)$ is infinitesimally rigid.

**Spider web:** A *spider web* is a tensegrity framework were some subset of the vertices are pinned and all edges are cables.

### 14.4.1 Examples

We now describe examples of super stable and thus universally rigid tensegrities.

Suppose we have an abstract tensegrity polygon in which the Hamilton cycle vertices are labelled $1,2,\ldots,n$. A Cauchy polygon is an abstract tensegrity polygon in which the struts are of the form $i(i+2)$ for $1 \leq i \leq n-2$. A Grunbaum polygon is an abstract tensegrity polygon in which the struts are 13 and $2i$ for $4 \leq i \leq n$. Figure 14.9 gives examples of Cauchy and Grunbaum polygons, respectively. Note also that Figure 14.1 is the smallest Cauchy and Grunbaum polygon.

Roth and Whiteley [49] proved that Generalized Grunbaum polygons are robust and stable. Geleji and Jordán [34] proved that, for an abstract tensegrity polygon $G$, the properties strong, robust and stable are equivalent.

**Theorem 14.13 (Connelly, 1982 [12])** *Suppose a tensegrity framework $(G,p)$ in $\mathbb{R}^2$ consists of a convex polygon, with cables on the boundary and struts inside. If there is a nonzero stress then the stress matrix has rank $n-3$ and is positive semidefinite (and hence universally rigid).*

The special class of tensegrity frameworks where every edge is a cable, known as spiderwebs, admits a more fully understood theory. In particular Connelly [12] proved the following theorem.

**Theorem 14.14** *Let $(G,p)$ be a pinned spider web framework in $\mathbb{R}^d$, where $G$ is connected, with a strict equilibrium stress. Then $(G,p)$ is globally rigid in $\mathbb{R}^d$.*

Observe that this theorem can be converted to a statement about the global rigidity of unpinned tensegrities consisting of a spider web plus a complete graph of struts on the previously pinned vertices. See, for example, Figure 14.10.

A number of further classes of super stable tensegrities are discussed in [16] such as tensegrities with a triangle of struts and then only cables inside with emphasis on the potential to use inductive constructions to show that all graphs in the class that are also 3-connected and redundantly rigid have super stable realizations.

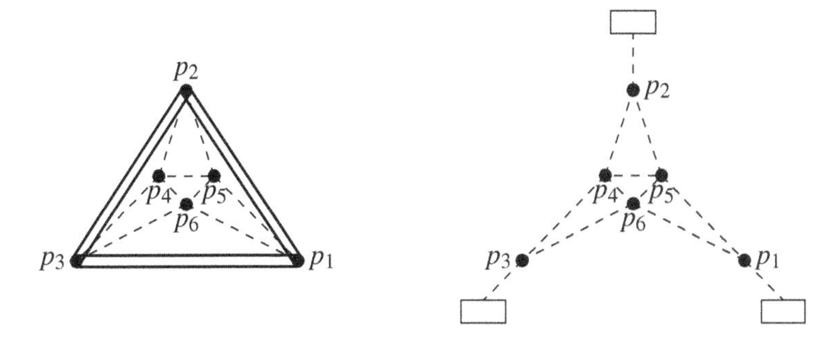


**Figure 14.10**
A spiderweb with a complete graph of struts around the outside and the pinned version of this spiderweb.

### 14.4.2 Applications

The chapter [13] discusses the use of global rigidity for tensegrities in one proof of Cauchy's arm lemma, which is of course used in the proof of Cauchy's famous theorem on convex polyhedra. See also Chapter 13. In particular the idea is to use a Cauchy polygon with struts "representing" angles and prove the universal rigidity of the Cauchy polygon using induction and Theorem 14.7. The problem is then translated back to Cauchy's arm problem by replacing the cables by bars and noticing that the increases in length of (pairs of incident) struts correspond to increases in the size of the corresponding angles. Further applications of tensegrities to distance geometry problems can be found in [2, 24, 54].

A wealth of further information on tensegrities can be found in the literature. For complete bipartite examples, see [23]; for highly symmetric tensegrities see the catalogue [11] (also [10]); for dihedrally symmetric examples including Snelson's example, see [67].

Pak and Vilenchik used Theorem 14.7 to show particular instances of the graph realization problem are tractable and deduced from their work a method of constructing uniquely $k$-colorable graphs [46]. Laurent and Varvitsiotis [44] studied tensegrity frameworks in the context of universal completability and semidefinite programming. Similarly Tanigawa studied spherical tensegrities [58] in the context of semidefinite programming and matrix completion.

Szabadka [57] conjectured that there exists an integer $k$ such that if $(G,p)$ is a generic tensegrity framework with at least $k$ cables and at least $k$ struts, $G$ is 3-connected and the underlying bar-joint framework $(\bar{G},p)$ is redundantly rigid, then $(G,p)$ is rigid in $\mathbb{R}^2$. Szabadka also proved several results about tensegrities relating to 1-extensions, and to complete graphs and wheels.

We finish by mentioning that the Carpenter's rule problem [18, 55] is an example of a fundamental problem in discrete geometry which can be interpreted as a tensegrity problem. Here the opening of a polygonal chain can be seen to be an 'expansive motion' where distances only increase so can be replaced by struts. Indeed any context with expansive motions fits this philosophy (e.g., [56]).

# References

[1] A. Y. Alfakih and Viet-Hang Nguyen. On affine motions and universal rigidity of tensegrity frameworks. *Linear Algebra Appl.*, 439(10):3134–3147, 2013.

[2] Maria Belk. Realizability of graphs in three dimensions. *Discrete Comput. Geom.*, 37(2):139–162, 2007.

[3] Maria Belk and Robert Connelly. Making contractions continuous: a problem related to the kneser-poulsen conjecture. Technical report, http://inside.bard.edu/academic/programs/math/mbelk/Contractions.pdf, 2007.

[4] A. Bezdek, K. Bezdek, and R. Connelly. Finite and uniform stability of sphere coverings. *Discrete Comput. Geom.*, 13(3-4):313–319, 1995.

[5] A. Bezdek, K. Bezdek, and R. Connelly. Finite and uniform stability of sphere packings. *Discrete Comput. Geom.*, 20(1):111–130, 1998.

[6] Károly Bezdek and Robert Connelly. Pushing disks apart—the Kneser-Poulsen conjecture in the plane. *J. Reine Angew. Math.*, 553:221–236, 2002.

[7] C. R. Calladine. Buckminster fuller's "trensegrity" structures and clerk maxwell's rules for the construction of stiff frameworks. *International Journal of Solids and Structures*, 14:161–172, 1978.

[8] C. R. Calladine and S. Pellegrino. First-order infinitesimal mechanisms. *Internat. J. Solids Structures*, 27(4):505–515, 1991.

[9] R. Connelly. Combining globally rigid frameworks. *Tr. Mat. Inst. Steklova*, 275(Klassicheskaya i Sovremennaya Matematika v Pole Deyatelnosti Borisa Nikolaevicha Delone):202–209, 2011.

[10] R. Connelly and M. Terrell. Tenségrités symétriques globalement rigides. *Structural Topology*, (21):59–78, 1995. Dual French-English text.

[11] Robert Connelly. Highly symmetric tensegrity structures. http://www.math.cornell.edu/ tens/.

[12] Robert Connelly. Rigidity and energy. *Invent. Math.*, 66(1):11–33, 1982.

[13] Robert Connelly. Rigidity. In *Handbook of convex geometry, Vol. A, B*, pages 223–271. North-Holland, Amsterdam, 1993.

[14] Robert Connelly. Generic global rigidity. *Discrete Comput. Geom.*, 33(4):549–563, 2005.

[15] Robert Connelly. Rigidity of packings. *European J. Combin.*, 29(8):1862–1871, 2008.

[16] Robert Connelly. Questions, conjectures and remarks on globally rigid tensegrities. http://www.math.cornell.edu/ connelly/09-Thoughts.pdf, 2009.

[17] Robert Connelly. Tensegrities and global rigidity. In *Shaping space*, pages 267–278. Springer, New York, 2013.

[18] Robert Connelly, Erik D. Demaine, and Günter Rote. Straightening polygonal arcs and convexifying polygonal cycles. *Discrete Comput. Geom.*, 30(2):205–239, 2003. U.S.-Hungarian Workshops on Discrete Geometry and Convexity (Budapest, 1999/Auburn, AL, 2000).

[19] Robert Connelly and William Dickinson. Periodic planar disc packings. *Philos. Trans. R. Soc. Lond. Ser. A Math. Phys. Eng. Sci.*, 372(2008):20120039, 17, 2014.

[20] Robert Connelly, Matthew Funkhouser, Vivian Kuperberg, and Evan Solomonides. Packings of equal disks in a square torus. arXiv:1512.08762.

[21] Robert Connelly and Steven Gortler. Prestress stability of triangulated convex polytopes and universal second order rigidity. arXiv:1510.04185.

[22] Robert Connelly and Steven Gortler. Iterative universal rigidity. *Discrete Comput. Geom.*, 53(4):847–877, 2015.

[23] Robert Connelly and Steven Gortler. Universal rigidity of complete bipartite graphs. arXiv: 1502.02278, 2016.

[24] Robert Connelly and Jean-Marc Schlenker. On the infinitesimal rigidity of weakly convex polyhedra. *European J. Combin.*, 31(4):1080–1090, 2010.

[25] Robert Connelly, Jeffrey D. Shen, and Alexander D. Smith. Ball packings with periodic constraints. *Discrete Comput. Geom.*, 52(4):754–779, 2014.

[26] Robert Connelly and Walter Whiteley. Second-order rigidity and prestress stability for tensegrity frameworks. *SIAM J. Discrete Math.*, 9(3):453–491, 1996.

[27] Miguel de Guzmán and David Orden. From graphs to tensegrity structures: geometric and symbolic approaches. *Publ. Mat.*, 50(2):279–299, 2006.

[28] Aleksandar Donev, Salvatore Torquato, Frank H. Stillinger, and Robert Connelly. A linear programming algorithm to test for jamming in hard-sphere packings. *J. Comput. Phys.*, 197(1):139–166, 2004.

[29] Amy C. Edmondson. *A Fuller explanation.* Design Science Collection. Birkhäuser Boston, Inc., Boston, MA, 1987. The synergetic geometry of R. Buckminster Fuller, A Pro Scientia Viva Title.

[30] Yaser Eftekhari. *Geometry of point-hyperplane and spherical frameworks.* PhD thesis, York University, 2017.

[31] Richard Buckminster Fuller. *Synergetics 2: Further Explorations in the Geometry of Thinking, Volume 2.* Macmillan, 1983.

[32] Richard Buckminster Fuller and E. J. Applewhite. *Synergetics: Explorations in the geometry of thinking.* Macmillan, 1982.

[33] Zsolt Gáspár, Tibor Tarnai, and Krisztián Hincz. Partial covering of a circle by equal circles. Part II: the case of 5 circles. *J. Comput. Geom.*, 5(1):126–149, 2014.

[34] János Geleji and Tibor Jordán. Robust tensegrity polygons. *Discrete Comput. Geom.*, 50(3):537–551, 2013.

[35] Igors Gorbovickis. Strict Kneser-Poulsen conjecture for large radii. *Geom. Dedicata*, 162:95–107, 2013.

[36] S. D. Guest. The stiffness of tensegrity structures. *IMA J. Appl. Math.*, 76(1):57–66, 2011.

[37] Bruce Hendrickson. Conditions for unique graph realizations. *SIAM J. Comput.*, 21(1):65–84, 1992.

[38] D.E. Ingber. Cellular tensegrity: defining new rules of biological design that govern the cytoskeleton. *J. Cell Sci.*, 104:613–627, 1993.

[39] Donald E. Ingber, Ning Wang, and Dimitrije Stamenović. Tensegrity, cellular biophysics, and the mechanics of living systems. *Rep. Progr. Phys.*, 77(4):046603, 21, 2014.

[40] Bill Jackson, Tibor Jordán, and Csaba Király. Strongly rigid tensegrity graphs on the line. *Discrete Appl. Math.*, 161(7-8):1147–1149, 2013.

[41] Tibor Jordán, András Recski, and Zoltán Szabadka. Rigid tensegrity labelings of graphs. *European J. Combin.*, 30(8):1887–1895, 2009.

[42] Yoshihiro Kanno. Exploring new tensegrity structures via mixed integer programming. *Struct. Multidiscip. Optim.*, 48(1):95–114, 2013.

[43] E. N. Kuznetsov. On immobile kinematic chains and a fallacious matrix analysis. *Trans. ASME J. Appl. Mech.*, 56(1):222–224, 1989.

[44] M. Laurent and A. Varvitsiotis. Positive semidefinite matrix completion, universal rigidity and the strong Arnold property. *Linear Algebra Appl.*, 452:292–317, 2014.

[45] Benjamin Nabet and Naomi Ehrich Leonard. Shape control of a multi-agent system using tensegrity structures. In *Lagrangian and Hamiltonian methods for nonlinear control 2006*, volume 366 of *Lect. Notes Control Inf. Sci.*, pages 329–339. Springer, Berlin, 2007.

[46] Igor Pak and Dan Vilenchik. Constructing uniquely realizable graphs. *Discrete Comput. Geom.*, 50(4):1051–1071, 2013.

[47] András Recski. Combinatorial conditions for the rigidity of tensegrity frameworks. In *Horizons of combinatorics*, volume 17 of *Bolyai Soc. Math. Stud.*, pages 163–177. Springer, Berlin, 2008.

[48] András Recski and Offer Shai. Tensegrity frameworks in one-dimensional space. *European J. Combin.*, 31(4):1072–1079, 2010.

[49] B. Roth and W. Whiteley. Tensegrity frameworks. *Trans. Amer. Math. Soc.*, 265(2):419–446, 1981.

[50] J. B. Saxe. Embeddability of weighted graphs in k-space is strongly np-hard. In *Proceedings of the 17th Allerton Conference in Communications, Control and Computing*, pages 480–489, 1979.

[51] M. Schenk, S. D. Guest, and J. L. Herder. Zero stiffness tensegrity structures. *Internat. J. Solids Structures*, 44(20):6569–6583, 2007.

[52] Meera Sitharam and Mavis Agbandje-Mckenna. Modeling virus self-assembly pathways: avoiding dynamics using geometric constraint decomposition. *J. Comput. Biol.*, 13(6):1232–1265, 2006.

[53] R.T. Skelton and C. Sultan. Controllable tensegrity: a new class of smart structures. *Proc. SPIE*, 3039:166–177, 1997.

[54] Anthony Man-Cho So and Yinyu Ye. A semidefinite programming approach to tensegrity theory and realizability of graphs. In *Proceedings of the Seventeenth Annual ACM-SIAM Symposium on Discrete Algorithms*, pages 766–775. ACM, New York, 2006.

[55] Ileana Streinu. Pseudo-triangulations, rigidity and motion planning. *Discrete Comput. Geom.*, 34(4):587–635, 2005.

[56] Ileana Streinu and Walter Whiteley. Single-vertex origami and spherical expansive motions. In *Discrete and computational geometry*, volume 3742 of *Lecture Notes in Comput. Sci.*, pages 161–173. Springer, Berlin, 2005.

[57] Zoltan Szabadka. *Globally rigid frameworks and rigid tensegrity graphs in the plane*. PhD thesis, Eotvos Lorand University, 2010.

[58] Shin-Ichi Tanigawa. The signed positive semidefinite matrix completion problem for odd-$k_4$ minor free signed graphs. arXiv: 1603.08370, 2016.

[59] G. Tibert. *Deployable tensegrity structures for space applications*. PhD thesis, Royal institute of technology, Stockholm, 2002.

[60] Neil L. White and Walter Whiteley. The algebraic geometry of stresses in frameworks. *SIAM J. Algebraic Discrete Methods*, 4(4):481–511, 1983.

[61] Walter Whiteley. Cones, infinity and 1-story buildings. *Structural Topology*, (8):53–70, 1983. With a French translation.

[62] Walter Whiteley. A correspondence between scene analysis and motions of frameworks. *Discrete Appl. Math.*, 9(3):269–295, 1984.

[63] Walter Whiteley. Infinitesimally rigid polyhedra. I. Statics of frameworks. *Trans. Amer. Math. Soc.*, 285(2):431–465, 1984.

[64] Walter Whiteley. Rigidity and polarity. II. Weaving lines and tensegrity frameworks. *Geom. Dedicata*, 30(3):255–279, 1989.

[65] Darrell Williamson, Robert E. Skelton, and Jeongheon Han. Equilibrium conditions of a tensegrity structure. *Internat. J. Solids Structures*, 40(23):6347–6367, 2003.

[66] J. Y. Zhang and M. Ohsaki. Self-equilibrium and stability of regular truncated tetrahedral tensegrity structures. *J. Mech. Phys. Solids*, 60(10):1757–1770, 2012.

[67] J.Y. Zhang, S.D. Guest, R. Connelly, and M. Ohsaki. Dihedral 'star' tensegrity structures. *International Journal of Solids and Structures*, 47(1):1 – 9, 2010.

# Chapter 15

## *Geometric Conditions of Rigidity in Nongeneric Settings*

**Oleg Karpenkov**
*University of Liverpool*

**CONTENTS**

## 15.1 Introduction

In this chapter we discuss geometric approach to infinitesimal rigidity in non-generic settings. Currently this approach is developed for the case of the plane and there is almost nothing known for three- and higher-dimensional spaces. Recall that a framework (i.e., a realization of a graph in the plane, or in the space) is *infinitesimally rigid* if every infinitesimal isometric deformation of the framework is an infinitesimal isometric deformation for the corresponding complete graph on the vertices. One of the fundamental characterizations of infinitesimal rigidity is via the linear independence of the rows of the rigidity matrix. Namely the configuration is infinitesimally rigid if and only if the corresponding rigidity matrix is of full rank.

Since the rank of the rigidity matrix is defined via algebraic equations (by the determinants of a given rank submatrices of the rigidity matrix), a rigidity matrix of a *generic* graph realization in the plane is always either full rank or not full rank. In this chapter we study the case of graphs when generic realizations of a given graph have a full rank rigidity matrix. In this situation infinitesimal flexibility is achieved only for specific frameworks, forming a positive codimension orbifold. As we see, such graphs are infinitesimally flexible only in *nongeneric settings.* A geometric description of such setting (conditions) is the main goal of this chapter.

First algebraic characterization of nongeneric settings were developed in 1983 by N. L. White and W. Whiteley in [31]. The authors proposed the techniques of tie-downs, where one supplements the rigidity matrix by additional lines up to a square matrix. Then the determinant of the obtained matrix gives an equation for infinitesimal flexibility configurations. Finally, one should rewrite it in terms of bracket ring and factorize (this is important for removing extra lines used to write the determinant). Several of the obtained factors would be independent on the choice of the supplementary lines. The locus of these factors is precisely the set of all infinitesimally flexible configurations. Here we should notice that the factorization of such bracket expressions is a hard open problem even in the two-dimensional case.

In this chapter we consider an alternative geometric approach where the conditions of infinitesimal flexibility are written in terms of *meet* and *join* operations and relations of extended Cayley algebra (see in [16]). It is based on the study of tensegrities, where the characterization of nongeneral settings is given by simple geometric conditions on the vertices of frameworks.

To familiarize the reader with the problem of geometric conditions of infinitesimal rigidity/flexibility in nongeneric settings we discuss one simple example. Recall that the *bracket* for three points in the plane $p_1, p_2, p_3$ is the following expression

$$[p_1, p_2, p_3] = \det(p_2 - p_1, p_3 - p_1),$$

where $\det(v, w)$ is the determinant of a matrix whose columns are $v$ and $w$ respectively. Here the bracket is considered as a degree 2 polynomial on 6 coordinates of the points involved.

**Example 41** *Consider a tensegrities on 6 distinct points with the graph $G$ as on Figure 15.1 on the left. What are all 6-point configurations $P = (p_1, \ldots, p_6)$ in the plane whose frameworks $G(P)$ are not infinitesimally rigid? In geometric terms we have the following conditions (later we describe such geometric conditions in terms of extended Cayley algebra):*

- *The lines $p_1p_2$, $p_3p_4$, $p_5p_6$ have a common point or parallel to each other (see Figure 15.1 at the middle);*
- *the points $p_1$, $p_4$, $p_5$ are in a line;*
- *the points $p_2$, $p_3$, $p_6$ are in a line.*

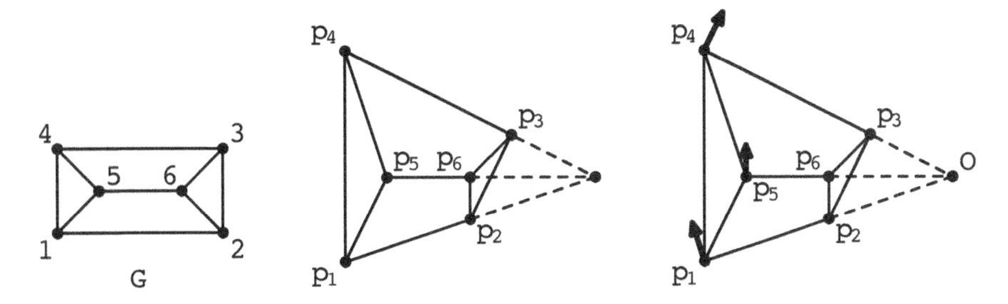


**Figure 15.1**
The graph $G$ (on the left); one of its realization with nonzero equilibrium stresses for $G$ (in the middle); a corresponding infinitesimally isometric flexion (on the right).

*Each of these conditions define a subset (strata) on the configuration space of all* 6*-point configurations. The corresponding bracket ring expression for this graph is:*

$$\big([p_4,p_1,p_2][p_3,p_6,p_5]-[p_3,p_1,p_2][p_4,p_6,p_5]\big)[p_1,p_4,p_5][p_2,p_3,p_6]=0.$$

*Here we have three factors: the first, the second, and the third factors correspond to the first, the second, and the third items, respectively.*

**Remark 15.1** Notice that the direction of infinitesimal flexion for the above example is very easy to see. Fix one of the triangles (say $p_2p_6p_3$) and infinitesimally isometrically rotate another triangle $p_1p_4p_5$ about the intersection $O$ (see Figure 15.1 on the right).

In fact if one of the triangles is obtained by a translation from another we have a finite flexion.

**Further organization of this chapter.** In Section 15.2 we recall the definition of tensegrity, define the configuration space of tensegrities and discuss its stratification; as an example we study here a simple case of tensegrities on 4 points in the plane. Further in Section 15.3 we study the extended Cayley algebra and introduce the corresponding geometric relations on the elements of the configuration spaces of lines and points in the plane. After that we study geometric conditions of infinitesimal rigidity in terms of extended Cayley algebra in Section 15.4. We continue in Section 15.5 with an introduction of the techniques of local surgeries on graphs that preserve geometric conditions. Finally in Section 15.6 we conclude with the algorithmic questions in the subject.

## 15.2 Configuration Space of Tensegrities and its Stratification

Let us start with some definitions and background. Within this chapter we will employ the techniques of tensegrities that is very useful in the study of infinitesimal rigidity. Nonzero tensegrities and infinitesimal flexions are closely related. In fact, the following statement holds:

*A realization of the graph is infinitesimally flexible if and only if it admits a non-zero tensegrity.*

One of the benefits of tensegrities is an additional structure of a force-load for the edges of the realizations. The force-loads are extremely helpful for understanding geometric conditions describing infinitesimally flexible configurations in nongeneric settings.

### 15.2.1 Background

The story of tensegrities goes back to J.C. Maxwell who studied equilibrium states for frames under the action of static forces, see [20]. In the second half of the twentieth century tensegrities reappeared in the art, K. Snelson constructed several surprising cable-bar sculptures that are actually tensegrities [26]. Recently *TensegriTree* was erected (see [7]) to celebrate the 50th anniversary of the University of Kent (see [21] for an overview and history of tensegrity constructions). The term "tensegrity" (a combination of words "tension" and "integrity") was introduced by R. Buckminster after erection of first such constructions. There are various applications of tensegrities in different branches of science. In particular they were used in the study of viruses [1, 24], cells [12, 13], and deployable mechanisms [25, 28].

In mathematics, tensegrities were studied in different contexts: rigidity and flexibility of tensegrities [2, 3, 22, 34]; minimal rigidity [29]; mechanical properties [14]. Classical tensegrities were extended to spherical and projective geometry [23]; to other non-Euclidean geometries and special normed spaces [15, 19]; to surfaces in $\mathbb{R}^3$ [15]. See also a general introductory paper on tensegrities in mathematics by R. Connelly [4].

In [31, 32] N.L. White and W. Whiteley studied the existence of non-zero tensegrities for a given $n$-tuple of points. They have introduced algebraic conditions for the existence of nontrivial tensegrities (see also [33]). Later in the papers [8, 9, 10] M. de Guzmán and D. Orden introduce atom decomposition techniques and describe several conditions for some further examples of graphs. Finally in [16] the existence of non-zero tensegrities was described in terms of extended Cayley algebra.

### 15.2.2 Definition of a Tensegrity

Let us bring together main notions and definitions in theory of tensegrities.

**Definition 15.1** Fix a positive integer $d$. Let $G = (V,E)$ be an arbitrary graph without loops and multiple edges. Let it have $n$ vertices.

- A *framework* $G(P)$ in $\mathbb{R}^d$ is a map of the graph $G$ with vertices $v_1,\dots,v_n$ on a finite point configuration $P = (p_1,\dots,p_n)$ in $\mathbb{R}^d$ with straight edges, such that $G(P)(v_i) = p_i$ for $i = 1,\dots,n$.

- A *stress* $w$ on a framework is an assignment of real scalars $w_{i,j}$ (called *tensions*) to its edges $p_ip_j$. We also put $w_{i,j} = 0$ if there is no edge between the corresponding vertices. Observe that $w_{i,j} = w_{j,i}$, since they refer to the same edge.

- A stress $w$ is called a *self-stress* if, in addition, the following equilibrium condition is fulfilled at every vertex $p_i$:
$$\sum_{\{j|j\neq i\}} w_{i,j}\overline{p_ip_j} = 0.$$
By $\overline{p_ip_j}$ we denote the vector from the point $p_i$ to the point $p_j$.

- A pair $(G(P),w)$ is called a *tensegrity* if $w$ is a self-stress for the framework $G(P)$.

- We say that $F$ is a *force-load* for a tensegrity $(G(P),w)$ if for every edge $p_ip_j$ there exists a force $F_{i,j}$ acting along this edge such that
$$F_{i,j} = w_{i,j}\overline{p_ip_j}.$$

### 15.2.3 Stratification of the Space of Tensegrities

Let us consider the configuration space of all ordered $n$-tuples of points in $\mathbb{R}^d$, i.e., $(\mathbb{R}^d)^n$. Let also $G$ be some graph on $n$ vertices. Denote by $\text{Tens}_G(P)$ the space of all tensegrities on $G(P)$. Note that $\text{Tens}_G(P)$ is a linear space whose dimension coincide with codimension of the rigidity matrix.

**Definition 15.2**

- A maximal connected component of $(\mathbb{R}^d)^n$ whose all points have the same dimension $\text{Tens}_G(*)$ is called a *stratum* for the graph $G$.
- A disjoint decomposition of $(\mathbb{R}^d)^n$ into strata with respect to $G$ is called the *stratification* of $(\mathbb{R}^d)^n$ related to $G$
- The *universal tensegrity stratification of* $(\mathbb{R}^d)^n$ is the intersection of all the stratifications for all possible graphs $G$ on $n$ vertices. The strata of universal tensegrity stratification are called the *universal tensegrity strata*.

It is known that all the strata are semialgebraic sets [5].

### 15.2.4 Tensegrities on 4 Points in the Plane

In this subsection we consider the universal stratification of the space of all tensegrities on 4 points in the plane. For a generic 4-tuple of points $P$ we have: *no three points in a line, no two points coincide*. A generic $P$ admits the unique up to a scalar multiplication tensegrity for a complete graph $K_4$ and no tensegrities for all other graphs on 4 points. There are exactly 14 connected components of generic points.

The universal strata of codimension 1 correspond to the configurations $P$ where *three out of four vertices of the graph lie in a line*. For such configurations there are some $K_3 \subset K_4$ graphs that admits a nonzero tensegrity. The number of such strata is 24.

Combinatorial adjacency structure of the full dimension and codimension 1 universal strata is shown on Figure 15.2. Each oval corresponds to the union of 6 universal strata of codimension 1 having the same triples of points in a line. The ovals divide the plane into 14 connected component representing universal strata of full dimension. Finally the large dots represent the universal strata of higher codimension (which is rather non-trivial to show on the picture).

All the universal strata of codimension greater than 1 are the intersections of the closures of the codimension 1 universal strata. The table of all these strata follows.

| Stratum description | codim | quantity |
|---|---|---|
| Four points in a line | 2 | 12 |
| Two points coincide | 2 | 12 |
| Four points in a line, two of which coincide | 3 | 18 |
| Three points coincide | 4 | 4 |
| Two pairs of points coincide | 4 | 3 |
| All points coincide | 6 | 1 |

For a more detailed description of the configuration spaces we refer to papers [5]. The case of planar tensegrities on 5 points is exhaustively studied in [17]. It has the following amount of universal strata. The universal stratum of codimension 8 is the stratum corresponding to the case when all the points coincide, it is of dimension 2.

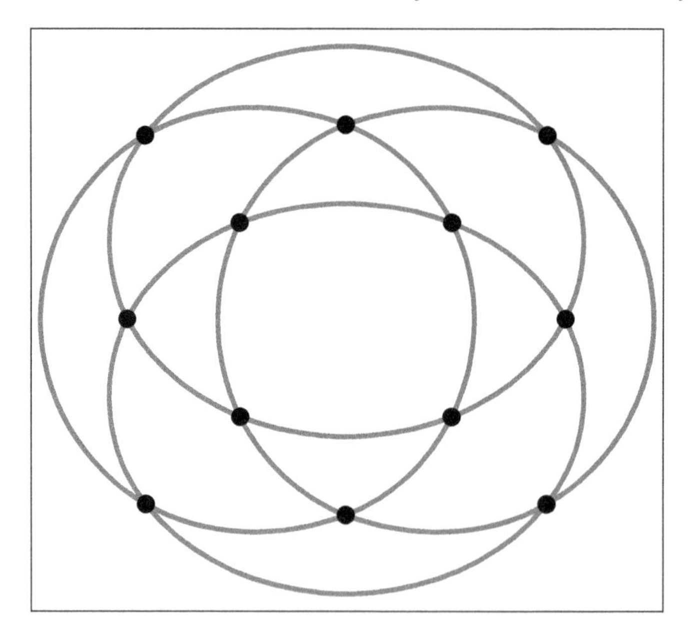

**Figure 15.2**
Adgacency of codimension 1 and full dimension strara of tensegrities on 4 points in the plane.

| Codimension of a stratum | 0 | 1 | 2 | 3 | 4 | 5 | 6 | 7 | 8 |
|---|---|---|---|---|---|---|---|---|---|
| Number of strata | 264 | 600 | 810 | 300 | 170 | 75 | 15 | 0 | 1 |

## 15.3 Extended Cayley Algebra and the Corresponding Geometric Relations

If the number of points is greater than 5 then a more interesting universal strata appear. They are defined by geometric conditions of the extended Cayley algebra which we study in this section.

The elements of the extended Cayley algebra are points, lines and a special element which we denote as "*true*." The element *true* is similar to element 1 in the algebra of logic. It is introduced in order not to consider multiple cases when certain non-generality of the configuration occur. Usually *true* means that either certain lines or points coincide. Whenever the element *true* is involved in an expression on Cayley algebra, this expression is *true* itself.

### 15.3.1 Extended Cayley Algebra

For simplicity we work in the projectivization of the Euclidean plane, i.e., in real projective plane $\mathbb{R}P^2$. First of all we define four elementary operations on points and lines in $\mathbb{R}P^2$. These operations have much in common with *join* and *meet* operations of Cayley algebra (i.e., $\vee$ and $\wedge$). For more information on Cayley algebras we refer to [6], [30], and [11]. Note that it is much more convenient to work in the projective settings since every two lines here intersect.

**Operation I (2-point operation).** Denote the first operation by $(*,*)$. This operation is a binary operation defined on the set of all points in the projective plane and the additional element *true*, as follows.

| $(*,*)$ | $\mathbf{p_1}$ | $\mathbf{p_2}(\neq \mathbf{p_1})$ | **true** |
|---|---|---|---|
| $\mathbf{p_1}$ | *true* | $p_1 \vee p_2$ | *true* |
| $\mathbf{p_2}(\neq \mathbf{p_1})$ | $p_1 \vee p_2$ | *true* | *true* |
| **true** | *true* | *true* | *true* |

Here $p_1$ and $p_2$ are arbitrary distinct points and $p_1 \vee p_2$ denotes the line through these points.
In case if there is no confusion we write $p_1 p_2$ instead of $(p_1, p_2)$.

**Operation II (2-line operation).** Similarly we define the binary operation $* \cap *$ on the the set of all lines in the projective plane and the additional element *true*.

| $* \cap *$ | $\ell_1$ | $\ell_2(\neq \ell_1)$ | **true** |
|---|---|---|---|
| $\ell_1$ | *true* | $\ell_1 \wedge \ell_2$ | *true* |
| $\ell_2(\neq \ell_1)$ | $\ell_1 \wedge \ell_2$ | *true* | *true* |
| **true** | *true* | *true* | *true* |

Here $\ell_1$ and $\ell_2$ are arbitrary distinct lines and $\ell_1 \vee \ell_2$ denotes the intersection point of these lines.

**Choice operations.** Further we need two choice operations:

| **Operation III**<br>**Point choice** | **Operation IV**<br>**Line choice** |
|---|---|
| Pick a point on a given line avoiding a given discrete subset of points | Pick a line through a given point avoiding a given discrete subset of lines |

Let us study the following curious example.

**Remark 15.2** Consider the following 4-tiple of points of projective plane:

$$p_1 = [0:0:1], \quad p_2 = [1:0:1], \quad p_3 = [0:1:1], \quad p_4 = [1:1:1]$$

and let $p = [a:b:c]$. *What can we obtain starting with the points $p_1, p_2, p_3$, and $p_4$ applying consequently Operations I and II?* The answer to this question is very surprising. One can construct precisely all points of $p \in \mathbb{Q}P^2$ (i.e., $p = (a,b,c)$ is proportional to a triple of rational numbers).

**Relations.** Finally we define several basic relations for points and lines in $\mathbb{R}P^2$ which will be used to generate equations in extended Cayley algebra:

| **Relation** | **Notation** | **It is fulfilled in the following cases** |
|---|---|---|
| **3-point** | $(p_1, p_2, p_3) = true$ | 1) One of the entries is *true*;<br>2) At least two points coincide;<br>3) $p_1$, $p_2$, $p_3$ are in a line. |
| **point-line** | $p \in \ell = true$ | 1) One of the entries is *true*;<br>2) $p$ is contained in $\ell$. |
| **3-line** | $\ell_1 \cap \ell_2 \cap \ell_3 = true$ | 1) One of the entries is *true*;<br>2) at least two lines coincide;<br>3) $\ell_1$, $\ell_2$, $\ell_3$ are concurrent. |

We conclude this subsection with an important remark.

**Remark 15.3** In fact every extended Cayley algebra expression is well defined for points and lines of the Euclidean plane (which is considered as an affine plane in the projective plane). For that reason we will use extended Cayley algebra in Euclidean settings as well. While operating with points and lines of the Euclidean plane, it is useful to keep in mind that the resulting object can be the line at infinity or a point at the line at infinity.

### 15.3.2 Geometric Relations on Configuration Spaces of Points and Lines

Let us consider geometric relations for special configuration spaces of lines. For a fix $n$-point configuration $P$ consider the configuration space of (non-fixed) lines passing through prescribed points, namely

$$\{(\ell_1,\dots,\ell_m)\big|\ell_i \text{ passes through } p_{j(i)},\, i=1,\dots,m\}$$

We denote it by $\Xi_P(R)$, where $R$ is the list of inclusion conditions defining the configuration space (i.e., $R=\{p_{j(i)}\in\ell_i\big|i=1,\dots,m\}$).

**Remark 15.4** One may think of lines $\ell_1,\dots,\ell_m$ to be variables of equations while points $p_1,\dots,p_n$ to be parameters on which equations depend.

Let us give the following general definitions.

**Definition 15.3** Consider $\Xi_P(R)$ as above.

- A *geometric condition* on $\Xi_P(R)$ is a composition of several geometric operations and one geometric relation on points $(p_1,\dots,p_n)$ and lines $(\ell_1,\dots,\ell_m)$ of $\Xi_P(R)$.
- We say that a system of geometric conditions on $\Xi_P(R)$ is *fulfilled* at $P$ if there exists a choice of $m$ lines satisfying all the conditions $R$ such that every geometric condition is "*true*" for this choice of lines.
- Two systems of geometric conditions on $n$-tuples of points and $m$-tuples of lines satisfying conditions $R$ are *equivalent* if for every configuration $P$ these systems are simultaneously either fulfilled or not fulfilled for $\Xi_P(R)$ at $P$.

Let us consider the following simple example.

**Example 42** *For the configuration space*

$$\Xi_{(p_1,\dots,p_6)}(p_5\in\ell_1)$$

*we consider the following system of geometric conditions:*

$$\begin{cases} p_1p_4\cap p_2p_3\cap\ell=true \\ p_6\in\ell=true \end{cases}.$$

*Then a generic configuration of 6 points does not fulfill this system, while the following configuration does.*

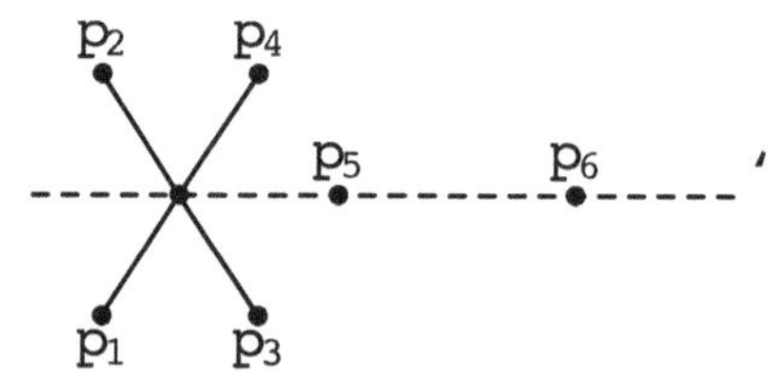

## 15.4 Geometric Conditions of Infinitesimal Flexibility in Terms of Extended Cayley Algebra

Recall that the geometric conditions of infinitesimal flexibility coincide with geometric conditions on realizability of non-zero tensegrities for the corresponding frameworks.

In Subsection 15.4.1 we study examples of such conditions for graphs of 6 and 7 vertices. Further in Subsections 15.4.2 and 15.4.3 we discuss the generality assumptions for frameworks and tensegrities respectively. In Subsection 15.4.4 we formulate the theorem which states that conditions on realizability of a generic (i.e., non-parallelizable) tensegrity for a given graph or framework are expressed in terms of extended Cayley algebra. This theorem is one of the central statements of this chapter. Finally in Subsection 15.4.5 we discuss a potential way to improve this theorem.

### 15.4.1 Examples in the Plane

In all the examples of this subsection the list of non-fixed lines is empty. So we have simply $\Xi_P()$. There are two interesting examples on 6 vertices, they are listed below.

| Graph (6 vert.) | Sufficient geometric conditions |
|---|---|
| 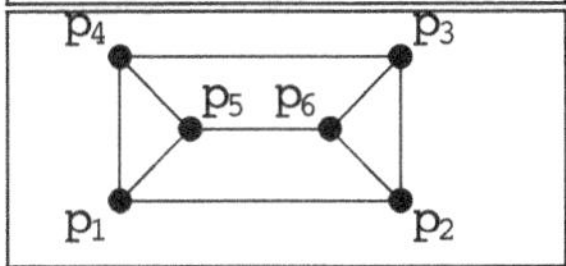 p4 p3 p5 p6 p1 p2 | $p_1p_2 \cap p_3p_4 \cap p_5p_6 = true$ |
| 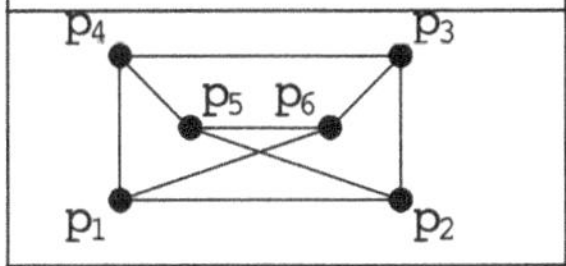 p4 p3 p5 p6 p1 p2 | $(p_1p_2 \cap p_4p_5, p_2p_3 \cap p_5p_6, p_3p_4 \cap p_6p_1) = true$<br>(Equivalently: the six points $p_1$, $p_2$, $p_3$, $p_4$, $p_5$, and $v_6$ are on a conic) |

In the right column of the above table we write a sufficient condition for a 6-tuple of points to admit a non-zero tensegrity (or, equivalently, to have a non-trivial infinitesimal isometric flexion) for the corresponding graph.

We have the following collection of graphs on 7 vertices together with geometric conditions for them.

| Graph (7 vert.) | Sufficient geometric conditions |
|---|---|
| 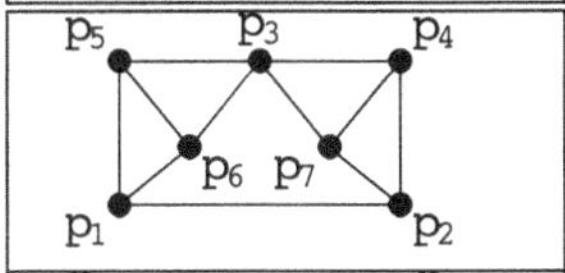 p5 p3 p4 p6 p7 p1 p2 | $(p_1, p_2, p_3) = true$ |
| 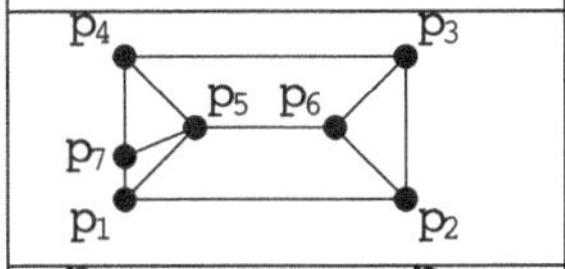 p4 p3 p5 p6 p7 p1 p2 | $p_1p_2 \cap p_3p_4 \cap p_5p_6 = true$ |
| 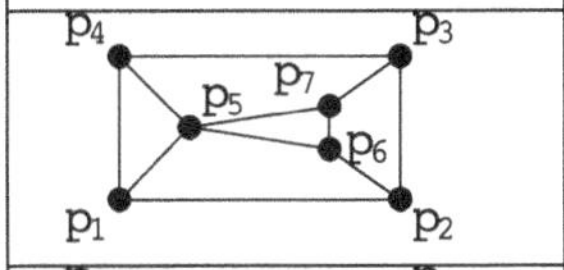 p4 p3 p5 p7 p6 p1 p2 | $p_1p_2 \cap p_3p_4 \cap (p_5, p_2p_6 \cap p_3p_7) = true$ |
| 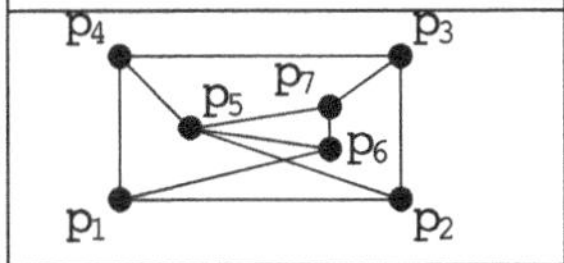 p4 p3 p5 p7 p6 p1 p2 | $(p_1p_2 \cap p_4p_5, p_2p_3 \cap (p_5, p_1p_6 \cap p_3p_7), p_3p_4 \cap (p_1, p_1p_6 \cap p_3p_7)) = true$ |

### 15.4.2 Frameworks in General Position

Recall that a *cycle* is a graph homeomorphic to the circle. A *simple cycle* in a graph $G$ is a subgraph of $G$ homeomorphic to the circle.

**Definition 15.4** Let us consider the following notions of general position.

- An $n$-tuple of lines in the Euclidean plane are said to be *in general position* if the following two conditions hold:

  — no three lines meet in a point;

  — no three lines are parallel.

- Let $C(P)$ be a realization of a cycle $C$ in the plane, where $P = (p_1, \ldots, p_n)$. We say that $C(P)$ is *in general position* if the lines passing through the edges of $C(P)$ are in general position.

- A framework $G(P)$ in the plane is said to be *in general position* if every simple cycle of at most $n-1$ vertex is in general position (recall that $G$ has $n$ vertices).

In particular if a cycle is in general position, then all its edges are of nonzero length.

### 15.4.3 Non-parallelizable Tensegrities

When study conditions of realisability we will assume some natural general position assumptions for the force-loads of tensegrities. These assumptions are collected in the following definition.

**Definition 15.5** Let $G$ be a graph on $n$ vertices, $(G(P), w)$ be a tensegrity and $F$ be an equilibrium force load for the tensegrity $(G(P), w)$. We say that the equilibrium force-load $F$ is *non-parallelizable* at vertex $p$, if the following two conditions are fulfilled.

Suppose that the forces of $F$ at all edges adjacent to $p$ are $F_1, \ldots, F_s$.

- Let $\varepsilon_i \in \{0, 1\}$ for $i = 1, \ldots, s$. Then

$$\sum_{i=1}^{s} \varepsilon_i F_i = 0 \quad \text{if and only if} \quad \varepsilon_1 = \ldots = \varepsilon_s.$$

- All the following forces

$$F_1 + \sum_{i=2}^{s} \varepsilon_i F_i, \quad \text{where} \quad (\varepsilon_2, \ldots \varepsilon_s) \in \{0,1\}^s \setminus \{(1, \ldots, 1)\}$$

  are nonproportional to each other. (Recall that here we consider precisely $2^{s-1} - 1$ forces.)

We sat that a tensegrity $(G(P), F)$ is *non-parallelizable* if it is non-parallelizable at every its vertex.

**Remark 15.5** If the first condition of Definition 15.5 is not fulfilled at $p$, then one of the following surgeries on the graph can be done:

— either some edge can be deleted

— or $p$ can be splitted in two vertices (each edge is adjacent either to one copy of $p$ or to another).

The resulting framework admits tensegrity with the same non-zero forces along the corresponding edges. So in some sense this tensegrity is realizable for a framework of a "simpler" graph (i.e., such graph has a smaller first Betti number).

### 15.4.4 Geometric Conditions for Existence Non-parallelizable Tensegrities

In this subsection we briefly describe the objects that are involved in the geometric conditions for non-parallelizable tensegrities. The actual algorithm to write them would be given in the last section of this chapter.

Consider an arbitrary graph $G$ on $n$ vertices, and let $G(P)$ be a framework with distinct points $P=(p_1,\dots,p_n)$. Consider the following arrangement of lines: *At each point $p_i \in P$ we choose* $\deg p_i - 3$ *ordered lines passing through $p_i$.* Denote by $\Xi_G(P)$ the configuration space of all such arrangements.

**Remark 15.6** We have

$$\dim(\Xi_G(P)) = \sum_{i=1}^{n} \left(\deg(p_i)-3\right),$$

In particular, if all vertices are of degree 3, then the configuration space $\Xi_G(P)$ is empty.

We use the configuration space $\Xi_G(P)$ for the detection non-parallelizable tensegrities. Let us formulate one of the central theorems of this chapter.

**Theorem 15.1** *A framework $G(P)$ in general position admits a non-parallelizable tensegrity if and only if this framework satisfies a certain system of geometric conditions on points of $P$ and lines of $\Xi_G(P)$ for all simple cycles of $G$.*

**Remark 15.7** The system of geometric conditions is explicitly described by the algorithm of Subsection 15.6.6, see also Theorem 15.3.

The proof of this theorem is rather technical, we refer an interested reader to [16].

### 15.4.5 Conjecture on Strong Geometric Conditions for Tensegrities

As we have seen in the examples of the tables above, all the geometric conditions are written entirely in terms of the vertices of the framework $P$, none of the lines of $\Xi_G(P)$ for them are involved. One might expect that this situation is general for planar tensegrities. Let us briefly discuss this here.

**Definition 15.6** Let $G$ be a graph and let $G(P)$ be one of the frameworks for $G$. We say that a geometric condition on points $P$ is a *strong geometric condition* for $G(P)$ if it does not involve choice operations (i.e., operations III and IV).

In many cases geometric conditions on points and lines passing through them are equivalent to certain strong geometric conditions on points only. However this is not always the case, we illustrate this with the following example.

**Example 43** *In this example we deal with the configuration space*

$$\Xi_{(p_1,p_2,p_3,p_4,p_5,p_6)}\left(p_1 \in \ell_1, p_2 \in \ell_2, p_3 \in \ell_3\right)$$

*and the following system of geometric conditions for it:*

$$\begin{cases} \ell_1 \cap \ell_2 \cap p_4p_5 = true \\ \ell_2 \cap \ell_3 \cap p_5p_6 = true \\ \ell_3 \cap \ell_1 \cap p_6p_4 = true \end{cases}$$

*Below is the example of a 9-point and 3-line configuration satisfying the above system of condition.*

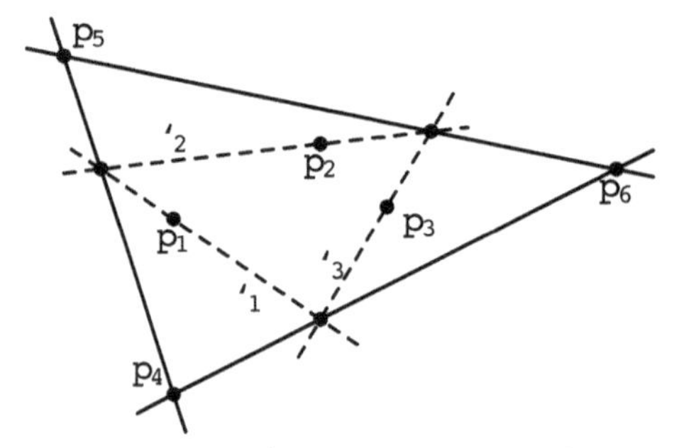


*From the one hand, this system is not equivalent to any system of strong geometric conditions on P. From the other hand this system is not related to any graph G.*

In the view of the last example it would be interesting to check if the following statement holds.

**Conjecture ([16]).** For every graph $G$ there exists a system of strong geometric conditions such that a framework $G(P)$ in general position admits a non-parallelizable tensegrity if and only if $P$ satisfies this system of strong geometric conditions.

In other words, this conjecture implies that all the non-parallelizable tensegrities are described in terms of Cayley algebra operations on the *vertices* of frameworks.

We would like to conclude this section with the following interesting question.

**Problem.** Develop a similar geometric approach to the study of infinitesimal flexibility in three- and higher-dimensional cases.

## 15.5 Surgeries on Graphs

Surgeries on graphs provide a techniques to obtain geometric conditions on graphs while knowing geometric conditions on some another graphs. Here we change the structure of the graph locally leaving most of the vertices and edges of the original graph unchanged. The more surgeries one knows, the smaller the set of initial graphs one needs to study in order to develop a list of geometric conditions for all graphs. This techniques is rather successful in the planar case and rather unstudied in higher dimensional cases. Below we describe most of the currently known graph surgeries.

In the diagrams we show a part of a graph. Black dots indicate vertices that might have some edges that we do not see on the corresponding diagram (i.e., the edges connecting such vertices with the vertices that are not in this part of the graph). The edges of white dots are precisely the edges shown in the diagram. Finally we would like to mention that surgeries nay be applied only under certain genericity conditions. Such conditions are indicated in the last column.

**Basic surgeries.** We start with the simplest possible type of surgeries. These surgeries remove the points of degree 1 and 2. Here is the complete list of them.

| Source | Target | Conditions |
|---|---|---|
| p1 ●——○ p2 | ● p1 | $p_1 \neq p_2$. |
| p1 ●——○ p2 ——● p3 | p1 ●    ● p3 | The points $p_1$, $p_2$ and $p_3$ are not in a line. |
| p1 ●——○ p2 ——● p3 | p1 ●————● p3 | The points $p_1$, $p_2$ and $p_3$ are in a line and distinct to each other. |

**Dimension 1 subgraph surgeries.** The second class of surgeries is rather wide. First of all we give the following definition.

**Definition 15.7** We say that a triple $(G, e_1, e_2)$ where $G$ is a graph and $e_1$ and $e_2$ are its edges *fulfills the condition* $\langle G, e_1, e_2 \rangle$ at a framework $(G, P)$ if the following two conditions hold.

— The graph $G$ has a unique (up to a scalar) non-zero tensegrity.

— The stresses at edges $e_1$ and $e_2$ for non-zero tensegrities on $(G, P)$ are non-zero.

Here we have the following general statement.

***Proposition 15.1***
*If a $(G, e_1, e_2)$ fulfills the condition $\langle G, e_1, e_2 \rangle$ then the tensegrities for the graph $G \cup e_1$ admits non-zero tensegrities at P if and only if the graph $G \cup e_2$ admits non-zero tensegrities at P.*

The last proposition give rise to the following list of surgeries.

| Source | Target | Conditions |
|---|---|---|
| 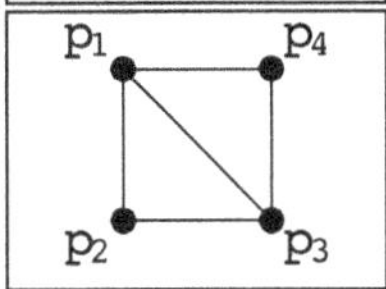 $p_1$, $p_4$, $p_2$, $p_3$ | 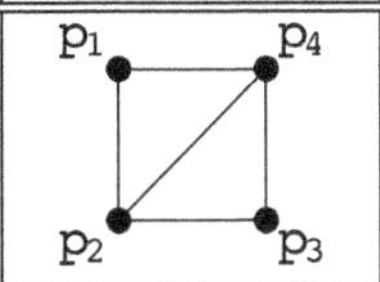 $p_1$, $p_4$, $p_2$, $p_3$ | All the triples of points are not in a line. |
| 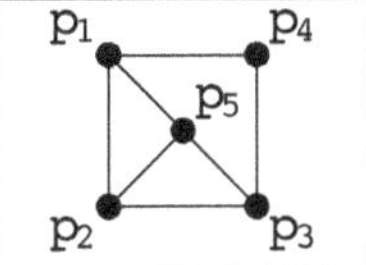 $p_1$, $p_4$, $p_5$, $p_2$, $p_3$ | 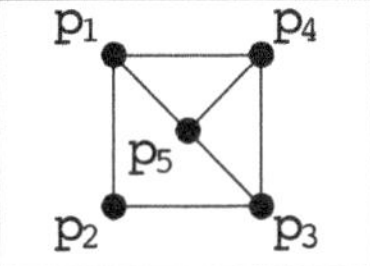 $p_1$, $p_4$, $p_5$, $p_2$, $p_3$ | The triples of points are not in a line: $(p_4, p_1, p_5)$ and $(p_i, p_{i+1}, p_5)$ where $i = 1, 2, 3$, and the points $(p_1, p_2, p_3, p_4)$ are not in one line. |
| ... | ... | ... |
| $H \setminus e_1$ | $H \setminus e_2$ | There exists $G \subset H$ with $e_1, e_2 \in G$ satisfying Consdition $\langle G, e_1, e_2 \rangle$. |

One should be careful while using these surgeries, since the condition $\langle G, e_1, e_2 \rangle$ might already contain nontrivial configuration with non-zero tensegrities. So while removing such cases one might remove possible realizations.

**Degree 3 vertex surgeries.** The next class of surgeries plays an important role in the planar case.

| Source | Target | Relations | Conditions |
|---|---|---|---|
| 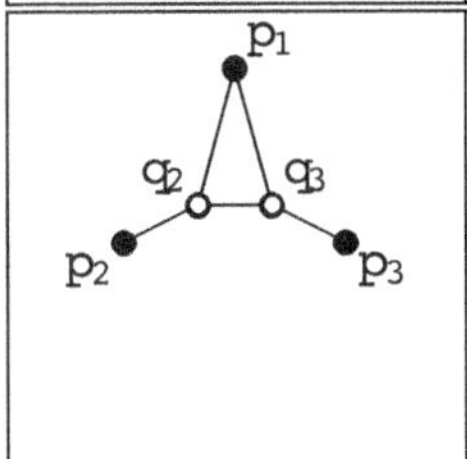 $p_1$, $q_2$, $q_3$, $p_2$, $p_3$ | 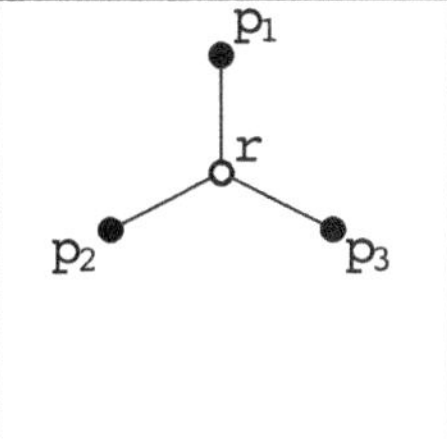 $p_1$, $r$, $p_2$, $p_3$ | $r = p_2q_2 \cap p_3q_3$ | the triples of points are not in a line: $(p_1, q_2, q_3)$, $(p_2, p_3, q_i)$, and $(p_1, p_i, q_i)$, where $i = 2, 3$. |
| 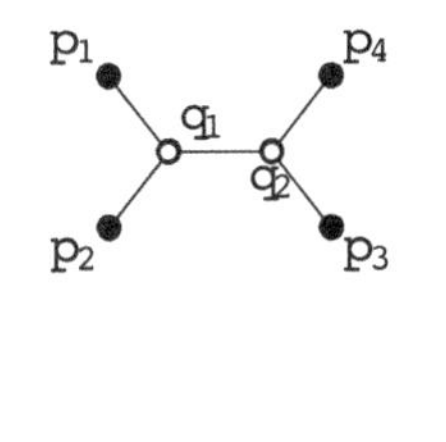 $p_1$, $p_4$, $q_1$, $q_2$, $p_2$, $p_3$ | 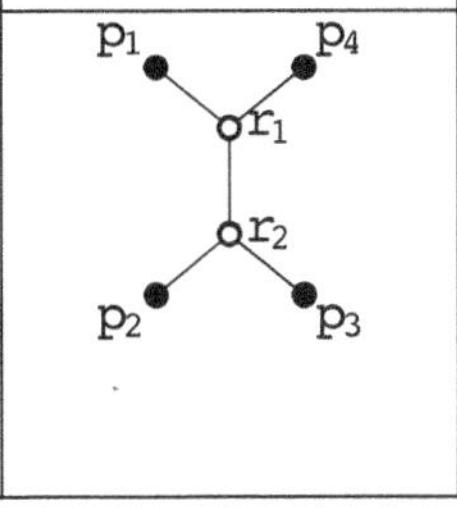 $p_1$, $p_4$, $r_1$, $r_2$, $p_2$, $p_3$ | $r_1 = p_1q_1 \cap p_4q_2$<br>$r_2 = p_2q_1 \cap p_3q_2$ | the triples of points are not in a line: $(p_1, p_2, q_1)$, $(p_3, p_4, q_2)$, and $(p_i, q_1, q_2)$ for $i = 1, \ldots, 4$. |

The last surgery is called an HΦ-surgery. It is essentially used in the study of planar tensegrities, see in [16].

*Remark on strong geometric conditions for graphs on small number of vertices.* Using these surgeries one might find strong geometric conditions of infinitesimal isometric flexibility to all graphs having 9 or less vertices (see in [16]). The complete list of codimension 1 graphs for 8 or less vertices can be found in [5]. The first example of a graph which we unable to reduce to 9-point graphs using the above surgeries is as follows:

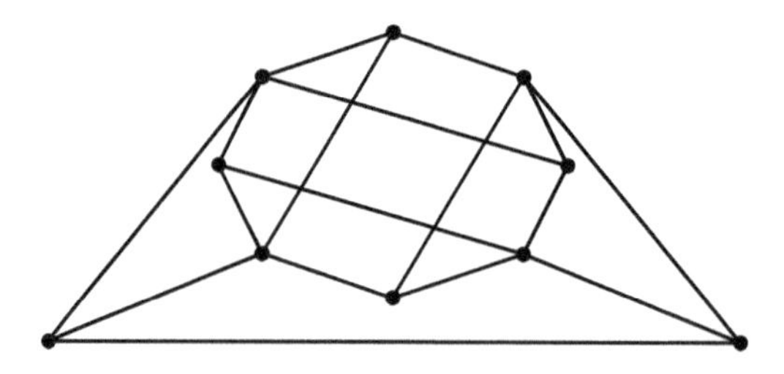

This graph is a good candidate as a counterexample to the conjecture of Subsection 15.4.5.

## 15.6 Algorithm to Write Geometric Conditions of Realizability of Generic Tensegrities

In this section we show how to construct geometric conditions for cycles mentioned in Theorem 15.1.

We start this section with the study of framed cycles, here for each framed cycle in general position we introduce a certain geometric condition related to it. This already allows us to define geometric conditions for trivalent graphs (see Subsection 15.6.1). For the graphs having vertices of degree greater than 3 we introduce resolution schemes for such vertices (see Subsection 15.6.2). Further in Subsections 15.6.3 and 15.6.4 using resolution schemes we construct the framings for simple cycles of a general graph $G$, which provides us with desired geometric conditions for $G$. Finally we summarize the construction techniques of geometric conditions detecting non-zero tensegrities in Subsection 15.6.5 (see also Theorem 15.3).

### 15.6.1 Framed Cycles in General Gosition

#### 15.6.1.1 Basic Definitions

Let us start with the following general definition.

**Definition 15.8** Let $P = (p_1, \dots, p_k)$.

- A realization of the cycle $C(P)$ in the Euclidean plane has a *framing* if every vertex $p_i$ is equipped with a line $\ell_i$ passing through it.
- The realization $C(P)$ together with its framing is called the *framed cycle*. Denote it by

$$C(P,L) = \big((p_1, \dots, p_k), (\ell_1, \dots, \ell_k)\big).$$

We use the following notion of genericity for framed cycles.

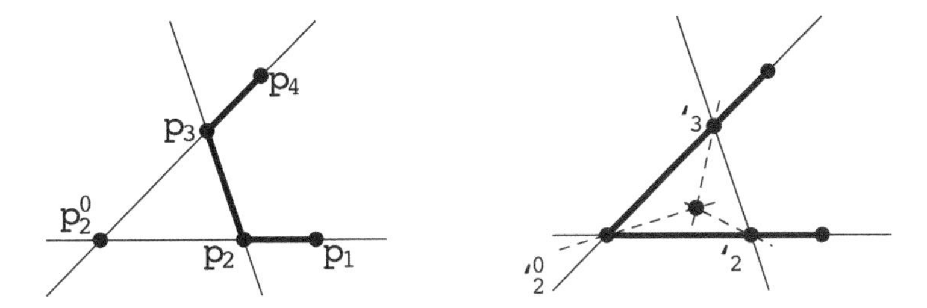


**Figure 15.3**
A projection operation $\omega_2$. On the left we have: $p'_2 = p_1p_2 \cap p_3p_4$; on the right: $\ell'_2 = (p'_2, \ell_2 \cap \ell_3)$.

**Definition 15.9** A framed cycle $C(P,L)$ is in *general position* if
— the cycle $C(P)$ is in general position;
— for every admissible $i$ we have: the line $\ell_i$ does not contain the points $p_{i-1}$ and $p_{i+1}$.

Let $C(P,L)$ be a cycle with $k \geq 4$ vertices. for an arbitrary $i \in \{1,2,\ldots,k\}$ we set

$$p'_i = p_{i-1}p_i \cap p_{i+1}p_{i+2};$$
$$\ell'_i = p'_i(\ell_i \cap \ell_{i+1}).$$

(Here we set $p_0 = p_k$, $p_{k+1} = p_1$, and $p_{k+2} = p_2$.)

The following surgery on the framed cycle $C(P,L)$ is called a *projection operation* of a cycle:

$$\omega_i(C(P,L)) = C(P',L'),$$

where

$$\begin{array}{ll} P' = (p_1, p_2, \ldots, p_{i-1}, p'_i, p_{i+2}, \ldots, p_k), & \text{and} \\ L' = (\ell_1, \ell_2, \ldots, \ell_{i-1}, \ell'_i, \ell_{i+2}, \ldots, \ell_k). & \end{array} \tag{15.1}$$

In other words the projection operation $\omega_i(C(P,L))$ removes the edge $p_ip_{i+1}$ by prolonging the edges $p_{i-1}p_i$ and $p_{i+1}p_{i+2}$ towards the intersection point of their lines (i.e., $p'_i$) and introduce a framing $\ell'_i$ to $p'_i$. See an example of $\omega_2(C(P,L))$ on Figure 15.3.

**Remark 15.8** It is clear that the projection operation is entirely expressed by the elementary operations (Operations II and I).

#### 15.6.1.2 Geometric Conditions for Framed Cycles

Consider a framed cycle

$$C(P,L) = C((p_1,\ldots,p_k),(\ell_1,\ldots,\ell_k))$$

in general position. Then the composition of any admissible $k-3$ projection operations applied to $C(P,L)$ will result in a framed triangular cycle in general position. Denote the resulting cycle by

$$C(\hat{P},\hat{L}) = C((\hat{p}_1,\hat{p}_2,\hat{p}_3),(\hat{\ell}_1,\hat{\ell}_2,\hat{\ell}_3)).$$

**Definition 15.10** Let $C(P,L)$ and $C(\hat{P},\hat{L})$ be as above. Then the condition

$$\hat{\ell}_1 \cap \hat{\ell}_2 \cap \hat{\ell}_3 = true \tag{15.2}$$

is the *geometric condition* defined by $C$.

Note the following.

- The geometric condition for $C(P,L)$ does not depend on the choice of projection operations. The resulting geometric relations are equivalent.
- The geometric condition for $C(P,L)$ is a combination of $9k-9$ Operations I; $6k-6$ Operations II; and one 3-line relation on points $P$ and lines $L$. (Hence it is a true geometric condition of extended Cayley algebra).

For simplicity one might always fix the following composition of projection operators:

$$\underbrace{\omega_1 \circ \ldots \circ \omega_1}_{k-3 \text{ times}}(C(P,L)).$$

The expression in terms of Operations I, II and one 3-line relation are written directly from expressions (15.1).

**Example 44** *For the cycles on 3, 4, and 5 vertices we have the following geometric relations:*

$$\begin{array}{ll} k{=}3: & \ell_1 \cap \ell_2 \cap \ell_3 = true; \\ k{=}4: & (\ell_1 \cap \ell_4, \ell_2 \cap \ell_3, p_1p_2 \cap p_3p_4) = true; \\ k{=}5: & (\ell_2 \cap \ell_3, p_1p_2 \cap p_3p_4) \cap \ell_1 \cap (\ell_4 \cap \ell_5, p_1p_3 \cap p_3p_4) = true. \end{array}$$

#### 15.6.1.3 Geometric Conditions for Trivalent Graphs

In the case of trivalent graphs we are in position write down all the geometric conditions for cycles mentioned in Theorem 15.1.

Let $G$ be a trivalent graph and $G(P)$ be a framework in general position. Consider a cycle $C$ in $G$ and and set the natural framing for $C(P(C))$ (here $P(C) \subset P$) as follows. Let the vertex $p_i \in C$ by adjacent to the edges $p_ip_{i,1}$, $p_ip_{i,2}$, and $p_ip_{i,3}$ of the framework $G(P)$. Without loss of generality we assume that the edges $p_ip_{i,1}$, $p_ip_{i,2}$ are edges of $C(P)$ at vertex $p_i$. Then we set $\ell_i(C) = p_i, p_{i,3}$. Set

$$L(C) = (\ell_1(C), \ldots, \ell_k(C)).$$

The geometric condition for the cycle $C \in G$ in Theorem 15.1 is precisely the geometric condition (15.2) of Definition 15.10 for $C(P(C), L(C))$. Now Theorem 15.1 for 3-valent graphs can be reformulated as follows.

**Theorem 15.2** *Let $G$ be a trivalent graph. A framework $G(P)$ in general position admits a non-parallelizable tensegrity if and only if every simple (framed) cycle $C(P(C), L(C))$ of $G$ satisfies the geometric condition (15.2) for $C(P(C), L(C))$.*

### 15.6.2 Resolution Schemes

Suppose now $G$ has vertices of degree greater than 3, so we cannot write geometric conditions for cycles using Theorem 15.2. The main idea here is to consider *resolutions* at vertices of degree $k > 3$ replacing them by unrooted full binary trees with $k$ leaves. Then one can define a similar geometric conditions for the resulting graph. We show the main steps to do this below.

#### 15.6.2.1 Definition of Resolution Schemes

Let us study how to replace a vertex of the framework by a unrooted full binary tree. Recall that an unrooted full binary tree is a tree without the root where the degree of every vertex of $T$ is either 1 or 3. Recall also that an edge of a tree is a *leaf* if one of its vertices is of degree 1. All other edges are *interior* edges of a tree.

Denote by $\mathrm{Gr}(1,\mathbb{R}P^2)$ the Grassmannian of 2-dimensional planes in $\mathbb{R}^3$ (i.e., $\mathrm{Gr}(1,\mathbb{R}P^2)$ is the set of all lines in the projective plane). Here as usual when we work in affine settings we consider an affine Grassmannian as a subset of the projective Grassmannian.

**Definition 15.11** Consider an unrooted full binary tree $T$ and let

$$\mathcal{L} : E(T) \to \mathrm{Gr}(1,\mathbb{R}P^2).$$

We say that a pair $(T,\mathcal{L})$ is a *resolution scheme* at point $p$ in the Euclidean plane if for every edge $e \in T$ it holds $p \in \mathcal{L}(e)$. Denote it by $(T,\mathcal{L})_p$.

#### 15.6.2.2 Resolution of a Framework

In what follows we restrict ourselves to graphs whose vertices are all of degree 3 or greater.

**Definition 15.12** Let $G$ be a graph on $n$ vertices and let $G(P)$ be its framework on $P=(p_1,\ldots,p_n)$. We say that the collection

$$(G(P),((T_1,\mathcal{L}_1)_{p_1},\ldots,(T_n,\mathcal{L}_n)_{p_n}))$$

is a *resolution of* $G(P)$ if for every $i$ we have:

- the resolution scheme $(T_i,\mathcal{L}_i)_{p_i}$ has exactly $\deg p_i$ leaves.
- the edges of $G$ adjacent to $p_i$ are enumerated by the leaves $(T_i,\mathcal{L}_i)_{p_i}$ (i.e., the one-to-one correspondence between the adjacent edges and the leaves is fixed).
- let $v$ be a leaf at of $T_i$ corresponding to an edge $p_ip_j$ then

$$\mathcal{L}_i(v) = (p_i,p_j).$$

We denote it by $G(P)^{\mathcal{T}}_{\mathcal{L}}$.

#### 15.6.2.3 HΦ-Surgeries on Completely Generic Resolution Schemes

In this section we describe a surgery for a certain class of generic resolution schemes. This surgery is similar to HΦ-surgery (the second degree 3 vertex surgery shown on page 329) for graphs.

Let $(T,\mathcal{L})_p$ be a resolution scheme. Consider the tree $T'$ obtained from $T$ by the following flip operation:

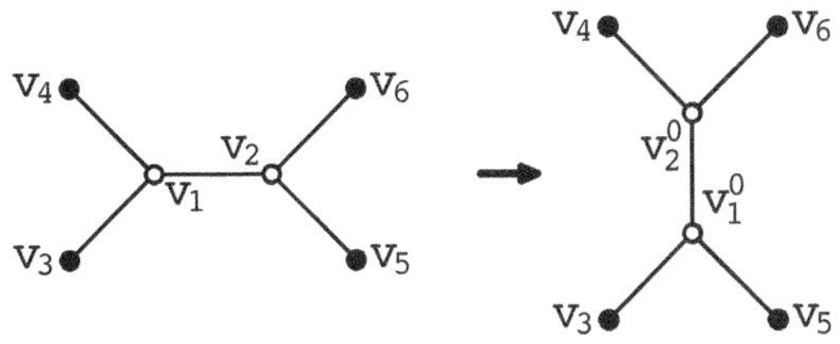


Here white vertices are interior while black vertices can be both interior or leaves.

Let us construct the line $\ell$ which we further associate to $v_1'v_2'$. This is done geometrically using the following sequence of Operations I–IV.

- Operation IV: Pick a point $p_\infty \neq p \in \mathcal{L}(v_1v_2)$;
- Operation III: Pick a line $\ell_\infty \neq \mathcal{L}(v_1v_2)$ through $p_\infty$;

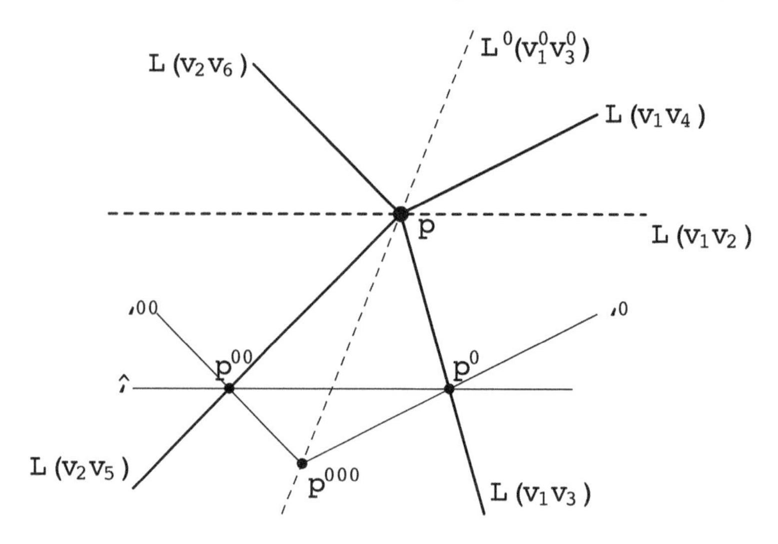


**Figure 15.4**
Geometric construction of $\ell = \mathcal{L}'(v_1'v_2')$.

- Operation III: Pick a point $p' \notin \{p, p_\infty\}$;
- Operations I and II: Define $\hat{\ell} = p(\mathcal{L}(v_1v_2) \cap \ell_\infty)$;
- Operations I: $p'' = \hat{\ell} \cap \mathcal{L}(v_2v_5)$;
- Operations I and II: $\ell' = p'(\mathcal{L}(v_1v_4) \cap \ell_\infty)$;
- Operations I and II: $\ell'' = p''(\mathcal{L}(v_2v_6) \cap \ell_\infty)$;
- Operations I: $p''' = \ell' \cap \ell''$;
- Operations II: $\ell = pp'''$.

See Figure 15.4.

**Remark 15.9** In this construction we recommend to consider the line $\ell_\infty$ to the the line at infinity and $p_\infty$ be one of its points. As we have already mentioned in Remark 15.3 the lines and points involved in our construction may be at infinity.

**Definition 15.13** An HΦ-*surgery on* $(T, \mathcal{L})_p$ *at the interior edge* $v_1v_2$ is the operation that replaces $(T, \mathcal{L})$ with $(T', \mathcal{L}')$ where $\mathcal{L}'$ is defined as follows:

$$\begin{array}{ll} \mathcal{L}'(v_1'v_2') = \ell; & \mathcal{L}'(v_1'v_3) = \mathcal{L}(v_1, v_3); \\ \mathcal{L}'(v_1'v_5) = \mathcal{L}(v_2, v_5); & \mathcal{L}'(v_2'v_4) = \mathcal{L}(v_1, v_4); \\ \mathcal{L}'(v_2'v_6) = \mathcal{L}(v_2, v_6); & \mathcal{L}'(e) = \mathcal{L}(e) \text{ for any other edge } e. \end{array}$$

In fact, the resulting resolution scheme is not always well-defined due to some non-genericity phenomena (e.g., when the lines $v_1v_3$ and $v_2v_5$ coincide). So the following definition is actual here.

**Definition 15.14** We say that a resolution scheme $(T, \mathcal{L})$ is *completely generic* if every composition of HΦ-surgeries is well-defined.

For a geometric description of completely generic resolution trees we refer to [16].

**Remark 15.10** Let $(T,\mathcal{L})_p$ be a completely generic resolution scheme (where $T$ has $n$ leaves). Then applying all possible different compositions of H$\Phi$-surgeries one gets precisely $(2n-1)!!$ distinct resolutions scheme, which is the number of the all unrooted binary full trees with $n$ marked leaves (e.g., see ex. 5.2.6 in [27]). In fact, these schemes are in a natural one-to-one correspondence with the set of all unrooted binary full trees with $n$ marked leaves ($(T,\mathcal{L}) \to T$).

**Definition 15.15** We say that a resolution of a graph is *generic* if every its resolution scheme is completely generic.

### 15.6.3 Construction of Framing for Pairs of Leaves in Completely Generic Resolution Schemes

Assume that we are given by a completely generic resolution scheme $(T,\mathcal{L})_p$. Let us define a special line for every pair of leaves $u,v$.

**Definition 15.16** Let $(T,\mathcal{L})$ be a completely generic resolution scheme. Consider a composition $\phi$ of H$\Phi$-surgeries and a resolution scheme $(T',\mathcal{L}')_p$ such that the following two conditions hold

- $\phi\big((T,\mathcal{L})_p\big) = (T',\mathcal{L}')_p$;
- the leaves $u$ and $v$ are adjacent edges in the tree $T'$ (note that any H$\Phi$-surgery does not affect leaves, so $u$ and $v$ are leaves for both trees $T$ and $T'$).

Assume that $w$ is the third edge of $T'$ adjacent to the common point of the leaves $u$ and $v$. Set

$$\ell_p(u,v) = \mathcal{L}'(w).$$

The following two statements clarify the correctness of Definition 15.16.

***Proposition 15.2***
*For a completely generic resolution scheme $(T,\mathcal{L})$ we have:*

- *there exists a pair $\big(\phi,(T',\mathcal{L}')_p\big)$ satisfying both conditions of Definition 15.16.*
- *the line $\ell_p(v,w)$ does not depend on the choice of $\phi$ and $(T',\mathcal{L}')_p$.*

We will skip the proof of the second statement and provide the algorithm to construct a certain pair $\big(\phi,(T',\mathcal{L}')_p\big)$ below.

**Remark 15.11** **(Construction of $\ell_p(v,w)$.)** Let $v_1\ldots v_s$ be a simple path connecting the leaf $u = v_1v_2$ where $\mathcal{L}(v_1v_2) = p_ip_j$ and the leaf $v = v_{s-1}v_s$ with framing to $\mathcal{L}(v_{s-1}v_s) = p_ip_k$. Then we consequently apply $s-3$ H$\Phi$-surgeries along the edges $v_2v_3, v_3v_4, \ldots, v_{s-2}v_{s-1}$. As a result we have a resolution scheme $(T_i',\mathcal{L}')_{p_i}$ whose leaves $\mathcal{L}'^{-1}(e_{ij})$ and $\mathcal{L}'^{-1}(e_{ik})$ share a common vertex.

### 15.6.4 Framed Cycles Associated to Generic Resolutions of a Graph

First, we define the framing for two adjacent edges of frameworks.

**Definition 15.17** Consider a generic resolution $G(P)_{\mathcal{L}}^{\mathcal{T}}$ of a framework $G(P)$. Let $p_ip_j$ and $p_ip_k$ be two edges in $G(P)$ with a common vertex $p_i$ and let $v$ and $w$ be the associated leaves in the resolution scheme $(T_i, \mathcal{L}_i)_{p_i}$. Then the line $\ell_{p_i}(v,w)$ introduced in Definition 15.16 is the *associated framing for the pair of edges* $(e_{ij}, e_{ik})$ *at* $p_i$. We denote it by $\ell_{j,i,k}$.

The above definition leads to the natural notion of framed cycles associated to generic resolutions of a graph.

**Definition 15.18** Consider a generic resolution $G(P)_{\mathcal{L}}^{\mathcal{T}}$ of a framework $G(P)$. Let $C = p_1 \dots p_s$ be a cycle in $G(P)$. Denote by $C(G, \mathcal{T}, P, \mathcal{L})$ the framed cycle with vertices $p_1 \dots p_s$ such that

$$\ell_{i-1,i,i+1} \text{ is a framing at } q_i \text{ (for } i = 1, \dots, s).$$

We say that this cycle is a *framed cycle associated to* $G(P)_{\mathcal{L}}^{\mathcal{T}}$.

**Definition 15.19** Any line $\ell_{j,i,k}$ in the framing associated to a $G(P)_{\mathcal{L}}^{\mathcal{T}}$ is explicitly expressed in terms of Operations I–IV on the lines of $\mathcal{L}(E(G_{\mathcal{T}}))$ (by Definition 15.13 and Remark 15.11). Let us fix one of the possible composition of Operations I–IV defining the framing $\ell_{j,i,k}$ and call it the *sequence of geometric operations* defining $\ell_{j,i,k}$.

### 15.6.5 Natural Correspondences Between $\Xi_G(P)$ and the Set of all Resolutions for $G(P)$

Given a framework $G(P)$ and the corresponding configuration space $\Xi_G(P)$. Let us fix the following data and notation:

- Fix a resolution tree $T_i$ at each vertex $p_i$ and denote $\mathcal{T} = (T_1, \dots, T_n)$.
- Denote by $G(P)_*^{\mathcal{T}}$ the configuration space of all resolutions for the framework $G(P)$ whose resolution schemes at $p_i$ has a tree $T_i$ for $i = 1 \dots, n$.
- Enumerate all interior edges for every tree $T_i$.
- Enumerate all the lines of $\Xi_G(P)$ passing through every point $p_i$.

Once the above is done, we have a natural isomorphism between $G(P)_*^{\mathcal{T}}$ and $\Xi_G(P)$. Here the line $\ell_j$ at point $p_i$ of the configuration in $\Xi_G(P)$ corresponds to the line $\mathcal{L}_i(v_j)$ of the $j$-th interior edge of the resolution scheme $(T_i, \mathcal{L}_i)_{p_i}$.

### 15.6.6 Techniques to Construct Geometric Conditions Defining Tensegrities

Finally let us show step by step how to write down the system of geometric conditions for the existence of non-parallelizable tensegrities.

**Input Data.** We start with a framework $G(P)$ in general position.

*Step 1.* Fix $\mathcal{T} = (T_1, \dots, T_n)$ in resolution schemes at all vertices and associate the configuration space $\Xi_G(P)$ with $G(P)_*^{\mathcal{T}}$ (see Subsection 15.6.5).

*Step 2.* Pick all simple cycles $C_1, \dots, C_N$ in $G$ that does not pass through all the points of $G$.

*Step 3.* Write all lines $\ell_{j,i,k}$ in terms of compositions of Operations I–IV on the points of $P$ and the lines corresponding to the interior edges of $G(P)_*^{\mathcal{T}}$. (See Definition 15.19.)

*Step 4.* Define framed cycles $C_i(G, \mathcal{T}, P, \mathcal{L})$ related to $C_i$ for $i = 1, \dots, N$. Here we use the lines obtained in Step 3 for framings.

*Step 5.* Write down geometric conditions for $C_i(G,\mathcal{T},P,\mathcal{L})$ for $i=1,\dots,N$ in terms of lines $\ell_{i,j,k}$. (See the construction of Section 15.6.1.2.)

*Step 6.* Combining together Step 3 and Step 5 we write down geometric conditions for framed cycles $C_i(G,\mathcal{T},P,\mathcal{L})$ for $i=1,\dots,N$ in terms of $P$ and the lines of $\Xi_G(P)$ (which is isomorphic to $G(P)_*^{\mathcal{T}}$, see Section 15.6.5).

**Output data.** As an output we get the system of geometric conditions on the space $\Xi_G$. By Theorem 15.1 this system is fulfilled if and only if there exists a non-parallelizable tensegrity at $\Xi_G$.

**Theorem 15.3** *The above algorithm produces geometric conditions for Theorem 15.1.*

**Remark 15.12** In fact, at Step 2 it is sufficient to pick only the simple cycles generating $H_1(G)$. In practice it is sufficient to choose even less cycles to get the corresponding geometric existence condition of a non-parallelizable tensegrity.

For further details and justification of the above algorithm we refer to [16].

# References


[1] D.L.D. Caspar and A. Klug, Physical principles in the construction of regular viruses, in *Proceedings of Cold Spring Harbor Symposium on Quantitative Biology*, vol. 27 (1962), pp. 1–24.

[2] R. Connelly and W. Whiteley, Second-order rigidity and prestress stability for tensegrity frameworks, *SIAM Journal of Discrete Mathematics*, vol. 9, no. 3 (1996), pp. 453–491.

[3] R. Connelly, Tensegrities and global rigidity, *Shaping space*, Springer, New York (2013), pp. 267-278.

[4] R. Connelly, What is ... a tensegrity?, *Notices Amer. Math. Soc.*, vol. 60, no. 1 (2013), pp. 78-80.

[5] F. Doray, O. Karpenkov, J. Schepers, Geometry of configuration spaces of tensegrities, *Disc. Comp. Geom.*, vol. 43, no. 2 (2010), pp. 436–466.

[6] P. Doubilet, G.-C. Rota, J. Stein, On the foundations of combinatorial theory. IX. Combinatorial methods in invariant theory, *Studies in Appl. Math.*, vol. 53 (1974), pp. 185-216.

[7] D. Gray, http://expedition.uk.com/projects/tensegritree-university-of-kent/

[8] M. de Guzmán, *Finding Tensegrity Forms*, preprint (2004).

[9] M. de Guzmán, D. Orden, Finding tensegrity structures: Geometric and symbolic aproaches, in *Proceedings of EACA-2004*, pp. 167–172 (2004).

[10] M. de Guzmán, D. Orden, From graphs to tensegrity structures: Geometric and symbolic approaches, *Publ. Mat.* 50 (2006), pp. 279–299.

[11] H. Li, *Invariant algebras and geometric reasoning.* With a foreword by David Hestenes. World Scientific Publishing Co. Pte. Ltd., Hackensack, NJ, 2008, xiv+518 pp.

[12] D. E. Ingber, Cellular tensegrity: defining new rules of biological design that govern the cytoskeleton, *Journal of Cell Science*, vol. 104 (1993), pp. 613–627.

[13] D.E. Ingber, N. Wang, D. Stamenović, Tensegrity, cellular biophysics, and the mechanics of living systems, *Rep. Progr. Phys.*, vol. 77, no. 4 (2014), 046603, 21 p.

[14] B. Jackson, T. Jordán, B. Servatius, H. Servatius, Henneberg moves on mechanisms, *Beitr. Algebra Geom.*, vol. 56, no. 2 (2015), pp. 587-591.

[15] B. Jackson, A. Nixon, Stress matrices and global rigidity of frameworks on surfaces, *Discrete Comput. Geom.* vol. 54, no. 3 (2015), pp. 586-609.

[16] O. Karpenkov, *The combinatorial geometry of stresses in frameworks*, preprint (2015), arXiv:1512.02563 [math.MG].

[17] O. Karpenkov, J. Schepers, B. Servatius, On stratifications for planar tensegrities with a small number of vertices, *ARS Mathematica Contemporanea*, vol. 6, no. 2 (2013), pp. 305–322.

[18] D. Kitson, S.C. Power, Infinitesimal rigidity for non-Euclidean bar-joint frameworks, *Bull. Lond. Math. Soc.*, vol. 46, no. 4, (2014), pp. 685-697.

[19] D. Kitson, B. Schulze Maxwell-Laman counts for bar-joint frameworks in normed spaces, *Linear Algebra and its Applications* vol. 481, (2015), pp. 313–329.

[20] J. C. Maxwell, On reciprocal figures and diagrams of forces, *Philos. Mag.* vol. 4, no. 27 (1864), pp. 250–261.

[21] R. Motro, *Tensegrity: Structural systems for the future*, Kogan Page Science, London, 2003.

[22] B. Roth, W. Whiteley, Tensegrity frameworks, *Trans. Amer. Math. Soc.*, vol. 265, no. 2 (1981), pp. 419–446.

[23] F.V. Saliola, W. Whiteley, *Some notes on the equivalence of first-order rigidity in various geometries*, arXiv:0709.3354 [math.MG], 15 p.

[24] C. Simona-Mariana, B. Gabriela-Catalina, Tensegrity applied to modelling the motion of viruses, *Acta Mech. Sin.* vol. 27, no. 1 (2011), pp. 125-129.

[25] R.E. Skelton, *Deployable tendon-controlled structure*, United States Patent 5642590, July 1, 1997.

[26] K. Snelson, http://www.kennethsnelson.net.

[27] R.P. Stanley, *Enumerative combinatorics. Vol. 2.* With a foreword by Gian-Carlo Rota and appendix 1 by Sergey Fomin. Cambridge Studies in Advanced Mathematics, 62. Cambridge University Press, Cambridge, 1999, xii+581 pp.

[28] A. G. Tibert. *Deployable tensegrity structures for space applications*, Ph.D. Thesis, Royal Institute of Technology, Stokholm 2002.

[29] M. Wang, M. Sitharam, Combinatorial Rigidity and Independence of Generalized Pinned Subspace-Incidence Constraint Systems, in *Automated Deduction in Geometry–10th International Workshop*, Coimbra, Portugal (2014), pp. 166–180, arXiv:1503.01837 [cs.CG].

[30] N.L. White, T. McMillan, Cayley factorization. Symbolic and algebraic computation, Rome, (1988), pp. 521-533, *Lecture Notes in Comput. Sci.*, 358, Springer, Berlin, 1989.

[31] N. L. White, W. Whiteley, The algebraic geometry of stresses in frameworks, *SIAM J. Alg. Disc. Meth.*, vol. 4, n. 4 (1983), pp. 481–511.

[32] N. L. White, W. Whiteley, The algebraic geometry of motions of bar-and-body frameworks, *SIAM J. Algebraic Discrete Methods*, vol. 8, no. 1 (1987), pp. 1-32.

[33] N. White, The bracket ring of a combinatorial geometry, I, *Trans. Amer. Math. Soc.*, vol. 202 (1975), pp. 79–95.

[34] W. Whiteley, Rigidity and scene analysis, in J.E. Goodman and J. O'Rourke, editors, *Handbook of Discrete and Computational Geometry*, chapt. 49, pp. 893–916, CRC Press, New York, 1997.

# Chapter 16

# *Generic Global Rigidity in General Dimension*

**Steven J. Gortler**

*sjg@cs.harvard.edu, SEAS, Harvard University*

## CONTENTS



## 16.1 Basic Setup

In this chapter we discuss the concept of global rigidity for bar and joint frameworks in an arbitrary fixed dimension, $d$. This is stronger than the plain "rigidity" property (also called local rigidity), in that it requires that, up to congruence, there is a unique framework satisfying the given length constraints. In particular, even if there are a only finite number of equivalent and noncongruent frameworks, if that number is greater than one, then such frameworks are not considered to be globally rigid.

In this chapter, we will always assume that $d$ is a fixed dimension, $G$ is a graph with $n$ vertices and $m$ edges, and $n \geq d+2$. A (bar and joint) framework in $\mathbb{R}^d$, denoted as $(G,\mathbf{p})$, is a graph $G$ together with a configuration $\mathbf{p} = (\mathbf{p}_1,\dots,\mathbf{p}_n)$ of points in $\mathbb{R}^d$.

**Globally rigid:** A bar and joint framework $(G,\mathbf{p})$ in $\mathbb{R}^d$ is *globally rigid* if any equivalent framework $(G,\mathbf{q})$ in $\mathbb{R}^d$ is, in fact, congruent to $(G,\mathbf{p})$. Otherwise, we say that the framework is *globally flexible*.

**Generic:** We say that a set of configuration $\mathbf{p}$ (or a framework $(G,\mathbf{p})$) is *generic* if its coordinates do not satisfy *any* nontrivial algebraic relations with rational coefficients.

**Generically globally rigid:** We say that a graph $G$ is *generically globally rigid* in $\mathbb{R}^d$, if all generic frameworks are globally rigid.

**Generically globally flexible:** We say that a graph $G$ is *generically globally flexible* in $\mathbb{R}^d$, if all generic frameworks are globally flexible. We can also call this property: *generically not globally rigid.*

**Generic property:** A property $P$ is *generic* if for every graph, either all generic frameworks satisfy $P$ or none do.

Global rigidity naturally arises in the context of attempting to find a $d$-dimensional framework from an input set of inter-point length constraints. Only when the underlying framework is globally rigid is such an inverse problem well posed. Note that even when the framework is globally rigid, it may not be computationally easy to solve such an inverse problem.

The bad news is that it is NP-Hard to determine if a framework $(G,\mathbf{p})$ (say with integer valued coordinates) is globally rigid [11]. For some initial insight into why this may be so, consider the case where the graph is an $n$-cycle and we have a framework in $\mathbb{R}^1$. Any one-dimensional framework of this graph gives rise to a set of signs on the edge lengths that solves a SUBSET-SUM problem (since the cycle must close up). The ability to efficiently determine global rigidity would give one the ability to efficiently determine if there is a second solution to a SUBSET-SUM problem that has one known satisfying set of signs. This is an NP-HARD problem.

This pessimism leads us to look instead on understanding the so-called "generic" case. Before defining this carefully, we first describe what we are after. Let us consider the case of a one-dimensional framework $(G,\mathbf{p})$ of some graph $G$. Suppose that this graph is not 2-connected, then we can find some disconnecting vertex, $i$, and "flip" the embedding of one of the components across the point $\mathbf{p}_i$ while maintaining all of the edge lengths. This will give us some equivalent framework $(G,\mathbf{q})$. This new framework will not be congruent to $(G,\mathbf{p})$ unless all but one of the components are collapsed into a single point. The special inputs, $\mathbf{p}$, that have such collapses can easily be described using a set of algebraic relations on the coordinates of $\mathbf{p}$.

On the other hand consider the case where $G$ is an $n$-cycle (modeling a SUBSET-SUM problem with real edge lengths). Unless the points (or lengths) are chosen very carefully, we would not expect to find a second way of solving the SUBSET-SUM problem. Indeed if such a solution existed, there must be some specific non-trivial algebraic relation on the $\mathbf{p}$ that allows for this unexpected second solution. If we were sufficiently dedicated, We could spend some effort to write down this algebraic relation, but we are not so inclined; we are content to know that inputs allowing such second solutions are very special and hard to come by. They certainly have measure 0, but as they arise from algebraic solutions, there are even more special than that.

With a bit more effort, we can maybe convince ourselves that if $G$ is 2-connected, then the existence of all of the overlapping cycles will ensure that $(G,\mathbf{p})$ is globally rigid, unless we choose our $\mathbf{p}$ very carefully (so that they satisfy some specify-able, but perhaps unspecified, algebraic equations with rational coefficients).

Doing this, we conclude that if $G$ is 2-connected then $(G,\mathbf{p})$ in $\mathbb{R}^1$ will be globally rigid, unless $\mathbf{p}$ is "special". Conversely, if $G$ is not 2-connected, then $(G,\mathbf{p})$ in $\mathbb{R}^1$ will be globally flexible, unless $\mathbf{p}$ is "special".

There are a bunch of ways of making this notion formal, and the one typically used in the rigidity literature is through the notion of a generic framework.

Given this notion, we can state

**Theorem 16.1** *A graph $G$ is generically globally rigid in $\mathbb{R}^1$ if it is 2-connected. A graph $G$ is generically globally flexible in $\mathbb{R}^1$ if it is not 2-connected.*

Note that our formal definition is a bit of nominal overkill, as it has ruled out saying anything about any $\mathbf{p}$ that satisfy *any* algebraic equations defined over the rationals. While in fact, as claimed above we really know about the behavior of $(G,\mathbf{p})$ outside of one specific algebraically definable set. But it is immediately clear that these singular sets are at the very least semi-algebraic, defined using real algebraic equalities and inequalities with rational coefficients. Thus if all generic points lie outside such a singular set, then the singular set must be of lower dimension, and indeed lie in some algebraic subset (defined using equalities only). Thus we could also state the theorem in the form

**Theorem 16.2** *If $G$ is 2-connected, then $(G,\mathbf{p})$ is globally rigid in $\mathbb{R}^1$ over a Zariski-open set of configurations. If $G$ is not 2-connected, then $(G,\mathbf{p})$ in $\mathbb{R}^1$ is globally flexible over a Zariski-open set of $\mathbf{p}$. Moreover, these Zariski open sets can be expressed over the rationals.*

And so when you read theorems in the form of Theorem 16.1 in this chapter, you can always think of them in the form of Theorem 16.2.

We now notice something important in this theorem. Since a graph $G$ is either 2-connected, or it is not-2-connected, we see that in $\mathbb{R}^1$, $G$ is either generically globally rigid, or it is generically globally flexible. In other words, if $G$ is not generically globally rigid in $\mathbb{R}^1$ then it must be generically globally flexible in $\mathbb{R}^1$.

Apriori, we might have imagined that there could be some graphs $G$ that are neither generically globally rigid nor generically globally flexible in $\mathbb{R}^1$. Perhaps, half the frameworks are globally rigid while the other half are globally flexible. Theorem 16.2 states that this does not happen.

We formalize this with the concept of a generic property, and we state the corollary

**Corollary 16.3** *Global rigidity in $\mathbb{R}^1$ is a generic property.*

The goal of this chapter is to explore the property of generic global rigidity in any arbitrary but fixed dimension, denoted as $d$. In fact, as we will describe below, global rigidity in $\mathbb{R}^d$ is a generic property as well.

## 16.2 Connelly's Sufficiency Theorem

In this section we will describe Connelly's sufficiency condition for generic global rigidity in $\mathbb{R}^d$ [3].

**Stress vector:** Given a graph $G$, a *stress vector* $\omega = (\ldots, \omega_{ij}, \ldots)$, is an assignment of a real scalar $\omega_{ij} = \omega_{ji}$ to each edge, $\{i,j\}$ in $G$. (We have $\omega_{ij} = 0$, when $\{i,j\}$ is not an edge of $G$).

**Equilibrium stress vector:** We say that $\omega$ is an *equilibrium stress vector* for $(G,\mathbf{p})$ if the vector equation

$$\sum_j \omega_{ij}(\mathbf{p}_i - \mathbf{p}_j) = 0 \tag{16.1}$$

holds for all vertices $i$ of $G$. The equilibrium stress vectors of $(G,\mathbf{p})$ form the co-kernel of its rigidity matrix $R(\mathbf{p})$.

**Stress matrix:** We associate an $n$-by-$n$ *stress matrix* $\Omega$ to a stress vector $\omega$, by saying that $i,j$ entry of $\Omega$ is $-\omega_{ij}$, for $i \neq j$, and the diagonal entries of $\Omega$ are such that the row and column sums of $\Omega$ are zero. The stress matrices of $G$ are simply the symmetric matrices with zeros associated to nonedge pairs, and such that the vector of all-ones is in its kernel.

**Equilibrium stress matrix:** If $\omega$ is an equilibrium stress vector for $(G,\mathbf{p})$ then we say that the associated $\Omega$ is an *equilibrium stress matrix* for $(G,\mathbf{p})$.

For each of the $d$ spacial dimensions, if we define a vector $v$ in $\mathbb{R}^n$ by collecting the the associated coordinate over all of the points in $\mathbf{p}$, we have $\Omega v = 0$. Thus if the dimension of the affine span of the vertices $\mathbf{p}$ is $d$, then the rank of $\Omega$ is at most $n-d-1$, but it could be less.

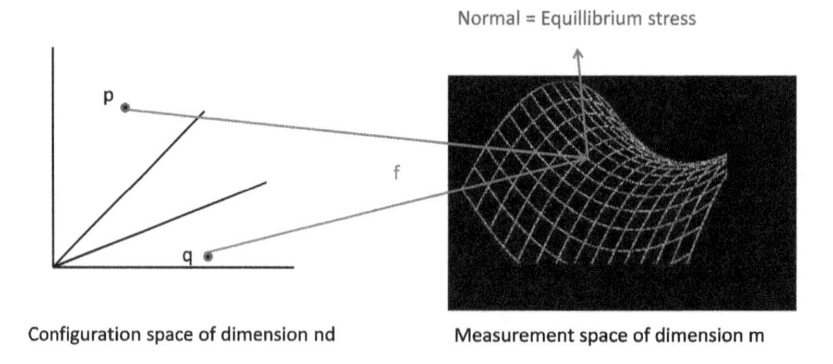


**Figure 16.1**
On the left we have the configuration space, $\mathbb{R}^{nd}$. The map $f$ measures the squared lengths along each of the $m$ edges and maps to a point in $\mathbb{R}^m$. The image of $f$ over all configurations is the measurement set $M_d$. A generic configuration maps to a smooth point of $M_d$ where it has a well defined tangent and normal space. At smooth points $f(\mathbf{p})$, normal vectors are equilibrium stress vectors for $(G,\mathbf{p})$. If $(G,\mathbf{p})$ and $(G,\mathbf{q})$ are equivalent, they map to the same point on the measurement set, and share the equilibrium stress matrices. If this matrix has rank $n-d-1$, then such $\mathbf{p}$ and $\mathbf{q}$ must be related by an affine transform.

**Conic at infinity:** We say that *the edge directions of* $(G,\mathbf{p})$ *are on a conic at infinity* in $\mathbb{R}^d$ if there exists some nonzero $d$-by-$d$ symmetric matrix $Q$, such that for all edges $ij$, we have $(\mathbf{p}_i-\mathbf{p}_j)^t Q(\mathbf{p}_i-\mathbf{p}_j)=0$.

With these definitions, we can now state Connelly's sufficiency condition [3].

**Theorem 16.4** *Suppose that there is a generic configuration* $\mathbf{p}$ *in* $\mathbb{R}^d$ *such that* $(G,\mathbf{p})$ *has a equilibrium stress matrix of rank* $n-d-1$*. Then* $G$ *is generically globally rigid in* $\mathbb{R}^d$.

**Remark 16.1** If one generic $(G,\mathbf{p})$ in $\mathbb{R}^d$ has such an equilibrium stress matrix, then so too must any other generic framework $(G,\mathbf{q})$ in $\mathbb{R}^d$.

In fact, due to the properties of matrix ranks the theorem can be slightly strengthened [5, 7] to say

**Theorem 16.5** *Suppose that there is an infinitesimally rigid configuration* $\mathbf{p}$ *in* $\mathbb{R}^d$ *such that* $(G,\mathbf{p})$ *has a nonzero equilibrium stress matrix of rank* $n-d-1$*. Then* $G$ *is generically globally rigid in* $\mathbb{R}^d$.

**Remark 16.2** Caveats [5]:

Even if a infinitesimally rigid framework $(G,\mathbf{p})$ has an equilibrium stress matrix of rank $n-d-1$, this does not imply that $(G,\mathbf{p})$ itself is globally rigid in $\mathbb{R}^d$. It just implies that $G$ is generically globally rigid in $\mathbb{R}^d$.

Even when $G$ is generically globally rigid, there can still be (nongeneric) infinitesimally rigid frameworks that are globally flexible.

Even if a generic framework $(G,\mathbf{p})$ is globally rigid, there may be another framework $(G,\mathbf{q})$ that is globally flexible, where $\mathbf{q}$ is obtained from $\mathbf{p}$ by applying an affine transform in $\mathbb{R}^d$.

It is worthwhile outlining the central ideas used in the proof of this theorem. (See Figure 16.1)

For a fixed graph $G$ and dimension $d$, let $f$ be the map that takes a configuration of $n$ points in $\mathbb{R}^d$, and measures the squared Euclidean lengths on the $m$ edges, thought of as a point in $\mathbb{R}^m$.

We will refer to the image of $f$, acting on all possible configurations in $\mathbb{R}^d$ as the *measurement set* $M_d$. The measurement set is a semi-algebraic set and so most of its neighborhoods look like smooth manifolds in $\mathbb{R}^m$. The nonsmooth points of $M_d$ must satisfy some additional algebraic equations in $\mathbb{R}^m$ that do not vanish identically on $M_d$, and these pull back through $f$ to some non-trivial algebraic equations on configuration space. Thus the image of $f$, of a small ball around some generic $\mathbf{p}$, must be a smooth manifold, with a well defined tangent and normal space in $\mathbb{R}^m$.

Recall that any equilibrium stress vector $\omega$ of $(G,\mathbf{p})$ is in the cokernel of the rigidity matrix of $(G,\mathbf{p})$ which represents the Jacobian of $f$. Thus $\omega$ is orthogonal in $\mathbb{R}^m$ to image of the differential of $f$ at $\mathbf{p}$. When $\mathbf{p}$ is generic, the image of the differential spans the tangent space of $M_d$ at $f(p)$, placing $\omega$ within its normal space.

With this picture in mind, if (the generic) $(G,\mathbf{p})$ and $(G,\mathbf{q})$ are equivalent, then $f(p) = f(q)$; both configurations map to the same smooth point in $M_d$. Thus $\omega$ (as a normal vector) must be orthogonal to the columns of the rigidity matrix of $(G,\mathbf{q})$. Thus $\omega$ must also be an equilibrium stress for $(G,\mathbf{q})$.

Suppose that the the rank of $\Omega$ is $n-d-1$, then this implies that the configuration $\mathbf{q}$ must arise as a $d$-dimensional affine transform of the configuration $\mathbf{p}$.

At this point we are almost done, all we need to rule out is the possibility that there is an equivalent $(G,\mathbf{q})$ which arises as an affine (but non Euclidean) transform of $(G,\mathbf{p})$. This possibility is formalized using the notion of a conic at infinity.

The following simple proposition is proved in [3].

***Proposition 16.1***
*Every affine transform of* $\mathbf{p}$ *preserving edge lengths of* $(G,\mathbf{p})$ *is a congruence, if and only if the edge directions of* $(G,\mathbf{p})$ *are not on a conic at infinity.*

And the last piece of the puzzle is then the following proposition from [3].

***Proposition 16.2***
*Suppose that* $(G,\mathbf{p})$ *is a generic framework in* $\mathbb{R}^d$ *and each vertex has degree at least d, then the edge directions of* $(G,\mathbf{p})$ *are not on a conic at infinity.*

Putting these pieces together, one can complete Connelly's proof.

Here is a useful variant of Proposition 16.2 that does not require genericity [1]:

***Proposition 16.3***
*Suppose that* $(G,\mathbf{p})$ *is a framework in* $\mathbb{R}^d$ *and for each vertex v, the affine span of v and its neighbors in* $(G,\mathbf{p})$ *span all of* $\mathbb{R}^d$*. Suppose also that* $(G,\mathbf{p})$ *has an equilibrium stress matrix of rank* $n-d-1$*. Then the edge directions of* $(G,\mathbf{p})$ *are not on a conic at infinity.*

## 16.3 Hendrickson's Necessary Conditions

In this section we describe Hendrickson's necessary conditions for generic global rigidity [9]. As described in Chapter 21, these condition are also sufficient for generic global rigidity in $\mathbb{R}^2$. In three dimensions and higher, though, these conditions are not sufficient.

**Redundantly rigid:** A graph $G$ is *redundantly rigid* in $\mathbb{R}^d$ if it remains rigid even after removing any single edge.

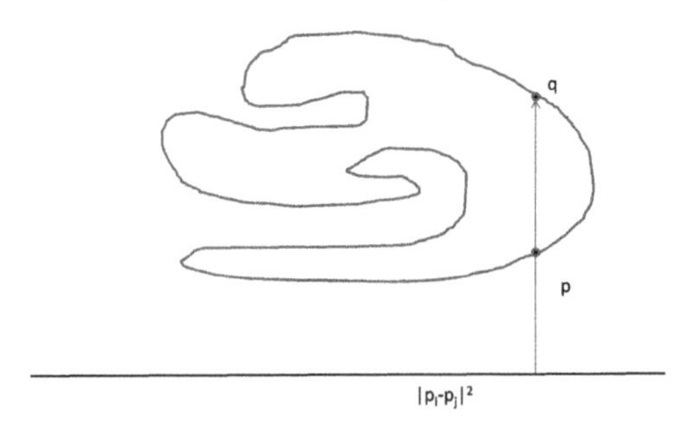


**Figure 16.2**
The closed curve (topological circle) represents a set of configurations that are equivalent once edge $e_{ij}$ is removed from $G$. The vertical projection, $\pi$, represents the measurement of the squared length between vertex $i$ and vertex $j$. A generic point in the range of this measurement, $\pi(\mathbf{p})$, must have an even number of points in the same fiber, proving the existence of the equivalent $(G, \mathbf{q})$.

**Theorem 16.6** *Let $(G, \mathbf{p})$ be a generic framework in $\mathbb{R}^d$. If $(G, \mathbf{p})$ is globally rigid then both of the following must hold*

*(a) $G$ is vertex $(d+1)$-connected.*

*(b) $G$ is redundantly rigid in $\mathbb{R}^d$.*

This means that if $G$ does not satisfy one of these two conditions, then $G$ is generically globally flexible in $\mathbb{R}^d$.

Again, it is worthwhile outlining the central ideas used in the proof of this theorem. (See Figure 16.2.)

The necessity of the first condition is something we already saw above when discussing the necessity of 2-connectivity for generic global rigidity in one dimension. This generalizes easily to $d$ dimensions. If the graph is not $d+1$ connected, then we can find $d$ vertices whose removal would disconnect the graph into multiple components. Generically, we can simply reflect one of these components across the hyperplane spanned by the cut vertices to obtain an equivalent but noncongruent second framework.

The second condition is more interesting. Suppose that $G$ is not redundantly rigid. We remove one edge $e_{ij}$ that makes the resulting graph $G'$ flexible. Henrickson then argues (using genericity, Sard's theorem and the implicit function theorem) that the set of equivalent frameworks (modulo congruence) to $(G', \mathbf{p})$ must form a smooth compact manifold, and thus must contain a circle. One then looks at the squared distance between the pair $\mathbf{p}_i$ and $\mathbf{p}_j$ (corresponding to the removed edge) as we travel along this circle in configuration space. This gives a map from a circle to the real line. Such a map must have an mod-2-degree of zero. Thus any generic point in its image must have at least one other point its fiber. Any such other point will be an equivalent but incongruent framework with respect to the original graph $G$.

### 16.3.1 Nonsufficiency

In three and higher dimensions, there are examples of graphs that are $d+1$ connected and redundantly rigid while not being generically globally rigid. Connelly found [2] a family of such examples that were complete bipartite graphs in various dimensions The smallest such instance is $K_{5,5}$ in $\mathbb{R}^3$. More kinds of examples have been found by Frank and Jiang [6] and more recently by Jordan et al. [10].

## 16.4 Necessity of Connelly's Condition

Connelly conjectured that the conditions of Theorem 16.4 were in fact necessary and this was proven by Gortler, Healy, and Thruston [7].

**Theorem 16.7** *Suppose that a generic configuration* $\mathbf{p}$ *in* $\mathbb{R}^d$ *is globally rigid in* $\mathbb{R}^d$*. Then* $(G,\mathbf{p})$ *must have a equilibrium stress matrix of rank* $n-d-1$.

Moreover, in light of Remark 16.1, one generic framework has an equilibrium stress matrix of rank $n-d-1$ iff all generic frameworks do. So the theorem tells us that if one generic framework does not have an equilibrium stress matrix of rank $n-d-1$ then $G$ is generically globally flexible in $\mathbb{R}^d$.

Together with Theorem 16.4, this gives us

**Corollary 16.8** *Global rigidity in* $\mathbb{R}^d$ *is a generic property.*

Additionally, using the corollary one can prove [5, 7]:

**Theorem 16.9** *Suppose that an infinitesimally rigid configuration* $\mathbf{p}$ *in* $\mathbb{R}^d$ *is globally rigid in* $\mathbb{R}^d$*. Then* $G$ *must be generically globally rigid. And conversely if* $G$ *is generically globally flexible, then all infinitesimally rigid frameworks must be globally flexible.*

**Remark 16.3** Caveat [5]: $G$ can be generically globally rigid, and $(G,\mathbf{p})$ can be infinitesimally rigid, while $(G,\mathbf{p})$ still does not have an equilibrium stress matrix of rank $n-d-1$.

The proof of theorem 16.7 is inspired by the mod-2 degree argument used in the proof of Hendrickson's theorem, though the details get a bit more complicated. In this case they look at the degree of a certain map with a certain range and certain domain: The domain is the be the set of frameworks $(G,\mathbf{q})$ that are in equilibrium under some chosen $\Omega$, an equilibrium stress of $(G,\mathbf{p})$. The range is $\mathbb{R}^m$ where $m$ is the number of edges. The map simply measures the squared length of each edge in a framework.

Under the assumption of genericity and that $(G,\mathbf{p})$ does not have an equilibrium stress matrix of rank $n-d-1$, they show that this map has a well defined degree-mod-2. The proof of this step uses some non-trivial properties of ruled algebraic varieties. We note, for further reference that it also uses the fact that if the graph is connected and one-vertex has been pinned down (to mod out translations), then the squared length map is *proper*.

## 16.5 Randomized Algorithm for Testing Generic Global Rigidity

The above theory leads naturally to a randomized efficient algorithm for checking if a graph $G$ is generically globally rigid in $\mathbb{R}^d$.

The basic idea is that if $G$ is generically globally rigid, then a randomly selected equilibrium stress matrix of a randomly selected configuration $\mathbf{p}$ will (with high probability) have rank $n-d-1$. (Keep in mind that a randomly select matrix from a linear space of matrices will, with high probability, exhibit the largest rank of any matrix in the space).

The algorithm is outlined as follows:

(a) Pick a random configuration $\mathbf{p}$ in $\mathbb{R}^d$.

(b) If $(G, \mathbf{p})$ is infinitesimally flexible, then output "generically globally flexible" and exit.

(c) Calculate the space of equilibrium stresses of $(G, \mathbf{p})$.

(d) Pick a random equilibrium stress from this space.

(e) Calculate the rank of this equilibrium stress.

(f) If this rank is less than $n-d-1$ then output "generically globally flexible" and exit.

(g) Output "generically globally rigid" and exit.

The test in step (2) is done by computing the rigidity matrix of $(G, \mathbf{p})$ and comparing its rank to $nd - \binom{d+1}{2}$ as in Chapter 18.

Step (3) is done by computing the co-kernel of the rigidity matrix of $(G, \mathbf{p})$. Any vector in the co-kernel is an equilibrium stress vector.

If the algorithm has reached step (7), then it must have found an equilibrium stress matrix of rank $n-d-1$, for an infinitesimally rigid $(G, \mathbf{p})$. In this case, $G$ is certifiably generically globally rigid using Theorem 16.5. Thus the algorithm is always correct when it outputs "generically globally rigid." On the other hand, when the algorithm outputs "generically globally flexible," there is (only) a small chance that it picked an exceptional framework $(G, \mathbf{p})$, or an exceptional equilibrium stress matrix $\Omega$.

In practice, one might perform the above steps numerically using floating point numbers. To obtain theoretical guarantees [7], one can perform the above steps using integers modulo some random prime larger than $4mn$.

## 16.6 Surgery

There are various combinatorial operations on graphs that preserve generic global rigidity. We summarize some of these results here.

Connelly and Whiteley [5] discuss generic global rigidity under the coning operation. Starting with a graph $G$, one new vertex $v$ is added and edges are added between $v$ and all of the other vertices.

**Theorem 16.10** *A graph $G$ is generically globally rigid in $\mathbb{R}^d$ iff its cone graph is generically globally rigid in $\mathbb{R}^{d+1}$.*

Connelly [4] showed the following

**Theorem 16.11** *Suppose that $G_1$ and $G_2$ are generically globally rigid in $\mathbb{R}^d$, each with at least $d+2$ vertices. Suppose that $d+1$ vertices of $G_1$ are identified with $d+1$ vertices of $G_2$. Next, Let $G$ be the union of $G_1$ and $G_2$. Next, up to one edge can be removed from $G$, if it is common to both $G_1$ and $G_2$. Then $G$ is generically globally rigid in $\mathbb{R}^d$.*

This theorem has been generalized by Tanigawa [13]. That paper also includes the following interesting theorem:

**Theorem 16.12** *If $G$ is vertex redundantly rigid in $\mathbb{R}^d$, then $G$ is generically globally rigid in $\mathbb{R}^d$,*

Szabadka [12] shows the following

**Theorem 16.13** *A graph obtained from a generically globally rigid graph in $\mathbb{R}^d$ by a 1-extension is generically globally rigid in $\mathbb{R}^d$.*

A 1-extension in $\mathbb{R}^d$ removes an existing edge $e$ and adds a new vertex with $d+1$ new incident edges so that the new vertex is incident with both the endpoints of $e$.

## 16.7 Other Spaces

Connelly and Whiteley [5] studied the relationship between generic global rigidity in $\mathbb{R}^d$ and on the sphere $S^d$.

**Theorem 16.14** *A graph $G$ is generically globally rigid in $\mathbb{R}^d$ iff $G$ is generically globally rigid in $S^d$. Additionally, generic global rigidity in $S^d$ is a generic property.*

Similarly, Gortler and Thurston [8] studied the relationship between generic global rigidity in various spaces.

**Theorem 16.15** *A graph $G$ is generically globally rigid in $\mathbb{R}^d$ iff $G$ is generically globally rigid in $\mathbb{C}^d$. Additionally, generic global rigidity in $\mathbb{C}^d$ is a generic property.*

In the above notion, the complex edge squared length is measured as the complex $(\mathbf{p}_i - \mathbf{p}_j)^2$, with no conjugation. (If conjugation is used, then this problem simply reduces to that of global rigidity in $\mathbb{R}^{2d}$.)

They also looked at generic global rigidity in pseudo-Euclidean spaces such as Minkowski space.

**Theorem 16.16** *A graph $G$ is generically globally rigid in $\mathbb{R}^d$ iff $G$ is generically globally rigid in any and all $d$-dimensional pseudo-Euclidean spaces and hyperbolic space.*

**Remark 16.4** It is not known if generic global rigidity in a pseudo-Euclidean space is a generic property. It is possible that there are graphs that are not generically globally rigid in $\mathbb{R}^d$ (equiv. $\mathbb{C}^d$) but are not generically globally flexible in a pseudo-Euclidean space (rather it has generic points exhibiting both behaviors).

## References

[1] A.Y. Alfakih and Viet-Hang Nguyen. On affine motions and universal rigidity of tensegrity frameworks. *Linear Algebra and its Applications*, 439(10):3134–3147, 2013.

[2] Robert Connelly. On generic global rigidity. In *Applied geometry and discrete mathematics*, volume 4 of *DIMACS Ser. Discrete Math. Theoret. Comput. Sci.*, pages 147–155. Amer. Math. Soc., Providence, RI, 1991.

[3] Robert Connelly. Generic global rigidity. *Discrete Comput. Geom*, 33(4):549–563, 2005.

[4] Robert Connelly. Combining globally rigid frameworks. *Proceedings of the Steklov Institute of Mathematics*, 275(1):191–198, 2011.

[5] Robert Connelly and WJ Whiteley. Global rigidity: the effect of coning. *Discrete & Computational Geometry*, 43(4):717–735, 2010.

[6] Samuel Frank and Jiayang Jiang. New classes of counterexamples to hendricksons global rigidity conjecture. *Discrete & Computational Geometry*, 45(3):574–591, 2011.

[7] Steven J. Gortler, Alexander D. Healy, and Dylan P. Thurston. Characterizing generic global rigidity. *American Journal of Mathematics*, 132(4):897–939, 2010.

[8] Steven J. Gortler and Dylan P. Thurston. Generic global rigidity in complex and pseudo-euclidean spaces. In *Rigidity and symmetry*, pages 131–154. Springer, 2014.

[9] Bruce Hendrickson. Conditions for unique graph realizations. *SIAM J. Comput.*, 21(1):65–84, February 1992.

[10] Tibor Jordán, Csaba Király, and Shin-Ichi Tanigawa. Generic global rigidity of body-hinge frameworks. Technical report, EGRES Technical Reports, TR2014-06, 2014.

[11] J. B. Saxe. Embeddability of weighted graphs in $k$-space is strongly NP-hard. In *Proc. 17th Allerton Conf. in Communications, Control, and Computing*, pages 480–489, 1979.

[12] Zoltán Szabadka. *Globally rigid frameworks and rigid tensegrity graphs in the plane.* PhD thesis, Department of Operations Research, Eötvös Loránd University, 2010.

[13] Shin-Ichi Tanigawa. Sufficient conditions for the global rigidity of graphs. *Journal of Combinatorial Theory, Series B*, 113:123–140, 2015.

# Chapter 17

## *Change of Metrics in Rigidity Theory*

**Anthony Nixon**

*Department of Mathematics and Statistics, Lancaster University, U.K.*

**Walter Whiteley**

*Department of Mathematics, York University, Canada*

### CONTENTS



## 17.1 Introduction

In this chapter we abstract the basic notions of rigidity and flexibility to apply in various geometries which live in a shared projective space. We show the equivalence of infinitesimal rigidity across the Cayley-Klein geometries using projective transformations.

Given rigidity properties of a framework $(G,p)$ in $\mathbb{R}^d$, we can often lift these properties to a coned framework in $\mathbb{R}^{d+1}$, and then re-project to a different hyperplane to obtain another $d$-dimensional realization $(G,q)$ as a slice of the cone, preserving key rigidity properties. This is a concrete form of a projective transformation from $(G,p)$ to $(G,q)$. Alternatively, there may be a direct analysis of the impact of a projective transformation on the various properties. These projective transformations preserve the infinitesimal rigidity of $(G,p)$, and have a clearly described impact on the coefficients of any equilibrium-stress of $(G,p)$. These processes also provide the tools to confirm the transfer of these basic properties among Euclidean, Spherical, Pseudo-Euclidean, and Hyperbolic metrics –all of which live in a common projective space. The geometric transfer, for a specific projective configuration $\tilde{p}$ in any of the metrics, then gives corresponding combinatorial transfers of generic properties.

We also consider global (and universal) rigidity in these spaces by analysing the stress matrix. In this setting a number of key transfer results are known generically, or combinatorially, but the transfers are less geometrically robust.

In Section 17.2 we present the equivalence of infinitesimal rigidity for Euclidean and Spherical frameworks by passing through cone frameworks and then in Section 17.3 we consider projective frameworks directly. Section 17.4 considers extensions to Minkowski and Hyperbolic geometries and analogues for symmetric frameworks are discussed in Section 17.5. In Section 17.6 we consider the more detailed transfer of global and universal rigidity before concluding by mentioning some related work and summarising the results of the chapter in Section 17.7.

## 17.2 Projective Transfer of Infinitesimal Rigidity

**Infinitesimal flex:** Let $G=(V,E)$. We say that $u: V \to \mathbb{R}^d$ is an *infinitesimal flex* of $(G,p)$ if $u$ satisfies

$$\langle p_i - p_j, u_i - u_j \rangle = 0 \qquad (\{i,j\} \in E).$$

**Infinitesimally rigid:** A framework $(G,p)$ in $\mathbb{R}^d$ is *infinitesimally rigid* if the dimension of the space of infinitesimal flexes of $(G,p)$ is equal to $\binom{d+1}{2}$ (assuming that the points $p(V)$ affinely span $\mathbb{R}^d$).

**Rigidity matrix:** The *rigidity matrix* $R_d(G,p)$ of a framework $(G,p)$ in $\mathbb{R}^d$ is the $|E| \times d|V|$ matrix with rows indexed by edges and columns by $d$-tuples of vertices with the following form:

$$R_d(G,p) = \begin{array}{c} \\ \\ \{i,j\} \\ \\ \end{array} \begin{pmatrix} & & & \overset{i}{\vdots} & & & & \overset{j}{\vdots} & & & \\ 0 & \dots & 0 & p_i - p_j & 0 & \dots & 0 & p_j - p_i & 0 & \dots & 0 \\ & & & \vdots & & & & \vdots & & & \end{pmatrix}.$$

**Cone graph:** Let $G=(V,E)$ be a graph. The *cone* of $G$ is the graph $G*o$ which has vertex set $V \cup o$ and edge set $E \cup \{\{x,o\} : x \in V\}$.

**Cone framework:** Given a framework $(G,p)$ in $\mathbb{R}^d$, the *cone framework* $(G*o, \bar{p}^*)$ is defined by the realization $\bar{p}^* : V \cup o \to \mathbb{R}^{d+1}$ such that $\bar{p}_i = (p_i, 1) \in \mathbb{R}^{d+1}$ and $\bar{p}^*(o) = (0,\dots,0)$.

**Spherical framework:** Define the unit sphere $\mathbb{S}^d$ in $\mathbb{R}^{d+1}$ by $\mathbb{S}^d = \{(x_1, x_2, \dots, x_{d+1}) \in \mathbb{R}^{d+1} : x_1^2 + x_2^2 + \dots + x_{d+1}^2 = 1\}$. Let $G=(V,E)$ be a graph, a *framework* $(G,p)$ on $\mathbb{S}^d$ is an ordered pair consisting of a graph $G$ and a realization $p$ such that $p_i \in \mathbb{S}^d$ for all $i \in V$.

**Spherical infinitesimal flex:** A map $u: V \to \mathbb{R}^{d+1}$ is said to be an *infinitesimal flex* of a Spherical framework $(G,p)$ if it satisfies

$$\langle p_i, u_j \rangle + \langle p_j, u_i \rangle = 0 \qquad (\{i,j\} \in E) \text{ and}$$

$$\langle p_i, u_i \rangle = 0 \qquad (i \in V).$$

**Projective transformation:** A $d$-dimensional framework $(G,q)$ is a *projective transformation* of a $d$-dimensional framework $(G,p)$ if there is an invertible $(d+1)\times(d+1)$ matrix $T$ for which $T(p_i,1) = \lambda_i(q_i,1)$ for all $1 \leq i \leq n$ (where $\lambda_i$ is a scalar and $(p_i,1)$ are the affine coordinates of $p_i$).

**Equilibrium stress:** An *equilibrium stress* $\omega$ of a framework $(G,p)$ in $\mathbb{R}^d$ is a vector $\omega$ in the cokernel of $R_d(G,p)$. Equivalently, $\omega$ is an equilibrium stress if

$$\sum_{j=1}^{n} \omega_{ij}(p_i - p_j) = 0 \text{ for all } 1 \leq i \leq n, \tag{17.1}$$

where $\omega_{ij}$ is taken to be equal to $\omega_e$ if $e = \{i,j\} \in E$ and to be equal to 0 if $\{i,j\} \notin E$.

**Sliding:** Given a cone framework $(G * o, \bar{p}^*)$ in $\mathbb{R}^{d+1}$, the process of *sliding* moves the points $\bar{p}_i, i \in V$, along the unique lines connecting them to the cone vertex (while avoiding the cone vertex itself). Algebraically, $\bar{p}_i$ goes to $\alpha_i \bar{p}$ for $\alpha \neq 0$.

### 17.2.1 Coning and Spherical Frameworks

We embed $\mathbb{R}^d$ into $\mathbb{R}^{d+1}$, in the hyperplane $x_{d+1} = 1$ (where $x_{d+1}$ is the last coordinate of vectors in $\mathbb{R}^{d+1}$). In this manner we consider a framework $(G,p)$ in $\mathbb{R}^d$ as a framework $(G,p)$ on the hyperplane $x_{d+1} = 1$ by setting $\bar{p}_i = (p_i,1)$ for $1 \leq i \leq |V|$.

With this embedding we then cone to $(G * o, \bar{p}^*)$ by adding a new vertex $o$ and setting $p^*(o) = (0,\ldots,0)$. Now slide the points $p_i$ along the line through the origin $o$ and $p_i$ such that the lengths of the edges $oi$ are all equal. See Figure 17.1 for an example. We get a new framework which is infinitesimally rigid if and only if the original was. If we assume the cone edges are all length 1 then we have a framework in $\mathbb{R}^{d+1}$ which we can view as a framework on the sphere.

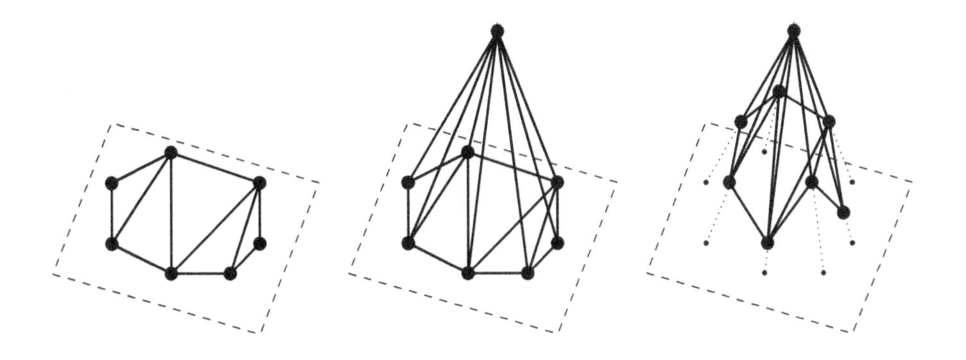

**Figure 17.1**
An infinitesimally rigid framework in $\mathbb{R}^2$ realized in a hyperplane in $\mathbb{R}^3$ then coned to an infinitesimally rigid framework in $\mathbb{R}^3$. Each point $x$ in the cone (minus its vertex), determines a unique line. Sliding $x$ along this line (minus the cone vertex) preserves infinitesimal rigidity.

This motivates the following presentation. Following [32] we focus on infinitesimal flexes. The transfer of infinitesimal rigidity can also be seen at the level of matrix operations on the rigidity matrices which we will subsequently define. Define the upper-hemisphere $\mathbb{S}^d_{>0} = \{x \in \mathbb{S}^d : \langle x,e\rangle > 0\}$ where $e = (0,\ldots,0,1) \in \mathbb{R}^{d+1}$. Suppose we have a framework $(G,p)$ on $\mathbb{S}^d_{>0}$. If $p(t)$ is a continuous flex of $(G,p)$ on $\mathbb{S}^d_{>0}$, then for all $t$ in some interval and for all $\{i,j\} \in E$, we have

$d_{\mathbb{S}^d_{>0}}(p_i(t) \cdot p_j(t)) = c_{ij}$, where $d_{\mathbb{S}^d_{>0}}$ denotes Spherical distance, $c_{ij}$ is constant for all $\{i,j\} \in E$, and for all $t$ and $k \in V$, $p_k(t) \cdot p_k(t) = 1$. Equivalently, for all $t$, $\{i,j\} \in E$ and $k \in V$, we have the system

$$p_i(t) \cdot p_j(t) = \cos c_{ij} \tag{17.2}$$

$$p_k(t) \cdot p_k(t) = 1. \tag{17.3}$$

If the flex $p(t)$ is differentiable at $t = 0$, then $p(t)$ must satisfy $p_i \cdot p_j(0) + p_i(0) \cdot p_j = 0$ and $p_k \cdot p_k(0) = 0$.

It follows that an infinitesimal flex of the framework $(G,p)$ on $\mathbb{S}^d_{>0}$ is a map $u : V \to \mathbb{R}^{d+1}$ satisfying, for each $\{i,j\} \in E$ and for each $k \in V$,

$$p_i \cdot u_j + p_j \cdot u_i = 0 \text{ and } p_k \cdot u_k = 0. \tag{17.4}$$

A trivial infinitesimal flex of $\mathbb{S}^d_{>0}$ is an infinitesimal flex which is also an infinitesimal isometry of $\mathbb{R}^{d+1}$. The framework $(G,p)$ is infinitesimally rigid on $\mathbb{S}^d_{>0}$ if all infinitesimal flexes of $(G,p)$ are restrictions of trivial infinitesimal flexes.

If $(G,p)$ is a bar-and-joint framework on $\mathbb{S}^d_{>0}$, then the cone framework $(G * o, \bar{p}^*)$ is infinitesimally rigid in $\mathbb{R}^{d+1}$ if and only if $(G,p)$ is infinitesimally rigid on $\mathbb{S}^{d+1}_{>0}$. That is, frameworks on $\mathbb{S}^d_{>0}$ can be modeled by the cone on the same framework in $\mathbb{R}^{d+1}$.

We now present two maps, a map carrying a framework $(G,p)$ on $\mathbb{S}^d_{>0}$ into a framework $(G,q)$ in $\mathbb{R}^d$, and a map carrying the infinitesimal flexes of $(G,p)$ into infinitesimal flexes of $(G,q)$. The latter map carries trivial infinitesimal flexes of $\mathbb{S}^d_{>0}$ to trivial infinitesimal flexes of $\mathbb{R}^d$, yielding the equivalence of infinitesimal rigidity in these spaces.

If $(G,p)$ is a framework on $\mathbb{S}^d_{>0}$, then $G(\phi \circ p)$ is a framework in $\mathbb{R}^d$, where $\phi : \mathbb{S}^d_{>0} \to \mathbb{R}^d$ is given by

$$\phi(x) = \frac{x}{e \cdot x}.$$

The inverse of $\phi$ is given by

$$\phi^{-1}(x) = \frac{x}{x \cdot x}.$$

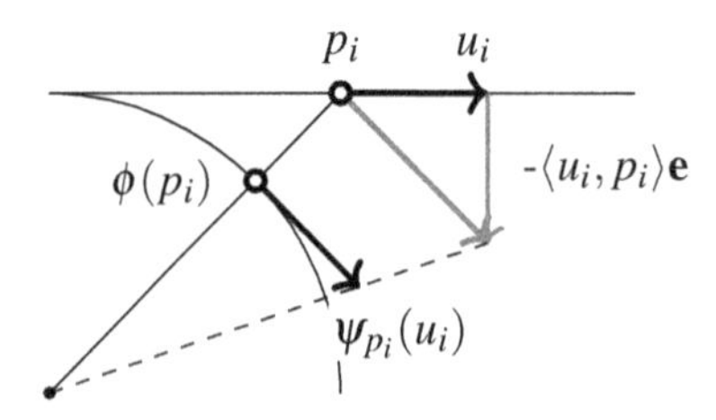


**Figure 17.2**
Transfer of infinitesimal flexes between $\mathbb{R}^d$ and $\mathbb{S}^d_{>0}$.

If $u$ is an infinitesimal flex of the framework $(G,p)$ on $\mathbb{S}^d_{>0}$, let $\psi$ denote the map

$$\psi : u_i \to \frac{1}{e \cdot p_i}(u_i - (u_i \cdot e)e).$$

If $(G,q)$ is a framework in $\mathbb{R}^d$ with infinitesimal flex $u$, then $\psi^{-1}$ is given by

$$\psi^{-1} : u_i \to \frac{1}{\sqrt{q_i \cdot q_i}}(u_i(u_i \cdot q_i)e).$$

These maps are illustrated in Figure 17.2. Observe that $\psi$ and $\psi^{-1}$ map into the appropriate tangent spaces: $\phi^{-1}(q_i) \cdot \psi^{-1}(u_i) = 0$ and $\psi(u_i) \cdot e = 0$. By direct computation we have the following theorem.

**Theorem 17.1** *A vector $u \in \mathbb{R}^d$ is an infinitesimal flex of the framework $(G,p)$ on $\mathbb{S}^d_{>0}$ if and only if $\psi \circ u$ is an infinitesimal flex of the framework $(G, \phi \circ p)$ in $\mathbb{R}^d$. Moreover, $u$ is a trivial infinitesimal flex if and only if $\psi \circ u \circ \phi^{-1}$ is a trivial infinitesimal flex.*

**Corollary 17.2** *A framework $(G,p)$ is infinitesimally rigid on $\mathbb{S}^d_{>0}$ if and only if $(G, \phi \circ p)$ is infinitesimally rigid in $\mathbb{R}^d$.*

### 17.2.2 Rigidity Matrices

One can take multiple viewpoints on the Spherical rigidity matrix. Consider the following matrix:

$$\begin{array}{c} \\ \{i,j\} \\ \\ i \\ \\ j \end{array} \begin{array}{c} \begin{array}{ccccccccccc} & & & i & & & & j & & & \end{array} \\ \begin{pmatrix} 0 & \dots & 0 & p_i - p_j & 0 & \dots & 0 & p_j - p_i & 0 & \dots & 0 \\ & & & \vdots & & & & \vdots & & & \\ 0 & \dots & 0 & p_i & 0 & \dots & 0 & 0 & 0 & \dots & 0 \\ & & & \vdots & & & & \vdots & & & \\ 0 & \dots & 0 & 0 & 0 & \dots & 0 & p_j & 0 & \dots & 0 \end{pmatrix} \end{array}.$$

Firstly, as above this is the rigidity matrix as a cone framework in $\mathbb{R}^{d+1}$ with the cone vertex at the origin and the columns for the cone vertex deleted. Secondly we can view this as the rigidity matrix for a framework on the sphere with vertex rows encoding the fact that infinitesimal flexes are tangent to the surface of the sphere (here we can interpret the edges as either Euclidean [25] or geodesic [43]). Thirdly, and we will take this viewpoint in what follows, the following row equivalent form of the rigidity matrix which is more natural for projective reasoning:

$$R_{\mathbb{S}^d}(G,p) = \begin{array}{c} \\ \{i,j\} \\ \\ i \\ \\ j \end{array} \begin{array}{c} \begin{array}{ccccccccccc} & & & i & & & & j & & & \end{array} \\ \begin{pmatrix} 0 & \dots & 0 & p_j & 0 & \dots & 0 & p_i & 0 & \dots & 0 \\ & & & \vdots & & & & \vdots & & & \\ 0 & \dots & 0 & p_i & 0 & \dots & 0 & 0 & 0 & \dots & 0 \\ & & & \vdots & & & & \vdots & & & \\ 0 & \dots & 0 & 0 & 0 & \dots & 0 & p_j & 0 & \dots & 0 \end{pmatrix} \end{array}.$$

A more subtle task is to extend this transfer of infinitesimal flexes to a transfer of finite flexes. This is more challenging because flexibility is not a projective invariant, unlike infinitesimal flexibility, see [30]. We illustrate this with an example in Figure 17.3.

However, rigidity does transfer for *regular* configurations, which form an open dense subset of all configurations. A configuration $p$ in $\mathbb{S}^d_{>0}$ is regular for $G$ if $R_{\mathbb{S}^d}(G,p)$ has the maximum rank over all configurations $x$ in $\mathbb{S}^d_{>0}$. Similarly $q$ in $\mathbb{R}^d$ is regular for $G$ if $R_d(G,p)$ has the maximum rank over all $x$ in $\mathbb{R}^d$. Note that $p$ is regular on $\mathbb{S}^d_{>0}$ if and only if $(G, \phi \circ p)$ is regular for $G$ in $\mathbb{R}^d$. This gives a strong transfer of rigidity and infinitesimal rigidity at regular configurations.

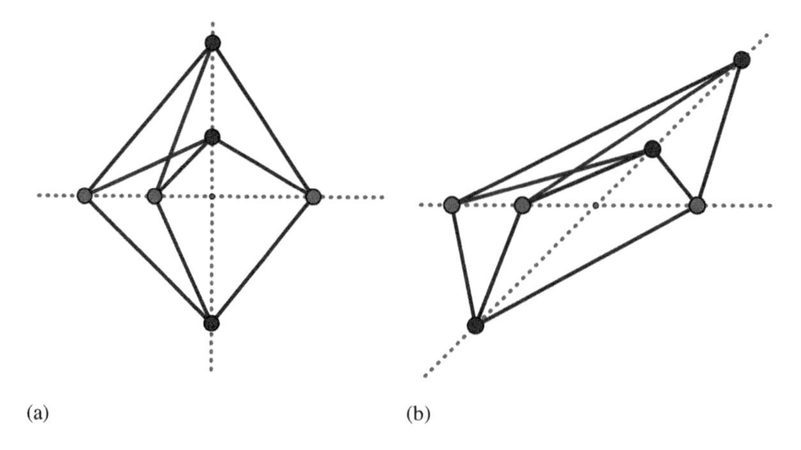


**Figure 17.3**
Two realizations of the complete bipartite graph $K_{3,3}$. Both realizations are infinitesimally flexible but only the realization on the left (with the lines perpendicular) has a continuous flex.

**Theorem 17.3** *If p is a regular point for $(G,p)$ in $\mathbb{S}^d_{>0}$ the the following are equivalent:*

*(a) $(G,p)$ is rigid on $\mathbb{S}^d_{>0}$;*

*(b) $(G,p)$ is infinitesimally rigid on $\mathbb{S}^d_{>0}$;*

*(c) $(G,\phi\circ p)$ is rigid in $\mathbb{R}^d$;*

*(d) $(G,\phi\circ p))$ is infinitesimally rigid in $\mathbb{R}^d$.*

Since flexibility is the negation of rigidity; we conclude that, at a regular point, $(G,p)$ is flexible on $\mathbb{S}^d_{>0}$ if and only if $(G,\phi\circ p)$ is flexible in $\mathbb{R}^d$. See also [34, Section 4.2] for an example at a non-regular point where flexibility is not transferred under coning.

We will consider affine space $\mathbb{A}^d$ as the hyperplane $x_{d+1}=1$ in $\mathbb{R}^{d+1}$. Recall $\phi^{-1}:\mathbb{A}^d\to\mathbb{S}^d_{>0}$ is defined by $\phi^{-1}(x)=\frac{x}{x\cdot x}$ (central projection).

For notational simplicity we will define the affine rigidity matrix only for $\mathbb{A}^2$. The rigidity matrix for $(G,p)$ in $\mathbb{A}^2$ has the form:

$$R_{\mathbb{A}^2}(G,\tilde{p})=\begin{array}{r}\textit{edge } \{i,j\}\\ \\ \textit{vertex } i\\ \\ \textit{vertex } j\end{array}\left(\begin{array}{cccccccccc} & \multicolumn{3}{c}{\tilde{p}_i} & & \multicolumn{3}{c}{\tilde{p}_j} & \\ \dots & \frac{x_i}{z_i}-\frac{x_j}{z_j} & \frac{y_i}{z_i}-\frac{y_j}{z_j} & 0 & \dots & \frac{x_j}{z_j}-\frac{x_i}{z_i} & \frac{y_j}{z_j}-\frac{y_i}{z_i} & 0 & \dots\\ & \vdots & \vdots & \vdots & & \vdots & \vdots & \vdots & \\ \dots & \frac{x_i}{z_i} & \frac{y_i}{z_i} & 1 & \dots & 0 & 0 & 0 & \dots\\ & \vdots & \vdots & \vdots & & \vdots & \vdots & \vdots & \\ \dots & 0 & 0 & 0 & \dots & \frac{x_j}{z_j} & \frac{y_j}{z_j} & 1 & \dots \end{array}\right).$$

**Theorem 17.4 (Saliola and Whiteley, 2007 [32])** *A bar-joint framework $(G,p)$ is infinitesimally rigid in $\mathbb{A}^d$ if and only if $(G,\phi\circ p)$ is infinitesimally rigid on $\mathbb{S}^d_{>0}$.*

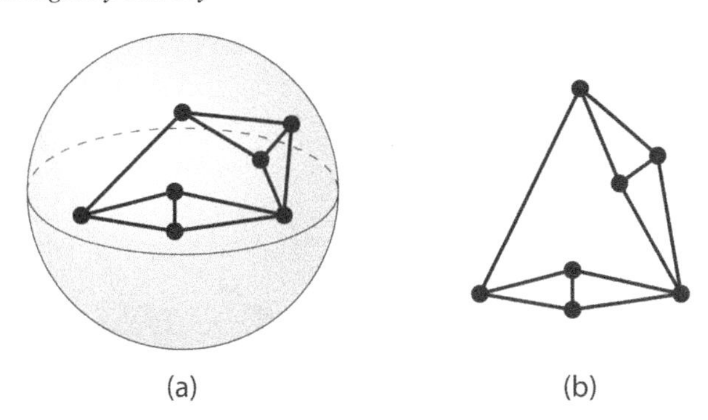


**Figure 17.4**
(a) A Spherical bar-joint framework $(G,p)$ on $\mathbb{S}^2_{>0}$ and (b) the corresponding bar-joint framework in $\mathbb{R}^2$.

As an example we present Figure 17.4. This gives an infinitesimally rigid framework on $\mathbb{S}^2_{>0}$ and the corresponding framework in $\mathbb{R}^2$.

If we focus on the framework $(G,p)$ on the upper hemisphere $\mathbb{S}^2_{>0}$ and scale each $p_i = (x_i, y_i, z_i)$ with $z_i \neq 0$ by $\frac{1}{z_i}$ to obtain $\tilde{p}_i = (\frac{x_i}{z_i}, \frac{y_i}{z_i}, 1)$ to create a framework $(G,\tilde{p})$ which is row equivalent to $R_{\mathbb{S}^2_{>0}}(G,p)$ (see, also, [34]). In particular they have isomorphic kernels. The corresponding matrix transformations result in the affine rigidity matrix of the framework.

Given a framework $(G,q)$ on $\mathbb{S}^d$ and $I \subseteq V$, the *inversion* $\iota$ (with respect to $I$) is an operator acting on $q$ such that $(\iota \circ q)_i = -q_i$ if $i \in I$ and $(\iota \circ q)_i = q_i$ otherwise. Note that $u$ is an infinitesimal flex of $(G,q)$ if and only if $\iota \circ u$, defined by $(\iota \circ u)_i = -u_i$ $(i \in I)$ and $(\iota \circ u)_i = u_i$ $(i \in V \setminus I)$, is an infinitesimal flex of $(G, \iota \circ q)$, which again means that $\iota$ preserves infinitesimal rigidity. See Figure 17.5. Inversion also takes a regular point $q$ to a regular point $(\iota \circ q)$. More generally, inversion takes a flexible framework $(G,q)$ to a flexible framework $(G, \iota \circ q)$.

We use inversion to flip points in $\mathbb{S}^d_{<0}$ to $\mathbb{S}^d_{>0}$ so that a framework $(G,q)$ on $\mathbb{S}^d$ (with no points on the equator) is transferred to a framework $(G, \iota \circ q)$ on $\mathbb{S}^d_{>0}$. It follows that a framework $(G,q)$ on $\mathbb{S}^d$ can be transformed to a framework $(G, \iota \circ \gamma \circ q)$ on $\mathbb{S}^d_{>0}$ by first applying a rotation $\gamma$ which moves all points off the equator, and then applying $\iota$ to flip points to $\mathbb{S}^d_{>0}$. In this way all of the results about $\mathbb{S}^d_{>0}$ extend to $\mathbb{S}^d$.

#### 17.2.2.1 Spherical to Affine Transfer

We can move from the Spherical rigidity matrix to the affine rigidity matrix by a sequence of row and column operations. These do not affect the dimension of the kernel. They are just multiplying the matrix on the left by an invertible matrix $T_{\mathbb{A},\mathbb{S}}$. The only column operation required is to multiply the three columns corresponding to the vertex $p_i$ by $z_i$ (and similar situation for each vertex $p_j$) which shows that velocities (elements of the kernel) will transfer by $u_i \to (\frac{1}{z_i})u_i = \tilde{u}_i$ from the Spherical matrix $R_{\mathbb{S}}(G,p)$ to the affine matrix $R_{\mathbb{A}}(G,\tilde{p})$ and conversely. The kernels are isomorphic, and a framework $(G,p)$ is infinitesimally rigid on the sphere if and only if the corresponding framework $G,\tilde{p})$ is infinitesimally rigid.

Suppose, for simplicity, that $d=2$. We can go a step further and isolate the framework on the plane $z=1$ as a plane framework, $(G,\hat{p})$. When we compare the matrix $R_{\mathbb{A}^2}(G,p)$ with the usual

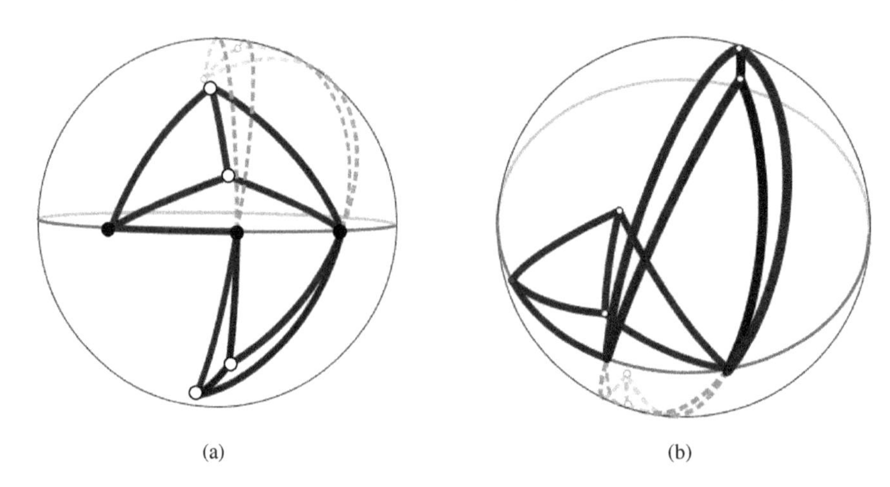


**Figure 17.5**
A framework on $\mathbb{S}^2$ with points in the lower and upper-hemispheres and on the equator (a). The dotted lines illustrate inversion applied to part of the Spherical framework to move to $\mathbb{S}^2_{\geq 0}$, and the result of inversion is shown in (b).

rigidity matrix $R_2(G,\hat{p}) =$

$$\begin{array}{c} \\ \\ \textit{edge } \{i,j\} \\ \\ \end{array} \begin{pmatrix} & \hat{p}_i & & & \hat{p}_j & & \\ & \vdots & \vdots & & \vdots & \vdots & \\ \dots & \hat{x}_i - \hat{x}_j & \hat{y}_i - \hat{y}_j & \dots & \hat{x}_j - \hat{x}_i & \hat{y}_j - \hat{y}_i & \dots \\ & \vdots & \vdots & & \vdots & \vdots & \end{pmatrix},$$

where $\hat{p}_i = (\hat{x}_i, \hat{y}_i) = (\frac{x_i}{z_i}, \frac{y_i}{z_i})$ for $i \in V$, we notice that they record equivalent information. That is, there exists a one-one correspondence between the solution spaces of the Spherical matrix $R_{\mathbb{S}^2}(G,p)$ and the plane matrix $R_2(G,\hat{p})$. Specifically, the kernels of the affine matrix $R_{\mathbb{A}}(G,\tilde{p})$ and the plane matrix $R_2(G,\hat{p})$ correspond in the following way. Define the maps: $\hat{u} : V \to \mathbb{R}^{2|V|}$ by $\hat{u}_i = (\hat{u}_{i,x}, \hat{u}_{i,y})$; $\tilde{u} : V \to \mathbb{R}^{3|V|}$ by $\tilde{u}_i = (\hat{u}_{i,x}, \hat{u}_{i,y}, -\langle \hat{u}_i, \hat{p}_i \rangle)$; and $u : V \to \mathbb{R}^{3|V|}$ by $u_i = \alpha_i \tilde{u}_i$ (where $\alpha_i = \|\tilde{p}_i\|$ is a scalar). Then $\hat{u}$ is an infinitesimal flex of $(G,\hat{p})$ if and only if $\tilde{u}$ is an infinitesimal flex of $(G,\tilde{p})$ which holds if and only if $u$ is an infinitesimal flex of $(G,p)$.

A summary of the transfer of infinitesimal rigidity between the different geometries is given in Figure 17.6.

### 17.2.3 Equilibrium Stresses

Having examined the theory of first-order flexes, we pause to present the flexes as the solutions to a matrix equation $R_d(G,p)x = 0$. In this setting, we have the equivalent theory of static rigidity working with the row space and row dependences (the equilibrium stresses) of these matrices, instead of the column dependencies (the infinitesimal flexes). The correspondence is immediate, but it takes a particularly nice form for the "projective" models in Euclidean space of the standard metrics [32]. In this setting, the rigidity correspondence is a simple matrix multiplication:

$$R_{\mathbb{S}^d_{>0}}(G,p)[T_{XY}] = R_d(G,p)$$

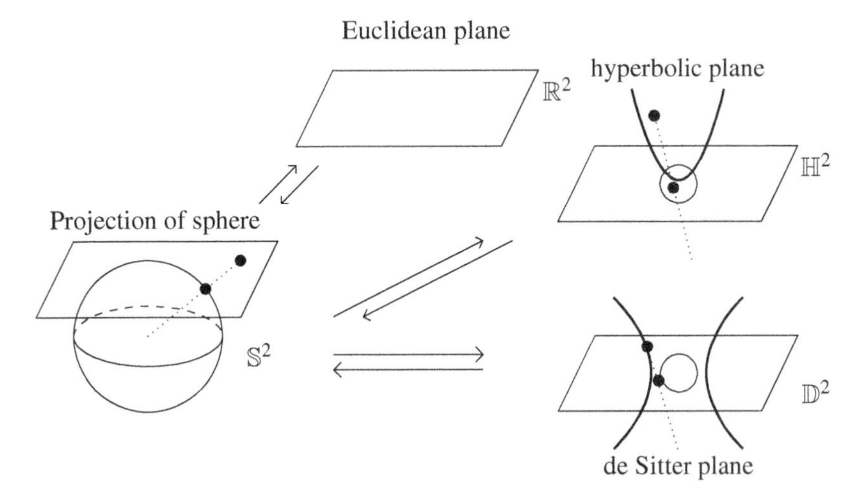


**Figure 17.6**
Visualization of the transfer between metrics.

for the same underlying configuration p, where $T_{XY}$ is a block diagonal matrix with a block entry for each vertex, based on how the sense of perpendicular is twisted at that location from one metric to the other. As a consequence of this simple correspondence of matrices, we see that row dependencies (the static equilibrium stresses) are completely unchanged by the switch in metric. As a biproduct of this static correspondence, there is a correspondence for the infinitesimal rigidity of the structures with inequalities, the tensegrity frameworks, which are well understood as a combination of first-order theory and equilibrium stresses of the appropriate signs for the edges with pre-assigned inequality constraints (see Chapter 14).

***Lemma 17.1 Saliola and Whiteley, 2007 [32]***
*There is an equilibrium stress $\omega$ for a framework $(G,p)$ in $\mathbb{R}^d$ if and only if there is an equilibrium stress $\bar{\omega}$ for the corresponding framework $(G,\bar{p})$ on $\mathbb{S}^d_{>0}$.*

We note that the static theory of rigidity was studied in detail by Crapo and Whiteley [12], in a projective setting. The projective invariance of statics was known back in the 1860's, at the time that projective geometry reached the U.K. [29].

### 17.2.4 Point-Hyperplane Frameworks

We now extend the transfer of infinitesimal rigidity to include points on the equator of the sphere and hence points at infinity in the corresponding affine plane. Here a surprising link to point-line frameworks (see Chapter 22) was developed in [14].

Let us denote the underlying graph of a point-hyperplane framework in $\mathbb{R}^d$ by $G = (V_P \cup V_L, E_{PP} \cup E_{PL} \cup E_{LL})$, where $V_P$ and $V_L$ represent the set of points and the set of hyperplanes, respectively. The edge set is partitioned into $E_{PP}, E_{PL}, E_{LL}$ according to the bipartition $\{V_P, V_L\}$ of the vertex set. Alongside the usual point configuration we have the line configuration $\ell = (a,r) : V_L \to \mathbb{S}^{d-1} \times \mathbb{R}$. Then each $i \in V_P$ is associated with $p_i \in \mathbb{R}^d$ while each $\ell_j \in V_L$ is associated with a hyperplane $\{x \in \mathbb{R}^d : \langle a_j, x\rangle + r_j = 0\}$ for some $a_j \in \mathbb{S}^{d-1}$ and $r_j \in \mathbb{R}$.

Point-hyperplane rigidity is then the natural generalisation of point-line rigidity to higher dimensions with constraints representing the distances between some pairs of points, the distances between some pairs of points and hyperplanes and angles between some pairs of hyperplanes. More formally the constraints, after differentiating, are determined by the following system of equations:

$$\langle p_i - p_j, u_i - u_j \rangle = 0 \qquad (\{i,j\} \in E_{PP}) \tag{17.5}$$
$$\langle p_i, \dot{a}_j \rangle + \langle u_i, a_j \rangle - \dot{r}_j = 0 \qquad (\{i,j\} \in E_{PL}) \tag{17.6}$$
$$\langle a_i, \dot{a}_j \rangle + \langle \dot{a}_i, a_j \rangle = 0 \qquad (\{i,j\} \in E_{LL}) \tag{17.7}$$
$$\langle a_i, \dot{a}_i \rangle = 0 \qquad (i \in V_L). \tag{17.8}$$

A map $(u, \dot{\ell})$, with $\dot{\ell} = (\dot{a}, \dot{r})$, is said to be an *infinitesimal flex* of $(G, p, \ell)$ if it satisfies the system (17.5)–(17.8), and $(G, p, \ell)$ is *infinitesimally rigid* if the dimension of the space of its infinitesimal flexes is equal to $\binom{d+1}{2}$, assuming the points $p(V_P)$ and hyperplanes $\ell(V_L)$ affinely span $\mathbb{R}^d$.

Translate the point-hyperplane framework $(G, p, \ell)$ to a point-hyperplane framework $(G, \hat{p}, \ell)$ in affine space $\mathbb{A}^d$ by taking $\hat{p}_i = (p_i, 1)$ for all $i \in V_p$. It becomes clear that the linear constraints for an infinitesimal flex of a point-hyperplane framework in $\mathbb{A}^d$ are almost identical to those for a bar-joint framework in $\mathbb{S}^d$. (See [14] for more details.)

**Theorem 17.5 (Eftekhari et al., 2017 [14])** *Let $(G, \hat{p}, \ell)$ be a point-hyperplane framework in $\mathbb{R}^d$ with $G = (V_P \cup V_L, E)$. Let $(G, p)$ be the corresponding framework on $\mathbb{S}^d$ with the points $p(v)$ for $v \in V_L$ lying on the equator* and let $(G, q)$ be the corresponding framework in $\mathbb{R}^d$ with the points $p(v)$ for $v \in V_L$ lying on a hyperplane. Then the following are equivalent:*

*(a) the point-hyperplane framework $(G, \hat{p}, \ell)$ is infinitesimally rigid in $\mathbb{R}^d$;*

*(b) the bar-joint framework $(G, p)$ on $\mathbb{S}^d$ is infinitesimally rigid; and*

*(c) the bar-joint framework $(G, q)$ in $\mathbb{R}^d$ with the points in $V_L$ lying on a hyperplane is infinitesimally rigid.*

To illustrate how this transfer can be useful, we note that it leads directly to a combinatorial characterisation of bar-joint frameworks in the plane in which a subset of points are collinear. This follows from the equivalence above because of a combinatorial characterisation of rigidity for point-line frameworks due to Jackson and Owen [21]. This result extends a theorem of Jackson and Jordán [20] dealing with the case of 3 collinear points.

In [14] a number of further combinatorial results are given for generic point-hyperplane frameworks with various constraints on the hyperplanes (fixed, fixed intercept and fixed normal).

### 17.2.5 Tensegrity Frameworks

As well as bar-joint frameworks one may consider the generalisation to tensegrity frameworks where some bars are allowed to increase but not decrease in length and some are allowed to decrease but not increase (see Chapter 14). We point out two theorems that show that infinitesimal rigidity for tensegrities can be understood by coning to the sphere.

**Theorem 17.6 (Schulze and Whiteley, 2012 [34])** *Given a tensegrity framework $(G, p)$ in $\mathbb{R}^d$ and a corresponding cone framework $(G * o, \bar{p}^*)$ in $\mathbb{R}^{d+1}$ (where all edges in $G$ remain as bars, cables and struts respectively and the cone edges are all bars). Then $(G, p)$ is an infinitesimally rigid tensegrity framework in $\mathbb{R}^d$ if and only if $(G * o, \bar{p}^*)$ is an infinitesimally rigid tensegrity framework in $\mathbb{R}^{d+1}$.*

*A calculation translates the system (17.5)–(17.8) into an equivalent linear system where $r_j$ does not appear [14]. In other words the last coordinate of $\ell_j$ does not affect the infinitesimal rigidity so we may assume that $\ell : V_L \to \mathbb{S}^{d-1} \times \{0\}$.

**Theorem 17.7 (Schulze and Whiteley, 2012 [34])** *Given a tensegrity framework $(G,p)$ in $\mathbb{R}^d$ and a corresponding framework $(G*o,q)$ on $\mathbb{S}^d$, then $(G,p)$ is an infinitesimally rigid tensegrity framework in $\mathbb{R}^d$ if and only if $(G*o,q)$ is an infinitesimally rigid tensegrity framework on $\mathbb{S}^d$.*

We note that Figure 17.7 (b) and (e) give a sample of how the cables and struts of a tensegrity framework are transformed in a projective transformation.

## 17.3 Projective Frameworks

When we look again at the matrix $R_{\mathbb{S}^d}(G,p)$ for a Spherical framework $(G,p)$, we can view it through a projective lens. We are now working in $\mathbb{P}^d$ and considering the rigidity matrix for a configuration $\check{p}_i \in \mathbb{P}^d$.

$$
R_{\mathbb{P}^d}(G,\check{p}) = \begin{array}{c} \\ \{i,j\} \\ \\ i \\ \\ j \end{array}
\begin{array}{c} \begin{array}{ccccccccccc} & & & i & & & & j & & & \end{array} \\
\begin{pmatrix}
0 & \dots & 0 & \check{p}_j & 0 & \dots & 0 & \check{p}_i & 0 & \dots & 0 \\
 & & & \vdots & & & & \vdots & & & \\
0 & \dots & 0 & \check{p}_i & 0 & \dots & 0 & 0 & 0 & \dots & 0 \\
 & & & \vdots & & & & \vdots & & & \\
0 & \dots & 0 & 0 & 0 & \dots & 0 & \check{p}_j & 0 & \dots & 0
\end{pmatrix} \end{array}
$$

In this projective space, we can apply scaling to individual vertices, replacing $\check{p}_i$ with $\alpha_i\check{p}_i$ where $\alpha_i \neq 0$, and preserve the rank of the rigidity matrix:
- multiply the columns for $i$ by $\frac{1}{\alpha_i}$;
- the rows for edges $\{i,j\}$ by $\alpha_i$; and
- the row for $\check{p}_i$ by $\alpha_i^2$.

The result is the desired matrix for the scaled framework:

$$
R_{\mathbb{P}^d}(G,\check{p}) = \begin{array}{c} \\ \{i,j\} \\ \\ i \\ \\ j \end{array}
\begin{array}{c} \begin{array}{ccccccccccc} & & & i & & & & j & & & \end{array} \\
\begin{pmatrix}
0 & \dots & 0 & \check{p}_j & 0 & \dots & 0 & \alpha_i\check{p}_i & 0 & \dots & 0 \\
 & & & \vdots & & & & \vdots & & & \\
0 & \dots & 0 & \alpha_i\check{p}_i & 0 & \dots & 0 & 0 & 0 & \dots & 0 \\
 & & & \vdots & & & & \vdots & & & \\
0 & \dots & 0 & 0 & 0 & \dots & 0 & \check{p}_j & 0 & \dots & 0
\end{pmatrix} \end{array}.
$$

Notice that this matrix operation includes inversion in the sphere for $\check{p}_i$, with $\alpha_i = -1$. However, our focus is currently on the shared projective framework.

We can trace the impact of an arbitrary projective transformation with an invertible matrix $T_{d+1\times d+1}$ by multiplying the entire rigidity matrix on the right by the block diagonal $|V|(d+1) \times |V|(d+1)$ matrix $T$ with block entries $T_{d+1\times d+1}$. This multiplication by an invertible matrix preserves the rank of the rigidity matrix. Combined with the scaling above, we have a matrix form for an arbitrary projective transformation.

In these combined operations, an equilibrium stress $\omega_{ij}$ is transformed to $\frac{1}{\alpha_i\alpha_j}\omega_{ij}$ and $\omega_i$ is transformed to $\frac{1}{\alpha_i^2}\omega_i$.

If we apply a Cayley algebra join $\vee p_i$ on the right to the columns associated to $\check{p}_i$, the equilibrium equation for an equilibrium stress becomes $\sum_{\{j|\{i,j\}\in E\}} \check{p}_j \vee \check{p}_i = 0$, since $\check{p}_i \vee \check{p}_i = 0$. This is the simple form of an equilibrium stress in terms of the 2-extensors for the forces at $\check{p}_i$ [12]. (See

Chapters 3 and 4 for Cayley algebra background.) Figure 17.7 (b) and (d) illustrates how the signs for an equilibrium stress change under a projective transformation in the plane. The idea is that when the projective transformation applies a negative weight to one end, but not the other, of an edge, the sign of the equilibrium stress (compression or tension) is changed.

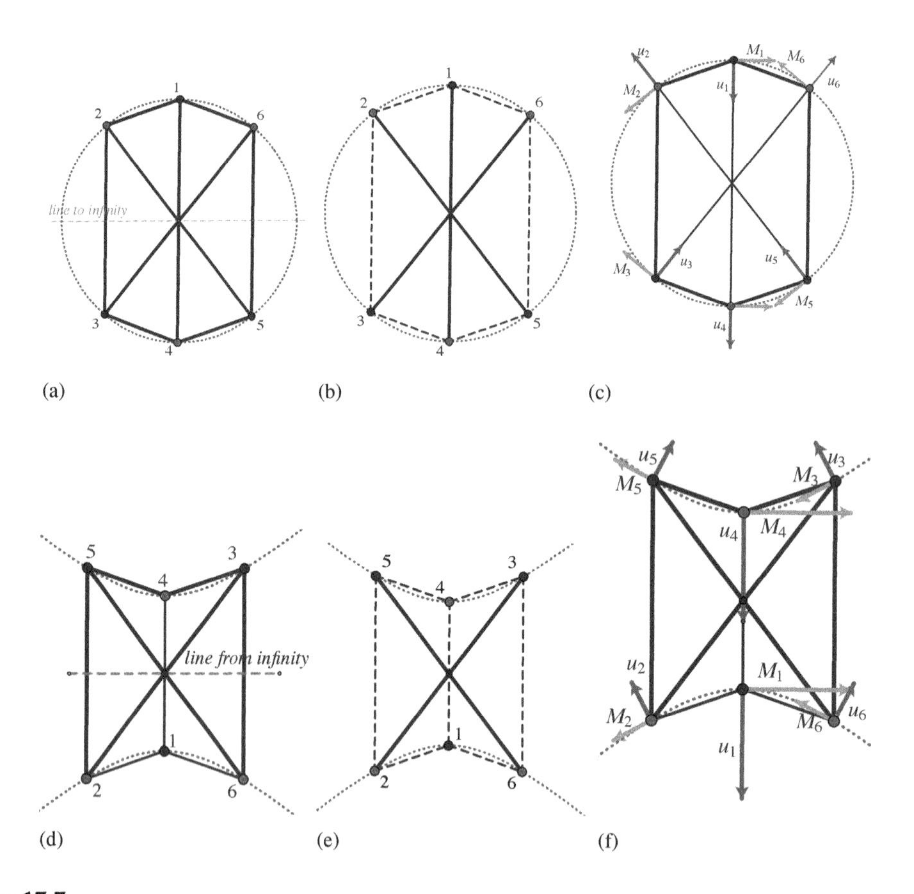


**Figure 17.7**
The framework with graph $K_{3,3}$ on a circle (a) has an equilibrium stress with tension on dotted bars and compression on solid bars (b). In (c) we see a non-trivial infinitesimal flex and the corresponding (perpendicular) projective momenta. After a projective transformation, the framework lies on a hyperbola (d) with the transformed equilibrium stress pattern (e) and the transformed projective momenta and corresponding perpendicular infinitesimal flexes (f).

The projective interpretation of the infinitesimal flexes is less well known but it provides significant geometric insight into how the infinitesimal flexes transform.

We begin with the projective re-presentation of infinitesimal flexes in $\mathbb{R}^d$ as solutions of the matrix equation $R_{\mathbb{P}^d}(G,\check{p})M = 0$. It is implicit in the coning theorems, that there is an isomorphism of the solutions to $R_{\mathbb{R}^d}(G,p)v = 0$ and $R_{\mathbb{P}^d}(G,\check{p})M = 0$ which fix the origin, We now give an explicit correspondence, and interpret the entries $M_i$, as hyperplane coordinates in $\mathbb{P}^d$ for each vertex. The equation

$$\langle \check{p}_i, M_i \rangle = \check{p}_i \cdot M_i = 0$$

says that $M_i$ represents a hyperplane through the projective point $\check{p}_i$.

***Proposition 17.1***
*For a framework $(G,p)$ in $\mathbb{R}^d$ and the corresponding projective (affine) framework $(G,p)$ with weight (last coordinate) 1 in $\mathbb{P}^d$, $(\ldots,u_i,\ldots)$ is an infinitesimal flex of $(G,p)$ if and only if $(\ldots,M_i,\ldots) = (\ldots,(u_i,-u_i\cdot p_i),\ldots)$ is a projective motion.*

**Proof.** For every edge $\{i,j\}$ we find the correspondence:

$$(p_i-p_j)\cdot(u_i)+(p_j-p_i)\cdot v_j = (p_i-p_j)\cdot(u_i)+(p_j-p_i)\cdot v_j = 0 \tag{17.9}$$
$$\iff [(p_j u_i - p_i u_i)+(p_i v_j - p_j v_j)] = \langle \check{p}_i, M_i\rangle + \langle \check{p}_i, M_j\rangle = 0. \tag{17.10}$$

Combined with the observation above, extended we confirm,

$$\langle \check{p}_i, M_i\rangle = (p_i,1)\cdot(u_i,-p_i u_i) = p_i u_i - p_i u_i = 0$$

This gives the full correspondence for configurations in $\mathbb{R}^d$. □

We call the hyperplanes $M_i$ the *momentum* of the point. See Figure 17.7 (c), (f) for the pairings of $u_i$ and $M_i$. If we change $\check{p}_i$ to $\alpha\check{p}_i$ then tracking the change through the matrix operations above, transforms $M_i$ to $\frac{1}{\alpha_i}M_i$. Above, a projective transformation is applied by right multiplication of the rigidity matrix by $T$. Left multiplication by $T^{-1}$ preserves the solution space of $M$.

In Figure 17.7, the original infinitesimal flexes for the circle are perpendicular to the circle, inward for $1,3,5$ and outward for $2,4,6$, and the momenta are therefore tangent to the circle, but with different "signs." After transformation, the momenta remain tangent to the conic and the rescaled infinitesimal flexes are perpendicular to the new conic. This is a sample of how this representation provides additional geometric insight.

When transforming frameworks and infinitesimal flexes to other metrics, we can continue to follow the momenta, and then use the inverse transformation to corresponding normals to the momenta –using the orthogonality of the new metric.

A valuable exploration of the underlying projective geometry of bar and joint frameworks is the study of *pure conditions* which present projective polynomials to capture which realizations of a generically isostatic framework are singular (lowering the rank of the rigidity matrix) [37]. These polynomials apply across all of the metrics being explored here (see Chapter 4).

## 17.4 Pseudo-Euclidean Geometries

**Pseudo-Euclidean space:** Consider $\mathbb{R}^d$ as equipped with a pseudo Euclidean metric. That is we define the length of a vector $w \in \mathbb{R}^d$ as

$$||w||^2 = -\sum_{i=1}^{s} w_i^2 + \sum_{i=s+1}^{d} w_i^2.$$

This defines a *Pseudo-Euclidean space* $\mathbb{F}^d$. Note also that the case when $s=0$ is the usual Euclidean case we have already discussed. These are sometimes called Minkowski Spaces, and written as $\mathbb{M}^d_{t,s}$, where $t=d-s$, see [34].

**Minkowski space:** *Minkowski space* $\mathbb{M}^d$ is the Pseudo-Euclidean space with $s=1$.

**Spheres in Pseudo Euclidean space:** Similarly, for $x,y \in \mathbb{R}^{d+1}$, let $\langle x,y\rangle_k$ denote the function

$$\langle x,y\rangle_k = x_1y_1 + \cdots + x_{d-k+1}y_{d-k+1} - x_{d-k+2}y_{d-k+2} - \cdots - x_{d+1}y_{d+1},$$

and let $X^d_{c,k}$ denote the set,

$$X^d_{c,k} = \{x \in \mathbb{R}^{d+1} : \langle x,x\rangle_k = c\},$$

for some constant $c \neq 0$ and $k \in \mathbb{N}$. The space $X^d_{-1,1}$ is the $d$-dimensional *Hyperbolic* space, which we denote by $\mathbb{H}^d$. The space $X^d_{1,1}$ is the $d$-dimensional *de Sitter space* which we denote by $\mathbb{D}^d$. (Note that Spherical space $\mathbb{S}^d$ is the case when $k = 0, c = 1$.) Some literature asks that $x_{d+1} > 0$, corresponding to the upper hemisphere [32], but we choose to be more general, using inversion to complete the picture [34].

### 17.4.1 Hyperbolic and Minkowski Spaces

Now we consider the transfer of rigidity results for Hyperbolic and Pseudo-Euclidean (in particular Minkowski) spaces. More details on non-Euclidean, or Cayley-Klein, geometries may be found in [36].

Define $\phi' : \mathbb{F}^d \to \mathbb{S}^d_{>0}$ by $\phi'(x) = \frac{x}{\|x\|}$.

**Theorem 17.8 (Saliola and Whiteley, 2007 [32])** *A bar-joint framework $(G,p)$ is infinitesimally rigid in $\mathbb{F}^d$ if and only if $(G, \phi' \circ p)$ is infinitesimally rigid on $\mathbb{S}^d_{>0}$.*

Recall Figure 17.6. This shows the map from a Spherical framework to Hyperbolic and corresponding infinitesimal flexes. Similarly to the Euclidean case, we can cone in Pseudo-Euclidean space to pass to any of spheres (in particular to Hyperbolic and de Sitter spheres).

**Theorem 17.9 (Saliola and Whiteley, 2007 [32])** *A bar-joint framework $(G,p)$ is infinitesimally rigid in $\mathbb{H}^d$ if and only if the corresponding Spherical framework is infinitesimally rigid on $\mathbb{S}^d_{>0}$.*

We may also consider "points at infinity" in the Hyperbolic geometry: points on the *absolute*, a $d$-sphere in the Klein model in $\mathbb{R}^{d+1}$. Since we are coning from the sphere, in principle through the affine plane $x_{n+1} = 1$ which has a Euclidean metric even in Minkowski space, the "absolute" is not special in Minkowski space, though the metric along rays of the cone has all distances $= 0$. This should not cause major changes in the rigidity of the framework.

It is, however, an open problem to extend the results of [14] to frameworks in the de Sitter space $\mathbb{D}^d$ with points on the "equator" $x_{n+1} = 0$. We anticipate that the methods and results of [14] will extend with appropriate adjustments.

We can also consider tensegrity frameworks in Hyperbolic space.

**Theorem 17.10 (Schulze and Whiteley, 2012 [34])** *Given a tensegrity framework $(G,p)$ in $\mathbb{R}^d$ and a corresponding framework $(G*o,q)$ in $\mathbb{H}^d$, then $(G,p)$ is an infinitesimally rigid tensegrity framework in $\mathbb{R}^d$ if and only if $(G*o,q)$ is an infinitesimally rigid tensegrity framework in $\mathbb{H}^d$.*

## 17.5 Transfer of Symmetric Infinitesimal Rigidity

***S*-symmetric graph:** An action of a group $S$ on $G$ is a group homomorphism $\theta : S \to \mathrm{Aut}(G)$, where $\mathrm{Aut}(G)$ denotes the automorphism group of the graph $G$. If $S$ acts on $G$ by $\theta$, then we say that the graph $G$ is *S-symmetric* (with respect to $\theta$).

**$S$-gain graph:** For an $S$-symmetric graph $G = (V, E)$ with a free action $\theta$, the *$S$-gain graph* is an oriented labelled quotient graph $G/S$. The quotient is the multi-graph (possibly with loops) which has the set $V/S = \{Si : i \in S\}$ of vertex orbits as its vertex set and the set $E/S = \{Se : e \in E\}$ of edge orbits as its edge set. An edge orbit connecting $Si$ and $Sj$ can be written as $\{(\theta(\gamma)(i), \theta(\gamma) \circ \theta(\alpha)(j)) : \gamma \in S\}$ for a unique $\alpha$. For each such edge orient it from $Si$ to $Sj$ and assign the label $\alpha$.

**$S$-symmetric framework:** Let $S$ be an abstract group and $G = (V, E)$ be an $S$-symmetric graph with respect to an action $\theta : S \to \mathrm{Aut}(G)$. Suppose also that $S$ acts on $\mathbb{R}^d$ via the homomorphism $\tau : S \to O(\mathbb{R}^d)$, where $O(\mathbb{R}^d)$ denotes the orthogonal group. Then, we say that a framework $(G, p)$ is *$S$-symmetric* (with respect to $\theta$ and $\tau$) if:

$$\tau(x)(p(v)) = p(\theta(x)v) \qquad \text{for all } x \in S \text{ and all } v \in V.$$

**Forced symmetric infinitesimal flex:** Let $S$ be a symmetry group. An infinitesimal flex $u : V \to \mathbb{R}^d$ on an $S$-symmetric graph $(G, p)$ is a *forced symmetric infinitesimal flex* if $x(u_i) = u_{x(i)}$ for all $i \in V$ and $x \in S$.

**Forced symmetrically rigid:** An $S$-symmetric framework is *forced symmetrically infinitesimally rigid* if every forced symmetric infinitesimal flex is trivial.

**$S$-regular:** A symmetric framework $(G, p)$ in $\mathbb{R}^d$ (resp. on $\mathbb{S}^d$) is *$S$-regular* if the rank of the rigidity matrix $R_d(G, p)$ (resp. $R_{\mathbb{S}^d}(G, p)$) is maximal over all $S$-symmetric configurations $q \in \mathbb{R}^{d|V|}$ (resp. $q \in \mathbb{S}^{d|V|}$).

### 17.5.1 Symmetric Frameworks

Let $S$ be a symmetry group, let $x \in S$ and let $F_x$ denote the linear subspace of $\mathbb{R}^d$ which consists of all points $a \in \mathbb{R}^d$ with $x(a) = a$. If $(G, p)$ is $S$-symmetric then $p_i$ lies in the subspace

$$U(p_i) = \bigcap_{x \in S : x(p_i) = p_i} F_x.$$

For given bases $B_i$ and $B_j$ of $U(p_i)$ and $U(p_j)$, let $M_i$ and $M_j$ be the matrices whose columns are the coordinate vectors of $B_i$ and $B_j$ relative to the canonical basis of $\mathbb{R}^d$.

Let $(G, p)$ be an $S$-symmetric framework in $\mathbb{R}^d$ which has no joint that is "fixed" by a non-trivial symmetry operation in $S$. Further, let $(G_0, \psi)$ be the quotient $S$-gain graph of $(G, p)$. For each edge $e \in E(G_0)$, the *orbit rigidity matrix* $O(G, p, S)$ of $(G_0, p)$ is the $|E(G_0)| \times d|V(G_0)|$ matrix of the form

$$\begin{array}{c} \\ \\ \{i,x(j)\} \\ \{i,x(i)\} \\ \\ \end{array}
\begin{pmatrix}
 & & & i & & & & & j & & & \\
 & & & & & \vdots & & & & & & \\
0 & \dots & 0 & (p_i - x(p_j))M_i & 0 & \dots & 0 & (p_j - x^{-1}(p_i))M_j & 0 & \dots & 0 \\
0 & \dots & 0 & (2p_i - x(p_i) - x^{-1}(p_i))M_i & 0 & \dots & 0 & 0 & 0 & \dots & 0 \\
 & & & & & \vdots & & & & &
\end{pmatrix}.$$

We refer the reader to Chapter 25 and [34] for more details. Similarly, for an $S$-symmetric Spherical framework $(G, p)$ with $p : V \to \mathbb{S}^d$, we can also write down a standard Spherical orbit matrix. This is the $(|E(G_0)| + |V(G_0)|) \times (d+1)|V(G_0)|$ matrix $O_{\mathbb{S}}(G, p)$:

$$\begin{array}{l} \\ \\ \{i,x(j)\} \\ \{i,x(i)\} \\ i \\ \\ j \\ \\ \end{array} \begin{pmatrix} & & & i & & & & j & & & \\ & & & & & \vdots & & & & & \\ 0 & \dots & 0 & (p_i - x(p_j))M_i & 0 & \dots & 0 & (p_j - x^{-1}(p_i))M_j & 0 & \dots & 0 \\ 0 & \dots & 0 & (2p_i - x(p_i) - x^{-1}(p_i))M_i & 0 & \dots & 0 & 0 & 0 & \dots & 0 \\ 0 & \dots & 0 & p_i & 0 & \dots & 0 & 0 & 0 & \dots & 0 \\ & & & & & & & & & & \\ 0 & \dots & 0 & 0 & 0 & \dots & 0 & p_j & 0 & \dots & 0 \\ & & & & & \vdots & & & & & \end{pmatrix}.$$

Our next theorem can be proved by symmetrising the coning transformation from the Euclidean to the Spherical setting we have already presented, see [34] for details.

**Theorem 17.11** *[34] Let $q$ be a configuration of points on $\mathbb{S}^d_{>0}$ such that the projection $\pi(q)$ from the centre of $\mathbb{S}^d$ onto $\mathbb{R}^d$ (via the affine plane) satisfies $\pi(q) = p$. Let $S^*$ be a symmetry group of $\mathbb{S}^d_{>0}$ and let $S$ be the corresponding symmetry group (under $\pi$) of $\mathbb{R}^d$. Then:*

- *the space of $S$-symmetric infinitesimal flexes of $(G,p)$ in $\mathbb{R}^d$ is isomorphic to the space of $S^*$-symmetric infinitesimal flexes of $(G,q)$ on $\mathbb{S}^d_{>0}$;*
- *the space of $S$-symmetric equilibrium stresses of $(G,p)$ in $\mathbb{R}^d$ is isomorphic to the space of $S^*$-symmetric equilibrium stresses of $(G,q)$ on $\mathbb{S}^d_{>0}$.*

We also note that the transfer takes an $S$-regular configuration $p$ in $\mathbb{R}^d$ to an $S^*$-regular configuration $q$ on $\mathbb{S}^d$. With this, symmetric flexes also transfer from $\mathbb{R}^d$ to $\mathbb{S}^d$ through coning. See [34, Section 4.1] for details and also cautionary examples.

We mention that the transfer described in Theorem 17.11 was geometric and hence, noting that the transfer operations preserve the symmetry group (as long as the group exists in both Spherical and Euclidean space), gives us the transfer for incidentally symmetric frameworks. One can also consider the transfer at a more detailed level. In particular Schulze and Tanigawa developed a block decomposition of the rigidity matrix (see Chapter 25 and [33]). With this, we expect that the rank preserving matrix operations can be applied to each block of the decomposition of the rigidity matrix for incidental symmetry.

In Figure 17.8 we give an example, adapted from Figure 17.7, to illustrate how the analysis works, even with fixed vertices and fixed edges.

In this subsection we explain how the transfer of infinitesimal rigidity extends to symmetric frameworks. When this is done, then the general transfer of infinitesimal rigidity can be thought of as the special case when the symmetry group is the identity.

In fact this symmetric transfer extends to Pseudo-Euclidean spaces and to Hyperbolic and de Sitter spaces [34] whenever the symmetry group exists. The basic examples of symmetry groups that exist throughout the range of Pseudo-Euclidean geometries are the reflection and half turn rotation groups. It is an open problem, currently being investigated in [4], to confirm that this transfer extends to include points on the equator of $\mathbb{S}^d$ (points at infinity). We also mention that the results in Theorem's 17.6 and 17.7 were extended to symmetric tensegrities in [34].

## 17.6 Global Rigidity

**Globally rigid framework:** A bar-joint framework $(G,p)$ in $\mathbb{R}^d$ is *globally rigid* if every equivalent framework $(G,q)$ arises from $(G,p)$ by a congruence of $\mathbb{R}^d$.

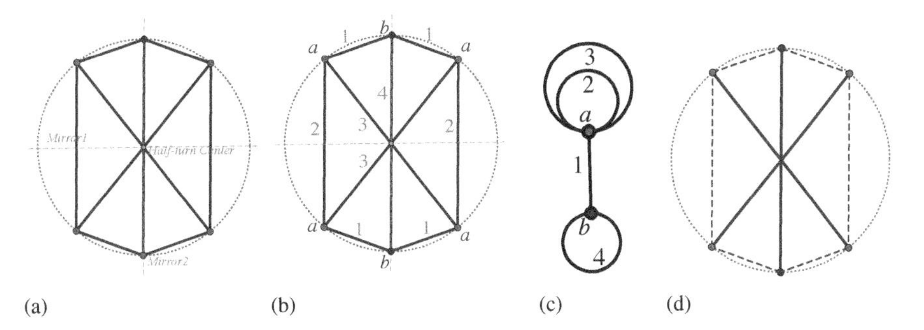


**Figure 17.8**
A framework, with underlying graph $K_{3,3}$, with mirrors and a half-turn (a) has two vertex orbits $a, b$ and four edges orbits (b). The corresponding gain graph (c) has the count of 3 columns (2 for a, 1 for b fixed on the mirror) and 4 rows (edge orbits). Since $4 > 3$ there is a fully symmetric equilibrium stress (d). The corresponding infinitesimal flex in Figure 17.7 is not symmetric, and the framework is rigid and globally rigid [11]. Note this analysis also applies without change to the projected version in Figure 17.7.

**Globally rigid graph:** A graph is *globally rigid* in $\mathbb{R}^d$ if all generic realizations $(G, p)$ are globally rigid in $\mathbb{R}^d$.

**Quasi-generic Spherical framework:** A framework on the sphere is *quasi-generic* if it is the central projection of a generic framework in $\mathbb{R}^{d+1}$.

**Universally rigid:** A framework $(G, p)$ in $\mathbb{R}^d$ is *universally rigid* if every equivalent framework $(G, q)$ in $\mathbb{R}^D$ for any $D \geq d$ arises from $(G, p)$ by a congruence.

**Dimensionally rigid:** A framework $(G, p)$ in $\mathbb{R}^d$ is *dimensionally rigid* if there are no equivalent frameworks with a higher dimensional affine span.

**Stress matrix:** The *stress matrix* $\Omega$ is the $n \times n$ symmetric matrix with off-diagonal entries $-\omega_{ij}$ and diagonal entries $\sum_j \omega_{ij}$.

**Conic at infinity:** A finite set of (non-zero) vectors in $\mathbb{R}^d$ lie on a *conic at infinity* if when regarded as points in projective $d-1$-dimensional space, they lie on a conic.

**Super stable:** A framework $(G, p)$ in $\mathbb{R}^d$ with a full dimensional affine span is called *super stable* if it has an equilibrium stress matrix $\Omega$ which is PSD with maximum rank and its edge directions do not lie on a conic at infinity.

**Slicing:** Given a cone framework $(G, p)$ in $\mathbb{R}^{d+1}$, the process of *slicing* means to slide all points of $p$ to the affine plane and then consider the resulting subframework (without the cone) as a framework in $\mathbb{R}^d$.

Let us now consider the effect of changing the metric on the global rigidity of the corresponding frameworks (see also [23]). We have already seen that equilibrium stresses transfer, and it is clear that these necessary conditions also transfer. Hendrickson gave necessary conditions for the global rigidity of a framework $G(p)$ in dimension $d$: (i) the graph is $(d+1)$-connected and (ii) $G$ is redundantly rigid in $\mathbb{R}^d$. In dimensions $d = 1$ and $d = 2$ these are also sufficient in $\mathbb{R}^d$. The conditions are well-known not to be sufficient for $d > 3$.

It follows from the main result of [16] that global rigidity in Euclidean spaces is a generic property. That is, if some generic realization $(G, p)$ of $G$ is globally rigid in $\mathbb{R}^d$ then all generic

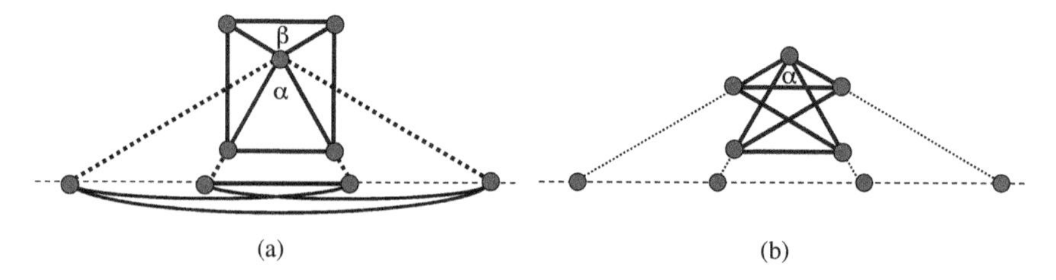


**Figure 17.9**
(a) A globally rigid realization of the 4-cycle on the line is coned to a framework in $\mathbb{R}^2$ that is not globally rigid. The angles $\alpha = 60°$, $\beta = 120°$, and the other two angles at the cone vertex are 90° induce a second realization. (b) An equivalent but non-congruent realization.

realizations $(G,q)$ in $\mathbb{R}^d$ are also globally rigid. A fundamental result of Connelly and Whiteley transfers this to Spherical frameworks, which also appears in the work of Pogorelov [27, 28].

**Theorem 17.12 (Connelly and Whiteley, 2010 [6])** *A graph G is generically globally rigid in $\mathbb{R}^d$ if and only if G is quasi-generically globally rigid on the sphere $\mathbb{S}^d$.*

To understand this theorem lets start by considering coning. It is not hard to see that this process takes a redundantly rigid framework in $\mathbb{R}^d$ to a redundantly rigid framework in $\mathbb{R}^{d+1}$ and takes a $d$-connected graph to a $d+1$ connected graph. Thus Hendrickson's necessary conditions transfer, and the sufficiency for $d = 1$ and $d = 2$ transfers to the sphere, and also to $\mathbb{H}^2$ and $\mathbb{D}^2$ (private communication from Stephen Gortler). One of the key ideas to the stronger transfer of global rigidity is the averaging technique which is implicit in the work of Pogorelov [27, 28], and appears in [35] .

**Theorem 17.13 (Connelly and Whiteley, 2010 [6])** *Given a globally rigid and infinitesimally rigid framework $(G,p)$ in $\mathbb{R}^d$, there exists an open neighborhood $N_p$ such that for any $q \in N_p$, the framework $(G,q)$ is globally rigid and infinitesimally rigid in $\mathbb{R}^d$.*

The other key idea is the stress matrix characterisation of generic global rigidity given in [7, 16] (see Chapter 16). In particular the fact that projective transformations preserve infinitesimal rigidity is extended in [6] to show that projective transformations also preserve the property of having a maximum rank stress matrix. For coning, in particular, we have the following result.

**Theorem 17.14 (Connelly and Whiteley, 2010 [6])** *Let $(G,p)$ and $(G*o,\bar{p}^*)$ be frameworks in $\mathbb{R}^d$ of full affine span. Then $(G,p)$ has a stress matrix $\Omega$ with rank $|V|-d-1$ if and only if $(G*o,\bar{p}^*)$ has a stress matrix $\Omega^*$ with rank $|V|-d-1$.*

By combining this result with [7, 16] we have the following.

**Corollary 17.15** *A framework $(G,p)$ in $\mathbb{R}^d$ is globally rigid if and only if the cone framework $(G*o,\bar{p}^*)$ in $\mathbb{R}^{d+1}$ is globally rigid.*

The example in Figure 17.9 is based on a cycle (generically globally rigid on the line) whose cone is (generically) globally rigid in the plane. However the non-generic realization of the cycle results in an equivalent but non-congruent realization.

### 17.6.1 Universal Rigidity

While global rigidity in Euclidean space is a generic property, universal rigidity is not. Even on the real line a cycle $C_k$, with $k \geq 4$, may be universally rigid or flexible in $\mathbb{R}^2$ depending on the ordering of the vertices [22].

In the next three results we present coning for dimensional rigidity, super stability and universal rigidity [10].

**Theorem 17.16** *The framework $(G, p)$ is dimensionally rigid in $\mathbb{R}^d$ if and only if the cone framework $(G * o, \bar{p}^*)$ is dimensionally rigid in $\mathbb{R}^{d+1}$.*

**Theorem 17.17** *The framework $(G, p)$ is super stable in $\mathbb{R}^d$ if and only if the cone framework $(G * o, \bar{p}^*)$ is super stable in $\mathbb{R}^{d+1}$.*

**Theorem 17.18** *If the framework $(G, p)$ is super stable in $\mathbb{R}^d$ then the cone framework $(G * o, \bar{p}^*)$ is universally rigid in $\mathbb{R}^{d+1}$. If the cone framework $(G * o, \bar{p}^*)$ is universally rigid in $\mathbb{R}^{d+1}$ then the framework $(G, p)$ is dimensionally rigid in $\mathbb{R}^d$.*

The only failure for projecting a universally rigid cone framework to a universally rigid framework can come from the appearance of affine flexes due to the projection having member directions on a conic at infinity. If, for example, the framework $(G, p)$ is in general position, this cannot happen, as was observed by Alfakih and Ye [2].

***Lemma 17.2 [8]***
*Let $(G * o, \bar{p}^*)$ be a cone framework in $\mathbb{R}^{d+1}$ with an equilibrium stress matrix $\Omega$. Then any framework obtained from $(G * o, \bar{p}^*)$ by sliding has an equilibrium stress matrix with the same rank and signature as $\Omega$.*

In fact sliding, as we illustrated in Figure 17.1, preserves global rigidity. One of the key aspects here is how the equilibrium stress changes under sliding the framework. When the framework is slid to a hyperplane then a new equilibrium stress emerges from the flatness, but the cone edges simultaneously become unstressed. In [8] it is also shown that super stability is invariant under coning and slicing. A consequence being that projective transformations that do not send points to infinity preserve super stability.

It is also worth pointing out explicitly that inversion is a special case of sliding so the above discussion implies that inversion preserves global rigidity, superstability and universal rigidity.

### 17.6.2 Projective Transformations

A general projection in $\mathbb{R}^d$ can be formed by coning to $\mathbb{R}^{d+1}$, rotating, and re-projecting, perhaps several times. As a result, the coning results guarantee the projective invariance of key properties.

**Theorem 17.19** *A framework $(G, p)$ is super stable (resp. dimensionally rigid) in $\mathbb{R}^d$ if and only if every invertible projective transformation which keeps vertices finite is super stable (resp. dimensionally rigid).*

This projective invariance is almost always true for universal rigidity. The key is whether the directions of the members meet the set $X$ which is going to infinity, in a conic.

**Theorem 17.20 ([9])** *If a framework $(G, p)$ is universally rigid in $\mathbb{R}^d$, and $X$ is a hyperplane avoiding all vertices, such that the members do not meet it in a conic, then any invertible projective transformation $T$ which takes $X$ to infinity makes $(G, T(p))$ universally rigid.*

An illustrative example is given in [9, Figures 8 and 9]. Note that this theorem also extends to tensegrity frameworks, [9, Theorem 13.1].

Alfakih and Nguyen [1] showed that having a maximum rank PSD stress matrix is a sufficient condition for dimensional rigidity. When a nested sequence of affine spaces, with a corresponding sequence of PSD stress matrices, is used to iteratively demonstrate the dimensional rigidity of a given framework [9], the projected framework is also dimensionally rigid, since the projected framework has the projected affine sequence to demonstrate its dimensional rigidity. It is well known that there are special positions for which global rigidity is not projectively invariant. In fact it is not even an affine invariant, see [6, Figure 4 and Example 8.3]. However, the rank of any stress matrix $\Omega$ on the framework is invariant under projective transformations [6].

### 17.6.3 Pseudo-Euclidean Metrics

Global rigidity has a natural extension to all the Pseudo-Euclidean metrics and their spheres. There are two distinct ways that global rigidity of $G$ can be "generic" within a metric: (i) there are some generic configurations at which $(G,p)$ is globally rigid; and (ii) if $(G,p)$ is globally rigid at one generic configuration, then $(G,q)$ is globally rigid for all generic $q$. We use the language *generically globally rigid* for the second sense. In arbitrary Pseudo-Euclidean spaces we have the following result of Gortler and Thurston [15].

**Theorem 17.21** *A graph $G$ is generically globally rigid in $\mathbb{R}^d$ if and only if it is globally rigid for some generic configurations in $\mathbb{F}^d$.*

However this does not complete the story. It is not clear if global rigidity is a generic property in Pseudo-Euclidean spaces. It is possible that there are graphs which are not generically globally rigid but nevertheless do have some generic realizations which are globally rigid. A partial result given in [15] resolves this positively when $G$ has a (generically) globally rigid subgraph on at least $d+1$ vertices.

Here Hyperbolic space was modelled as the "'-1" sphere in Minkowski space. More formally Minkowski space $\mathbb{M}^{d+1}$ is the $(d+1)$-dimensional Pseudo-Euclidean space with one negative coordinate in its signature. We then model Hyperbolic space as the vectors $x \in \mathbb{M}^{d+1}$ such that $|x|^2 = -1$ in the Minkowski metric.

It is an open problem to understand the transfer of universal rigidity for Pseudo-Euclidean spaces. Generic universal rigidity was characterised in Euclidean spaces in [17] as graphs which admit maximum rank PSD stress matrices. We have already seen how maximum rank equilibrium stresses transfer.

## 17.7 Summary and Related Topics

We conclude the chapter with Table 17.1 summarizing the key results we have presented and then by briefly mentioning a number of related pieces of work. The table reveals a number of problems that have not been studied in detail.

As we assembled the table we recognized some gaps that have not been carefully explored for a transfer of metric. One example is the concepts of *second-order rigidity* and *pre-stress stability* [11]. Second-order rigidity is not projectively invariant, and will only be of interest in the presence of some equilibrium stress. There is some evidence these results will extend geometrically to the Spherical metric through coning, but we are not aware of work transferring these concepts carefully to the Spherical metric or to other metrics. Even dimensional rigidity, and universal rigidity, have not been carefully defined for other metrics. Though they appear amendable to generalizations and transfers, with appropriate definitions.

We have discussed that infinitesimal rigidity, as a projective invariant, is equivalent in the Cayley-Klein metrics (Euclidean, Spherical, Minkowski, Hyperbolic spaces, etc.). Let us make two comments on the limitations of this approach. First, one may wonder if the alternative viewpoint

**Table 17.1**
We compare the properties of Euclidean frameworks with frameworks in other geometries. "Yes" refers to the equivalence of the property in the given geometry with the Euclidean case and "no" refers to the non-equivalence.

| **Type of rigidity** | **Nature of transfer** | **Cone** | **Spherical** $\mathbb{S}^d$ | **Pseudo-Euclidean** $\mathbb{P}^d$ | **Hyperbolic** $\mathbb{H}^d$ |
|---|---|---|---|---|---|
| Infinitesimal | geometric | Yes, [32] | Yes, [32] | Yes, [32] | Yes, [32] |
| Rigidity | generic | Yes | Yes | Yes | Yes |
| $S$-symmetric | geometric | Yes, [34] | Yes, [34] | Yes, [34] | Yes, [34] |
| Super stability | geometric | Yes, [10] | ? | ? | ? |
| Dimensional | geometric | Yes, [10] | ? | ? | ? |
| Global | generic | Yes, [6] | ? | ? | Yes $d = 2$ |
| Universal | geometric | No, [23] | No, [23] | ? | ? |

these other geometries provide can help with conjectures about Euclidean frameworks. We know of no results in this direction. Second, one may consider more abstract metric spaces. In general the transfer processes we have described break down since such spaces have different numbers of "trivial" infinitesimal flexes and hence different combinatorics required for rigidity. Rigidity in various other metric space contexts have been considered [24, 25, 26], see also Chapter 24.

We now mention some relevant work in progress [4]. A half turn symmetric framework in $\mathbb{R}^d$ is projectively equivalent to a mirror reflection, in all dimensions, which guarantees all infinitesimal and static properties are preserved. This same projective transformation applies in all Pseudo Euclidean spaces and will preserve all infinitesimal rigidity properties. This projective transformation also shows that forced-symmetric rigidity is preserved under the transformation. The reader may like to compare and contrast this with combinatorial results for symmetric frameworks given in Chapter 25. In [4] more general pairings of symmetry groups with equivalent rigidity properties are explored.

Saliola and Whiteley [31] analysed arrangements of lines and circles in the plane, and more generally arrangements of hyperplanes and hyperspheres in any dimension $d$. This resulted in a direct, geometric correspondence to points and distances in $\mathbb{D}^{d+1}$, which preserves infinitesimal rigidity and flexibility. This transformation relies on stereographic projection and lifting between the Euclidean space and the sphere, which takes spheres and hyperplanes into spheres on $\mathbb{S}^{d+1}$ and preserves angles. This correspondence is interesting, as it embeds the entire infinitesimal theory of de Sitter space into a set of geometric constraints in a lower dimensional Euclidean space, as well as giving an example where we have a clear analysis for angles. See [31] for further results and comments, including the connections of Cauchy's Theorem on the infinitesimal rigidity for convex polyhedra in all of these spaces, along with Andreev's Theory [3] for angles in polyhedra in Hyperbolic space.

Since the basic infinitesimal rigidity is projectively invariant, this invites exploration of the impact of polarity. The situation is very different for Spherical, Hyperbolic, and de Sitter spaces, where points and distances polarize to hyperplanes and angles, and the setting of Euclidean space. In $\mathbb{R}^3$ polarity connects bar-joint frameworks to a form of infinitesimal rigidity for interesting structures call sheet works [41]. It is now natural to ask whether this same correspondence appears in $\mathbb{M}^3$? There is a related special correspondence of first-order rigidity in $\mathbb{R}^2$, with a lifting matroid of lines in $\mathbb{R}^2$ [44]. Again, we now propose extending this to $\mathbb{M}^2$ and liftings into $\mathbb{M}^3$.

Izmestiev [18] provides an alternative viewpoint, with new proofs, on the projective theory of rigidity and in [19] constructed examples of infinitesimally flexible Hyperbolic cone manifolds.

Much of the work on body-hinge frameworks is presented in projective form (e.g. [12]). In addition the work on the algebraic geometry of infinitesimal body-bar frameworks [38] is presented in a completely projective form, that will transfer across metrics. In addition, the inductive proof of rigidity for body-bar frameworks in $\mathbb{R}^d$ carries over directly to the Spherical metric, and these appear in a chapter on CAD frameworks in [13]. We note that for body-bar frameworks there is an efficient combinatorial algorithm for generic global rigidity in all dimensions [5], something that does not exist for bar and joint frameworks.

Further relevant work on cones and buildings was considered in [40], where coning is presented with projective algebra, and the cone-point at infinity. An extensive early presentation of both the statics and infinitesimal theory, with many worked examples, is available on the web at [39].

Turning the infinitesimal velocities 90 degrees in $\mathbb{R}^2$ has a second interpretation as the vectors of a parallel drawing in $\mathbb{R}^2$ [35]. This theory of parallel drawings has a number of analogous properties to first-order rigidity: (i) it generalizes to all dimensions with a rigidity-like matrix; (ii) the rank of the matrix is projectively invariant; and (iii) it transfers to all metrics which have the full space of translations: the Pseudo-Euclidean metrics, including Minkowski space. In addition, this theory has a dual theory of projections and liftings, and stronger combinatorial algorithms [42]. Since these theories are already fully linear, infinitesimal rigidity and global rigidity coincide for parallel drawing and scene analysis.

## References

[1] A. Y. Alfakih and Viet-Hang Nguyen. On affine motions and universal rigidity of tensegrity frameworks. *Linear Algebra Appl.*, 439(10):3134–3147, 2013.

[2] A. Y. Alfakih and Yinyu Ye. On affine motions and bar frameworks in general position. *Linear Algebra Appl.*, 438(1):31–36, 2013.

[3] E. M. Andreev. Convex polyhedra in Lobačevskiĭ spaces. *Mat. Sb. (N.S.)*, 81 (123):445–478, 1970.

[4] Katharine Clinch, Anthony Nixon, Bernd Schulze, and Walter Whiteley. Pairing symmetries for projective and spherical frameworks. in preparation, 2018.

[5] R. Connelly, T. Jordán, and W. Whiteley. Generic global rigidity of body-bar frameworks. *J. Combin. Theory Ser. B*, 103(6):689–705, 2013.

[6] R. Connelly and W. J. Whiteley. Global rigidity: the effect of coning. *Discrete Comput. Geom.*, 43(4):717–735, 2010.

[7] Robert Connelly. Generic global rigidity. *Discrete Comput. Geom.*, 33(4):549–563, 2005.

[8] Robert Connelly, Steven Gortler, and Louis Theran. Affine rigidity and conics at infinity. arXiv: 1605.07911, 2016.

[9] Robert Connelly and Steven J. Gortler. Iterative universal rigidity. *Discrete Comput. Geom.*, 53(4):847–877, 2015.

[10] Robert Connelly and Steven J. Gortler. Universal rigidity of complete bipartite graphs. *Discrete Comput. Geom.*, 57(2):281–304, 2017.

[11] Robert Connelly and Walter Whiteley. Second-order rigidity and pre-stress stability for tensegrity frameworks. *SIAM J. on Discrete Methods*, 9:453–492, 1986.

[12] Henry Crapo and Walter Whiteley. Statics of frameworks and motions of panel structures, a projective geometric introduction. *Structural Topology*, 6:43–82, 1982. With a French translation.

[13] Yaser Eftekhari. *Geometry of point-hyperplane and spherical frameworks.* PhD thesis, York University, 2017.

[14] Yaser Eftekhari, Bill Jackson, Anthony Nixon, Bernd Schulze, Shin-Ichi Tanigawa, and Walter Whiteley. Point-hyperplane frameworks, slider joints, and rigidity preserving transformations. arXiv: 1703.06844, 2017.

[15] Steven Gortler and Dylan Thurston. Generic global rigidity in complex and pseudo-euclidean spaces. In *Rigidity and Symmetry*. Fields Institute Communications, 2014.

[16] Steven J. Gortler, Alexander D. Healy, and Dylan P. Thurston. Characterizing generic global rigidity. *Amer. J. Math.*, 132(4):897–939, 2010.

[17] Steven J. Gortler and Dylan P. Thurston. Characterizing the universal rigidity of generic frameworks. *Discrete Comput. Geom.*, 51(4):1017–1036, 2014.

[18] Ivan Izmestiev. Projective background of the infinitesimal rigidity of frameworks. *Geom. Dedicata*, 140:183–203, 2009.

[19] Ivan Izmestiev. Examples of infinitesimally flexible 3-dimensional hyperbolic cone-manifolds. *J. Math. Soc. Japan*, 63(2):581–598, 2011.

[20] Bill Jackson and Tibor Jordán. Rigid two-dimensional frameworks with three collinear points. *Graphs Combin.*, 21(4):427–444, 2005.

[21] Bill Jackson and J. C. Owen. A characterisation of the generic rigidity of 2-dimensional point-line frameworks. *J. Combin. Theory Ser. B*, 119:96–121, 2016.

[22] Tibor Jordán and Viet-Hang Nguyen. On universally rigid frameworks on the line. *Contrib. Discrete Math.*, 10(2):10–21, 2015.

[23] Tibor Jordan and Walter Whiteley. Global rigidity. In Csaba D Toth, Joseph O'Rourke, and Jacob E Goodman, editors, *Handbook of discrete and computational geometry 3rd Edition.* Chapman and Hall/CRC, 2018.

[24] D. Kitson and S. C. Power. Infinitesimal rigidity for non-Euclidean bar-joint frameworks. *Bull. Lond. Math. Soc.*, 46(4):685–697, 2014.

[25] A. Nixon, J. C. Owen, and S. C. Power. Rigidity of frameworks supported on surfaces. *SIAM J. Discrete Math.*, 26(4):1733–1757, 2012.

[26] Anthony Nixon and Stephen Power. Double-distance frameworks and mixed sparsity graphs. arXiv: 1709.06349, 2017.

[27] A. V. Pogorelov. *Topics in the theory of surfaces in elliptic space.* Translated by Royer and Roger, Inc. Edited and with a preface by Richard Sacksteder. Russian Tracts on Advanced Mathematics and Physics, Vol. I. Gordon and Breach, New York, 1961.

[28] A. V. Pogorelov. *Extrinsic geometry of convex surfaces.* American Mathematical Society, Providence, R.I., 1973. Translated from the Russian by Israel Program for Scientific Translations, Translations of Mathematical Monographs, Vol. 35.

[29] W. J. M. Rankine. On the application of barycentric perspective to the transformation of structures. *Phil. Mag.*, 26:387–388, 1863.

[30] B. Roth and W. Whiteley. Tensegrity frameworks. *Trans. Amer. Math. Soc.*, 265(2):419–446, 1981.

[31] Franco Saliola and Walter Whiteley. Constraining plane configurations in CAD: circles, lines, and angles in the plane. *SIAM J. Discrete Math.*, 18(2):246–271, 2004.

[32] Franco Saliola and Walter Whiteley. Some notes on the equivalence of first-order rigidity in various geometries. arXiv: 0709.3354, 2007.

[33] Bernd Schulze and Shin-ichi Tanigawa. Infinitesimal rigidity of symmetric bar-joint frameworks. *SIAM J. Discrete Math.*, 29(3):1259–1286, 2015.

[34] Bernd Schulze and Walter Whiteley. Coning, symmetry and spherical frameworks. *Discrete Comput. Geom.*, 48(3):622–657, 2012.

[35] Bernd Schulze and Walter Whiteley. Rigidity and scene analysis. In Csaba D Toth, Joseph O'Rourke, and Jacob E Goodman, editors, *Handbook of discrete and computational geometry 3rd Edition*. Chapman & Hall/ CRC, 2018.

[36] Horst Struve and Rolf Struve. Non-Euclidean geometries: the Cayley-Klein approach. *J. Geom.*, 98(1-2):151–170, 2010.

[37] Neil White and Walter Whiteley. Algebraic geometry of stresses in frameworks. *SIAM J. Alg. Disc. Math.*, 4:53–70, 1983.

[38] Neil White and Walter Whiteley. The algebraic geometry of motions of bar and body frameworks. *SIAM J. Alg. Disc. Math.*, 8:1–32, 1987.

[39] Walter Whiteley. Introduction to structual topology. http://wiki.math.yorku.ca/index.php/Resources_in_Rigidity_Theory, 1978.

[40] Walter Whiteley. Cones, infinity and 1-story buildings. *Structural Topology*, 8:53–70, 1983. With a French translation.

[41] Walter Whiteley. Rigidity and polarity. I. Statics of sheet structures. *Geom. Dedicata*, 22(3):329–362, 1987.

[42] Walter Whiteley. A matroid on hypergraphs, with applications in scene analysis and geometry 75-95. *Disc. and Comp. Geometry*, 4:75–95, 1988.

[43] Walter Whiteley. The union of matroids and the rigidity of frameworks. *SIAM J. Discrete Math.*, 1(2):237–255, 1988.

[44] Walter Whiteley. Rigidity and polarity. II. Weaving lines and tensegrity frameworks. *Geom. Dedicata*, 30(3):255–279, 1989.

# Part IV

# Combinatorial Rigidity

# Chapter 18

# *Planar Rigidity*

**Brigitte Servatius**

*Mathematics Department, Worcester Polytechnic Institute, Worcester, MA*

**Herman Servatius**

*Mathematics Department, Worcester Polytechnic Institute, Worcester, MA*

## CONTENTS



## Introduction

The consideration of geometric constraint systems as a general subject is a relatively recent development, although many specific embodiments do have a long history. The oldest examples can be expressed in the form of a bar and joint framework, considered by Watt, Maxwell, and Cremona in the nineteenth century [10, 37]. Frameworks which have been most successfully studied are those which lie in the plane, such as the mechanisms studied by Kempe [19].

In fact, many of the ruler and compass constructions of classical geometry, which go back to antiquity, can be viewed in the context of bar and joint frameworks in the plane.

## 18.1 Rigidity of Bar and Joint Frameworks

We want to consider the rigidity of a framework $F = ((V,E),\mathbf{p})$ in $d$-space with nodes corresponding to the vertices $V$, length constraints corresponding to the edges $E$, and a placement of the vertices into Euclidean space by $\mathbf{p} : V \to \mathbb{R}^{|V|d}$, particularly the case $d = 2$, the plane. For this consideration, it is important to distinguish the different and related forms of rigidity. The most natural and straightforward notion is local rigidity, or simply, rigidity. A framework is *rigid* if every placement $\mathbf{q} : V \to \mathbb{R}^d$ sufficiently close to $\mathbf{p}$ and preserving the distances between the placements of adjacent vertices, is related to $\mathbf{p}$ by a congruence of $\mathbb{R}^d$.

Regarding $\mathbf{p}$ as a point in real $|V|d$-space, the *rigidity function* corresponding to the graph $G = (V,E)$ is the map $\rho : \mathbb{R}^{|V|d} \to \mathbb{R}^{|E|}$ where the coordinate of $\rho(\mathbf{p})$ corresponding to $e = (v,w) \in E$ is $\|\mathbf{p}(v) - \mathbf{p}(w)\|^2$. We can now describe the set of all frameworks on $G$ with edge lengths equivalent to $F$ as

$$\mathcal{C}(F) = \{(G,\mathbf{q}) \mid \rho(\mathbf{q}) = \rho(\mathbf{p})\} \tag{18.1}$$

Placements in $\mathcal{C}(F)$ correspond to the solutions of the constraint system consisting of $|E|$ quadratic equations

$$(\mathbf{p}(v) - \mathbf{p}(w))^2 = (\mathbf{q}(v) - \mathbf{q}(w))^2 \qquad (v,w) \in E, \tag{18.2}$$

so $\mathcal{C}(F)$ is an algebraic set and it is called the *configuration space of* $F$. Clearly $\mathbf{p} \in \mathcal{C}(F)$ and we may describe a physical movement of the framework in $\mathbb{R}^d$, that is, a movement of the vertices which preserves the lengths of the edges, by a continuous path in $\mathcal{C}$ starting at $\mathbf{p}$. Moreover, since $\mathcal{C}(F)$ is an algebraic set, it follows that if the framework $F$ is not rigid, then there is in fact a motion of $F$, see [38]. For example, an equilateral triangle framework is rigid in the plane, while a square framework is non-rigid, deformable through a family of rhombi.

In the 1970s, Bolker and Crapo [3, 4] developed a criterion for how to rigidify a grid of squares in the plane using diagonal braces along a selection of the square diagonals. Given a braced grid, they defined a bipartite graph whose vertex set is the set of rows and columns of the grid, and let row $i$ be adjacent to column $j$ if there is a brace in the square $(i, j)$. See Figure 18.1 for some examples of braced grids and their brace graphs.

**Theorem 18.1 (Bolker)** *A braced grid is rigid if and only if the bipartite graph of braces is connected.*

In [3] the circuits in the structure matroid of face diagonal braces for a grid of cubes in space are characterized. In [49] necessary conditions for the bases are given and it is pointed out that these conditions may be checked in polynomial time using Edmonds' matroid partition algorithm [12].

A strictly stronger notion than rigidity is *global rigidity*, which requires that all edge distance preserving placements of $(V,E)$ be related to $\mathbf{p}$ by a congruence of $\mathbb{R}^d$. For a simple example, a square framework with a diagonal brace is rigid, but not globally rigid, since it has an alternate placement folded into an isosceles right triangle with two vertices placed together. The square framework with two diagonal braces, whose underlying graph is the complete graph on 4 vertices, is globally rigid. Every placement of a complete graph is globally rigid, so every rigid framework can be made globally rigid by adding a sufficient number of braces.

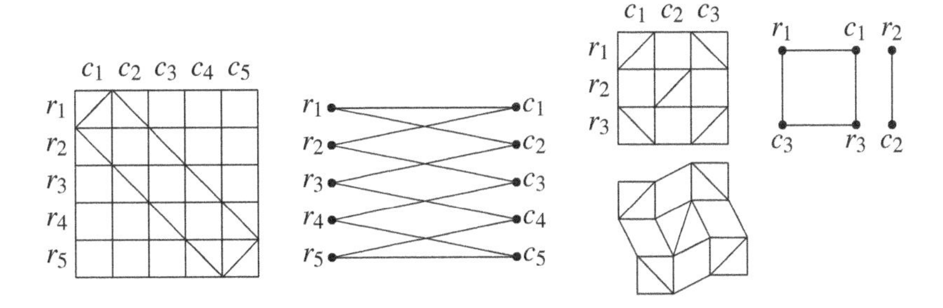


**Figure 18.1**
A rigid braced $4 \times 4$ grid with its 2-connected brace graph, so deleting any brace still leaves a rigid grid. A braced $3 \times 3$ grid with disconnected brace graph and a deformation derived from its components.

A more theoretically tractable notion is *infinitesimal rigidity*, which forbids not actual motions of the vertices, but infinitesimal ones. Given a framework $F$, hence a fixed placement $\mathbf{p}$, we have seen that the constraint system for rigidity and global rigidity is the system of $|E|$ quadratic equations (18.2) whose derivative, evaluated at $\mathbf{p}$, gives

$$(\mathbf{p}(v) - \mathbf{p}(w)) \cdot (\mathbf{p}'(v) - \mathbf{p}'(w)) = 0 \qquad (v,w) \in E, \tag{18.3}$$

a linear system of constraints on the initial velocities $\{\mathbf{p}'(v)\}$ of any motion of $F$, geometrically requiring that the velocities of an adjacent pair of vertices must have equal projections onto the edge between them, see Figure 18.2a.

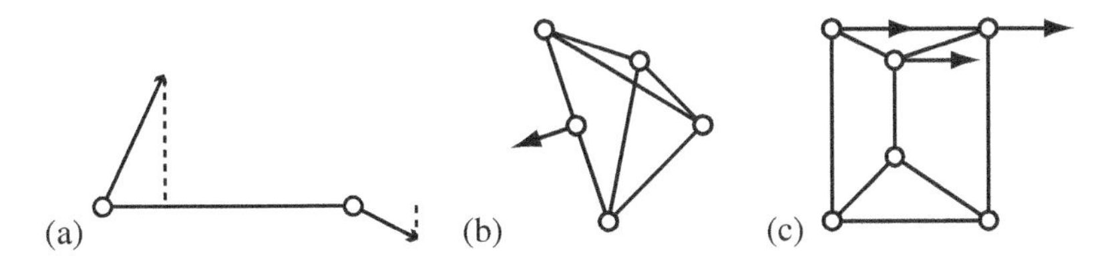


**Figure 18.2**
(a) The infinitesimal constraint on each edge. (b) and (c) Two globally rigid frameworks which are not infinitesimally rigid.

It is common practice when moving to infinitesimal considerations to assume that the placement is an embedding, a one-to-one map on the vertex set, so that none of the equations in (18.3) is trivial; and that the framework contains a set of at least $d$ vertices in general position, so that the system (18.3) always has a solution space of rank at least $\binom{d+1}{2}$, corresponding to the linear space $I_0$ of infinitesimal isometries of $d$-space, the *trivial* solutions. A convenient basis for $I_0$ consists of the $d$ vectors $(\mathbf{e}_i, \mathbf{e}_i, \ldots, \mathbf{e}_i)$, where $\mathbf{e}_i$ is the $i$'th standard basis element of $\mathbb{R}^d$ (written as a row vector), and the $\binom{d}{2}$ vectors $(M_{i,j}\mathbf{p}(v_1), M_{i,j}\mathbf{p}(v_2), \ldots, M_{i,j}\mathbf{p}(v_{|V|}))$, where $M_{i,j}$, $1 \le i < j \le d$, is a $d \times d$ skew symmetric matrix which has exactly two non-zero entries, a 1 in the $(i, j)$ position, and a $-1$ in position $(j,i)$, $i = 1, \ldots d$. If the vectors $\mathbf{e}_1 \ldots \mathbf{e}_d$ are the principle axes of the point set $\mathbf{p}(v_1) \ldots \mathbf{p}(v_{|V|})$ and their centroid is at the origin then these $\binom{d+1}{2}$ vectors form an orthogonal basis for $I_0$. A framework for which the elements of $I_0$ are the only solutions is said to be *infinitesimally rigid.*

It is not true that every framework which has a non-trivial motion, that is, a motion changing the distance between at least one pair of non-adjacent vertices, has a motion whose initial velocities are also non-trivial, see [7]. Nevertheless, since the solutions to (18.2) form an algebraic set, any motion may be assumed to be analytic, and by considering higher derivatives the following key result follows, see [17].

**Theorem 18.2** *If a framework in d-space is infinitesimally rigid, then it is rigid.*

The converse is not true, see Figures 18.2b and c in which the existence of the non-trivial infinitesimal motion of a rigid framework correlates to the placement being in a special position. For b, the placement is not in general position, and for c there is a separating set of three collinear edges.

In [2] Bolker and Roth prove (among many other things) that a placement of $K_{3,3}$ in $\mathbb{R}^2$ is infinitesimally rigid unless its six vertices lie on a conic. Actual motions of planar bipartite frameworks were studied in [36, 45].

In [15] it is shown that there is an open dense set of placements, the *generic embeddings*, for which neither rigidity nor infinitesimal rigidity depend on the placement, and, in addition, the converse of Theorem 18.2 is true.

**Theorem 18.3** *If $F = (G, \mathbf{p})$ is a rigid framework in d-space with a generic placement $\mathbf{p}$, then $F$ is infinitesimally rigid.*

For simplicity, one can take the generic embeddings to be those for which the $|V|d$ coordinates are algebraically independent from one another, or, practically, a "random" embedding. If a framework has a rigid generic embedding, we say its graph is generically rigid. Table 18.1 illustrates the connections between these concepts. It is often stated that the framework $F = ((V,E),\mathbf{p})$ is generically rigid if for any generic embedding $\mathbf{q}$ of $V$, the framework $((V,E),\mathbf{q})$ is rigid, however much confusion can be avoided if one reserves the concept of generic rigidity to graphs, so that the first three columns in Table 18.1 are referring to the frameworks $F = ((V,E),\mathbf{p})$ as drawn, while for the fourth column, only the graph $(V,E)$ is relevant.

## 18.1.1 Rigidity Matrix and Augmented Rigidity Matrices

Given an ordering on the vertices and the edges, the linear constraint system of the equations in (18.3) for infinitesimal rigidity is encoded in a $|V|d \times |E|$ matrix, $R(F)$, called the *rigidity matrix.* The infinitesimal motions of the framework, also called *flexes*, are the elements of the kernel of this matrix.

The placement $\mathbf{p}(u) = (-2,0)$, $\mathbf{p}(v) = (1,3)$, $\mathbf{p}(w) = (1,-3)$ corresponds to the rigidity matrix R(F),

$$R((\{u,v,w\},\{(u,v),(u,w)\}),\mathbf{p}) = \begin{matrix} \\ (u,v) \\ (u,w) \end{matrix} \begin{matrix} \quad u \quad & \quad v \quad & \quad w \quad \\ \left[\begin{matrix} 3 & 3 \\ 3 & -3 \end{matrix}\right. & \begin{matrix} -3 & -3 \\ 0 & 0 \end{matrix} & \left.\begin{matrix} 0 & 0 \\ -3 & 3 \end{matrix}\right] \end{matrix},$$

a basis for whose flexes is illustrated in Figure 18.4, where the first three flexes are a basis for $I_0$, and the fourth flex spans the orthogonal complement of $I_0$. So the fourth flex is non-trivial and corresponds to a displacement which does not change the centroid of $\mathbf{p}(V)$ and has zero moment. If the centroid of $\mathbf{p}(V)$ is at the origin, the space of such representatives for the non-trivial flexes can be found by computing the kernel of $R(F)$ augmented by rows corresponding to the $\binom{d+1}{2}$ normalization equations

$$0 = \sum_{k=1}^{|V|} \mathbf{e}_i \cdot \mathbf{p}'(v_k),\ 1 \le i \le d, \quad 0 = \sum_{k=1}^{|V|} M_{ij}\mathbf{p}(v_k) \cdot \mathbf{p}'(v_k),\ 1 \le i < j \le d. \tag{18.4}$$

**Table 18.1**
A comparison of types of rigidity. All frameworks are in the plane.

| | rigid | globally rigid | infinitesimally rigid | generically rigid |
|---|---|---|---|---|
| | No | No | No | No |
| | No | No | No | Yes |
| | Yes | No | No | No |
| | Yes | No | No | Yes |
| | Yes | No | Yes | Yes |
| | Yes | Yes | No | No |
| | Yes | Yes | No | Yes |
| | Yes | Yes | Yes | Yes |

For the example above, the augmented rigidity matrix is

$$\begin{array}{r} \\ (u,v) \\ (u,w) \\ (e_1,e_1,e_1) \\ (e_2,e_2,e_2) \\ M_{1,2}(u,v,w) \end{array} \begin{array}{c} \begin{array}{cccccc} \quad u & & \quad v & & \quad w & \end{array} \\ \left[\begin{array}{cccccc} 3 & 3 & -3 & -3 & 0 & 0 \\ 3 & -3 & 0 & 0 & -3 & 3 \\ 1 & 0 & 1 & 0 & 1 & 0 \\ 0 & 1 & 0 & 1 & 0 & 1 \\ 0 & 2 & 3 & -1 & -3 & -1 \end{array}\right] \end{array}.$$

Its kernel has dimension 1 containing the vector $(-2,0,1,-3,1,3)$ corresponding to a non-trivial infinitesimal motion leaving the centroid at the origin.

An augmentation of the constraint system eliminating the trivial motions is often called *pinning* the framework.

To apply the above method of pinning to a given framework, first compute the centroid and translate the framework so that the centroid is at the origin.

A more direct method of pinning is to choose a set $\{v_1,\ldots,v_d\}$ of vertices placed in general position and to constrain any flex to require that $\mathbf{p}'(v_1) = \mathbf{0}$ and also for $2 \le i \le d$ that $\mathbf{p}'(v_i)$ be

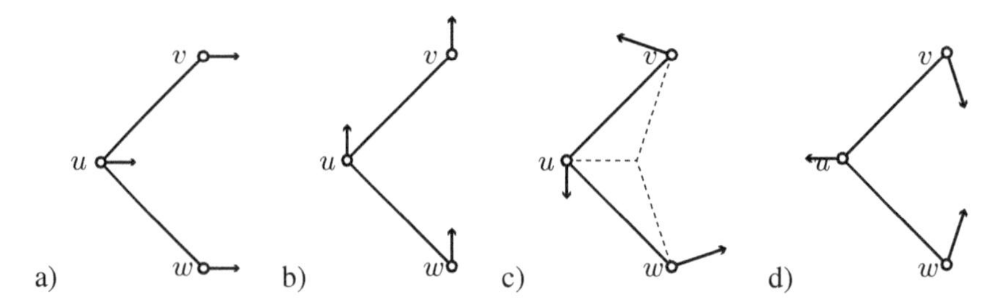


**Figure 18.3**
A basis for the flexes of a two-bar framework.

contained in the affine span of $\{\mathbf{p}(v_1), \ldots \mathbf{p}(v_{i-1})\}$. If we assume, without loss of generality, that $\mathbf{p}(v_i) \in \langle \mathbf{e}_1, \ldots, \mathbf{e}_{i-1} \rangle$, for $i = 1, \ldots, d$, these restricted flexes can be found by augmenting $R(F)$ by rows corresponding to the system

$$\mathbf{e}_j \cdot \mathbf{p}'(v_i) = 0, \ 1 \leq i \leq j \leq d, \tag{18.5}$$

This constrains any flex to satisfy $\mathbf{p}'(v_1) = \mathbf{0}$ and for $2 \leq i \leq d$ to require $\mathbf{p}'(v_i)$ to be contained in the affine span of $\{\mathbf{p}(v_1), \ldots \mathbf{p}(v_{i-1})\}$. See Figure 18.4 in which the second example illustrates pinning

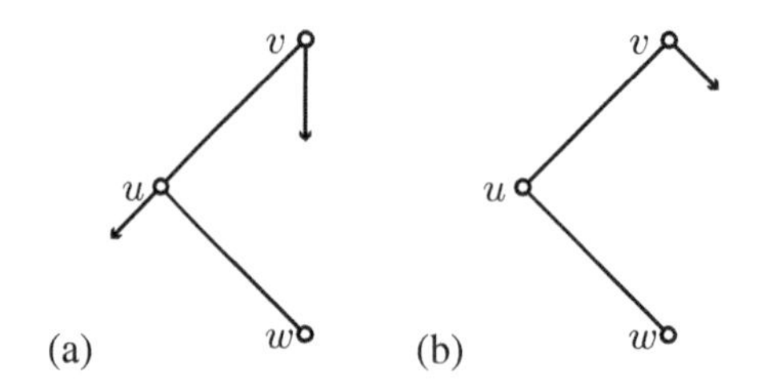


**Figure 18.4**
Normalized flexes for frameworks pinned with the equations in (18.5) using (a) $v_1 = w$ and $v_2 = v$ and (b) $v_1 = w$ and $v_2 = u$.

the vertices of a complete subgraph $K_d$ of $G$, in the plane just an edge. This forces $\mathbf{p}'(v_i) = \mathbf{0}$ for $1 \leq i \leq d$, and in this case values of the normalized flexes on the remaining vertices can be found using the *reduced rigidity matrix*, obtained from $R(F)$ by deleting the $d^2$ columns corresponding to $\{v_1, \ldots, v_d\}$, and the $\binom{d}{2}$ rows corresponding to the edges between them. The reduced rigidity matrix in the above example for pinning $u$ and $w$ is just $[3,3]$.

### 18.1.2 Rigidity Matrix as a Transformation

It is very useful when considering the abstract structures which follow to regard the rigidity matrix of a framework, $R(F)$, as a linear transformation from $\mathbb{R}^{d|V|}$ to $\mathbb{R}^{|E|}$. The domain is the set $\mathbb{R}^{d|V|}$ of *vertex displacements*, $\Delta\mathbf{p} \in \mathbb{R}^{d|V|}$, written as $1 \times |V|d$ column vector. The range is the space of *edge elongations*, $\Delta\mathbf{e} \in \mathbb{R}^{|E|}$. A $1 \times |E|$ column vector $\Delta\mathbf{e}$ of edge elongations is in the image of $R(F)$ if it is the linear approximation to the result of a vertex displacement, $\Delta\mathbf{e} = R(F)\mathbf{p}$, in which case $\Delta\mathbf{e}$ is called a *strain*. Note that here strains have units of square length, which differs from common engineering practice. Of course, a displacement $\Delta\mathbf{p}$ corresponds to an infinitesimal motion if its strain is the zero vector, $R(F)\Delta\mathbf{p} = \mathbf{0}$.

The dual transformation carries just as much information about the framework, although the physical interpretation is different. Since each edge constrains the motion of its incident vertices, it is natural to regard the role of that edge as to exert force on its endpoints, equal in magnitude but opposite in direction, depending on whether the edge is under tension or compression. The state of these edge forces in the framework is given by a row vector $\omega \in \mathbb{R}^{|E|}$, whose coordinates have units of force per unit length, called a *stress* vector. One easily computes that the $d$ coordinates of $\mathbf{f} = \omega R(F)$ corresponding to vertex $v$ form the *resultant force* at vertex $v$ from the forces at its incident edges. We think of a resultant $\mathbf{f}$ as being available to balance, or resist, an external force $-\mathbf{f}$ on the vertices of the framework. In short, the domain of the dual transformation is the space of edge stresses, the codomain is the space of vertex forces, and the image is the space of *resolvable forces*.

Of course the framework will be unable to resist forces which tend to translate or rotate the framework as a whole, so we are only concerned with vertex forces which are *equilibriated*, that is, orthogonal to $I_0$. The framework $F$ is rigid in this context, we say *statically rigid*, if any applied equilibriated force can be balanced by vertex forces arising from a stress assignment to the edges. In other words, the rank of $R(F)$ must be $|V|d - \binom{d+1}{2}$, which is the same rank that $R(F)$ must have in order for every flex to be trivial, so these dual notions of rigidity, static and infinitesimal rigidity, are the same. The kernel of the dual transformation consists of those stresses which resolve to the zero vector on all vertices. These stresses are called variously, *equilibrium stresses*, *prestresses*, *resolvable stresses*, and, unfortunately, *stresses*. (This ambiguity is usually resolved by the context –if it is stated that a framework has a stress, or the rank of the space of stresses is computed, equilibrium stresses are meant). Table 18.2 summarizes the dual terminology, kinematic and static, surrounding the rigidity matrix.

**Table 18.2**
The rigidity nomenclature for the linear algebra of the rigidity matrix.

| linear algebra | framework |
|---|---|
| domain $\mathbb{R}^{|V|d}$ | placements |
| codomain $\mathbb{R}^{|E|}$ | stresses |
| range | edge elongations |
| corange | vertex forces |
| row | infinitesimal constraint<br>resultant force from a single stressed edge |
| column | edge strain |
| columns for $v \in V$ | vector star at $\mathbf{p}(v)$ |
| column space | strains |
| row space | resolvable forces |
| kernel | flexes |
| kernel dimension | $\binom{d+1}{2}+$ degree of freedom |
| cokernel | resolvable stresses<br>constraint dependence |
| rank | strain dimension<br>dimension of resolvable forces |

The connection between these dual points of view is in the consideration of the work done by a displacement of the vertices under the vertex forces,

$$W = \mathbf{f} \cdot \Delta\mathbf{p} = (\omega R(F))\,\Delta\mathbf{p} = \omega\,(R(F)\Delta\mathbf{p}) = \omega \cdot \Delta\mathbf{e} \tag{18.6}$$

which will be examined more thoroughly in a later Chapter 14. Here we only want to note the roles played by the various components of the linear transformation $R(F)$ in terms of infinitesimal rigidity.

**Theorem 18.4** *A framework $F$ is infinitesimally rigid if and only if any one of the following conditions is satisfied:*

*(a) $F$ has no non-trivial flexes.*

*(b) The rank of $R(F)$ is $d|V| - \binom{d+1}{2}$.*

*(c) Every set of equilibriated forces on the vertices is equal to the resultant forces from some stress on the edges.*

*(d) The number of independent resolvable stresses is $|E| + \binom{d+1}{2} - d|V|$.*

*(e) For each pair of vertices $v$ and $w$, there is a stress $\omega$ whose resultant force $\mathbf{f} = \omega R(F)$ is zero except at vertices $v$ and $w$, where the forces are equal and opposite, directed along the line between $\mathbf{p}(v)$ and $\mathbf{p}(w)$.*

### 18.1.3 The Infinitesimal Rigidity Matroid of a Framework

Given a framework $F = ((V,E),\mathbf{p})$, the rows of $R(F)$ correspond to the infinitesimal constraints on the vertices. Independence of these constraints is a key feature of the system. If one row is linearly dependent on the remaining rows, that row may be deleted without changing the flexes of the framework, and the corresponding bar is *implied* by the other bars, or *redundant*, and the framework *overbraced.* The proper combinatorial structure for studying independence is the matriod. The rows of $R(F)$ comprise a matroid, the *infinitesimal rigidity matroid of $F$*, $\mathcal{F}(F)$, whose independent sets are those subsets of rows which are linearly independent. The following lists the important structures in this matroid, contrasted with language of linear algebra, and that of frameworks.

*Independence:* A set $E'$ is independent in $\mathcal{F}(F)$. — The rows in $E'$ are linearly independent. — The subframework $((V(E'),E'),\mathbf{p})$ is nowhere overbraced.

*Dependence:* A set $E'$ is dependent in $\mathcal{F}(F)$. — The rows in $E'$ are linearly dependent. — The subframework $((V(E'),E'),\mathbf{p})$ is somewhere overbraced.

*Cycle:* A set $E'$ is a cycle in $\mathcal{F}(F)$, a minimally dependent set. — The rows in $E'$ are minimally linearly dependent. — The subframework $((V(E'),E'),\mathbf{p})$ is overbraced, but all of its proper subframeworks are not overbraced.

*Basis:* A set $E'$ is a basis in $\mathcal{F}(F)$, a maximal independent set. — The rows in $E'$ are linearly independent and span the rowspace. — The subframework $((V,E'),\mathbf{p})$ is a minimal subframework with the same flexes as $F$.

*Rank:* The rank of $E'$, $\text{rank}(E')$ in $\mathcal{F}(F)$, the cardinality of the largest independent subset of $E'$. — The rank of the row vectors in $E'$. — The dimension of the space of resultant forces from stresses on $E'$.

*Closure:* The closure of $E'$ in $\mathcal{F}(F)$ is the set of all $e \in E$ with $\text{rank}(E'+e) = \text{rank}(E')$. — The set of all rows in $E$ which lie in the span of $E'$. — The set of bars in $E$ which are implied by the bars of $E'$.

## Glossary

**Framework, linkage:** A pair $F = (G, \mathbf{p})$ where $G = (V, E)$ is a graph and $\mathbf{p} : V \to \mathbb{R}^d$ is a placement of the vertices into $d$-dimensional Euclidean space. It is also common to notate the framework $((V, E), \mathbf{p})$ as $(V, E, \mathbf{p})$.

**Joint, node:** A point of the framework, represented by a vertex of the graph, where bars may meet, and whose motion is constrained by the bars.

**Bar, link, brace:** A connection between two joints, represented by an edge of the graph, whose role is to fix the distance between that pair of nodes.

**Configuration Space:** The configuration space of the framework $F$ is the space $\mathcal{C}(F)$ of all placements $\mathbf{q} : V \to \mathbb{R}^d$ such that $\|\mathbf{p}(v) - \mathbf{p}(w)\| = \|\mathbf{q}(v) - \mathbf{q}(w)\|$ for every edge $(v, w) \in E$. The configuration space $\mathcal{C}(F)$ inherits a natural metric and differentiable structure from $\mathbb{R}^{|V|d}$.

**Pinned framework:** A framework normalized to exclude trivial motions and flexes.

**Rigid framework, locally rigid framework:** A framework $(G, \mathbf{p})$ such that $\mathbf{p}$ has a neighborhood in $\mathcal{C}(\mathbf{p})$ containing only embeddings $\mathbf{q}$ such that $\mathbf{q} = \mathbf{p}I$, for some isometry $I$ of $d$-space.

**Rigidity function, length function:** The rigidity function of the graph $G$ is $\rho : \mathbb{R}^{|V|d} \to \mathbb{R}^{|E|}$ which takes each placement $\mathbf{p}$ to the vector $\rho(\mathbf{p})$ of square bar lengths; the coordinate of $\mathbb{R}^{|E|}$ corresponding to the edge $(i, j) \in E$ is $(\mathbf{p}(i) - \mathbf{p}(j))^2$.

**Rigidity matrix:** The rigidity matrix $R(F)$ of a framework is an $|E| \times d|V|$ matrix, the Jacobian matrix of the rigidity function of the graph, evaluated at the embedding $\mathbf{p}$, divided by 2.

**Motion of the framework $F$, deformation:** A continuous one parameter family $F_t$ of frameworks with $F_0 = F$, not all of which are congruent to $F$.

**Infinitesimal motion, first order motion, flex:** Solution $\mathbf{p}'(v)$ to the linear system

$$(\mathbf{p}(v) - \mathbf{p}(w)) \cdot (\mathbf{p}'(v) - \mathbf{p}'(w)) = 0 \qquad (v, w) \in E, \tag{18.3}$$

encoded by the rigidity matrix.

**Infinitesimally rigid framework, first order rigid framework:** A framework $F = ((V, E), \mathbf{p})$ whose rigidity matrix $R(F)$ has rank $d|V| - \binom{d+1}{2}$.

**General position embedding:** Embedding $\mathbf{p}$ such that no $n$ vertices, $n \geq 2$, are placed on any $n-2$ dimensional affine subspace.

**Generic embedding:** Embedding $\mathbf{p} : V \to \mathbb{R}^d$ is generic if no sub-determinant of the rigidity matrix is zero unless it is identically equal to zero.

**Generically rigid graph:** Graph with an (infinitesimally) rigid generic embedding.

**(Internal) degree of freedom:** Number of independent flexes of the framework. Rank of the kernel of the rigidity matrix, $\mathrm{df}(F)$,

$$\mathrm{df}(E) = r(E) - m|V(E)| + \binom{d+1}{2}.$$

## 18.2 Abstract Rigidity Matroids

Let $K = K(V)$ denote the edge set of the complete graph on the set $V$, $|V| = n$. A matroid $\mathcal{A}_d$ on $K$ with closure operator $\langle \cdot \rangle$ is called a *d-dimensional abstract rigidity matroid for V* if, besides the usual closure axioms for a matroid:

**C1** $T \subseteq \langle T \rangle$;

**C2** If $R \subseteq T$, then $\langle R \rangle \subseteq \langle T \rangle$;

**C3** $\langle \langle T \rangle \rangle = \langle T \rangle$;

**C4** If $s, t \in (E - \langle T \rangle)$, then $s \in \langle T \cup \{t\} \rangle$ if and only if $t \in \langle T \cup \{s\} \rangle$;

it satisfies the additional conditions C5 and C6 below:

**C5** If $E, F \subseteq K$ and $|V(E) \cap V(F)| < d$, then $\langle E \cup F \rangle \subseteq (K(V(E)) \cup K(V(F)))$.

**C6** If $\langle E \rangle = K(V(E))$, $\langle F \rangle = K(V(F))$ and $|V(E) \cap V(F)| \geq d$, then $\langle E \cup F \rangle = K(V(E \cup F))$.

Note that Axiom C5 implies $\langle E \rangle \subseteq K(V(E))$ so, motivated by the infinitesimal case, we say $E \subseteq K$ is *rigid* if $\langle E \rangle = K(V(E))$. With this definition C5 expresses the idea that any separating set of codimension less than $d$ allows a hinge motion, and C6 reads: if two rigid sets have $d$ or more vertices in common then their union is rigid.

All infinitesimal rigidity matroids satisfy C6. If $\mathbf{p}$ embeds the vertices of a framework in general position, i.e., no $m$ vertices, $m \leq d+1$, lie on any $m-2$ dimensional affine subspace, then $\mathcal{F}(p)$ satisfies C6 also and is an abstract rigidity matroid. There exist, however, abstract rigidity matroids which are not infinitesimal, [17], and abstract rigidity may also arise in other contexts, see [26] or [64].

Let $\mathcal{A}_d$ be a $d$-dimensional abstract rigidity matroid. If $E$ is any edge set such that $|V(E)| \leq d+1$ then, by C5, $E$ is independent. In particular, $K_{d+1}$ is independent. If $v \notin V(E)$ and $F$ is a set of $d$ edges joining $v$ to vertices in $V(E)$, then $E \cup F$ is called a *0-extension* of $E$. Again by C5, 0-extensions of independent sets are independent. In particular, $K_{d+2} - e$ is a 0-extension of $K_{d+1}$, and hence independent. Moreover, $K_{d+2} - e$ is the union of two $K_{d+1}$'s along a $K_d$, thus $\langle K_{d+2} - e \rangle = K_{d+2}$ by C6, and we have that $K_{d+2}$ is a cycle in $\mathcal{A}_d$. The same kind of argument shows that, in fact, if $|V(E)| \geq d+1$, then

$$r(E) \leq m|V(E)| - \binom{d+1}{2}$$

with equality when $\langle E \rangle$ is complete. This inequality is fundamental for inductive constructions of independent sets, which are treated in Chapter 19. If $\langle E \rangle$ is not complete, then the discrepancy measures the *(internal) degree of freedom*, df$(E)$, of the edge set $E$,

$$\mathrm{df}(E) = r(E) - m|V(E)| + \binom{d+1}{2}$$

.

Abstract rigidity matroids were introduced by Graver in [16], and further investigated in [17, 44, 41].

In dimension 1 the generic rigidity matroids are the only examples of abstract rigidity matroids and they are known as connectivity matroids, or graphic matroids.

There are 2-dimensional infinitesimal rigidity matroids which are not generic, for example see Bolker and Roth [2], where it is shown that if $V = \{1, \ldots, 6\}$ is embedded on a conic in $\mathbb{R}^2$, then each copy of $K_{3,3}$ in this 2-dimensional infinitesimal rigidity matroid is dependent.

There also exist 2-dimensional abstract rigidity matroids which are not infinitesimal rigidity matroids. Take 6 generically embedded vertices, but remove one particular $K_{3,3}$ subgraph from the set of bases. It is easy to verify, that this yields another abstract rigidity matroid, which, according to the Bolker-Roth result is not infinitesimal. A second example is based on the observation that, in the 2-dimensional infinitesimal rigidity matroid for $V = \{1,\ldots,6\}$, not all prisms can be dependent. For example in Figure 18.2 there are two prisms on the same symmetrically embedded vertex set. Figure 18.2a has an infinitesimal motion, hence its edge set is dependent in the corresponding infinitesimal rigidity matroid, while the prism in Figure 18.2b is isostatic. One can show that starting with the bases for the 2-dimensional generic rigidity matroid of $V$ and deleting all prisms results in the bases for a 2-dimensional abstract rigidity matroid for $V$ which is clearly not infinitesimal.

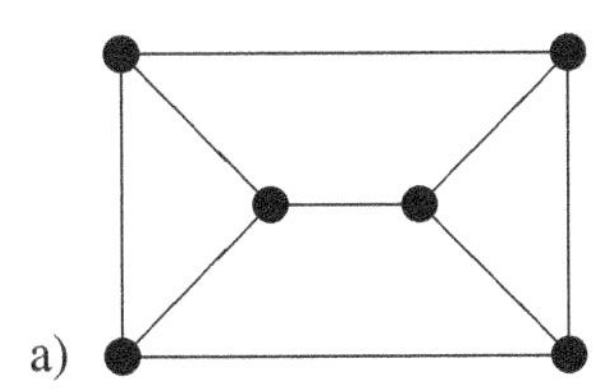


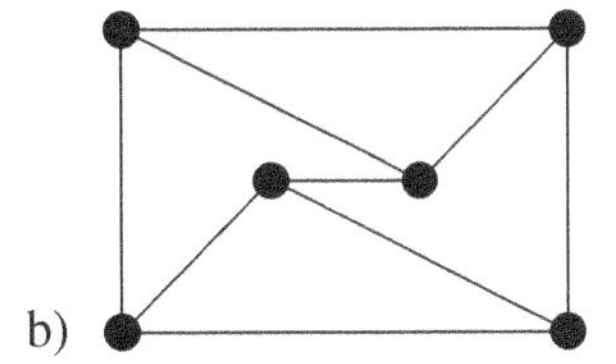


**Figure 18.5**
Two prisms on the same vertex set.

We know that all 2-dimensional abstract rigidity matroids on $n$ vertices have rank $2n-3$ and that those edge sets obtained from a single edge by a sequence of 0-extensions must therefore be bases. Any 2-dimensional abstract rigidity matroid on $n$ vertices, and hence any minimal 2-dimensional abstract rigidity matroid on $n$ vertices, must include these edge sets as bases. In general, however, additional bases are necessary, as illustrated in Figure 18.2. The sets $B_1$ and $B_2$ are bases in any

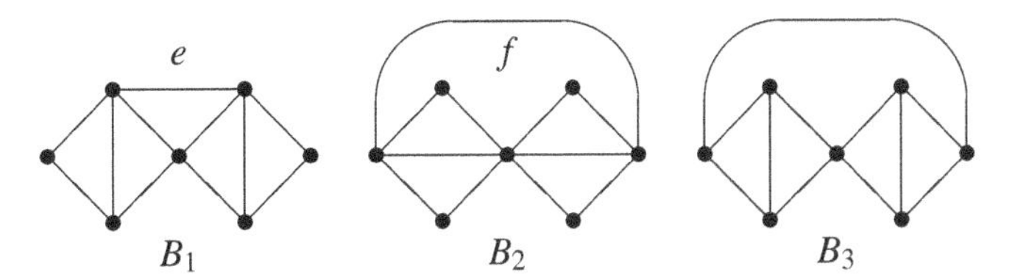


**Figure 18.6**
$B_3$ is obtained by basis exchange.

abstract rigidity matroid on seven vertices since they may be obtained by five 0-extensions on a single edge. The basis axioms assert that $B_1 - e$ can be augmented to a basis with an edge from $B_2$, which, since tetrahedra are dependent, must be $B_3 = B_1 - e + f$. Thus $B_3$ is also a basis in any 2-dimensional abstract rigidity matroid, but $B_3$ is not the result of a sequence of 0-extensions of an edge since it has no vertex of valence 2.

### 18.2.1 Characterizations of $\mathcal{A}_2$ and $(\mathcal{A}_2)^\perp$

The dual matroid, although it contains the same mathematical information, has structures with interesting and important interpretations. The abstract rigidity matroid generalizes the infinitesimal rigidity matroid, which encodes the independence structure of the rigidity transformation. The dual matroid does not directly generalize the dual transformation, but the duality in the sense of homology versus cohomology.

For every matroid on groundset $E$, complements of bases are the bases of the dual matroid. Matroid cocycles, minimally dependent sets in the dual matroid, are the minimal sets which intersect every basis of the matroid. Using the rigidity matrix it is easy to verify that in the 2-dimensional infinitesimal rigidity matroid on the edge set of a complete graph on at least 4 vertices, the set of edges, $S(v)$, incident to a vertex $v \in V$ is dependent in the dual matroid and deleting any element $e \in S(v)$ yields a cocycle. We call these cocycles the *vertex cocycles* of $v$; and, for each edge $e \in S(v)$, we denote the vertex cocycle $S(v) - e$ by $S_e(v)$

$S_e(v)$ is actually a cocycle in any 2-dimensional abstract rigidity matroid and, in fact, these cocycles are of use in characterizing 2-dimensional abstract rigidity matroids.

**Theorem 18.5** *Let $(V,K)$ be the complete graph on $n$ vertices. If $\mathcal{M}$ is a 2-dimensional abstract rigidity matroid on $K$, then each of the following three conditions hold:*

a. *For each $v \in V$ and each $e \in S(v)$, $S_e(v)$ is a cocycle of $\mathcal{M}$.*

b. *No cycle of $\mathcal{M}$ contains a vertex of valence less than three.*

c. *Each* 2*-valent* 0*-extension of an independent set of $\mathcal{M}$ is also an independent set of $\mathcal{M}$.*

*Conversely, if any one of the conditions a, b or c hold then $\mathcal{M}$ is a 2-dimensional abstract rigidity matroid on $K$.*

Theorem 18.5, proved in [15], can be used to construct abstract rigidity matroids that are not infinitesimal. It was extended to higher dimensions by Nguyen in [41].

Observe that the existence of the vertex cocycles in Theorem 18.5 forces the rank of the matroid to be at least $2n-3$. Also all $K_4$'s must be cycles, bounding the rank from above.

Both of these conditions are necessary. The uniform matroid on $K$ of rank 5 has each $K_4$ as a cycle but, when $|V| > 4$, it is not a 2-dimensional abstract rigidity matroid; and the uniform matroid on $K$ of rank $\binom{|V|}{2} - (|V|-3)$ has each vertex star minus an edge as a cocycle but, when $|V| > 3$, it is not a 2-dimensional abstract rigidity matroid.

We now turn our attention to an arbitrary rigidity matroid $\mathcal{A}_2$ for $V = \{1,\ldots,n\}$. The complete set of cocycles can be described. Recall that we say that a spanning edge set $E \subseteq K(V)$ has degree of freedom 1 when $r(E) = 2n-4 = r(K(V)) - 1$.

**Theorem 18.6** *Let $V = \{1,\ldots,n\}$ be given. The cocycles of $\mathcal{A}_2$ are the complements of closed spanning subsets with degree of freedom 1 in $\mathcal{A}_2$.*

It is easily verified that all cocycles of $K_5$ are isomorphic to one of the following four graphs. The first graph is a vertex cocycle.

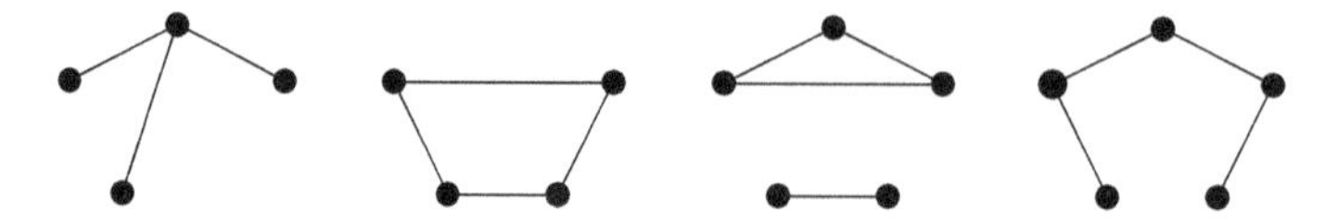

**Figure 18.7**
Cocycles of $\mathcal{G}_2(K_5)$.

The reason for studying cocycles, and in particular vertex cocycles, is to get information about the graph from the matroid defined on the edge set. In dimension 1 we have the following nice result by Whitney [66] that the cycle matroid of a 3-connected graph uniquely determines the graph. In this case the vertex cocycles span the cocycle space, which allows us to draw a graph from a set of cocycles with the property that every edge is contained in exactly two of them. Jordán and Kaszanitzky extended this result to rigidity matroids in [25]. They show that if G is 7-vertex-connected then it is uniquely determined by its two-dimensional rigidity matroid, and if a two-dimensional rigidity matroid is $(2k-3)$-connected then its underlying graph is k-vertex-connected.

### 18.2.2 The 2-Dimensional Generic Rigidity Matroid

If $v \notin V(E)$, $e \in E$ and $F$ is a set of $d+1$ edges joining $v$ to vertices in $V(E)$, including both endpoints of $e$, then $E \cup F - e$ is called a 1*–extension* of $E$.

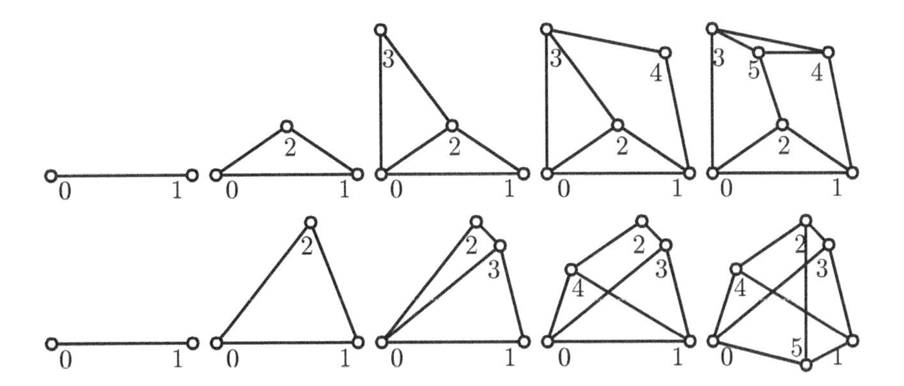


**Figure 18.8**
Constructing frameworks in the plane on the triangular prism and $K_{3,3}$ from a single edge via a sequence of 0-extensions and 1-extensions.

In dimension two, we can re-state the rank bound (18.2) using the 1-extension. If $E \subseteq K$ is independent, then $|F| \leq 2|V(F)|-3$, for each nonempty subset $F$ of $E$. We say that independent sets satisfy *Laman's condition.* An abstract rigidity matroid is said to have the 1*-extendability property* if, given an independent set $E$ on vertex set $V$, every 1–extension of $E$ yields an independent edge set on $V \bigcup v$.

**Theorem 18.7 (Laman's Theorem)** *Let $\mathcal{A}_2$ be an abstract rigidity matroid on the complete graph on n vertices. The following are equivalent:*

*(a) $\mathcal{A}_2$ satisfies Laman's condition,*

*(b) $\mathcal{A}_2$ has the* 1*-extendability property,*

*(c) $\mathcal{A}_2 = \mathcal{G}_2(n)$.*

This characterization of $\mathcal{G}_2(n)$ yields the following.

**Theorem 18.8** *$\mathcal{G}_2(n)$ is the unique maximal* 2*-dimensional abstract rigidity matroid on n vertices.*

The Laman characterization of the isostatic sets in $\mathcal{G}_2(n)$ is a powerful tool in the study of isostatic sets and, in view of the maximality of $\mathcal{G}_2(n)$, tells us much about the structure of isostatic sets in an arbitrary 2-dimensional abstract rigidity matroid.

Given a graph $G$, with edge set $E$, one can show directly that the collection of subsets of $E$ which satisfy Laman's condition in dimension 2 are the independent sets of a matroid on $E$. Laman's famous 1970 paper [33] used linear algebra techniques to prove Theorem 18.7. A paper by Pollaczek-Geiringer [46] in 1927 gave an equivalent theorem to Laman's theorem, where an induction proof using 0–extensions and 1–extensions is provided. Moreover, Pollaczek-Geiringer extends Theorem 18.7 to mechanisms: A framework on $n$ vertices and $e$ edges has $2n-3-e$ generic degrees of freedom if no subframework is overbraced. She also formulates an algebraic interpretation of Theorem 18.7 in the spirit of Frobenius [14] and points out interesting references to special position frameworks, see [31].

It is natural to ask if the edge sets that satisfy Laman's inequalities for dimension $m>2$ also form the collection of independent sets of some matroid, since such a matroid would be a logical candidate for the generic rigidity matroid in dimension $m$. Unfortunately we have the following negative result.

**Theorem 18.9** *The collection of edge sets that satisfy Laman's condition for dimension m cannot be the collection of independent sets of any matroid on K if $m \geq 3$.*

In the case $d=3$, one may observe that the proposed rank bound, $r(E) \leq 3|V(E)|-6$ fails for single edges. On the other hand, the bound $r(E) \leq 3|V(E)|-5$ works for single edges.

Whiteley, in [65], presents an important survey on an array of matroids drawn from three sources in discrete applied geometry: (i) static (or first-order) rigidity of frameworks and higher skeletal rigidity; (ii) parallel drawings (or, equivalently, polyhedral pictures); and (iii) $C_r^{r-1}$-cofactors abstracted from multivariate splines in all dimensions. In dimension 2 all of these three matroids are the same, in fact all equal $\mathcal{G}_2(n)$.

### 18.2.3 Cycles in $\mathcal{G}_2(n)$

Laman's characterization of $\mathcal{G}_2(n)$ allows us to combinatorially identify the independent sets, from which it is easy to deduce a characterization for the cycles of $\mathcal{G}_2(n)$.

**Theorem 18.10** *An edge set C is a cycle in $\mathcal{G}_2(n)$ if and only if $|E(C)| = 2|V(C)|-2$ and $|F| \leq 2|V(F)|-3$ for every proper subset F of $E(C)$.*

In Figure 18.9, all cycles in $\mathcal{G}_2(n)$ having fewer than 7 vertices are described.

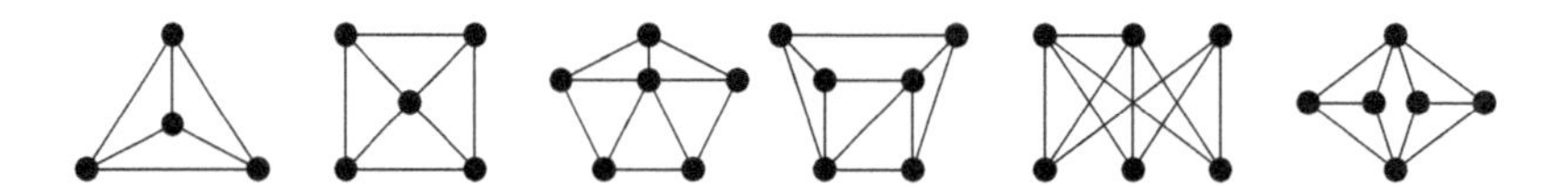

**Figure 18.9**
The generic rigidity cycles on up to 6 vertices.

Note that $\mathcal{A}_2$ admits no nongeneric cycles on fewer than six vertices. The only possible nongeneric cycles on six vertices are $K_{3,3}$ and the triangular prism.

Also note that the wheel $W_n$, the graph obtained from an $n$-gon by attaching a single new vertex which is adjacent to all the vertices of the $n$-gon, is a cycle in any abstract rigidity matroid in which it is contained. In Figure 18.9 the first three graphs are wheels.

It follows immediately from Theorem 18.10 that generic cycles are rigid, in fact over braced in a homogeneous way, specifically, the removal of any edge leaves a rigid graph whose edge set is independent in $\mathcal{G}_2(n)$. In fact, the rigidity of the cycles characterizes $\mathcal{G}_2(n)$.

**Theorem 18.11** *Let $\mathcal{A}_2$ be a 2-dimensional abstract rigidity matroid on n vertices. Then $\mathcal{A}_2 = \mathcal{G}_2(n)$ if and only if all cycles in $\mathcal{A}_2$ are rigid.*

The following lemma lists some simple but useful properties of cycles in $\mathcal{G}_2(n)$.

***Lemma 18.1***
*Let C be a cycle in $\mathcal{G}_2(n)$. Then:*

*a. $(V(C),C)$ is 2-connected.*

*b. $(V(C),C)$ is 3-edge-connected.*

*c. If the removal of 3 edges disconnects $(V(C),C)$, then the 3 edges have a common endpoint v, and v is of valence 3 in C.*

It is easy to verify that a 1-extension of a cycle is a cycle. The question, posed by R. Connelly, whether or not all 3-connected cycles in $\mathcal{G}_2(n)$ can be obtained from the tetrahedron by a sequence of 1-extensions was answered affirmatively by Berg and Jordán in [1].

**Theorem 18.12 (Berg-Jordán)** *All 3-connected cycles in $\mathcal{G}_2(n)$ can be obtained from the tetrahedron by a sequence of 1-extensions.*

The Berg-Jordán result is a very useful tool in investigating the structure of generic cycles. For independent sets, all 1-extensions yield independent sets, so one might suspect that all 1-extensions of generic cycles are generic cycles, but this turns out to be false: Consider $C$, the fifth cycle in Figure 18.9. Let $v$ be the upper right-hand vertex of $C$. There is no 1-extension giving $C$ with $v$ as the new vertex being attached, since $C - v$ has 3 vertices of degree 2. Connelly in [6] has intriguing examples with many adjacent vertices of degree 3 which he calls spider webs.

Let $C_1$ and $C_2$ be cycles in $\mathcal{G}_d(n)$ such that $E(C_1) \cap E(C_2) = \{e\}$ and $|V(E_1) \cap V(E_2)| = 2$, i.e., $C_1$ and $C_2$ have exactly one edge and no vertices other than the endpoints of this edge in common. Then the symmetric difference of $C_1$ and $C_2$ is an $d$-cycle for all $d$. This means that for $d > 2$ there must be nonrigid cycles, in fact cycles with arbitrarily large degree of freedom.

If a cycle in $\mathcal{G}_2(n)$ is not 3-connected, then it is the symmetric difference of two cycles in $\mathcal{G}_2(n)$ which have exactly one edge and its endpoints in common. This operation is called the *2-sum*. So all cycles in $\mathcal{G}_2(n)$ are obtained by the operations of 1-extension and 2-sum from a set of disjoint tetrahedra. However, $\mathcal{G}_2(n)$ is not closed under 2-sum decomposition [52].

In [20] Jackson and Jordán proved that in the plane a graph $G$ is globally rigid if $G$ is vertex 3-connected and $\mathcal{G}_2(G)$ is connected (i.e., for any two edges of $G$ there is a cycle of $\mathcal{G}_2(G)$ containing both of them), so the obvious two necessary conditions for a unique embedding of $G$ are also sufficient in the plane.

While the rigidity of a graph and the connectivity of a graph are closely related, planarity and rigidity seem quite unrelated. However, we have the following curious property in the plane, see [51].

**Theorem 18.13** *Let C be a cycle in $\mathcal{G}_2(n)$ such that $(V(C),C)$ is planar. Then the edge set of the geometric dual of $(V(C),C)$ is also a generic cycle.*

Wheels, $W_n$ are self-dual planar generic cycles.

### 18.2.4 Rigid Components of $\mathcal{G}_2(G)$

We regard a graph $G$ as a subgraph of $K_n$ for some $n > |V(G)|$ and denote the restriction of $\mathcal{G}_2(n)$ to $E(G)$ by $\mathcal{G}_2(G)$. Since $\mathcal{G}_2(n)$ is a submatroid of $\mathcal{G}_2(m)$ for $m > n$, this definition is independent of $n$, and we may define a set $F$ in $\mathcal{G}_2(G)$ to be *rigid* if $r(F) = 2|V(F)| - 3$, where $r$ is the rank function of $\mathcal{G}_2(n)$ restricted to $E(G)$.

A maximal rigid subgraph of $G$ is called an *r-component.* It is an immediate consequence of Axiom C6 that two r-components have at most one vertex in common, hence the r-components may be regarded as a partition of the edges of $G$. A non-rigid graph has at least two r-components.

Since the cycles of $\mathcal{G}_2(G)$ consist of those cycles of $\mathcal{G}_2(n)$ which are completely contained in $E(G)$, and since cycles are rigid by Theorem 18.11, it follows that every cycle of $\mathcal{G}_2(n)$ is completely contained in some rigid component. This gives the following theorem.

**Theorem 18.14** *Let $G$ be a graph and let $E_1, \ldots, E_k$ be the rigid components of $G$. Then $\mathcal{G}_2(G) = \mathcal{G}_2(G_1) \oplus \ldots \oplus \mathcal{G}_2(G_k)$, where $G_i = (V(E_i), E_i)$.*

If $G' = (V', E')$ is a subgraph of $G$, and $G'_i = (V'_i, E'_i)$ are the rigid components of $G'$, then the closure of $E'$ in $\mathcal{G}_2(G)$ is $E'$ together with all edges in $E$ both of whose endpoints lie in some $G'_i$, and the closure of $E'$ in $\mathcal{G}_2(n)$ is the union of the cliques $K(V(E'_1))$. This leads to the following characterization of $\mathcal{G}_2(n)$.

**Theorem 18.15** *Let $\mathcal{A}_2$ be a 2-dimensional abstract rigidity matroid on $n$ vertices. Then $\mathcal{A}_2 = \mathcal{G}_2(n)$ if and only if for any closed set $E$ with cliques $E_1, \ldots, E_k$, and $E = \cup_{i=1}^k E_i$ we have $r(E) = r(E_1) + \cdots + r(E_k)$.*

It follows from this result that, if $\mathcal{A}_2$ is a 2-dimensional abstract rigidity matroid which is not generic, then it must contain a closed set violating the the equality in this theorem. Look, for example, at Figure 18.2a, where any additional edge would kill the existing infinitesimal motion, so the edge set is closed, the rigid components are the two triangles and the three parallel edges, but the rank is only 8.

Consider $V = \{1, \ldots, 6\}$ embedded on a conic. Then, as we have noted, each copy of $K_{3,3}$ in $K(V)$ is a cycle. $K_{3,3}$ is closed and its r-components are simply its edges, so the equality in the theorem is violated for $K_{3,3}$.

In many ways, the r-components are analogous to the connected components of a graph. However, the two concepts do have some important differences: The removal of any edge from a graph increases the number of connected components by at most one. By contrast, the removal of an edge from a rigid graph can result in a graph with many r-components. For example, if $G$ is a quadrilateral with 1 diagonal, then the removal of the diagonal results in a graph with four r-components, all of which are single edges. Removing any other edge of $G$ leaves two r-components, a triangle and a single edge. It is not difficult to show, however, that deleting an edge from a minimally rigid graph in the plane always yields an even number of r-components.

The number of r-components does not give us any information on the rank of a given set. Of course, we could compute the rank by summing the ranks of the rigid components. Sometimes it is of interest to know how many additional edges are needed to achieve rigidity of $(V(E), E)$. Recall that the *degree of freedom* of an edge set $E$, $\mathrm{df}(E)$, in $\mathcal{G}_2(n)$ is defined by

$$\mathrm{df}(E) + r(E) = 2|V(E)| - 3.$$

A graph $G = (V, E)$ with degree of freedom 1 and $2k$ r-components, each of which has $n_i$ vertices, has

$$|V| = \sum_{i=1}^{2k} n_i - 3k + 2.$$

Using the direct sum decomposition of $\mathcal{G}_2(G)$ induced by the r-components we can give a combinatorial description of $\mathcal{G}_2(G)$ via the rank function.

**Theorem 18.16** *Let $G = (V,E)$ be a graph and let $r$ denote the rank function of $\mathcal{G}_2(G)$. Then*

$$r(E) = \min \sum_{i=1}^{k} (2|V(E_i)| - 3),$$

*where the minimum is taken over all collections $\{E_i\}$ such that $E = \cup E_i$.*

The minimum in (18.16) is achieved if $\{E_i\}$ coincides with the collection of rigid components of $G$, however other collections may also give the minimum, as seen in Figure 18.10, in which the

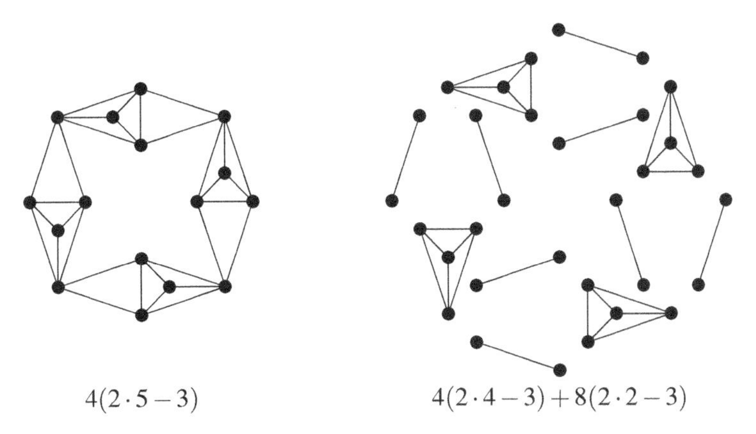


**Figure 18.10**
Computing $\sum(2|V(E_i)| - 3)$ for different edge partitions.

count at the left is via the rigid components. From Theorem 18.16 there follows the following result first due to Lovász and Yemini, see [35].

**Corollary 18.17**

$$df(E) = 2|V(E)| - 3 - \min \sum_{i=1}^{k} (2|V(E_i)| - 3),$$

*where the minimum extends over all systems $\{E_1, \ldots, E_k\}$ of subsets of $E$ such that $E_1 \cup \ldots \cup E_k = E$.*

### 18.2.5 Representability of $\mathcal{G}_2(n)$

The graphic matroid, in our notation $\mathcal{G}_1(n)$, has the nice property of being binary, in fact, regular, see [43, 63]. We know that $\mathcal{G}_2(n)$ is representable over the real numbers, since a representation is obtained from a generic embedding $\mathbf{p}$ of the vertices $V$ of a complete graph $K$, by the rigidity matrix $R(\mathbf{p}) = R(K, \mathbf{p})$.

**Theorem 18.18** *No 2-dimensional abstract rigidity matroid on more than 4 vertices is binary.*

**Proof 18.1** Recall that a matroid is binary if and only if the symmetric difference of two cycles is the disjoint union of cycles. If $n > 4$, we can find in $K_n$ two $K_4$'s having 3 vertices in common so that their disjoint union is independent. See Figure 18.11.

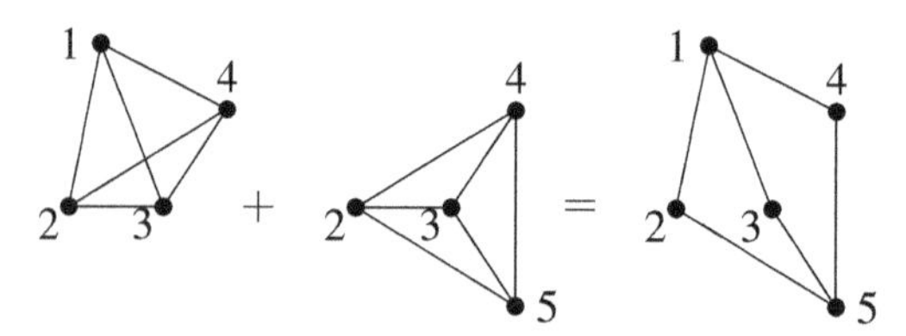


**Figure 18.11**
Two cycles whose symmetric difference is independent.

In any matroid $\mathcal{M}$ on $E$ representable over a finite field $\mathbb{F}$, the number of cycles is bounded by $\frac{|\mathbb{F}|^k-1}{|\mathbb{F}|-1}$, where $k=|E|-r(E)$. So if $k$ is small, there are relatively few cycles.

The next theorem uses this fact to show that no finite field is large enough to represent all generic rigidity matroids in dimension 2.

**Theorem 18.19** *There is no finite field $\mathbb{F}$ such that $\mathcal{G}_2(n)$ is representable over $\mathbb{F}$ for all $n$.*

**Proof 18.2** The graph $(V,E)$ in Figure 18.12 is rigid and has $k=|E|-r(E)=2$. The removal

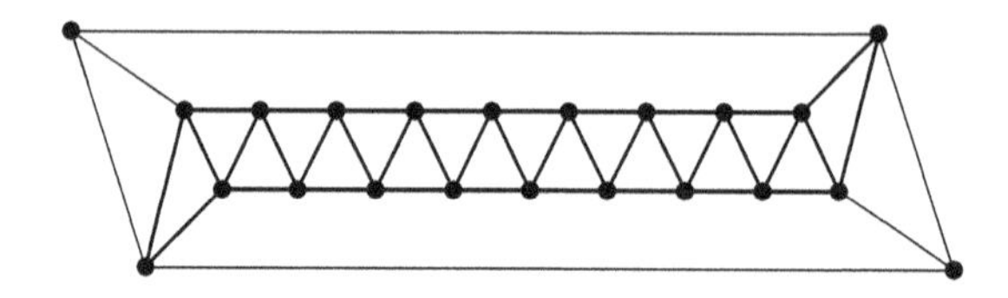

**Figure 18.12**
A rigid graph with $k=|E|-r(E)=2$.

of any edge with two endpoints of valence 4, as well as removal of a vertex of valence 3, yields a cycle. Clearly, we can make the string of triangles as long as we wish, so if a graph $G$ of this form has $2n$ vertices, $\mathcal{G}_2(n)$ contains $2(n+1)$ cycles. Now assume that $\mathcal{G}_2(n)$ is representable over $\mathbb{F}$. We have:

$$|\mathbb{F}|+1=\frac{|\mathbb{F}|^2-1}{|\mathbb{F}|-1}\geq 2(n+1).$$

However, $\mathcal{G}_2(n)$ is representable over the rational numbers. Representability of abstract rigidity matroids over fields of prime characteristic has not yet been studied.

## Glossary

**0-extension, Henneberg-1 move:** The 0-extension of a framework $((V,E),\mathbf{p})$ in dimension $d$ is a new framework consisting of $((V,E),\mathbf{p})$ augmented by a new vertex $v$ together $d$ edges joining $v$ to vertices in $V$.

**1-extension, Henneberg-2 move, edge split:** The 1-extension of a framework $((V,E),\mathbf{p})$ at an edge $e\in E$ in dimension $d$ is a new framework consisting of $((V,E-e),\mathbf{p})$ augmented by a new vertex $v$ together with $d+1$ edges joining $v$ to vertices in $V$, including the endpoints of the deleted (split) edge $e$.

**1-Extendability property:** An abstract rigidity matroid has the 1-*extendability property* if, given an independent set $E$ on vertex set $V$, every 1–extension of $E$ yields an independent edge set on $V \bigcup v$.

**Abstract rigidity matroid:** A matroid $\mathcal{A}_d$ on the edges of a complete graph $K$ with closure operator $\langle \cdot \rangle$ is called a *d-dimensional abstract rigidity matroid for* $V$ if it satisfies the usual four closure axioms for a matroid as well as

**C5** If $E, F \subseteq K$ and $|V(E) \cap V(F)| < d$, then $\langle E \cup F \rangle \subseteq (K(V(E)) \cup K(V(F)))$.

**C6** If $\langle E \rangle = K(V(E))$, $\langle F \rangle = K(V(F))$ and $|V(E) \cap V(F)| \geq d$, then $\langle E \cup F \rangle = K(V(E \cup F))$.

**Cocycles:** Minimally dependent sets in the $\mathcal{M}^*$, i.e., minimal sets which intersect every basis of $\mathcal{M}$.

**Generic rigidity matroid:** the matroid $\mathcal{G}_d(n)$, the infinitesimal rigidity matroid on the edges of the complete graph on $n$ vertices placed generically in $\mathbb{R}^d$.

**Infinitesimal rigidity matroid:** Matroid on the edges of a framework in which the independent sets are those edge sets whose corresponding rows in the rigidity matrix are linearly independent.

**Isostatic framework (infinitesimally, generically):** (Infinitesimally, generically) rigid framework such that the removal of any bar destroys the rigidity.

**Laman's condition:** An edge set satisfies Laman's condition if $|F| \leq 2|V(F)| - 3$ for all all non-empty subsets $F \subseteq E$. Independent sets of $\mathcal{G}_2(n)$ satisfy Laman's condition.

**Rigid component, *r*-component:** A maximal rigid subgraph of $G$. The $r$ components of an abstract rigidity matroid partition $E$.

**Vertex cocycles:** Cocycles of an abstract rigidity matroid on $(V, E)$ consisting only of edges incident at one vertex. In $\mathcal{A}_2$, the vertex cocycles are sets of the form $S_e(v)$ consisting of all edges incident to $v$ except $e$.

## 18.3 Rigidity and Connectivity

The matroid $\mathcal{G}_1(G)$ is the connectivity matroid of the graph $G$, and is closely related to the connectivity of the graph. As such, generic rigidity in higher dimensions has often been regarded as a generalization of connectivity. Both concepts require a certain number of edges to be well distributed over the graph. The purpose of this section is to collect results on the relation between the generic rigidity of graphs, the connectivity of graphs, and the connectivity of the relevant matroids.

The connectivity of a graph may be measured in many ways. In general, a graph is (vertex) $k$-connected if it has at least $k+1$ vertices and the removal of any $k-1$-vertices results in a connected graph. A graph is edge $k$-connected if it has at least $k$ edges and the removal of any $k-1$ edges results in a connected graph. Except for trivial examples, if $k=1$ the edge and vertex versions coincide and connectivity is characterized by a counting condition completely analogous to Laman's Theorem:

**Theorem 18.20** *A graph $G = (V,E)$ is connected if and only if there is a subset $F \subseteq E$ such that*

*(a) $|F| = |V| - 1$, and*

*(b) $|F'| \leq |V(F')| - 1$ for all subsets $F' \subseteq F$.*

The relation between vertex and edge $k$-connectivity is well studied, see for example the fundamental work of Tutte [61], or the modern text by Diestel [11].

The higher the vertex connectivity of the graph, the more likely it is that the graph is generically rigid in dimension $d$. However, even in dimension 2, this heuristic is often violated: there are generically rigid graphs which are 2-connected but not 3-connected, for instance the rightmost graph in Figure 18.9, yet in Figure 18.13a we have a graph which is 5-connected but generically flexible.

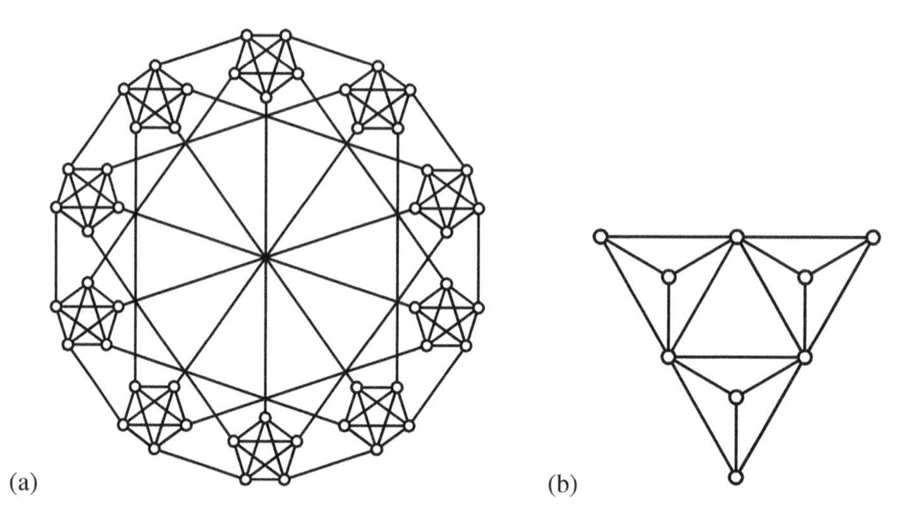


**Figure 18.13**
(a) A 5-regular graph whose vertices are incident vertex-edge pairs $(v,e)$ in $K_{5,5}$, with two pairs being adjacent if they have an element in common.(b) An edge-birigid graph which is the union of three cycles.

Each of the copies of $K_5$ is over-braced by 3 edges, and removing these 30 redundant edges leaves the remainder under braced. The next result shows that the region of uncertainty for generic rigidity in the plane is from 2-connected to 5-connected.

**Theorem 18.21 (Lovász and Yemini [35])** *Every vertex 6-connected graph is generically rigid in* $\mathbb{R}^2$.

There is no similar result for edge connectivity alone, since taking the union of two copies of $K_{n+1}$ at a vertex yields a graph which is $n$-edge-connected but generically flexible in $\mathbb{R}^2$.

### 18.3.1 Birigidity

A graph is *vertex k-rigid* if it has at least $k+1$ vertices and the removal of any $k-1$ vertices, together with their incident edges, leaves a rigid graph A graph is *edge k-rigid* if the removal of any $k-1$ edges leaves a rigid graph. For $k = 2$ the term birigidity is used. Every edge in an edge birigid graph, that is a 2-edge rigid graph, is redundant, and the term *redundantly rigid* is also used for $G$. With these definitions, the generic (edge) $k$-rigidity of a graph in $\mathbb{R}^1$ is exactly its $k$-connectivity. A cycle in $\mathcal{G}_2(n)$ is edge-birigid, but Figure 18.13 illustrates the general case.

Parallel to vertex- and edge connectivity in graphs, Tutte [62] also introduced the concept of connectivity in matroids. A matroid $\mathcal{M}$ on $S$ is *connected* if $r(F)+r(E-F) > r(E)$ holds for every non-empty proper subset $F$ of $E$. If there is a subset $F$ such that $r(F)+r(E-F) \leq r(E)$, then necessarily $r(F)+r(E-F) = r(E)$ and $F$ and $E-F$ are said to *separate* the matroid, and we write $\mathcal{M} = F \oplus (E-F)$. The connectivity matroid, $\mathcal{G}_1(G)$, of a graph $G$ is connected if and only if $G$ is 2-connected.

The situation in the plane is more complex. Suppose that $G = (V,E)$ is vertex birigid in the plane. Let $v \in V$ and let $v$ be incident to edges $e_1,\ldots,e_m$. If $G$ is not simply a triangle, we must have $3 \geq m$, otherwise deleting one of the neighbors of $v$ would leave a non-trivial motion of $v$. Since $G$ is birigid, $E' = E - \{e_1,\ldots,e_m\}$ is rigid, as well as $E' \cup \{e_i, ej\}$ for any pair $1 \leq i < j \leq m$, and $E' \cup \{e_i, e_j, e_k\}$ must be dependent, with a cycle of $\mathcal{G}_2(G)$ containing both $e_i$ and $e_j$. So every pair of edges incident to $v$ belong to a cycle of $\mathcal{G}_2(G)$, and hence to the same direct summand of $\mathcal{G}_2(G)$. Since the graph $G$ is connected, $\mathcal{G}_2(G)$ can have only one direct summand, so $\mathcal{G}_2(G)$ is a connected matroid.

Suppose, on the other hand, that $\mathcal{G}_2(G)$ is connected. Then by Theorem 18.14 it has only one rigid component, so $G$ is rigid, and $\langle E \rangle = 2|V|-3$. If any edge $e$ of $G = (V,E)$ were not redundant then $\langle E-e \rangle = 2|V|-4$, and $E-e$ and $\{e\}$ would be a separating partition of the matroid $\mathcal{G}_2(G)$, $\langle E-e \rangle + \langle e \rangle = \langle E \rangle$, so $G$ is edge birigid.

**Theorem 18.22 (Graver, Servatius, and Servatius, 1993 [15])** *If $G$ is vertex birigid in the plane, then $\mathcal{G}_2(G)$ is connected.*

*If $\mathcal{G}_2(G)$ is connected then $G$ is edge-birigid.*

The argument above implies this birigidity result is also true in any abstract rigidity matroid, but neither implication can be reversed. The graph of Figure 18.13b is edge birigid but its generic rigidity matroid is the direct sum of three connected components. The generic rigidity matroid on the wheel $W_n$, $n \geq 4$ is connected, since $\mathcal{G}_2(W_n)$ is just a cycle, but removing the center vertex yields a non-rigid graph.

In reviewing the proof of the Lovász Yemini Theorem, we see that, if a graph is 6-connected, then it is not only rigid but over-braced. In [22] 6-connectivity is replaced by 6-mixed connectivity, where a graph $G$ is 6-mixed connected if $G-U-D$ is connected for all vertex subsets $U$ and edge subsets $D$ satisfying $2|U|+|D| \leq 5$.

The connectivity of the rigidity matroid together with the fact that cycles are rigid was used by Jackson and Jordàn in [20] to characterize global rigidity in the plane.

**Theorem 18.23 (Jackson and Jordán, 2005 [20])** *A graph $G$ is generically globally rigid in the plane if and only if $G$ is 3-connected and edge birigid.*

### 18.3.2 Tree Decomposition Theorems

Tutte, resp. Nash-Williams, see [39, 40, 60] proved (1961–1964) that a graph $G = (V,E)$ is decomposable into $k$ spanning forests if and only if $|F| \leq k(|V(F)|-1)$ for all nonempty subsets $F$ of $E$. For $k = 2$ this looks remarkably similar to Laman's characterization of independent sets in $\mathcal{G}_2(n)$. The graphs described by Tutte/Nash-Williams have the same number of edges, relative to the number of vertices, as a cycle in $\mathcal{G}_2(n)$.

Two spanning trees of $G$ which are edge disjoint and have $G$ as their union are called a *2-tree decomposition* of $G$.

A 2-tree decomposition of $G$ is said be a *proper 2-tree decomposition* if no pair of proper subtrees, excepting single vertices, have the same span.

Choosing two paths of length $n$ randomly on the same vertex set of size $n$ will yield a graph with 4 vertices of degree three (the respective endpoints of the two paths, and all other vertices of degree 4. By construction this graph has a proper two tree (= path) decomposition, so it will be a cycle in $\mathcal{G}_2(n)$.

However, it is not true that every cycle with vertices of valence 3 and 4 only, has a two tree decomposition into two paths. Kijima and Tanigawa, [29] construct infinitely many planar counterexamples. See also [30].

If an edge set $E$ is isostatic, $x,y \in V(E)$, $(x,y) \notin E$, then $E+(x,y)$ contains exactly one cycle of $\mathcal{G}_2(n)$ and admits a 2-tree decomposition. To test isostaticity by using tree decompositions it is more economical to use the following result of Recski [47, 48, 50].

**Theorem 18.24** *Let $G=(V,E)$ be a graph. Then $E$ is isostatic in $\mathcal{G}_2(n)$ if and only if, doubling any edge $e \in E$, results in a multigraph which admits a 2-tree decomposition.*

Theorem 18.24 implies that an isostatic set is the union of three trees. One of the trees is spanning, the other two arise from removing the extra edge from the second tree in the decomposition, so one of the small trees could actually be an isolated vertex. Crapo [9] noted that many other *3-tree decompositions* exist and went on to prove the 3-tree decomposition theorem below. In Figure 18.14 we illustrate his result by listing three 3-tree decompositions for $K_{3,3}$.

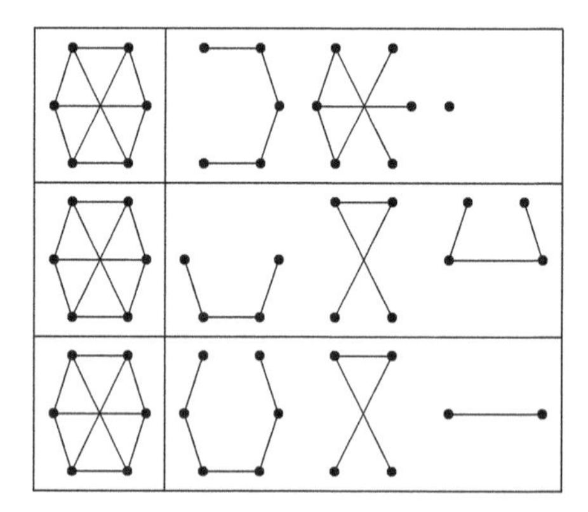

**Figure 18.14**
Three 3-tree decompositions of $K_{3,3}$.

**Theorem 18.25** *A graph $G=(V,E)$ is isostatic if and only if $G$ is the edge disjoint union of three trees such that each vertex of $G$ is contained in exactly two of the trees, and no two subtrees have the same span.*

#### 18.3.2.1 Computation of Independence in $\mathcal{G}_2(n)$

Given a graph $G=(V,E)$, to decide whether $E$ is independent in $\mathcal{G}_2(G)$ we can do better than naïvely check the inequality of Laman's Theorem for all subsets of $E$. The overbrace function $\phi(F)=2|V(F)|-3-|F|$ is *submodular* on the subsets $F$ of $E$, $\phi(F_1 \cup F_2)+\phi(F_1 \cap F_2) \leq \phi(F_1)+\phi(F_2)$ so we may use any one of the polynomial time algorithms which can find the minimum of a submodular set function, see [18]. If this minimum is non-negative, then $E$ is independent. The following is a modification of an algorithm by Edmonds, [12], and makes use directly of the tree decomposition of Theorem 18.25.

**INPUT** A graph $G = (V, E)$

**OUTPUT** A 2-forest decomposition of $G$ when $G$ is independent and the message "DEPENDENT" when $G$ is dependent.

**Step 1** Set $F_0 = F_1 = \emptyset$.

**Step 2** If $F_0 \cup F_1 = E$, GOTO Step 7; otherwise, let $e$ be the edge of least index in $E - (F_0 \cup F_1)$.

**Step 3** If $e \notin \langle F_0 \rangle$, replace $F_0$ by $F_0 + e$ and GOTO Step 2.

**Step 4** If $e \notin \langle F_1 \rangle$, replace $F_1$ by $F_1 + e$ and GOTO Step 2.

**Step 5** Construct the sequences of nested sets:

$$F_0 \supseteq F_2 = F_0 \cap \langle F_1 \rangle \supseteq F_4 = F_0 \cap \langle F_3 \rangle \supseteq \cdots;$$

$$F_1 \supseteq F_3 = F_1 \cap \langle F_2 \rangle \supseteq F_5 = F_1 \cap \langle F_4 \rangle \supseteq \cdots;$$

until the sequences become stationary. If the stationary sets are not empty, STOP and output "DEPENDENT".

**Step 6** Let $j$ be the first index so that $e \notin \langle F_j \rangle$. Let $C$ be the unique cycle contained in $F_{j \pmod 2} + e$ and let $j'$ be the first index so that $C - e \not\subseteq \langle F_{j'} \rangle$. Next let $e'$ be the edge in $C - e$ with smallest index such that $e' \notin \langle F_{j'} \rangle$. Replace $F_{j \pmod 2}$ by $F_{j \pmod 2} + e - e'$, replace $e$ by $e'$ and GOTO Step 3.

**Step 7** Construct the sequences of nested sets:

$$F_0 \supseteq F_2 = F_0 \cap \langle F_1 \rangle \supseteq F_4 = F_0 \cap \langle F_3 \rangle \supseteq \cdots;$$

$$F_1 \supseteq F_3 = F_1 \cap \langle F_2 \rangle \supseteq F_5 = F_1 \cap \langle F_4 \rangle \supseteq \cdots;$$

until the sequences become stationary. If the stationary sets are not empty, STOP and output "DEPENDENT", otherwise STOP and output $F_0$ and $F_1$.

We apply algorithm to the dependent graph in Figure 18.15.

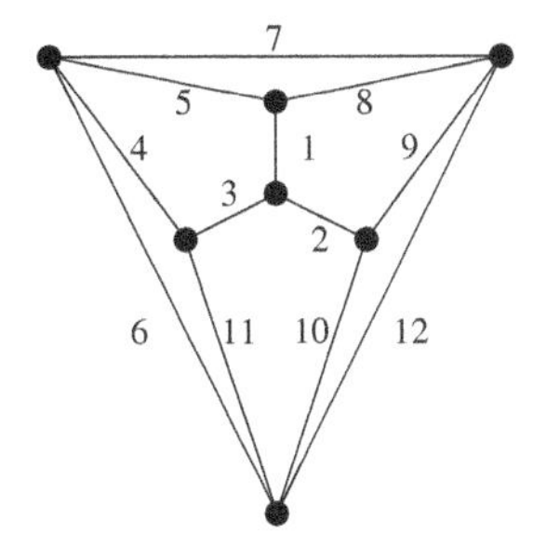


**Figure 18.15**
A 3-connected planar cycle in $\mathcal{G}_2(n)$.

**Example 45** *Starting with $F_0 = F_1 = \emptyset$, the algorithm iterates Step 2 eleven times before it can no longer simply add an edge to either $F_0$ or $F_1$. At this point, we have $F_0 = \{1,2,3,4,6,7\}$ and $F_1 = \{5,8,9,10,11\}$. (See the table at the end of this example.) Edge 12 is in the closures of both $F_0$ and $F_1$; so, the algorithm moves on to Step 5 (for the first time). The closures of the sets it generates are:*

$$\begin{aligned}
\langle F_0 \rangle &= E, \\
\langle F_1 \rangle &= \{4,5,6,7,8,9,10,11,12\}, \\
\langle F_2 \rangle &= \langle F_0 \cap \langle F_1 \rangle \rangle = \{4,6,7,11,12\}, \\
\langle F_3 \rangle &= \langle F_1 \cap \langle F_2 \rangle \rangle = \{11\}, \\
\langle F_4 \rangle &= \langle F_0 \cap \langle F_3 \rangle \rangle = \emptyset.
\end{aligned}$$

*In step 6, we note $j = 3$; we add edge 12 to $F_1$ and find the cycle $C = \{9,10,12\}$; we note $j' = 2$: we then replace edge 9 by edge 12 in $F_1$ and return to Step 3 with e equal to edge 9. Since edge 9 is in the closures of both $F_0$ and the new $F_1$, we again proceed to Step 5. This time we get the sequence:*

$$\begin{aligned}
\langle F_0 \rangle &= E, \\
\langle F_1 \rangle &= \{4,5,6,7,8,9,10,11,12\}, \\
\langle F_2 \rangle &= \langle F_0 \cap \langle F_1 \rangle \rangle = \{4,6,7,11,12\}, \\
\langle F_3 \rangle &= \langle F_1 \cap \langle F_2 \rangle \rangle = \{11,12\}, \\
\langle F_4 \rangle &= \langle F_0 \cap \langle F_3 \rangle \rangle = \emptyset.
\end{aligned}$$

*This time j equals the "old" $j'$ or 2. Edge 9 is added to $F_0$ creating the cycle $C = \{2,3,4,7,9\}$. Then $j' = 1$ and edge 9 replaces edge 2 in $F_0$. We move to Step 3 with e equal to edge 2 and then to Step 4 where edge 2 is added to $F_1$. The algorithm then returns to Step 2 and moves directly to Step 7. We have a 2-forest decomposition of E. But, since $\langle F_0 \rangle = \langle F_1 \rangle = E$, the algorithm terminates with the message "DEPENDENT".*

*In Table 18.3 we list the sequence of sets $F_0$ and $F_1$ which occur.*

The algorithm can be altered to check for cycles by noting that in a graph $G = (V,E)$ with $|E| = 2|V| - 2$ edges any proper subcycle must be contained in one of the subgraphs of $G$ obtained by deleting a vertex.

Jacobs and Hendrickson, see [24], use yet another version, similar to Recski's Theorem, of Laman's Theorem, namely the fact that an edge set $E$ is independent in $\mathcal{G}_2(G(E))$ if quadrupling any one of its edges produces no induced subgraph of average degree larger than 2. They show that the existence of a pebble covering is equivalent to the independence condition in the quadrupled edge formulation, where a pebble covering is the result of the *Pebble game*: Each vertex is given two pebbles. A vertex can use its pebbles to cover any two edges which are incident to that vertex. An assignment covering all edges is a *pebble covering*.

Pebble games were further investigated in [34, 56] and generalized to hypergraphs in [55] and Cad systems [13].

### 18.3.3 Pinned Frameworks and Assur Decomposition

#### 18.3.3.1 Isostatic Pinned Framework

A *mechanism* is a framework with one internal degree of freedom. Given a mechanism, we are interested in its internal motions, not the trivial ones, so following the mechanical engineers we pin the framework by prescribing, for example, the coordinates of the endpoints of an edge, or in general by fixing the position of the vertices of some rigid subgraph, see Figure 18.16. Vertices with fixed positions are called *pinned*, the others *inner*. (Inner vertices are sometimes called *free* or

**Table 18.3**
Steps of the modified Edmonds Algorithm

| iteration of Step 2 | $F_0$ | $F_1$ |
|---|---|---|
| 0 | $\emptyset$ | $\emptyset$ |
| 1 | $\{1\}$ | $\emptyset$ |
| 2 | $\{1,2\}$ | $\emptyset$ |
| 3 | $\{1,2,3\}$ | $\emptyset$ |
| 4 | $\{1,2,3,4\}$ | $\emptyset$ |
| 5 | $\{1,2,3,4\}$ | $\{5\}$ |
| 6 | $\{1,2,3,4,6\}$ | $\{5\}$ |
| 7 | $\{1,2,3,4,6,7\}$ | $\{5\}$ |
| 8 | $\{1,2,3,4,6,7\}$ | $\{5,8\}$ |
| 9 | $\{1,2,3,4,6,7\}$ | $\{5,8,9\}$ |
| 10 | $\{1,2,3,4,6,7\}$ | $\{5,8,9,10\}$ |
| 11 | $\{1,2,3,4,6,7\}$ | $\{5,8,9,10,11\}$ |
| 12 | $\{1,2,3,4,6,7\}$ | $\{5,8,10,11,12\}$ |
| $12'$ | $\{1,3,4,6,7,9\}$ | $\{5,8,10,11,12\}$ |
| $12''$ | $\{1,3,4,6,7,9\}$ | $\{2,5,8,10,11,12\}$ |

*unpinned* in the literature.) Edges among pinned vertices are irrelevant to the analysis of a pinned framework. We will denote a pinned graph by $G(I,P;E)$, where $I$ is the set of inner vertices, $P$ is the set of pinned vertices, and $E$ is the set of edges, where each edge has at least one endpoint in $I$.

A pinned graph $G(I,P;E)$ is said to satisfy the *pinned framework conditions* if $|E| = 2|I|$ and for all subgraphs $G'(I',P';E')$ the following conditions hold:

(a) $|E'| \leq 2|I'|$ if $|P'| \geq 2$,

(b) $|E'| \leq 2|I'| - 1$ if $|P'| = 1$ , and

(c) $|E'| \leq 2|I'| - 3$ if $P' = \emptyset$.

We call a pinned graph $G(I,P;E)$ *pinned isostatic* if $E = 2|I|$ and $G \cup K_P$ is rigid as an unpinned graph, where $K_P$ is a complete graph on a vertex set containing all pins (but no inner vertices). In other words, we "replace" the pinned vertex set by a complete graph containing the pins and call $G(I,P;E)$ isostatic, if choosing any basis in that replacement produces an (unpinned) isostatic graph.

A pinned graph $G(I,P;E)$ realized in the plane, with $\mathbf{P}$ for the pins, and $\mathbf{p}$ for all the vertices, is a *pinned framework.* A pinned framework is *rigid* if the matrix $R(G \cup K_P)$ has rank $2|I|$, with the columns corresponding to the vertex set of $K_p$ removed, *independent* if the rows of $R(G \cup K_P)$ corresponding to $E$ are independent, and *isostatic*, if it is rigid and independent. The vertices $I$ of a pinned framework are in *generic position* if any submatrix of the rigidity matrix is zero only if it is identically equal to zero with the coordinates of the inner vertices as variables. The coordinates of the pins are prescribed constants.

Figure 18.16 shows an example of a pinned isostatic $G$ and a corresponding basis of $R(G \cup K_P)$.

It is common in engineering to choose pins in advance and their placement $\mathbf{P}$ might not be generic, in fact not even in general position, as it is sometimes necessary to have all pins on a line. The following result shows that this is not a problem.

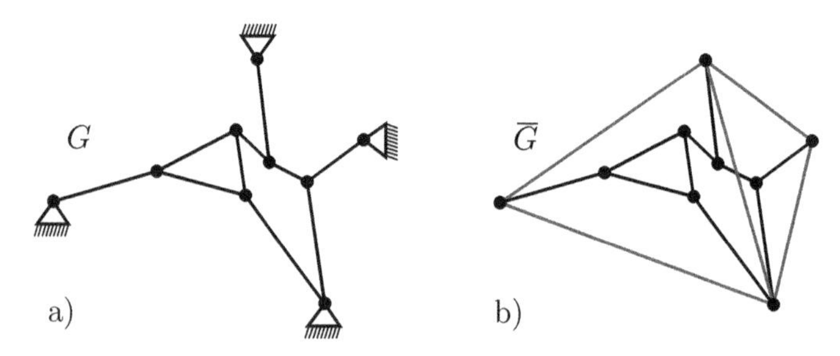


**Figure 18.16**
Framework (a) is pinned isostatic because framework (b) is isostatic.

**Theorem 18.26** *Given a pinned graph $G(I,P;E)$, the following are equivalent:*
*(i) There exists an isostatic realization of G.*
*(ii) The Pinned Framework Conditions are satisfied.*
*(iii) For all placements* **P** *of P with at least two distinct locations and all generic positions of I the resulting pinned framework is isostatic.*

A pinned graph $G(I,P;E)$ satisfying the Pinned Framework Conditions must have at least two pins and in every isostatic realization of $G$ there must be at least two distinct pin locations. Placing all pins in the same location never yields an isostatic framework, but we can make an important observation about the degree of freedom of such a "pin collapsed" framework.

**Theorem 18.27** *Let $G(I,P;E)$ be a pinned graph satisfying the Pinned Framework Conditions. Identifying the pinned vertices to one vertex $p^*$ yields a graph $G^*(V,E)$, $V = I \bigcup \{p^*\}$ and the degree of freedom of $G^*$ is one less than the number of circuits contained in $\mathcal{G}_((G)^*)$.*

**Theorem 18.28** *Assume $G = (I,P;E)$ is a pinned isostatic graph. Then the following are equivalent:*
*(i) $G = (I,P;E)$ is minimal as a pinned isostatic graph: that is for all proper subsets of vertices $I' \cup P'$, $I' \cup P'$ induces a pinned subgraph $G' = (I' \cup P', E')$ with $|E'| \leq 2|I'| - 1$.*
*(ii) If the set P is contracted to a single vertex $p^*$, inducing the unpinned graph $G^*$ with edge set E, then $G^*$ is a cycle.*
*(iii) Either the graph has a single inner vertex of degree 2 or each time we delete a vertex, the resulting pinned graph has a motion of all inner vertices (in generic position).*
*(iv) Deletion of any edge from G results in a pinned graph that has a motion of all inner vertices (in generic position).*

Condition (i) is a refinement of the Grübler count [42], in a form which is now necessary and sufficient.

Condition (ii) translates the minimality condition to minimal dependence in $\mathcal{G}_2(n)$ and thus serves as a purely combinatorial description of Assur graphs and may be checked for example by the modified Edmonds algorithm described in Section 18.3.2.1.

Conditions (iii) and (iv) are similar in nature. Condition (iii) provides the engineer with a quicker check for the Assur property for smaller graphs than (iv), since there are fewer vertices than edges to delete. However, condition (iv) tells the engineer that a driver inserted for an arbitrary edge will (generically) move all inner vertices.

Some examples of Assur graphs are drawn in Figure 18.17 and their corresponding generic cycles in Figure 18.18.

A general isostatic framework can be decomposed into a partially ordered set of Assur graphs. This partial order can be represented in an *Assur scheme* as in Figure 18.22.

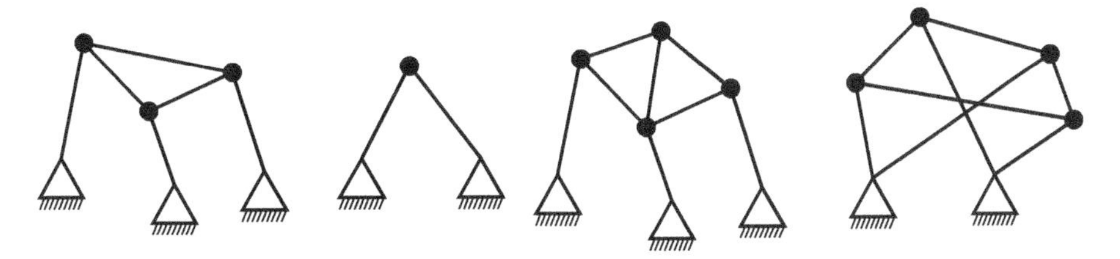

**Figure 18.17**
Assur graphs.

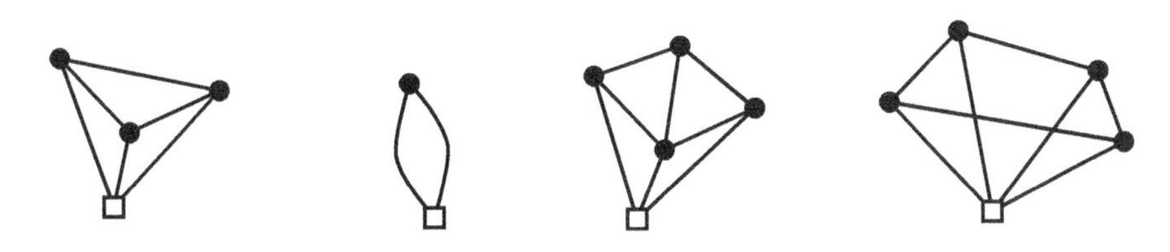

**Figure 18.18**
Corresponding cycles for Assur graphs.

Figure 18.19 shows isostatic pinned frameworks and Figure 18.21 indicates their decomposition.

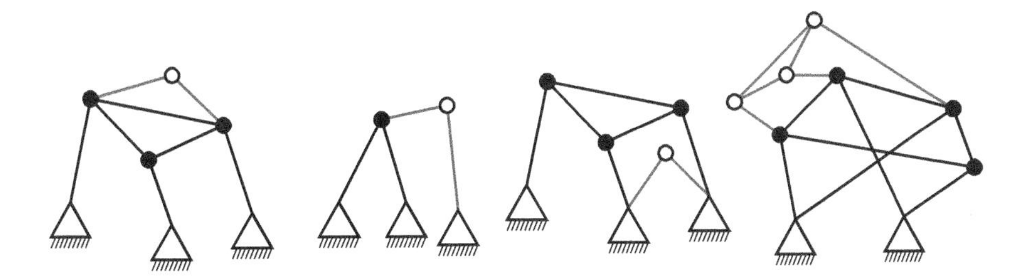

**Figure 18.19**
Decomposable (not Assur) graphs.

Under this operation, the Assur graphs will be the minimal, indecomposable graphs. Every pinned isostatic graph $G$ is a unique composition of Assur graphs, called *Assur components* of $G$. Decomposition by identifying the pins, then deleting the resulting generic cycle and pinning the vertices of attachment naturally induces a partial order on the Assur components of an isostatic graph: component $A \leq B$ if $B$ occurs at a higher level, and $B$ has at least one vertex of $A$ as a pinned vertex. The algorithm for decomposing the graph guarantees that $A \leq B$ means that $B$ occurs at a later stage than $A$.

This partial order, with the identifications needed for linkage composition, can be used to reassemble the graph from its Assur components.

Deleting any edge in an isostatic framework produces a mechanism. Its decomposition into Assur components permits the analysis of this mechanism in layers. In fact, we can delete an edge in each Assur component to obtain a pinned framework with several degrees of freedom whose complex behavior can be simply described by analyzing the individual Assur components. The

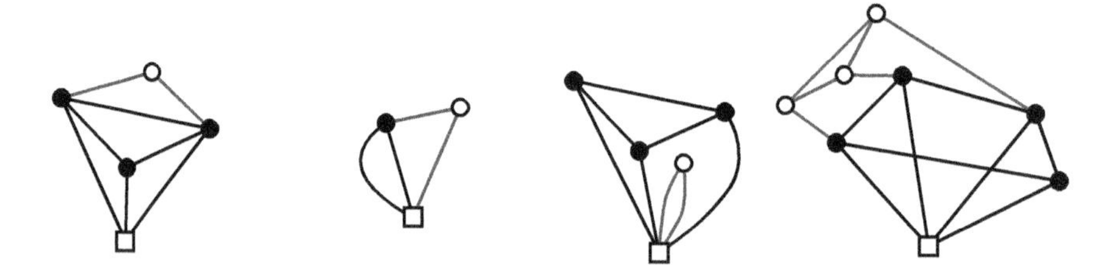

**Figure 18.20**
The first step of a decomposition for isostatic frameworks in 18.19 –with identified subcircuit(s).

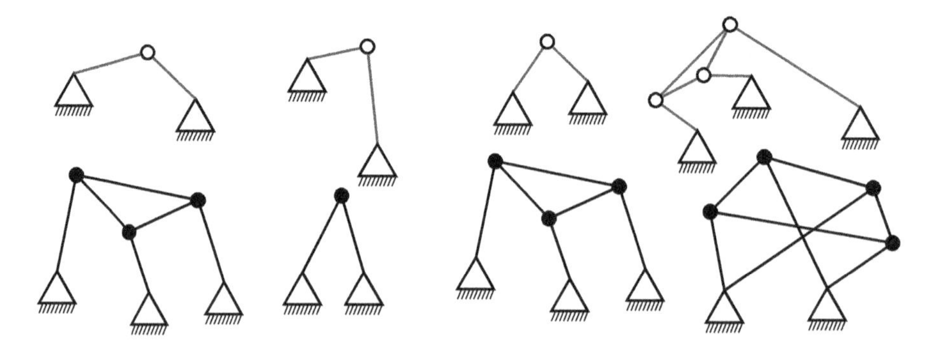

**Figure 18.21**
Recomposing the pinned isostatic graphs in Figure 18.19 from their Assur components.

engineer thinks of edge deletion as replacing an edge by a *driver*. This process of adding drivers is studied in [53].

The dyad is the only Assur graph on three vertices. There is no Assur graph on four vertices. An Assur graph, whose corresponding generic cycle is $K_4$ is called a *basic* Assur graph.

To generate all Assur graphs (on five or more vertices) we use Theorem 18.28(ii) together with Theorem 18.12 to generate all rigidity circuits. To get from a rigidity circuit $C$ to an Assur graph, we choose a vertex $p^*$ of $C$ and split it into two or more pins. The choice of $p^*$, the splitting of $p^*$ into a set $P$ of pins ($2 \leq |P| \leq val(p^*)$), and choosing for each edge incident to $p^*$ an endpoint from $P$ allows us to construct several Assur graphs from one generic cycle, see Figure 18.23. We say that $G(I,P;E)$ and $G'(I,P',E)$ are related by *pin rearrangement* if $G^* = G'^*$ (see Figure 18.23).

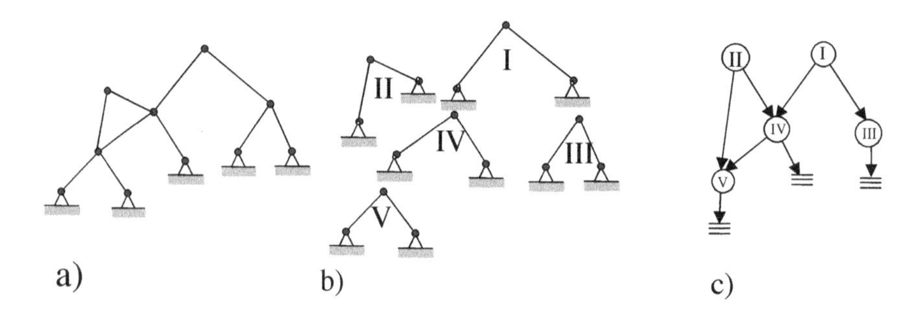


**Figure 18.22**
An isostatic pinned framework (a) has a unique decomposition into Assur graphs (b) which is represented by a partial order or Assur scheme (c).

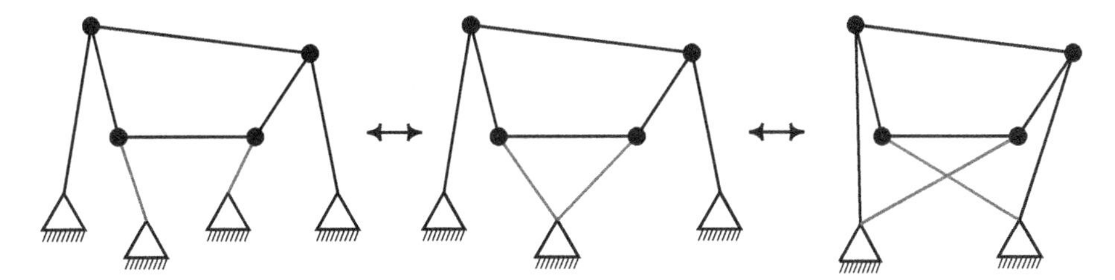

**Figure 18.23**
Pin rearrangement (maintaining at least two pins).

**Theorem 18.29** *All Assur graphs on 5 or more vertices can be obtained from basic Assur graphs by a sequence of edge-splits, pin-rearrangements and 2-sums of smaller Assur graphs.*

There are additional operations under which the class of Assur graphs is closed, which are of interest to the mechanical engineer, for example vertex-split (Figure 18.24).

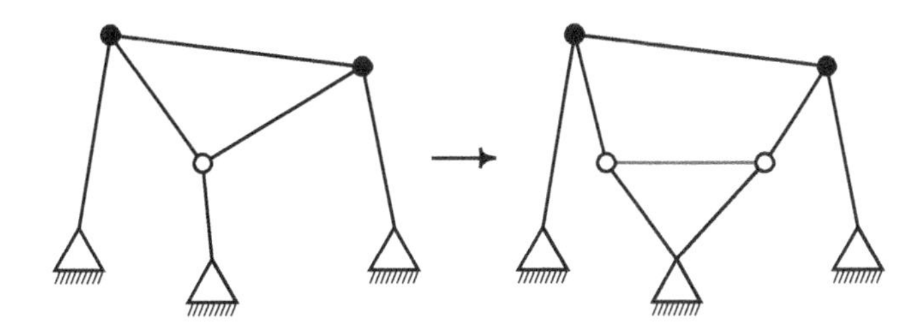

**Figure 18.24**
Vertex split taking an Assur graph to an Assur graph.

The inductive constructions for Assur graphs can be used to provide a visual *certificate sequence* for an Assur graph. If we are given a sequence of edge-splits and 2-sums starting from a dyad and ending with $G$, see Figure 18.25, it is trivial to verify that $G$ is an Assur graph. It is well known, see [59], that there are exponential algorithms to produce such a certificate. However, all algorithms mentioned in section 18.3.2.1 can be adapted to verify the Assur property.

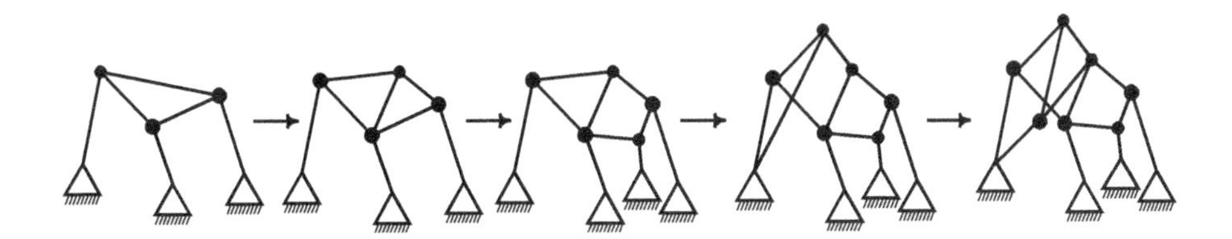

**Figure 18.25**
Certificate sequence for the final Assur graph.

### 18.3.4 Body and Pin Structures

Given a simple graph $G=(V,E)$, we may regard $G$ as a body-and-pin graph of a structure in the plane: The vertices are bodies, edges denote pins. Each pin connects just two bodies. As realizations consider amorphous bodies. An embedding specifies the location of the pins. One possibility for bodies in the plane are (long enough) rods, i.e., line segments. In this case all the pins on a body are collinear and we speak of a pin-collinear body-and-pin-structure. Note that if $G=(V,E)$ is simple, then a pin collinear structure exists. And can be constructed from any generic embedding of the structure graph $G=(V,E)$ in $\mathbb{R}^2$. by forming the polar of that embedding, see Figure 18.26.

Jackson and Jordán show in [21] that the body-and-pin and rod-and-pin 2-polymatroids of a graph are identical.

**Theorem 18.30** *Let $G(V,E)$ be a multigraph. Then the following statements are equivalent:*
*(a) G has a realization as an infinitesimally rigid body-and-hinge framework in $\mathbb{R}^2$.*
*(b) G has a realization as an infinitesimally rigid body-and-hinge framework $(G,q)$ in $\mathbb{R}^2$ with each of the sets of points $\{q(e) : e \in E_G(v)\}, v \in V$, collinear.*
*(c) $2G$ contains 3 edge disjoint spanning trees.*

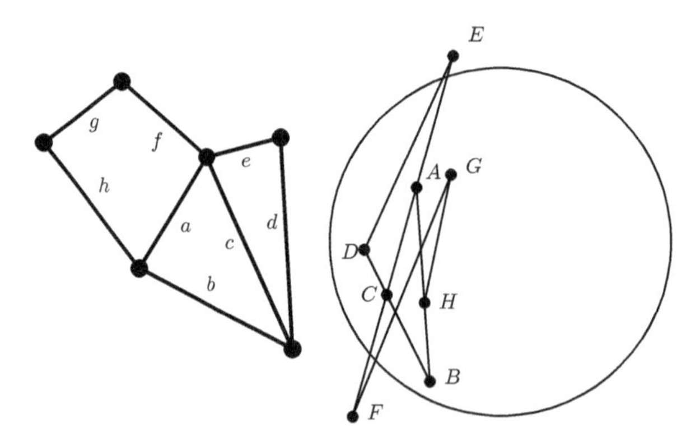


**Figure 18.26**
A graph and its polar with respect to a circle.

Figure 18.26 shows a simple graph. We want to interpret here the vertices as bodies and the edges prescribing incidences between bodies. If the bodies are to be realized as line segments, the polar of the graph $G$ with respect to some conic yields the desired realization. It is rigid, because the graph contains three spanning trees on the edge sets $\{a,b,d,f,h\},\{a,c,e,f,g\},\{b,c,d,g,h\}$, collectively using any edge of $G$ at most twice, which means that $2G$ contains 3 edge disjoint spanning trees.

Theorem 18.30 is a special case of the Tay-Whiteley theorem, [58], where the molecular conjecture was formulated: A multigraph has a generically rigid realization as a hinged structure in $n$-space if and only it has a rigid realization as a hinged structures in n-space with all hinges of body $v_i$ in a hyperplane $H_i$ of the space. The molecular conjecture was proved by Katoh and Tanigawa [28] in all dimensions. Tay, [57], extended the Tay-Whiteley Theorem to the case where each hinge can be shared by more than two bodies. It would be interesting to know if the Katoh-Tanigawa Theorem can also be extended in the same way.

### 18.3.5 Rigidity of Random Graphs

Let $G_{n,d}$ denote the probability space of all $d$-regular graphs on $n$ vertices with the uniform probability distribution. A sequence of graph properties $A_n$ holds asymptotically almost surely, or a.a.s. for short, in $G_{n,d}$ if $\lim_{n\to\infty} \Pr_{G_{n,d}}(A_n) = 1$. Graphs in $G_{n,d}$ are known to be a.a.s. highly connected. It was shown by Bollobás [5] and Wormald [68] that if $G \in G_{n,d}$ for any fixed $d \geq 3$, then $G$ is a.a.s. $d$-connected. This result was extended to all $3 \leq d \leq n-4$ by Cooper et al. [8] and Krivelevich et al. [32]. Stronger results hold if we discount "trivial" cutsets. In [67], Wormald shows that if if $G \in G_{n,d}$ for any fixed $d \geq 3$, then $G$ is a.a.s. cyclically $(3d-6)$-edge-connected. Together with the fact, shown in [23] that if $G = (V,E)$ is a cyclically 5-edge-connected 4-regular graph, then $G$ is globally rigid, this immediately gives:

**Theorem 18.31 ( [23])** *If $G \in G_{n,d}$ and $d \geq 4$ then $G$ is a.a.s. globally rigid.*

For a random $d$-regular graph $G$ on $n$ vertices the expected number, $\mathbb{E}X_j$, of subgraphs $H$ of minimum degree three of order $j$ and size $2j-k$ in an $m$-edged subgraph of $G$ was calculated in [54].

Considering $\log(\mathbb{E}X_j)/n$ as a function of $m/n$ and $j/n$ the graphs in Figure 18.27 for $d = 5$ and $d = 6$ were obtained. To see for which $m/n$ and $j/n$ the expected number $\mathbb{E}X_j$ is roughly one, or $\log(\mathbb{E}X_j)/n = 0$, we have inserted the plane $z = 0$ in the graphs.

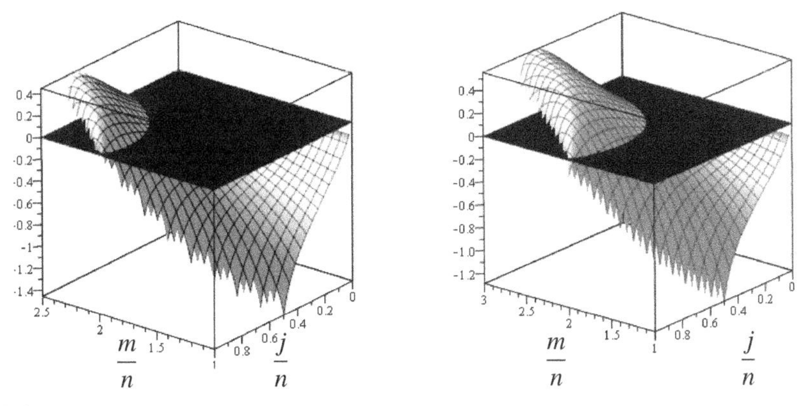


**Figure 18.27**
$\log(\mathbb{E}X_j)$ as function of $m/n$ and $j/n$; $d = 5$ (left), $d = 6$ (right).

From Figure 18.27 one might predict that as edges are percolated randomly into a graph to obtain a $d$-regular graph, that the rigidity phase transition will be first order, since small rigid subgraphs are unlikely.

Let $G(n,p)$ denote the probability space of all graphs on $n$ vertices in which each edge is chosen independently with probability $p$.

**Theorem 18.32 (Jackson, Servatius, and Servatius, 2007 [23])** *Let $G \in G(n,p)$, where $p = (\log n + k \log\log n + w(n))/n$, and $\lim_{n\to\infty} w(n) = \infty$.*
*(a) If $k = 2$ then $G$ is a.a.s. rigid.*
*(b) If $k = 3$ then $G$ is a.a.s. globally rigid.*

Rigidity percolation in this Erdős-Rényi model was studied in [27], where it is shown that there exists a sharp threshold for a giant rigid component to emerge.

## Glossary

**2-tree decomposition:** Partition of the edge set of $G$ into two spanning trees.

**Body pin graph:** A graph in which the vertices represent rigid bodies and an edge indicates that the two bodies are pinned together at a point.

**Grübler count:** An engineering rule of thumb to measure the degree of freedom (ignoring constraint independence) of a system. The Grübler count of a body pin framework in the plane adds 3 for each body and subtracts 2 for each pin, and subtracts 3 for the isometries of the plane.

**Mechanism:** a framework with one internal degree of freedom, ($k$'th order mechanism, with $k$ degrees of freedom, is also sometimes used.)

**Pinned Framework:** A framework with at least two vertices of fixed position.

**proper 2-tree decomposition:** A 2-tree decomposition of $G$ such that no pair of proper subtrees, excepting single vertices, have the same span.

**edge $k$-rigid:** A rigid graph such that the removal of any $k-1$ edges yields a framework with the same type of rigidity.

**edge birigid, edge 2-rigid, redundantly rigid:** A $k$-rigid framework (graph) with $k=2$.

**(vertex) $k$-rigid:** $G$ has at least $k+1$ vertices such that the removal of any $k-1$ vertices, together with their incident edges, leaves a rigid graph.

**(vertex) birigid, vertex 2-rigid:** A vertex 2-rigid framework (graph).

**spanning forest of $G$:** A subgraph of a graph $G$ which is cycle free and contains all the vertices of $G$. Its connected components are trees and isolated vertices.

**submodular function:** Function defined on $\mathcal{P}(E)$ satisfying $\phi(F_1 \cup F_2) + \phi(F_1 \cap F_2) \leq \phi(F_1) + \phi(F_2)$.

## References


[1] Alex R. Berg and Tibor Jordán. A proof of Connelly's conjecture on 3-connected circuits of the rigidity matroid. *J. Combin. Theory Ser. B*, 88(1):77–97, 2003.

[2] E. D. Bolker and B. Roth. When is a bipartite graph a rigid framework? *Pacific J. Math.*, 90(1):27–44, 1980.

[3] Ethan D. Bolker. Bracing rectangular frameworks. II. *SIAM J. Appl. Math.*, 36(3):491–508, 1979.

[4] Ethan D. Bolker and Henry Crapo. Bracing rectangular frameworks. I. *SIAM J. Appl. Math.*, 36(3):473–490, 1979.

[5] Béla Bollobás. *Random graphs*, volume 73 of *Cambridge Studies in Advanced Mathematics*. Cambridge University Press, Cambridge, second edition, 2001.

[6] Robert Connelly. Rigidity. In *Handbook of convex geometry, Vol. A, B*, pages 223–271. North-Holland, Amsterdam, 1993.

[7] Robert Connelly and Herman Servatius. Higher-order rigidity—what is the proper definition? *Discrete Comput. Geom.*, 11(2):193–200, 1994.

[8] Colin Cooper, Alan Frieze, and Bruce Reed. Random regular graphs of non-constant degree: connectivity and Hamiltonicity. *Combin. Probab. Comput.*, 11(3):249–261, 2002.

[9] Henry Crapo. On the generic rigidity of plane frameworks. Research Report RR-1278, INRIA, August 1990. Projet ICSLA.

[10] Luigi Cremona. Le figure reciproche. *Civilità delle Macchine*, 4(5):55–62, 1956.

[11] Reinhard Diestel. *Graph theory*, volume 173 of *Graduate Texts in Mathematics*. Springer, Heidelberg, fourth edition, 2010.

[12] Jack Edmonds. Minimum partition of a matroid into independent subsets. *J. Res. Nat. Bur. Standards Sect. B*, 69B:67–72, 1965.

[13] James Farre, Helena Kleinschmidt, Jessica Sidman, Audrey St. John, Stephanie Stark, Louis Theran, and Xilin Yu. Algorithms for detecting dependencies and rigid subsystems for CAD. *Comput. Aided Geom. Design*, 47:130–149, 2016.

[14] F. G. Frobenius. Über zerlegbare Determinanten. *Sitzungsberichte der Berl. Akademie*, XVIII, 1917.

[15] Jack Graver, Brigitte Servatius, and Herman Servatius. *Combinatorial rigidity*, volume 2 of *Graduate Studies in Mathematics*. American Mathematical Society, Providence, RI, 1993.

[16] Jack E. Graver. Rigidity matroids. *SIAM J. Discrete Math.*, 4(3):355–368, 1991.

[17] Jack E. Graver, Brigitte Servatius, and Herman Servatius. Abstract rigidity in $m$-space. In *Jerusalem combinatorics '93*, volume 178 of *Contemp. Math.*, pages 145–151. Amer. Math. Soc., Providence, RI, 1994.

[18] Martin Grötschel, László Lovász, and Alexander Schrijver. *Geometric algorithms and combinatorial optimization*, volume 2 of *Algorithms and Combinatorics: Study and Research Texts*. Springer-Verlag, Berlin, 1988.

[19] E. W. Hobson, H. P. Hudson, A. N. Singh, and A. B. Kempe. *Squaring the Circle and Other Monographs*. Chelsea Publishing Company, New York, NY, 1953.

[20] Bill Jackson and Tibor Jordán. Connected rigidity matroids and unique realizations of graphs. *J. Combin. Theory Ser. B*, 94(1):1–29, 2005.

[21] Bill Jackson and Tibor Jordán. On the rigidity of molecular graphs. *Combinatorica*, 28(6):645–658, 2008.

[22] Bill Jackson and Tibor Jordán. A sufficient connectivity condition for generic rigidity in the plane. *Discrete Appl. Math.*, 157(8):1965–1968, 2009.

[23] Bill Jackson, Brigitte Servatius, and Herman Servatius. The 2-dimensional rigidity of certain families of graphs. *J. Graph Theory*, 54(2):154–166, 2007.

[24] Donald J. Jacobs and Bruce Hendrickson. An algorithm for two-dimensional rigidity percolation: the pebble game. *J. Comput. Phys.*, 137(2):346–365, 1997.

[25] Tibor Jordán and Viktória E. Kaszanitzky. Highly connected rigidity matroids have unique underlying graphs. *European J. Combin.*, 34(2):240–247, 2013.

[26] Gil Kalai. Hyperconnectivity of graphs. *Graphs Combin.*, 1(1):65–79, 1985.

[27] Shiva Prasad Kasiviswanathan, Cristopher Moore, and Louis Theran. The rigidity transition in random graphs. In *Proceedings of the Twenty-Second Annual ACM-SIAM Symposium on Discrete Algorithms*, pages 1237–1252. SIAM, Philadelphia, PA, 2011.

[28] Naoki Katoh and Shin-ichi Tanigawa. A proof of the molecular conjecture. *Discrete Comput. Geom.*, 45(4):647–700, 2011.

[29] Shuji Kijima and Shin-ichi Tanigawa. Sparsity and connectivity of medial graphs: concerning two edge-disjoint Hamiltonian paths in planar rigidity circuits. *Discrete Math.*, 312(16):2466–2472, 2012.

[30] Csaba Király and Ferenc Péterfalvi. Balanced generic circuits without long paths. *Discrete Math.*, 312(15):2262–2271, 2012.

[31] E. Kötter. Über die Möglichkeit $n$ Punkte in der Ebene oder im Raum durch weniger als $2n-3$ oder $3n-6$ Stäbe von ganz unveränderlicher Länge unverschieblich miteinander zu verbinden. *Festschrift für H. Mueller-Breslau*, 1912.

[32] Michael Krivelevich, Benny Sudakov, Van H. Vu, and Nicholas C. Wormald. Random regular graphs of high degree. *Random Structures Algorithms*, 18(4):346–363, 2001.

[33] G. Laman. On graphs and rigidity of plane skeletal structures. *J. Engrg. Math.*, 4:331–340, 1970.

[34] Audrey Lee and Ileana Streinu. Pebble game algorithms and sparse graphs. *Discrete Math.*, 308(8):1425–1437, 2008.

[35] L. Lovász and Y. Yemini. On generic rigidity in the plane. *SIAM J. Algebraic Discrete Methods*, 3(1):91–98, 1982.

[36] H. Maehara and N. Tokushige. When does a planar bipartite framework admit a continuous deformation? *Theoret. Comput. Sci.*, 263(1-2):345–354, 2001. Combinatorics and computer science (Palaiseau, 1997).

[37] J. Clerk Maxwell. On Reciprocal Diagrams in Space, and their relation to Airy's Function of Stress. *Proc. London Math. Soc.*, S1-2(1):58.

[38] John Milnor. *Singular points of complex hypersurfaces*. Annals of Mathematics Studies, No. 61. Princeton University Press, Princeton, NJ, 1968.

[39] C. St. J. A. Nash-Williams. Edge-disjoint spanning trees of finite graphs. *J. London Math. Soc.*, 36:445–450, 1961.

[40] C. St. J. A. Nash-Williams. Decomposition of finite graphs into forests. *J. London Math. Soc.*, 39:12, 1964.

[41] Viet-Hang Nguyen. On abstract rigidity matroids. *SIAM J. Discrete Math.*, 24(2):363–369, 2010.

[42] Robert L. Norton. *Design of Machinery: An Introduction To The Synthesis and Analysis of Mechanisms and Machines*. McGraw Hill, New York, 2004.

[43] James G. Oxley. *Matroid theory*. Oxford Science Publications. The Clarendon Press, Oxford University Press, New York, 1992.

[44] Sachin Patkar, Brigitte Servatius, and K. V. Subrahmanyam. Abstract and generic rigidity in the plane. *J. Combin. Theory Ser. B*, 62(1):107–113, 1994.

[45] Kevin Peterson. The stress spaces of bipartite frameworks. *Pacific J. Math.*, 197(1):173–182, 2001.

[46] H. Pollaczek-Geiringer. Über die Gliederung ebener Fachwerke. *ZAMM - Journal of Applied Mathematics and Mechanics / Zeitschrift für Angewandte Mathematik und Mechanik*, 7(1):58–72, 1927.

[47] András Recski. A network theory approach to the rigidity of skeletal structures. I. Modelling and interconnection. *Discrete Appl. Math.*, 7(3):313–324, 1984.

[48] András Recski. A network theory approach to the rigidity of skeletal structures. II. Laman's theorem and topological formulae. *Discrete Appl. Math.*, 8(1):63–68, 1984.

[49] András Recski. Bracing cubic grids—a necessary condition. In *Proceedings of the Oberwolfach Meeting "Kombinatorik" (1986)*, volume 73, pages 199–206, 1989.

[50] András Recski. *Matroid theory and its applications in electric network theory and in statics*, volume 6 of *Algorithms and Combinatorics*. Springer-Verlag, Berlin; Akadémiai Kiadó (Publishing House of the Hungarian Academy of Sciences), Budapest, 1989.

[51] Brigitte Servatius and Peter R. Christopher. Construction of self-dual graphs. *Amer. Math. Monthly*, 99(2):153–158, 1992.

[52] Brigitte Servatius and Herman Servatius. On the 2-sum in rigidity matroids. *European J. Combin.*, 32(6):931–936, 2011.

[53] Brigitte Servatius, Offer Shai, and Walter Whiteley. Geometric properties of Assur graphs. *European J. Combin.*, 31(4):1105–1120, 2010.

[54] Brigitte Servatius and Nicholas Wormald. On the size and number of rigid subgraphs of d-regular graphs. *Preprint.*

[55] Ileana Streinu and Louis Theran. Sparse hypergraphs and pebble game algorithms. *European J. Combin.*, 30(8):1944–1964, 2009.

[56] Ileana Streinu and Louis Theran. Sparsity-certifying graph decompositions. *Graphs Combin.*, 25(2):219–238, 2009.

[57] Tiong-Seng Tay. Linking $(n-2)$-dimensional panels in $n$-space. II. $(n-2,2)$-frameworks and body and hinge structures. *Graphs Combin.*, 5(3):245–273, 1989.

[58] Tiong-Seng Tay and Walter Whiteley. Recent advances in the generic rigidity of structures. *Structural Topology*, (9):31–38, 1984. Dual French-English text.

[59] Tiong-Seng Tay and Walter Whiteley. Generating isostatic frameworks. *Structural Topology*, (11):21–69, 1985. Dual French-English text.

[60] W. T. Tutte. On the problem of decomposing a graph into $n$ connected factors. *J. London Math. Soc.*, 36:221–230, 1961.

[61] W. T. Tutte. *Connectivity in graphs*. Mathematical Expositions, No. 15. University of Toronto Press, Toronto, Ont.; Oxford University Press, London, 1966.

[62] W. T. Tutte. Connectivity in matroids. *Canad. J. Math.*, 18:1301–1324, 1966.

[63] D. J. A. Welsh. *Matroid theory*. Academic Press [Harcourt Brace Jovanovich, Publishers], London-New York, 1976. L. M. S. Monographs, No. 8.

[64] Walter Whiteley. Matroids and rigid structures. In *Matroid applications*, volume 40 of *Encyclopedia Math. Appl.*, pages 1–53. Cambridge Univ. Press, Cambridge, 1992.

[65] Walter Whiteley. Some matroids from discrete applied geometry. In *Matroid theory (Seattle, WA, 1995)*, volume 197 of *Contemp. Math.*, pages 171–311. Amer. Math. Soc., Providence, RI, 1996.

[66] Hassler Whitney. 2-Isomorphic Graphs. *Amer. J. Math.*, 55(1-4):245–254, 1933.

[67] N. C. Wormald. Models of random regular graphs. In *Surveys in combinatorics, 1999 (Canterbury)*, volume 267 of *London Math. Soc. Lecture Note Ser.*, pages 239–298. Cambridge Univ. Press, Cambridge, 1999.

[68] Nicholas C. Wormald. The asymptotic connectivity of labelled regular graphs. *J. Combin. Theory Ser. B*, 31(2):156–167, 1981.

# Chapter 19

# *Inductive Constructions for Combinatorial Local and Global Rigidity*

**Anthony Nixon**

*Department of Mathematics and Statistics, Lancaster University, U.K.*

**Elissa Ross**

*MESH Consultants Inc., Fields Institute, Toronto, Canada*

**CONTENTS**



## 19.1 Introduction

A bar-joint framework is a geometric realization of a graph. Formally, a *framework* is a pair $(G,p)$ where $G=(V,E)$ is a (finite, simple) graph and $p:V\to\mathbb{R}^d$. The framework $(G,p)$ is *rigid* if every edge-length-preserving continuous deformation of the vertices arises from an isometry of $\mathbb{R}^d$ (see Chapter 18). A stronger condition is *global rigidity*, where $(G,p)$ is the only realization of $G$ in $\mathbb{R}^d$, up to isometries, with the edge lengths prescribed by $p$ (see Chapter 21).

Determining the rigidity, or global rigidity, of a given framework is NP-hard [1, 69]. However, the situation improves for generic frameworks where one can linearize the problem and characterize generic rigidity via the rank of the rigidity matrix [3]. A framework $(G,p)$ is *generic* if the coordinates of $p$ form an algebraically independent set over $\mathbb{Q}$.

A key topic in rigidity theory, perhaps the fundamental topic, is to characterize generic rigidity, and generic global rigidity, in purely combinatorial terms. That is, we seek to characterize generic rigidity as a property of the graph alone. Cornerstone theorems in rigidity theory are such characterizations of rigidity [51] and global rigidity [8, 28] for generic frameworks in $\mathbb{R}^2$. In both cases the proof was obtained via inductive, or recursive, constructions.

Our focus will be on such constructions both for these fundamental results and for related problems in various areas of rigidity theory. Since the first of these results in 1970 [51], there has been a multitude of papers using inductive constructions for a variety of problems. We will briefly describe a number of these. (In a number of the results we state, the geometric operations do not need the full strength of generic; general position, where no $d+1$ points define a $(d-1)$-dimensional hyperplane often suffices.) One of our aims is to show the versatility of proofs via inductive construction. In particular, how similar arguments can be adapted to seemingly disparate areas of rigidity theory.

An inductive construction has two distinct aspects to it. The first aspect is to prove a constructive characterization of the class of graphs. That is, to show that a short list of (typically) local operations is sufficient to characterize a class of graphs, in the sense that every graph in the class, and no others, can be built recursively from a (or a few) "base graph(s)" using only these local operations. A key starting point in such characterizations is often to establish what the minimum degree is for such a graph and for each possibility provide a way of adding a vertex with such a degree. The second aspect then considers these graph operations as framework operations. Here the goal is to show that the operation preserves generic rigidity. This is done by induction, first by showing directly that a generic realization of the base graph is rigid and then by showing that applying any of the operations to a rigid framework results in a larger rigid framework. A typical technique for this is to use the matrix characterization of rigidity and often involves finding a special, non-generic, realization that is rigid and then using the fact that generic frameworks have maximal rank to conclude the rigidity of a generic realization.

Alongside the other chapters in this book, we direct the reader to [19, 39, 62, 84, 85] for a wealth of further information and references about rigidity and to [16, 49] for details on related constructive characterizations in combinatorial optimization. Since we will not put any focus on 1-dimensional frameworks we direct the interested reader to [27].

We close the introduction by briefly outlining what follows. In Section 19.2 we describe results relevant to the rigidity and global rigidity of generic (bar-joint) frameworks in $\mathbb{R}^d$. We separate these results into three kinds: geometric operations on frameworks; combinatorial results about graphs; and characterizations of rigidity in combinatorial terms (for this last point we can only discuss $\mathbb{R}^2$). In Section 19.3 we repeat this 3-step analysis for special classes of frameworks, such as body-bar frameworks in $\mathbb{R}^d$ where we have a much fuller combinatorial understanding of rigidity. In Section 19.4 we try to survey all corners of the rigidity literature that have been influenced by inductive construction techniques. Necessarily this means that, in most cases, we give little more than references for the interested reader to follow.

*Notation:* $G=(V,E)$ is a finite simple graph. When loops and multiple edges are permitted we will explicitly use the term multigraph.

## 19.2 Rigidity in $\mathbb{R}^d$

In this section we will describe results relevant to inductive constructions used in the study of rigidity of bar-joint frameworks in Euclidean spaces. We split our analysis into two sections: geometric and combinatorial.

**0-extension:** The operation of ($d$-dimensional) *0-extension* forms a new graph $G'$ from $G$ by adding a new vertex $v_{n+1}$ with $d$ edges incident to distinct vertices of $G$.

**1-extension:** The operation of ($d$-dimensional) *1-extension* forms a new graph $G'$ from $G$ by deleting an edge $v_iv_j$ and adding a new vertex $v_{n+1}$ and $d+1$ new edges incident to $v_i, v_j$ and $d-1$ further distinct vertices of $G$.

**0-reduction:** The operation of ($d$-dimensional) *0-reduction* forms a new graph $G'$ from $G$ by deleting a degree $d$ vertex $v$ from $G$.

**1-reduction:** The operation of ($d$-dimensional) *1-reduction* forms a new graph $G'$ from $G$ by deleting a degree $d+1$ vertex $v$ from $G$ and adding an edge between two vertices formerly adjacent to $v$.

**$X$-replacement:** The operation of ($d$-dimensional) *$X$-replacement* forms a new graph $G'$ from $G$ by deleting two non-adjacent edges $v_iv_j, v_kv_\ell$ and adding a new vertex $v_{n+1}$ and $d+2$ new edges incident to $v_i, v_j, v_k, v_\ell$ and $d-2$ further distinct vertices of $G$.

**vertex splitting:** Let $v \in V$ have $N(v) = \{u_1, \ldots, u_m\}$. A *vertex splitting* operation (in $d$-dimensions) on $v$ removes $v$ and its incident edges and adds vertices $v_0, v_1$ and edges $u_1v_0, u_2v_0, u_1v_1, u_2v_1, \ldots, u_{d-1}v_0, u_{d-1}v_1, v_0v_1$ and rearranges the edges $u_dv, \ldots, u_mv$ in some way into edges $u_iv_j$ for $i \in \{d, \ldots, m\}$ and $j \in \{0, 1\}$.

For each of the above definitions, the dimension corresponds to the dimension we intend to use these graph operations as operations on frameworks. Since this is usually clear from the context we typically drop "$d$-dimensional" for brevity when referring to these operations.

**$V$-replacement:** The operation of *$V$-replacement* (note only applicable in 3-dimensions) forms a new graph $G'$ from $G$ by deleting two adjacent edges $v_iv_j, v_jv_k$ and adding a new vertex $v_{n+1}$ and 5 new edges incident to $v_i, v_j, v_k$ and 2 further distinct vertices of $G$.

**vertex-to-4-cycle:** Let $v \in V$ have $N(v) = \{u_1, \ldots, u_m\}$. A *vertex-to-4-cycle* operation (note only applicable in 2-dimensions) on $v$ removes $v$ and its incident edges and adds vertices $v_0, v_1$ and edges $u_1v_1, v_1u_2, u_2v_0, v_0u_1$ and rearranges the edges $u_3v, \ldots, u_mv$ in some way into edges $u_iv_j$ for $i \in \{3, \ldots, m\}$ and $j \in \{0, 1, \}$.

Our remaining definitions concern graphs.

**$(2,3)$-sparse:** $G$ is *$(2,3)$-sparse* if $|E'| \leq 2|V'| - 3$ for all subgraphs $(V', E')$ of $G$ with $|E'| > 0$.

**$(2,3)$-tight:** $G$ is *$(2,3)$-tight* if $|E| = 2|V| - 3$ and $G$ is $(2,3)$-sparse.

**$(k,\ell)$-sparse:** $G$ is *$(k,\ell)$-sparse* if $|E'| \leq k|V'| - \ell$ for all subgraphs $(V', E')$ of $G$ with $|V'| \geq k$.

**$(k,\ell)$-tight:** $G$ is *$(k,\ell)$-tight* if $|E| = k|V| - \ell$ and $G$ is $(k,\ell)$-sparse.

**$(2,3)$-circuit:** $G$ is a *$(2,3)$-circuit* if $|E| = 2|V| - 2$ and every proper subgraph of $G$ is $(2,3)$-sparse.

**admissible:** A degree 3 vertex $v$ in a $(2,3)$-circuit is *admissible* if there is a 1-reduction at $v$ which results in a $(2,3)$-circuit.

**feasible:** A degree 3 vertex $v$ in a 3(-vertex)-connected $(2,3)$-circuit is *feasible* if there is a 1-reduction at $v$ which results in a 3-connected $(2,3)$-circuit.

**redundantly rigid graph:** $G$ is *redundantly rigid* if $G - e$ contains a spanning $(2,3)$-tight subgraph for all $e \in E$.

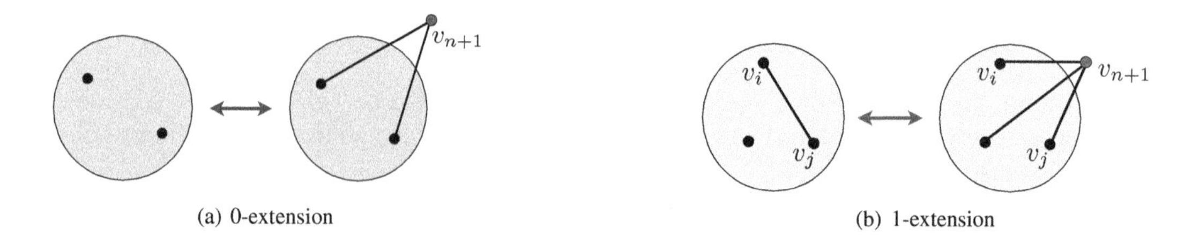


(a) 0-extension

(b) 1-extension

**Figure 19.1**
The fundamental moves in two dimensions.

**planar 1-extension:** A 1-extension is *planar* if $G$ and $G'$ are both planar.

**planar vertex splitting:** A vertex split is *planar* if $G$ and $G'$ are both planar.

**Laman plus one graph:** $G$ is a *Laman plus one graph* if there exists $e \in E$ such that $G - e$ is $(2,3)$-tight.

**2-sum:** A *2-sum* of two graphs $G_i = (V_i, E_i)$, for $i = 1,2$, over the pair $x, y$ forms the graph $G = (V, E)$ where $V = V_1 \cup V_2$ with $V_1 \cap V_2 = \{x, y\}$, $xy \in E_i$ for $i = 1,2$ and $E = (E_1 \cup E_2) - xy$. A 2-separation is the inverse of a 2-sum.

***M*-connected:** A graph $G$ is *M-connected* if the rigidity matroid of a generic framework $(G, p)$ is connected.

### 19.2.1 Inductive Operations on Frameworks

In this subsection we will consider a variety of local operations on graphs and when they are known to preserve the rigidity or global rigidity of frameworks. Our focus will be on rigidity (and global rigidity) of bar-joint frameworks in $\mathbb{R}^d$.

0- and 1-extensions (see Figure 19.1) are so fundamental they have been described in the literature under a number of different names, such as *Henneberg*-1 *and* -2 moves [24], or *vertex addition* and *edge-splitting* [84]. Together they may be called *Henneberg moves* [24]. Both are well known to preserve rigidity in all dimensions. The first is trivially proved via considering the rank of a matrix. The second has been proved in several different ways, see, for example, [51, 79].

***Lemma 19.1***
*Let $(G, p)$ be a rigid framework in $\mathbb{R}^d$. Let $G'$ be formed from $G$ by a 0-extension. Then $(G', p')$ is generically rigid in $\mathbb{R}^d$.*

***Lemma 19.2***
*Let $(G, p)$ be a generically rigid framework in $\mathbb{R}^d$. Let $G'$ be formed from $G$ by a 1-extension. Then $(G', p')$ is generically rigid in $\mathbb{R}^d$.*

Several proofs of Lemma 19.2 can be extended to prove our next result, which is only known in $\mathbb{R}^2$. Indeed an example from [19] confirms the analogue in dimension at least 4 is false.

***Lemma 19.3***
*Let $(G, p)$ be a generically rigid framework in $\mathbb{R}^2$. Let $G'$ be formed from $G$ by $X$-replacement. Then $(G', p')$ is generically rigid in $\mathbb{R}^2$.*

**Figure 19.2**
$X$-replacement (a) and $V$-replacement (b) in three dimensions.

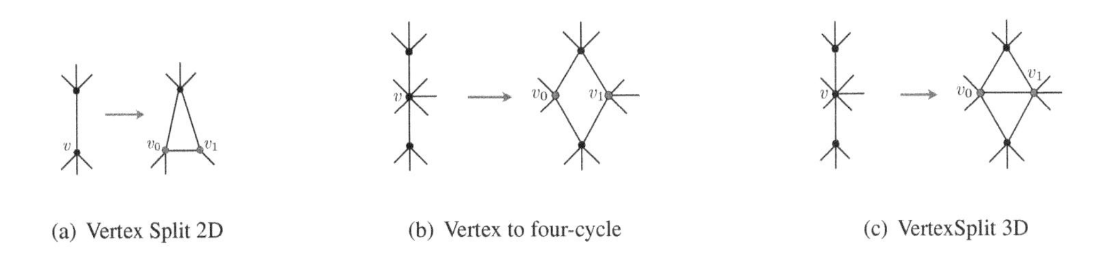


**Figure 19.3**
Vertex splitting in two and three dimensions, and the two-dimensional vertex-to-four-cycle move.

The 3-dimensional version of $X$-replacement (see Figure 19.2(a)) is unclear.

**Conjecture 3 ([79])** *Let $(G,p)$ be a generically rigid framework in $\mathbb{R}^3$. Let $G'$ be formed from $G$ by $X$-replacement. Then $(G',p')$ is generically rigid in $\mathbb{R}^3$.*

A number of special cases have now been proved [10, 19, 26], however the general case still seems to be difficult. A resolution to this conjecture would shed great light on the 3D rigidity problem. The sister operation, $V$-replacement (see Figure 19.2(b)), is known not to preserve rigidity but for purely graph theoretical reasons (for example it may turn a degree 3 vertex at the point of the $V$ into a degree 2 vertex). Tay and Whiteley [79] proposed a "double $V$" conjecture which gets around the obvious combinatorial defect and leaves a geometric conjecture analogous to Conjecture 3. Positive resolution of both of these conjectures would lead to a combinatorial characterization of generic rigidity in $\mathbb{R}^3$.

We mention one more fundamental operation which is known to preserve rigidity in all dimensions, Whiteley's vertex splitting operation, Figure 19.3(a).

***Proposition 19.1 Whiteley [83]***
*Let $(G,p)$ be a generically rigid framework in $\mathbb{R}^d$. Let $G'$ be formed from $G$ by vertex splitting. Then $(G',p')$ is generically rigid in $\mathbb{R}^d$.*

Vertex splitting was also crucial in work of Finbow-Singh, Ross and Whiteley analyzing block and hole polyhedra [15]. There has been recent work on triangulated surfaces in [11] using inductive techniques (in particular vertex splitting) to extend this. Also recently global rigidity has been characterized for a number of triangulated surfaces (sphere, torus, projective plane) again using vertex splitting [44].

A variant of the 2-dimensional version of vertex splitting, which we call a vertex-to-4-cycle move (see Figure 19.3(b)), is also known to preserve rigidity.

***Proposition 19.2 Lomeli, Moshe and Whiteley [50]***
*Let $(G,p)$ be a generically rigid framework in $\mathbb{R}^2$. Let $G'$ be formed from $G$ by a vertex-to-4-cycle move. Then $(G',p')$ is generically rigid in $\mathbb{R}^2$.*

This move has also been extended to other rigidity contexts [58].

We now move on to consider global rigidity. It is quickly apparent that 0-extension does not preserve global rigidity. However 1-extension does.

**Theorem 19.1 (Connelly [8])** *Let $(G,p)$ be a generically globally rigid framework in $\mathbb{R}^d$. Let $G'$ be formed from $G$ by 1-extension. Then $(G',p')$ is generically globally rigid in $\mathbb{R}^d$.*

Here the proof is far more intricate than the proof of Lemma 19.2. Connelly [8] proved this result as a corollary of his sufficient condition for a generic framework to be globally rigid. He did this by utilizing equilibrium stresses (also known as self-stresses) and, in particular, a natural stress matrix whose rank being maximal, along with the framework being rigid, guarantees global rigidity. An alternative, more direct proof, was later given by Jackson, Jordán, and Szabadka [33] in their work on globally linked pairs. Both proofs rely essentially on the coordinates of $p$ being algebraically independent.

It is easy to check that $(d+1)$-connectivity can fail when we apply $X$-replacement and, this being a necessary condition for global rigidity [22], $X$-replacement does not preserve global rigidity.

Vertex splitting can create a degree $d$ vertex so clearly does not, in general, preserve global rigidity. Ruling this out is known to be sufficient for vertex splitting to preserve generic global rigidity in $\mathbb{R}^2$. However the proof, found in [43], is as a consequence of combinatorial characterizations of rigidity and global rigidity (see Subsection 19.2.3). In higher dimensions the problem is open.

**Conjecture 4 (Connelly and Whiteley [7])** *Let $(G,p)$ be a generically globally rigid framework in $\mathbb{R}^d$. Let $G'$ be formed from $G$ by a vertex splitting operation such that $G'$ has minimum degree $d+1$. Then $(G',p')$ is generically globally rigid in $\mathbb{R}^d$.*

Insight into why this is difficult is given by Connelly in [9]. In particular he proves a partial result [9, Theorem 29] which reduces the problem to proving that $(G',p')$ is generically redundantly rigid (when $(G,p)$ is generically globally rigid). There has been very recent progress toward solving Conjecture 4. Jordán and Tanigawa [44] have used a "non-degenerate stress" idea to prove that every graph generated from $K_{d+2}$ by a sequence of vertex splitting operations (that preserve the minimum degree requirement) is globally rigid.

Finally let us mention that Connelly proved a method of combining two generically globally rigid frameworks in dimension $d$, by identifying $d+1$ common vertices, that results in a generically globally rigid framework [5].

### 19.2.2 Recursive Characterizations of Graphs

In this subsection we consider our operations graph theoretically. The key aim being to understand intricate classes of graphs by building any member using these simple operations. The majority of this subsection focuses on two dimensions, corresponding results on the line are typically easy and analogues for dimension $\geq 3$ are only partially understood. We do discuss higher dimensions: for bar-joint frameworks briefly at the end of this subsection; and for special classes of frameworks in the next section.

Our first result shows that 0- and 1-extensions are sufficient to characterize the class of graphs relevant to rigidity in $\mathbb{R}^2$.

**Theorem 19.2 (Henneberg, Laman [24, 51])** *$G$ is $(2,3)$-tight if and only if $G$ can be generated recursively from $K_2$ using only the operations of 0- and 1-extension.*

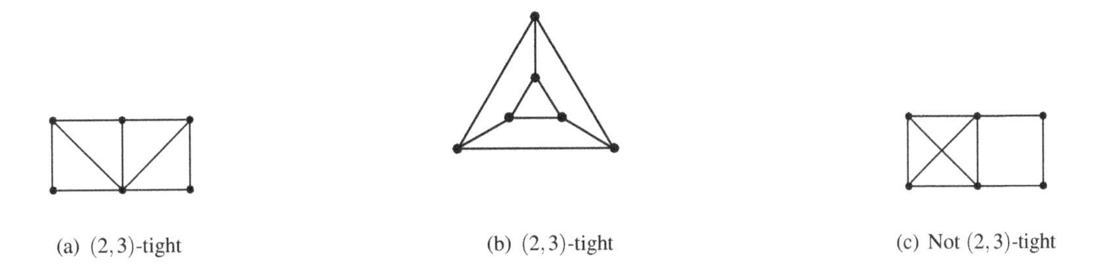


**Figure 19.4**
Examples (and non-examples) of $(2,3)$-tight graphs.

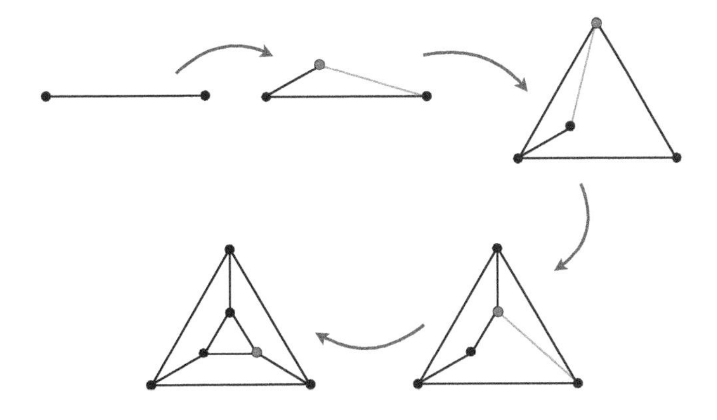

**Figure 19.5**
Sample construction sequence of the triangular prism: 0-extension followed by three 1-extensions.

Jackson and Jordán [28] mildly strengthened this result by proving that a $(2,3)$-tight graph can be generated from any $(2,3)$-tight subgraph using 0- and 1-extensions.

The proof of Theorem 19.2, starts with the simple fact that every $(2,3)$-tight graph, with $|V| \geq 3$, has minimum degree in the set $\{2,3\}$. Consider a vertex of minimum degree. If it is degree 2, apply a 0-reduction and the result is always $(2,3)$-tight. If it is degree 3, we need a modicum of care: there are three possible new edges in a 1-reduction and it may be that only one of those results in a $(2,3)$-tight graph. Nevertheless every degree 3 vertex is reducible; for example if every possible new edge already exists, there is a copy of $K_4$ and hence the graph is not $(2,3)$-tight, (Figure 19.4(c)). Figure 19.5 constructs the triangular prism from $K_2$.

This result has been extended to prove the stronger fact that we can generate all planar $(2,3)$-tight graphs using these operations: insisting that every intermediate graph is also planar. (To do this we can no longer apply arbitrary 1-extensions, but 1-extensions that preserve planarity suffice.)

**Theorem 19.3 (Haas et al [21])** *G is planar and $(2,3)$-tight if and only if G can be generated recursively from $K_2$ using only the operations of planar 0- and planar 1-extension.*

For planar $(2,3)$-tight graphs there is an alternative characterization using vertex splitting instead. Here the reduction relies on the guarantee of a triangular face (in fact two triangular faces are guaranteed) in a planar $(2,3)$-tight graph distinct from $K_2$.

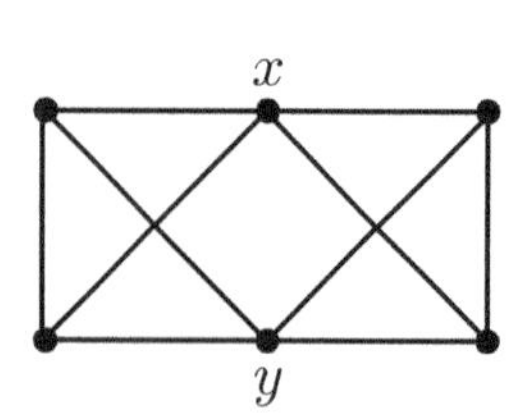

(a) A $(2,3)$-circuit with no admissible degree 3 vertex.

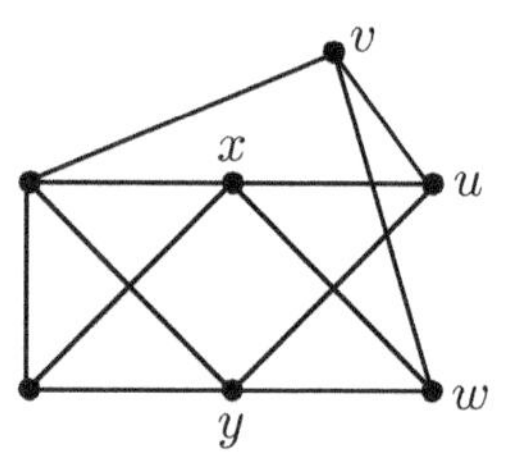

(b) In this $(2,3)$-circuit, $v$ is admissible but not feasible whereas $u$ and $w$ are both feasible.

**Figure 19.6**
Examples of $(2,3)$-circuits.

**Theorem 19.4 (Fekete, Jordan and Whiteley [13])** *G is planar and* $(2,3)$*-tight if and only if G can be generated recursively from* $K_2$ *using only planar vertex splitting.*

This result was also implicit in Owen and Power's [65] study of 3-connected, planar $(2,3)$-tight graphs; which they showed could be generated using either vertex splitting or one other move. In [2] an analogue is given for planar $(2,2)$-tight graphs (using the vertex-to-$K_4$ move, the inverse of the operation of contracting a copy of $K_4$, in addition to vertex splitting) and used to prove results about contacts of circular arcs in the plane.

The 0- and 1-extension operations are even sufficient to characterize the following special class of $(2,2)$-tight graphs.

**Theorem 19.5 (Haas et al [21])** *G is Laman plus one if and only if G can be generated recursively from* $K_4$ *using only the operations of 0- and 1-extension.*

For a related class, the $(2,3)$-circuits, the minimum degree is 3 so 0-extension is of no use. Instead we have the following theorem, which was the first combinatorial step towards understanding globally rigid graphs in the plane. Note that it is not true that every degree 3 vertex can be reduced. For example the graph in Figure 19.6(a) has no admissible vertex. The key to the theorem is isolating when some degree 3 vertex can be reduced.

**Theorem 19.6 (Berg and Jordán [4])** *Every* $(2,3)$*-circuit, distinct from* $K_4$*, contains at least 3 admissible degree 3 vertices.*

In the following the 2-sum operation, as is consistent with its usage in matroid theory, glues two $(2,3)$-circuits together along an edge and deletes the common edge. Servatius and Servatius [73] showed further that the rigidity matroid is not closed under 2-sum decomposition.

**Theorem 19.7 (Berg and Jordán [4])** *G is a* $(2,3)$*-circuit if and only if G can be generated recursively from disjoint copies of* $K_4$ *by applying 1-extensions within connected components and taking 2-sums of connected components.*

Berg and Jordán also extended their result to show that 1-extension alone is sufficient to generate all 3-connected $(2,3)$-circuits. The key being that they extended Theorem 19.6 to guarantee at least 2 feasible nodes. (Figure 19.6(b) gives an example with an admissible node $v$ which is not feasible; the example does, though, contain feasible nodes.)

**Theorem 19.8 (Berg and Jordán [4])** *Every 3-connected* $(2,3)$*-circuit can be generated from* $K_4$ *by 1-extensions.*

A detailed argument using ear decompositions of the $(2,3)$-sparse matroid extended this to the following crucial characterization.

**Theorem 19.9 (Jackson and Jordán [28])** *Let $G$ be $M$-connected and let $G'$ be formed from $G$ by a 1-extension. Then $G'$ is $M$-connected. Moreover if $G$ is $M$-connected and 3-connected then $G$ can be generated from $K_4$ by 1-extensions and edge additions.*

This quickly implies the following result.

**Theorem 19.10 (Jackson and Jordán [28])** *$G$ is 3-connected and redundantly rigid if and only if $G$ can be generated recursively from $K_4$ using only the operations of 1-extension and edge addition.*

To understand the difficulty in extending Theorem 19.8 to Theorem 19.9, it is instructive to note, first, that there are 3-connected redundantly rigid graphs which do not contain a spanning $(2,3)$-circuit and, second, that it is not immediately clear that every 3-connected and "minimally" $M$-connected graph even contains a vertex of degree 3.

It seems to be hard to find a recursive construction for the class of redundantly rigid graphs. Weaken this a little and say a graph $G$ (with an associated generic framework $(G,p)$) is redundant if every edge is in a $(2,3)$-circuit. Then an inductive construction of redundant graphs was given in [42].

### 19.2.3 Combinatorial Characterizations of Rigidity

For the two main rigidity theorems in $\mathbb{R}^2$ let us briefly discuss how the inductive proof proceeds. In most similar results, a proof by inductive construction has the same style.

**Theorem 19.11 (Laman [51])** *A generic framework $(G,p)$ in $\mathbb{R}^2$ is rigid if and only if it contains a spanning subgraph which is $(2,3)$-tight.*

The inductive proof is for the deeper direction $(2,3)$-tight implies minimally rigid. We proceed by induction on $n$ and the base case is the trivial check that $K_2$ is minimally rigid. Suppose every $(2,3)$-tight graph on at most $n$ vertices is minimally rigid. Consider a $(2,3)$-tight graph $G$ on $n+1$ vertices. By Theorem 19.2 we can reduce $G$ to a smaller $(2,3)$-tight graph using a 0-reduction or a 1-reduction. By induction this smaller graph is minimally rigid. Now apply Lemma 19.1 or 19.2 to deduce that $G$ was minimally rigid.

**Theorem 19.12 (Hendrickson [22], Connelly [8], Jackson and Jordán [28])** *Let $G$ be a graph with $|V| \geq 4$. A generic framework $(G,p)$ in $\mathbb{R}^2$ is globally rigid if and only if $G$ is 3-connected and $(G,p)$ is redundantly rigid.*

Hendrickson [22] proved that every globally rigid generic framework is 3-connected and redundantly rigid. The sufficiency direction is proved by induction, combining Theorems 19.10 and 19.1.

Tanigawa [76] gave a new sufficient condition for global rigidity using "vertex-redundant rigidity" which allowed him to simplify some parts of the proof of Theorem 19.12.

Let $(G,p)$ be a framework and let $u,v \in V$. Then the pair $u,v$ is globally linked if the distance between $u$ and $v$ is the same in $(G,p)$ as it is in any other generic realization $(G,q)$. In [33] globally linked pairs were studied. In particular 1-extension preserves global linkedness when the edge being deleted is redundant. Subsequently they showed [34] that 1-extensions on non-redundant edges preserve the property of being not globally linked.

For 3- and higher-dimensional frameworks one of the key combinatorial difficulties in proving a 3D rigidity theorem is that 0-extensions and 1-extensions are not sufficient. However Jackson and Jordán proved an analogue of Theorem 19.2 for degree bounded graphs, in arbitrary dimension, using just these two operations. We say that a framework $(G,p)$ is *independent* if its rigidity matrix has maximum possible rank over all realizations of $G$.

**Theorem 19.13 (Jackson and Jordán [29])** *Let $G$ be a connected graph with minimum degree $d+1$ and maximum degree $d+2$. Then $G$ is $(d, \binom{d+1}{2})$-sparse if and only if $(G,p)$ is generically independent.*

They also used 0- and 1-extensions to prove a similar result for sparse graphs (see [29, Corollary 4.3].)

In [59] the rigidity of frameworks on concentric $d$-spheres whose radii are allowed to vary continuously was considered. Here the combinatorial objects are vertex colored graphs and analogues of Laman's theorem for circles with arbitrarily many radii varying independently were proved using inductive techniques, in particular 0- and 1-extensions on colored graphs. There results were also extended to 2-spheres with 1 or 3 radii varying independently. By the equivalence of rigidity in $\mathbb{R}^d$ with rigidity on the $d$-sphere (see Chapter 17, Section 17.2), these results give interpolation theorems between rigidity characterizations in $\mathbb{R}$ and $\mathbb{R}^2$ and from $\mathbb{R}^2$, a small step, in the direction of $\mathbb{R}^3$.

Penne [66] showed that triangle free 1-extensions preserve the property of having an irreducible pure condition (in the sense of White-Whiteley [81]). He conjectured that an isostatic graph (in the plane) has an irreducible pure condition if and only if the only proper rigid subgraphs are single edges. He called such graphs minimally isostatic graphs (MIGs). There are countably many MIG's that cannot be generated from $K_{3,3}$ by triangle free 1-extensions. Proving an inductive construction for this class of graphs is an interesting open problem. A closely related class of graphs was considered in [34] in the study of globally loose pairs. These are isostatic graphs in which the only proper rigid subgraphs are complete graphs. Again finding an inductive construction is an open problem.

## 19.3 Body-Bar, Body-Hinge, Molecular, etc.

While we have a good understanding of rigidity in $\mathbb{R}^2$ it is still true that little is known for bar-joint frameworks in higher dimensions. For example, combinatorially, no analogue of Theorem 19.2 is known for $(3,6)$-tight graphs with, or without, any consideration of excluding known problematic examples like the double banana.

However, a number of special classes of frameworks have been studied extensively. In the previous section we saw some special cases (for example, Theorem 19.13). Here we will focus on frameworks in $\mathbb{R}^d$ with special geometry and show that in these cases a lot can be said combinatorially about rigidity.

**Body-bar framework:** A $d$-dimensional *body-bar framework* is a structural model consisting of rigid bodies and stiff bars. The bodies are free to move continuously in $\mathbb{R}^d$ subject to the constraints imposed by the bars.

**Body-hinge framework:** A $d$-dimensional *body-hinge framework* is a structural model consisting of rigid bodies and hinges. Each hinge is a $(d-2)$-dimensional affine subspace that joins some pair of bodies. The bodies are free to move continuously in $\mathbb{R}^d$ subject to the constraint that the relative motion of any two bodies joined by a hinge is a rotation about the hinge.

**Molecular framework:** A *molecular framework* is a body-hinge framework in which the hinges are coplanar. That is, all the $(d-2)$-dimensional affine subspaces incident to a body are contained in a common $(d-1)$-dimensional affine subspace.

**edge pinch:** Let $0 \le j \le m \le k$. An edge pinch $K(k,m,j)$ pinches $j$ edges of $G$ into a new node $z$, places $m-j$ loops on $z$ and connects $z$ to other vertices of $G$ by $k-m$ new edges.

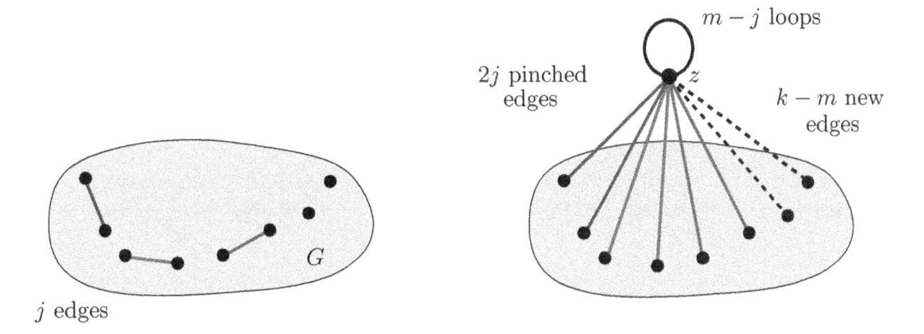


**Figure 19.7**
Edge pinch $K(k,m,j)$.

### 19.3.1 Geometry and Combinatorics

The simplest geometrically special class is that of rigid bodies attached by bars. Since such body-bar frameworks are special classes of bar-joint frameworks, Lemmas 19.1 and 19.2 clearly apply. Similar proofs show that the multigraph variants also preserve rigidity. Similarly 0-extension does not preserve global rigidity for body-bar frameworks, while Theorem 19.1 implies that 1-extension does. However, Connelly, Jordán, and Whiteley [6], in their study of body-bar global rigidity, had to work significantly harder including use of a grafting operation (to combine graphs) and a triangle exchange operation (swapping edges within a collinear triangle).

The following theorem extends the 0- and 1-extension construction to $(k,k)$-tight graphs for arbitrary $k$.

**Theorem 19.14 (Tay [78])** *A loopless multigraph G is $(k,k)$-tight if and only G can be formed from the single vertex by a series of loopless edge pinches ($K(k,m,m)$, where $0 \leq m \leq k$).*

As an aside we note that this was then extended to $(k,\ell)$-tight multigraphs by Fekete and Szegő [14]. Let $P_n$ be the multigraph on one vertex with $n$ incident loops.

**Theorem 19.15 (Fekete and Szegő [14])** *Let $1 \leq k \leq \ell$. A multigraph G is $(k,\ell)$-tight if and only if G can be obtained from $P_{k-\ell}$ by edge pinches $K(k,m,j)$ where $j \leq m \leq k-1$ and $m-j \leq k-\ell$. A multigraph G is $(k,0)$-tight if and only if G can be obtained from $P_k$ by edge pinches $K(k,m,j)$ where $j \leq m \leq k$ and $m-j \leq k$.*

A classical result of Nash-Williams [54] expresses $(k,k)$-tightness in terms of packing of $k$ edge-disjoint spanning trees. This leads to analogues of $(2,3)$-tight graphs in terms of nearly $k$-tree connected graphs and of redundantly rigid graphs in terms of highly $k$-tree connected graphs. For these classes of graphs, Frank and Szegő [17] proved constructive characterizations using analogues of pinching. There are also a number of constructive characterizations of related classes of graphs, see [16] and the references therein.

In their work on the molecular conjecture, Katoh and Tanigawa [45] proved an inductive construction of graphs $G$ such that the graph $\tilde{G}$ formed from taking $\binom{d+1}{2}-1$ copies of each edge contains $\binom{d+1}{2}$ edge-disjoint spanning trees and deleting any edge from $G$ destroys the spanning tree property. Their characterization, for such graphs $G$ which are 2-edge-connected uses contraction of proper rigid subgraphs and the inverse of subdivision.

In [48] an inductive construction of minimally rigid body-hinge graphs was given.

### 19.3.2 Characterizations

For body-bar frameworks, rigidity can be elegantly characterized via tree packing in arbitrary dimension.

**Theorem 19.16 (Tay [77])** *A multigraph $G$ is generically minimally rigid in $\mathbb{R}^d$ as a body-bar framework if and only if $G$ is $(D,D)$-tight, where $D = \binom{d+1}{2}$ is the dimension of the Euclidean group $Euc(\mathbb{R}^d)$.*

Tay [77] and Whiteley [82] extended this result to body-hinge frameworks, using non-inductive means. Tay and Whiteley [79] conjectured that it could be extended to "hinge-concurrent" frameworks and this became known as the molecular conjecture and was used by material scientists for many years [38, 80]. Jackson and Jordán [31] used 0-extensions, 1-extensions and vertex splitting in their proof of the conjecture for 2-dimensions. (Since they worked in dimension 2 the hinges were actually pins.)

**Theorem 19.17 (Jackson and Jordán [31])** *A multigraph $G$ has an infinitesimally rigid pin-collinear body-pin realization if and only if $2G$ (the graph obtained from $G$ by doubling all edges) contains three edge-disjoint spanning trees.*

The following theorem, proved by Katoh and Tanigawa [45], extended this to $d$-dimensions and hence turned the molecular conjecture into a theorem. This is one of the most powerful results proved using an inductive construction.

**Theorem 19.18 (Katoh and Tanigawa [45])** *A multigraph $G$ can be realised as an infinitesimally rigid body-and-hinge framework in $\mathbb{R}^d$ if and only if $G$ can be realised as an infinitesimally rigid panel-and-hinge framework in $\mathbb{R}^d$.*

Global rigidity is also well understood for body-bar frameworks.

**Theorem 19.19 (Connelly, Jordan, and Whiteley [6])** *A body-bar framework is generically globally rigid in $\mathbb{R}^d$ if and only if it is generically redundantly rigid in $\mathbb{R}^d$.*

An interesting aspect of Connelly, Jordan and Whiteley's proof is that they were able to go outside the inductive class, using looped graphs in intermediate steps of the induction process.

Very recently this has been extended to body-hinge frameworks by Jordán, Kiraly and Tanigawa [41] (using non-inductive techniques). A global rigidity version of the molecular theorem is an open problem, although a conjecture was presented in [6].

## 19.4 Further Rigidity Contexts

**Gain graph:** A $\Gamma$-*gain graph* is a multigraph with directions whose edges are labelled by elements of a group $\Gamma$.

**Balanced:** A subgraph $H$ of a $\Gamma$-gain graph $G$ is *balanced* if, for every cycle in $H$, the product of the edge labels equals the identity (making the appropriate choice of group element or its inverse depending on the direction of the edge).

**$(k,\ell,m)$-gain-sparse/tight:** A $\Gamma$-gain graph $G$ is $(k,\ell,m)$-*gain-sparse* if $|F| \leq k|V(F)| - \ell$ for every balanced subset $F \subset E$ and $|F| \leq k|V(F)| - ml$ for every unbalanced subset $F \subset E$. Moreover a $\Gamma$-gain graph $G$ is $(k,\ell,m)$-*gain-tight* if $|E| = k|V| - m$ and $G$ is $(k,\ell,m)$-gain-sparse.

**mechanism:** A framework with exactly one degree of freedom.

**direction-length graph:** A loopless multigraph $G = (V, D \cup L)$ whose edge set is partitioned into *direction edges D* and *length edges L*. Neither $D$ nor $L$ contains parallel edges.

**1-extension on direction-length graphs:** These are 1-extensions, as defined above, that may be *direction-pure*, *length-pure* (involving only direction or length edges, respectively) or *mixed.*

In this final section we briefly describe a number of areas of current research in rigidity theory where inductive constructions have been useful.

### 19.4.1 Frameworks with Symmetry

There has been significant use of inductive constructions for the study of symmetric frameworks. These are frameworks, typically in $\mathbb{R}^2$, which admit an action of a symmetry group but are otherwise as generic as possible (see Chapter 25). Symmetry analysis in rigidity splits into two distinct strands. First, there is incidental symmetry where a framework happens to possess some symmetry but deformations which break the symmetry are permitted. The key sources for such considerations are [70, 71, 72]. Known characterizations of incidental rigidity are, currently, limited to very small groups but make significant use of inductive constructions.

Second, there is forced symmetry, where the framework is symmetric and only deformations which preserve that symmetry are considered. We mention here one sample theorem of Jordan, Kaszanitsky, and Tanigawa [40] which uses symmetrized versions of 0-extension, 1-extension and $X$-replacement to characterize symmetry-forced rigidity. They first use these operations to give an inductive construction of $(2,3,1)$-gain-tight graphs and then show that these operations preserve symmetric rigidity. See Figures 19.8 and 19.9 for samples of gain graphs, symmetric frameworks, and the 1-extension operations in these cases.

Moreover of interest across several rigidity contexts are $(2,\ell,m)$-gain-tight graphs in the range $0 \leq m \leq \ell < 4$. For the range $1 \leq m \leq \ell < 4$, inductive constructions are actually known in each case (see [64]). However when $m = 0$ proving analogues is open. Two hints of the additional difficulty arises from the facts that the minimum degree may be 4 (contrary to the case when $m \geq 1$) and that $(2,3,0)$-gain-sparsity does not, in general, induce a matroid (see [72]).

### 19.4.2 Infinite Frameworks

Infinite frameworks which are periodic with respect to some lattice have been considered from an inductive construction perspective. In particular using the viewpoint of considering periodic frameworks as frameworks on a torus, Ross [68] proved that, with a fixed lattice, periodic rigidity was equivalent to the graph being $(2,3,2)$-gain-tight. The proof extended Theorem 19.14, in the case $k = 2$, to deal with the group elements in subgraphs. This has been extended inductively in [63] to a partially variable lattice, in [67] to body-bar periodic frameworks in $\mathbb{R}^3$ and extended non-inductively by Malestein and Theran [53] to a fully variable lattice.

Whiteley [86] has also considered inductions for geometrically special types of repetitive structures (fragments, substrates, etc.) and suggested potential applications of inductive techniques such as an extended form of vertex splitting.

Theorem 19.2 has been generalized to countably infinite graphs (with bounded degree vertices), still using 0- and 1-extensions, by Kitson and Power [46]. Using this, a countably infinite version of Laman's theorem has been derived. Their combinatorial object is a containment tower of finite $(2,3)$-tight subgraphs.

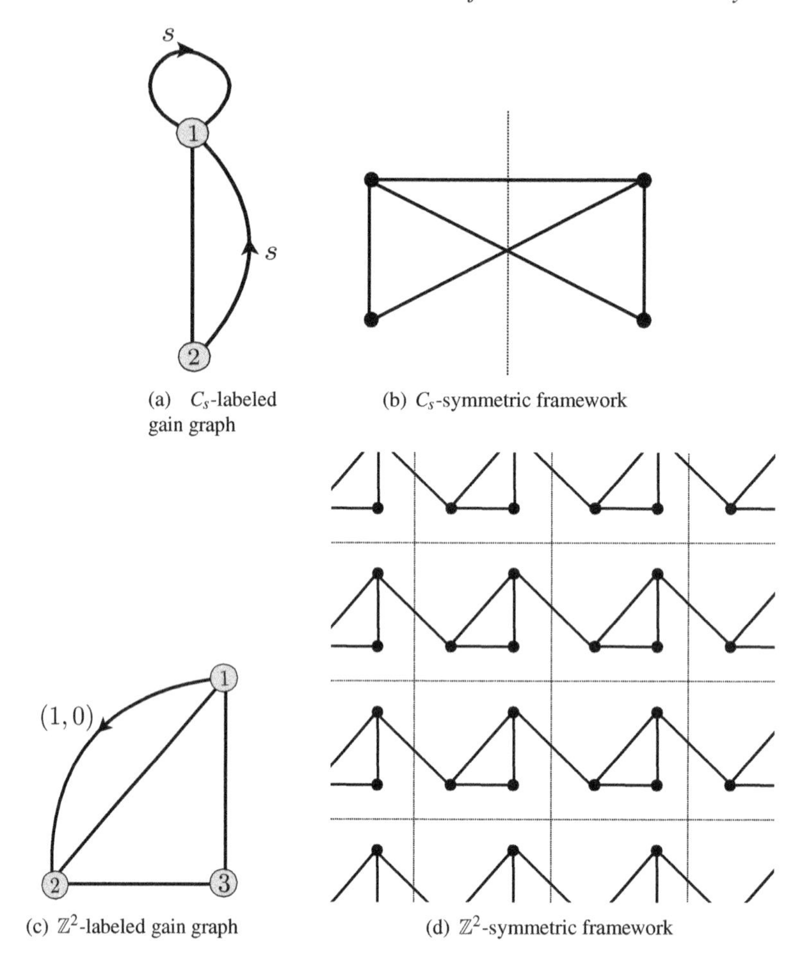


**Figure 19.8**
A mirror-symmetric framework (b) and its gain graph (a). A plane-periodic framework (d) and its gain graph (c).

### 19.4.3 Surfaces

The rigidity (and global rigidity) of frameworks in $\mathbb{R}^3$ which are forced to lie on a 2-dimensional manifold has also been studied in [25, 37, 57, 58]. Characterizations were given for the rigidity of generic frameworks on concentric spheres and concentric cylinders [57] and this was extended to surfaces with exactly one isometry in [58]. These theorems were proved inductively requiring characterizations of (simple) $(2,2)$-tight and $(2,1)$-tight graphs, see also [56]. (The simplicity requirement preventing these results from being deducable from Theorems 19.14 and 19.15.) The operations used were 0-extension, 1-extension, vertex-to-4-cycle, vertex-to-$K_4$, and in the $(2,1)$-tight case, joining two graphs in the class by a bridge.

For example, we briefly describe the $(2,2)$-tight characterization given in [58]. The minimum degree is in the set $\{2,3\}$. By using 0-extensions we may assume that any graph that we cannot generate using these moves has minimum degree 3. The problem with extending Theorem 19.2 is that $K_4$ is $(2,2)$-sparse so it can happen that every vertex of degree 3 cannot be reduced to a (simple) $(2,2)$-tight graph using a 1-reduction; all degree 3 vertices may be contained in subgraphs isomorphic to $K_4$. In such a case we can contract a copy of $K_4$ to a single vertex unless there is a

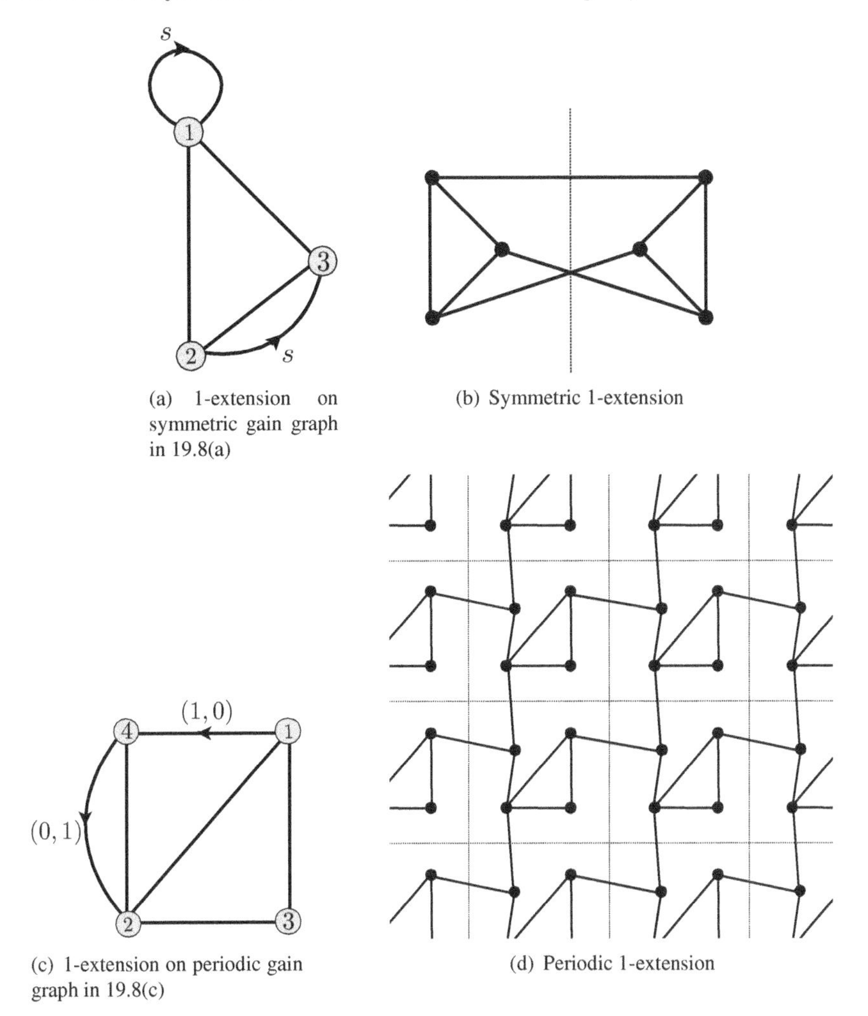


(a) 1-extension on symmetric gain graph in 19.8(a)

(b) Symmetric 1-extension

(c) 1-extension on periodic gain graph in 19.8(c)

(d) Periodic 1-extension

**Figure 19.9**
1-extensions on the graphs of Figure 19.8.

copy of $K_3$ which intersects the $K_4$ in a single edge. However this structure gives a 4-cycle on the vertex set of the $K_3$ along with the degree 3 vertex of the $K_4$ to which we can always apply an inverse vertex-to-4-cycle move.

A recursive construction of (simple) $(2,0)$-tight graphs was recently obtained in [20]. The case of rigidity on a specific surface admitting no isometries such as the ellipsoid remains open due to the difficulty of understanding X-replacement on such a surface.

Considering global rigidity in this context leads to the study of $(2,2)$-circuits, for which a characterization was given in [60]. Also an analogue of the stress matrix for frameworks on surfaces was developed, using which, the 1-extension operation was shown to preserve global rigidity on surfaces [25]. In [37] these results were extended to completely characterize global rigidity on the cylinder. As well as building on these earlier results the proof used a generalized form of vertex splitting and two further recursive construction results for cylindrical frameworks. Finally, symmetric frameworks on surfaces were considered giving rise to several inductive constructions for various classes of $(k,\ell,m)$-gain-tight graphs (see [64]).

The above-mentioned construction of $(2,2)$-tight graphs was also used to prove an analogue of Laman's theorem for frameworks in $\mathbb{R}^2$ equipped with a non-Euclidean metric (an $\ell_p$ norm for

$p \in (1,\infty)$ with $p \neq 2$). See [47] and Chapter 24 for details. Generalizing these contexts to metric spaces with multiple types of distance constraints with their combinatorics handled by sparsity constraints on edge-colored subgraphs has also recently been considered [61].

### 19.4.4 Mechanisms

Servatius, Shai, and Whiteley [74] use 0- and 1-extensions in their study of pinned frameworks in the plane and Assur graphs. They give an analogue of Theorem 19.2 in the pinned setting and present a recursive construction of Assur graphs using 1-extensions and 2-sums. Assur graphs arise, in engineering, in the synthesis and analysis of mechanisms.

Gao and Sitharam [18] studied the class of graphs constructable by 0-extensions in their analysis of geometric properties (connectivity, convexity, etc.) of Cayley configuration spaces.

### 19.4.5 Applications of Rigidity Techniques

Jackson, Jordán, and Tanigawa [35] used inductive construction techniques, including 0-extensions, 1-extensions and what they call double 1-extensions, to prove results about matrix completion problems for low rank semidefinite matrices.

Algorithmic aspects of rigidity have been considered using inductive constructions. In particular inductive moves give efficient ways of constructing families of rigid [52] and globally rigid [43] graphs in the plane.

Recently rigidity has found applications in control of robotic formations. Here inductive techniques have been used, for example, to understand rigidity with direction constraints (known to control theorists as bearing rigidity) [87]. Versions of the 0- and 1-extension operations were also used in a similar context to analyse persistence [23], which concerns rigidity for directed graphs where constraints are "one-way."

### 19.4.6 Direction-Length Frameworks and CAD

Rigidity and inductions have also been applied to various problems in computer aided design (CAD) [75]. Building on a result of Jackson and Jordán on the global rigidity of 2-dimensional direction-length graphs [32], Nguyen proved that $d$-dimensional 1-extensions preserve global rigidity of direction-length graphs [55]. These results are significant because it is not known whether global rigidity is a generic property of direction-length frameworks even in two dimensions. Therefore, it is helpful to have inductive methods to generate families of globally rigid direction-length graphs, and this also provides a tool to verify the global rigidity of certain frameworks.

Owen and Power [65] showed that vertex splitting preserves the property of being radically solvable; a property, intimately linked to rigidity, of importance in CAD. It seems to be a challenging open problem to determine if the 1-extension operation preserves the property of being radically solvable.

### 19.4.7 Nearly Generic Frameworks

Since genericity is a very strong assumption there is substantial interest in weakening this assumption. We have already discussed such contexts such as symmetric frameworks and frameworks on surfaces. Here we mention combinatorial results for frameworks in the plane with some specified non-genericity. Jackson and Jordán [30] characterized rigidity for frameworks in the plane where three designated vertices are collinear but are otherwise generic. To do this they extended the Henneberg-Laman technique by avoiding these 3 special vertices in their 0- and 1-extension arguments. Second, Fekete, Jordán, and Kaszanitzky [12] proved a delete-contract characterization of rigidity in the plane where two designated vertices are coincident but the framework is otherwise generic. Again the technique was to use 0- and 1-extensions but the proofs required analyzing a more complicated count matroid. Further use of these recursive operations extended the result to a triple of coincident points [20]. These ideas were very recently extended to frameworks on the sphere and on the cylinder with two coincident points [36] using similar operations to those described in Subsection 19.4.3.

## 19.5 Summary Tables

Tables 19.1 and 19.2 summarize the use of inductive moves for local and global rigidity.

**Table 19.1**

Summary of inductive moves for local rigidity.

| Operation | Figure | Type of framework | Rigidity | Class of graphs | Characterization |
|---|---|---|---|---|---|
| 0-, 1-extension | Fig. 19.1 | 2-dim bar-joint | Lemmas 19.1, 19.2 | $(2,3)$-tight, Theorem 19.2<br>Laman-plus-one, Theorem 19.5 | 2-dim rigidity |
| | | $d$-dim bar-joint | Lemmas 19.1, 19.2 | | Degree bounded |
| | Fig. 19.9(a), (b) | Symmetric | Yes [71] | $(2,3,1)$-gain-tight, [40] | Forced rotational<br>Forced reflectional<br>Odd-order dihedral, [40] |
| | Fig. 19.9(c), (d) | Periodic | Yes, [68] | $(2,3,2)$-gain-tight, [68] | Fixed lattice<br>Partial lattice, [63] |
| | | Surfaces | Yes, [58] | $(2,2)$-tight, [58]<br>$(2,1)$-tight, [56] | cylinder, [57]<br>1 isometry surfaces, [58] |
| | | Symmetry and surfaces | Yes, [64] | $(2,3,3)$-gain-tight, [64]<br>$(2,2,2)$-gain-tight, [64]<br>$(2,2,1)$-gain-tight, [64]<br>$(2,1,1)$-gain-tight, [64] | Spherical inversion, [64]<br>Cylinder with<br>Various groups, [64]<br>Cone, [64] |
| | | Infinite bar-joint | Yes, [46] | Sequential $(2,3)$-tight, [46] | |
| X-replacement | Fig. 19.2(a) | 2-dim bar-joint | Lemma 19.3 | | |
| | Fig. 19.2(a) | 3-dim bar-joint | Conjecture 3 | | |
| | Fig. 19.2(a) | d-dim bar-joint | No, [19] | | |
| V-replacement | Fig. 19.2(b) | 3-dim bar-joint | No | | |
| Vertex splitting | Fig. 19.3(a), (c) | d-dim bar-joint | Proposition 19.1 | | |
| Vertex-to-4-cycle | Fig. 19.3(b) | 2-dim bar-joint | Yes, [50] | | |
| | | Surfaces | Yes, [58] | $(2,2)$-tight<br>$(2,1)$-tight | Cylinder, [57]<br>1 isometry surfaces |
| 2-sum | | 2-dim bar-joint | Yes | | |
| Pinching | Fig. 19.7 | d-dim body-bar | Yes | $(D,D)$-tight, [78] | Body-bar rigidity, [77] |

**Table 19.2**

Summary of inductive moves for global rigidity.

| Operation | Type of framework | Global rigidity | Class(es) of graphs | Characterization |
|---|---|---|---|---|
| 0-extension | $d$-dim bar-joint | No | | |
| 1-extension | 2-dim bar-joint | Theorem 19.1 | $(2,3)$-circuits, [4]<br>3-connected and<br>redundantly rigid graphs, [28] | 2-dim global, [8, 28] |
| | $d$-dim bar-joint | Theorem 19.1 | | |
| X-replacement | 2-dim bar-joint | No, see, for example, [62, Figure 6] | | |
| Vertex splitting | 2-dim bar-joint | Yes, [13] | | |
| | $d$-dim bar-joint | Conjecture 4 | | |
| 2-sum | 2-dim bar-joint | No | $(2,3)$-circuits | |
| Pinching | d-dim body-bar | Yes, [6] | Highly $k$-tree connected, [17] | Body-bar global, [6] |
| 1-extension | Surfaces | Yes, [25] | $(2,2)$-circuits, [60] | |
| 1-extension | Direction-length | Yes, [32] | | |

# References

[1] T. Abbott. Generalizations of Kempe's universality theorem. Master's thesis, 2008.

[2] M. Alam, D. Eppstein, M. Kaufmann, S. Kobourov, S. Pupyrev, A. Schulz, and T. Ueckerdt. Contact representations of sparse planar graphs. *arXiv:1501.00318*, 2015.

[3] L. Asimow and B. Roth. The rigidity of graphs. *Trans. Amer. Math. Soc.*, 245:279–289, 1978.

[4] Alex R. Berg and Tibor Jordán. A proof of Connelly's conjecture on 3-connected circuits of the rigidity matroid. *J. Combin. Theory Ser. B*, 88(1):77–97, 2003.

[5] R. Connelly. Combining globally rigid frameworks. *Tr. Mat. Inst. Steklova*, 275(Klassicheskaya i Sovremennaya Matematika v Pole Deyatelnosti Borisa Nikolaevicha Delone):202–209, 2011.

[6] R. Connelly, T. Jordán, and W. Whiteley. Generic global rigidity of body-bar frameworks. *J. Combin. Theory Ser. B*, 103(6):689–705, 2013.

[7] R. Connelly and W. J. Whiteley. Global rigidity: the effect of coning. *Discrete Comput. Geom.*, 43(4):717–735, 2010.

[8] Robert Connelly. Generic global rigidity. *Discrete Comput. Geom.*, 33(4):549–563, 2005.

[9] Robert Connelly. Questions, conjectures and remarks on globally rigid tensegrities. Technical Report, November 2009.

[10] James Cruickshank. On spaces of infinitesimal motions and three dimensional Henneberg extensions. *Discrete Comput. Geom.*, 51(3):702–721, 2014.

[11] James Cruickshank, Derek Kitson, and Stephen C. Power. The generic rigidity of triangulated spheres with blocks and holes. *J. Combin. Theory Ser. B*, 122:550–577, 2017.

[12] Zsolt Fekete, Tibor Jordán, and Viktória E. Kaszanitzky. Rigid two-dimensional frameworks with two coincident points. *Graphs Combin.*, 31(3):585–599, 2015.

[13] Zsolt Fekete, Tibor Jordán, and Walter Whiteley. An inductive construction for plane Laman graphs via vertex splitting. In *Algorithms—ESA 2004*, volume 3221 of *Lecture Notes in Comput. Sci.*, pages 299–310. Springer, Berlin, 2004.

[14] Zsolt Fekete and László Szegő. A note on $[k,l]$-sparse graphs. In *Graph theory in Paris*, Trends Math., pages 169–177. Birkhäuser, Basel, 2007.

[15] Wendy Finbow-Singh and Walter Whiteley. Isostatic block and hole frameworks. *SIAM J. Discrete Math.*, 27(2):991–1020, 2013.

[16] András Frank. *Connections in combinatorial optimization*, volume 38 of *Oxford Lecture Series in Mathematics and Its Applications*. Oxford University Press, Oxford, 2011.

[17] András Frank and László Szegő. Constructive characterizations for packing and covering with trees. *Discrete Appl. Math.*, 131(2):347–371, 2003.

[18] Heping Gao and Meera Sitharam. Characterizing 1-dof henneberg-i graphs with efficient configuration spaces. In *Proceedings of the 2009 ACM symposium on Applied Computing*, pages 1122–1126. ACM, 2009.

[19] Jack Graver, Brigitte Servatius, and Herman Servatius. *Combinatorial rigidity*, volume 2 of *Graduate Studies in Mathematics*. American Mathematical Society, Providence, RI, 1993.

[20] Hakan Guler. *Rigidity of Frameworks.* PhD thesis, Queen Mary, University of London, 2018.

[21] Ruth Haas, David Orden, Günter Rote, Francisco Santos, Brigitte Servatius, Herman Servatius, Diane Souvaine, Ileana Streinu, and Walter Whiteley. Planar minimally rigid graphs and pseudo-triangulations. *Comput. Geom.*, 31(1-2):31–61, 2005.

[22] Bruce Hendrickson. Conditions for unique graph realizations. *SIAM J. Comput.*, 21(1):65–84, 1992.

[23] Julien M. Hendrickx, Brian D. O. Anderson, Jean-Charles Delvenne, and Vincent D. Blondel. Directed graphs for the analysis of rigidity and persistence in autonomous agent systems. *Internat. J. Robust Nonlinear Control*, 17(10-11):960–981, 2007.

[24] L. Henneberg. *Die Graphische Statik der starren Systeme.* (Johnson Reprint), 1911.

[25] B. Jackson and A. Nixon. Stress matrices and generic global rigidity of frameworks on surfaces. *Discrete Comput. Geom.*, 54(3):586–609, 2015.

[26] B. Jackson and J. Owen. Notes on henneberg moves. 2013.

[27] Bill Jackson. Notes on the rigidity of graphs. *Levico Conference Notes*, 2007. `http://www.science.unitn.it/cirm/JacksonLectures.pdf`.

[28] Bill Jackson and Tibor Jordán. Connected rigidity matroids and unique realizations of graphs. *J. Combin. Theory Ser. B*, 94(1):1–29, 2005.

[29] Bill Jackson and Tibor Jordán. The $d$-dimensional rigidity matroid of sparse graphs. *J. Combin. Theory Ser. B*, 95(1):118–133, 2005.

[30] Bill Jackson and Tibor Jordán. Rigid two-dimensional frameworks with three collinear points. *Graphs Combin.*, 21(4):427–444, 2005.

[31] Bill Jackson and Tibor Jordán. Pin-collinear body-and-pin frameworks and the molecular conjecture. *Discrete Comput. Geom.*, 40(2):258–278, 2008.

[32] Bill Jackson and Tibor Jordán. Operations preserving global rigidity of generic direction-length frameworks. *International Journal of Computational Geometry & Applications*, 20(06):685–706, 2010.

[33] Bill Jackson, Tibor Jordán, and Zoltán Szabadka. Globally linked pairs of vertices in equivalent realizations of graphs. *Discrete Comput. Geom.*, 35(3):493–512, 2006.

[34] Bill Jackson, Tibor Jordán, and Zoltán Szabadka. Globally linked pairs of vertices in rigid frameworks. In Robert Connelly, Walter Whiteley, and Asia Weiss, editors, *Rigidity and Symmetry*. Fields Institute, 2014.

[35] Bill Jackson, Tibor Jordán, and Shin-ichi Tanigawa. Combinatorial conditions for the unique completability of low-rank matrices. *SIAM J. Discrete Math.*, 28(4):1797–1819, 2014.

[36] Bill Jackson, Viktoria Kaszanitsky, and Anthony Nixon. Rigid cylindrical frameworks with two coincident points. arXiv: 1607.02039, 2016.

[37] Bill Jackson and Anthony Nixon. Global rigidity of generic frameworks on the cylinder. arXiv: 1610.07755, 2017.

[38] Donald J Jacobs, Leslie A Kuhn, and Michael F Thorpe. Flexible and rigid regions in proteins. In *Rigidity theory and applications*, pages 357–384. Springer, 2002.

[39] Tibor Jordán. Combinatorial rigidity: graphs and matroids in the theory of rigid frameworks. Technical Report TR-2014-12, Egerváry Research Group, Budapest, 2014. `www.cs.elte.hu/egres`.

[40] Tibor Jordán, Viktória E. Kaszanitzky, and Shin-ichi Tanigawa. Gain-sparsity and symmetry-forced rigidity in the plane. *Discrete Comput. Geom.*, 55(2):314–372, 2016.

[41] Tibor Jordán, Csaba Király, and Shin-ichi Tanigawa. Generic global rigidity of body-hinge frameworks. *J. Combin. Theory Ser. B*, 117:59–76, 2016.

[42] Tibor Jordán, András Recski, and Zoltán Szabadka. Rigid tensegrity labelings of graphs. *European J. Combin.*, 30(8):1887–1895, 2009.

[43] Tibor Jordán and Zoltán Szabadka. Operations preserving the global rigidity of graphs and frameworks in the plane. *Comput. Geom.*, 42(6-7):511–521, 2009.

[44] Tibor Jordan and Shin-Ichi Tanigawa. Global rigidity of triangulations with braces. *EGRES TR-2017-06*, 2017.

[45] Naoki Katoh and S.-I. Tanigawa. A proof of the molecular conjecture. *Discrete & Computational Geometry*, 45(4):647–700, 2011.

[46] D. Kitson and S. Power. The rigidity of infinite graphs. *arXiv:1310.1860*, 2013.

[47] D. Kitson and S. C. Power. Infinitesimal rigidity for non-Euclidean bar-joint frameworks. *Bull. Lond. Math. Soc.*, 46(4):685–697, 2014.

[48] Yuki Kobayashi, Yuya Higashikawa, Naoki Katoh, and Naoyuki Kamiyama. An inductive construction of minimally rigid body–hinge simple graphs. *Theoretical Computer Science*, 556(0):2 – 12, 2014. Combinatorial Optimization and Applications.

[49] Erika R. Kovács and László A. Végh. Constructive characterization theorems in combinatorial optimization. In *Combinatorial optimization and discrete algorithms*, RIMS Kôkyûroku Bessatsu, B23, pages 147–169. Res. Inst. Math. Sci. (RIMS), Kyoto, 2010.

[50] L. Moshe L. Lomeli and W. Whiteley. Bases and circuits for 2-rigidity: constructions via tree partitions.

[51] G. Laman. On graphs and rigidity of plane skeletal structures. *J. Engrg. Math.*, 4:331 – 340, 1970.

[52] Audrey Lee and Ileana Streinu. Pebble game algorithms and sparse graphs. *Discrete Math.*, 308(8):1425–1437, 2008.

[53] Justin Malestein and Louis Theran. Generic combinatorial rigidity of periodic frameworks. *Adv. Math.*, 233:291–331, 2013.

[54] C. St. J. A. Nash-Williams. Decomposition of finite graphs into forests. *J. London Math. Soc.*, 39:12, 1964.

[55] Viet-Hang Nguyen. 1-extensions and global rigidity of generic direction-length frameworks. *International Journal of Computational Geometry & Applications*, 22(06):577–591, 2012.

[56] A. Nixon and J. Owen. An inductive construction of (2,1)-tight graphs. *Contributions to Discrete Math*, 9(2):1–16, 2014.

[57] A. Nixon, J. C. Owen, and S. C. Power. Rigidity of frameworks supported on surfaces. *SIAM J. Discrete Math.*, 26(4):1733–1757, 2012.

[58] A. Nixon, J. C. Owen, and S. C. Power. A characterization of generically rigid frameworks on surfaces of revolution. *SIAM J. Discrete Math.*, 28(4):2008–2028, 2014.

[59] A. Nixon, B. Schulze, S.I. Tanigawa, and W. Whiteley. Rigidity of frameworks on expanding spheres. *arXiv:1501.01391*, 2015.

[60] Anthony Nixon. A constructive characterisation of circuits in the simple $(2,2)$-sparsity matroid. *European J. Combin.*, 42:92–106, 2014.

[61] Anthony Nixon and Stephen Power. Double-distance frameworks and mixed sparsity graphs. *arXiv: 1709.06349*, 2017.

[62] Anthony Nixon and Elissa Ross. One brick at a time: a survey of inductive constructions in rigidity theory. In Robert Connelly, Walter Whiteley, and Asia Weiss, editors, *Rigidity and Symmetry*. Fields Institute, 2014.

[63] Anthony Nixon and Elissa Ross. Periodic rigidity on a variable torus using inductive constructions. *Electron. J. Combin.*, 22(1), 2015.

[64] Anthony Nixon and Bernd Schulze. Symmetry-forced rigidity of frameworks on surfaces. *Geom. Dedicata*, 182:163–201, 2016.

[65] J. C. Owen and S. C. Power. The non-solvability by radicals of generic 3-connected planar Laman graphs. *Trans. Amer. Math. Soc.*, 359(5):2269–2303, 2007.

[66] Rudi Penne. Isostatic bar and joint frameworks in the plane with irreducible pure conditions. *Discrete Appl. Math.*, 55(1):37–57, 1994.

[67] Elissa Ross. The rigidity of periodic body-bar frameworks on the three-dimensional fixed torus. *Philos. Trans. R. Soc. Lond. Ser. A Math. Phys. Eng. Sci.*, 372(2008):20120112, 23, 2014.

[68] Elissa Ross. Inductive constructions for frameworks on a two-dimensional fixed torus. *Discrete & Computational Geometry*, 54(1):78–109, 2015.

[69] J. Saxe. Embeddability of weighted graphs in k-space is strongly np-hard. In *Proceedings of the 17th Allerton conference in communications, control and computing*, pages 480–489, 1979.

[70] B. Schulze. *Combinatorial and geometric rigidity with symmetry constraints*. PhD thesis, York University, 2009.

[71] Bernd Schulze. Symmetric versions of Laman's theorem. *Discrete Comput. Geom.*, 44(4):946–972, 2010.

[72] Bernd Schulze and Shin-ichi Tanigawa. Infinitesimal rigidity of symmetric bar-joint frameworks. *SIAM J. Discrete Math.*, 29(3):1259–1286, 2015.

[73] Brigitte Servatius and Herman Servatius. On the 2-sum in rigidity matroids. *European J. Combin.*, 32(6):931–936, 2011.

[74] Brigitte Servatius, Offer Shai, and Walter Whiteley. Combinatorial characterization of the Assur graphs from engineering. *European J. Combin.*, 31(4):1091–1104, 2010.

[75] Brigitte Servatius and Walter Whiteley. Constraining plane configurations in computer-aided design: combinatorics of directions and lengths. *SIAM J. Discrete Math.*, 12(1):136–153 (electronic), 1999.

[76] Shin-ichi Tanigawa. Sufficient conditions for the global rigidity of graphs. *J. Combin. Theory Ser. B*, 113:123–140, 2015.

[77] Tiong-Seng Tay. Rigidity of multigraphs I: linking rigid bodies in $n$-space. *J. Combinatorial Theory B*, 26:95 – 112, 1984.

[78] Tiong-Seng Tay. Henneberg's method for bar and body frameworks. *Structural Topology*, 17:53–58, 1991.

[79] Tiong-Seng Tay and Walter Whiteley. Generating isostatic frameworks. *Structural Topology*, (11):21–69, 1985. Dual French-English text.

[80] Michael Thorpe. http://flexweb.asu.edu/.

[81] Neil L. White and Walter Whiteley. The algebraic geometry of stresses in frameworks. *SIAM J. Algebraic Discrete Methods*, 4(4):481–511, 1983.

[82] Walter Whiteley. The union of matroids and the rigidity of frameworks. *SIAM J. Discrete Math.*, 1(2):237–255, 1988.

[83] Walter Whiteley. La division de sommet dans les charpentes isostatiques. *Structural Topology*, (16):23–30, 1990. Dual French-English text.

[84] Walter Whiteley. Some matroids from discrete applied geometry. In *Matroid theory (Seattle, WA, 1995)*, volume 197 of *Contemp. Math.*, pages 171–311. Amer. Math. Soc., Providence, RI, 1996.

[85] Walter Whiteley. Rigidity and scene analysis. In *Handbook of discrete and computational geometry*, CRC Press Ser. Discrete Math. Appl., pages 893–916. CRC, Boca Raton, FL, 1997.

[86] Walter Whiteley. Fragmentary and incidental behaviour of columns, slabs and crystals. *Philosophical Transactions of the Royal Society of London A: Mathematical, Physical and Engineering Sciences*, 372(2008), 12 2013.

[87] Shiyu Zhao and Daniel Zelazo. Bearing rigidity and almost global bearing-only formation stabilization. *IEEE Trans. Automat. Control*, 61(5):1255–1268, 2016.

# Chapter 20

# *Rigidity of Body-Bar-Hinge Frameworks*

**Csaba Király**

*Department of Operations Research, ELTE Eötvös Loránd University, and MTA-ELTE Egerváry Research Group on Combinatorial Optimization, Pázmány Péter sétány 1/C, Budapest, Hungary*

**Shin-ichi Tanigawa**

*Department of Mathematical Informatics, The University of Tokyo, Hongo, Bunkyo-ku, Tokyo, Japan*

## CONTENTS



This chapter discusses body-bar-hinge frameworks which are abstract models of structures consisting of rigid bodies connected by bars and/or hinges. A body-bar-hinge framework can be regarded as a special case of bar-joint frameworks, however, its underlying combinatorics is well understood even in higher dimensional spaces. In particular its generic rigidity has an exact connection to the tree packing problem, one of the fundamental problems in graph theory, and this link enables us to analyze the higher dimensional rigidity by graph theoretical approach.

# 20.1 Rigidity of Body-Bar-Hinge Frameworks

## 20.1.1 Body-Bar Frameworks

A body-bar framework models a structure consisting of full-dimensional rigid bodies connected by stiff bars. The bodies are free to move subject to the constraint that the distance between the points linked by bars must be constant. We shall summarize the basics of body-bar frameworks.

**Body-bar framework:** A $d$-dimensional *body-bar framework* is a pair $(G,\boldsymbol{b})$ of a multi-graph $G=(V,E_B)$ and $\boldsymbol{b}:\hat{E}_B \ni (e,v) \mapsto b_{e,v} \in \mathbb{R}^d$, where $\hat{E}_B$ denotes the set of pairs $(e,v)$ of an edge $e$ in $E_B$ and its endvertex $v$. Here $G$ represents the underlying graph, which is obtained by regarding each body as a vertex and each bar as an edge. See Figure 20.1 and Figure 20.2(a). $\boldsymbol{b}$ represents the bar-configuration, where the bar corresponding to $e=uv$ is the line segment $[b_{e,u},b_{e,v}]$ between $b_{e,u}$ and $b_{e,v}$.

**Body-bar realization:** A body-bar framework $(G,\boldsymbol{b})$ with the underlying graph $G$ is called a *body-bar realization* of $G$.

**Replacement:** A *replacement* of a $d$-dimensional body-bar framework $(G,\boldsymbol{b})$ is a map $\boldsymbol{r}:V\ni v\mapsto r_v \in \mathrm{Euc}(d)$ that satisfies

$$\|r_v(b_{e,v})-r_u(b_{e,u})\| = \|b_{e,v}-b_{e,u}\| \qquad (e=uv\in E_B),$$

where $\mathrm{Euc}(d)$ denotes the set of isometries of the $d$-dimensional Euclidean space $\mathbb{R}^d$ and $\|\cdot\|$ denotes the Euclidean norm. A replacement is *nontrivial* if $r_u \neq r_v$ for some pairs of vertices $u,v$.

**Global rigidity:** A $d$-dimensional body-bar framework $(G,\boldsymbol{b})$ is *globally rigid* if it has no nontrivial replacement.

**Finite motion:** A *finite motion* of a body-bar framework $(G,\boldsymbol{b})$ is a continuous family $\boldsymbol{r}^t$ $(t\in[0,1])$ of replacements of $(G,\boldsymbol{b})$ parameterized by $t$ such that $r_v^0$ is the identity for every $v\in V$. A finite motion is *nontrivial* if $r_v^t \neq r_u^t$ for some $u,v\in V$ and $t\in(0,1]$.

**Rigidity:** A $d$-dimensional body-bar framework $(G,\boldsymbol{b})$ is *rigid* if $(G,\boldsymbol{b})$ has no nontrivial finite motion.

**Infinitesimal isometry:** An *infinitesimal isometry* $i$ of $\mathbb{R}^d$ is $i:\mathbb{R}^d\ni p\mapsto Sp+t\in\mathbb{R}^d$ for some skew-symmetric matrix $S$ of size $d$ and $t\in\mathbb{R}^d$.

**Infinitesimal motion:** An *infinitesimal motion* $\boldsymbol{m}$ of a $d$-dimensional body-bar framework $(G,\boldsymbol{b})$ is a map $\boldsymbol{m}:V\ni v\mapsto m_v\in\mathrm{Inf}(d)$ that satisfies

$$\langle m_v(b_{e,v})-m_u(b_{e,u}), b_{e,v}-b_{e,u}\rangle = 0 \qquad (e=uv\in E_B), \tag{20.1}$$

where $\mathrm{Inf}(d)$ denotes the set of infinitesimal isometries of $\mathbb{R}^d$. An infinitesimal motion $\boldsymbol{m}$ is *trivial* if $m_u=m_v$ for every $u,v\in V$.

**Infinitesimal rigidity:** A body-bar framework $(G,\boldsymbol{b})$ is *infinitesimally rigid* if every infinitesimal motion of $(G,\boldsymbol{b})$ is trivial.

**Degree of freedom:** The *degree of freedom* of a body-bar framework is the dimension of the space of infinitesimal motions modulo trivial motions.

**Generic framework:** A body-bar framework $(G, \boldsymbol{b})$ is *generic* if the set of coordinates in $\boldsymbol{b}$ is algebraically independent over $\mathbb{Q}$.

$(k,\ell)$**-sparsity:** A multigraph $G$ is $(k,\ell)$*-sparse* if $|F| \leq k|V(F)| - \ell$ for any nonempty $F \subseteq E(G)$, and a $(k,\ell)$-sparse graph $G$ is $(k,\ell)$*-tight* if $|E(G)| = k|V(G)| - \ell$.

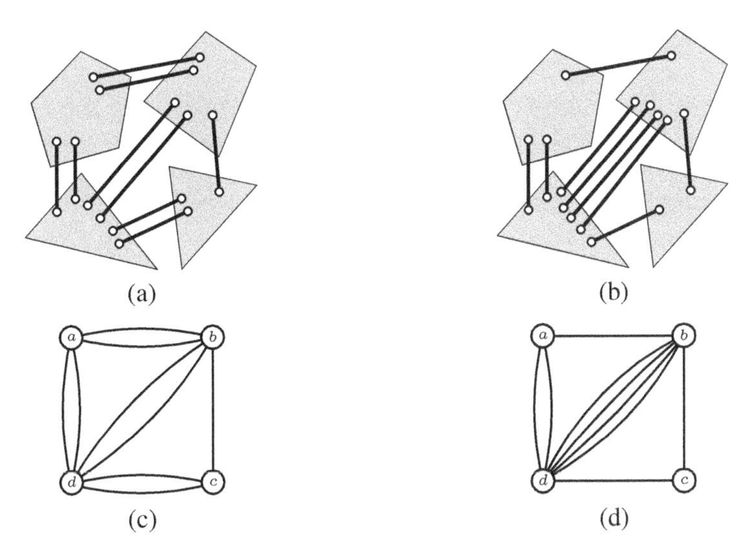


**Figure 20.1**
Body-bar frameworks in $\mathbb{R}^2$ and the underlying graphs.

**Equivalent bar-joint frameworks.** A body-bar framework can be defined in an alternative way by regarding each body as a globally rigid bar-joint subframework. The underlying graph of the resulting bar-joint framework is called a body-bar graph. Specifically, for a graph $G$, the ($d$-dimensional) *body-bar graph induced by* $G$, denoted by $G^{\mathrm{B}}$, is defined as follows:

- $G^{\mathrm{B}}$ consists of $(d+1)|V(G)| + 2|E(G)|$ vertices; for each $v \in V(G)$ we have $d+1$ vertices $x_{v,1}, \ldots, x_{v,d+1}$ and for each $e = uv \in E(G)$ we have two vertices $x_{e,u}$ and $x_{e,v}$;
- the *core* $C(v)$ *of* $v \in V(G)$ is the complete graph on $\{x_{v,1}, \ldots, x_{v,d+1}\}$;
- the *body* $B(v)$ *of* $v \in V(G)$ is the complete graph on $V(C(v)) \cup \{x_{e,v} : e \in \delta_G(v)\}$, where $\delta_G(v)$ denotes the set of edges in $G$ incident to $v$;
- $E(G^B) := \bigcup_{v \in V(G)} E(B(v)) \cup \{\{x_{e,u}, x_{e,v}\} : e = uv \in E(G)\}$.

See Figure 20.2 for an example.

Given a body-bar framework $(G, \boldsymbol{b})$, the above construction of $G^{\mathrm{B}}$ can be extended at the level of frameworks to obtain an equivalent bar-joint framework $(G^{\mathrm{B}}, \boldsymbol{p})$. Here the joint configuration $\boldsymbol{p} : V(G^{\mathrm{B}}) \to \mathbb{R}^d$ is taken such that $\boldsymbol{p}(x_{e,v}) = b_{e,v}$ for every $(e,v) \in \hat{E}_B$ and $\boldsymbol{p}(C(v))$ is an affinely independent set of points whose coordinates are algebraically independent from the coordinates

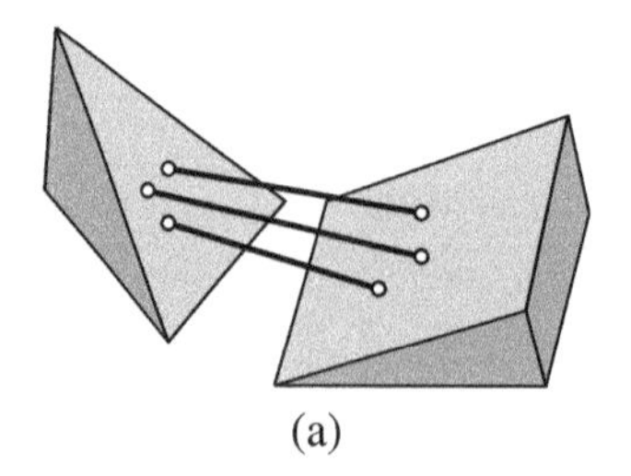
(a)

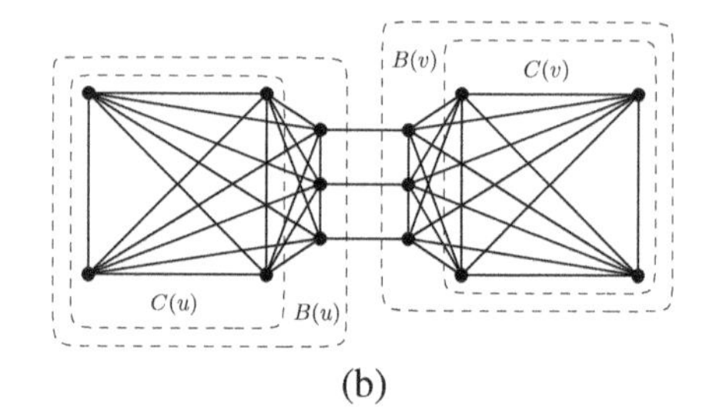

(b)

**Figure 20.2**
A body-bar framework in $\mathbb{R}^3$ and the body-bar graph induced by the underlying graph.

in $\boldsymbol{b}$. Then $(G,\boldsymbol{b})$ is infinitesimally rigid/rigid/globally rigid if and only if $(G^{\mathrm{B}},\boldsymbol{p})$ is infinitesimally rigid/rigid/globally rigid. In this sense, body-bar frameworks form a special class of bar-joint frameworks, and several basic facts about bar-joint frameworks can be translated in terms of body-bar frameworks. An example is the following connection between rigidity and infinitesimal rigidity.

***Proposition 20.1***
*If a body-bar framework $(G,b)$ is infinitesimally rigid, then it is rigid.*

**Rigidity matrices.** The study of body-bar frameworks was initiated by Tay [39], where he proposed a rigidity matrix written in terms of the Plücker coordinates of bars.

Let $\bigvee^2\mathbb{R}^{d+1}$ be a $\binom{d+1}{2}$-dimensional vector space each of whose coordinate is indexed by a pair $(i,j)$ with $1\leq i<j\leq d+1$, and for $p,q\in\mathbb{R}^{d+1}$, let $p\vee q$ be a vector in $\bigvee^2\mathbb{R}^{d+1}$ whose $(i,j)$-th entry is the determinant of the $2\times 2$-matrix consisting of the $i$-th and the $j$-th rows of $(d+1)\times 2$-matrix $(p\ q)$.

For a vector $x\in\mathbb{R}^d$, let $\hat{x}=(x^\top,1)^\top\in\mathbb{R}^{d+1}$ be a vector in $\mathbb{R}^{d+1}$ obtained from $x\in\mathbb{R}^d$ by appending one at the last coordinate. Let $[x,y]$ be the (oriented) line segment from $x$ to $y$ in $\mathbb{R}^d$. Then the *Plücker coordinate vector* of $[x,y]$ is defined by $\hat{x}\vee\hat{y}$. For example, if $x=(x_i)_{1\leq i\leq 3}$ and $y=(y_i)_{1\leq i\leq 3}$, then

$$\hat{x}\vee\hat{y}=\begin{pmatrix}x_1y_2-x_2y_1 & x_2y_3-x_3y_2 & x_3y_1-x_1y_3 & x_1-y_1 & x_2-y_2 & x_3-y_3\end{pmatrix}^\top.$$

Let $m\in\mathrm{Inf}(d)$ be an infinitesimal isometry that can be represented as $m(x)=Sx+t$ for some skew-symmetric matrix $S$ and $t\in\mathbb{R}^d$ for every $x\in\mathbb{R}^d$. We can alternatively represents $m$ by using the skew-symmetric matrix $\hat{S}:=\begin{pmatrix}S & t\\ -t & 0\end{pmatrix}$ of size $(d+1)\times(d+1)$. Then $\begin{pmatrix}m(x)\\ -\langle m,t\rangle\end{pmatrix}=\hat{S}\hat{x}$ for every $x\in\mathbb{R}^d$. By regarding $\hat{S}$ as a $\binom{d+1}{2}$-dimensional vector $s\in\bigvee^2\mathbb{R}^{d+1}$ whose $(i,j)$-the entry is equal to the $(i,j)$-th entry of $\hat{S}$, one can canonically represent each infinitesimal isometry $m\in\mathrm{Inf}(d)$ as a vector $s$ in $\bigvee^2\mathbb{R}^{d+1}$. Then for any $x,y,z\in\mathbb{R}^d$ we have

$$\langle x-y,m(z)\rangle=\langle\hat{x}-\hat{y},\hat{S}\hat{z}\rangle=(\hat{x}-\hat{y})^\top\hat{S}\hat{z}=\langle s,(\hat{x}-\hat{y})\vee\hat{z}\rangle. \tag{20.2}$$

Consider a body-bar framework $(G,\boldsymbol{b})$ and an infinitesimal motion $\boldsymbol{m}$ of $(G,\boldsymbol{b})$. If we represent each $m_v$ $(v\in V(G))$ by $s_v\in\bigvee^2\mathbb{R}^{d+1}$, then it follows from (20.2) and anticommutativity of $\vee$ that

$$\langle m_v(b_{e,v})-m_u(b_{e,u}),b_{e,v}-b_{e,u}\rangle=\langle s_v-s_u,\hat{b}_{e,v}\vee\hat{b}_{e,u}\rangle.$$

In view of this relation, one can alternatively define an infinitesimal motion of $(G,\boldsymbol{b})$ as $\boldsymbol{s}:V(G)\to\bigvee^2\mathbb{R}^{d+1}$ satisfying

$$\langle s_v-s_u,\hat{b}_{e,v}\vee\hat{b}_{e,u}\rangle=0\quad(e=uv\in E_B). \tag{20.3}$$

Then an infinitesimal motion $\boldsymbol{s}$ is trivial if and only if $s_u = s_v$ for every $u, v \in V$.

Note that (20.3) is a system of linear equations in variable $\boldsymbol{s}$, where the solution space is the space of infinitesimal motions of $(G, \boldsymbol{b})$. The *rigidity matrix* $R^{\mathrm{B}}(G, \boldsymbol{b})$ is a matrix of size $|E(G)| \times \binom{d+1}{2}|V(G)|$ representing the system (20.3), i.e.,

$$e = uv \begin{pmatrix} 0 & \dots & 0 & \overset{u}{\hat{b}_{e,u} \vee \hat{b}_{e,v}} & 0 & \dots & 0 & \overset{v}{\hat{b}_{e,v} \vee \hat{b}_{e,u}} & 0 & \dots & 0 \end{pmatrix}.$$

Since the set of trivial infinitesimal motions forms a $\binom{d+1}{2}$-dimensional linear space, we have the following.

***Proposition 20.2***
*A body-bar framework $(G, \boldsymbol{b})$ in $\mathbb{R}^d$ is infinitesimally rigid if and only if* $\operatorname{rank} R^{\mathrm{B}}(G, \boldsymbol{b}) = \binom{d+1}{2}|V(G)| - \binom{d+1}{2}$.

**Maxwell-type condition.** By Proposition 20.2,

$$|E(G)| \geq \binom{d+1}{2}|V(G)| - \binom{d+1}{2}$$

if a body-bar framework $(G, \boldsymbol{b})$ is infinitesimally rigid. This condition can be strengthened as in the form of Maxwell's condition for bar-joint frameworks.

***Proposition 20.3***
*If a body-bar framework $(G, \boldsymbol{b})$ is infinitesimally rigid, then $G$ contains $\binom{d+1}{2}$ edge-disjoint spanning trees, or equivalently a spanning $(\binom{d+1}{2}, \binom{d+1}{2})$-tight subgraph.*

The equivalence between the two conditions in Proposition 20.3 is due to Nash-Williams' theorem [32]. (See Section 20.5 for more details.)

Proposition 20.3 is a sufficient condition for the existence of nontrivial infinitesimal motions, and thus it can be used for proving that a body-bar framework is not infinitesimally rigid. However, the form of Proposition 20.3 (which is the most popular form in the literature) is not convenient for this purpose as it is not clear how to show that $G$ does not contain $\binom{d+1}{2}$ edge-disjoint spanning trees. The Tutte-Nash-Williams theorem explained in Theorem 20.18 converts Proposition 20.3 to the following useful form.

***Proposition 20.4***
*A body-bar framework $(G, \boldsymbol{b})$ is not infinitesimally rigid if there is a partition $\mathcal{P}$ of $V(G)$ satisfying $e_G(\mathcal{P}) < \binom{d+1}{2}(|\mathcal{P}| - 1)$, where $e_G(\mathcal{P})$ denotes the number of edges connecting different sets in $\mathcal{P}$.*

For example, in graph $G$ in Figure 20.1(d), take partition $\mathcal{P} = \{\{a\}, \{b, d\}, \{c\}\}$ of $V(G)$. Then $e_G(\mathcal{P}) = 5 < 6 = 3(|\mathcal{P}| - 1)$. Hence this partition certifies that the body-bar framework in Figure 20.1(b) is not infinitesimally rigid.

In fact the connection between infinitesimal rigidity and tree-packings in Proposition 20.3 is explicit in the combinatorial zero/nonzero pattern of the rigidity matrix $R^{\mathrm{B}}(G, \boldsymbol{b})$. For $d = 1$, $R^{\mathrm{B}}(G, \boldsymbol{b})$ is exactly the incidence matrix of (an orientation) $G$, and in general $R^{\mathrm{B}}(G, \boldsymbol{b})$ can be considered as the "union" of $\binom{d+1}{2}$ copies of the incidence matrix, where the meaning of "union" can be formalized using *matroid union* in matroid theory, see [36] for more details. Such a combinatorial pattern of the rigidity matrix can be exploited even for analyzing singular cases [46]. The details can be found in Chapter 4.

### 20.1.2 Body-Hinge Frameworks

A $d$-dimensional body-hinge framework models a structure consisting of rigid bodies connected by hinges. Each hinge is a $(d-2)$-dimensional simplex that connects a pair of bodies. The bodies are free to move in $\mathbb{R}^d$ subject to the constraint that the relative motion of any two bodies joined by a hinge is a rotation about the hinge. We shall summarize basic terminology of body-hinge frameworks.

**Body-hinge framework:** A $d$-dimensional *body-hinge framework* is a pair $(G,\boldsymbol{h})$ of a multigraph $G=(V,E_H)$ and $\boldsymbol{h}: E_H \to \Delta^d_{d-2}$, where $\Delta^d_{d-2}$ denotes the set of $(d-2)$-dimensional simplices in $\mathbb{R}^d$. Here $G$ represents the underlying graph, where each vertex corresponds to a body and each edge corresponds to a hinge. $\boldsymbol{h}$ represents the hinge-configuration. See Figure 20.3 and Figure 20.4(a).

**Body-hinge realization:** A body-hinge framework $(G,\boldsymbol{h})$ with the underlying graph $G$ is called a *body-hinge realization* of $G$.

**Replacement:** A *replacement* of a $d$-dimensional body-hinge framework $(G,\boldsymbol{h})$ is a map $\boldsymbol{r}: V \ni v \mapsto r_v \in \mathrm{Euc}(d)$ that satisfies

$$r_u(x) = r_v(x) \qquad (e = uv \in E_H, x \in h_e).$$

A replacement $\boldsymbol{r}$ is *nontrivial* if $r_u \neq r_v$ for some pairs of vertices $u,v$.

**Global rigidity:** A $d$-dimensional body-hinge framework $(G,\boldsymbol{h})$ is *globally rigid* if it has no nontrivial replacements.

**Finite motion:** A *finite motion* is a continuous family $\boldsymbol{r}^t$ $(t \in [0,1])$ of replacements of $(G,\boldsymbol{h})$ such that $r^0_v$ is the identity for every $v \in V$. A finite motion $\boldsymbol{r}^t$ is nontrivial if $r^t_v \neq r^t_u$ for some $t \in (0,1]$ and some $u,v \in V$.

**Rigidity:** A $d$-dimensional body-hinge framework $(G,\boldsymbol{h})$ is *rigid* if $(G,\boldsymbol{h})$ has no nontrivial finite motion.

**Infinitesimal motion:** An *infinitesimal motion* $m$ of a $d$-dimensional body-hinge framework $(G,\boldsymbol{h})$ is a map $\boldsymbol{m}: V \ni v \mapsto m_v \in \mathrm{Inf}(d)$ that satisfies

$$m_u(x) = m_v(x) \qquad (e = uv \in E_H, x \in h_e). \tag{20.4}$$

An infinitesimal motion $\boldsymbol{m}$ is *trivial* if $m_u = m_v$ for every $u,v \in V$.

**Infinitesimal rigidity:** $(G,\boldsymbol{h})$ is *infinitesimally rigid* if every infinitesimal motion of $(G,\boldsymbol{h})$ is trivial.

**Generic framework:** A body-hinge framework $(G,\boldsymbol{h})$ is *generic* if the set of coordinates of the vertices of the simplices in $\boldsymbol{h}$ is algebraically independent over $\mathbb{Q}$.

**Equivalent bar-joint frameworks.** As in the case of body-bar frameworks, the rigidity property of body-hinge frameworks can be captured by looking at equivalent bar-joint frameworks, obtained by replacing each body by a bar-joint realization of a large enough complete graph. For a graph $G=(V,E_H)$, the ($d$-dimensional) *body-hinge graph induced by* $G$, denoted by $G^{\mathrm{H}}$, is defined as follows:

- $G^{\mathrm{H}}$ consists of $(d+1)|V(G)|+(d-1)|E(G)|$ vertices; for each $v \in V(G)$ we have $d+1$ vertices $x_{v,1},\ldots,x_{v,d+1}$ and for each $e \in E(G)$ we have $d-2$ vertices $x_{e,1},\ldots,x_{e,d-1}$;

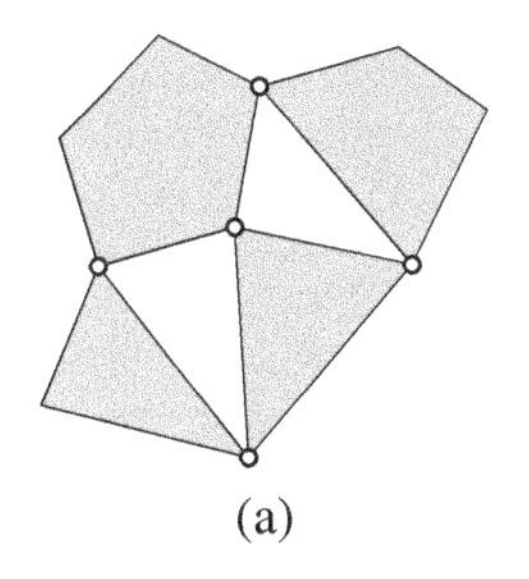
(a)

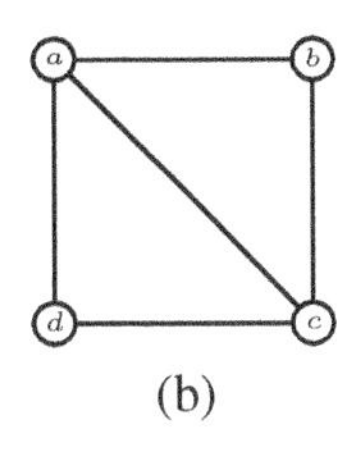

(b)

**Figure 20.3**
A body-hinge framework in $\mathbb{R}^2$ and the underlying graph.

- the *hinge* $H(e)$ *of* $e$ is the complete graph on $\{x_{e,1},\dots,x_{e,d-1}\}$ for each $e \in E$;
- the *core* $C(v)$ *of* $v$ is the complete graph on $\{x_{v,1},\dots,x_{v,d+1}\}$ for each $v \in V$;
- the *body* $B(v)$ *of* $v$ is the complete graph on $V(C(v)) \cup \bigcup_{e\in\delta_G(v)} V(H(e))$;
- $E(G^{\mathrm{H}}) := \bigcup_{v\in V} E(B(v))$.

See Figure 20.4.

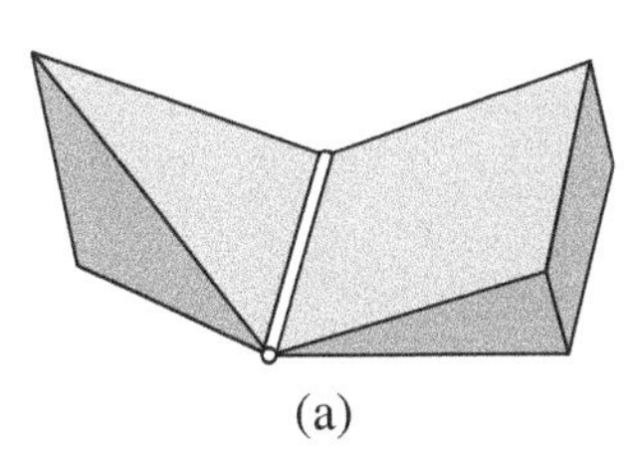
(a)

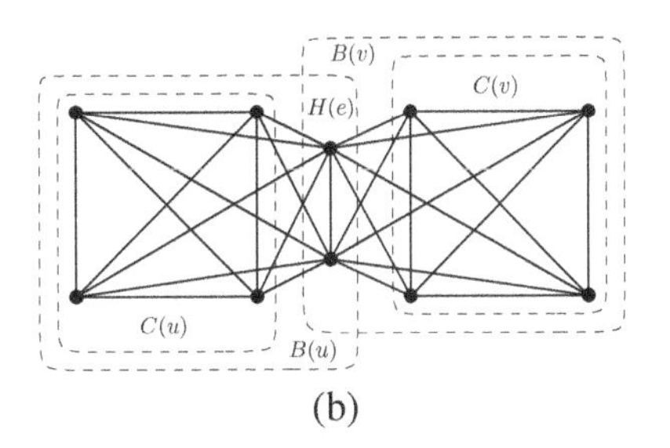

(b)

**Figure 20.4**
A body-hinge framework in $\mathbb{R}^3$ and the body-hinge graph induced by the underlying graph.

Given a body-hinge framework $(G,\boldsymbol{h})$, the above construction of $G^{\mathrm{H}}$ can be extended at the level of frameworks to obtain an equivalent bar-joint framework $(G^{\mathrm{H}},\boldsymbol{p})$. Here $\boldsymbol{p}: V(G^H) \to \mathbb{R}^d$ is taken such that $h_e = \mathrm{conv}\{\boldsymbol{p}(x_{e,i}) : 1 \le i \le d-1\}$ for every $e \in E_H$ and $\boldsymbol{p}(C(v))$ is an affinely independent set of points independent from the coordinates of the vertices in $\boldsymbol{h}$. Then $(G,\boldsymbol{h})$ is infinitesimally rigid/rigid/globally rigid if and only if $(G^{\mathrm{H}},\boldsymbol{p})$ is infinitesimally rigid/rigid/globally rigid.

**Maxwell-type condition.** Suppose that there are two bodies indexed by $u$ and $v$ and connected by a hinge $h_{uv}$ in $\mathbb{R}^d$. The hinge restricts the possible infinitesimal motions $m_u, m_v \in \mathrm{Inf}(d)$ assigned to those bodies by

$$m_u(x) = m_v(x) \qquad (x \in h_{uv}). \tag{20.5}$$

This is equivalent to

$$\langle y - x, m_u(x) - m_v(x)\rangle = 0 \qquad (x \in h_{uv}, y \in \mathbb{R}^d). \tag{20.6}$$

Letting $s_u, s_v \in \bigvee^2 \mathbb{R}^{d+1}$ to be the $\binom{d+1}{2}$-dimensional vectors representing $m_u$ and $m_v$ as in the last section, (20.6) can be written as

$$\langle s_u - s_v, \hat{x} \vee \hat{y}\rangle = 0 \qquad (x \in h_{uv}, y \in \mathbb{R}^d). \tag{20.7}$$

The set of (oriented) segments in $\mathbb{R}^d$ is said to be *independent* if the corresponding Plücker coordinate vectors form a linearly independent set. For any independent set of $(\binom{d+1}{2}-1)$ segments $[a_i,b_i]$ intersecting $h_{uv}$, we have $\text{span}\{\hat{x}\vee\hat{y} : x\in h_{uv}, y\in\mathbb{R}^d\} = \text{span}\{\hat{a}_i\vee\hat{b}_i : 1\leq i\leq \binom{d+1}{2}-1\}$. Thus (20.7) is further simplified to a system of $(\binom{d+1}{2}-1)$ equations,

$$\langle s_u - s_v, \hat{a}_i\vee\hat{b}_i\rangle = 0 \qquad \left(1\leq i\leq \binom{d+1}{2}-1\right). \tag{20.8}$$

Each equation in (20.7) is nothing but the linear constraint imposed by a bar between $a_i$ and $b_i$. Therefore the hinge constraint is equivalently regarded as length constraints of $(\binom{d+1}{2}-1)$ bars intersecting the hinge.

This replacement of a hinge with bars enables us to regard body-hinge frameworks as a special case of body-bar frameworks, and a rigidity matrix of a body-hinge framework $(G,\boldsymbol{h})$ can be defined as that of a resulting body-bar framework.

For a graph $G$ and a positive integer $k$, let $kG$ be the graph obtained by replacing each edge with $k$ parallel ones. By the above technique of replacing each hinge with $\left(\binom{d+1}{2}-1\right)$ bars, the following Maxwell-type condition for body-hinge frameworks follows from Proposition 20.3.

***Proposition 20.5***
*If a body-hinge framework $(G,\boldsymbol{h})$ is infinitesimally rigid, then $(\binom{d+1}{2}-1)G$ contains $\binom{d+1}{2}$ edge-disjoint spanning trees.*

### 20.1.3 Body-Bar-Hinge Frameworks

The combined notion of body-bar frameworks and body-hinge frameworks are useful in some applications.

**Body-bar-hinge framework:** A *body-bar-hinge framework* is a triple $(G,\boldsymbol{b},\boldsymbol{h})$ of a 2-edge-colored graph $G=(V,E_B\cup E_H)$, $\boldsymbol{b}:\hat{E}_B\to\mathbb{R}^d$, and $\boldsymbol{h}:E_H\to\Delta^d_{d-2}$. In the underlying graph $G=(V,E_B\cup E_H)$, each vertex corresponds to a body, each edge in $E_B$ corresponds to a bar, and each edge in $E_H$ is corresponding to a hinge.

The rigidity/global rigidity/infinitesimal rigidity of a body-bar-hinge framework is defined in an obvious manner by unifying the definitions of body-bar frameworks and body-hinge frameworks. The constructions of equivalent bar-joint frameworks are also extendable, and hence the rigidity question for body-bar-hinge frameworks can be answered by looking at corresponding equivalent bar-joint frameworks.

For the analysis of infinitesimal rigidity, one may convert each hinge with $\binom{d+1}{2}-1$ independent bars intersecting the hinge, and the analysis can be carried out in an equivalent body-bar framework. This in particular implies the following.

***Proposition 20.6***
*If a body-bar-hinge framework $(G=(V,E_B\cup E_H),\boldsymbol{b},\boldsymbol{h})$ is infinitesimally rigid, then $H$ contains $\binom{d+1}{2}$ edge-disjoint spanning trees, where $H$ is the graph obtained from $G$ by replacing each edge in $E_H$ with $\binom{d+1}{2}-1$ parallel edges.*

## 20.2 Generic Rigidity

In this section we give a characterization of the rigidity of generic body-bar-hinge frameworks.

### 20.2.1 Body-Bar Frameworks

A remarkable feature of body-bar frameworks is that, in the generic case, the converse direction of Proposition 20.3 is true in any dimension. This is in contrast to bar-joint frameworks, whose combinatorial characterization is a central open question in rigidity theory.

**Theorem 20.1 (Tay [39])** *Let $G$ be a multi-graph. Then a generic $d$-dimensional body-bar realization of $G$ is rigid if and only if $G$ contains $\binom{d+1}{2}$ edge-disjoint spanning trees.*

For example, for the body-bar framework in Figure 20.1(a), three edge-disjoint spanning trees in the underlying graph in Figure 20.1(c) can be used for a certificate of the infinitesimal rigidity. Depending on the context, different combinatorial conditions are used, see Theorem 20.20.

The necessity of Theorem 20.1 follows from Proposition 20.3. Tay's original proof for the sufficiency is based on a constructive characterization of $(k,k)$-tight graphs, see Theorem 20.24. Several different proofs are known in the literature, and one of the simplest proofs is the following. Consider proving a slightly generalized statement by induction on the lexicographical ordering of $(V(G),E(G))$: a generic $d$-dimensional body-bar realization of $G$ has $k$ degrees of freedom if $G$ is $(\binom{d+1}{2},\binom{d+1}{2})$-sparse with $|E(G)| = \binom{d+1}{2}|V(G)| - \binom{d+1}{2} - k$. We assume $E(G) \neq \emptyset$ since otherwise the claim is trivial. Take any edge $e = uv$ and consider a generic realization $(G-e,\boldsymbol{b})$ of $G-e$, which has $k+1$ degrees of freedom by induction. If a generic realization of $G$ has more than $k$ degrees of freedom, then every infinitesimal motion $\boldsymbol{m}$ of $(G-e,\boldsymbol{b})$ satisfies $m_u = m_v$. Now, consider the graph $G/e$ obtained by contracting $e$. Then, as $m_u = m_v$ holds for every infinitesimal motion $m$ of $(G-e,\boldsymbol{b})$, the degree of freedom of a generic realization of $G/e$ should be at least that of $(G-e,\boldsymbol{b})$, which is $k+1$. However, one can show that $G/e$ contains a spanning $(\binom{d+1}{2},\binom{d+1}{2})$-sparse subgraph $H$ with $|E(H)| \geq \binom{d+1}{2}|V(H)| - \binom{d+1}{2} - k$, which is a contradiction since a generic realization of $H$ has degree of freedom at most $k$ by induction.

Another simple proof was due to Whiteley [47] using tree-decompositions, and the idea is used in various different places [21, 22, 26, 35]. Suppose that $G$ can be decomposed into $\binom{d+1}{2}$ edge-disjoint spanning connected subgraphs $T_{i,j}$ $(1 \leq i < j \leq d+1)$. Take an affinely independent set $x_1,\ldots,x_{d+1}$ of points in $\mathbb{R}^d$ and set $\boldsymbol{b}$ to be $b_{e,u} = x_i$ and $b_{e,v} = x_j$ if $e = uv \in T_{i,j}$. Then due to the specialty of $\boldsymbol{b}$ it can be easily shown that the rigidity matrix $R^{\mathrm{B}}(G,\boldsymbol{b})$ is equivalent to the direct sum of the adjacency matrices of $T_{i,j}$ over all pairs $i,j$. Thus the rigidity matrix has rank equal to $\binom{d+1}{2}(|V(G)|-1)$, and $(G,\boldsymbol{b})$ is infinitesimally rigid.

### 20.2.2 Body-Hinge Frameworks

Tay [40, 41] and Whiteley [47] independently observed that a similar combinatorial condition as in Theorem 20.1 can be used to characterize the generic rigidity of body-hinge frameworks.

**Theorem 20.2 (Tay [40, 41], Whiteley [47])** *Let $G$ be a multi-graph. Then a generic $d$-dimensional body-hinge realization of $G$ is rigid if and only if $(\binom{d+1}{2}-1)G$ contains $\binom{d+1}{2}$ edge-disjoint spanning trees.*

Whiteley's original proof [47] is simpler although Tay [40, 41] proved a more general statement (see Subsection 20.3.2). In fact, a proof of Theorem 20.1 given in the last section is applicable for Theorem 20.2. To see this, suppose that $(\binom{d+1}{2}-1)G$ can be decomposed into $\binom{d+1}{2}$ edge-disjoint spanning connected subgraphs $T_{i,j}$ $(1 \leq i < j \leq d+1)$. As in the last section, we construct a body-bar realization $(((\binom{d+1}{2})-1)G, \boldsymbol{b})$ of $((\binom{d+1}{2})-1)G$ by taking an affinely independent set $x_1, \ldots, x_{d+1}$ of points in $\mathbb{R}^d$ and setting $\boldsymbol{b}$ to be $b_{e,u} = x_i$ and $b_{e,v} = x_j$ if $e \in T_{i,j}$. Then due to the specialty of $\boldsymbol{b}$, one can check that $(((\binom{d+1}{2})-1)G, \boldsymbol{b})$ is infinitesimally rigid. Moreover the set of the $\binom{d+1}{2}-1$ bars associated with an edge $f \in E(G)$ intersects a $(d-2)$-dimensional simplex, denoted by $h_f$. Thus by setting $\boldsymbol{h}$ to be $\boldsymbol{h}(f) = h_f$ for each $f \in E(G)$ we get an infinitesimally rigid body-hinge realization $(G, \boldsymbol{h})$, as $(G, \boldsymbol{h})$ is infinitesimally rigid if and only if $(((\binom{d+1}{2})-1)G, \boldsymbol{b})$ is infinitesimally rigid.

As noted by Jackson and Jordán [21], it is possible to extend the characterization to body-bar-hinge frameworks by applying Whiteley's proof. Moreover the rank of the rigidity matrix can be combinatorially described.

**Theorem 20.3 (Jackson and Jordán [21])** *Let $G = (V, E_B \cup E_H)$ be a 2-edge-colored multigraph, and let $H$ be a graph obtained from $G$ by replacing each edge in $E_H$ by $\binom{d+1}{2}-1$ parallel copies. Then the rank of the rigidity matrix of a generic body-bar-hinge realization of $G$ in $\mathbb{R}^d$ is equal to*

$$\min\left\{\binom{d+1}{2}(|V(G)| - |\mathcal{Q}|) + e_H(\mathcal{Q}) : \mathcal{Q} \text{ is a partition of } V(H)\right\}.$$

## 20.3 Other Related Models

In this section we summarize known results on variants of body-bar-hinge frameworks.

**$k$-plate:** A *$k$-plate* is a compact connected set in $\mathbb{R}^d$ whose affine span is $k$-dimensional.

**$k$-plate-bar framework:** A *$k$-plate-bar framework* is a structure consisting of $k$-plates connected by bars. Each bar can connect two distinct plates. The underlying graph is defined by associating each $k$-plate with a vertex and each bar with an edge. See Figure 20.5 for an example.

**$k$-plate-bar realization:** A $k$-plate-bar framework with the underlying graph $G$ is called a *$k$-plate-bar realization* of $G$.

**Identified body-hinge framework:** An *identified body-hinge framework* is a generalized model of a body-hinge framework, where each hinge may connect more than two bodies.

**Panel-hinge framework:** A *panel-hinge framework* is a structure consisting of rigid panels connected by hinges, where a *panel* means a $(d-1)$-plate in $\mathbb{R}^d$.

**Square of a graph:** The *square* $G^2$ of a graph $G$ is the graph on $V(G)$ which contains an edge $uv$ if and only if the distance between $u$ and $v$ in $G$ is at most two. See Figure 20.6.

**Molecular framework:** A *molecular framework* is a 3-dimensional body-hinge framework in which all the hinges incident to each body pass through a point.

**Body-pin framework:** A *body-pin framework* is a structure consisting of bodies connected at points.

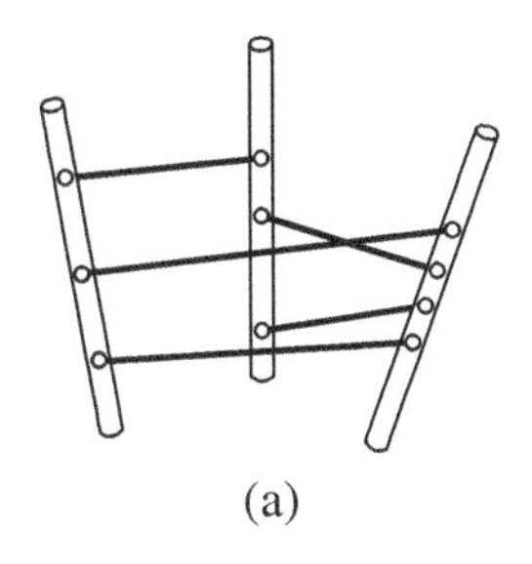
(a)

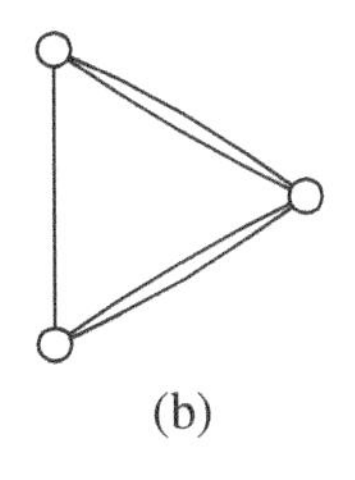
(b)

**Figure 20.5**
A 1-plate-bar framework in $\mathbb{R}^3$ and the underlying graph.

### 20.3.1 Plate-Bar Frameworks

The rigidity/infinitesimal rigidity/global rigidity of a $k$-plate-bar framework is defined in the identical manner as those of body-bar framework. The only difference is that for a $k$-plate-bar framework the space of replacements of each plate is defined by factoring the isometries fixing the plate.

Although plate-bar frameworks appear in real applications, understanding the rigidity property is also theoretically important. When $k = d$, those are body-bar frameworks, whose generic rigidity is characterized combinatorially in Theorem 20.1. When $k = 0$, those are bar-joint frameworks, whose generic rigidity has not yet been characterized when $d \geq 3$. The concept of $k$-plate-bar frameworks interpolates between those two well-studied models.

It follows from the definition that the same combinatorial condition can be used to characterize the generic rigidity for $k = d - 1$. The following result by Tay [40, 41] solves the case when $k = d - 2$.

**Theorem 20.4 (Tay [40, 41])** *Let $G$ be a multigraph. Then $G$ has an infinitesimally rigid $(d-2)$-plate-bar realization in $\mathbb{R}^d$ if and only if $G$ contains a $(\binom{d+1}{2} - 1, \binom{d+1}{2})$-tight spanning subgraph.*

Tanigawa [36] gave an alternative inductive proof starting from Theorem 20.1. A constructive characterization of $(k, k+1)$-tight graphs is known by Frank and Szegő [10].

It seems that going beyond Theorem 20.4 is less explored. Understanding $(d-3)$-plate-bar frameworks is certainly difficult for $d = 3$, but for higher dimensions the Maxwell-type condition may be necessary and sufficient.

### 20.3.2 Identified Body-Hinge Frameworks

The body-hinge model given in Section 20.1.2 can capture identified body-hinge frameworks by locating several hinges at the same position. However the resulting body-hinge framework is no longer generic and Theorem 20.2 cannot be applied to identified body-hinge frameworks.

In order to capture its combinatorics correctly, the underlying graph of an identified body-hinge framework is defined by a hypergraph by associating each body with a vertex and each hinge with a hyperedge, or equivalently a bipartite graph $G = (V_B \cup V_H, E)$ whose one side $V_B$ corresponds to the set of bodies and the other side $V_H$ corresponds to the set of hinges. There is an edge between $u \in V_B$ and $v \in V_H$ if the hinge of $v$ is incident to the body of $u$.

Since each hinge is a $(d-2)$-plate, each identified body-hinge framework can be considered as a framework consisting of bodies and $(d-2)$-plates linked by bars. Hence the following extension of Theorem 20.2 follows as a corollary of Theorem 20.4.

**Theorem 20.5 (Tay [40, 41])** *Let $G = (V_B \cup V_H, E)$ be a bipartite graph. Then a generic identified body-hinge realization of $G$ in $\mathbb{R}^d$ is rigid if and only if $(\binom{d+1}{2} - 1)G$ contains a $(\binom{d+1}{2}, \binom{d+1}{2} -$

$1, \binom{d+1}{2})$*-tight spanning subgraph* $H$*, i.e.,* $|E(H)| = \binom{d+1}{2}|V_B| + (\binom{d+1}{2} - 1)|V_H| - \binom{d+1}{2}$ *and* $|E(I)| \leq \binom{d+1}{2}|V_B \cap V(I)| + (\binom{d+1}{2} - 1)|V_H \cap V(I)| - \binom{d+1}{2}$ *for any nonempty subgraph* $I$ *of* $H$*.*

### 20.3.3 Panel-Hinge Frameworks

Panel-hinge frameworks are another important variants of body-hinge frameworks that frequently appear in structural engineering. Theoretically they also have a rich historical background dating back to Cauchy. Cauchy's theorem says that the 2-skeleton of a strictly convex polytope is rigid in $\mathbb{R}^3$ if it is regarded as a panel-hinge framework by identifying each face as a panel and each edge as a hinge. The rigidity of polytopes is explained in detail in Chapter 13.

A panel-hinge framework can be regarded as a body-hinge framework with *hinge-coplanarity condition*, i.e., the set of hinges incident to a body lies on a $(d-1)$-dimensional subspace. Hence the Maxwell-type necessary condition for infinitesimal rigidity (Proposition 20.5) is still valid for any panel-hinge framework. On the other hand the combinatorial characterization for generic body-hinge frameworks (Theorem 20.2) cannot be applied because of the hinge-coplanarity condition. Tay and Whiteley [42] conjectured that the same combinatorial condition is still valid for panel-hinge frameworks. This conjecture was confirmed by Jackson and Jordán [20] for $d = 2$, and then by Katoh and Tanigawa [25] for general $d$.

**Theorem 20.6 (Katoh and Tanigawa [25])** *Let* $G$ *be a graph. Then a generic* $d$*-dimensional panel-hinge realization of* $G$ *is rigid if and only if* $\left(\binom{d+1}{2} - 1\right) G$ *contains* $\binom{d+1}{2}$ *edge-disjoint spanning trees.*

Extending Theorem 20.6 to identified panel-hinge frameworks is an important unsolved problem.

### 20.3.4 Molecular Frameworks

A molecular framework is a special class of body-hinge frameworks which can be used to understand the rigidity of ideal molecules. Consider a simplest model of ideal molecules in $\mathbb{R}^3$, where atoms are connected only by covalent bonds. Each bond constrains the distance between two atoms while two consecutive bonds fixes the angle between the bonds. Hence the rigidity property of an ideal molecule with the underlying graph $G$ is analyzed by looking at its square $G^2$. See Figure 20.6.

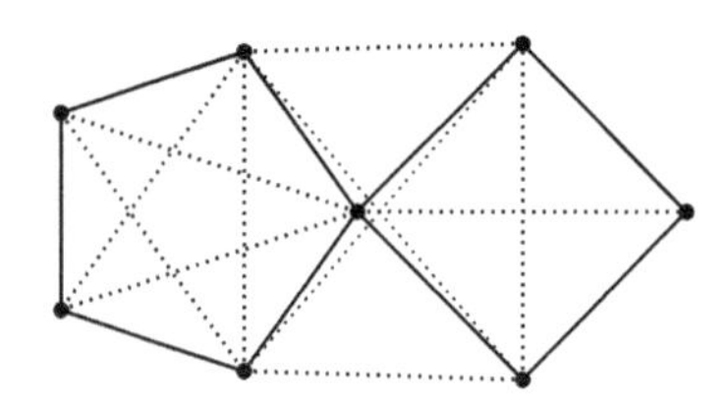

**Figure 20.6**
$G$ and $G^2$, where $G$ consists of the bold edges and $G^2$ consists of the bold edges and the dotted edges.

Observe that for each vertex $v \in V(G)$ the neighbors of $v$ and $v$ itself induce a complete subgraph in $G^2$. By regarding such a complete subgraph as a body associated with $v$, we see that two bodies share a hinge if and only if the associated two vertices are adjacent in $G$. Thus, with this identification between complete subgraphs and bodies, a bar-joint framework $(G^2, \boldsymbol{p})$ of $G^2$ leads to a body-hinge framework $(G, \boldsymbol{h})$ with the underlying graph $G$, where $\boldsymbol{h}(ij) = [\boldsymbol{p}(i), \boldsymbol{p}(j)]$ for each edge $ij \in E(G)$.

Formally we have the following:

***Proposition 20.7***
*Let $(G^2, \boldsymbol{p})$ be a bar-joint framework in general position for a graph $G$ with minimum degree at least two, and let $(G, \boldsymbol{h})$ be the 3-dimensional body-hinge framework with $\boldsymbol{h}(ij) = [\boldsymbol{p}(i), \boldsymbol{p}(j)]$ for each edge $ij \in E(G)$. Then $(G^2, \boldsymbol{p})$ is infinitesimally rigid in $\mathbb{R}^3$ if and only if $(G, \boldsymbol{h})$ is infinitesimally rigid in $\mathbb{R}^3$.*

Note that the body-hinge framework $(G, \boldsymbol{h})$ in Proposition 20.7 is hinge-concurrent.

Tay and Whiteley [42] observed that a molecular framework in $\mathbb{R}^3$ can be converted to a panel-hinge framework through polarity. Recall that in a three-dimensional projective space and its polar space a point is associated with a hyperplane while a line is associated with a line keeping the incidence structure. Hence for each hinge-concurrent body-hinge framework in $\mathbb{R}^3$ one can associate a panel-hinge framework in the polar space. A classical observation by Crapo and Whiteley [6] is that the infinitesimal rigidity is preserved under projective polarity, which implies the following.

**Theorem 20.7 (Crapo and Whiteley [6])** *Let $G$ be a graph. Then a molecular framework $(G, \boldsymbol{h})$ is infinitesimally rigid in $\mathbb{R}^3$ if and only if the polar panel-hinge framework $(G, \boldsymbol{h}^*)$ is infinitesimally rigid in $\mathbb{R}^3$.*

Combining Theorem 20.6, Proposition 20.7, and Theorem 20.7, we have the following, which was known as the *molecular rigidity conjecture* before [25]:

**Theorem 20.8 (Katoh and Tanigawa[25])** *For a graph $G$, the following are equivalent:*

**(a)** *A generic molecular framework with the underlying graph $G$ is rigid in $\mathbb{R}^3$;*

**(b)** *A generic panel-hinge realization of $G$ is rigid in $\mathbb{R}^3$;*

**(c)** *A generic bar-joint realization of $G^2$ is rigid in $\mathbb{R}^3$;*

**(d)** *$5G$ contains six edge-disjoint spanning trees.*

Jackson and Jordán [18, 19] gave a number of important consequences of the molecular conjecture.

**Theorem 20.9 (Jackson and Jordán [19])** *Let $G = (V, E)$ be a graph with minimum degree at least two. Then the rank of the generic 3-dimensional rigidity matroid of $G^2$ is equal to* $\min\{3|V| - 6|\mathcal{P}| - 3 + 5e_G(\mathcal{P}) : \mathcal{P}$ *is a partition of* $V(G)\}$.

See Chapter 1 for the definition of the generic rigidity matroid.

**Theorem 20.10 (Jackson and Jordán [19])** *Let $G = (V, E)$ be a graph. Then $G^2$ is independent in the 3-dimensional rigidity matroid if and only if $G^2$ satisfies Maxwell's condition, i.e., for any $F \subseteq E(G^2)$ with $|V(F)| \geq 3$, $|F| \leq 3|V(F)| - 6$.*

Recall that a *rigid component* of a graph $H = (V, E)$ is a rigid subgraph of $H$ which is maximal with respect to inclusion. In several applications, the task is to identify rigid components rather than checking just rigidity, and hence characterizing rigid components is important.

**Theorem 20.11 (Jackson and Jordán [18])** *Let $G = (V, E)$ be a graph of minimum degree at least two. Then for each rigid component $C$ of $G^2$ there is a $6/5$-brick $B = G[X]$ of $G$ with $C = G[X \cup N_G(X)]^2$. Conversely for each $6/5$-brick $B = G[X]$ of $G$ the subgraph $G[X \cup N_G(X)]^2$ is a rigid component of $G^2$.*

For the definition of 6/5-brick, see Section 20.5.

As explained in Section 20.5.4 there is an efficient algorithm for computing the collection of 6/5-bricks of $G$, and Theorem 20.11 leads to an efficient algorithm for computing the rigid component decomposition of $G^2$.

It was also shown in [18] that Theorem 20.8 implies the rank formula of $G^2$ in terms of the rigid component decomposition of $G^2$.

Understanding the mechanisms of proteins is a challenging question in molecular biology, and the theory of statics forms a basics for attacking the problem. Combinatorial approaches based on Theorem 20.8 are implemented in several places, where the popular pebble game algorithm or its variant is commonly used for checking the condition of Theorem 20.8. See, e.g., Section 20.5.4 and [44, 49] for more details. A geometric analysis of various special cases can be found in [48, 49].

### 20.3.5 Body-Pin Frameworks

Body-pin frameworks also form a natural class of structures in real applications. If a pin may connect more than two bodies, understanding rigidity properties of body-pin frameworks turns out to be challenging as any rigidity question of bar-joint frameworks can be formulated in this body-pin model. Thus in this subsection we shall focus on the model where each pin connects exactly two bodies. Under this assumption, the underlying graph of a body-pin framework becomes a graph, where each vertex represents a body and each edge represents a pin connecting the two bodies associated with the endvertices. A body-pin framework is then denoted by a pair $(G,\boldsymbol{p})$ of the underlying graph $G=(V,E_P)$ and the pin-configuration $\boldsymbol{p}: E_P \to \mathbb{R}^d$. A replacement of a body-pin framework $(G,\boldsymbol{p})$ is defined as $\boldsymbol{m}: V \ni v \mapsto m_v \in \mathrm{Euc}(d)$ that satisfies

$$m_u(p_e) = m_v(p_e) \qquad (e = uv \in E_P).$$

$(G,\boldsymbol{p})$ is said to be rigid if it has no nontrivial finite motion, i.e., a continuous family $\boldsymbol{r}^t$ $(t \in [0,1])$ of replacements of $(G,\boldsymbol{p})$ such that $m_v^0$ is identity and $r_v^t \neq r_u^t$ for some $t \in (0,1]$.

Dress gave a conjecture for characterizing the rigidity of generic bar-joint frameworks. Although Dress' conjecture was disproved by Jackson and Jordán [17], one may wonder if it still holds for some sub-classes of graphs. Jackson, Jordán and Tanigawa conjecture that Dress' conjecture is true for generic body-pin frameworks. The conjecture has the following attractive form for body-pin frameworks.

**Conjecture 5** *Let G be a graph. Then a generic three-dimensional body-pin realization of G is rigid if and only if*

$$\sum_{\{X,X'\}\in\binom{\mathcal{P}}{2}} h_G(X,X') \geq 6(|\mathcal{P}|-1)$$

*for every partition $\mathcal{P}$ of $V$, where $\binom{\mathcal{P}}{2}$ denotes the set of pairs of subsets in $\mathcal{P}$*

$$h_G(X,X') = \begin{cases} 6 & \text{if } d_G(X,X') \geq 3 \\ 5 & \text{if } d_G(X,X') = 2 \\ 3 & \text{if } d_G(X,X') = 1 \\ 0 & \text{if } d_G(X,X') = 0 \end{cases}$$

*and $d_G(X,X)$ denotes the number of edges in G connecting X and X.*

Note that, if $h_G$ were defined to be $h_G(X,X')=6$ for $d_G(X,X')=2$, then the combinatorial condition is equivalent to the Tutte-Nash-Williams condition in Theorem 20.18 for $3G$ to contain six edge-disjoint spanning trees.

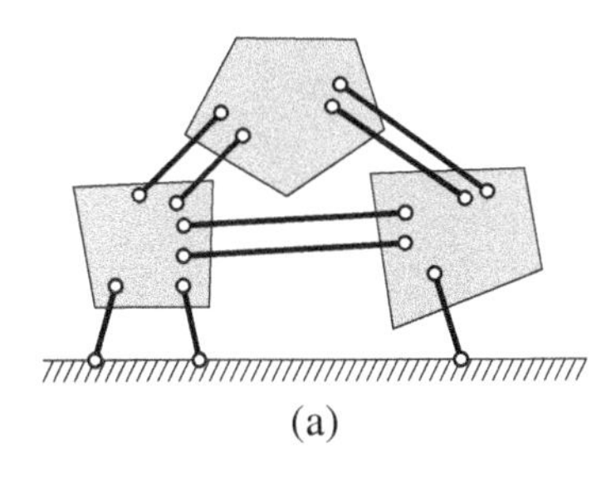
(a)

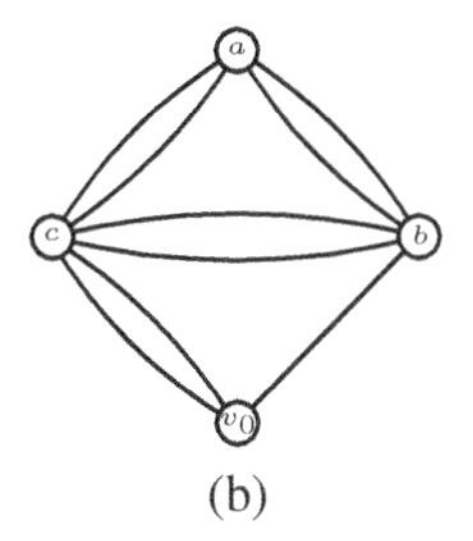

(b)

**Figure 20.7**
A body-bar framework with bar-boundary and the underlying graph.

### 20.3.6 Body-Bar Frameworks with Boundaries

In engineering applications a structure has some links to the external environment such as the ground or walls. In rigidity theory, such a structure is commonly understood by using a *pinned* framework, where pinned points are fixed in the ambient space. Its rigidity property can be captured by the conventional rigidity of frameworks (without pinned points) by linking every pair of fixed points by a bar. However, in the body-bar model, the resulting body-bar frameworks are no longer generic, and Theorem 20.1 cannot be applied. Katoh and Tanigawa [26] gave an extension of Theorem 20.1 to pinned body-bar frameworks.

The underlying combinatorics of a pinned body-bar framework is captured by a graph $G = (V, E_B)$ along with a multisubset $P$ of $V$. Then a pinned body-bar framework is a tuple $(G, P, \boldsymbol{b}, \boldsymbol{p})$ with $\boldsymbol{b} : \hat{E}_B \to \mathbb{R}^d$ and $\boldsymbol{p} : P \to \mathbb{R}^d$, where $\boldsymbol{b}$ and $\boldsymbol{p}$ encode the locations of bars and pins, respectively.

An infinitesimal motion of $(G, P, \boldsymbol{b}, \boldsymbol{p})$ is an infinitesimal motion of the ordinary body-bar framework $(G, \boldsymbol{b})$ with an extra pinning constraint:

$$m_v(\boldsymbol{p}(v)) = 0 \qquad (v \in P).$$

We say that $(G, P, \boldsymbol{b}, \boldsymbol{p})$ is infinitesimally rigid if $(G, P, \boldsymbol{b}, \boldsymbol{p})$ has no nonzero infinitesimal motion.

By extending Tutte-Nash-Williams' tree-packing theorem, Katoh and Tanigawa [26] proved the following.

**Theorem 20.12 (Katoh and Tanigawa [26])** *Let $G = (V, E_B)$ be a graph, $P$ be a multiset $P$ of vertices in $V$, and $\boldsymbol{p} : P \to \mathbb{R}^d$. Then there exists a bar-configuration $b : \hat{E}_B :\to \mathbb{R}^d$ such that the pinned body-bar framework $(G, P, \boldsymbol{b}, \boldsymbol{p})$ is infinitesimally rigid if and only if*

$$e_G(\mathcal{P}) \geq \binom{d+1}{2}|\mathcal{P}| - \sum_{X \in \mathcal{P}} \sum_{i=1}^{d_X+1} (d - i + 1)$$

*for every partition $\mathcal{P}$ of $V$, where $d_X$ denotes the dimension of the affine span of $p(X \cap P)$, which is $-1$ if $X \cap P = \emptyset$.*

It should be noted that $\boldsymbol{p}$ is given in advance in Theorem 20.12 and it may not be generic. Theorem 20.12 is a corollary of the following result on body-bar frameworks with bar-boundary. A body-bar framework with bar-boundary is a body-bar framework, some of whose bodies are linked to the ground by bars. By regarding the ground as a body fixed in the ambient space, the underlying combinatorics can be captured by the augmented graph $G \cup \{v_0\}$ by adding vertex $v_0$ representing the fixed body. See Figure 20.7.

**Theorem 20.13 (Katoh and Tanigawa [26])** *Let $G \cup \{v_0\}$ be a graph with a designated vertex $v_0$, $E_0$ be the set of edges in $G \cup \{v_0\}$ incident to $v_0$, and $\boldsymbol{b}^0 : \hat{E}_0 \to \mathbb{R}^d$. Then there exists an extension*

*$\boldsymbol{b}: \hat{E}(G \cup \{v_0\}) \to \mathbb{R}^d$ of $\boldsymbol{b}^0$ such that the body-bar framework $(G \cup \{v_0\}, \boldsymbol{b})$ is infinitesimally rigid if and only if*

$$e_G(\mathcal{P}) \geq \binom{d+1}{2}|\mathcal{P}| - \sum_{X \in \mathcal{P}} \dim \text{span}\{b^0_{e,u} \vee b^0_{e,v_0} : e \in E_0(X)\}$$

*for every partition $\mathcal{P}$ of $V(G)$, where $E_0(X)$ denotes the set of edges in $E_0$ incident to $X$.*

It should be noted that $\boldsymbol{b}^0$ is given and may not be generic in Theorem 20.13. Theorem 20.13 also has a nice application to *constrained* body-bar frameworks, where each body is allowed to move only specific directions. For example, if a body is a horizontally flat shaped disc, then it is natural to restrict its motion to those keeping in a horizontal position. Note that body-bar frameworks with bar-boundary capture such constrained body-bar frameworks. Even if allowable directions of motions of bodies are non-generic, Theorem 20.13 gives a combinatorial characterization of infinitesimal rigidity.

### 20.3.7 Other Variants

There are several further extensions of Theorem 20.1 and Theorem 20.2. Jackson and Nguyen [22] considered the length-direction rigidity of body-bar frameworks, where some of the edges represent direction-constraints between points on bodies. They gave an extension of Theorem 20.1 in terms of sparsity condition of the underlying graphs. Haller et al. [14] discussed further different types of geometric constraints among bodies. See Chapter 23 for this general model.

Theorem 20.1 and Theorem 20.2 assume generic placements of bars and hinges, but such generic assumptions sometimes limit applications. In [13] Guest, Schulze, and Whiteley discussed Maxwell-type condition for body-bar frameworks under point group symmetry. In [35, 37] extensions of Theorem 20.1 and Theorem 20.2 were given for body-bar frameworks with point group symmetry. In [2, 34, 37] extensions of Theorem 20.1 to periodic infinite body-bar frameworks were discussed. See Chapter 25 for more details.

## 20.4 Generic Global Rigidity

One of the most challenging questions in rigidity theory is to give a combinatorial characterization of the global rigidity of generic bar-joint frameworks. This was solved for $d = 2$ by Jackson and Jordán [16] and it remains open for $d \geq 3$ (see Chapter 21). Again, in contrast to bar-joint frameworks, a succinct characterization is known for body-bar-hinge frameworks in general dimension.

### 20.4.1 Body-Bar Frameworks

As mentioned in Section 20.1, the global rigidity question for body-bar-hinge frameworks can be reduced to that of equivalent bar-joint frameworks. This conversion allows us to use several related results from the theory of global rigidity of bar-joint frameworks. The most important result in our context is Hendrickson's theorem [15], which states that if a generic bar-joint framework $(G, \boldsymbol{p})$ on at least $d+2$ vertices is globally rigid in $\mathbb{R}^d$, then $G$ is $(d+1)$-connected and the framework is redundantly rigid, that is, $(G-e, \boldsymbol{p})$ is rigid for every edge of $G$. If we apply this result for the equivalent bar-joint frameworks to body-bar frameworks we immediately get the following proposition.

***Proposition 20.8***
*If a generic body-bar framework $(G, \boldsymbol{b})$ is globally rigid in $\mathbb{R}^d$, then $G - e$ contains $\binom{d+1}{2}$ edge-disjoint spanning trees for every edge $e$ of $G$.*

The global rigidity of generic body-bar frameworks was characterized by Connelly, Jordán and Whiteley [4] by showing the sufficiency of the condition in Proposition 20.8.

**Theorem 20.14 (Connelly, Jordán, and Whiteley [4])** *Let $G$ be a multi-graph. Then a generic $d$-dimensional body-bar realization of $G$ is globally rigid if and only if $G-e$ contains $\binom{d+1}{2}$ edge-disjoint spanning trees for every edge $e$ of $G$.*

The original proof of [4] used the constructive characterization of Frank and Szegő [10] (see Theorem 20.25). A different, simplified proof was given by Tanigawa [38]. It was shown in [38] that a generic bar-joint framework $(G, \boldsymbol{p})$ is globally rigid in $\mathbb{R}^d$ if $(G, \boldsymbol{p})$ is *vertex-redundantly rigid*, i.e., $(G-v, \boldsymbol{p})$ is rigid for every $v \in V(G)$. It turns out that if $G$ satisfies the combinatorial condition in Theorem 20.14 then a generic realization of the body-bar graph $G^{\mathrm{B}}$ is vertex-redundantly rigid.

### 20.4.2 Body-Hinge Frameworks

Tanigawa [38] observed that if $\left(\binom{d+1}{2}-1\right)G-2e$ contains $\binom{d+1}{2}$ edge-disjoint spanning trees for every edge $e$ of $G$, then a generic $d$-dimensional body-hinge realization of $G$ is vertex-redundantly rigid and hence globally rigid. Especially, if a body-hinge framework is hinge-redundantly rigid (that is, it remains rigid after a deletion of an arbitrary hinge), then it is globally rigid. With some further ideas Jordán, Király and Tanigawa [23] characterized the global rigidity of generic body-hinge frameworks.

**Theorem 20.15 (Jordán, Király, and Tanigawa [23, 24])** *Let $G$ be a multigraph and $d \geq 3$. Then a generic $d$-dimensional body-hinge realization of $G$ is globally rigid if and only if $(\binom{d+1}{2}-1)G-e$ contains $\binom{d+1}{2}$ edge-disjoint spanning trees for every edge $e$ of $(\binom{d+1}{2}-1)G$.*

Unlike body-bar frameworks, the necessity of Theorem 20.15 is not a direct consequence of Hendrickson's theorem (yet the proof in [23] relies on Hendrickson's theorem). The key idea is that if $d \geq 3$ and the degree is at least 2 in $G$, then the rigidity/global rigidity of a generic body-hinge framework $(G, \boldsymbol{h})$ is equivalent to the rigidity/global rigidity of a bar-joint framework that arises from the equivalent bar-joint framework $(G^{\mathrm{H}}, \boldsymbol{p})$ by removing each core $C(v)$ for every $v \in V(G)$. This latter bar-joint framework $(S_G^{\mathrm{H}}, \boldsymbol{p}')$ is called a *skeleton*. It was shown that $(S_G^{\mathrm{H}}, \boldsymbol{p}')$ is not redundantly rigid if $G$ does not satisfy the combinatorial condition of Theorem 20.15 if $d = 3$. For $d \geq 4$, a slightly modified bar-joint framework is used for applying the same argument.

For $d = 2$, a much simpler condition characterizes global rigidity. (See also Corollary 20.19 for a connection to redundant rigidity.)

**Theorem 20.16 (Jordán, Király, and Tanigawa [23])** *Let $G$ be a multigraph and $d \geq 3$. Then a generic 2-dimensional body-hinge realization of $G$ is globally rigid if and only if $G$ is 3-edge-connected.*

The proof technique in Theorem 20.15 can be extended to body-bar-hinge frameworks.

**Theorem 20.17 (Jordán, Király, and Tanigawa [23])** *Let $G = (V, E_B \cup E_H)$ be a 2-edge-colored multigraph, $d \geq 3$, and let $H$ be a graph obtained from $G$ by replacing each edge in $E_H$ by $\binom{d+1}{2}-1$ parallel copies. Then a generic $d$-dimensional body-bar-hinge realization of $G$ is globally rigid if and only if $H-e$ contains $\binom{d+1}{2}$ edge-disjoint spanning trees for every edge $e$ of $H$.*

Hinge-redundant rigidity/global rigidity of body-hinge frameworks were discussed by Kobayashi et al. [29].

In Section 20.3 we have seen various extensions of Tay's theorem (Theorem 20.1) for body-bar frameworks and Tay-Whiteley's theorem (Theorem 20.2) for body-hinge frameworks. To the best

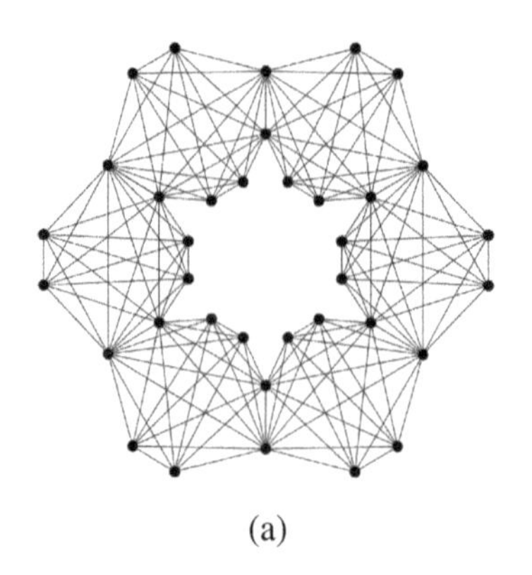


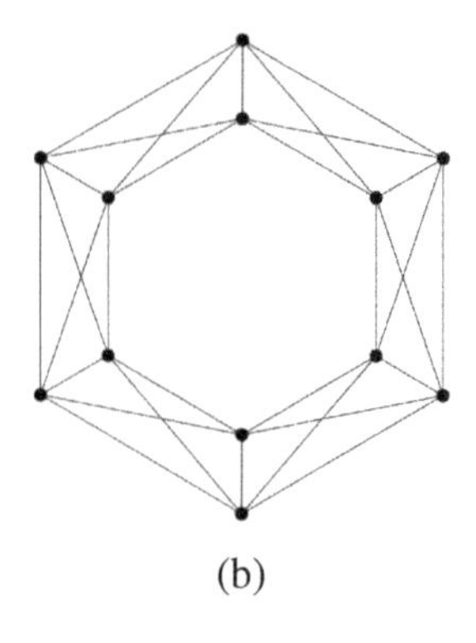


**Figure 20.8**
(a) $C_6^{\mathrm{H}}$ is not globally rigid because (b) its skeleton is minimally rigid.

of our knowledge the corresponding global rigidity questions have not been addressed. The most important and interesting question would be the characterization of the global rigidity of molecular frameworks or generic bar-joint realizations of the square of graphs in $\mathbb{R}^3$. It was shown in [23] that, contrary to the rigidity case, the combinatorial condition in Theorem 20.15 is no longer sufficient. Figure 20.6 is an example that satisfies the condition in Theorem 20.15 but its square is not globally rigid as it is not 4-connected.

### 20.4.3 Counterexamples to Hendrickson's Conjecture

Hendrickson [15] conjectured that his necessary conditions are also sufficient for the global rigidity of bar-joint frameworks. Although the conjecture is true for $d \leq 2$ [16], Connelly gave counterexamples for $d \geq 3$ [3]. Connelly and Whiteley [5] then conjectured that there exist no counterexample in $\mathbb{R}^d$ containing $K_{d+1}$ as a subgraph. They also conjectured that the number of counterexamples is finite in $\mathbb{R}^d$, for all $d \geq 3$. Later Frank and Jiang [11] disproved the latter conjecture for $d \geq 5$, and then Jordán, Király and Tanigawa [23] disproved both conjectures for all $d \geq 3$.

The counterexamples by Jordán, Király, and Tanigawa [23] were built based on Theorem 20.15. By Theorem 20.2, if $(G, \boldsymbol{b})$ is rigid, then $(\binom{d+1}{2} - 1)G$ contains $\binom{d+1}{2}$ edge-disjoint spanning trees, which in turn implies that a generic bar-joint realization $(G^{\mathrm{H}}, \boldsymbol{p})$ is $(d+1)$-connected and redundantly rigid. In other words, $G^{\mathrm{H}}$ satisfies Hendrickson's condition. On the other hand, if $(\binom{d+1}{2} - 1)G - e$ does not contain $\binom{d+1}{2}$ edge-disjoint spanning trees for some edge $e$, then $(G^{\mathrm{H}}, \boldsymbol{p})$ is not globally rigid by Theorem 20.15. Since there are infinitely many graphs $G$ satisfying the above two combinatorial properties, we get an infinite family of counterexamples.

The simplest example is the case when $G$ is the cycle $C_6$ of size six: a generic realization of $C_6^{\mathrm{H}}$ is 4-connected and redundantly rigid but it is not globally rigid by Theorem 20.15. See Figure 20.8.

## 20.5 Graph Theoretical Aspects

In this section we summarize some graph theoretical results concerning body-bar-hinge frameworks.

**$k$-edge-connectivity:** An undirected graph $G = (V, E)$ is *$k$-edge-connected* if there are $k$ edge-disjoint paths between every pair of vertices in $V$.

**$k$-connectivity:** $G=(V,E)$ is *$k$-connected* if there are $k$ internally vertex-disjoint paths between every pair of vertices in $V$.

**Tree-connectivity:** $G=(V,E)$ is *$(k,\ell)$-tree-connected* if $G-F$ contains $k$ edge-disjoint spanning trees for all $F\subseteq E$ with $|F|\leq \ell$. When $\ell=0$ or 1 a $(k,\ell)$-tree-connected graph is also *$k$-tree-connected* or *highly $k$-tree-connected*, respectively.

**Rooted $(k,\ell)$-arc-connectivity:** For positive integers $k$ and $\ell$, a digraph $D=(V,A)$ with a root vertex $r_0\in V$ is *$r_0$-rooted $(k,\ell)$-arc-connected* if there are $k$ arc-disjoint (one-way) paths from $r_0$ to every other vertex and there are $\ell$ arc-disjoint paths from every vertex to $r_0$. A digraph $D$ is *rooted $(k,\ell)$-arc-connected* if it is $r_0$-rooted $(k,\ell)$-arc-connected for some $r_0\in V$.

**Brick:** Let $q=b/c$ with positive integers $b$ and $c$. A *$q$-brick* of a graph $G$ is a maximal subgraph $H$ such that $cH$ is $b$-tree-connected.

**Superbick:** Let $q=b/c$ with positive integers $b$ and $c$. A *$q$-superbrick* of a graph $G$ is a maximal subgraph $H$ such that $cH$ is highly $b$-tree-connected.

**Pinching:** Let $G$ be an undirected graph. By *pinching a set $F$ of edges* in $G$, we mean an operation that inserts a new vertex and subdivides the edges in $F$ by the new vertex.

**$(k,\ell)$-sparsity matroid:** A matroid on the edge set of a graph $G$ is a *$(k,\ell)$-sparsity matroid* (or *$(k,\ell)$-count matroid*) if its independent set family is the family of the edge sets of the $(k,\ell)$-sparse subgraphs in $G$.

### 20.5.1 Tree Packing and Connectivity

Connectivity is one of the most fundamental concepts in graph theory and various variants have been investigated in the literature. The most relevant one in our context is tree-connectivity. By Menger's theorem, $k$-edge-connectivity (resp., $k$-connectivity) is equivalent to the property that the graph remains connected after the removal of at most $k-1$ edges (resp., $k-1$ vertices). The following theorem by Tutte [45] and Nash-Williams [31] gives a Menger-type characterization for $(k,\ell)$-tree-connected graphs.

**Theorem 20.18 (Tutte [45] and Nash-Williams [31])** *A graph $G=(V,E)$ is $(k,\ell)$-tree-connected if and only if*

$$e_G(\mathcal{P})\geq k(|\mathcal{P}|-1)+\ell \tag{20.9}$$

*holds for every partition $\mathcal{P}$ of $V$.*

Theorem 20.18 gives a simple certificate for a graph not being $(k,\ell)$-tree-connected. Such a certificate is useful even in the rigidity context as demonstrated in Proposition 20.4.

In the rigidity context Theorem 20.18 implies a robust version of Theorem 20.1. Namely $G$ is $\left(\binom{d+1}{2},\ell\right)$-tree-connected if and only if a generic body-bar realization of $G$ is *$\ell$-edge-redundantly rigid* in $\mathbb{R}^d$, i.e., it is rigid after the removal of any $\ell$ edges.

In several papers a graph $G=(V,E)$ satisfying the latter condition of Theorem 20.18 is called *$(k,\ell)$-partition-connected.* A $(k,0)$-partition-connected graph is called *$k$-partition-connected.* Observe that $(k,\ell)$-partition-connectivity coincides with $(k+\ell)$-edge-connectivity when $\ell\geq k$, which means that a generic body-bar realization of $G$ is $\ell$-edge-redundantly rigid if and only if $G$ is $(k+\ell)$-edge-connected. Similarly, by a simple edge-counting argument, Theorem 20.18 implies the following

**Corollary 20.19** *$G$ is $k$-edge-connected if and only if $2G$ is $(k,k)$-tree-connected.*

Graph orientations are important subjects even in rigidity theory as they appear in the pebble game algorithm for checking $(k,\ell)$-sparsity. As pointed in [1], the pebble game can be seen as a variant of Hakimi's classical orientation theorem. A far reaching generalization of Hakimi's orientation theorem was shown by Frank [8], which leads to a characterization of $(k,\ell)$-tree-connectivity in terms of graph orientations.

By Menger's theorem, the $r_0$-rooted $(k,\ell)$-arc-connectivity of a digraph $D=(V,E)$ is equivalent to the following: $\delta_D(X)\geq k$ and $\rho_D(X)\geq \ell$ for every set $X\subsetneq V$ with $r_0\in X$ where $\delta_D(X)$ and $\rho_D(X)$ denote the numbers of out-going arcs from $X$ and in-coming arcs to $X$ in $D$, respectively. From a general orientation result of Frank [8], it was shown that $(k,\ell)$-arc-connected orientability is equivalent to $(k,\ell)$-partition-connectivity.

We summarize the characterizations we have seen.

**Theorem 20.20** *For positive integers $k,\ell,d$ with $k\geq \ell$ and a graph $G=(V,E)$, the following are equivalent:*

- *$G$ is $(k,\ell)$-tree-connected;*
- *$G$ is $(k,\ell)$-partition-connected (Tutte [45] and Nash-Williams [31]);*
- *$G$ contains a $(k,k)$-tight spanning graph after removing any $\ell$ edges (Nash-Williams [32]);*
- *$G$ has a rooted $(k,\ell)$-arc-connected orientation (Frank [8]);*
- *$G$ has an $r$-rooted $(k,\ell)$-arc-connected orientation for any $r\in V$ (Frank [8]).*

*Moreover, if $k=\binom{d+1}{2}$, then the above conditions are equivalent to the following:*

- *A generic $d$-dimensional body-bar realization of $G$ is $\ell$-edge-redundantly rigid (Tay [39]).*

*If $k=\binom{d+1}{2}$ and $\ell\geq 1$, then the above conditions are equivalent to the following:*

- *A generic $d$-dimensional body-bar realization of $G$ is $(\ell-1)$-edge-redundantly globally rigid (Connelly, Jordán, and Whiteley [4]).*

Frank's characterization in terms of orientability leads to the most compact certificate for $(k,\ell)$-tree-connectivity, and thus for $\ell$-edge-redundant rigidity.

Nash-Williams' result [32] in Theorem 20.20 implies that $G$ is $(k,k)$-tree-connected if and only if the rank of the $(k,k)$-sparsity matroid is equal to $k|V|-k$. Hence one can investigate the properties of tree-connectivity through the $(k,k)$-sparsity matroid.

It is important to understand which rigid frameworks are globally rigid. In view of Theorem 20.1 and Theorem 20.14, the following corollary gives a matroidal view for such a question.

**Corollary 20.21** *Suppose that a graph $G=(V,E)$ is $k$-tree-connected. Then $G$ is highly $k$-tree-connected if and only if every $e\in E$ is covered by a circuit of the $(k,k)$-sparsity matroid of $G$, that is, the $(k,k)$-sparsity matroid of $G$ is connected.*

A stronger version of Corollary 20.21 was observed by Király [28].

**Theorem 20.22 (Király [28])** *Let $G=(V,E)$ be a $k$-tree connected graph and let $T=(V,F)$ be a spanning $(k,k)$-tight subgraph of $G$. Then $G-e$ is $k$-tree-connected if and only if $e\in E-F$ or $e$ is induced by a minimal $(k,k)$-tight subgraph of $T$ covering both $u$ and $v$ for some edge $uv\in E-F$.*

### 20.5.2 Brick Partitions

In order to find the maximal rigid or globally rigid components of a body-bar-hinge framework, one would like to determine maximal $k$-tree-connected and highly $k$-tree-connected subgraphs. This can be done using a result of Jackson and Jordán [18]. It follows from Theorem 20.18 that the definitions of $q$-bricks and $q$-superbricks for a rational number $q$ are independent from the representation of $q$ as $b/c$. As the union of two $(k,\ell)$-tree-connected graphs with at least one common vertex is also $(k,\ell)$-tree-connected, the vertex sets of $q$-bricks (resp., $q$-superbricks) of a graph $G$ form a partition of $V(G)$, called the *$q$-brick-partition* (resp., *$q$-superbrick-partition*) of $G$.

For a partition $\mathcal{P}$ of $V$ and a rational number $q$, $\mathrm{def}_{G,q}(\mathcal{P}) = q(|\mathcal{P}|-1) - e_G(\mathcal{P})$ denotes the *deficiency* of $\mathcal{P}$ in $G$ (with respect to $q$). Furthermore, let $\mathrm{def}_q(G) = \max\{\mathrm{def}_{G,q}(\mathcal{P}) : \mathcal{P}$ is a partition of $V\}$. A partition $\mathcal{P}$ is *$q$-tight* if $\mathrm{def}_{G,q}(\mathcal{P}) = \mathrm{def}_q(G)$. Jackson and Jordán [18] characterized the brick- and superbrick-partitions of a graph as follows.

**Theorem 20.23 (Jackson and Jordán [18])** *Let $G = (V,E)$ be a graph and let $q$ be a rational number. Let $\mathcal{P}$ and $\mathcal{Q}$ be $q$-tight partitions of $V$ such that $|\mathcal{P}|$ is as small as possible and $|\mathcal{Q}|$ is as large as possible. Then $\mathcal{P}$ is the $q$-brick-partition of $G$ and $\mathcal{Q}$ is the $q$-superbrick-partition of $G$.*

### 20.5.3 Constructive Characterizations

Tay [39] used the following constructive characterization of $k$-tree-connected graphs to prove Theorem 20.1.

**Theorem 20.24 (Tay [39])** *An undirected graph $G = (V,E)$ is $k$-tree-connected if and only if $G$ can be built up from a single vertex graph by the following operations:*

(i) *add a new edge,*

(ii) *pinch $i$ $(0 \leq i \leq k-1)$ existing edges with a new vertex $v$, and add $k-i$ new edges connecting $v$ with existing vertices.*

A constructive characterization of highly $k$-tree-connected graphs was given by Frank and Szegő [10], which was used by Connelly, Jordán, and Whiteley [4] to prove Theorem 20.14.

**Theorem 20.25 (Frank and Szegő [10])** *An undirected graph $G = (V,E)$ is highly $k$-tree-connected if and only if $G$ can be built up from a single vertex graph by the following operations:*

(i) *add a new edge (that can be a loop),*

(ii) *pinch $i$ $(1 \leq i \leq k-1)$ existing edges with a new vertex $v$, and add $k-i$ new edges connecting $v$ with existing vertices.*

In some applications in rigidity, graphs are required to be *simple*. A constructive characterization of *simple* $k$-tree-connected graphs is known for $k = 2$ by Nixon, Owen and Power [33], but the general case is still open.

### 20.5.4 Algorithms

We collect here some algorithmic results that are useful in the investigation of body-bar-hinge frameworks.

**Algorithms for testing $k$-tree-connectivity.** There are several algorithms for deciding whether a graph is $k$-tree-connected. The first group of these algorithms is based on the fact that $k$-tree-connectivity can be decided by computing the rank of the $(k,k)$-sparsity matroid. This matroid is the union of $k$ copies of the graphic matroid, which immediately provides an algorithm to calculate the rank function (see Gabow and Westerman [12]). There is also an algorithm to calculate the rank in any $(k,\ell)$-sparsity matroid based on degree-constrained orientations, widely called the pebble game algorithm [30]. This algorithm is a bit less efficient but is much easier to implement and extend to further applications. For example, it can be used to output the maximal $(k,k)$-tight subgraphs of a graph, that is, the $k$-brick partition. We refer to [30, 43] for more details.

Another way for deciding $k$-tree-connectivity is due to Frank [7], and is based on rooted $(k,0)$-arc-connected orientations (see Theorem 20.20). A further efficient implementation of Frank's algorithm was given by Király [27].

**Algorithms for testing high $k$-tree-connectivity.** By exploiting the $(k,k)$-sparsity matroid, one can also design an efficient algorithm for testing high $k$-tree-connectivity. Again, the pebble game algorithm can be modified to determine the connected components of the $(k,k)$-sparsity matroid of $G$ (see [30, 43]), which gives the $k$-superbrick partition of the graph. Hence this also provides an algorithm to compute the $q$-superbrick partition for every rational $q$.

The original proof of Frank's theorem in [8] is algorithmic, and hence it provides an algorithm for testing $(k,\ell)$-tree-connectivity by using orientations. This algorithm is more involved than that for $k$-tree-connectivity. For $\ell = 1$, Király [27] gave a simplified algorithm. The main observation in [27] is that if we take an arbitrary rooted $k$-arc-connected orientation and take the set of vertices from which the root is reachable in a one-way path, then these vertices induce a rooted $(k,1)$-arc-connected sub-digraph. Hence the algorithm is the following. Take an $r_0v$ edge and orient it towards $r_0$ and take a rooted $k$-arc-connected orientation of $G - r_0v$. Take the set $R$ (containing $r_0$ and $v$) of vertices from which $r_0$ is reachable by this orientation, shrink it to a new root and restart the procedure with this graph.

**Augmenting to a (highly) $k$-tree-connected graph.** Augmenting a given framework to a rigid or globally rigid framework with minimum cost is a natural engineering problem, and due to the combinatorial characterizations we may focus on augmenting a graph to a (highly) $k$-tree-connected graph when the framework is generic. By the matroid structure behind, augmenting a graph to a $k$-tree-connected graph can be solved easily. Moreover, we can also handle the weighted version of this problem. However, the augmentation to high tree-connectivity is only solved for the unweighted case.

Frank and T. Király [9] gave a polynomial algorithm to augment a graph to a $(k,\ell)$-tree-connected graph with minimum number of edges. A rather simple algorithm can be given for the important subcase when the input is $k$-tree-connected and $\ell = 1$ by an algorithm of Király [28]. This algorithm uses several results and algorithms mentioned in this section. It first takes the $k$-superbrick partition and observes that shrinking the members of this partition to single vertices results in a $(k,k)$-tight graph. To augment this latter graph, we need to cover its edge-set by minimal $(k,k)$-tight subgraphs by Theorem 20.22. This is done with a two phase greedy algorithm that uses the pebble game algorithm as a subroutine.

**How to decide whether a bar-joint realization of graph is a body-bar-hinge framework.** In order to use the above algorithms to decide the rigidity or global rigidity of some specific generic bar-joint frameworks, one needs to discover its body-bar-hinge structure. In [23] a simple algorithm for testing whether a graph is a body-hinge graph is given. This algorithm can be modified for testing whether a graph has a body-bar-hinge or some other related structure (e.g. a skeleton).

# References

[1] A.R. Berg and T. Jordán. Algorithms for graph rigidity and scene analysis. In U. Zwick G. Di Battista, editor, *Proc. 11th Annual European Symposium on Algorithms (ESA)*, Springer Lecture Notes in Computer Science 2832, pages 78–89, 2003.

[2] C. Borcea, I. Streinu, and S. Tanigawa. Periodic body-and-bar frameworks. *SIAM J. Discrete Math.*, 29(93–112), 2015.

[3] R. Connelly. On generic global rigidity. In P. Gritzmann and B. Sturmfels, editors, *Applied geometry and discrete mathematics*, volume 4 of *DIMACS Ser. Discrete Math. Theoret. Comput. Sci*, pages 147–155. AMS, 1991.

[4] R. Connelly, T. Jordán, and W. Whiteley. Generic global rigidity of body-bar frameworks. *Journal of Combinatorial Theory Series B*, 103:689–705, 2013.

[5] R. Connelly and W. Whiteley. Global rigidity: The effect of coning. *Discrete & Computational Geometry*, 43(4):717–735, 2010.

[6] H. Crapo and W. Whiteley. Statics of frameworks and motions of panel structures, a projective geometric introduction. *Structural Topology*, 6(43–82), 1982.

[7] A. Frank. On disjoint trees and arborescences. In *Algebraic Methods in Graph Theory*, 25, pages 59–169. Colloquia Mathematica Soc. J. Bolyai, Norh-Holland, 1978.

[8] A. Frank. On the orientation of graphs. *J. Comb. Theory, Ser. B*, 28(3):251–261, 1980.

[9] A. Frank and T. Király. Combined connectivity augmentation and orientation problems. *Discrete Appl. Math.*, 131(2):401–419, 2003.

[10] A. Frank and L. Szegő. Constructive characterizations for packing and covering with trees. *Discrete Applied Mathematics*, 131(2):347–371, 2003.

[11] S. Frank and J. Jiang. New classes of counterexamples to Hendrickson's global rigidity conjecture. *Discrete & Computational Geometry*, 45:574–591, 2011.

[12] H.N. Gabow and H.H. Westermann. Forests, frames, and games: Algorithms for matroid sums and applications. *Algorithmica*, 7(5&6):465–497, 1992.

[13] S.D. Guest, B. Schulze, and W. Whiteley. When is a symmetric body-bar structure isostatic? *Internat. J. Solids Structures*, 47:2745–2754, 2010.

[14] K. Haller, A. Lee-St. John, M. Sitharam, I. Streinu, and N. White. Body-and-cad geometric constraint systems. *Comput. Geom. Theory Appl.*, (45):385–405, 2012.

[15] B. Hendrickson. Conditions for unique graph realizations. *SIAM J. Comput.*, 21(1):65–84, 1992.

[16] B. Jackson and T. Jordán. Connected rigidity matroids and unique realizations of graphs. *J. Comb. Theory, Ser. B*, 94:1–29, 2005.

[17] B. Jackson and T. Jordán. The Dress conjectures on rank in the 3-dimensional rigidity matroid. *Advances in Applied Mathematics*, 35(4):355–367, 2005.

[18] B. Jackson and T. Jordán. Brick partitions of graphs. *Discrete Mathematics*, 310(2):270–275, 2008.

[19] B. Jackson and T. Jordán. On the rigidity of molecular graphs. *Combinatorica*, 28(6):645–658, 2008.

[20] B. Jackson and T. Jordán. Pin-collinear body-and-pin frameworks and the molecular conjecture. *Discrete and Computational Geometry*, 40(2):258–278, 2008.

[21] B. Jackson and T. Jordán. The generic rank of body–bar-and-hinge frameworks. *European Journal of Combinatorics*, 31(2):574–588, 2009.

[22] B. Jackson and V.H. Nguyen. Graded sparse graphs and body-length-direction frameworks. *European Journal of Combinatorics*, 46:51–67, 2015.

[23] T. Jordán, Cs. Király, and S. Tanigawa. Generic global rigidity of body-hinge frameworks. *Journal of Combinatorial Theory, Series B*, 117:59–76, 2016.

[24] T. Jordán, Cs. Király, and S. Tanigawa. Generic global rigidity of body-hinge frameworks: a correction. Technical Report TR-2017-11, Egerváry Research Group, Budapest, 2017. `www.cs.elte.hu/egres`.

[25] N. Katoh and S. Tanigawa. A proof of the Molecular conjecture. *Discrete & Computational Geometry*, 45:647–700, 2011.

[26] N. Katoh and S. Tanigawa. Rooted-tree decompositions with matroid constraints and the infinitesimal rigidity of frameworks with boundaries. *SIAM Journal on Discrete Mathematics*, 27:155–185, 2013.

[27] Cs. Király. Algorithms for finding a rooted $(k,1)$-edge-connected orientation. *Discrete Appl. Math.*, 166:263–268, March 2014.

[28] Cs. Király. Rigid graphs and an augmentation problem. Technical Report TR-2015-03, Egerváry Research Group, Budapest, 2015. `www.cs.elte.hu/egres`.

[29] Y. Kobayashi, Y. Higashikawa, N. Katoh, and A. Sljoka. Characterizing redundant rigidity and redundant global rigidity of body-hinge graphs. *Information Processing Letters*, 116(2):175 – 178, 2016.

[30] A. Lee and I. Streinu. Pebble game algorithms and sparse graphs. *Discrete Mathematics*, 308(8):1425–37, 2008.

[31] C.St.J.A. Nash-Williams. Edge-disjoint spanning trees of finite graphs. *J. London Math. Soc.*, 36:445–450, 1961.

[32] C.St.J.A. Nash-Williams. Decomposition of finite graphs into forests. *Journal of the London Mathematical Society*, 1(1):12, 1964.

[33] A. Nixon, J.C. Owen, and S.C. Power. Rigidity of frameworks supported on surfaces. *SIAM J. Discrete Math.*, 26(4):1733–1757, 2012.

[34] E. Ross. The rigidity of periodic body-bar frameworks on the three-dimensional fixed torus. *Phil. Trans. Royal Soc. A*, 372:2008, 2014.

[35] B. Schulze and S. Tanigawa. Linking rigid bodies symmetrically. *European Journal of Combinatorics*, 42:145–166, 2014.

[36] S. Tanigawa. Generic rigidity matroids with Dilworth truncations. *SIAM Journal on Discrete Mathematics*, 26:1412–1439, 2012.

[37] S. Tanigawa. Matroids of gain graphs in applied discrete geometry. *Tran. Amer. Math. Soc.*, 367:8597–8641, 2015.

[38] S. Tanigawa. Sufficient conditions for globally rigidity of graphs. *Journal of Combinatorial Theory Series B*, 113:123–140, 2015.

[39] T.S. Tay. Rigidity of multi-graphs. I: Linking rigid bodies in $n$-space. *Journal of Combinatorial Theory. Series B*, 36(1):95–112, 1984.

[40] T.S. Tay. Linking $(n-2)$-dimensional panels in $n$-space II:$(n-2,2)$-frameworks and body and hinge structures. *Graphs and Combinatorics*, 5(1):245–273, 1989.

[41] T.S. Tay. Linking $(n-2)$-dimensional panels in $n$-space I:$(k-1,k)$-graphs and $(k-1,k)$-frames. *Graphs and Combinatorics*, 7(3):289–304, 1991.

[42] T.S. Tay and W. Whiteley. Recent advances in the generic rigidity of structures. *Structural Topology*, 9:31–38, 1984.

[43] L. Theran. Sparsity matroids for combinatorial rigidity. in this handbook.

[44] M.F. Thorpe, M. Chubynsky, B. Hespenheide, S. Menor, D.J. Jacobs, L.A. Kuhn, M.I. Zavodszky, M. Lei, A.J. Rader, and W. Whiteley. Flexibility in biomolecules. In *Current topics in physics*, pages 97–112 (Chapter 6). Imperial College Press, 2005.

[45] W.T. Tutte. On the problem of decomposing a graph into $n$ connected factors. *Journal of the London Mathematical Society*, 36:221–230, 1961.

[46] N. White and W. Whiteley. The algebraic geometry of motions of bar-and-body frameworks. *SIAM J. Alg. Disc. Math.*, 8:1–32, 1987.

[47] W. Whiteley. The union of matroids and the rigidity of frameworks. *SIAM Journal on Discrete Mathematics*, 1(2):237–255, 1988.

[48] W. Whiteley. Rigidity of molecular structures: geometric and generic analysis. In M.F. Thorpe and P.M. Duxbury, editors, *Rigidity theory and applications*, pages 21–46. Kluwer, 1999.

[49] W. Whiteley. Counting out to the flexibility of molecules. *Physical Biology*, 2:S116–S126, 2005.

# Chapter 21

# *Global Rigidity of Two-Dimensional Frameworks*

**Bill Jackson**
*School of Mathematical Sciences, Queen Mary University of London, Mile End Road, London, England*

**Tibor Jordán**
*Department of Operations Research, Eötvös University, Pázmány Péter sétány 1/C, Budapest, Hungary, and MTA-ELTE Egerváry Research Group on Combinatorial Optimization, Budapest, Hungary*

**Shin-Ichi Tanigawa**
*Department of Mathematical Informatics, The University of Tokyo, Hongo, Bunkyo-ku, Tokyo, Japan*

## CONTENTS

## 21.1 Introduction

**Framework:** A $d$-dimensional *framework* is a pair $(G, p)$, where $G = (V, E)$ is a graph and $p$ is a map from $V$ to $\mathbb{R}^d$. We also call $(G, p)$ a *realization* of $G$ in $\mathbb{R}^d$.

**Generic framework:** A framework $(G, p)$ is *generic* if the set of coordinates of the points $p(v)$, $v \in V(G)$, is algebraically independent over the rationals.

**Equivalent frameworks:** Two $d$-dimensional frameworks $(G, p)$ and $(G, q)$ are *equivalent* if $||p(u) - p(v)|| = ||q(u) - q(v)||$ holds for all pairs $u, v$ with $uv \in E$, where $||.||$ denotes the Euclidean norm in $\mathbb{R}^d$.

**Congruent frameworks:** Two $d$-dimensional frameworks $(G, p)$ and $(G, q)$ are *congruent* if $||p(u) - p(v)|| = ||q(u) - q(v)||$ holds for all pairs $u, v$ with $u, v \in V$. This is the same as saying that $(G, q)$ can be obtained from $(G, p)$ by an isometry of $\mathbb{R}^d$.

**Globally rigid:** A $d$-dimensional framework $(G, p)$ is *globally rigid* if every $d$-dimensional framework which is equivalent to $(G, p)$ is congruent to $(G, p)$. The graph $G$ is *globally rigid in* $\mathbb{R}^d$ if every generic realization of $G$ in $\mathbb{R}^d$ is globally rigid. See Figure 21.1.

**Rigid:** A $d$-dimensional framework $(G, p)$ is *rigid* if there exists an $\varepsilon > 0$ such that, if a $d$-dimensional framework $(G, q)$ is equivalent to $(G, p)$ and $||p(u) - q(u)|| < \varepsilon$ for all $v \in V$, then $(G, q)$ is congruent to $(G, p)$. The graph $G$ is *rigid in* $\mathbb{R}^d$ if every generic realization of $G$ in $\mathbb{R}^d$ is rigid.

It is a hard problem to decide if a given framework is rigid or globally rigid. Indeed Saxe [46] showed that it is NP-hard to decide if even a 1-dimensional framework is globally rigid and Abbot [1] showed that the rigidity problem is NP-hard for 2-dimensional frameworks.

The analysis and characterization of rigid and globally rigid frameworks become more tractable, however, if we consider generic frameworks. The main reason for this is that rigidity and global rigidity of frameworks in $\mathbb{R}^d$ are both *generic properties* in the sense that they depend only on the

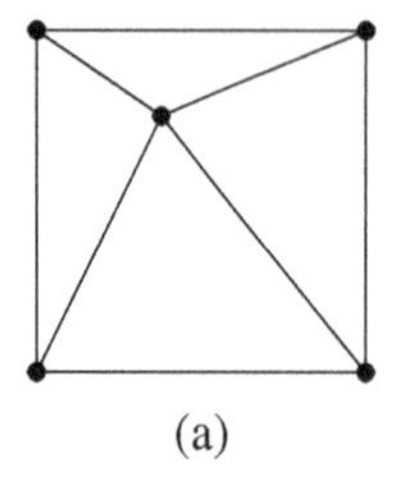
(a)

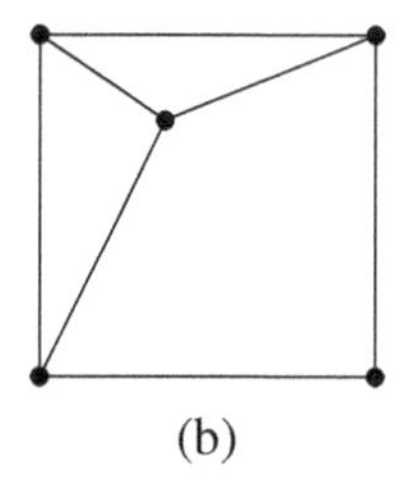
(b)

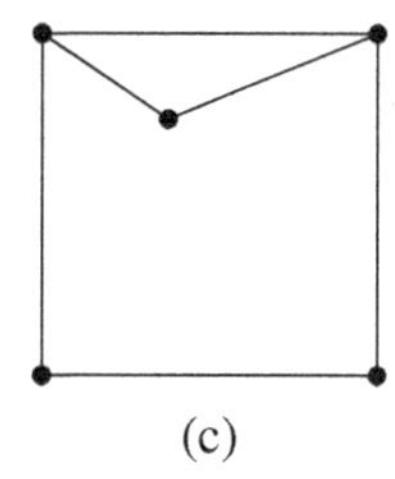
(c)

**Figure 21.1**
Frameworks in the plane: (a) globally rigid, (b) rigid but not globally rigid, (c) not rigid.

graph $G$ and not the particular realization $p$, when $(G,p)$ is generic. This follows from results of Asimow and Roth [2] and Gortler, Healy and Thurston [23], respectively. The problems of finding a polynomially verifiable characterization for graphs which are rigid or globally rigid in $\mathbb{R}^d$ have been solved for $d = 1,2$, but are major open problems for $d \geq 3$.

Global rigidity has several real life applications. One such application occurs in the *network localization problem*, in which the locations of some nodes (called anchors) of a network as well as the distances between some pairs of nodes are known, and the goal is to determine the location of all nodes. This is one of the fundamental algorithmic problems in the theory of wireless sensor networks and has been the focus of a number of research articles and survey papers, see for example [3, 19, 48]. When constructing a sensor network, we would like the solution to the localization problem to be unique, in this case we say that the network is *uniquely localizable*. It is not difficult to see that the unique localizability of a generic network is equivalent to the global rigidity of its underlying distance graph (in which two nodes are adjacent if and only if the distance between them is known). We will return to this problem in Sections 21.7 and 21.10.

We will mostly be concerned with global rigidity in $\mathbb{R}^2$ but will state results for $\mathbb{R}^d$ when they hold in all dimensions. Some of the material in this chapter is discussed in more detail in the survey by Jordán [37].

## 21.2 Conditions for Global Rigidity

**Equilibrium stress:** An *equilibrium stress* for a framework $(G,p)$ in $\mathbb{R}^d$ is an assignment $\omega : E \to \mathbb{R}$ such that, for each vertex $v_i \in V$:

$$\sum_{j:v_iv_j \in E} \omega_{i,j}(p(v_i) - p(v_j)) = 0. \tag{21.1}$$

**Stress matrix:** The *stress matrix* $\Omega$ associated to an equilibrium stress $\omega$ is the $|V| \times |V|$ symmetric matrix in which the entries are defined so that $\Omega[i,j] = -\omega_{i,j}$ for all edges $v_iv_j \in E$, $\Omega[i,j] = 0$ for all non-adjacent vertex pairs $v_i, v_j \in V$, and $\Omega[i,i]$ is chosen so that each row and column sum is equal to zero.

**Redundantly rigid graph:** A graph $G$ is *redundantly rigid in* $\mathbb{R}^d$ if $G$ has at least two vertices and $G - e$ is rigid in $\mathbb{R}^d$ for all $e \in E(G)$.

**$k$-connected graph:** A graph $G$ is *$k$-connected* for some positive integer $k$ if $|V(G)| \geq k+1$ and $G - X$ is connected for all $X \subset V(G)$ with $|X| \leq k-1$.

### 21.2.1 Stress Matrix Characterization in $\mathbb{R}^d$

Connelly [13] built a theoretical foundation for the analysis of global rigidity based on stress matrices. He observed that the rank of any stress matrix for $(G,p)$ is at most $|V| - d - 1$ when $p(V)$ affinely spans $\mathbb{R}^d$. Subsequently, Connelly [15] (sufficiency) and Gortler, Healy and Thurston [23] (necessity) showed that the property of having a stress matrix with this maximum possible rank characterizes global rigidity for generic frameworks.

**Theorem 21.1** *Let $(G,p)$ be a generic framework in $\mathbb{R}^d$ on at least $d+2$ vertices. Then $(G,p)$ is globally rigid in $\mathbb{R}^d$ if and only if $(G,p)$ has an equilibrium stress $\omega$ for which the rank of the associated stress matrix $\Omega$ is $|V| - d - 1$.*

We will say that $(G, p)$ has a *maximum rank* stress matrix when its rank is $|V| - d - 1$.

We can use Theorem 21.1 to deduce that global rigidity in $\mathbb{R}^d$ is a generic property. It also gives rise to a randomized algorithm for deciding if a given graph is globally rigid in $\mathbb{R}^d$, but it does not give a deterministic polynomial algorithm or a polynomially verifiable characterization for global rigidity. See Chapter 16 for more details.

### 21.2.2 Hendrickson's Necessary Conditions for Global Rigidity

Hendrickson [24] showed that redundant rigidity and $(d+1)$-connectivity are both necessary conditions for a graph to be globally rigid in $\mathbb{R}^d$.

**Theorem 21.2** *Let G be a globally rigid graph in $\mathbb{R}^d$. Then either G is a complete graph on at most $d+1$ vertices, or G is $(d+1)$-connected and redundantly rigid in $\mathbb{R}^d$.*

These necessary conditions are also sufficient to imply the global rigidity of $G$ in $\mathbb{R}^d$ when $d = 1,2$, as we shall see below. This is not the case, however, when $d \geq 3$. We say that a graph $G$ is a *Hendrickson graph* in $\mathbb{R}^d$ if it satisfies the necessary conditions for global rigidity given in Theorem 21.2, but it is *not* globally rigid in $\mathbb{R}^d$. For $d = 3$, Connelly [14] showed that the complete bipartite graph $K_{5,5}$ is a Hendrickson graph. He also constructed similar examples (specific complete bipartite graphs on $\binom{d+2}{2}$ vertices) for all $d \geq 3$. Frank and Jiang [21] found two more (bipartite) Hendrickson graphs in $\mathbb{R}^4$ as well as infinite families in $\mathbb{R}^d$ for $d \geq 5$. Jordán, Király, and Tanigawa [38] constructed infinite families of Hendrickson graphs for all $d \geq 3$. Further examples can be obtained by using the observation that the cone of a $d$-dimensional Hendrickson graph is a $(d+1)$-dimensional Hendrickson graph.

## 21.3 Graph Operations

**1-extension operation:** Given a graph $H$ with a designated edge $e = v_1v_2$, and $d-1$ additional vertices $v_3, \ldots, v_{d+1}$, the *d-dimensional* 1*-extension operation on e* (sometimes called an *edge d-split*) adds a new vertex $v_0$, removes $e$, and inserts $d+1$ new edges $v_0v_1, v_0v_2, \ldots, v_0v_{d+1}$. See Figure 21.2.

**Cone operation:** Given a graph $H$, the *cone* of $H$ is the graph obtained by adding a new vertex $v$ and joining $v$ to every vertex of $H$.

**Vertex splitting operation:** Given a graph $G = (V, E)$, a vertex $v \in V$, and a tripartition $F_0, F_1, F_2$ of the edges incident to $v$ with $|F_0| = d-1$, the *d-dimensional vertex splitting operation at* $v$ replaces the vertex $v$ by two new vertices $v_1$ and $v_2$ and an edge $v_1v_2$, replaces each edge $uv \in F_0$ by two new edges $uv_1, uv_2$, and replaces each edge $wv \in F_i$ by an edge $wv_i$, $i = 1,2$, see Figure 21.3. The vertex splitting operation is said to be *non-trivial* if $F_1, F_2$ are both non-empty, or equivalently, if each of the split vertices $v_1, v_2$ has degree at least $d+1$.

Inductive constructions for graphs which use global rigidity preserving operations are frequently used, both to prove that certain families of graphs are globally rigid and to analyze the global rigidity of a particular graph.

It is easy to see that adding an edge or a vertex of degree $d+1$ to a graph will preserve global rigidity in $\mathbb{R}^d$. By applying the degree $d+1$ vertex addition operation recursively to $K_{d+1}$ we obtain the family of $(d+1)$*-lateration graphs*, which are examples of sparse globally rigid graphs in $\mathbb{R}^d$.

The 1-extension operation is a finer and more useful graph operation that preserves global rigidity.

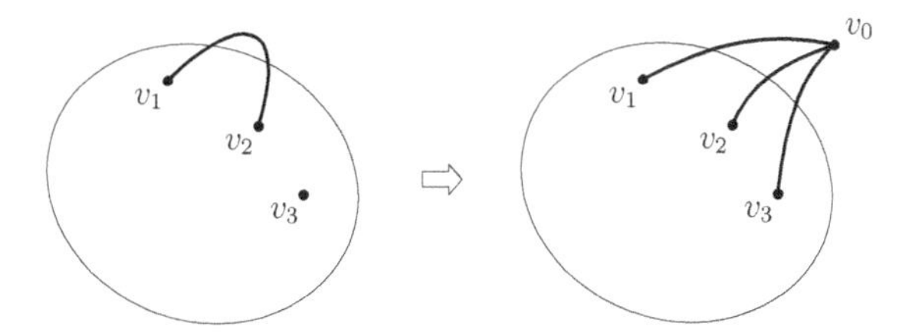


**Figure 21.2**
A 2-dimensional 1-extension.

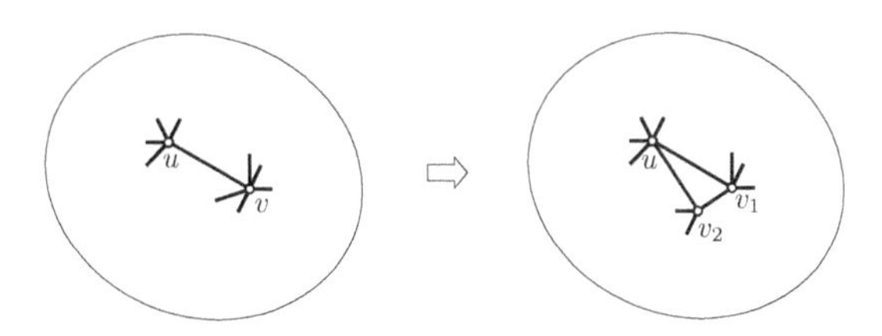


**Figure 21.3**
2-dimensional vertex splitting.

**Theorem 21.3** *[15] Suppose that G can be obtained from $K_{d+2}$ by a sequence of 1-extensions and edge additions. Then G is globally rigid in $\mathbb{R}^d$.*

Figure 21.4 shows that every wheel can be constructed from $K_4$ by a sequence of 2-dimensional 1-extensions and hence is globally rigid in $\mathbb{R}^2$.

Connelly proved Theorem 21.3 by showing that $K_{d+2}$ has an equilibrium stress with a maximum rank stress matrix and that this property is preserved by the 1-extension operation. He then used (the sufficiency part of) Theorem 21.1 to deduce that $G$ is globally rigid. Since the necessity part of Theorem 21.1 is now known to be true, we can deduce the following stronger result using Connelly's result on 1-extensions.

**Theorem 21.4** *Suppose G is obtained from a graph H by the d-dimensional 1-extension operation. If H is globally rigid in $\mathbb{R}^d$ and $|V(H)| \geq d+2$ then G is globally rigid in $\mathbb{R}^d$.*

A direct proof of Theorem 21.4 for the case when $d = 2$ is given in [31]. This proof is extended to all $d$ in [49].

Connelly and Whiteley used Theorem 21.1 to show that the cone operation transfers global rigidity from $\mathbb{R}^d$ to $\mathbb{R}^{d+1}$.

**Theorem 21.5** *[16] Suppose that G is the cone of H. Then G is globally rigid in $\mathbb{R}^{d+1}$ if and only if H is globally rigid in $\mathbb{R}^d$.*

We have seen that wheeels are globally rigid in $\mathbb{R}^2$. Since a wheel on $n+1$ vertices is the cone of a cycle on $n$ vertices, Theorem 21.5 implies that cycles are globally rigid in $\mathbb{R}^1$. (This observation could also be deduced directly from Theorem 21.4.)

Cheung and Whiteley [10] conjecture that the non-trivial $d$-dimensional vertex splitting operation preserves generic global rigidity in $\mathbb{R}^d$ for all $d \geq 1$. Jordán and Szabadka [39] used the characterization of generic global rigidity in $\mathbb{R}^2$, see Theorem 21.10 below, to verify this conjecture when $d = 2$.

**Theorem 21.6** *[39] Suppose G is obtained from H by a non-trivial 2-dimensional vertex splitting operation. If H is globally rigid in $\mathbb{R}^2$ then G is also globally rigid in $\mathbb{R}^2$.*

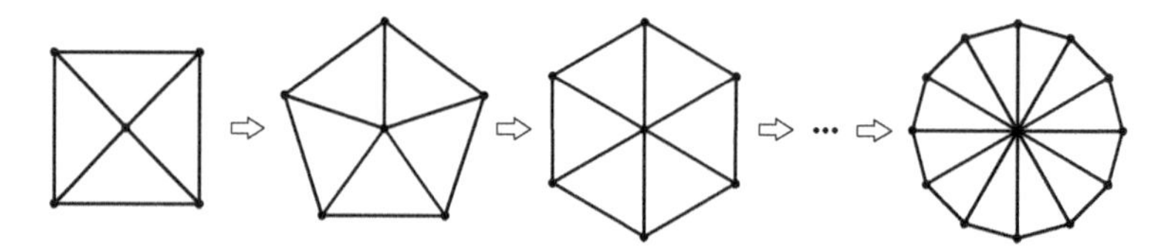

**Figure 21.4**
A sequence of 2-dimensional 1-extensions.

Two other graph operations which are sometimes used are the *$X$-replacement* and *diamond split* operations, see Figure 21.5. In general an $X$-replacement may not preserve 3-connectivity. It was pointed out in [39] that a diamond split may not preserve redundant rigidity in $\mathbb{R}^2$. Hence neither operation is guaranteed to preserve global rigidity in $\mathbb{R}^2$.

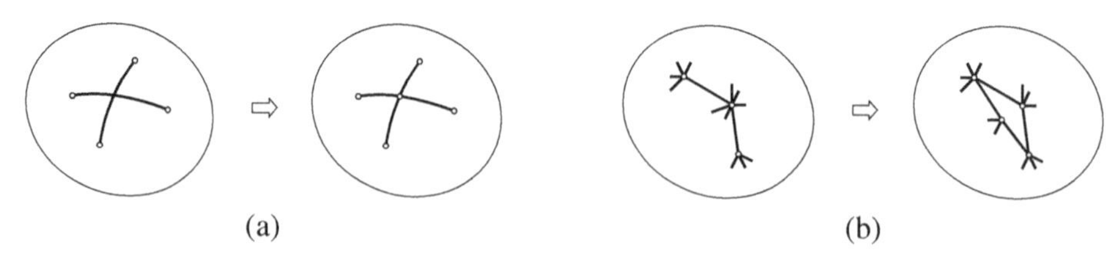


**Figure 21.5**
(a) $X$-replacement, and (b) diamond split in $\mathbb{R}^2$.

## 21.4 Characterization of Global Rigidity in $\mathbb{R}^1$ and $\mathbb{R}^2$

Generic global rigidity has been completely characterized in $\mathbb{R}^1$ and $\mathbb{R}^2$. The 1-dimensional result is folklore.

**Theorem 21.7** *A graph is globally rigid in $\mathbb{R}^1$ if and only if it is a complete graph on at most two vertices or is 2-connected.*

Note that Hendrickson's condition of redundant rigidity is implied by 2-connectivity since connected graphs are rigid in $\mathbb{R}^1$. One way to prove Theorem 21.7 is to use the well-known ear-decompositions of 2-connected graphs to obtain the following inductive construction for 2-connected graphs and then apply Theorem 21.3.

**Theorem 21.8** *A graph is 2-connected if and only if it can be obtained from $K_3$ by a sequence of (1-dimensional) 1-extensions and edge additions.*

A similar proof technique works for global rigidity in $\mathbb{R}^2$. A key step is the following inductive construction for 3-connected redundantly rigid graphs.

**Theorem 21.9** *[26] Every 3-connected graph which is redundantly rigid in $\mathbb{R}^2$ can be obtained from $K_4$ by a sequence of (2-dimensional) 1-extensions and edge additions.*

The 2-dimensional rigidity matroid plays a pivotal role in extending the proof technique of Theorem 21.8 to obtain Theorem 21.9. More details will be given in Section 21.5 below.

Theorems 21.2, 21.3 and 21.9 combine to give the following characterization of globally rigid graphs in $\mathbb{R}^2$.

**Theorem 21.10** *[26] A graph G is globally rigid in $\mathbb{R}^2$ if and only if G is a complete graph on at most three vertices or G is 3-connected and redundantly rigid.*

Theorem 21.9 gives rise to an efficient deterministic algorithm for testing whether a graph is globally rigid in $\mathbb{R}^2$. More details will be given in Section 21.9.

## 21.5 The Rigidity Matroid

**Rigidity matrix:** The *rigidity matrix* $R(G,p)$ of a $d$-dimensional framework $(G,p)$ is a matrix of size $|E| \times d|V|$, where, for each edge $e = v_iv_j \in E$, in the row corresponding to $e$, the entries in the $d$ columns corresponding to vertices $v_i$ and $v_j$ contain the $d$ coordinates of $(p(v_i) - p(v_j))$ and $(p(v_j) - p(v_i))$, respectively, and the remaining entries are zeros.

**Rigidity matroid:** The rigidity matrix of $(G,p)$ defines the *rigidity matroid* of $(G,p)$ on the ground set $E$, where a set of edges $F \subseteq E$ is independent if and only if the rows of the rigidity matrix indexed by $F$ are linearly independent. (For more details on matroids and related combinatorial results the reader is referred to [20, 45].) Since the entries of the rigidity matrix are polynomial functions of the coordinates of the points $p(v_i)$ with integer coefficients, any two generic $d$-dimensional frameworks $(G,p)$ and $(G,q)$ have the same rigidity matroid. We call this the *$d$-dimensional rigidity matroid* of the graph $G$. We denote the rigidity matroid of $G$ by $\mathcal{R}_d(G)$ and its rank by $r_d(G)$.

**$\mathcal{R}_d$-independent graph:** A graph $G = (V,E)$ is *$\mathcal{R}_d$-independent* if $E$ is independent in $\mathcal{R}_d(G)$.

**$\mathcal{R}_d$-circuit:** A subgraph $H = (W,C)$ of a graph $G$ is an *$\mathcal{R}_d$-circuit* of $G$ if $C$ is a circuit (i.e., a minimal dependent set) in $\mathcal{R}_d(G)$. In particular, $G$ is an *$\mathcal{R}_d$-circuit* if $E$ is a circuit in $\mathcal{R}_d(G)$.

**2-sum operation:** Given two graphs $H_1 = (V_1,E_1)$ and $H_2 = (V_2,E_2)$ with $V_1 \cap V_2 = \emptyset$ and two designated edges $u_1v_1 \in E_1$ and $u_2v_2 \in E_2$, the *2-sum* of $H_1$ and $H_2$ (along the edge pair $u_1v_1$, $u_2v_2$) is the graph $H_1 \oplus_2 H_2$ obtained from $H_1 - u_1v_1$ and $H_2 - u_2v_2$ by identifying $u_1$ with $u_2$ and $v_1$ with $v_2$. See Figure 21.6.

**Components of a matroid :** Given a matroid $\mathcal{M} = (E,\mathcal{I})$, we define a relation on $E$ by saying that $e, f \in E$ are related if $e = f$ or if there is a circuit $C$ in $\mathcal{M}$ with $e, f \in C$. It is well-known that this is an equivalence relation. The equivalence classes are called the *components* of $\mathcal{M}$. If $\mathcal{M}$ has at least two elements and only one component then $\mathcal{M}$ is said to be *connected.* If $\mathcal{M}$ has components $E_1, E_2, \ldots, E_t$ and $\mathcal{M}_i$ is the matroid restriction of $\mathcal{M}$ onto $E_i$ then $\mathcal{M} = \mathcal{M}_1 \oplus \mathcal{M}_2 \ldots \oplus \mathcal{M}_t$, where $\oplus$ denotes the *direct sum* of matroids, see [45].

**$\mathcal{R}_d$-components of a graph:** A graph $G = (V,E)$ is *$\mathcal{R}_d$-connected* if its $d$-dimensional rigidity matroid $\mathcal{R}_d(G)$ is connected. The *$\mathcal{R}_d$-components* of $G$ are the subgraphs of $G$ induced by the components of $\mathcal{R}_d(G)$, see Figure 21.8. We say that an $\mathcal{R}_d$-component is *trivial* if it has only one edge, that is, if it is induced by an edge which belongs to no $\mathcal{R}_d$-circuit. Such an edge is also called an *$\mathcal{R}_d$-bridge.*

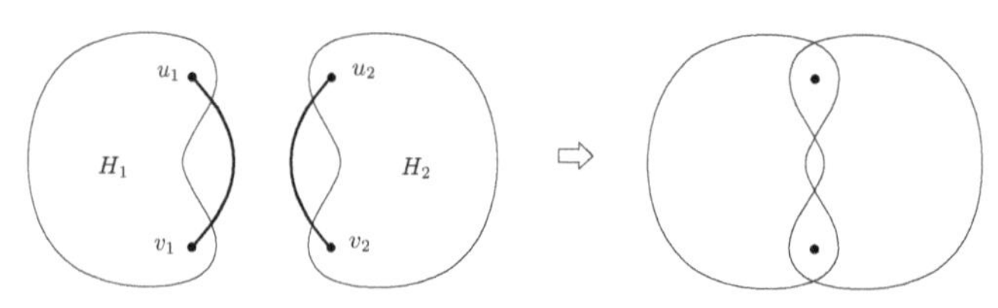


**Figure 21.6**
The 2-sum operation.

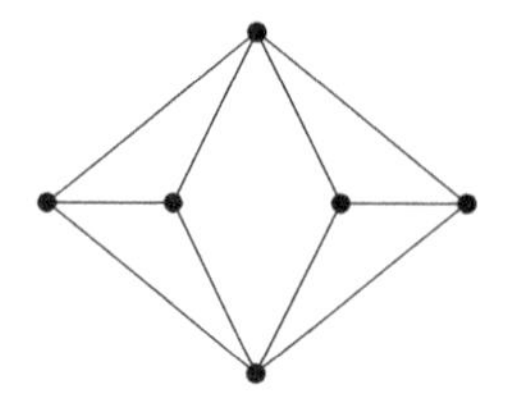

**Figure 21.7**
An $\mathcal{R}_2$-circuit that is not 3-connected. This graph can be constructed by a 2-sum of two copies of $K_4$.

### 21.5.1 $\mathcal{R}_d$-Independent Graphs

It is not difficult to see that a graph $G$ is $\mathcal{R}_1$-independent if and only if it is a forest and hence $\mathcal{R}_1(G)$ is the well known cycle matroid of $G$. A fundamental theorem of Laman [41] characterizes $\mathcal{R}_2$-independent graphs. It remains an open problem to find good characterizations for $\mathcal{R}_d$-independence or, more generally, the rank function of $\mathcal{R}_d(G)$ when $d \geq 3$.

### 21.5.2 $\mathcal{R}_d$-Circuits

Wheels, $K_{3,3}$ plus an edge, and $K_{3,4}$ are all examples of 3-connected $\mathcal{R}_2$-circuits. An example of an $\mathcal{R}_2$-circuit that is not 3-connected is given in Figure 21.7. Laman's characterization of independence in $\mathcal{R}_2$ implies that $\mathcal{R}_2$-circuits are redundantly rigid in $\mathbb{R}^2$. It follows that a graph $G$ is redundantly rigid in $\mathbb{R}^2$ if and only if $G$ is rigid in $\mathbb{R}^2$ and each edge of $G$ belongs to an $\mathcal{R}_2$-circuit of $G$. (This is not true in higher dimensions, since $\mathcal{R}_d$-circuits may not be rigid in $\mathbb{R}^d$ when $d \geq 3$.)

Berg and Jordán [5] obtained the following inductive construction for $\mathcal{R}_2$-circuits.

**Theorem 21.11** *A graph $G$ is an $\mathcal{R}_2$-circuit if and only if $G$ is a connected graph obtained from disjoint copies of $K_4$'s by taking 2-sums and 2-dimensional 1-extensions.*

They then used Theorem 21.11 to obtain a simpler construction for 3-connected $\mathcal{R}_2$-circuits.

**Theorem 21.12** *[5] A graph $G$ is a 3-connected $\mathcal{R}_2$-circuit if and only if $G$ can be constructed from $K_4$ by a sequence of 2-dimensional 1-extensions.*

Theorem 21.12 is an important step in the characterization of generic global rigidity in $\mathbb{R}^2$. Together with Theorem 21.3, it immediately implies that 3-connected $\mathcal{R}_2$-circuits are globally rigid.

### 21.5.3 $\mathcal{R}_d$-Connected Graphs

By definition the $\mathcal{R}_d$-components of a graph $G$ are pairwise edge-disjoint subgraphs of $G$ and that the rigidity matroid $\mathcal{R}_d(G)$ can be expressed as the direct sum of the rigidity matroids of its $\mathcal{R}_d$-components.

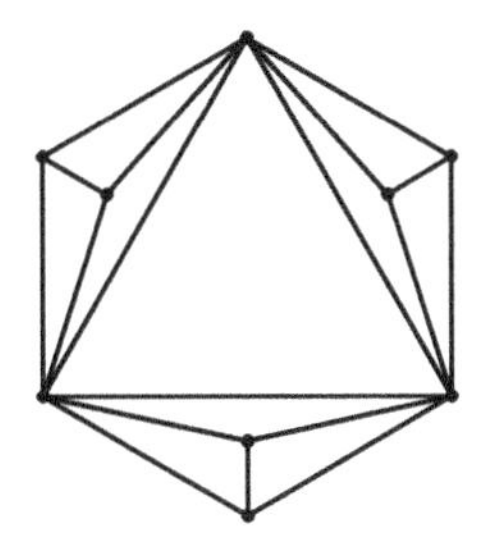

**Figure 21.8**
A graph that is redundantly rigid in $\mathbb{R}^2$ but it not $\mathcal{R}_2$-connected. It has three $\mathcal{R}_2$-components given by the three copies of $K_4$.

When $d = 2$, we can also define the *rigid components* of $G$ to be the maximal subgraphs which are rigid in $\mathbb{R}^2$, and the *redundantly rigid components* to be the maximal subgraphs which are redundantly rigid in $\mathbb{R}^2$, together with the subgraphs which are induced by an $\mathcal{R}_2$-bridge. We refer to the second type of redundantly rigid component as *trivial*. Thus the trivial $\mathcal{R}_2$-components and the trivial redundantly rigid components are the same. Since the non-trivial $\mathcal{R}_2$-components of $G$ are redundantly rigid, the $\mathcal{R}_2$-components are pairwise edge-disjoint vertex-induced subgraphs of $G$ and the partition of $E(G)$ given by the $\mathcal{R}_2$-components is a refinement of the partition given by the redundantly rigid components and hence a further refinement of the partition given by the rigid components. In addition, the rigidity matroid $\mathcal{R}_2(G)$ can be expressed as the direct sum of the rigidity matroids of either the $\mathcal{R}_2$-components of $G$, the redundantly rigid components of $G$, or the rigid components of $G$.

We say that a graph $G$ is *nearly 3-connected* if $G$ can be made 3-connected by adding at most one new edge.

**Theorem 21.13** *[26] If $G$ is nearly 3-connected and every edge of $G$ is in some $\mathcal{R}_2$-circuit, then $G$ is $\mathcal{R}_2$-connected.*

As a corollary we obtain:

**Theorem 21.14** *[26] If $G$ is a 3-connected graph and is redundantly rigid in $\mathbb{R}^2$, then $G$ is $\mathcal{R}_2$-connected.*

Figure 21.8 shows an example of a graph that is redundantly rigid in $\mathbb{R}^2$ but is not $\mathcal{R}_2$-connected.

The main result of this subsection is the following inductive construction. Its proof is inductive and uses Theorem 21.12 as a base case. The inductive step uses an ear-decomposition of the rigidity matroid of an $\mathcal{R}_2$-connected graph, in a similar way that an ear decomposition of a 2-connected graph can be used to prove Theorem 21.8.

**Theorem 21.15** *[26] A graph is 3-connected and $\mathcal{R}_2$-connected if and only if it can be obtained from $K_4$ by a sequence of (2-dimensional) 1-extensions and edge additions.*

Theorem 21.9 follows from Theorems 21.14 and 21.15.

An example of this inductive construction is given in Figure 21.9.

By using the above results one can also obtain an inductive construction for $\mathcal{R}_2$-connected graphs using 1-extensions, edge-additions and 2-sums [26], cf. Theorem 21.11.

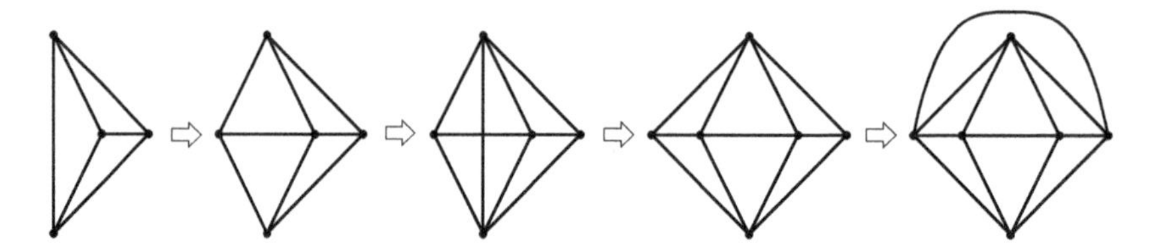

**Figure 21.9**
A sequence of 1-extensions and edge additions.

## 21.6 Special Families of Graphs

**Mixed connectivity:** In a graph $G = (V,E)$, a pair $(U,D)$ with $U \subseteq V$ and $D \subseteq E$ is a *mixed cut* if $G - U - D$ is not connected. The graph $G$ is *6-mixed-connected* if $|V| \geq 4$ and $2|U| + |D| \geq 6$ for all mixed cuts $(U,D)$ in $G$.

**Cyclic edge-connectivity:** A graph $G = (V,E)$ is *cyclically k-edge-connected* if, for all $X \subseteq V$ such that $G[X]$ and $G[V - X]$ both contain cycles, we have at least $k$ edges from $X$ to $V - X$.

**Vertex redundant rigidity:** A graph $G$ is *vertex-redundantly rigid* in $\mathbb{R}^d$ if $G - v$ is rigid in $\mathbb{R}^d$ for all $v \in V(G)$.

**$k$-rigid graph:** A graph $G$ is *$k$-rigid* in $\mathbb{R}^d$ if $G - U$ is rigid in $\mathbb{R}^d$ for all $U \subseteq V(G)$ with $|U| \leq k - 1$.

**Unit disk framework:** A *unit disk framework (with sensing radius R)* is a framework $(G, p)$ in $\mathbb{R}^2$ for which $uv \in E(G)$ if and only if $||p(u) - p(v)|| \leq R$.

**Unit disk graph:** A graph $G$ is a *unit disk graph* if it can be realized as a unit disk framework in $\mathbb{R}^2$.

**Square of a graph:** The *square* $G^2$ of a graph $G$ is obtained from $G$ by adding a new edge $uv$ for each pair $u, v \in V(G)$ of distance two in $G$.

### 21.6.1 Highly Connected Graphs

The proof technique used by Lovász and Yemini [44] to show that 6-connected graphs are rigid in $\mathbb{R}^2$ yields the stronger result that these graphs are in fact redundantly rigid. We can combine this result with Theorem 21.10 to obtain:

**Theorem 21.16** *[26] Every* 6*-connected graph is globally rigid in* $\mathbb{R}^2$.

An infinite family of 5-connected non-rigid graphs given in [44] shows that the hypothesis on vertex connectivity in Theorem 21.16 cannot be reduced from six to five. On the other hand, Jackson, Servatius, and Servatius showed in [34] that the connectivity hypothesis can be replaced by a slightly weaker hypothesis of "essential-6-vertex-connectivity" which allows vertex cuts of size four or five as long as they only separate one or at most three vertices, respectively, from the rest of the graph.

It was shown in [28] that the connectivity hypothesis can be weakened in a more substantial way and still guarantee the rigidity and global rigidity of a graph.

**Theorem 21.17** *[28] If $G$ is a* 6*-mixed-connected graph, then $G - e$ is globally rigid in* $\mathbb{R}^2$ *for all* $e \in E$.

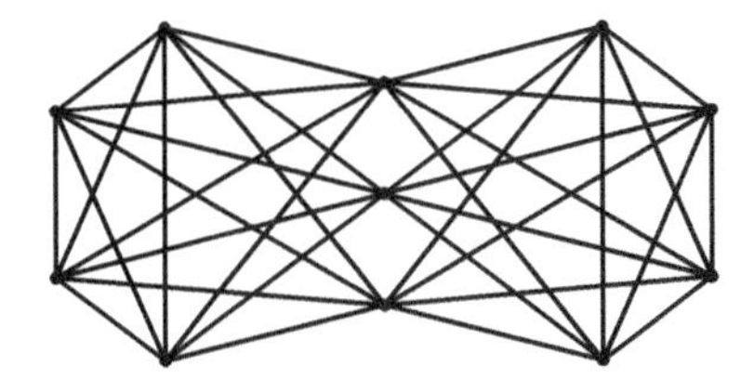

**Figure 21.10**
A graph that is mixed-6-connected but is not 4-connected.

Note that $G$ is 6-mixed-connected if and only if $G$ is 6-edge-connected, $G-v$ is 4-edge-connected for all $v \in V$, and $G-\{u,v\}$ is 2-edge-connected for all $u,v \in V$. It follows that 6-vertex-connected graphs are 6-mixed-connected and 6-mixed-connected graphs are 3-vertex-connected. See Figure 21.10 for an example.

The final result of this subsection observes that for 4-regular graphs an even weaker connectivity condition implies global rigidity.

**Theorem 21.18** *[34] Every cyclically* 5*-edge-connected* 4*-regular graph is globally rigid in* $\mathbb{R}^2$.

Examples of 4-regular 4-connected graphs and 5-regular 5-connected graphs which are not globally rigid are given in Theorem 21.20 (c),(d) below. See also Figure 21.11.

### 21.6.2 Vertex-Redundantly Rigid Graphs

Tanigawa [50] showed that vertex-redundant rigidity is a sufficient condition for global rigidity in all dimensions.

**Theorem 21.19** *Every graph which is vertex-redundantly rigid in* $\mathbb{R}^d$ *is globally rigid in* $\mathbb{R}^d$.

Kaszanitzky and Király [40] solved a number of extremal problems related to $k$-rigidity. One can also consider higher degrees of redundant rigidity with respect to edge removal as well as similar notions for global rigidity. It is an open problem to decide whether any of these graph properties can be tested in polynomial time for $d \geq 2$.

### 21.6.3 Vertex Transitive Graphs

Vertex transitive graphs which are rigid or globally rigid were characterized by Jackson, Servatius, and Servatius [34].

**Theorem 21.20** *Let $G$ be a connected $k$-regular vertex transitive graph on $n$ vertices. Then $G$ is not globally rigid in* $\mathbb{R}^2$ *if and only if one of the following holds:*
*(a)* $k=2$ *and* $n \geq 4$.
*(b)* $k=3$ *and* $n \geq 6$.
*(c)* $k=4$ *and $G$ has a* 3*-factor $F$ consisting of $s$ disjoint copies of* $K_4$ *where* $s \geq 3$.
*(d)* $k=5$ *and $G$ has a* 4*-factor $F$ consisting of $s$ disjoint copies of* $K_5$ *where* $s \geq 6$.

As a corollary they determine all vertex transitive graphs which are rigid but not globally rigid in $\mathbb{R}^2$. See Figure 21.11 for an example of such a graph.

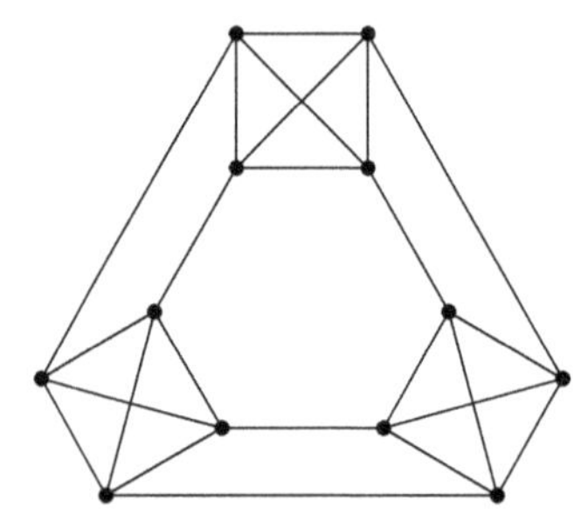

**Figure 21.11**
A 4-regular 4-connected vertex transitive graph that is rigid but not globally rigid in $\mathbb{R}^2$.

### 21.6.4 Graphs of Large Minimum Degree

Jackson and Jordán obtained a sufficient condition for global rigidity in terms of minimum degree.

**Theorem 21.21** *[27] Let $G$ be a graph on $n \geq 4$ vertices with minimum degree at at least $\frac{n+1}{2}$. Then $G$ is globally rigid in $\mathbb{R}^2$.*

The graph consisting of two complete graphs of equal size with two vertices in common shows that the bound on the minimum degree in Theorem 21.21 is best possible.

### 21.6.5 Random Graphs

We consider three different models of random graphs. Throughout this subsection, we assume that all logarithms are natural.

Our first model is the Erdős-Rényi model of random graphs. Let $G(n,p)$ denote the probability space of all graphs on $n$ vertices in which each pair of vertices is joined by an edge with independent probability $p$, see [6]. A sequence of graph properties $A_n$ holds asymptotically almost surely, or a.a.s. for short, in $G(n,p)$ if $\lim_{n\to\infty} \Pr_{G(n,p)}(A_n) = 1$.

**Theorem 21.22** *[34] Let $G \in G(n,p)$, where $p = (\log n + k \log\log n + w(n))/n$, and $\lim_{n\to\infty} w(n) = \infty$.*
*(a) If $k = 2$ then $G$ is a.a.s. rigid in $\mathbb{R}^2$.*
*(b) If $k = 3$ then $G$ is a.a.s. globally rigid in $\mathbb{R}^2$.*

The bounds on $p$ given in Theorem 21.22 are best possible since if $G \in G(n,p)$ and $p = (\log n + k \log\log n + c)/n$ for any constant $c$, then $G$ a.a.s. does not have minimum degree at least $k$, see [6].

Our second model is of random regular graphs. Let $G_{n,d}$ denote the probability space of all $d$-regular graphs on $n$ vertices chosen with the uniform probability distribution. (We refer the reader to [6] for a mathematical procedure for generating the graphs in $G_{n,d}$.) Since globally rigid graphs on at least four vertices are redundantly rigid, the only globally rigid graphs in $G(n,d)$ for $d \leq 3$ are $K_2$, $K_3$, and $K_4$. The situation changes drastically for $d \geq 4$.

**Theorem 21.23** *[34] If $G \in G_{n,d}$ and $d \geq 4$ then $G$ is a.a.s. globally rigid in $\mathbb{R}^2$.*

Our third model is of geometric random graphs. Let $Geom(n,r)$ denote the probability space of all graphs on $n$ vertices in which the vertices are distributed uniformly at random in the unit square and all pairs of vertices of distance at most $r$ are joined by an edge. Suppose $G \in Geom(n,r)$. Li et al. [43] have shown that if $n\pi r^2 = \log n + (2k-3)\log\log n + w(n)$ for $k \geq 2$ a fixed integer and $\lim_{n\to\infty} w(n) = \infty$, then $G$ is a.a.s. $k$-connected. As noted by Eren et al. [19], this result can be combined with Theorem 21.16 to deduce that if $n\pi r^2 = \log n + 9\log\log n + w(n)$ then $G$ is a.a.s. globally rigid. We do not know if this result is best possible. However, it is also shown in [43] that if $n\pi r^2 = \log n + (k-1)\log\log n + c$ for any constant $c$, then $G$ is a.a.s. not $k$-connected. Thus, if $n\pi r^2 = \log n + 2\log\log n + c$ for any constant $c$, then $G$ is a.a.s. not 3-connected, and hence is a.a.s. not globally rigid.

### 21.6.6 Unit Disk Graphs

The family of unit disk graphs is a natural model for sensor networks in which the distance between two nodes is known if and only if this distance is at most the sensing radius $R$ of the nodes, (c.f. Section 21.7.4). It is NP-hard to test whether a graph is a unit disk graph [8], and it is also NP-hard to test whether a unit disk framework is globally rigid [3]. However, it may be possible to use the unit disk property of a unit disk framework $(G,p)$ and bounds on the sensing radius to deduce necessary or sufficient conditions which imply that $(G,p)$ is globally rigid in the sense that it is a unique realization of $G$ as a unit disk framework, with the given edge lengths. This is a largely unexplored area of research.

### 21.6.7 Squares of Gaphs, Line Graphs, and Zeolites

The square $G^2$ of a graph $G$ is used in several applications. For example, if we double the sensing radius of a unit disk framework $(G,p)$, the underlying graph of the new unit disk framework will contain $G^2$. It follows from the next result that this doubling operation makes a generic unit disk framework, whose underlying graph is 2-edge-connected, globally rigid in $\mathbb{R}^2$.

**Theorem 21.24** *[10] Let $G$ be a connected graph. Then $G^2$ is globally rigid in $\mathbb{R}^2$ if and only if one endvertex of every bridge of $G$ has degree one.*

A $d$-dimensional *body-pin framework* is a different structural model consisting of full dimensional rigid bodies and pins. Each pin is a joint that connects two bodies. The bodies are free to move continuously in $\mathbb{R}^d$ subject to the constraint that the relative motion of any two bodies joined by a pin is a rotation about the pin. The framework is rigid if every such motion preserves the distances between all pairs of points belonging to different rigid bodies, i.e., the motion extends to an isometry of $\mathbb{R}^d$. The underlying graph of the framework has the bodies as its vertices and the pins as its edges. We can obtain an equivalent bar-joint framework to a given body-pin framework by taking a bar-joint realization of the line graph of the underlying graph of the body-pin framework (this operation replaces each body by a bar-joint realization of a large complete graph in such a way that two bodies joined by a pin share a joint.

The special case when $d = 2$ and the underlying graph is 3-regular gives rise to the (2-dimensional) *combinatorial zeolites*. The investigation of these structures is motivated by the interest in the flexibility properties of real zeolites, which are molecules formed by corner-sharing tetrahedra. Their rigidity can be understood by considering their equivalent bar-joint realizations.

**Theorem 21.25** *[35] Let $G = (V,E)$ be a 3-regular graph. Then the line graph of $G$ is globally rigid in $\mathbb{R}^2$ if and only if $G$ is 3-edge-connected.*

It was further shown in [38] that the underlying graphs of globally rigid generic body-pin frameworks in $\mathbb{R}^2$ can be characterized by the same connectivity condition. A generalization to higher dimensions can also be found in [38], see Chapter 20.

## 21.7 Related Properties

In this section we consider properties of generic frameworks which are weaker than global rigidity. We will assume that all frameworks are 2-dimensional unless explicitly stated otherwise.

**Globally linked pair of vertices:** A pair of vertices $\{u,v\}$ in a framework $(G,p)$ is *globally linked* in $(G,p)$ if, in all equivalent frameworks $(G,q)$, we have $||p(u)-p(v)|| = ||q(u)-q(v)||$. The pair $\{u,v\}$ is *globally linked* in the graph $G$ if it is globally linked in all generic frameworks $(G,p)$.

**Globally rigid cluster:** A *globally rigid cluster* of $G$ is a maximal subset of $V$ in which all pairs of vertices are globally linked in $G$.

**Globally loose pair of vertices:** A pair of vertices $\{u,v\}$ in a graph $G$ is *globally loose* if $\{u,v\}$ is not globally linked in all generic realizations of $G$.

**Uniquely localizable vertex:** Let $(G,p)$ be a generic framework with a designated set of vertices $P$. A vertex $v \in V(G)$ is *uniquely localizable in $(G,p)$ with respect to $P$* if whenever $(G,q)$ is equivalent to $(G,p)$ and $p(b) = q(b)$ for all vertices $b \in P$, then we also have $p(v) = q(v)$. We say that $v$ is *uniquely localizable in the graph $G$ with respect to $P$*, if $v$ is uniquely localizable with respect to $P$ in all generic frameworks $(G,p)$.

### 21.7.1 Globally Linked Pairs of Vertices

It follows from the above definition that a graph $G$ is globally rigid if and only if all pairs of vertices of $G$ are globally linked. Unlike global rigidity, however, 'global linkedness' is not a generic property. Figure 21.12 shows an example of a pair of vertices in a rigid graph $G$ which is globally linked in one generic realization, but not in another. (Note, however that global linkedness is a generic property for 1-dimensional frameworks since, in this case, $\{u,v\}$ is globally linked in $G$ if and only if $G$ has two openly disjoint $uv$-paths.)

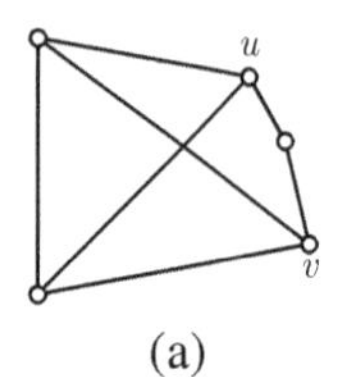

(a)

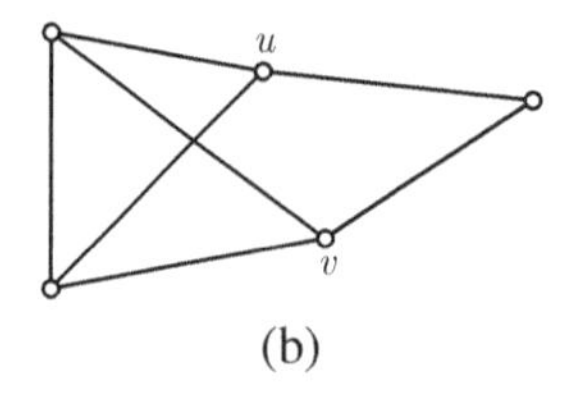

(b)

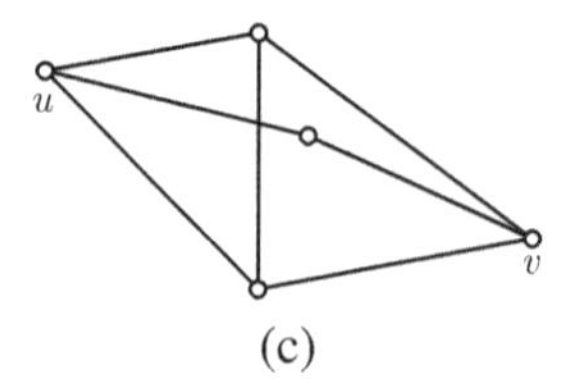

(c)

**Figure 21.12**
The pair $\{u,v\}$ is globally linked in the framework (a), but not in (b) since the distance between $u$ and $v$ is different in the equivalent framework (c).

No complete characterization of globally linked pairs in graphs is known when $d \geq 2$. We will describe several results and conjectures in the 2-dimensional case. The first shows that the 1-extension operation preserves the property that a pair of vertices is globally linked as long as the deleted edge is redundant.

**Theorem 21.26** *[31] Let $G,H$ be graphs such that $G$ is obtained from $H$ by a 1-extension operation on the edge $xy$ and vertex $w$. Suppose that $H-xy$ is rigid and that $\{u,v\}$ is globally linked in $H$. Then $\{u,v\}$ is globally linked in $G$.*

Let $G = (V, E)$ be a graph and $x, y \in V$. We use $\kappa_G(x,y)$ to denote the maximum number of pairwise openly disjoint $xy$-paths in $G$. Note that if $xy \notin E$ then, by Menger's theorem, $\kappa_G(x,y)$ is equal to the size of a smallest set $S \subseteq V - \{x,y\}$ for which there is no $xy$-path in $G - S$. It is easy to see that:

**Lemma 21.7.1** *[31] Let $(G, p)$ be a generic framework, $x, y \in V(G)$, $xy \notin E(G)$, and suppose that $\kappa_G(x,y) \leq 2$. Then $\{x,y\}$ is not globally linked in $(G, p)$.*

This observation, together with Theorem 21.26, can be used to characterize globally linked pairs in $\mathcal{R}_2$-connected graphs.

**Theorem 21.27** *[31] Let $G = (V, E)$ be an $\mathcal{R}_2$-connected graph and $x, y \in V$. Then $\{x,y\}$ is globally linked in $G$ if and only if $\kappa_G(x,y) \geq 3$.*

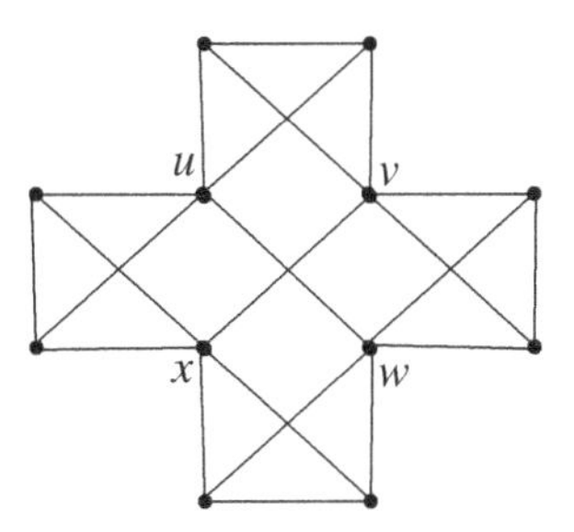


**Figure 21.13**
An $\mathcal{R}_2$-circuit $G$. The pairs $(u,v),(v,w),(w,x),(x,u)$ are globally linked in $G$ since each pair of vertices is connected by three pairwise openly disjoint paths.

Theorem 21.27 is illustrated in Figure 21.13. It has the following immediate corollary.

**Corollary 21.28** *[31] Let $G = (V, E)$ be a graph and $x, y \in V$. If either $xy \in E$, or there is an $\mathcal{R}_2$-component $H$ of $G$ with $\{x,y\} \subseteq V(H)$ and $\kappa_H(x,y) \geq 3$, then $\{x,y\}$ is globally linked in $G$.*

It is conjectured in [31] that the sufficient condition for global linkedness given in Corollary 21.28 is also necessary:

**Conjecture 6** *The pair $\{x,y\}$ is globally linked in a graph $G = (V, E)$ if and only if either $xy \in E$ or there is an $\mathcal{R}_2$-component $H$ of $G$ with $\{x,y\} \subseteq V(H)$ and $\kappa_H(x,y) \geq 3$.*

The following two closely related conjectures are also given in [31]. It is shown that together, these conjectures are equivalent to Conjecture 6.

**Conjecture 7** *Suppose that $\{x,y\}$ is a globally linked pair in a graph $G$. Then there is a redundantly rigid component $R$ of $G$ with $\{x,y\} \subseteq V(R)$.*

**Conjecture 8** *Let $G$ be a graph. Suppose that there is a redundantly rigid component $R$ of $G$ with $\{x,y\} \subseteq V(R)$ and $\{x,y\}$ is globally linked in $G$. Then $\{x,y\}$ is globally linked in $R$.*

The fact that Conjecture 6 implies both Conjectures 7 and 8 follows from Theorem 21.13.

It is straightforward to show that the 0-*extension operation* (which adds a vertex of degree 2 to a graph) preserves the property that a pair of vertices is not globally linked, see [31]. Our next result (which is a counterpart to Theorem 21.26) shows that a similar result holds for 1-extension.

**Theorem 21.29** *[32] Let $H = (V,E)$ be a rigid graph and let $G$ be a 1-extension of $H$ on some edge $uw \in E$. Suppose that $H - uw$ is not rigid and that $\{x,y\}$ is not globally linked in $H$ for some $x,y \in V$. Then $\{x,y\}$ is not globally linked in $G$.*

This result was used in [32] to deduce:

**Theorem 21.30** *Let $G = (V,E)$ be a rigid graph, $u,v \in V$, and $R = (U,F)$ be a redundantly rigid component of $G$. Suppose that $G - e$ is not rigid for all $e \in E - F$. Then $\{u,v\}$ is globally linked in $G$ if and only if $uv \in E$ or $\{u,v\}$ is globally linked in $R$.*

The special case of Theorem 21.30 when $G$ has no non-trivial redundantly rigid components characterizes globally linked pairs in minimally rigid graphs.

**Theorem 21.31** *[32] Let $G = (V,E)$ be a minimally rigid graph and $u,v \in V$. Then $\{u,v\}$ is globally linked in $G$ if and only if $uv \in E$.*

### 21.7.2 Globally Rigid Clusters

We can identify globally rigid clusters in $G$ by using the following procedure. We first construct the $\mathcal{R}_2$-components $H$ of $G$. For each component $H$, we construct the *augmented graph* $H^+$ by adding an edge between all pairs of non-adjacent vertices $u,v$ such that $H - \{u,v\}$ is disconnected. We then construct the *cleavage units* of $H$ by taking the maximal subgraphs of $H^+$ which are 3-connected or are equal to $K_3$, see [26, Section 3] for more details. By Corollary 21.28, each pair of vertices in a 3-connected cleavage unit of an $\mathcal{R}_2$-component of $G$ is globally linked in $G$. This can be illustrated by the graph $G$ in Figure 21.13. It is $\mathcal{R}_2$-connected and has five cleavage units which are equal to the five $K_4$-subgraphs in $G \cup \{uv, vw, wx, xu\}$. The vertex sets of these five subgraphs are the globally rigid clusters of $G$.

The truth of Conjecture 6 would imply that the vertex sets of the 3-connected cleavage units of the $\mathcal{R}_2$-components of $G$ (and the copies of $K_2$ or $K_3$ not included in any such cleavage unit) are precisely the globally rigid clusters of $G$. Note that the vertices of a globally rigid cluster of $G$ need not induce a globally rigid subgraph in $G$. For example, in Figure 21.13 none of the globally rigid clusters induce a globally rigid subgraph.

### 21.7.3 Globally Loose Pairs

It follows from Lemma 21.7.1 and Theorem 21.27 that if $G$ is $\mathcal{R}_2$-connected then each pair of vertices $\{u,v\}$ is either globally linked or globally loose in $G$, and that $\{u,v\}$ is globally loose if and only if $\kappa_G(u,v) = 2$. On the other hand, there exist graphs $G$ with a pair of vertices $\{u,v\}$ that are neither globally linked nor globally loose in $G$.

Our next result gives a sufficient condition for a pair $\{u,v\}$ to be globally loose in a graph $G$.

**Theorem 21.32** *[31] Let $G = (V,E)$ be a globally rigid graph and $e = uv \in E$. Suppose that $G - e$ is not globally rigid. Then $\{u,v\}$ is globally loose in $G - e$.*

A minimally rigid graph $G$ is said to be *special* if its only rigid proper subgraphs are complete graphs on at most three vertices. Examples of special graphs are $K_{3,3}$ and the prism (i.e., the complement graph of a cycle of length six). It can be seen that special graphs are 3-connected. It follows that, if $G$ is special and $uv \notin E(G)$, then $G + uv$ is a 3-connected $\mathcal{R}_2$-circuit. Thus $G + uv$ is globally rigid by Theorem 21.10. Since $G - uv$ is not globally rigid, Theorem 21.32 implies:

**Theorem 21.33** *[31] Let $G$ be a special minimally rigid graph and suppose that $u,v \in V$. Then $\{u,v\}$ is globally loose in $G$ if and only if $uv \notin E$.*

The following stronger result was proved in [32]: if $G$ is minimally rigid and $G+uv$ is an $\mathcal{R}_2$-circuit for two non-adjacent vertices $u,v$ of $G$, then $\{u,v\}$ is globally loose. The special case when $G+uv$ is a 3-connected $\mathcal{R}_2$-circuit follows from Theorem 21.32.

### 21.7.4 Uniquely Localizable Vertices

Unique localizability has direct applications in sensor network localization. We can think of $P$ as a set of *pinned vertices* (or *anchor nodes*) in a sensor network. Vertices in $P$ are, by definition, uniquely localizable. It is easy to observe that, if some vertex $v \in V-P$ is uniquely localizable, then $|P| \geq 3$ and there exist three openly disjoint paths from $v$ to $P$ (cf. Lemma 21.7.1). As was the case for global linkedness, unique localizability is not a generic property.

Let $G+K(P)$ denote the graph obtained from $G$ by adding all edges $bb'$ for which $bb' \notin E$ and $b,b' \in P$. The following lemma from [31] is easy to prove.

**Lemma 21.7.2** *[31] Let $G=(V,E)$ be a graph, $P \subseteq V$ and $v \in V-P$. Then $v$ is uniquely localizable in $G$ with respect to $P$ if and only if $|P| \geq 3$ and $\{v,b\}$ is globally linked in $G+K(P)$ for all (or equivalently, for at least three) vertices $b \in P$.*

Lemma 21.7.2 and Theorem 21.27 imply the following characterization of uniquely localizable vertices when $G+K(P)$ is $\mathcal{R}_2$-connected.

**Theorem 21.34** *[31] Let $G=(V,E)$ be a graph, $P \subseteq V$ and $v \in V-P$. Suppose that $G+K(P)$ is $\mathcal{R}_2$-connected. Then $v$ is uniquely localizable in $G$ with respect to $P$ if and only if $|P| \geq 3$ and $\kappa(v,b) \geq 3$ for all $b \in P$.*

Similarly, Lemma 21.7.2 and Conjecture 6 would imply the following characterization of uniquely localizable vertices in an arbitrary graph.

**Conjecture 9** *Let $G=(V,E)$ be a graph, $P \subseteq V$ and $v \in V-P$. Then $v$ is uniquely localizable in $G$ with respect to $P$ if and only if $|P| \geq 3$ and there is an $\mathcal{R}_2$-component $H$ of $G+K(P)$ with $P+v \subseteq V(H)$ and $\kappa_H(v,b) \geq 3$ for all $b \in P$.*

Theorems 21.27 and 21.34 imply that the sets of globally linked pairs and uniquely localizable vertices can be determined for $\mathcal{R}_2$-connected graphs. Conjectures 6 and 9 would extend this to all graphs.

### 21.7.5 The Number of Non-Equivalent Realizations

An infinitesimally rigid framework $(G,p)$ is *regular valued* if all equivalent frameworks are infinitesimally rigid. (See Chapter 18 for the definition of infinitesimal rigidity.) It is known that an infinitesimally rigid, regular valued framework $(G,p)$ has only finitely many equivalent and pairwise non-congruent realizations. We denote this number by $\mathbb{R}(G,p)$. This parameter is related to global linkedness by the fact that two vertices $u,v$ are globally linked in an infinitesimally rigid, regular valued framework $(G,p)$ if and only if $\mathbb{R}(G,p)=\mathbb{R}(G+uv,p)$. As for global linkedness, the value taken by $\mathbb{R}(G,p)$ is not a generic property: it is not difficult to see that $\mathbb{R}(G,p)=2$ and $\mathbb{R}(G,q)=4$ for the frameworks $(G,p)$ and $(G,q)$ shown in Figure 21.12(a) and (b), respectively.

On the other hand the following result shows that $\mathbb{R}(G,p)$ is constant for all generic realizations of an $\mathcal{R}_2$-connected graph $G=(V,E)$. For each $\{u,v\} \subset V$, let $w_G(u,v)$ denote the number of connected components of $G-\{u,v\}$ and put $b(G)=\sum(w_G(u,v)-1)$, where the summation is over all $\{u,v\} \subset V$.

**Theorem 21.35** *[31] Let $(G,p)$ be a generic realization of a $\mathcal{R}_2$-connected graph $G$. Then $\mathbb{R}(G,p)=2^{b(G)}$.*

For example, in the graph $G$ of Figure 21.13, $b(G) = 4$ and hence $\mathbb{R}(G,p) = 16$ for all generic realizations $(G,p)$.

Theorem 21.35 enables us to obtain a representative of each distinct congruence class of frameworks which are equivalent to a given generic framework $(G,p)$ when $G$ is $\mathcal{R}_2$-connected by iteratively applying the following operation to $(G,p)$. Choose a 2-vertex-cut $\{u,v\}$ of $G$ and reflect some, but not all, of the components of $G-\{u,v\}$ in the line through the points $p(u)$ and $p(v)$. Thus, even if a sensor network with an $\mathcal{R}_2$-connected grounded graph is not uniquely localizable, we may still obtain all possible sets of locations from one set of feasible locations in a straightforward manner.

It would be of interest to find tight bounds on $\mathbb{R}(G,p)$ when $G$ belongs to a given family of graphs and $p$ is generic. To this end we let $\mathbb{R}(G)$ be the maximum value taken by $\mathbb{R}(G,p)$ over all generic realizations of $G$. Borcea and Streinu [7] show that $\mathbb{R}(G) \leq \frac{1}{2}\binom{2n-4}{n-2} \approx 4^n$ for all rigid graphs $G$ with $n$ vertices and construct an infinite family of rigd graphs $G$ with $\mathbb{R}(G) = 12^{(n-3)/3} \approx 2.29^n$. A slightly better construction due to Emiris and Moroz [17], see also [18], gives an infinite family of rigid graphs $G$ with $\mathbb{R}(G) = 28^{(n-3)/4} \approx 2.3^n$. Capco et al [9] have recently obtained a recurrence formula which gives an upper bound on $\mathbb{R}(G)$ when $G$ is minimally rigid (more precisely their recurrence formula determines the number of frameworks in $\mathbb{C}^2$ which are equivalent to a given generic realization of $G$).

We may also consider lower bounds on $\mathbb{R}(G)$. Jackson and Owen [33] conjecture that $\mathbb{R}(G) \geq 2^{n-3}$ when $G$ is minimally rigid and verify their conjecture in the special case when $G$ is planar.

### 21.7.6 Stability Lemma and Neighborhood Results

Connelly and Whiteley [16, Theorem 13] showed that if a $d$-dimensional framework $(G,p)$ is both infinitesimally rigid and globally rigid, then the same properties hold for all frameworks $(G,q)$ which are close enough to $(G,p)$:

**Theorem 21.36** *Given a framework $(G,p)$ which is globally rigid and infinitesimally rigid in $\mathbb{R}^d$, there is an open neighborhood $U$ of $p$ in $\mathbb{R}^{d|V(G)|}$ such that for all $q \in U$ the framework $(G,q)$ is globally rigid and infinitesimally rigid.*

We will describe analogues of this result for globally linked pairs and the number of non-equivalent realizations. The first result shows that, for an infinitesimally rigid, regular valued framework framework $(G,p)$, $\mathbb{R}(G,p)$ does not increase in some open neighborhood of $p$.

**Theorem 21.37** *[32] Suppose that $(G,p)$ is an infinitesimally rigid, regular valued framework. Then there exists an open neighborhood $U$ of $p$ such that, for all $q \in U$, $(G,q)$ is an infinitesimally rigid, regular valued framework with $\mathbb{R}(G,q) \leq \mathbb{R}(G,p)$.*

Note that Theorem 21.37 generalizes (the 2-dimensional version of) Theorem 21.36 since an infinitesimally rigid, globally rigid framework $(G,p)$ is regular valued and has $\mathbb{R}(G,p) = 1$. The following example shows that we can have $\mathbb{R}(G,q) < \mathbb{R}(G,p)$ for a framework $(G,p)$ satisfying the hypotheses of Theorem 21.37 and $q$ arbitrarily close to $p$. Consider the realization $(G,p)$ of a wheel in which the central vertex and two nonconsecutive rim vertices are collinear. Then $\mathbb{R}(G,p) = 2$ but $\mathbb{R}(G,q) = 1$ for all generic $(G,q)$ since wheels are globally rigid.

In contrast, our next results shows that $\mathbb{R}(G,p)$ is constant in some open neighborhood of $p$ if either $G$ is minimally rigid or $p$ is generic.

**Theorem 21.38** *[32] Suppose that $(G,p)$ is an infinitesimally rigid, regular valued realization of a minimally rigid graph $G = (V,E)$. Then there exists an open neighborhood $U$ of $p$ such that, for all $q \in U$, $(G,q)$ is infinitesimally rigid, regular valued, and has $\mathbb{R}(G,q) = \mathbb{R}(G,p)$.*

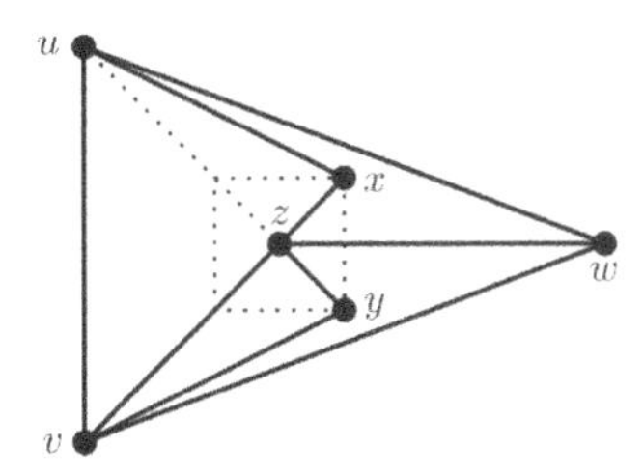


**Figure 21.14**
The framework $(G,p)$ is an infinitesimally rigid, regular valued realization of a minimally rigid graph. There are exactly four equivalent realizations which keep the triangle $p_u p_v p_w$ fixed, and they can be obtained by reflecting $x$ in the line through $p_u p_z$ and/or reflecting $y$ in the line through $p_v p_z$. The distance between $x$ and $y$ is the same in all such realizations so $(x,y)$ is globally linked in $(G,p)$. On the other hand, $\{x,y\}$ is not globally linked in any generic realization $(G,q)$.

**Corollary 21.39** *Suppose that $(G,p)$ is an infinitesimally rigid, regular valued realization of a minimally rigid graph $G=(V,E)$. Then there exists an open neighborhood $U$ of $p$ such that, for all $u,v \in V$ and all $q \in U$, $\{u,v\}$ is not globally linked in $(G,q)$ if $\{u,v\}$ is not globally linked in $(G,p)$.*

The analogous result for the property that $\{u,v\}$ *are* globally linked does not hold in general, see Figure 21.14.

We next show that $\mathbb{R}(G,p)$ remains constant in an open neighborhood of $p$ for any rigid graph $G$ when $p$ is generic.

**Theorem 21.40** *[32] Suppose that $G=(V,E)$ is rigid and $(G,p)$ is generic. Then there exists an open neighborhood $U$ of $p$ such that, for all $q \in U$, $(G,q)$ is infinitesimally rigid, regular valued, and has $\mathbb{R}(G,p)=\mathbb{R}(G,q)$.*

**Corollary 21.41** *Suppose that $G=(V,E)$ is rigid and $(G,p)$ is generic. Then there exists an open neighborhood $U$ of $p$ such that, for all $u,v \in V$ and all $q \in U$, $\{u,v\}$ is globally linked in $(G,p)$ if and only if $\{u,v\}$ is globally linked in $(G,q)$.*

The realization $(G,p)$ of a wheel in which the central vertex and two nonconsecutive rim vertices are collinear shows that Corollaries 21.39 and 21.41 become false if we remove the respective hypotheses that $G$ is minimally rigid or $p$ is generic. The problem is that there are pairs of vertices which are not globally linked in $(G,p)$ but are globally linked in $(G,q)$ for $q$ arbitrarily close to $p$. The example in Figure 21.14 shows that we can also have pairs of vertices which are globally linked in $(G,p)$ but are not globally linked in $(G,q)$ for $q$ arbitrarily close to $p$, if we remove the hypothesis that $(G,p)$ is generic.

## 21.8 Direction Constraints

**Parallel drawings:** Two $d$-dimensional frameworks $(G,p)$ and $(G,q)$ are *parallel* if $p(u)-p(v)$ is a scalar multiple of $q(u)-q(v)$ for all $uv \in E$.

**Direction congruence:** Two $d$-dimensional frameworks $(G,p)$ and $(G,q)$ are *direction congruent* if there exists a scalar $\lambda$ and a vector $t$ such that $q(v)=\lambda p(v)+t$ for all $v \in V$. This is equivalent to saying that $(G,q)$ can be obtained from $(G,p)$ by a translation and dilation.

**Tight framework:** A $d$-dimensional framework $(G,p)$ is *tight* if every $d$-dimensional framework which is parallel to $(G,p)$ is direction congruent to $(G,p)$. The graph $G$ is in $\mathbb{R}^d$ if every generic realization of $G$ in $\mathbb{R}^d$ is tight.

**Mixed graph:** A *mixed graph* is a graph together with a bipartition $D \cup L$ of its edge set. We refer to edges in $D$ as *direction edges* and edges in $L$ as *length edges.*

**Direction-length framework:** A $d$-dimensional *direction-length framework* $(G,p)$ is a mixed graph $G = (V;D,L)$ together with a map $p : V \to \mathbb{R}^d$.

**Direction-length equivalence:** Two $d$-dimensional direction-length frameworks $(G,p)$ and $(G,q)$ are *equivalent* if $p(u) - p(v)$ is a scalar multiple of $q(u) - q(v)$ for all $uv \in D$ and $||p(u) - p(v)|| = ||q(u) - q(v)||$ for all $uv \in L$.

**Direction-length congruence:** Two $d$-dimensional direction-length frameworks $(G,p)$ and $(G,q)$ are *congruent* if there exists a vector $t \in \mathbb{R}^d$ and $\lambda \in \{-1,1\}$ such that $q(v) = \lambda p(v) + t$ for all $v \in V$. This is equivalent to saying that $(G,q)$ can be obtained from $(G,p)$ by an inversion and a translation.

**Direction-length global rigidity:** A $d$-dimensional direction-length framework $(G,p)$ is *globally rigid* if every $d$-dimensional direction-length framework which is equivalent to $(G,p)$ is congruent to $(G,p)$.

**Direction-length rigidity:** A $d$-dimensional direction-length framework $(G,p)$ is *rigid* if there exists an $\varepsilon > 0$ such that every $d$-dimensional direction-length framework $(G,q)$ which is equivalent to $(G,p)$ and satisfies $\|p(v) - q(v)\| < \varepsilon$ for all $v \in V$, is congruent to $(G,p)$. The mixed graph $G$ is *rigid in* $\mathbb{R}^d$ if every generic realization of $G$ in $\mathbb{R}^d$ is rigid.

**Direction-balanced mixed graph:** A 2-*separation* $G_1, G_2$ of a mixed graph $G$ is is a pair of subgraphs $G_1, G_2$ such that $G = G_1 \cup G_2$, $|V(G_1) \cap V(G_2)| = 2$ and $V(G_1) - V(G_2) \neq \emptyset \neq V(G_2) - V(G_1)$. It is *direction-balanced* if both $G_1$ and $G_2$ contain a direction edge. The mixed graph $G$ is *direction balanced* if all 2-separations of $G$ are direction balanced.

### 21.8.1 Parallel Drawings

The fact that direction constraints are linear implies that the local and global versions of tightness are equivalent properties of frameworks (if two $d$-dimensional frameworks $(G,p)$ and $(G,q)$ are parallel then we can continuously move each vertex $v$ of $G$ on the line joining $p(v)$ to $q(v)$ to transform $(G,p)$ to $(G,q)$ in such a way that the intermediate frameworks remain parallel). Whiteley [51] characterized generic tightness in all dimensions.

**Theorem 21.42** *[51] Let $(G,p)$ be a $d$-dimensional generic framework. Then $(G,p)$ is tight if and only if*

$$\sum_{H \in \mathcal{H}} (d|V(H)| - d - 1) \geq d|V(G)| - d - 1$$

*for all families $\mathcal{H}$ of subgraphs which have at least two vertices and cover $E(G)$.*

This implies, in particular, that tightness is a generic property of frameworks.

When $d = 2$, the characterization of generic tightness given in Theorem 21.42 is the same as the characterization of generic rigidity given by Lovász and Yemini [44]. This implies that generic tightness and generic rigidity are equivalent when $d = 2$.

### 21.8.2 Direction-Length Global Rigidity

Since the problem of characterizing global rigidity for direction-length frameworks is at least as hard as the corresponding problem when there are only length constraints, we will restrict our attention to the 2-dimensional case, unless explicitly stated otherwise.

Servatius and Whiteley [47] developed a rigidity theory for direction-length frameworks in $\mathbb{R}^2$ analogous to that for length constrained frameworks. One may construct a $|D \cup L| \times 2|V|$ *(direction-length) rigidity matrix* for a direction-length framework $(G,p)$ and use its rows to define the *(direction-length) rigidity matroid* $\mathcal{R}^{DL}(G,p)$ of $(G,p)$. A generic direction-length framework is rigid if and only if its rigidity matrix, or matroid, has rank $2|V|-2$. It follows that the rigidity of direction-length frameworks is a generic property. All generic realizations of a mixed graph $G$ have the same rigidity matroid and this matroid is defined to be the *rigidity matroid* $\mathcal{R}^{DL}(G)$ of $G$. The main result of Servatius and Whiteley [47] characterizes the generic rigidity of mixed graphs in $\mathbb{R}^2$.

The problem of characterizing when a generic direction-length framework $(G,p)$ is globally rigid is still an open problem, even though some recent results get very close to this target. More precisely, global rigidity has been characterized for generic frameworks when the rigidity matroid of the underlying mixed graph is connected. Mixed graphs with the property that *all* their generic realizations are globally rigid have also been characterized. (It is not yet known whether global rigidity of direction-length frameworks is a generic property i.e., we do not know whether there exist mixed graphs which have both a globally rigid generic realization and a non-globally rigid generic realization in $\mathbb{R}^2$.)

We first give a necessary condition for global rigidity, which is analogous to the "3-connectedness condition" of Theorem 21.2.

**Theorem 21.43** *[29] Suppose that $(G,p)$ is a generic globally rigid direction-length framework. Then $G$ is 2-connected and direction balanced.*

Rigidity is also a necessary condition for global rigidity. Redundant rigidity, however, is no longer necessary. To see this consider a minimally rigid mixed graph $G$ with exactly one length edge $e$. The 2-dimensional version of Theorem 21.42 and the above mentioned characterization of generic direction-length rigidity imply that $G-e$ is tight, and this in turn implies that every generic realization of $G$ is globally rigid. On the other hand, $G-f$ is not rigid for all edges $f$ of $G$.

We next give a result on 1-extensions which is analogous to Theorem 21.4. For a mixed graph $G$, the operation 1-*extension* (on edge $uw$ and vertex $z$) deletes the edge $uw$ and adds a new vertex $v$ and new edges $vu, vw, vz$ for some vertex $z \in V(G)$, with the provisos that at least one of the new edges has the same type as the deleted edge and, if $z=u$, then the two edges from $z$ to $u$ are of different type.

**Theorem 21.44** *[30] Let $H$ be a mixed graph with at least three vertices and $G$ be obtained from $H$ by a 1-extension on an edge $uw$. Suppose that $(G,p)$ is a generic realization of $G$. If $H-uw$ is rigid and $(H,p|_H)$ is globally rigid, then $(G,p)$ is globally rigid.*

In mixed graphs, a special kind of 0-extension also preserves global rigidity.

**Theorem 21.45** *[30] Let $G$ and $H$ be mixed graphs with $|V(H)| \geq 2$ and $(G,p)$ be a generic realization of $G$. Suppose that $G$ can be obtained from $H$ by adding a vertex $v$ incident to two direction edges. Then $(G,p)$ is globally rigid if and only if $(H,p|_H)$ is globally rigid.*

Note that the graph $G$ we obtain in Theorem 21.45 will not be redundantly rigid since it will contain a vertex of degree two.

We can use Theorems 21.44 and 21.45 to show that a special family of generic direction-length frameworks are globally rigid. We say that a mixed graph $G=(V;D,L)$ is a *mixed $\mathcal{R}^{DL}$-circuit* if $D \neq \emptyset \neq L$ and $D \cup L$ is a circuit in $\mathcal{R}^{DL}(G)$. It is easy to show that mixed $\mathcal{R}^{DL}$-circuits are 2-connected. Jackson and Jordán [29] showed that the necessary condition for global rigidity given in Theorem 21.43 is also sufficient to imply that mixed $\mathcal{R}^{DL}$-circuits are globally rigid.

**Theorem 21.46** *Let $(G, p)$ be a generic realization of a mixed $\mathcal{R}^{DL}$-circuit. Then $(G, p)$ is globally rigid if and only if $G$ is direction balanced.*

Clinch [11] extended Theorem 21.46 to $\mathcal{R}^{DL}$*-connected mixed graphs*, i.e., mixed graphs $G$ for which the matroid $\mathcal{R}^{DL}(G)$ is connected.

**Theorem 21.47** *Let $(G, p)$ be a generic realization of an $\mathcal{R}^{DL}$-connected mixed graph. Then $(G, p)$ is globally rigid if and only if $G$ is direction balanced.*

A recent result of Clinch, Jackson, and Keevash [12] characterizes mixed graphs with the property that *all* their generic realizations in $\mathbb{R}^2$ are globally rigid.

**Theorem 21.48** *Suppose $G = (V, D \cup L)$ is a mixed graph. Then every generic realization of $G$ is globally rigid if and only if $G$ is rigid and either $|L| = 1$ or $G$ has a direction-balanced $\mathcal{R}^{DL}$-connected subgraph which contains $L$.*

We close this section by noting that we can characterize global rigidity in all dimensions for mixed graphs in which every pair of adjacent vertices is connected by both a length and a direction edge.

**Theorem 21.49** *[29] Let $G$ be a mixed graph in which every pair of adjacent vertices is connected by both a length and a direction edge, and $(G, p)$ be a generic realization of $G$ in $\mathbb{R}^d$. Then $(G, p)$ is globally rigid if and only if $G$ is 2-connected.*

## 21.9 Algorithms

The structural results presented in this chapter give rise to efficient combinatorial algorithms for testing different global rigidity properties of generic frameworks and for solving a number of related algorithmic problems in the plane. The major problems are as follows. How can we decide whether a given graph $G$ is rigid, redundantly rigid, or $\mathcal{R}_2$-connected? More generally, how can we identify the rigid, redundantly rigid, and $\mathcal{R}_2$-connected components of $G$?

The key ingredient in solving these problems is an efficient subroutine for checking if a set of edges is independent in the 2-dimensional rigidity matroid of $G$. For this matroid, which is known to be a so-called count matroid, we have fast combinatorial algorithms for testing independence. This subroutine can be implemented in $O(|V|^2)$ time by using various alternating path algorithms: methods from matching theory [24], network flows [25], and graph orientations [4, 42] have been used to do this. By using additional algorithmic techniques, each of the problems mentioned above can be solved in $O(|V|^3)$ time.

## 21.10 Optimization Problems

The *globally rigid pinning problem* is motivated by a natural question from network localization: the goal is to find a smallest set $P$ of vertices in a graph $G$ for which $G + K(P)$ is globally rigid in $\mathbb{R}^2$ (where $K(P)$ denotes a complete graph on the vertex set $P$). Although the complexity of this problem is still open, an efficient approximation algorithm is given by Jordán [36].

The *$\mathcal{R}_2$-connected pinning problem* is a variant of the globally rigid pinning problem in which the goal is to find a smallest set $P \subseteq V$ such that $G + K(P)$ is $\mathcal{R}_2$-connected. This problem can be formulated as finding a largest matroid matching in the *hypergraphic matroid* defined on $V$ by the $\mathcal{R}_2$-components of $G$. Hypergraphic matroids are known to be linear, but it is not known how to find a suitable linear representation. The complexity status of the matroid matching problem in hypergraphic matroids is still open. Nevertheless, this formulation can be used to design a $\frac{3}{2}$-approximation algorithm (which works for the minimum cost version as well). This approximation algorithm can be used as a subroutine to design a $\frac{5}{2}$-approximation algorithm for finding a smallest (or minimum cost) subset $P$ for which $G + K(P)$ is globally rigid in $\mathbb{R}^2$ [36].

The above methods can be used to design a constant factor approximation algorithm for the corresponding *augmentation problem* in which the goal is to add a smallest set $F$ of new edges to $G$ so that $G + F$ is globally rigid. It was shown by García and Tejel [22] that the related problem of augmenting a rigid graph to a redundantly rigid graph by a smallest set of new edges is NP-hard.

## Acknowledgment

This work was supported by the Hungarian Scientific Research Fund grant no. K109240 and K115483. The third author was supported by JSPS KAKENHI Grant Number JP15K15942.

## References

[1] T.G. Abbot. Generalizations of Kempe's universality theorem. Master's thesis, MIT, `http://web.mit.edu/tabbott/www/papers/mthesis.pdf`, 2008.

[2] L. Asimow and B. Roth. The rigidity of graphs. *Trans. Amer. Math. Soc.*, 245:279–289, 1978.

[3] J. Aspnes, T. Eren, D. K. Goldenberg, A. S. Morse, and W. Whiteley. A theory of network localization. *IEEE Transactions on Mobile Computing*, 5:1663–1678, 2006.

[4] A.R. Berg and T. Jordán. Algorithms for graph rigidity and scene analysis. In U. Zwick G. Di Battista, editor, *Proc. 11th Annual European Symposium on Algorithms (ESA)*, Springer Lecture Notes in Computer Science 2832, pages 78–89, 2003.

[5] A.R. Berg and T. Jordán. A proof of Connelly's conjecture on 3-connected circuits of the rigidity matroid. *J. Combinatorial Theory Ser. B.*, 88:77–97, 2003.

[6] B. Bollobás. *Random graphs*. Academic Press, New York, 1985.

[7] C. Borcea and I. Streinu. The number of embeddings of minimally rigid graphs. *Discrete Comput. Geom.*, 31:287–303, 2004.

[8] H. Breu and D.G. Kirkpatrick. Unit disk graph recognition is NP-hard. *Comput. Geom.*, 9:3–24, 1998.

[9] J. Capco, M. Gallet, G. Grassegger, C. Koutschan, N. Lubbes, and J. Schicho. The number of realizations of a Laman graph. Technical report, arXiv 1701.05500v2, 2017.

[10] M. Cheung and W. Whiteley. Transfer of global rigidity results among dimensions: graph powers and coning. Technical report, York University, July 2008.

[11] K. Clinch. Global rigidity of 2-dimensional direction-length frameworks with connected rigidity matroids. Technical report, arXiv:1608.08559, 2016.

[12] K. Clinch, B. Jackson, and P. Keevash. Global rigidity of 2-dimensional direction-length frameworks. Technical report, arXiv:1607.00508v2, 2018.

[13] R. Connelly. Rigidity and energy. *Invent. Math.*, 66:11–33, 1982.

[14] R. Connelly. On generic global rigidity. In *Applied Geometry and Discrete Mathematics*, volume 4 of *DIMACS Ser. Discrete Math, Theoret. Comput. Sci.*, pages 147–155. Amer. Math. Soc., 1991.

[15] R. Connelly. Generic global rigidity. *Discrete Comput. Geom.*, 33:549–563, 2005.

[16] R. Connelly and W. Whiteley. Global rigidity: the effect of coning. *Discrete Comput. Geom.*, 43:717–735, 2010.

[17] I.Z. Emiris and G. Moroz. The assembly modes of 11-bar linkages. In *Proc. IFToMM World Cong. Mechanism and Machine Science*, Guanajuato, Mexico, 2011.

[18] I.Z. Emiris and I. D. Psarros. Counting euclidean embeddings of rigid graphs. Technical report, arXiv:1402.1484v1, 2014.

[19] T. Eren, W. Whiteley, A. Morse, P.N. Belhumeur, and B.D.O. Anderson. Sensor and network topologies of formations with direction, bearing, and angle information between agents. In *Proc. of the 42nd IEEE Conference on Decision and Control*, pages 3064–3069, Maui, HI, USA, December 2003.

[20] A. Frank. *Connections in combinatorial optimization*. Oxford University Press, 2011.

[21] S. Frank and J. Jiang. New classes of counterexamples to Hendrickson's global rigidity conjecture. *Discrete Comput. Geom.*, 45(3):574–591, 2011.

[22] A. García and J. Tejel. Augmenting the rigidity of a graph in $R^2$. *Algorithmica*, 59:145–168, 2011.

[23] S. Gortler, A. Healy, and D. Thurston. Characterizing generic global rigidity. *American Journal of Mathematics*, 132(4):897–939, 2010.

[24] B. Hendrickson. Conditions for unique graph realizations. *SIAM J. Comput.*, 21:65–84, 1992.

[25] H. Imai. Network-flow algorithms for lower-truncated transversal polymatroids. *J. Operations Res. Soc. Japan*, 26:186–210, 1983.

[26] B. Jackson and T. Jordán. Connected rigidity matroids and unique realizations of graphs. *J. Combinatorial Theory Ser B*, 94:1–29, 2005.

[27] B. Jackson and T. Jordán. Graph theoretic techniques in the analysis of uniquely localizable sensor networks. In G. Mao and B. Fidan, editors, *Localization Algorithms and Strategies for Wireless Sensor Networks*. IGI Global, 2009.

[28] B. Jackson and T. Jordán. A sufficient connectivity condition for generic rigidity in the plane. *Discrete Appl. Math.*, 157:1965–1968, 2009.

[29] B. Jackson and T. Jordán. Globally rigid circuits of the direction-length rigidity matroid. *J. Combinatorial Theory Ser B*, 100:1–22, 2010.

[30] B. Jackson and T. Jordán. Operations preserving global rigidity of generic direction-length frameworks. *International Journal on Computational Geometry and Applications*, 20:685–706, 2010.

[31] B. Jackson, T. Jordán, and Z. Szabadka. Globally linked pairs of vertices in equivalent realizations of graphs. *Discrete Comput. Geom.*, 35:493–512, 2006.

[32] B. Jackson, T. Jordán, and Z. Szabadka. Globally linked pairs of vertices in rigid frameworks, in: Rigidity and symmetry. *Fields Institute Communications*, 70:177–203, 2014.

[33] B. Jackson and J.C. Owen. The number of equivalent realisations of a rigid graph. Technical report, arXiv:1204.1228, 2012.

[34] B. Jackson, B. Servatius, and H. Servatius. The 2-dimensional rigidity of certain families of graphs. *J. Graph Theory*, 54:154–166, 2007.

[35] T. Jordán. Generically globally rigid zeolites in the plane. *Information Processing Letters*, 110:841–844, 2010.

[36] T. Jordán. Rigid and globally rigid graphs with pinned vertices. In G.O.H. Katona, A. Schrijver, and T. Szőnyi, editors, *Fete of Combinatorics and Computer Science*, Bolyai Society Mathematical Studies. Springer, 2010.

[37] T. Jordán. Combinatorial rigidity: graphs and matroids in the theory of rigid frameworks. In *Discrete Geometric Analysis*, volume 34 of *MSJ Memoirs*. The Mathematical Society of Japan, 2016.

[38] T. Jordán, C. Király, and S. Tanigawa. Generic global rigidity of body-hinge frameworks. *J. Combinatorial Theory Ser B*, 117:59–76, 2016.

[39] T. Jordán and Z. Szabadka. Operations preserving the global rigidity of graphs and frameworks in the plane. *Computational Geometry*, 42:511–521, 2009.

[40] V.E. Kaszanitzky and C. Király. On minimally highly vertex-redundantly rigid graphs. *Graphs and Combinatorics*, 32:225–240, 2016.

[41] G. Laman. On graphs and rigidity of plane skeletal structures. *J. Engineering Math.*, 4:331–340, 1970.

[42] A. Lee and I. Streinu. Pebble game algorithms and sparse graphs. *Discrete Math.*, 308:1425–1437, 2008.

[43] X-Y. Li, P-J. Wan, Y. Wang, and C-W. Yi. Fault tolerant deployment and topology control in wireless networks. In *Proceedings of the ACM Symposium on Mobile Ad Hoc Networking and Computing (MobiHoc)*, pages 117–128, Annapolis, MD, June 2003.

[44] L. Lovász and Y. Yemini. On generic rigidity in the plane. *SIAM J. Algebraic Discrete Methods*, 3:91–98, 1982.

[45] J.G. Oxley. *Matroid theory*. Oxford University Press, 2nd edition, 2011.

[46] J.B. Saxe. Embeddability of weighted graphs in $k$-space is strongly NP-hard. Technical report, Computer Science Department, Carnegie-Mellon University, 1979.

[47] B. Servatius and W. Whiteley. Constraining plane configurations in computer-aided design: combinatorics of directions and lengths. *SIAM J. Discrete Math.*, 12:136–153, 1999.

[48] A. M. So and Y. Ye. Theory of semidefinite programming for sensor network localization. *Math. Program.*, 109:367–384, 2007.

[49] Z. Szabadka. *Globally rigid frameworks and rigid tensegrity graphs in the plane*. PhD thesis, Institue of Mathematics, Eötvös Loránd University, Hungary, 2010.

[50] S. Tanigawa. Sufficient conditions for the global rigidity of graphs. *J. Combinatorial Theory Ser B*, 113:123–140, 2015.

[51] W. Whiteley. Some matroids from discrete applied geometry. In J. Bonin, J. Oxley, and B. Servatius, editors, *Matroid Theory*, volume 197 of *Contemp. Math.* Amer. Math. Soc., 1996.

# Chapter 22

## *Point-Line Frameworks*

**Bill Jackson**

*School of Mathematical Sciences, Queen Mary University of London, London, England.*

**J.C. Owen**

*Siemens, Cambridge, England.*

**CONTENTS**



## 22.1 Introduction

In this chapter we will describe the rigidity properties of frameworks in which some of the points (representing joints) in a bar-joint framework are replaced by lines. The graph of the framework thus becomes a vertex-labelled graph in which some of the vertices are labelled as points and some are labelled as lines. This gives rise to three different types of edges – those connecting a pair of points, those connecting a point and a line and those connecting two lines. The geometry of the framework corresponds to a collection of points and a collection of unbounded lines. The three classes of edges correspond to a constrained distance between a pair of points, a constrained distance between a point and a line (where the distance is defined as the Euclidean distance from the point to the nearest point on the line) and a constrained angle between two lines.

### 22.1.1 Motivation from CAD

Our intial interest in point-line frameworks arises from applications in computer-aided design (CAD). The first CAD systems typically represented all of the component geometric elements by a number of points on the element [16, 11]. For example a line can be represented by two points on the line or a plane by three points and then the distance between a point and a line is a trivalent constraint and the angle between two lines a four-valent constraint. However it became clear [3, 12, 19] that it is simpler and more efficient to represent the geometric elements more directly so that for example a line has coordinates corresponding to its position and tangent direction and a plane has coordinates corresponding to its position and normal direction.

### 22.1.2 Motivation from Automated Deduction in Geometry (ADG) and Theorem Proving

Many theorems of Euclidean geometry and projective geometry can be described in terms of geometric elements such as points, lines and circles together with distance relations between pairs of them [1, 5, 15]. For example the coincidence of a point and line implies that their separation distance is zero and similarly for a line-circle tangency. Many theorems can be interpreted as a statement that some of these incidence relations can be deduced from the existence of other relations and recognising this by some automated procedure is an active area of research.

As a specific example the well-known theorem of Desargues on the plane can be represented as a point-line framework in which certain points are incident to certain lines. If the points and lines are subject to these constraints but are otherwise generic then the dependency of the rigidity matrix would imply the theorem. The existence of such theorems shows that charactarising the rigidity properties of such quasi-generic point-line frameworks is a hard (and unsolved) problem.

### 22.1.3 Constraint Graphs and Frameworks

In CAD and ADG we are interested in many different types of geometric elements. For example the representation of a simple machined part may entail representing points, lines, planes, circles, cylinders, tori, and cones [13, 20]. If we also specify relationships between them, such as the fact that some pair are tangent or have a constrained separation distance between them, then this is commonly represented by a constraint-graph in which the nodes in the graph are labelled with the type of geometric element which they represent and the edges are labelled with the type of relationship which exists between its incident vertices [9].

If we assign coordinates to the geometric elements then the constraint graph becomes a constraint framework. It is rigid if every continuous motion of the geometric elements which preserves the constraints results in a constraint framework which can be obtained from the initial framework by an isometry of the space we are working in. In this view the bar-joint frameworks which are the subject of many of the chapters in this book are constraint frameworks in which all of the geometric elements are points and all of the constraints are distances between pairs of points.

Figure 22.1 shows all complete point-line graphs on three vertices. In these graphs we have adopted our usual convention that vertices which represent points are drawn as filled circles and labelled $u_i$ and vertices which represent lines are drawn as unfilled circles and labelled $v_i$. Figure 22.2 shows point-line frameworks which represent a geometric realization of each graph in Figure 22.1. The dotted lines in these frameworks represent either a distance constraint or an angle constraint between the two geometries which they connect. The distance and angle constraints in frameworks

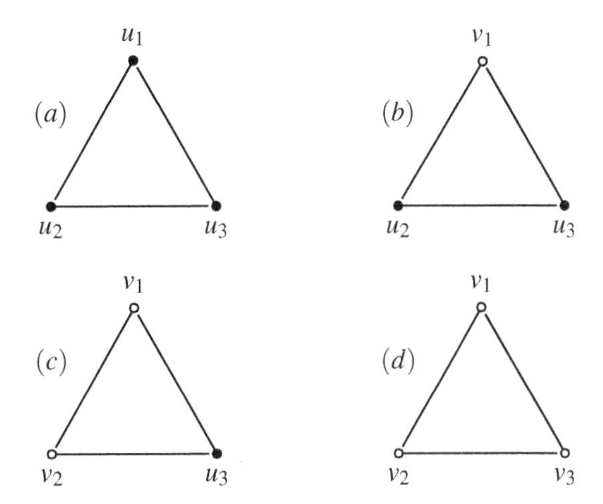


**Figure 22.1**
The four distinct, complete point-line graphs on three vertices. Filled circles are point-vertices labelled $u_i$ and unfilled circles are line-vertices labelled $v_i$.

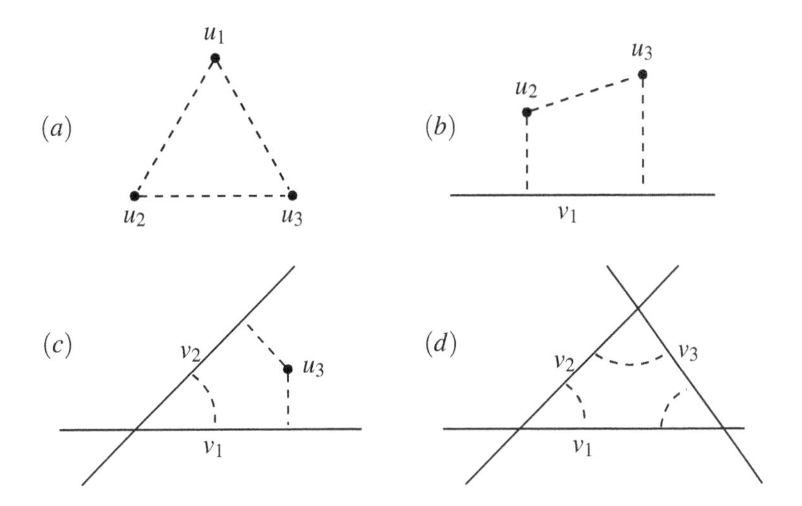


**Figure 22.2**
Four frameworks corresponding to the four point-line graphs in Figure 22.1.

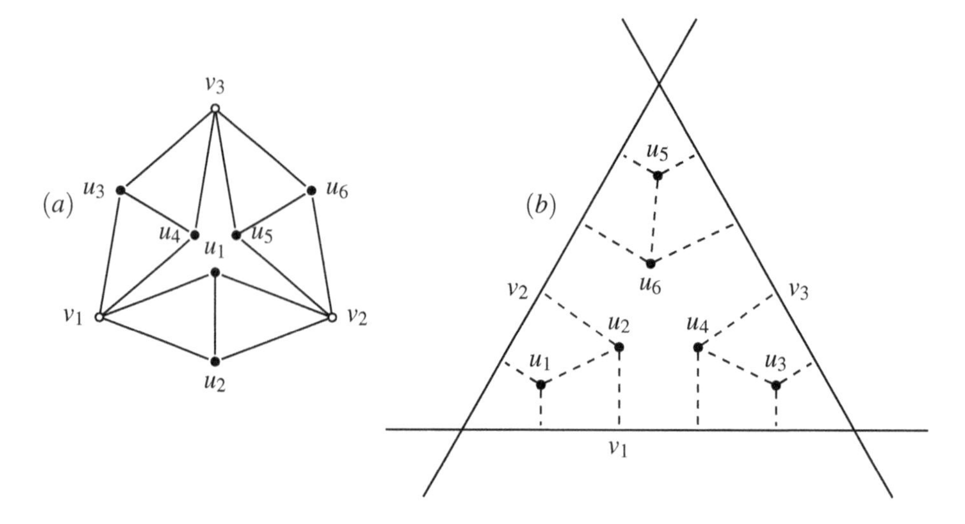


**Figure 22.3**
A point-line graph (a) and a corresponding point-line framework (b) in $\mathbb{R}^2$.

(a), (b) and (c) can be specified independently. However the three angle constraints in framework (d) must sum to 180 degrees and so cannot be specified independently.

Figure 22.3 shows a graph and corresponding framework which is derived from the 3-cycle of line-vertices shown in Figure 22.1(d) by substituting rigid subgraphs in place of the angle constraints. Although this graph has only distance constraints (between points and lines or between pairs of points), we will see that the distances in this graph cannot be specified independently.

As we observed in Section 22.1.1, a line can also be represented by a pair of points and so we may ask what is to be gained by allowing additional geometric elements. It is not difficult to see that many of the constraints can also be represented by bars in an appropriate bar-joint framework. For example a point may be constrained to be a fixed distance from a line which is represented by a pair of points using a Peaucellier mechanism which is itself a bar-joint framework.

However, the result of this substitution is a very specialized bar-joint framework and if there are several such substitutions it is difficult to get any intuition for the rigidity properties of the framework. Many of the important results for bar-joint frameworks refer to specific types of frameworks and these may not include those specialized bar-joint frameworks which can represent more general constraint frameworks [17]. The properties of such frameworks may only be obtained by studying the constraint framework directly. For example Laman's theorem describes the rigidity properties of a bar-joint framework for which the point coordinates are generic (algebraically independent over the rationals). A point-line framework for which the line positions and slopes are also generic would give a corresponding bar-joint framework in which the point coordinates are not generic and so Laman's theorem would not apply. We will describe below how we can obtain a theorem which corresponds to Laman's theorem by considering the point-line framework directly.

## 22.2 Point-Line Graphs and Frameworks

**Point-line graph:** A *point-line graph* $G = (V_P, V_L, E)$ is a graph $G = (V, E)$ without loops together with an ordered pair $(V_P, V_L)$ of, possibly empty, disjoint sets whose union is $V$ [9].

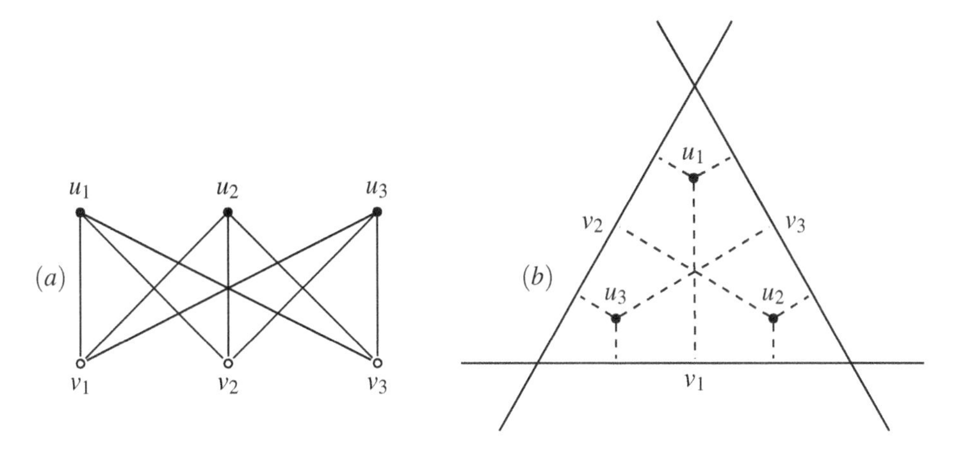


**Figure 22.4**
(a) The naturally bipartite point-line graph K(3,3) and (b) a framework on this graph.

We refer to vertices in $V_P$ and $V_L$ as *point-vertices* and *line-vertices*, respectively. We label the vertices as $V_P = \{u_1, \dots, u_s\}$ and $V_L = \{v_1, \dots, v_t\}$, and the edges as $E = \{e_1, e_2, \dots, e_m\}$. We use $E_{PP}$, $E_{PL}$ and $E_{LL}$ to denote the sets of edges incident to two point-vertices, to a point-vertex and a line-vertex and to two line-vertices respectively. For $e \in E$, we write $e = xy$ to mean that the end-vertices of $e$ are $x$ and $y$.

We supplement the above notation when it is not obvious which graph we are referring to by using $V(G), E(G)$, etc.

We will assume that a point-line graph is simple (without parallel edges) unless it is explicitly described as a point-line multigraph. Sometimes we will want to generate a multigraph from a simple graph $G$ by introducing a second copy of an edge which is already in the graph. We call this doubling an edge. If $e = ij \in E(G)$ then we write $G + ij$ for the graph with $e$ doubled and denote the new edge by $e'$.

**Naturally bipartite:** A point-line graph $G = (V_P, V_L, E)$ is *naturally bipartite* if $G$ is a bipartite graph with bipartition $V_P, V_L$.

In this case $E_{PP} = E_{LL} = \emptyset$ and $E = E_{PL}$. For example the point-line graph on $K(3,3)$ shown in Figure 22.4(a) is naturally bipartite.

Naturally bipartite point-line graphs are especially important. In particular we can derive a corresponding naturally bipartite point-line graph from any point-line graph $G$ by replacing any edge $e = ab$ in $E_{PP}$ or $E_{LL}$ with a copy of the naturally bipartite $K(3,3)$ shown in Figure 22.4(a) where either a pair of point-vertices or a pair of line-vertices in $K(3,3)$ are identified with the vertices $a$ and $b$ in $G$. Figure 22.5(b) shows the result of this substitution into the graph in Figure 22.5(a). We shall see that this substitution preserves infinitesimal rigidity and this will allow us to restrict to naturally bipartite point-line graphs.

### 22.2.1 Point-Line Frameworks and the Rigidity Map

A point-line framework is a realization of a point-line graph which is obtained by assigning appropriate coordinates to each point- and line-vertex in the graph. We will henceforth fix our attention to point-line frameworks in $\mathbb{R}^2$, unless explicitly stated otherwise. In this context, a point-vertex $u_i$ gets coordinates $(x_i, y_i) \in \mathbb{R}^2$. A line may be described in the $(x,y)$-plane by the equation $x = ay + b$ where $b$ gives the intersection of the line with the $x$-axis and $a$ gives its slope. Hence a line-vertex $v_i$

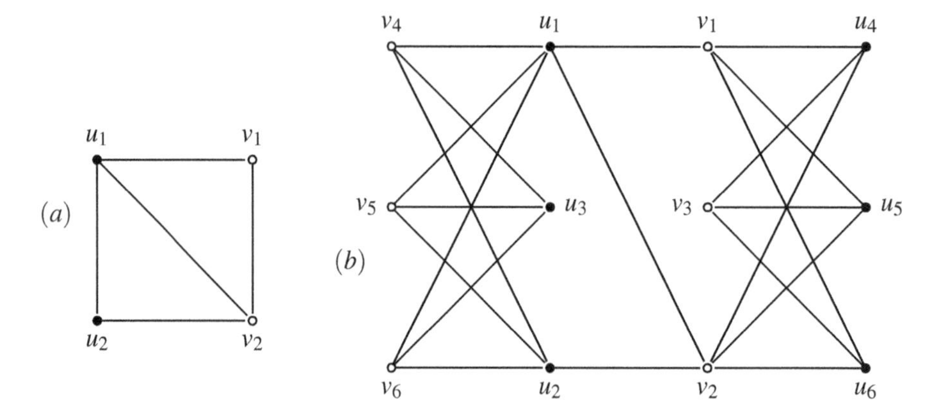


**Figure 22.5**
The point-line graph (a) converts into the naturally bipartite graph (b).

gets coordinates $(a_i, b_i) \in \mathbb{R}^2$ (assuming we choose a coordinate system such that none of the lines is parallel to the $x$-axis).

**Point-line framework:** A *point-line framework* is a pair $(G, p)$ where $G$ is a point-line graph and $p: V \to \mathbb{R}^2$.

We put $p(u_i) = (x_i, y_i)$ for each $u_i \in V_P$ and $p(v_i) = (a_i, b_i)$ for each $v_i \in V_L$. This gives a geometric realization of $(G, p)$ by taking the point corresponding to $u_i$ to have cartesian coordinates $(x_i, y_i)$ and the line corresponding to $v_i$ to be $x = a_i y + b_i$.

**Rigidity map:** The *rigidity map* of a point-line framework $(G, p)$ is the map $f_G : \mathbb{R}^{2|V|} \to \mathbb{R}^{|E|}$ defined as follows.

For each $p \in \mathbb{R}^{2|V|}$ we take $f_G(p) = (f_1(p), f_2(p), \ldots, f_m(p))$ where the components of $f_G(p)$ are indexed by the edges $e_i \in E$ and

$$f_i(p) = \begin{cases} (x_j - x_k)^2 + (y_j - y_k)^2 & \text{if } e_i = u_j u_k \in E_{PP} \\ (x_j - y_j a_k - b_k)(1 + a_k^2)^{-\frac{1}{2}} & \text{if } e_i = u_j v_k \in E_{PL} \\ \tan^{-1} a_j - \tan^{-1} a_k & \text{if } e_i = v_j v_k \in E_{LL} \text{ and } j < k. \end{cases}$$

These expressions for $f_i(p)$ are: the squared distance between the points represented by $u_j$ and $u_k$ when $e_i = u_j u_k \in E_{PP}$; the signed distance between the point represented by $u_j$ and the line represented by $v_k$ when $e_i = u_j v_k \in E_{PL}$; the angle between the lines represented by $v_j$ and $v_k$ when $e_i = v_j v_k \in E_{LL}$.

### 22.2.2 The Rigidity Matrix

**Jacobian matrix:** The *Jacobian matrix* $J(G, p)$ of the rigidity map is the $|E| \times 2|V|$-matrix of partial derivatives of $f_G$ evaluated at some point $p \in \mathbb{R}^{2|V|}$.

The rows of $J(G, p)$ are indexed by $E$ and pairs of columns of $J(G, p)$ are indexed by $V$ as follows: the two columns indexed by a vertex $u_i \in V_P$ are labelled $u_{i,x}$ and $u_{i,y}$ and the two columns indexed by a vertex $v_i \in V_L$ are labelled $v_{i,a}$ and $v_{i,b}$. The entries in $J(G, p)$ are as follows.

- A row indexed by an edge $e_i = u_j u_k \in E_{PP}$ has entries $2(x_j - x_k)$, $2(y_j - y_k)$, $2(x_k - x_j)$ and $2(y_k - y_j)$ in the columns indexed by $u_{j,x}$, $u_{j,y}$, $u_{k,x}$ and $u_{k,y}$, respectively.

- A row indexed by an edge $e_i = u_j v_k \in E_{PL}$ has entries $(1+a_k^2)^{-\frac{1}{2}}$, $-a_k(1+a_k^2)^{-\frac{1}{2}}$, $(-x_j a_k - y_j + a_k b_k)(1+a_k^2)^{-\frac{3}{2}}$ and $-(1+a_k^2)^{-\frac{1}{2}}$ in the columns indexed by $u_{j,x}$, $u_{j,y}$, $v_{k,a}$ and $v_{k,b}$, respectively.

- A row indexed by an edge $e_i = v_j v_k \in E_{LL}$ with $j < k$ has entries $(1+a_j^2)^{-\frac{1}{2}}$ and $-(1+a_k^2)^{-\frac{1}{2}}$ in the columns indexed by $v_{j,a}$ and $v_{k,a}$, respectively.

- All other entries are zero.

**Rigidity matrix:** The *rigidity matrix* of a point-line framework $(G,p)$ is the $|E| \times 2|V|$-matrix $R(G,p)$ with the following entries.

- A row indexed by an edge $e_i = u_j u_k \in E_{PP}$ has entries $x_j - x_k$, $y_j - y_k$, $x_k - x_j$ and $y_k - y_j$ in the columns indexed by $u_{j,x}$, $u_{j,y}$, $u_{k,x}$ and $u_{k,y}$ respectively.
- A row indexed by an edge $e_i = u_j v_k \in E_{PL}$ has entries $1$, $-a_k$, $-x_j a_k - y_j$ and $-1$ in the columns indexed by $u_{j,x}$, $u_{j,y}$, $v_{k,a}$ and $v_{k,b}$, respectively.
- A row indexed by an edge $e_i = v_j v_k \in E_{LL}$ with $j < k$ has entries 1 and $-1$ in the columns indexed by $v_{j,a}$ and $v_{k,a}$, respectively.
- All other entries are zero.

The Jacobian matrix $J(G,p)$ can be constructed from $R(G,p)$ using the following row and column operations. For each $v_i \in V_L$, multiply the column of $R(G,p)$ indexed by $v_{i,b}$ by $a_i b_i$ and subtract it from the column indexed by $v_{i,a}$, then divide the resulting column by $(1+a_i^2)$. For each $e \in E_{PP}$, multiply the row indexed by $e$ by 2. For each $e = u_j v_k \in E_{PL}$, divide the row indexed by $e$ by $(1+a_k^2)^{\frac{1}{2}}$. This construction shows that $J(G,p)$ and $R(G,p)$ are similar matrices and we have

**Lemma 22.2.1** *Let $(G,p)$ be a point-line framework. Then* $\operatorname{rank} J(G,p) = \operatorname{rank} R(G,p)$.

**Rigid:** A point-line framework is *rigid* if every continuous motion of the points and lines which satisfies the constraints results in a framework which can be obtained by a continuous isometry of $\mathbb{R}^2$ (i.e., a translation or rotation).

**Degenerate:** A framework is *degenerate* if there is an isometry which leaves all the points and lines in the framework unchanged.

Hence a degenerate framework has only parallel lines (and no points) and is unchanged by a translation along the lines, or has only coincident points (and no lines) and is unchanged by a rotation around the points. All other frameworks are nondegenerate.

If $(G,p)$ is nondegenerate then the rotations and translations of $\mathbb{R}^2$ generate a 3-dimensional subspace of the null space of $J(G,p)$. This gives the following result from [9].

**Lemma 22.2.2** *Let $(G,p)$ be a nondegenerate point-line framework. Then*
*(a)* $\operatorname{rank} R(G,p) \leq 2|V| - 3$.
*(b) If* $\operatorname{rank} R(G,p) = 2|V| - 3$ *then $(G,p)$ is rigid.*
*(c) If $p$ is a point where $rank(R(G,p))$ is maximum (i.e. a regular point of $f_G$) and* $\operatorname{rank} R(G,p) < 2|V| - 3$ *then $(G,p)$ is not rigid.*

**Generic:** A framework $(G,p)$ is *generic* if the coordinates in $p$ are algebraically independent over $\mathbb{Q}$.

It is straightforward to show that rank $R(G,p)$ is maximum for any generic point-line framework $(G,p)$.

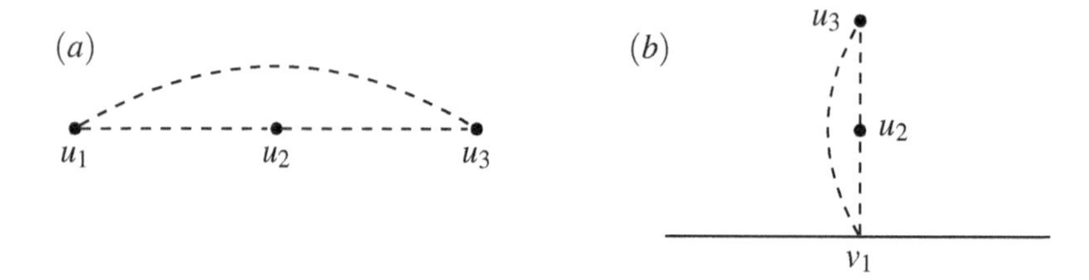


**Figure 22.6**
(a) A framework on the graph with three point-vertices in which the points $u_1$, $u_2$, and $u_3$ are collinear and (b) a framework on the graph with a line-vertex and two point-vertices in which the points $u_2$, $u_3$ are on an axis orthogonal to the line $v_1$.

### 22.2.3 The Rigidity Matroid

**Row matroid:** The *row matroid* of a matrix $R$ is the matroid $\mathcal{M}(R)$ on the set $E$ of rows of $R$, in which a set $F \subseteq E$ is independent in $\mathcal{M}(R)$ if and only if the corresponding rows of $R$ are linearly independent.

**Ridigity matroid:** The *rigidity matroid* $\mathcal{M}_{PL}(G,p)$ of a framework $(G,p)$ is the row matroid of its rigidity matrix $R(G,p)$. When $(G,p)$ is generic, its rigidity matroid depends only on the graph $G$. We refer to this matroid as the *rigidity matroid* of $G$ and denote it by $\mathcal{M}_{PL}(G)$.

Although the generic rigidity matroid $\mathcal{M}_{PL}(G)$ is not sufficient to fully describe the rigidity properties of all point-line frameworks $(G,p)$ it does describe the properties of a typical framework on $G$, that is $\mathcal{M}_{PL}(G,p) = \mathcal{M}_{PL}(G)$ for almost all $p$. In addition the matroid $\mathcal{M}_{PL}(G)$ does give some information on the rigidity properties of all frameworks $(G,p)$. For example, a set of rows in $R(G,p)$ is independent only if the corresponding set of edges are independent in $\mathcal{M}_{PL}(G)$ (although the converse is not necessarily true). For these reasons a charactarization of the rigidity matroid $\mathcal{M}_{PL}(G)$ is important.

### 22.2.4 Affine Properties of the Point-Line Rigidity Matrix

It is well known that the rank of the rigidity matrix of a bar-joint framework is invariant under general affine transformations of the underlying Euclidean space [23]. This is also true for a point-line framework with $V_P = \emptyset$ because the rigidity matrix does not depend on any of the point or line coordinates. However the rank of the rigidity matrix of a point-line framework is not invariant under general affine transformations of space when $V_P \neq \emptyset$ and $V_L \neq \emptyset$ as the following example shows.

Figure 22.6(a) shows a framework on the complete graph with three point-vertices and Figure 22.6(b) shows a framework on the complete graph with a line-vertex and two point-vertices. In Figure 22.6(a) the three points are colinear and in Figure 22.6(b) the line $v_1$ is orthogonal to the direction from $u_2$ to $u_3$.

The framework in 22.6(a) has an infinitesimal flex and the rank of its rigidity matrix is 2. Although the rank of the rigidity matrix of a generic framework on the same graph is 3 these two frameworks are not related by an affine transformation because any affine transform preserves the collinearity of points. Similarly the framework in 22.6(b) has an infinitesimal flex and the rank of its rigidity matrix is 2 whereas the rank of the rigidity matrix of a generic framework on the same graph is 3. However in this case there is an affine transformation connecting the two frameworks and hence this transform does not preserve the rank of the rigidity matrix.

### 22.2.5 Fixed-Slope Point-Line Frameworks

For some applications we may consider the slopes of the lines in a framework to be fixed to given values and not subject to variation when the framework is flexed [13, 14, 22]. For example if all the lines are joined by a connected network of angle constraints then there can be no variation of the relative angles of the lines and the rigidity matrix can be simplified by deleting the rows corresponding to the angle constraints and the columns corresponding to the line slopes. In addition we will see that if the graph is naturally bipartite and the coordinates of the points in a point-line framework are generic then the rigidity properties of the framework are fully determined by the properties of the corresponding fixed-slope point-line matroid.

**Fixed-slope rigidity matrix:** The *fixed-slope rigity matrix* $R_{fixed}(G,p)$ of a point-line framework $(G,p)$ is the $(|E_{PP}|+|E_{PL}|)\times(2(|V_P|+|V_L|)$ matrix in which the rows are indexed by $E_{PP}\cup E_{PL}$ and columns or pairs of columns are indexed by $V_L$ and $V_P$. The pairs of columns indexed by $u_i\in V_P$ are labelled as $u_{i,x}$ and $u_{i,y}$ and the column indexed by $v_i\in V_L$ is labelled as $v_{i,a}$. The entries are defined as follows.

- A row in $R_{fixed}(G,p)$ indexed by an edge $e_i=u_ju_k\in E_{PP}$ has entries $x_j-x_k$, $y_j-y_k$, $x_k-x_j$, $y_k-y_j$ in the columns indexed by $u_{j,x}$, $u_{j,y}$, $u_{k,x}$ and $u_{k,y}$, respectively.
- A row in $R_{fixed}(G,p)$ indexed by an edge $e_i=u_jv_k\in E_{PL}$ has entries $1$, $-a_k$, $-1$ in the columns indexed by $u_{j,x}$, $u_{j,y}$ and $v_{k,b}$, respectively.
- All other entries are zero.

**Fixed-slope rigid:** A point-line framework $(G,p)$ is *fixed-slope rigid* if every continuous motion of the points and lines which satisfies the constraints determined by the edges of $G$ and does not change the slope of the lines results in a framework which is a translation of $(G,p)$.

Corresponding to Lemma 22.2.2 we have following

**Lemma 22.2.3** *Let $(G,p)$ be a point-line framework with $|V_L|\neq\emptyset$. Then*
*(a)* $\operatorname{rank} R_{fixed}(G,p)\leq 2|V_P|+|V_L|-2$.
*(b) If* $\operatorname{rank} R_{fixed}(G,p)=2|V_P|+|V_L|-2$ *then $(G,p)$ is rigid.*
*(c) If $p$ is a point where rank $R_{fixed}(G,p)$ is maximum and rank $R_{fixed}(G,p)<2|V_P|+|V_L|-2$ then $(G,p)$ is not rigid.*

We denote the row matroid of $R_{fixed}(G,p)$ by $\mathcal{M}_{fixed}(G,p)$, and simplify this to $\mathcal{M}_{fixed}(G)$ when $p$ is generic.

If $G$ is naturally bipartite then $E_{PP}=\emptyset$ and $R_{fixed}(G,p)$ does not depend on the coordinates of the point-vertices. Then $R_{fixed}(G,p)=R_{fixed}(G,a)$ where $a=a_1,\dots,a_{|V_L|}$ are the line-slope coordinates. Whiteley [22] observed that in this case $R_{fixed}(G,a)$ is the same matrix as he obtained from the condition that an incidence graph on the line corresponds to the projection of a nontrivial scene in the plane. In this case, when $a$ is generic, he obtained the following charactarization of independence in the matriod $\mathcal{M}_{fixed}(G)$. A set $F\subseteq E$ is independent if and only if $|F'|\leq 2|V_P(F')|+|V_L(F')|-2$ for all nonempty subsets $F'$ of $F$, where $V_P(F')$ and $V_L(F')$ denote the sets of point- and line-vertices which are incident with $F'$.

In the general case, when $E_{PP}\neq\emptyset$, Owen and Power [14] obtained the following charactarization of independence in $\mathcal{M}_{fixed}(G)$.

**Theorem 22.2.4** *Let $G=(V_P,V_L,E)$ be a point-line graph and $F\subseteq E$. Then $F$ is independent $\mathcal{M}_{fixed}(G)$ if and only if $|F'|\leq 2|V_P(F')|+|V_L(F')|-2$ for all non-empty subsets $F'$ of $F$, with strict inequality whenever $F'\subseteq E_{PP}$.*

This charactarization may also be deduced from the charactarization of independence in $\mathcal{M}_{PL}(G)$ which we describe below.

## 22.3 Characterization of the Generic Rigidity Matroid for Point-Line Frameworks in $\mathbb{R}^2$

### 22.3.1 A Count Matroid for Point-Line Graphs

**Set function:** A *set function* is a map $f : 2^E \to \mathbb{Z}$ which assigns an integer to every subset of a set $E$.

**Non-decreasing:** The set function $f$ is *nondecreasing* if $f(S+e) \geq f(S)$ for all $e \in E$ and all $S \subseteq E$.

**Intersecting submodular:** The set function $f$ is *intersecting submodular* if $f(S_1)+f(S_2) \geq f(S_1 \cup S_2)+f(S_1 \cap S_2)$ for all $S_1, S_2 \subseteq S$ with $S_1 \cap S_2 \neq \emptyset$.

Let $f$ be a nondecreasing intersecting submodular set function which is nonnegative on the nonempty subsets of $E$.

**Induced matroid:** The *matroid induced by* $f$ is the matroid $\mathcal{M}(f)$ on $E$ in which a set $I \subseteq E$ is independent if and only if $|J| \leq f(J)$ for all $\emptyset \neq J \subseteq I$.

We will use the term *count matroid* to loosely describe a matroid on the edge set of a graph which is induced by a function which 'counts' the numbers of vertices incident with each set of edges in the graph.

Given a point-line graph $G = (V_P, V_L, E)$ we will use three nondecreasing, submodular set functions , $\nu_P$, $\nu_L$ and $\nu$, on $E$. These functions are defined by putting $\nu_P(F) = |V_P(F)|$, $\nu_L(F) = |V_L(F)|$ and $\nu(F) = |V(F)|$ for all $F \subseteq E$. They can be used to define several well known count matroids:

- $\mathcal{M}(\nu - 1)$ is the graphic matroid of a graph $G$ (which is the same as the 1-dimensional bar-joint rigidity matroid of $G$).

- $\mathcal{M}(2\nu - 3)$ is the 2-dimensional bar-joint rigidity matroid of a graph $G$. This implies that $\mathcal{M}_{PL}(G) = \mathcal{M}(2\nu - 3)$ when $G$ is a point-line graph and $V_L = \emptyset$.

- $\mathcal{M}(2\nu_P + \nu_L - 2)$ is the fixed slope rigidity matroid of a point-line graph $G$ whenever $G$ is naturally bipartite. (This follows from the results of Whiteley or Owen and Power mentioned at the end of Section 22.2.5.)

Since both points and lines have two degrees of freedom in $\mathbb{R}^2$ we might expect that the properties of $\mathcal{M}_{PL}(G)$ are not changed significantly when some of the points in a bar-joint framework are replaced with lines. The graphs on three vertices shown in Figure 22.1 show that this is not the case: the 3-cycle of point-vertices is generically rigid; the graphs in which one or two of these point-vertices are replaced by line-vertices are also generically rigid; but no realization of the 3-cycle of line-vertices in $\mathbb{R}^2$ is rigid since the three angles in the triangle cannot be specified independently.

The last example easily generalizes to include any graph which has $V_P = \emptyset$. In this case any cycle of edges corresponds to the lines in a geometric polygon in which all of the angles are given and are therefore dependent. Hence if $V_P(G) = \emptyset$, $\mathcal{M}_{PL}(G)$ is the graphic matroid $\mathcal{M}(\nu_L - 1)$.

We have seen that $\mathcal{M}_{PL}(G) = \mathcal{M}(2\nu - 3)$ when $V_L(G) = \emptyset$ and $\mathcal{M}_{PL}(G) = \mathcal{M}(\nu - 1)$ when $V_P(G) = \emptyset$. The other graphs on three vertices suggest the possibility that the first case may hold even when $V_L(G) \neq \emptyset$, provided that we have $V_P(G) \neq \emptyset$. This suggests that $\mathcal{M}_{PL}$ might be determined by the set function $f_t : 2^E \to \mathbb{Z}$ given by $f_t(F) = \nu(F) - 1$ if $V_P(F) = \emptyset$ and $f_t(F) = 2\nu(F) - 3$ otherwise. The graph and framework shown in Figure 22.7 show that this is not the case. It is easy

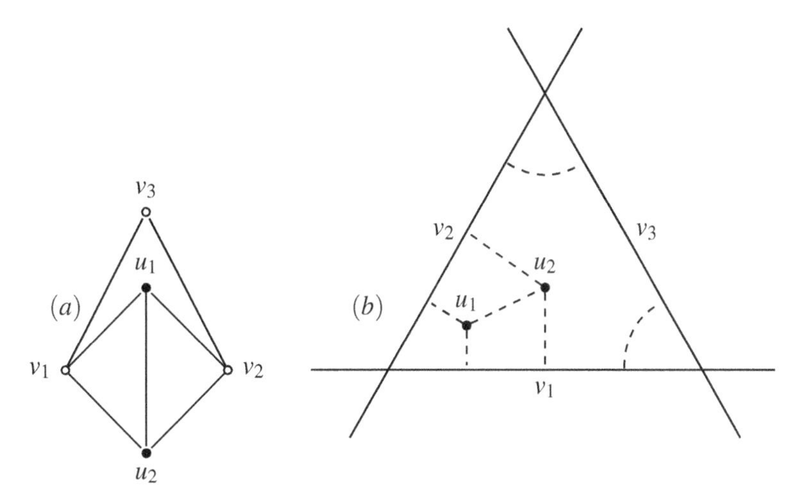


**Figure 22.7**
The graph (a) is a point-line graph with $|V_P| = 2$ and $|V_L| = 3$. (b) Shows a framework on this graph.

to check that $G$ has $2|V| - 3$ vertices and that $E$ is independent in $\mathcal{M}(2v-3)$. On the other hand, the line labelled $v_3$ has only angle constraints and hence its position is undetermined. This shows that $G$ is not generically rigid and hence that $\mathcal{M}_{PL}(G)$ is not determined by the set function $f_t$.

Another way to see that the edge set of the graph of figure 22.7(a) is not independent in $\mathcal{M}_{PL}(G)$ is to consider the graph $G + v_1v_2$. The edge $v_1v_2$ is contained in two distinct dependent sets $S_1 = \{v_1v_2, v_2v_3, v_3v_1\}$ and $S_2 = \{v_1v_2, u_1v_1, u_1v_2, u_2v_1, u_2v_2, u_1u_2\}$ of the point-line rigidity matroid. The matroid circuit axiom now implies that $E(G) = (S_1 \cup S_2) - v_1v_2$ is dependent in the point-line rigidity matroid.

We can also use the graph $G + v_1v_2$ to show that the set function $f_t$ does not have the submodularity property required to define a count matroid. We have $E(G + v_1v_2) = S_1 \cup S_2$ and $S_1 \cap S_2 = \{v_1v_2\}$. By considering the subgraphs induced by $S_1$, $S_2$, $S_1 \cup S_2$ and $S_1 \cap S_2$ shown in 22.8 we may deduce that $f_t(S_1) = 2$, $f_t(S_2) = 5$, $f_t(S_1) \cup S_2) = 7$ and $f_t(S_1 \cap S_2) = 1$. This implies that $f_t$ is not an intersecting submodular function.

Notice however that the set function $f_t$ does provide the following necessary condition for a set of edges in a graph $G$ to be independent in $\mathcal{M}_{PL}(G)$.

**Lemma 22.3.1** *Let $G = (V_P, V_L, E)$ be a point-line graph and $F \subset E$. If $F$ is independent in $\mathcal{M}_{PL}(G)$ then $|F'| \leq f_t(F')$ for all $\emptyset \neq F' \subseteq F$.*

We will use the idea of introducing new edges between pairs of line-vertices which are not already connected by an edge to obtain a stronger necessary condition for independence in $\mathcal{M}_{PL}(G)$, Lemma 22.3.2 below. This condition is given in terms of a set function which is submodular and hence is a candidate to induce $\mathcal{M}_{PL}(G)$ as a count matroid.

Suppose $G = (V_P, V_L, E)$ is a point-line graph. If a subgraph $H_i \subseteq G$ is generically rigid then we may add a spanning tree of edges on $V_L(H_i)$ to $G$ and remove the same number of edges in $E_{PP}(H_i) \cup E_{PL}(H_i)$ from $G$ in such a way that we do not change the maximum number of independent edges in $\mathcal{M}_{PL}(G)$. We may do this simultaneously on all subgraphs $H_i$ corresponding to a partition of $E$ to obtain an upper bound on the rank of $\mathcal{M}_{PL}(G)$. The fact that some of the subgraphs $H_i$ may not be generically rigid means that we may incorrectly increase the count of the maximum number of independent edges and hence in this way we get only an upper bound on the maximum number of independent edges in $\mathcal{M}_{PL}(G)$. Minimizing this bound over all partitions gives an even better upper bound. We used this idea in [9] to obtain the following necessary condition for independence in $\mathcal{M}_{PL}(G)$.

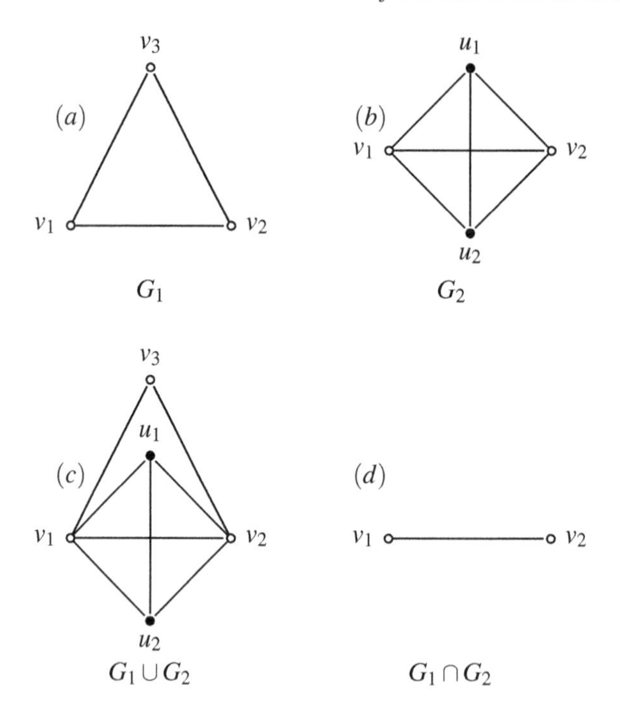


**Figure 22.8**
Point-line graphs $G_1$ and $G_2$ together with the graphs $G = G_1 \cup G_2$ and $G_1 \cap G_2$ induced respectively by the union and intersection of their edge sets.

**Lemma 22.3.2** *Let $G = (V_P, V_L, E)$ be point-line graphs. For $F \subseteq E$ let*

$$\rho(F) = \min\left\{\sum_{i=1}^{s}(2v_P(F_i)| + v_L(F_i) - 2)\right\}$$

*where the minimum is taken over all partitions $\{F_1, \ldots, F_s\}$ of $F$, and let*

$$f(F) = \rho(F) + v_L(F) - 1.$$

*If $F \neq \emptyset$ and $F$ is independent in $\mathcal{M}_{PL}(G)$ then $|F| \leq f(F)$.*

It is straightforward to check that, if $V_P(F) = \emptyset$, then the optimal partition has $s = |F|$ which gives $\rho(F) = 0$ and $f(F) = v(F) - 1$. Otherwise the partition with $s = 1$ shows that $\rho(F) \leq 2v_P(F) + v_L(F) - 2$ and hence $f(F) \leq 2v(F) - 3$. Thus $f(F) \leq f_t(F)$ for all $F \subseteq E$.

For the graph in Figure 22.7(a) we consider $F = E$ and the partition $E = \{F_1, F_2, F_3\}$ where $F_1 = \{v_1v_3\}$, $F_2 = \{v_2v_3\}$ and $F_3 = \{u_1v_1, u_2v_1, u_1v_2, u_2v_2, u_1u_2\}$. This partition gives $\rho(E) \leq 4$ and $f(E) \leq 6$. Since $|E| = 7$ this shows that $E$ is not independent in $\mathcal{M}_{PL}(G)$.

Similar calculations for the four graphs in Figure 22.8 give $f(S_1) = 2$, $f(S_2) = 5$, $f(S_1 \cap S_2) = 1$ and $f(S_1 \cup S_2) = 6$. Hence $f$ is submodular on these edge sets. Indeed, a theorem of Dunstan [7] implies that the function $\rho$ is submodular on $E$ and hence $f$ is also submodular on $E$. We can also show that $f$ is nondecreasing, and nonnegative on the nonempty subsets of $E$, see [9, Lemma 3.5]. Thus $f$ induces a count matroid $\mathcal{M}(f)$ on $E$, and Lemma 22.3.2 tells us that independence in $\mathcal{M}(f)$ is a necessary condition for independence in $\mathcal{M}_{PL}(G)$.

We will see that $\mathcal{M}_{PL}(G) = \mathcal{M}(f)$. The following properties of $\mathcal{M}(f)$ will be useful. They can be derived using standard matroid arguments, see [9].

Let $\mathcal{M}_1$ and $\mathcal{M}_2$ be matroids on the same ground set $E$.

**Matroid union:** The *matroid union* of $\mathcal{M}_1$ and $\mathcal{M}_2$ is the matroid $\mathcal{M}_1 \vee \mathcal{M}_2$ on $E$ defined by the condition that a set $I \subseteq E$ is independent in $\mathcal{M}_1 \vee \mathcal{M}_2$ if and only if $E$ is the disjoint union of $I_1$ and $I_2$ where $I_1$ is independent in $\mathcal{M}_1$ and $I_2$ is independent in $\mathcal{M}_2$.

**Lemma 22.3.3** *Let $G = (V_P, V_L, E)$ be a point-line graph. Then*
*(a) a set $S \subseteq E$ is independent in $\mathcal{M}(\rho + \nu_L - 1)$ if and only if $S + ij$ is independent in $\mathcal{M}(\rho + \nu_L)$ for all $ij \in S$,*
*(b) $\mathcal{M}(\rho + \nu_L) = \mathcal{M}(2\nu_P + \nu_L - 2) \vee \mathcal{M}(\nu_L)$.*

**Example** We can use Lemma 22.3.3 to show that the edge set of the naturally bipartite point-line graph $K(3,3)$ is independent in $\mathcal{M}(\rho + \nu_L - 1)$. By symmetry, it suffices to show that $E + ij$ is independent in $\mathcal{M}(2\nu_P + \nu_L - 2) \vee \mathcal{M}(\nu_L)$ for any fixed edge $ij \in E$. Let $S$ be any 1-factor of $K(3,3)$ which contains $ij$ and $T = (E + ij) - S$. Then it is straightforward to check that $T$ is independent in $\mathcal{M}(2\nu_P + \nu_L - 2)$ and $S$ is independent in $\mathcal{M}(\nu_L)$. Hence $E + ij = S \cup T$ is independent in $\mathcal{M}(2\nu_P + \nu_L - 2) \vee \mathcal{M}(\nu_L)$.

We will use matroid theory and some elementary linear algebra to characterize independence in $\mathcal{M}_{PL}(G)$ in the next section, in the special case when $G$ is naturally bipartite. This will allow us to use Lemma 22.3.3 to deduce that $\mathcal{M}_{PL}(G) = \mathcal{M}(\rho + \nu_L - 1)$ for any naturally bipartite point-line graph $G$. Then in Section 22.3.3 we will use the substitution of edges in $E_{PP}$ and $E_{LL}$ with copies of naturally bipartite $K(3,3)$ graphs to extend this result to all point-line graphs.

### 22.3.2 A Characterization of Independence in $\mathcal{M}_{PL}(G)$ when $G$ is Naturally Bipartite

Let $G = (V_P, V_L, E)$ be a naturally bipartite point-line graph. Let $(G, p)$ be a corresponding point-line framework and let $a = (a_1, \ldots, a_{|V_L|})$ denote the vector of line slope coordinates. We will first show that, if the point coordinates $p(u_i)$ for $u_i \in V_P$ are algebraically independent over $\mathbb{Q}(a)$, then the linear independence of the rows of $R(G, p)$ is closely related to the linear independence of the rows of a simpler matrix $A(G, a)$.

Let $c = (c_1, \ldots, c_{|E|}) \in \mathbb{R}^{|E|}$ be a vector of generic edge weights and let $A(G, a, c)$ be the matrix obtained by replacing the entries $-x_j a_k - y_j$ in the columns for $v_{k,a}$ in $R(G, p)$ by $c_{jk}$ for each $e = u_j v_k \in E$. We will refer to $A(G, a, c)$ as the *frame matrix corresponding to* $R(G, p)$. Note that this frame matrix does not depend on the coordinates of the point-vertices but only on the slopes of the lines (and on the parameters $c$).

**Lemma 22.3.4** *Let $G = (V_P, V_L, E)$ be a naturally bipartite point-line graph, let $(G, p)$ be a corresponding point-line framework, let $a = (a_1, \ldots, a_{|V_L|})$ denote the vector of line slope coordinates in $p$. Suppose that the coordinates $p(u_i) = (x_i, y_i)$ of the point-vertices $u_i \in V_P$ are algebraically independent over $\mathbb{Q}(a)$. Then the following statements are equivalent.*
*(a) The rows of $A(G + e', a, c)$ are independent for all $e' = u_i v_j$ with $e = u_i v_j \in E$.*
*(b) The rows of $R(G, p)$ are independent.*

**Proof.** $(a) \implies (b)$. Suppose that (a) holds. We can represent each vector in the null space of $A(G, a, c)$ as $(q, h)$ where $q : V_P \to \mathbb{R}^2$ and $h : V_L \to \mathbb{R}^2$. Let $q(u_k) = (q_{k,1}, q_{k,2})$ for all $u_k \in V_P$ and $h(v_k) = (h_{k,1}, h_{k,2})$ for all $v_k \in V_L$. Choose $e = u_i v_j \in E_{PL}$. Then $\operatorname{rank} A(G, a, c) = \operatorname{rank} A(G + e', a, c) - 1$ so we can find a $(q, h) \in \text{Null}\, A(G, a, c)$ such that $h_{j,1} \neq 0$.

Repeating this argument for each $e \in E$ and taking a suitable linear combination of the vectors we obtain, we can construct a $(q, h) \in \text{Null}\, A(G, a, c)$ such that $h_{j,1} \neq 0$ for all $v_j \in V_L$. The fact that $(q, h) \in \text{Null}\, A(G, a, c)$ gives

$$q_{i,1} - a_j q_{i,2} + c_e h_{j,1} - h_{j,2} = 0 \quad \text{for all} \quad e = u_i v_j \in E. \tag{22.1}$$

Construct a point-line framework $(G,p)$ by putting $p(u_i)=(-q_{i,2},q_{i,1})$ for all $u_i \in V_P$ and $p(v_j)=(a_j,0)$ for all $v_j \in V_L$. Equation (22.1) enables us to transform the rigidity matrix $R(G,p)$ to the $A$-matrix $A(G,a,c)$ by subtracting $h_{j,2}$ times column $v_{j,2}$ from column $v_{j,1}$ for all $v_j \in V_L$, and then dividing column $v_{j,1}$ by $h_{j,1}$. This implies that $\text{rank}\, R(G,p)=\text{rank}\, A(G,a,c)=|E|$. It follows that the rows of the rigidity matrix of any realization of $G$ as a point-line framework with generic coordinates for the point-vertices will be linearly independent. Hence (b) holds.

$(b) \implies (a)$. Suppose (b) holds. The specialization $c'_e=-x_j a_k - y_j$ for $e=jk \in E_{PL}$ shows that the rows of $A(G,a,c)$ are independent. The vector $(q,h)$ with $q_{k,1}=-y_k$, $q_{k,2}=x_k$, $h_{k,1}=-1$ and $h_{k,2}=0$, which corresponds to an infinitesimal rotation of $p$ about the origin, is in the nullspace of $R(G,p)$. This vector is also in the nullspace of $A(G,a,c')$ for the specialization $c'$ given above. However $(q,h)$ is not in the nullspace of $A(G+e',a,c')$ for any $e \in E$ provided we choose $c'_{e'} \neq c'_e$. Hence rank $A(G+e',a,c')>$ rank $(A(G,a,c'))$ for all $e \in E$ which implies (a). •

We next show that the row matroid of $A(G,a,c)$ can be expressed as the matroid union of two count matroids, one of which is the fixed slope rigidity matroid of $G$. We will use the following result of Brylawski [4, Lemma 7.6.14(1)], see also [9, Lemma 4.1], which gives a linear representation for the matroid union of two row matroids.

**Lemma 22.3.5** *Let $M_i$ be an $m \times n_i$-matrix for $i=1,2$. Let $X$ the $m \times m$ diagonal matrix with entries $x_i$ in row $i$ where $x_i$ are algebraically independent over $\mathbb{Q}(M_1,M_2)$ and let $(M_1,XM_2)$ be the $m \times (n_1+n_2)$-matrix whose columns are the combination of the columns of $M_1$ and the columns of $XM_2$. Then $\mathcal{M}(M_1,XM_2)=\mathcal{M}(M_1) \vee \mathcal{M}(M_2)$.*

**Lemma 22.3.6** *Let $G$ be a naturally bipartite graph, let $(G,p)$ be a framework with line slopes $a$ and let $A(G,a,c)$ be the corresponding frame matrix. Then $\mathcal{M}(A(G,a,c))=\mathcal{M}_{fixed}(G,a)) \vee \mathcal{M}(v_L)$.*

**Proof.** Let $R_L$ be the $|E| \times |V_L|$-matrix with entry 1 in the row for $u_i v_j$ and column for $v_j$ and 0 elsewhere. Then $R_L$ is the $(0,1)$-incidence matrix for a bipartite graph and it is well-known that $\mathcal{M}(R_L)=\mathcal{M}(v_L)$.

Let $X$ be the $|E| \times |E|$ diagonal matrix with generic parameters $c_e$ for the diagonal entry corresponding to $e \in E$. After a reordering of rows and columns, the matrix $A$ can be written as $A(G,a,c)=(R_{fixed}(G,a),XR_L(G))$. Then $\mathcal{M}(A(G,a,c))=\mathcal{M}_{fixed}(G,a) \vee \mathcal{M}(v_L)$ by Lemma 22.3.5. •

Lemmas 22.3.4 and 22.3.6 together with the result that $\mathcal{M}_{fixed}(G)=\mathcal{M}(2v_p+v_L-2)$ stated at the end of Section 22.2.5 give the following characterization of independence in $\mathcal{M}_{PL}(G)$ when $G$ is naturally bipartite.

**Lemma 22.3.7** *Let $G=(V_P,V_L,E)$ be a naturally bipartite point-line graph, let $(G,p)$ be a framework on $G$ and let $a=(a_1,\ldots,a_{|V_L|})$ be the vector of line-slopes. Suppose that the coordinates $p(u_i)=(x_i,y_i)$ of the point-vertices $u_i \in V_P$ are algebraically independent over $\mathbb{Q}(a)$. Let $S \subseteq E$. Then*

*(a) $S$ is independent in $\mathcal{M}_{PL}(G,p)$ if and only if $S+u_iv_j$ is independent in $\mathcal{M}_{fixed}(G+u_iu_j,a)) \vee \mathcal{M}(v_L)$ for all $u_iv_j \in E$.*

*(b) $S$ is independent in $\mathcal{M}_{PL}(G)$ if and only if $S+u_iv_j$ is independent in $\mathcal{M}(2v_P+v_L-2) \vee \mathcal{M}(v_L)$ for all $u_iv_j \in E$.*

### 22.3.3 A Characterization of Independence in $\mathcal{M}_{PL}(G)$

We described in Section 22.2 how any point-line graph $G$ can be converted to a naturally bipartite point-line graph $G^+$ by replacing all edges in $E_{PP}$ and $E_{LL}$ with copies of the naturally bipartite point-line graph $K(3,3)$. We showed in the example following lemma 22.3.3 that the edge set of $K(3,3)+ij$ is independent in $\mathcal{M}(2\nu_P+\nu_L-2))\vee\mathcal{M}(\nu_L)$ for any edge $ij$ of $K(3,3)$. Lemma 22.3.7 now implies that the edge set of $K(3,3)$ is independent in the generic point-line rigidity matroid.

It is straightforward to check that $2|V(G)|-|E(G)|=2|V(G^+)|-|E(G^+)|$. A more detailed analysis given in [9] shows that $E(G)$ is independent in $\mathcal{M}_{PL}(G)$ if and only if $E(G^+)$ is independent in $\mathcal{M}_{PL}(G)$, and $E(G)$ is independent in $\mathcal{M}(\rho+\nu_L-1)$ if and only if $G^+$ is independent in $\mathcal{M}(\rho+\nu_L-1)$.

Together with lemmas 22.3.3 and 22.3.7(b), this gives the following charactarization for independence in the rigidity matroid of any point-line graph

**Theorem 22.1** *Let $G=(V_P,V_L,E)$ be a point-line graph and $F\subseteq E$. The following statements are equivalent:*
*(a) $F$ is independent in $\mathcal{M}(\rho+\nu_L-1)$;*
*(b) for all $ij\in F$, $F+ij$ is independent in $\mathcal{M}(2\nu_P+\nu_L-2))\vee\mathcal{M}(\nu_L)$;*
*(c) $F$ is independent in $\mathcal{M}_{PL}(G)$.*

Property (b) in Theorem 22.1 leads to a polynomial-time algorithm to determine a maximal independent subset of $\mathcal{M}_{PL}(G)$, for any point-line graph $G$, see [9]. An implementation of the algorithm as a two-color pebble game is given in an arXiv version of this paper [10].

### 22.3.4 The Rank Function for $\mathcal{M}_{PL}(G)$

We used Theorem 22.1 to obtain the rank function $r_{PL}$ of $\mathcal{M}_{PL}(G)$ for any point-line graph $G$ in [9]. For any $F\subseteq E$ and any partition $\mathcal{F}=\{F_1,\dots,F_{|\mathcal{F}|}\}$ of F we define a graph $G_{\mathcal{F}}$ as follows. The vertices $v_i$ of $G_{\mathcal{F}}$ correspond to the partition sets $F_i$ and $v_iv_j\in E(G_{\mathcal{F}})$ if and only if $V_L(F_i)\cap V_L(F_j)\neq\emptyset$. We let $c_{\mathcal{F}}$ denote the number of connected components of $G_{\mathcal{F}}$.

**Theorem 22.2** *Let $G=(V_P,V_L,E)$ be a point-line graph and $F\subseteq E$. Then*

$$r_{PL}(F)=\nu_L(F)+min_{\mathcal{F}}\left\{\sum_{F_i\in\mathcal{F}}(2\nu_P(F_i)+\nu_L(F_i)-2)-c_{\mathcal{F}}\right\},$$

*where the minimum is taken over all partitions $\mathcal{F}$ of $F$.*

When $|V_P|=0$, the minimum in the above expression for $r_{PL}(F)$ occurs when the sets $F_i$ are all single edges. In this case $c_{\mathcal{F}}=c_F$ is just the number of connected components in the subgraph induced by $F$ and $r_{PL}(F)=\nu(F)-c_F$ is the rank function of the graphic matroid of $G$.

When $|V_L|=0$, we have $c_L(\mathcal{F})=|\mathcal{F}|$ for all $\mathcal{F}$ and

$$r_{PL}(F)=min_{\mathcal{F}}\left\{\sum_{F_i\in\mathcal{F}}(2\nu(F_i)-3)\right\}$$

is the rank function of the 2-dimensional bar-joint rigidity matroid of $G$.

## 22.4 Extensions to $\mathbb{R}^d$

### 22.4.1 Point-Line Frameworks in $\mathbb{R}^d$

A line in $\mathbb{R}^d$ is uniquely determined by two points and each of these two points can be anywhere on the line. Hence a line is represented by $2(d-1)$ coordinates. If we choose these coordinates to be the $d-1$ components of the line direction and the $d-1$ coordinates of the intersection of the line with a $d-1$ dimensional linear subspace (a hyperplane) we can obtain a Jacobean matrix with $|E|$ rows and $d|V_P|+2(d-1)|V_L|$ columns which has a similar form to the matrix $J(G,p)$ given above for 2-dimensional frameworks. An additional consideration is that we may specify more than one type of edge between two lines. For example if $d=3$ we can specify both the angle between two lines and the distance between them.

If $V_P = \emptyset$ then $E = E_{LL}$. If in addition $d=3$ and there are no line-line distance constraints then $J(G,p)$ corresponds to the rigidity matrix for points on the surface of a sphere [25] and the generic rigidity matroid is the same as the generic rigidity matroid for points in the plane.

If $d=3$ and we apply the restictions that (a) all vertices occur in rigid subgraphs and (b) each vertex has at most one neighbor which is not in its rigid subgraph, then the Jacobian matrix $J(G,p)$ corresponds to the rigidity matrix for a body and CAD framework [20].

A characterization of the generic rigidity matroid for point-line frameworks in $\mathbb{R}^3$ on all point-line graphs would give a characterization for bar-joint frameworks as a special case, which is a significant unsolved problem. It is conceivable however that the study of naturally bipartite point-line frameworks in $\mathbb{R}^3$ may be more tractable.

### 22.4.2 Point-Hyperplane Frameworks in $\mathbb{R}^d$

The Jacobean matrix descibed in Section 22.4.1 does not reduce to a simpler form like $R(G,p)$ when $d>2$. We do obtain a simpler rigidity matrix, however, if we replace the lines with planes when $d=3$ or more generally with subspaces of codimension 1 (hyperplanes) for any $d$. A hyperplane can be represented by $d$ parameters which we may take to be a unit normal vector $n$ and a distance $b$. Then the signed distance from a point with coordinates $p_j$ to a hyperplane with coordinates $(n_k,b_k)$ is $p_j.n_k - b_k$ and the rigidity matrix has a form similar to $R(G,p)$. Hence point-hyperplane frameworks are a natural extension of point-line frameworks when $d>2$.

Let $V_P$ and $V_H$ be the sets of point and hyperplane vertices, respectiviely. For any point-hyperplane graph $G$, we can define both a generic rigidity matroid and a generic fixed-slope rigidity matroid. Since both cases include the generic rigidity matroid for bar-joint frameworks as a special case, we may conclude that a characterization of indepedence in either case will be difficult when $d>2$. If $G$ is a naturally bipartite graph, however, then the fixed slope rigidity matrix is the same as the incidence matrix for points and hyperplanes that is studied in scene analysis, see [22]. In particular, it follows from [22, Theorem 4.1] that the generic $d$-dimensional fixed-slope rigidity matroid for a naturally bipartite point-hyperplane graph $G$ with vertex partitions $V_P$ and $V_H$ is $\mathcal{M}(d(v_P-1)+v_H)$ for all $d \geq 1$, see [6] for more details.

The main result of Eftekhari et al. [6] gives a rigidity preserving transformation between point-hyperplane frameworks and bar-joint frameworks. They show that, for a graph $G=(V,E)$ and $X \subseteq V$, $G$ has an infinitesimally rigid realization as a bar-joint framework in $\mathbb{R}^d$ in which the vertices in $X$ lie on a common hyperplane if and only if $G$ has an infinitesimally rigid realization as a point-hyperplane framework in $\mathbb{R}^d$ in which the vertices in $X$ are represented as hyperplanes and the vertices in $V \setminus X$ are represented as points. Combining this result with Theorem 22.1, we obtain a good characterization for when $G$ can be realized as an infinitesimally rigid bar joint framework in $\mathbb{R}^2$ in which the vertices in $X$ are colinear. The special case when $|X| = 3$ had previously been characterized in [8].

## 22.5 Direction-Length Frameworks

Instead of introducing lines as new objects in constraint graphs, we may adopt the approach described in Section 22.1.2 and extend the usual bar-joint frameworks by introducing directional constraints between pairs of points in the framework. For a pair of points $p_1$, $p_2$ this introduces a constraint of the form $(p_1 - p_2)/|p_1 - p_2| = \hat{a}_{12}$ where $\hat{a}_{12}$ is a unit vector describing a fixed direction. If all of the constraints are direction constraints we have a direction framework. These frameworks occur naturally in the theory of parallel redrawings and Whiteley [24] showed that the matroid for a generic direction framework in $\mathbb{R}^2$ is isomorphic to the matroid for a generic distance (bar-joint)framework on the same graph. Under this isomorphism the rotational isometry which is satisfied by the distance framework is replaced by a scaling isometry for the direction framework.

A constraint configuration which allows both distance and direction constraints between pairs of points is represented by an edge labelled graph for which all of the vertices represent points and the edges are partitioned into two sets $D$ and $L$ which represent the direction and distance (length) constraints, respectively. If the coordinates of the vertices are generic the rigidity of the framework is determined by a rigidity matroid on the direction-length graph. Servatius and Whiteley [18] characterized independence in this rigidity matroid by the following condition. A set $F \subseteq D \cup L$ is independent if and only if $|F'| \leq 2v(F') - 2$ for all nonempty subsets $F'$ of $F$, with strict inequality whenever $F' \subseteq D$ or $F' \subseteq L$.

Although direction-length frameworks constrain the direction between pairs of points they do this with respect to a fixed coordinate direction and cannot be used to constrain only the relative angle between two pairs of points. In particular, there is no structure in the rigidity matroid for a direction-length framework which represents the geometric fact that the three angles in a triangle are dependent and sum to 180 degrees. It follows that a point-line framework cannot be represented by a specialized direction-length framework.

On the other hand, any direction-length framework can be represented by a fixed-slope point-line framework by adding a fixed slope line for each pair of points which are constrained by a direction constraint, constraining this line to be at distance zero from the two points and deleting the direction constraint. Note, however, that this substitution will not give a generic point-line framework even if the initial direction-length framework is generic. Instead it gives rise to the type of quasi-generic point-line framework discussed in Section 22.1.3. The fact that independence can be characterized for the generic direction-length rigidity matroid indicates that we may also be able to characterize independence for such quasi-generic point-line frameworks when we add the condition that each line is incident with at most two points.

## References

[1] *Proceedings of Workshops on Automated Deduction in Geometry* 1996-2016.

[2] L. Asimow and B. Roth, The rigidity of graphs, *Trans. Amer. Math. Soc.* 245 (1978), 279-289.

[3] W. Bouma, I. Fudos, C. M. Hoffmann, J. Cai, and R. Paige, A geometric constraint solver, *Computer-Aided Design*, 27(1995) 487–501.

[4] T. Brylawski, Constructions, in *Theory of Matroids*, ed. N. White, CUP, London, 1986, 127–223.

[5] S. C. Chou, Mechanical geometry theorem proving, D. Reidel Publishing company (1988).

[6] Y. Eftekhari, B. Jackson, N. Nixon, B. Schulze, S.-I. Tanigawa and W. Whiteley, Point-hyperplane frameworks, slider joints, and rigidity preserving transformations, https://arxiv.org/pdf/1703.06844

[7] A. Frank, *Connections in combinatorial optimization*, Oxford Lecture Series in Mathematics and its Applications, 38, Oxford University Press, Oxford 2011.

[8] B. Jackson and J. Jordán, Rigid two-dimensional frameworks with three collinear points, *Graphs and Combinatorics* 21 (2005) 427–444.

[9] B. Jackson and J. C. Owen, A characterization of the generic rigidity of 2-dimensional point-line frameworks, *Journal of Combinatorial Theory, Series B* 119 (2016) 96–121.

[10] B. Jackson and J. C. Owen, A characterization of the generic rigidity of 2-dimensional point-line frameworks, http://arxiv.org/abs/1407.4675v1

[11] V. C. Lin, D. C. Gossard and R. A. Light, Variational geometry in computer aided design, *Proceedings of Siggraph* (1981) 171–177.

[12] J. C. Owen, Algebraic solution for geometry from dimensional constraints, *ACM Symposium on Foundations in Solid Modeling* (1991), 397–407.

[13] J. C. Owen, Constraints on simple geometry in two and three dimensions, *J. Comput. Geom. Appl.* 6 (1996) 421.

[14] J. C. Owen and S.C.Power, Independence conditions for point-line-position frameworks, preprint (2006).

[15] T. Pisanski, B. Servatius, Configurations from a graphical viewpoint, *Birkhauser*, 2013.

[16] I. E. Sutherland, Sketchpad: A man-machine graphical communication system, *Phd Thesis*, M.I.T., Cambridge, MA., 1963.

[17] B. Servatius and H. Servatius, Combinatorial local rigidity of 2 dimensional bar and joint frameworks, chapter in this book.

[18] B. Servatius and W. Whiteley, Constraining plane configurations in CAD:Combinatorics of directions and lengths, *SIAM J. Disc. Math.* 12 (1999) 136–153.

[19] P. Todd, A k-tree generalisation that characterises consistency of dimensioned engineering drawings, *SIAM Journal of Disc. Math.* 2 (1989) 255–261.

[20] A. St.John, Generic rigidity of body and cad frameworks, chapter in this book.

[21] W. Whiteley, The union of matroids and the rigidity of frameworks, *SIAM J. Disc. Math.* 1 (1988) 237–255.

[22] W. Whiteley, A matroid on hypergraphs with applications in scene analysis and geometry, *Disc. Comput. Geom.* 4 (1989) 75–95.

[23] W. Whiteley, Matroids and rigid structures, in Matroid Applications ed. Neil White, *Encyclopedia of Mathematics and Its Applications* 40 (1992) 1–51.

[24] W. Whiteley, Some Matroids from discrete applied geometry, in *Matroid Theory AMS Contemporary Mathematics* 197 (1996) 171–313.

[25] A. Nixon and W. Whiteley, Change of *Metrics in Rigidity Theory*, chapter in this book.

# Chapter 23

## *Generic Rigidity of Body-and-Cad Frameworks*

**Audrey St. John**

*Department of Computer Science, Mount Holyoke College, South Hadley, MA*

**CONTENTS**



## 23.1 Overview

Popular computer-aided design (CAD) software, such as SolidWorks or OnShape, enables users to build designs for 2D and 3D systems, typically by describing and constraining geometric elements such as points and lines. Underlying solvers check when a system is *well-constrained* (or *minimally rigid*), under-constrained (or *flexible*) or *over-constrained* (containing unnecessary dependencies that may or may not be consistent) to provide feedback to the user. These solvers generally rely on combinatorial and numerical techniques, but are susceptible to performance issues, prompting research questions related to speed and instability (especially near special positions). In addition, capturing *design intent* is a key goal for CAD software and there is room to improve the feedback given to the user. For example, simply being told a system is over-constrained is not as useful as being told how to resolve the dependencies. Users are not experts in geometric constraint systems, and more intuitive feedback can help users capture and refine their design intent.

In this chapter, we describe the body-and-cad rigidity model, defined to capture the majority of systems specified by this type of CAD software: a *body-and-cad framework* is composed of rigid bodies with pairwise *c*oincidence, *angular* or *distance* constraints between them [3]. The derivation of the theory follows the work of [10, 12] and similarly considers frameworks in dimension $d$; we denote the number of degrees of freedom availabe to a $d$-dimensional rigid body by $D = \binom{d+1}{2}$.

We start with the algebraic setting for rigidity in Section 23.2, walking through examples in 2D and 3D. The rigidity matrix and pure condition defined in Section 23.3 allow us to analyze the first-order *infinitesimal* behavior. Section 23.4 describes the *combinatorial* counting property that characterizes generic infinitesimal rigidity of a subset of body-and-cad frameworks [6]. Finally, we conclude in Section 23.5 with discussion around open questions and related applications.

*A note to the reader: This chapter is intended to be a "quick reference" guide with an emphasis on building intuition for body-and-cad rigidity; the more technical underpinnings required for the results are left in the References for conciseness. The theory relies on: linear algebra, combinatorics, matroid theory (see, e.g., [8]), sparsity counts (see Chapter 19), and pebble game algorithms [5].*

## 23.2 Algebraic Body-and-Cad Rigidity Theory

### 23.2.1 Glossary

**bicolored (multi)graph:** A multigraph $G = (V, E = R \sqcup B)$ with vertex set $V = [1..n]$ and edge set partitioned into two colors: red edges in $R$ and black edges in $B$.

**body-and-cad framework:** $F = (G, c, L_1, \ldots, L_{|\mathcal{C}|})$ defined on a cad graph $(G, c)$ and a family of length functions $L_1, \ldots, L_{|\mathcal{C}|}$ ($\mathcal{C}$ depends on the embedding dimension), where $L_i$ specifies the geometry of constraints associated with edges assigned color $c_i$.

**cad constraint:** A constraint of type (*c*oincidence, *a*ngle, or *d*istance); imposed on geometric elements identified on a pair of bodies; the set of possible constraints is denoted by $\mathcal{C}$ and depends on the embedding dimension.

**cad graph:** A pair $(G = (V,E), c)$ of a multigraph along with a coloring function $c : E \to \mathcal{C}$ ($\mathcal{C}$ depends on the embedding dimension); vertices represent bodies, edges represent cad constraints and $c(e)$ colors edge $e$ by its cad constraint label.

**congruent:** Two realizations of a framework are *congruent* if one can be obtained from the other by multiplying each $\mathcal{T}_i$ by some $\mathcal{T}' \in SE(d)$.

**flexible:** A framework is (locally) *flexible* with respect to a realization $\mathbf{T}$ if there exists another realization $\mathbf{T}'$ in some neighborhood of $\mathbf{T}$ not congruent to $\mathbf{T}$.

**geometric element:** A linear subspace (e.g., point, line, plane) rigidly affixed to a body.

**internal degrees of freedom:** The space of motions available to the system, modulo the trivial degrees of freedom.

**primitive angular constraint:** A constraint that may block at most one rotational degree of freedom; corresponds to one row in the rigidity matrix of Section 23.3 with $2\binom{d}{2}$ nonzero entries (in the "rotational" columns).

**primitive blind constraint:** A constraint that may block at most one (rotational or translational) degree of freedom; corresponds to one row in the rigidity matrix of Section 23.3 with $2D$ nonzero entries.

**primitive cad graph:** A bicolored multigraph $G = (V, E = R \sqcup B)$ associated to a body-and-cad framework, where $V$ is the set of rigid bodies, $R$ is the set of red edges corresponding to primitive angular constraints and $B$ is the set of black edges corresponding to primitive blind constraints.

**primitive constraint:** A constraint that may block at most one degree of freedom; corresponds to one row in the rigidity matrix of Section 23.3.

**realization:** An assignment $\mathbf{T} \in SE(d)^n$ of reference frames for each body such that the constraint equations determined by $L_1, \ldots, L_{|C|}$ are satisfied, where $F = (G, c, L_1, \ldots, L_{|C|})$ is a framework with $n$ bodies in dimension $d$.

**rigid:** A framework is (locally) *rigid* with respect to a realization $\mathbf{T}$ if all realizations in any neighborhood of $\mathbf{T}$ are congruent.

**trivial degrees of freedom:** The $D$-dimensional space of trivial Euclidean motions (for the ambient space $\mathbb{R}^d$).

### 23.2.2 Getting to Know Body-and-Cad Frameworks

The building blocks of a body-and-cad framework are fully dimensioned rigid bodies (see also Chapter 20) with no internal degrees of freedom, such as a chair leg in 3D. Geometric constraints are placed between pairs of rigid bodies by specifying geometric elements (e.g., points, lines or planes) affixed to each body. We consider *c*oincidence, *a*ngular and *d*istance (abbreviated *cad*) constraints, which can be used to model the majority of constraints used in CAD software and refer to constraints by the elements involved and the type of constraint; for instance, we could place a **plane-plane coincidence** constraint to force the top surface of a chair leg to coincide with the bottom of the chair's seat. Note that, in CAD software, users often work in a 3D "assembly" environment, which can be modeled as a body-and-cad framework with rigid bodies for the "parts" and geometric constraints for the "mates."

We now walk through a simple example to develop the foundations and terminology for body-and-cad rigidity theory. We follow the same process experienced by a CAD user of adding constraints individually to help the user gain intuition about the main rigidity questions.

**Example 46** *Refer to Figure 23.1(a) throughout this example. Consider a book and a clock placed on a table. We model this system in 2D, so each body has 3* degrees of freedom*: horizontal translation, vertical translation and rotation. The framework as a whole has 6 degrees of freedom. We are interested in the relative motion of bodies in the framework, so we fix the book's position at the origin to eliminate the framework's 3* trivial degrees of freedom*. We refer to the 3 remaining degrees of freedom as the* internal degrees of freedom *and say that the framework is* flexible*.*

- *We specify constraint I to be a* **line-line perpendicular** *constraint. The clock can no longer rotate relative to the book, and the framework now has two internal degrees of freedom. Refer to Figure 23.1(b).*
- *Constraint II, a* **point-line coincidence** *constraint, results in the framework having one internal degree of freedom; refer to Figure 23.1(c).*

- *To block this remaining internal degree of freedom, we can specify constraint III to be a* **point-line distance** *constraint so that the resulting framework, depicted in Figure 23.1(d), is* well-constrained *or* minimally rigid.

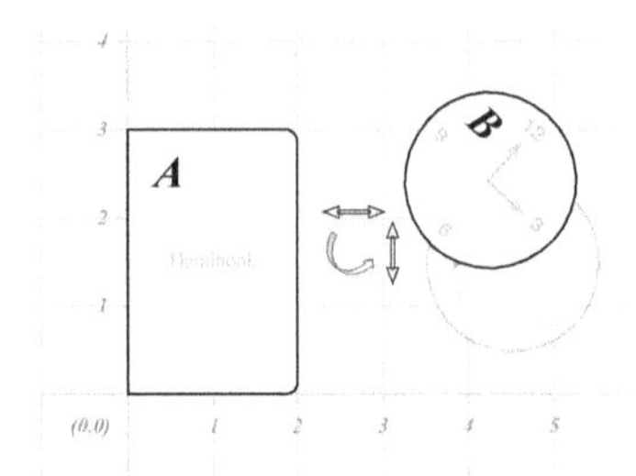


(a) A book and a clock each have 3 degrees of freedom in the plane: horizontal translation, vertical translation and rotation. Fixing the book at the origin eliminates the framework's trivial degrees of freedom, leaving 3 internal degrees of freedom: the clock can move freely.

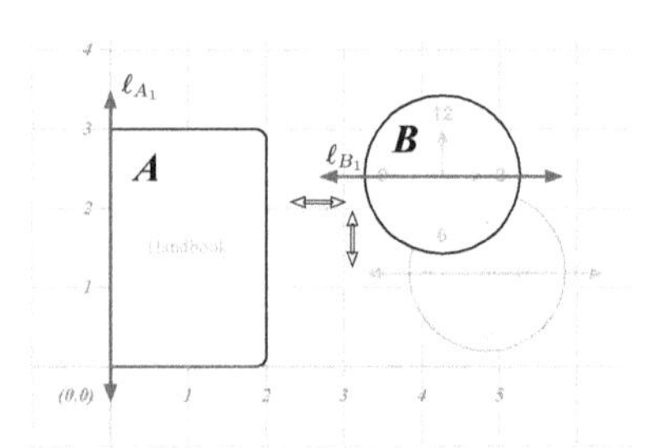


(b) Constraint I requires a line $\ell_{B_1}$ on the clock to be perpendicular to a line $\ell_{A_1}$ on the book. The resulting framework has 2 internal degrees of freedom: the clock can translate vertically and horizontally.

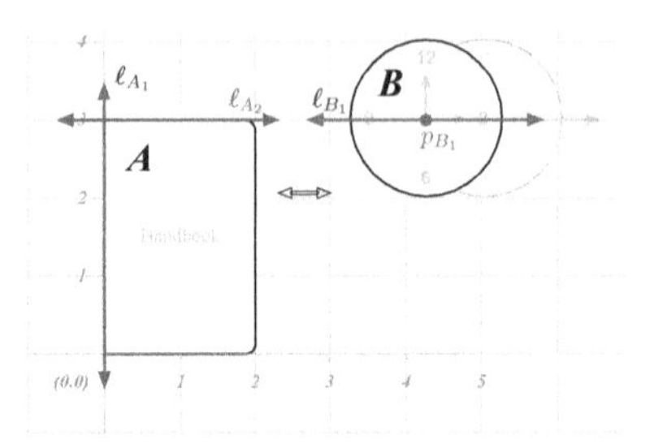


(c) Adding constraint II specifying the center point $p_{B_1}$ of the clock to lie on the short edge $\ell_{A_2}$ of the book removes a degree of freedom. The resulting framework has 1 internal degree of freedom: the clock can translate horizontally.

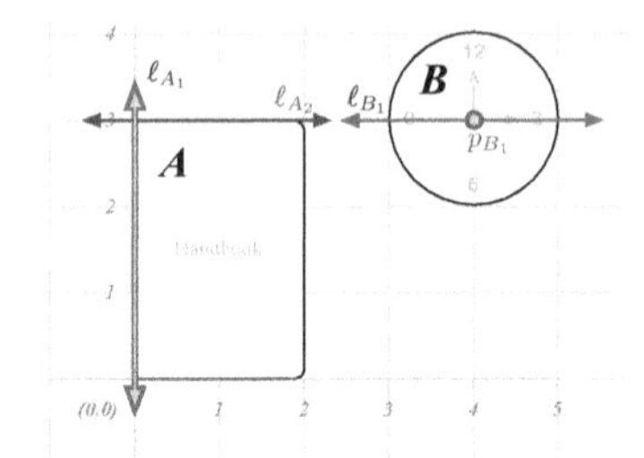


(d) A final constraint III requires the distance between the center point $p_{B_1}$ of the clock and the book's line $\ell_{A_1}$ to be 4 and removes the last internal degree of freedom, resulting in a minimally rigid framework.

**Figure 23.1**
A 2D body-and-cad framework has two rigid bodies (a book $A$ and a clock $B$) with pairwise constraints.

*We capture the combinatorics of this framework in a* cad graph, *as depicted in Figure 23.2.*

*Now consider constraint IV, a* **line-line coincidence** *constraint. A framework with only this constraint has one internal degree of freedom: the clock can translate horizontally (see Figure 23.3(a)). This highlights behavior that cad constraints may exhibit; unlike Constraints I, II and III, which each block a single degree of freedom, Constraint IV blocks two degrees of freedom. To capture this behavior, we associate a set of* primitive constraints *with each cad constraint; a primitive cad constraint blocks (at most) one degree of freedom and corresponds to a single row in the rigidity matrix discussed in Section 23.3. We further refine the classification of a primitive constraint as a* primitive angular constraint, *which blocks a rotational degree of freedom, or a* primitive blind constraint, *which may block a rotational or translational degree of freedom. We make this distinction as angular constraints require special consideration; notice that, while the framework defined with only Constraint IV has 1 remaining degree of freedom, it cannot be eliminated with angular constraints. The* primitive cad graph *associated with a cad graph is bicolored, with red edges for primitive angu-*

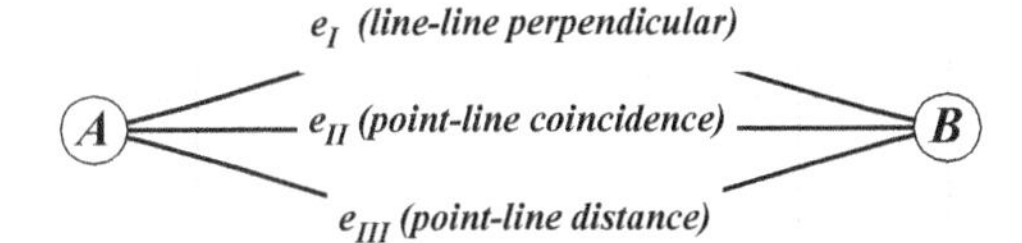


(a) The cad graph has a vertex for each body and a (labeled) edge for each cad constraint.

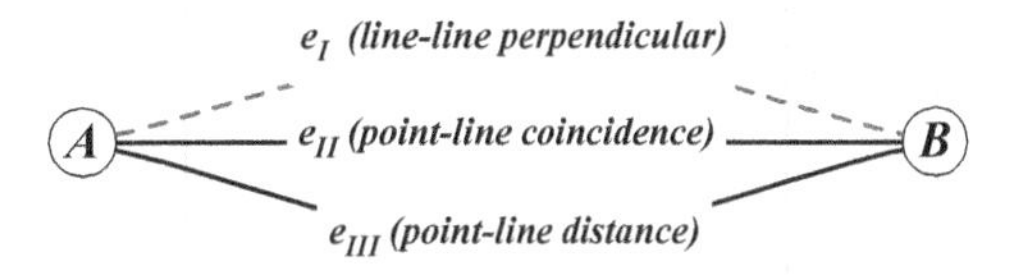


(b) The associated primitive cad graph: the **line-line perpendicular** cad constraint is associated to one primitive angular constraint, while the **point-line coincidence** and **distance** constraints are each associated to one primitive blind constraint.

**Figure 23.2**
The combinatorics of the body-and-cad framework with Constraints I, II and III.

*lar constraints and black edges for primitive blind constraints; refer to Figures 23.2(b) and 23.3(b).*

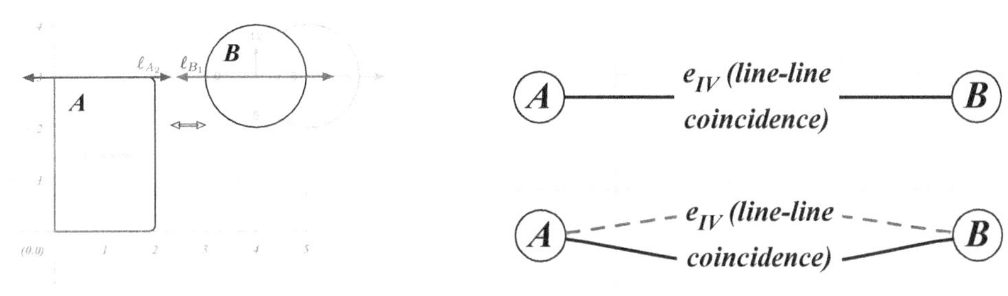


(a) Constraint IV requires the line $\ell_{B_1}$ on the clock to be coincident to the line $\ell_{A_2}$ on the book. The resulting framework has 1 internal degree of freedom: the clock can translate horizontally.

(b) The cad graph (top) and primitive cad graph (bottom) for a framework with Constraint IV; a **line-line coincidence** cad constraint is associated to one primitive angular constraint and one primitive blind constraint.

**Figure 23.3**
A **line-line constraint** highlights the need to further refine cad constraints.

*A framework with Constraints I, II, III and IV is* over-constrained *and* consistent, *as Constraint IV is already implied by Constraints I and II; indeed, a framework with constraints III and IV is minimally rigid. If, however, we specified constraint V to be a* **line-line coincidence** *constraint between $\ell_{B_1}$ and $\ell_{A_1}$ in the framework with constraints I, II, and III, we would obtain an* over-constrained *and* inconsistent *framework; there is no position that satisfies all constraints.*

### 23.2.3 Formalization of the Algebraic Setting

We now state the formal definitions for the algebraic rigidity concepts introduced by this example. Throughout this chapter, we work in dimensions 2 and 3, though the theory can be generalized further. A body-and-cad framework in dimension $d$ is composed of a set of rigid bodies with pairwise cad constraints specified between geometric elemeents (e.g., a point, line or plane).

While we might intuitively picture a rigid body as a solid object, we associate a frame of reference (local coordinate system) to each body, as is standard in robotics and graphics approaches. Then we can formally represent every rigid body $i$ by assigning an element $\mathcal{T}_i \in SE(d)$, where $SE(d)$ is the special Euclidean group for dimension $d$ containing rigid body motions; we represent elements of $SE(d)$ as $(d+1)\times(d+1)$ matrices. Since $\mathcal{T}_i$ gives the location and orientation of the reference frame in the global coordinate system, we can compute global coordinates for any geometric element: left-multiplying a point in homogeneous coordinates by a $\mathcal{T}\in SE(d)$ gives transformed coordinates in homogeneous coordinates (see, e.g., [9]). For a point $\mathbf{p}=(p_1,\ldots,p_d)\in\mathbb{R}^d$, denote its homogeneous coordinates by $\widehat{\mathbf{p}}=(p_1:p_2:\cdots:p_d:1)$; to convert back, we denote the extracted Cartesian coordinates by $\widetilde{\widehat{\mathbf{p}}}=\mathbf{p}$. Then, if a point on body $i$ has local coordinates $\mathbf{p}_i\in\mathbb{R}^d$, $\widetilde{\mathcal{T}_i\widehat{\mathbf{p}}_i}\in\mathbb{R}^d$ computes its global coordinates.

The specific sets of geometric elements and cad constraints that we consider depend on the dimension; to clarify definitions, we will specify the general theory to dimension 2 in this section. Each geometric element is rigidly affixed to a body $i$ and is represented with coordinates local to body $i$'s reference frame: $\mathbf{p}_i$ for a point or $(\mathbf{p}_i,\mathbf{q}_i)$ for the point-direction form of a line, with $\mathbf{p}_i,\mathbf{q}_i\in\mathbb{R}^2$. We denote the set of cad constraints by $\mathcal{C}$ and label each by the geometric elements involved. In 2D, a geometric element is either a point or a line, and $\mathcal{C}=\{$**point-point distance**, **point-point coincidence**, **point-line distance**, **point-line coincidence**, **line-line distance**, **line-line coincidence**, **line-line perpendicular**, **line-line parallel** and **line-line fixed angle**$\}$. We assume an ordering of the elements of $\mathcal{C}$ and abuse notation by identifying $\mathcal{C}$ with $[1..|\mathcal{C}|]$. For example, assuming the ordering of 2D body-and-cad constraints just given, the **point-line distance** constraint type is identified with index 3.

A *cad graph* is defined as a multigraph $G=(V,E)$ along with a coloring function $c:E\rightarrow\mathcal{C}$, where vertices represent rigid bodies and edges represent cad constraints. The coloring function $c(e)$ for $e\in E$ "colors" the edge $e$ with its cad constraint type. Denote by $E_i=\{e\in E|c(e)=i\}$ the edges of type $i$; e.g., $E_3=\{e_{III}\}$ in Example 46.

A *body-and-cad framework* $F=(G,c,L_1,\ldots,L_{|\mathcal{C}|})$ is a cad graph $(G,c)$ along with a family of length functions $L_1,L_{|\mathcal{C}|}$, where $L_i$ specifies the geometry of constraints associated with edges assigned color $i$. For instance, in 2D, the (squared) distance function $L_3:E_3\rightarrow\mathbb{R}^2\times(\mathbb{R}^2\times\mathbb{R}^2)\times\mathbb{R}$ expresses a **point-line distance** constraint $L_3(e)=(\mathbf{p}_i,(\mathbf{p}_j,\mathbf{q}_j),\delta)$ by requiring a point with local coordinates $\mathbf{p}_i$ on body $i$ to be a distance $\sqrt{\delta}$ from a line defined by point $\mathbf{p}_j$ and direction $\mathbf{q}_j$ in local coordinates on body $j$. In Example 46, if we assume the local reference frame for both bodies is aligned with the global reference frame (i.e., the local frame's origin is at global coordinates $(0,0)$ and is not rotated), then $L_3(e_{III})=((4,3),((0,0),(0,1)),16)$.

A *realization* of $F$ in $\mathbb{R}^d$ is a specification $\mathbf{T}=(\mathcal{T}_1,\ldots,\mathcal{T}_n)\in SE(d)^n$ of reference frames such that the length functions are satisfied. That is, a realization is a solution to the system of equations expressing the constraints; the (typically quadratic) equations themselves depend on the type of constraint.

For example, the (squared) length function $L_3$ for **point-line distance** constraints requires any realization $\mathbf{T}$ to place the bodies in such a way that the point on one body is a specified distance from the line on the other body. The distance can be expressed by projecting a vector $\mathbf{v}$ from the point to the line onto a vector $\mathbf{w}$ perpendicular to the line: $\frac{\mathbf{v}\cdot\mathbf{w}}{||\mathbf{w}||}$.

We provide this 2D constraint equation in full detail for concreteness. Let $e\in E_3$ and $L(e)=(\mathbf{p}_i,(\mathbf{p}_j,\mathbf{q}_j),\delta)$, where $\mathbf{p}_i$ gives the coordinates of the point relative to the reference frame of body $i$, $\mathbf{p}_j$ and $\mathbf{q}_j$ the coordinates of a point on the line and its direction relative to the reference frame of body $j$ and $\delta$ the squared distance to be satisfied. If $\mathbf{q}_i=(a,b)$, let $\mathbf{q}_j^{\perp}=(b,-a)$ be a vector perpendicular to $\mathbf{q}_j$. Thus, computing the global coordinates for a vector $\widetilde{\mathcal{T}_j\widehat{\mathbf{p}}_j}-\widetilde{\mathcal{T}_i\widehat{\mathbf{p}}_i}$ from the point to the line and for a vector $\widetilde{\mathcal{T}_j\widehat{\mathbf{q}_j^{\perp}}}$ perpendicular to the line, any realization $\mathbf{T}$ must satisfy:

$$\frac{((\widetilde{\mathcal{T}_j\widehat{\mathbf{p}}_j}-\widetilde{\mathcal{T}_i\widehat{\mathbf{p}}_i})\cdot\widetilde{\mathcal{T}_j\widehat{\mathbf{q}_j^{\perp}}})^2}{||\widetilde{\mathcal{T}_j\widehat{\mathbf{q}_j^{\perp}}}||^2}=\delta \tag{23.1}$$

Two realizations in $\mathbb{R}^d$ are *congruent* if they are related by a Euclidean motion, e.g., if one can be obtained from the other by multiplying each $\mathcal{T}_i$ by some $\mathcal{T}' \in SE(d)$. If all realizations of a framework are congruent, the framework is *globally rigid*. We are interested in the more intuitive concept of local rigidity, defined by assuming the framework is given along with a realization $\mathbf{T}$. If all realizations of $F$ in any neighborhood* of $\mathbf{T}$ are congruent to $\mathbf{T}$, then $F$ is (locally) *rigid* (with respect to $\mathbf{T}$); otherwise, it is *flexible*.

For example, let $F$ be a body-and-cad framework with the single **point-line distance** constraint (Constraint III) from Example 46, i.e., $L_3(e_{III}) = ((4,3),((0,0),(0,1)),16)$. Then $\mathbf{T} = (I_3, I_3)$, where $I_3$ is the 3x3 identity matrix, is a realization of $F$, depicted in Figure 23.1(d). For some parameter $t \in \mathbb{R}$, let $\mathbf{T}'(t) = (I_3, \begin{pmatrix} 1 & 0 & 0 \\ 0 & 1 & t \\ 0 & 0 & 1 \end{pmatrix})$; $\mathbf{T}'(t)$ also gives realizations of $F$, corresponding to translating the position of the clock vertically by $t$. For instance, $\mathbf{T}'(-2)$ is the realization obtained by vertically translating the clock so that its center is at position $(4,1)$. Since, for any neighborhood of $\mathbf{T}$, there exists a realization $\mathbf{T}'(t')$ (for some $t' \in \mathbb{R}$) that is not congruent to $\mathbf{T}$, the framework is flexible.

Note that, for the classical distance-based rigidity models of bar-and-joint or body-and-bar, minimal rigidity is defined in terms of removing any bar constraint, which block at most one degree of freedom. However, as highlighted by Constraint IV in Example 46, cad constraints may block more than one degree of freedom. Therefore, we define the analogous notion for body-and-cad in terms of primitive constraints. For a cad graph $G = (V,E)$, let the bicolored multigraph $H_G = (V, E' = R \sqcup B)$ denote the *primitive cad graph* associated to $G$, where each $e \in E$ is associated to a set of primitive angular constraints $R_e \in R$ and primitive blind constraints $B_e \in B$. The exact numbers of primitive angular and primitive blind constraints associated to a cad constraint depend on the dimension; for dimensions 2 and 3, refer to Tables 23.1 and 23.2. A rigid body-and-cad framework is *minimally rigid* if the removal of any primitive constraint results in a flexible framework.

**Table 23.1**
Association of 2D body-and-cad (*coincidence, angular, distance*) constraints with the number of primitive angular and blind constraints.

| | **Point** | | **Line** | |
|---|---|---|---|---|
| | Angular | Blind | Angular | Blind |
| **Point** | | | | |
| Coincidence | 0 | 2 | 0 | 1 |
| Distance | 0 | 1 | 0 | 1 |
| **Line** | | | | |
| Coincidence | | | 1 | 1 |
| Distance | | | 1 | 1 |
| Parallel | | | 1 | 0 |
| Perpendicular | | | 1 | 0 |
| Fixed angular | | | 1 | 0 |

*The definition of the metric space in which this neighborhood is defined is outside the scope of this chapter.

**Table 23.2**
Association of 3D body-and-cad (*coincidence, angular, distance*) constraints with the number of primitive angular and blind constraints.

| | Point | | Line | | Plane | |
|---|---|---|---|---|---|---|
| | Angular | Blind | Angular | Blind | Angular | Blind |
| **Point** | | | | | | |
| Coincidence | 0 | 3 | 0 | 2 | 0 | 1 |
| Distance | 0 | 1 | 0 | 1 | 0 | 1 |
| **Line** | | | | | | |
| Coincidence | | | 2 | 2 | 1 | 1 |
| Distance | | | 0 | 1 | 1 | 1 |
| Parallel | | | 2 | 0 | 1 | 0 |
| Perpendicular | | | 1 | 0 | 2 | 0 |
| Fixed angular | | | 1 | 0 | 1 | 0 |
| **Plane** | | | | | | |
| Coincidence | | | | | 2 | 1 |
| Distance | | | | | 2 | 1 |
| Parallel | | | | | 2 | 0 |
| Perpendicular | | | | | 1 | 0 |
| Fixed angular | | | | | 1 | 0 |

### 23.2.4 Building a 3D Body-and-Cad Framework

We conclude with the final example for this section: building a dimmer wall plate with slider controls.

**Example 47** *In 3D, each rigid body has 6 degrees of freedom (3 rotational and 3 translational), and we can have planes as geometric elements. Figure 23.4 contains a CAD system for a dimmer wall plate with 3 slider controls. The body-and-cad framework is composed of 4 parts, each modeled as a rigid body: the plate A and slider controls B, C and D. There are 6 cad constraints, depicted in Figures 23.5 and 23.6. We consider the process of adding the parts and constraints one at a time, as a CAD user would do.*

- *With the plate A added and fixed in place with its lower left corner at the origin and its base in the first quadrant of the xy-plane, we add the first slider control B. The framework has 6 internal degrees of freedom, as B can rotate and translate freely.*

- *After adding Constraint I, a* **plane-plane coincidence** *between A and B, the framework has 3 internal degrees of freedom: B can translate in the x- and y- directions and rotate about the z-axis. Refer to Figure 23.5(a).*

- *Specifiying Constraint II as a* **line-plane coincidence** *between an edge on A and a face on B achieves the intended design of a framework with 1 internal degree of freedom: B can only translate in the y-direction. Refer to Figure 23.5(b).*

- *The second slider control C is added, initially free to move with its 6 degrees of freedom. Constraint III, a* **plane-plane coincidence** *between B and C, is intended to be analogous to Constraint I, resulting in C having 3 degrees of freedom: C can translate in the x- and y- directions and rotate about the z-axis. However, due to the overhead of rotating the user*

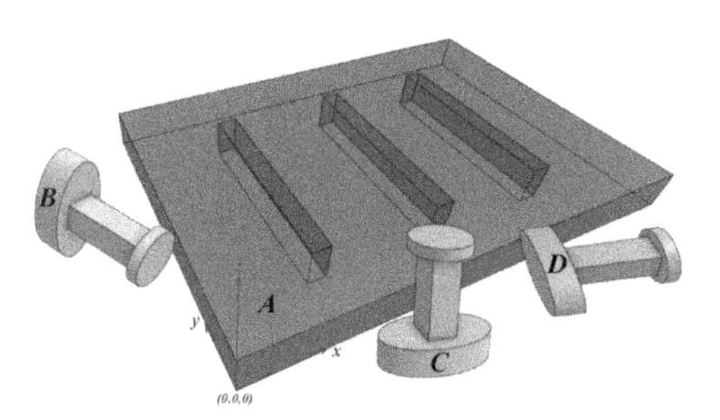


(a) Assume that the base $A$ is fixed at the origin. Then the framework has 18 (internal) degrees of freedom, as each of the controls can rotate and translate freely with 6 degrees of freedom.

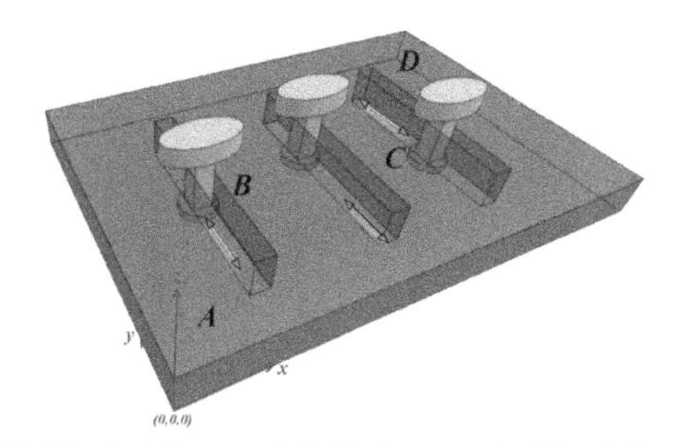


(b) The final framework (with six constraints) has 3 degrees of freedom, as each control can slide in the $y$-direction.

**Figure 23.4**
A dimmer wall plate can be designed with 4 parts: a base $A$ and three slider controls $B, C$, and $D$.

*interface, the constraint is placed between B and C, as the tops of the sliders are easy to select. The entire framework has 4 internal degrees of freedom. Refer to Figure 23.6(a).*

- *Constraint IV is analogous to Constraint II, specifying a* **line-plane coincidence** *between an edge on A and a face on C. This achieves the intended design of a framework with 2 internal degrees of freedom: B and C can only translate in the y-direction. Refer to Figure 23.6(a).*
- *The last slider control D is added, initially free to move with its 6 degrees of freedom. Constraint V, a* **plane-plane coincidence** *between B and D is specified similarly to Constraint III, resulting in D having the 3 degrees of freedom for translating in the x- and y- directions and rotating about the z-axis. The entire framework has 5 internal degrees of freedom. Refer to Figure 23.6(b).*
- *The last constraint is again chosen based on what happened to be easiest to select based on the CAD user interface. Constraint VI places a* **plane-plane parallel** *constraint between a face on C and a face on D. This achieves the final intended design of a framework with 3 internal degrees of freedom: each slider control B, C and D can only translate in the y-direction. Refer to Figure 23.6(b).*

*The combinatorics (cad and primitive cad graphs) of this framework are depicted in Figure 23.7.*

## 23.3 Infinitesimal Body-and-Cad Rigidity Theory

### 23.3.1 Glossary

**(generically) dependent:** A framework is (generically) *dependent* if its (generic) rigidity matrix does not have full rank.

**general:** A realization of a generically independent framework is *general* if it does not lie on the variety given by the pure condition.

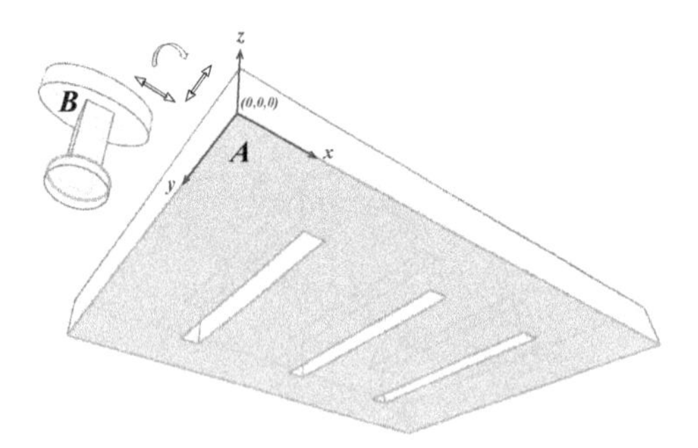


(a) The **plane-plane coincidence** Constraint *I* specifies that the bottom plane of $A$ is coincident to the top plane of the base of $B$. The view of the CAD software must be rotated to select the bottom plane of $A$. The resulting framework has 3 degrees of freedom, as $B$ can translate in the $x$- and $y$-directions and rotation about $z$.

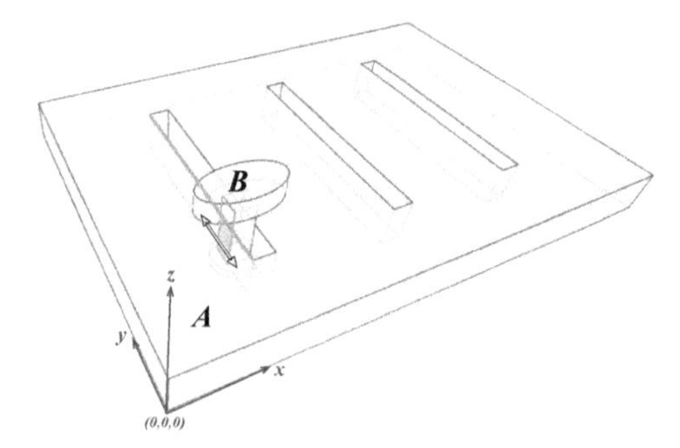


(b) Constraint II is a **line-plane coincidence** constraint restricting a line on $A$ to lie in a plane on $B$.

**Figure 23.5**
Two constraints result in a framework with 1 degree of freedom: $B$ can slide in the $y$-direction.

**generic rigidity matrix:** The matrix $M(G,\mathbf{x})$ on $(a+b)|V|$ columns and $|E|$ rows, where $\mathbf{x}$ maps $E$ to vectors of length $a+b$ with indeterminates as entries; if $e \in R$, $\mathbf{x}$ maps to a vector with indeterminates in the first $a$ entries and 0 in the remaining $b$, where $G=(V,E=B\sqcup R)$ is a bicolored graph, and $a$ and $b$ are integers.

**(generically) independent:** A framework is (generically) *independent* if its (generic) rigidity matrix has full rank.

**(generically) infinitesimally flexible:** A framework is (generically) *infinitesimally flexible* if its (generic) rigidity matrix has rank $< Dn-d$.

**infinitesimally minimally rigid:** An infinitesimally rigid framework is *infinitesimally minimally rigid* if the removal of any primitive constraint results in an infinitesimally flexible framework.

**infinitesimal motion:** A vector of length $Dn$ in $\ker(M(G,\mathbf{r}))$, where $M(G,\mathbf{r})$ is the rigidity matrix for a framework $(G,\mathbf{r})$, assigning an instantaneous motion for each body infinitesimally preserving the constraints.

**(generically) infinitesimally rigid:** A framework is (generically) *infinitesimally rigid* if its (generic) rigidity matrix has rank exactly $Dn-d$.

**instantaneous motion:** A vector $\boldsymbol{s} \in \mathbb{R}^D$, representing an element from $se(d)$, the Lie algebra associated with $SE(d)$; $\boldsymbol{s}=(\boldsymbol{\omega},\mathbf{v})$ with $\boldsymbol{\omega} \in \binom{d}{2}$ (the rotational component) and $\mathbf{v} \in \mathbb{R}^{D-\binom{d}{2}}$.

**pure condition:** A polynomial expressing the determinant of a tied-down generic rigidity matrix for a primitive cad graph with $Dn-D$ edges.

**rigidity matrix:** The Jacobian of the algebraic system of cad constraint equations; for a framework primitive cad graph $G=(V,E)$, the $|E|\times D|V|$ matrix $M(G,\mathbf{r})$ has entries encoded by $\mathbf{r}: E \to D$, where the last $D-\binom{d}{2}$ entries of $\mathbf{r}(e)$ are 0 if $e \in R$.

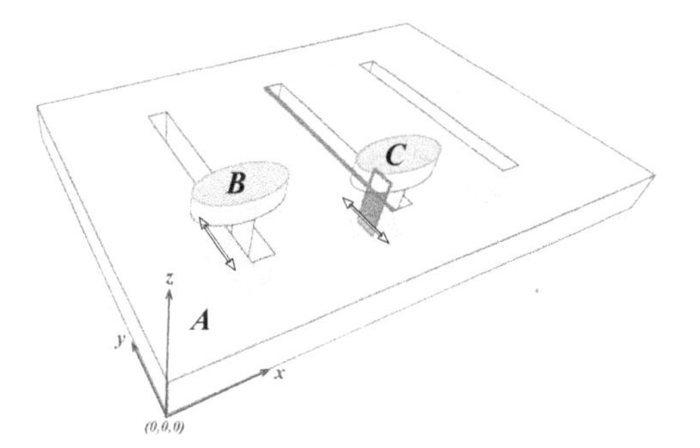


(a) Constraint *III* (orange) is a **plane-plane coincidence** constraint between the tops of $B$ and $C$. Selecting these planes is easier than following the design intent of a constraint analogous to Constraint *I* between $A$ and $C$. Constraint *IV* (green) is a **line-plane coincidence** constraint similar to Constraint *II* restricting a line on $A$ to lie in a plane on $C$. The resulting framework has 2 degrees of freedom: $B$ and $C$ can each slide in the $y$-direction.

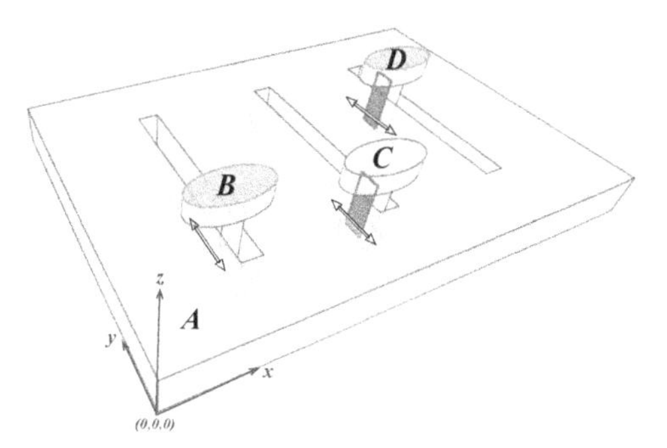


(b) Similar to Constraint *III*, Constraint *V* (orange) is a **plane-plane coincidence** constraint between the tops of $B$ and $D$. Constraint *VI* (green) is a **plane-plane distance** constraint between a plane on $C$ and a plane on $D$; again, due to the CAD user interface, this constraint was chosen instead of a constraint similar to Constraints *II* and *IV*, which would have better captured the design intent. In fact, Constraint *VI* overconstrains the system in a consistent way, as it blocks $D$'s rotation about $y$, already blocked by Constraint *V*.

**Figure 23.6**
Four additional constraints complete the design.

### 23.3.2 The Pattern of the Rigidity Matrix

As with other rigidity models, the *infinitesimal* rigidity theory captures the first-order behavior of a body-and-cad framework $F$. The equations describing the algebraic rigidity, of the type shown in Equation 23.1, are typically quadratic; solutions to the system are realizations $\mathbf{T} = (\mathcal{T}_1, \ldots, \mathcal{T}_n)$ that assign an element from $SE(d)$ to each of the $n$ bodies in the framework. We refer to the Jacobian of this system as the *rigidity matrix*.

For the technical setup of the infinitesimal rigidity theory, assume that $\mathbf{T}(t) : n \to SE(d)$ is a differentiable motion of the bodies and, for simplicity of analysis, assume that $\mathbf{T}(t_0) = (I_D, \ldots, I_D)$, where $D = \binom{d+1}{2}$. Then elements in the kernel of the rigidity matrix can be interpreted as assigning an *instantaneous motion*represented as an element from $se(d)$, the Lie algebra associated with $SE(d)$, to each body. We coordinatize $se(d)$ with vectors $\boldsymbol{s} \in \mathbb{R}^D$. An instantaneous motion $\boldsymbol{s}$ contains a rotational component $\boldsymbol{\omega} \in \mathbb{R}^{\binom{d}{2}}$;† let $\boldsymbol{s} = (\boldsymbol{\omega}, \mathbf{v})$ with $\mathbf{v} \in \mathbb{R}^{D-\binom{d}{2}}$. If $\mathbf{S} = (\boldsymbol{s}_1, \ldots, \boldsymbol{s}_n) \in \mathbb{R}^{Dn}$ is in the kernel of the rigidity matrix of $F$, we call $\mathbf{S}$ an *infinitesimal motion* of $F$.

Every framework will always have infinitesimal motions corresponding to the instantaneous *trivial motions* that assign the same element $\boldsymbol{s}'$ to each body. Thus, the kernel has dimension at least $D$. If the only infinitesimal motions of $F$ are the trivial ones, i.e., if the dimension of the kernel is exactly $D$, we say the framework is *infinitesimally rigid*; otherwise, it is *infinitesimally flexible*. Analogous to classical bar-and-joint rigidity [1], infinitesimal rigidity implies local rigidity.

#### 23.3.2.1 Primitive Angular and Blind Constraints

A framework is infinitesimally rigid exactly when its rigidity matrix has rank $Dn - D$, prompting closer analysis of the pattern of the matrix. Assume that the $Dn$ columns are ordered with $D$ for each body $i$, corresponding to the coordinates of $\boldsymbol{s}_i$. The rows correspond to the linearized constraints of the framework; each expresses a linearized primitive cad constraint between bodies $i$ and $j$. If the constraint is blind, there may be $D$ non-zero entries in the columns for body $i$, with their

†The component $\omega$ is drawn from the Lie algebra $so(d)$ associated with the special orthogonal group $SO(d)$ of rotations.

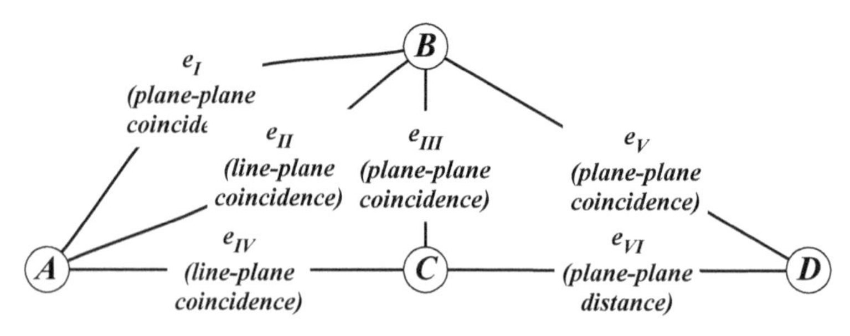


(a) The cad graph has 4 vertices and 6 edges.

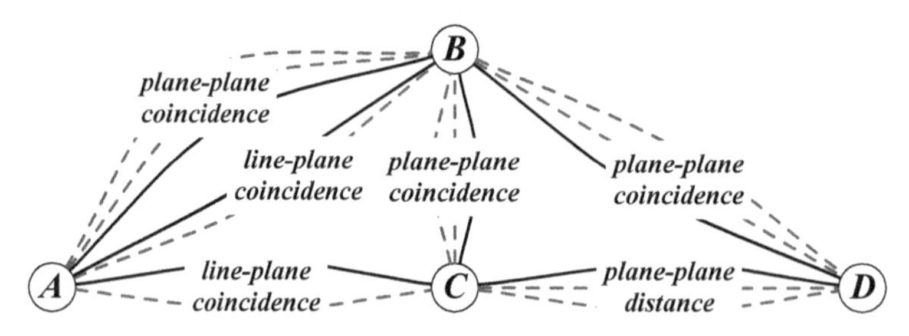


(b) The associated primitive cad graph has 4 vertices and 16 edges, one of which is redundant.

**Figure 23.7**
The combinatorics of the 3D body-and-cad framework for a dimmer wall plate with 3 slider switches.

negation in the columns for body $j$ and zeroes elsewhere. If the constraint is angular, there may be $\binom{d}{2}$ non-zero entries in the columns corresponding to the rotational component $\boldsymbol{\omega}_i$ with their negation in the columns corresponding to $\boldsymbol{\omega}_j$ and zeroes elsewhere (in particular, the columns for $\mathbf{v}_i$ and $\mathbf{v}_j$ contain zeroes). A schematic for the pattern of the rigidity matrix highlighting the distinction between angular and blind constraints is depicted below. For 3D body-and-cad constraints, the equations expressing the constraints can be developed directly in the infinitesimal setting using the Grassmann-Cayley algebra; for full details, refer to [3]. The analogous development can be carried through for 2D body-and-cad constraints.

The primitive cad graph $G = (V, E = R \sqcup B)$ captures the combinatorics of the rigidity matrix, with edges in $R$ corresponding to red primitive angular constraint rows and edges in $B$ corresponding to black primitive blind constraint rows. We can represent the rigidity matrix via a function $\mathbf{r} : E \to \mathbb{R}^D$ that labels an edge $e$ between vertices $i$ and $j$ of the primitive cad graph with a vector $\mathbf{r}(e) \in \mathbb{R}^D$; the corresponding row contains $\mathbf{r}$ in the $D$ columns for $i$ and $-\mathbf{r}$ in the columns for $j$. If $e \in R$, the last $D - \binom{d}{2}$ entries of $\mathbf{r}(e)$ are 0.

For $e \in R$ and $f \in B$ with endpoints $i$ and $j$, $\mathbf{r}(e) = (a_1, \ldots, a_{\binom{d}{2}}, 0, \ldots, 0)$ and $\mathbf{r}(f) = (b_1, \ldots, b_D)$, a schematic of the rigidity matrix rows corresponding to $e$ and $f$ follows.

| | $\cdots$ | $\boldsymbol{\omega}_i$ | $\mathbf{v}_i$ | $\cdots$ | $\boldsymbol{\omega}_j$ | $\mathbf{v}_j$ | $\cdots$ |
|---|---|---|---|---|---|---|---|
| Primitive angular $e \in R$ | $\cdots$ | $a_1 \cdots a_{\binom{d}{2}}$ | $\mathbf{0}$ | $\cdots$ | $-a_1 \cdots -a_{\binom{d}{2}}$ | $\mathbf{0}$ | $\cdots$ |
| Primitive blind $f \in B$ | $\cdots$ | $b_1 \cdots b_D$ | | $\cdots$ | $-b_1 \cdots b_D$ | | $\cdots$ |

The pair $(G, \mathbf{r})$ is sufficient to represent the infinitesimal behavior of a body-and-cad framework, and we denote the rigidity matrix by $M(G, \mathbf{r})$.

To summarize terminology, if the rank of the rigidity matrix $M(G, \mathbf{r})$ is:

- exactly $Dn - D$, the framework is *infinitesimally rigid*;

- $< Dn - D$, the framework is *infinitesimally flexible*;
- exactly $|E|$, the framework is *independent*;
- $< |E|$, the framework is *dependent*

### 23.3.3 Generic Rigidty

We are interested in the "typical" behavior of a framework, behavior that almost all realizations share. This is captured by the notion of *generic infinitesimal rigidity*, formally developed in [6] with a subsequent presentation in [2]. To reduce terminology for this chapter, we do not provide the full details here, instead summarizing the salient contributions. The main object of study is a polynomial called the *pure condition* expressing the determinant of a tied-down generic rigidity matrix; a framework is generically infinitesimally minimally rigid exactly when the pure condition is not identically zero.

Given a primitive cad graph $G = (V,E)$, the *generic rigidity matrix* $M(G,\mathbf{x})$ is obtained by replacing nonzero coordinates with indeterminates. Specifically, for each edge $e \in E$, $\mathbf{x}(e)$ maps to a vector with $D$ formal indeterminates if $e \in B$ and $\binom{d}{2}$ indeterminates if $e \in R$. The analogous definitions from above follow. If the rank of the generic rigidity matrix $M(G,\mathbf{r})$ is:

- exactly $Dn - D$, the framework is *generically infinitesimally rigid*;
- $< Dn - D$, the framework is *generically infinitesimally flexible*;
- exactly $|E|$, the framework is *generically independent*;
- $< |E|$, the framework is *generically dependent*

To remove the subspace of the kernel of a rigidity matrix corresponding to the $D$-dimensional space of "trivial motions," we can *tie down* a body $i$ by appending $D$ rows whose only non-zero entries are specified by embedding the $D \times D$ identity matrix in the columns corresponding to body $i$. Tying down a body [‡] in the generic rigidity matrix of a framework with primitive cad graph $G = (V,E)$, where $|E| = Dn - D$ gives a square matrix, whose determinant expresses the body-and-cad *pure condition* for all frameworks with the same underlying $G$. The pure condition is a polynomial in the indeterminates used to generalize the entries of these frameworks' rigidity matrices; if it is identically zero, every framework with the combinatorics of $G$ is infinitesimally flexible (and dependent). For frameworks with a nonzero pure condition, we refer to a realization as *general* (exhibiting generic behavior) if it does not lie on the variety given by the pure condition. An approach for analyzing special (e.g., non-general) realizations is given in Chapter 4.

## 23.4 Combinatorial Body-and-Cad Rigidity Theory

### 23.4.1 Glossary

**$[a,b]$-pebble game:** An $O(mn^2)$-time algorithm for determining independence in the $[a,b]$-sparsity matroid.

**$[a,b]$-sparse:** A bicolored graph $G = (V, R \sqcup B)$ is *$[a,b]$-sparse*, where $a$ and $b$ are non-negative integers, if there exists $B' \subseteq B$ such that (1) $R \cup B'$ is $(a,a)$-sparse and (2) $B \setminus B'$ is $(b,b)$-sparse.

[‡] The choice of tie-down does not matter [2].

**$[a,b]$-tight:** An $[a,b]$-sparse graph on $n$ vertices is *tight* if has $kn-k$ edges, where $k=a+b$.

**$(k,\ell)$-sparse:** A graph is $(k,\ell)$-*sparse*, where $k$ and $\ell$ are non-negative integers with $0 \le \ell < 2k$, if every set of $n'$ vertices spans at most $\max(0, kn'-\ell)$ edges.

**$(k,\ell)$-tight:** A $(k,\ell)$-sparse graph on $n$ vertices is *tight* if it has exactly $kn-\ell$ edges.

**(generic) rigidity matroid for $G$:** A set of edges is independent if the submatrix of $M(G,\mathbf{r})$ (or $M(G,\mathbf{x})$ for the generic setting) given by the corresponding rows has ful rank.

### 23.4.2 The Rigidity Matroid and Sparsity

As with other rigidity models, the *body-and-cad rigidity matroid* of a framework is defined on the edge set of its primitive cad graph $G$ and infinitesimal constraint function $\mathbf{r}$: a set of edges is independent if the submatrix of $(G,\mathbf{r})$ given by the corresponding rows has full rank. The *generic rigidity matroid* for a framework is defined in terms of the combinatorics of the framework's primitive cad graph $G$: a set of is independent if the submatrix of $M(G,\mathbf{x})$ given by the corresponding rows has full rank.

*Notation.* Throughout this section, there are several definitions that are modified with parameters, e.g., $[a,b]$-sparsity and $[a,b]$-pebble game; the main term is what follows the parametrization. Note also that both square brackets and parentheses appear in the parameters; square brackets (e.g., $[a,b]$-sparsity) are specific to behavior specific to the body-and-cad setting, while parentheses (e.g., $(k,\ell)$-sparsity) are consistent with notation standard in the field.

The pattern of $M(G,\mathbf{x})$ leads to a *sparsity* counting condition, as for other rigidity models (refer to Chapters 1, 18 and 20). A graph is $(k,\ell)$-*sparse* if every set of $n'$ vertices spans at most $\max(0, kn'-\ell)$ edges; a sparse graph is *tight* if it has exactly $kn-\ell$ edges. Results of Nash-Williams [7] and Tutte [11] show that a graph is $(k,k)$-tight if and only if it is the edge-disjoint union of $k$ spanning trees. For more on $(k,\ell)$-sparse graphs, see Chapter 19.

$M(G,\mathbf{x})$ has full rank if $G$ is $(D,D)$-sparse. In addition, since the red edge set $R$ corresponds to rows in $M(G,\mathbf{x})$ with additional zero entries, $G' = (V,R)$ (with only red edges) must satisfy $(\binom{d}{2},\binom{d}{2})$-sparsity. This *nested sparsity* condition, while necessary, is not sufficient for rigidity, as observed in [3]. Intuitively, one can consider independence in the rigidity matroid in terms of partitioning the rotational and translational degrees of freedom in a framework. This corresponds to a partitioning of the primitive constraints as well, with angular constraints only assigned to blocking rotational degrees of freedom. This perspective leads to the more general notion of $[a,b]$-*sparsity*; note the square brackets make a distinction from the $(k,\ell)$-sparsity just discussed.

A bicolored graph $G=(V,E=R\sqcup B)$ is $[a,b]$-*sparse* if there exists a $B' \subseteq B$ such that $(V,R\cup B')$ is $(a,a)$-sparse and $(V,B\setminus B')$ is $(b,b)$-sparse; it is *tight* if, in addition, it has exactly $(a+b)n-(a+b)$ edges. By the results of Nash-Williams and Tutte [7, 11], his implies that $G$ is tight if and only if $\exists B' \subseteq B$ such that $(V,R\cup B')$ is the edge-disjoint union of $a$ spanning trees and $(V,B\setminus B')$ is the edge-disjoint union of $b$ spanning trees. Figure 23.8 depicts a $[1,2]$-tight graph; in Figure 23.8(b), the solid lines include both red edges and form a spanning tree, while the remaining black edges form two edge-disjoint (dotted and dashed) spanning trees.

### 23.4.3 Characterizing Generic Body-and-Cad Rigidity

Independence of $M(G,\mathbf{x})$ can be analyzed via the pure condition defined in Section 23.3, leading to the following characterization for body-and-cad rigidity in terms of $[a,b]$-sparsity.

**Theorem 23.1 ([6])** *A body-and-cad framework $F$ with underlying primitive cad graph $G = (V,E=R\sqcup B)$ is independent if and only if:*

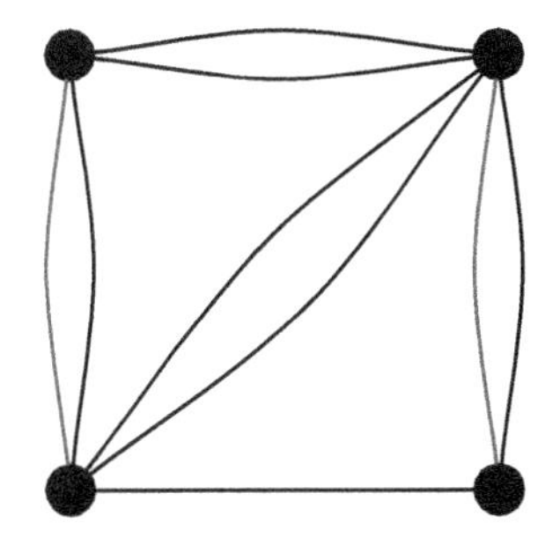

(a) The graph has two red edges and 8 black edges.

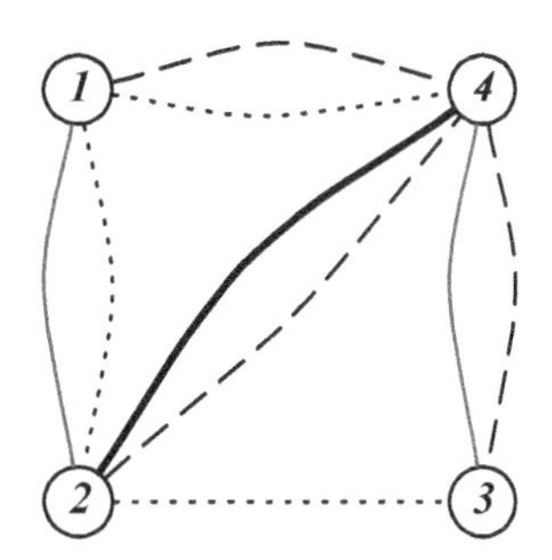


(b) The solid edges (including both red edges) form a spanning tree; the dotted and dashed edges form two other spanning trees.

**Figure 23.8**
A $1,2$-tight graph. Adding the bold black edge between vertices 2 and 4 to the red edge set results in a $(1,1)$-tight subgraph and a $(2,2)$-tight subgraph.

- *G is* $[1,2]$*-sparse and F is embedded in dimension 2*
- *G is* $[3,3]$*-sparse and F is embedded in dimension 2 with no point-point coincidence constraints*

The omission of point-point coincidence constraints in 3D is due to the corresponding rows in the rigidity matrix not following the generic pattern of $M(G,\mathbf{x})$. In particular, placing two point-point coincidence constraints between a pair of bodies results in a drop in the rank that does not appear generically; the resulting framework is essentially the "double banana" (see Chapter 1, Figure 1.4).

The primitive cad graph for the minimally rigid 2D body-and-cad framework in Figure 23.2(b) is $[1,2]$-tight. The primitive cad graph for the flexible, but dependent, 3D body-and-cad framework in Figure 23.7(b) is not $[3,3]$-sparse; the graph with only red edges is not $(3,3)$-sparse as it is a circuit in the $(3,3)$-sparsity matroid.

### 23.4.4 Algorithms

The $[a,b]$-sparsity condition is matroidal, leading to a greedy algorithm called the $[a,b]$-pebble game for determining independence [2]. This pebble game uses the approach of Knuth for matroid union [4], which generally requires "oracle" calls for determining independence in the $(a,a)$- and $(b,b)$-sparsity matroids. By integrating and maintaining the $(k,\ell)$-pebble game algorithm for $(k,\ell)$-sparsity [5], the $[a,b]$-pebble game achieves a faster time complexity of $O(mn^2)$. Since the $[a,b]$-sparsity condition is matroidal, the pebble game algorithm finds maximum-sized independent sets, allowing it to determine if the input graph contains a spanning tight graph as well as detecting dependencies and outputting $[a,b]$-components (vertex-maximal induced $[a,b]$-tight subgraphs in a graph that is not spannning a tight graph).

For containment, Algorithm 9 (reproduced from [2]) describes the $[a,b]$-pebble game algorithm. As with other pebble games, one can view the pebbles as tracking degrees of freedom: aqua pebbles for "angular" degrees of freedom and tan pebbles for "translational." Each edge is covered by a pebble, representing the degree of freedom "blocked" by that constraint. Observe that the black edges may be covered by either an aqua or tan pebble, generalizing the notion of blind constraints blocking either rotational or translational degrees of freedom, while red edges may only be covered by aqua pebbles (angular constraints can block only rotational degrees of freedom).

**Algorithm 9** The $[a,b]$-pebble game algorithm.

**Input:** A bi-colored graph $G = (V, E = B \sqcup R)$, with black $B$ and red $R$ edges.
**Output:** $[a,b]$-sparsity property **tight, sparse, dependent and contains spanning tight**, or **dependent**.
**Setup:** *Initialize an empty directed graph $H$ on vertex set $V$. On each vertex, place a aqua pebbles and b tan pebbles.*
**Allowed moves:**

**Add red edge** $ij$ [Precondition: $\geq a+1$ aqua pebbles on $i$ and $j$.]
*– Add the new edge, cover it with an aqua pebble from i (there is one by the precondition).*
*– Orient $ij$ out of i.*
**Add black edge** $ij$ [Precondition: $\geq a+1$ aqua pebbles on $i$ and $j$ or $\geq b+1$ tan pebbles on $i$ and $j$.]
*– Add the new edge; cover it with a pebble from i using aqua (if there are $a+1$ aqua) or tan (if there are $b+1$ tan).*
*– Orient $ij$ out of i.*
**Edge reversal** [Precondition: vertex $j$ has a pebble on it and an in-edge $ij$ covered by the same color.]
*– Reverse the edge by orienting it as ji out of j, covering with the pebble from j and returning the (same color) pebble originally covering $ij$ to i.*
**Aqua exchange edge reversal** [Precondition: vertex $j$ has an aqua pebble on it and a black in-edge $ij$ covered by a tan pebble; $i$ and $j$ do not belong to the same $(a,a)$-component of aqua pebble covered edges.]
*– Reverse the edge by orienting it as ji out of j, covering with the aqua pebble from j and returning the tan pebble originally covering $ij$ to i.*
**Tan exchange edge reversal** [Precondition: vertex $j$ has a tan pebble on it and a black in-edge $ij$ covered by an aqua pebble; $i$ and $j$ do not belong to the same $(b,b)$-component of tan pebble covered edges.]
*– Reverse the edge by orienting it as ji out of j, covering with the tan pebble from j and returning the aqua pebble originally covering $ij$ to i.*

**Method:**

(a) For each edge $e \in E$

  (a) If $e$ is black: attempt to collect $b+1$ tan pebbles on its endpoints with Alg. 10.
  (b) If Alg. 10 returns **true**: insert $e$ with an **add black edge** move.
  (c) Else, or if $e$ is red: attempt to collect $a+1$ aqua pebbles on its endpoints with Alg. 10.
  (d) If Alg. 10 returns **true**: insert $e$ with an **add black/red edge** move.
  (e) Else: *reject* it and highlight the edges returned by Alg. 10 as the fundamental circuit of the edge (if $e$ is black, this is the union of both calls to Alg. 10).

(b) If every edge is added: output **tight** if there are $a+b$ pebbles left and **sparse** otherwise.

(c) Else, there were rejected edges: output **dependent and contains spanning tight** if there are $a+b$ pebbles left and **dependent** otherwise.

**Algorithm 10** The subroutine for finding pebbles for the $[a,b]$-pebble game.

**Input:** An $[a,b]$-pebble game configuration (a directed bi-colored graph), an edge $e$, and a desired additional pebble color $c_e$ (**aqua** or **tan**).
**Output: true** if $a+1$ aqua (if $c_e$ is aqua) or $b+1$ tan (if $c_e$ is tan) pebbles can be collected on the endpoints of $e$ or **false** otherwise, along with the set of visited edges.
**Method:**

(a) Initialize set $F = \emptyset$.

(b) Initialize queue $Q = \emptyset$. Entries of $Q$ will be of the form $(f,c)$, recording an edge on which to cover with a pebble of color $c$.

(c) Set $e.predecessor =$ NIL.

(d) Enqueue $(e, c_e)$ into Q.

(e) While $Q$ is not empty

  (a) Dequeue $(f,c)$.

  (b) If $f \neq e$ and $f$ is red, continue to the next iteration of the loop.

  (c) Use the basic pebble game rules to try to collect $a+1$ (if $c$ is **aqua**) or $b+1$ (if $c$ is **tan**) pebbles on the endpoints of $f$; let $F'$ be the set of edges visited by that search.

  (d) If the pebbles were collected

    i. Let $g = f$.

    ii. While $g.predecessor \neq$ NIL

      A. Let $d$ be the color of the pebble covering $g$, $\overline{d}$ be the opposite color, $u$ and $v$ the source and target of $g$.

      B. Collect a pebble of color $\overline{d}$ on $v$ using the basic pebble game rules with **edge reversal** moves.

      C. Perform a $\overline{d}$ **exchange edge reversal** move to reverse the edge from $v$ to $u$, covering it with the $\overline{d}$-colored pebble and releasing a $d$-colored pebble back onto $u$.

      D. Set $g = g.predecessor$.

    iii. Collect $a+1$ (if $c$ is **aqua**) or $b+1$ (if $c$ is **tan**) pebbles on the endpoints of $g (= e)$.

    iv. Output **true** and $F \cup F'$.

  (e) Otherwise

    i. For each edge $g \in F'$ that is not in $F$

      A. Set $g.predecessor = f$; let $\overline{c}$ be the opposite of color $c$.

      B. Enqueue $(g, \overline{c})$ into $Q$.

    ii. Assign $F = F \cup F'$.

(f) Output **false** and $F$.

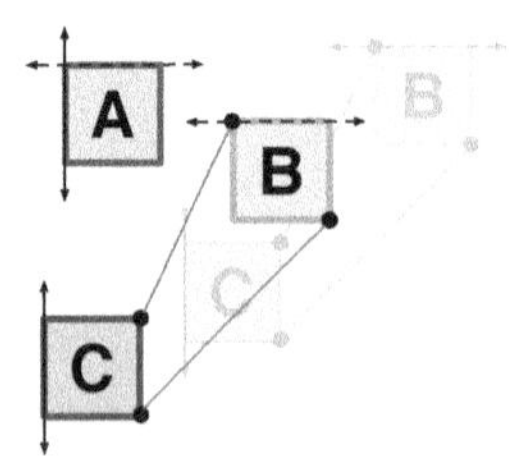


(a) A flexible 2D body-and-cad framework consisting of 3 bodies with the following 4 constraints: dashed lines on $A$ and $B$ must be parallel; solid lines on $A$ and $C$ must be parallel; two bars between bodies $B$ and $C$ fix the distance between pairs of points. The contextually rigid block {B,C} is flexible as an induced framework.

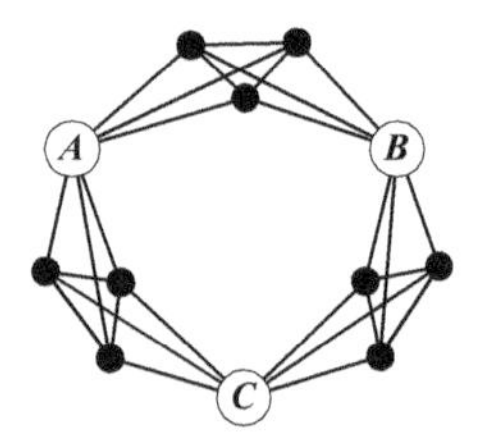


(b) The well-known "triple banana" 3D bar-and-joint framework spanning 12 joints is flexible as each "banana" can rotate. This framework has a contextually rigid block $\{A, B, C\}$ that is flexible as an induced framework, since there are no constraints among $A, B$ and $C$.

**Figure 23.9**
Contextually rigid blocks highlight behavior that does not appear in 2D bar-and-joint and $d$-dimensional body-and-bar rigidity models.

## 23.5 Open Questions

We conclude this chapter with a few open questions for body-and-cad rigidity that would improve feedback for users of CAD software.

**Contextually rigid systems.** A *rigid block* is defined to be a set of bodies that do not contain any relative degrees of freedom. Detecting rigid blocks can improve feedback when CAD users add a dependent (and inconsistent) constraint. Most CAD software attempts to find a minimal set of dependent (primitive) constraints (i.e., a circuit in the rigidity matroid) to give the user guidance on how to resolve the inconsistency. Detecting such dependent sets can be achieved if rigid blocks can be identified. In the well-understood rigidity models of 2D bar-and-joint and $d$-dimensional body-and-bar frameworks, rigid blocks have the property that they remain rigid as induced frameworks. However, a body-and-cad rigid component may be flexible as an induced framework; see the example in Figure 23.9(a). We refer to such a component as *contextually rigid*, as its rigidity depends on other constraints in the framework. This behavior is also exhibited by 3D bar-and-joint frameworks; the "triple banana" contains a contextually rigid component on the three connecting joints (see Figure 23.9(b)). The $[a,b]$-pebble game of Section 23.4.4 can detect rigid blocks that are rigid as induced frameworks. However, it remains an open question to find an efficient algorithm that detects contextually body-and-cad rigid blocks.

**Dependent systems.** Understanding these contextually rigid components may lead towards a complete combinatorial characterization of the circuits of the $[a,b]$-sparsity matroid. While the structure of some circuits is described in [2], it remains open to determine if the categorization is complete. This has an impact on the algorithm for factoring the pure condition, which is the starting point for developing approaches towards giving more intuitive feedback for overconstrained systems. Current CAD software typically highlights a (not necessarily minimal) set of constraints that cannot be resolved due to an dependency, without giving additional guidance or information to the user. Using Cayley factorization to find a geometric interpretation of the dependency may lead to more useful feedback for the user (e.g., "the design is overconstrained due to faces on Part 1 being parallel"); for a case study, refer to Section 3 of [2].

**Capturing design intent with equivalent systems.** While CAD software strives to accurately capture the design intent of the user, the often cumbersome user interface can prevent the designer from easily specifying constraints that match their intent. For example, the design intent for the 3D dimmer wall plate of Example 47 is that each slider control should move with respect to (only) the base plate. However, the combinatorics depicted in Figure 23.7 make it clear that the constraints do not effectively capture this intent. Indeed, the tree-like combinatorics shown in 23.10, which correspond to an equivalent body-and-cad framework, would better capture this design intent. This highlights an open challenge for CAD software. *Given a design, can the software suggest an equivalent one that better captures design intent?* For body-and-cad rigidity theory, this could be posed as an open question to find a way of generating equivalent frameworks. Further, we seek "simple" frameworks, which may better capture design intent, and may also lead to more efficient systems for the embedded numerical solvers to process.

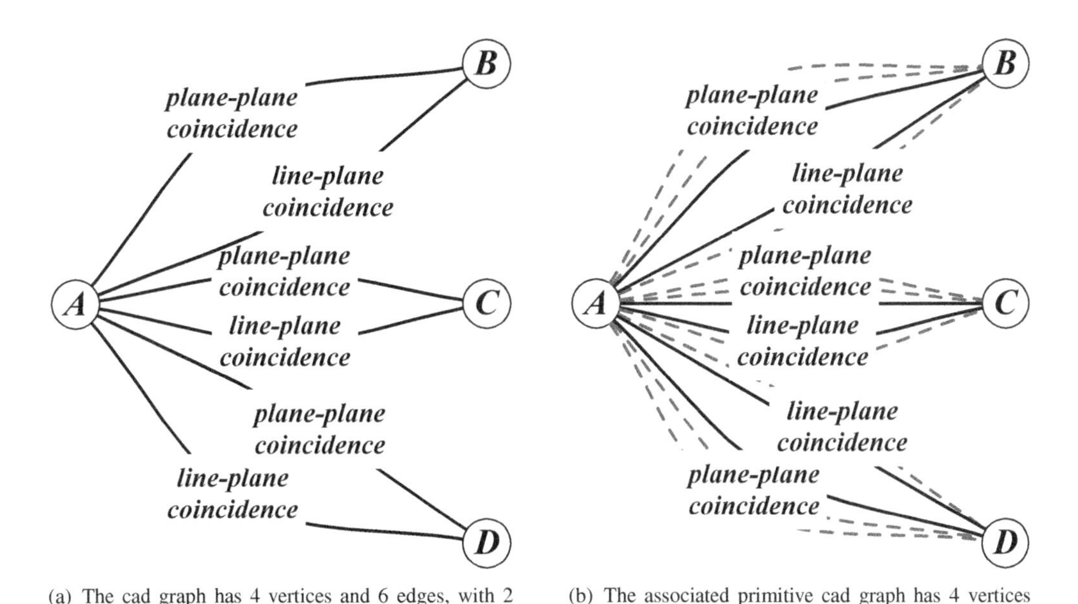


(a) The cad graph has 4 vertices and 6 edges, with 2 edges constraining each slider switch to the base.

(b) The associated primitive cad graph has 4 vertices and 15 edges with no redundancies.

**Figure 23.10**
The combinatorics of an equivalent 3D body-and-cad framework that better captures the "tree-like" design intent of a dimmer wall plate with 3 slider switches.

**Acknowledgments.** *Research partially supported by NSF IIS-1253146. Table 23.2 reproduced from [3], Figure 23.8 from [6], and Figure 23.9 from [2]. The author would like to thank Jessica Sidman for her valuable and insightful feedback on this chapter.*

# References

[1] Leonard Asimow and Ben Roth. The rigidity of graphs II. *Journal of Mathematical Analysis and Applications*, 68:171–190, March 1979.

[2] James Farre, Helena Kleinschmidt, Jessica Sidman, Audrey St. John, Stephanie Stark, Louis Theran, and Xilin Yu. Algorithms for detecting dependencies and rigid subsystems for cad. *Computer Aided Geometric Design*, 47:130–149, 10 2016.

[3] Kirk Haller, Audrey Lee-St.John, Meera Sitharam, Ileana Streinu, and Neil White. Body-and-cad geometric constraint systems. *Computational Geometry: Theory and Applications*, 45(8):385–405, 2012.

[4] Donald E. Knuth. Matroid partitioning. Technical report, Stanford, CA, 1973.

[5] Audrey Lee and Ileana Streinu. Pebble game algorithms and sparse graphs. *Discrete Math.*, 308(8):1425–1437, 2008.

[6] Audrey Lee-St.John and Jessica Sidman. Combinatorics and the rigidity of cad systems. *Computer-Aided Design*, 45(2):473–482, 2013.

[7] C. St. J. A. Nash-Williams. Edge-disjoint spanning trees of finite graphs. *Journal London Math. Soc.*, 36:445–450, 1961. Characterization graphs containing k edge-disjoint spanning trees with counting properties (including (k, k)-sparsity) and partition results).

[8] J.G. Oxley. *Matroid Theory*. Oxford graduate texts in mathematics. Oxford University Press, 2011.

[9] J. M. Selig. *Geometric Fundamentals of Robotics*. Springer Publishing Company, Incorporated, 2nd edition, 2010.

[10] Tiong-Seng Tay. Rigidity of multi-graphs. I. Linking rigid bodies in n-space. *Combinatorial Theory Series*, B(26):95–112, 1984.

[11] W. T. Tutte. On the problem of decomposing a graph into $n$ connected factors. *J. London Math. Soc.*, 36:221–230, 1961.

[12] Neil White and Walter Whiteley. The algebraic geometry of motions of bar-and-body frameworks. *SIAM J. Algebraic Discrete Methods*, 8(1):1–32, 1987.

# Chapter 24

# *Rigidity with Polyhedral Norms*

**Derek Kitson**

*Dept. Math. Stats., Lancaster University, Lancaster, U.K.*

## CONTENTS



## 24.1 Introduction

The goal of this chapter is to present a streamlined introduction to rigidity theory for bar-joint frameworks in $\mathbb{R}^d$ where the underlying metric is governed by a polyhedral norm (as opposed to the Euclidean norm). Non-Euclidean rigidity theory, in which an alternative metric or quadratic form is used to set geometric constraints, appears to be a relatively new topic. It has been considered by various authors in the contexts of spherical and hyperbolic geometry and in pseudo-Euclidean spaces (see [1, 3, 11, 12, 13]). However, the techniques used in each of these cases are different from those required when the constraints are derived from a polyhedral norm. The theory presented here is developed in [6, 4, 7, 8, 5]. For the purposes of exposition, proofs are omitted and the results are not presented in their most general form. To give a broader context, note that we are working within the geometry of finite dimensional normed linear spaces over $\mathbb{R}$, known in the literature as Minkowski spaces (not to be confused with Minkowski space-time). One benefit of working in this setting is that we retain much of the interplay between real analysis and linear algebra that underpins Euclidean rigidity theory. For further reading on Minkowski geometry see [9, 14] and references therein.

A norm on $\mathbb{R}^d$ is *polyhedral* if its unit ball is a convex polytope. The most familiar examples of polyhedral norms are the $\ell^1$ norm and its dual the $\ell^\infty$ norm. To build a picture of how a framework can "flex" in such a setting consider a two-dimensional connected framework with one designated *control* node pinned at the origin. Suppose that all other nodes are constrained to move in a straight line following one of only four directions: up, down, left and right. Note that these directions of motion correspond to the four extreme points of the $\ell^1$ unit ball. If the nodes of the framework undergo a continuous motion, with the control node remaining fixed and all other nodes following one of the four allowed directions, then Euclidean distances between adjacent nodes will in general not be preserved. However, there may exist a continuous motion which preserves $\ell^\infty$ distances and so it is natural to ask whether there is a rigidity theory for the $\ell^\infty$ norm which could be applied in this context.

More generally, we could suppose that the nodes of the framework are constrained to move in some finite number $n$ of pre-determined directions. If each direction is represented by a vector of Euclidean norm 1 then the absolutely convex hull of these vectors is the unit ball for a unique norm on $\mathbb{R}^2$. This norm, and its dual, are examples of polyhedral norms and play roles analogous to the $\ell^1$ and $\ell^\infty$ norms above. This motivates the development of a rigidity theory for general polyhedral norms on $\mathbb{R}^d$.

### 24.1.1 Glossary

In the following, $G = (V,E)$ is a finite simple graph and $\|\cdot\|$ is a norm on $\mathbb{R}^d$. The automorphism group of $G$ is denoted $\text{Aut}(G)$ and the group of linear isometries of $(\mathbb{R}^d, \|\cdot\|)$ is denoted $\text{Isom}(\mathbb{R}^d, \|\cdot\|)$.

**Polyhedral norm:** A norm on $\mathbb{R}^d$ with the property that the closed unit ball $B = \{x \in \mathbb{R}^d : \|x\| \leq 1\}$ is a convex polytope.

**Bar-joint framework:** A pair $(G, p)$ where $p = (p_v)_{v\in V} \in (\mathbb{R}^d)^V$ and $p_v \neq p_w$ for each edge $vw$ in $G$.

**Rigidity map:** The map $f_G : (\mathbb{R}^d)^V \to \mathbb{R}^E$, $(x_v)_{v\in V} \mapsto (\|x_v - x_w\|)_{vw\in E}$.

**Well-positioned framework:** A bar-joint framework $(G, p)$ with the property that the rigidity map $f_G$ is differentiable at $p$.

**Continuous flex:** For a bar-joint framework $(G, p)$, a continuous flex is a continuous path in $f_G^{-1}(f_G(p))$ which passes through $p$.

**Infinitesimal flex:** For a well-positioned bar-joint framework $(G, p)$, an infinitesimal flex is a vector which lies in the kernel of the differential $df_G(p)$.

**Rigid motion:** A family of continuous paths $\alpha_x : (-\delta, \delta) \to \mathbb{R}^d$, $x \in \mathbb{R}^d$, with the property that $\alpha_x(t)$ is differentiable at $t = 0$, $\alpha_x(0) = x$ for each $x \in \mathbb{R}^d$ and $\|x - y\| = \|\alpha_x(t) - \alpha_y(t)\|$ for all $x, y \in \mathbb{R}^d$ and all $t \in (-\delta, \delta)$.

**Infinitesimal rigid motion:** A vector field $\eta : \mathbb{R}^d \to \mathbb{R}^d$ derived from a rigid motion of $\mathbb{R}^d$ by the formula $\eta(x) = \alpha_x'(0)$.

**Rigid framework:** A bar-joint framework $(G, p)$ with the property that every continuous flex arises as the restriction of a rigid motion to $\{p_v : v \in V\}$.

**Flexible framework:** A bar-joint framework which is not rigid.

**Isostatic framework:** A rigid bar-joint framework $(G,p)$ with the property that every bar-joint framework obtained from $(G,p)$ by removing a single edge from $G$ is flexible.

**Symmetric framework:** A bar-joint framework $(G,p)$ with a group action $\theta:\Gamma\to \mathrm{Aut}(G)$ and a faithful group representation $\tau:\Gamma\to \mathrm{Isom}(X,\|\cdot\|)$ such that $\tau(\gamma)p_v = p_{\theta(\gamma)v}$ for all $\gamma\in\Gamma$ and all $v\in V$.

### 24.1.2 Statement of the Problem

Let $X$ be a finite dimensional real vector space and let $\|\cdot\|$ be a norm on $X$.

(a) Given a framework $(G,p)$ in $(X,\|\cdot\|)$ with symmetry group $\Gamma$, determine whether the framework is rigid, isostatic or flexible.

(b) Given a graph $G$ and a group action $\theta:\Gamma\to \mathrm{Aut}(G)$, determine whether there exists $p\in X^V$ and a representation $\tau:\Gamma\to \mathrm{Isom}(X,\|\cdot\|)$ such that $(G,p)$ is isostatic and $\Gamma$-symmetric with respect to $\theta$ and $\tau$.

While these problems are stated for general normed spaces we will restrict our attention here to *polyhedral norms* on $\mathbb{R}^d$. The first problem regarding the rigidity of frameworks is considered in Section 24.2. Initially it will be assumed the symmetry group $\Gamma$ is trivial (in which case no prior knowledge of symmetry is required). We will then consider non-trivial symmetry groups and for this the reader may wish to refer to Chapter 25. The second problem regarding the existence of rigid placements for graphs is considered in Section 24.3.

## 24.2 Rigidity of Frameworks

In this section we present methods for determining whether a given bar-joint framework is rigid with respect to a polyhedral norm.

### 24.2.1 Points of Differentiability

A key feature of the Euclidean norm, from the perspective of rigidity theory, is that given any bar-joint framework $(G,p)$, the corresponding rigidity map $f_G$ is *always* differentiable at $p$. This allows us to pass to the differential $df_G(p)$ and to detect *infinitesimal* flexibility for any framework $(G,p)$. In contrast, for polyhedral norms there exist valid framework placements $p$ which are points of non-differentiability for the rigidity map $f_G$ and so to begin we make the following observation.

**Theorem 24.1** *Let $(G,p)$ be a bar-joint framework in $\mathbb{R}^d$ and let $\|\cdot\|_{\mathcal{P}}$ be a polyhedral norm on $\mathbb{R}^d$. The following statements are equivalent.*

*(i) $f_G$ is differentiable at $p$.*

*(ii) For each edge $vw\in E$ the vector $p_v - p_w$ lies in the conical hull of exactly one facet of the polytope $\mathcal{P}$.*

A framework $(G,p)$ may be regarded as *well-positioned* for a particular choice of polyhedral norm if the corresponding rigidity map $f_G$ is differentiable at $p$. In this case, the elements of the kernel of the differential $df_G(p)$ represent *infinitesimal flexes* of $(G,p)$ and it can be shown that (continuous) rigidity is equivalent to *infinitesimal rigidity*. Two convenient methods for detecting rigidity emerge from this equivalence. The first is based on computing the rank of the differential $df_G(p)$ and for this we require an analogue of the Euclidean rigidity matrix. The second method is to consider properties of an induced edge-labeling of the graph. This method is made possible by the fact that the polyhedral rigidity matrix has only finitely many possible entries, derived from the finitely many facets of the polytope $\mathcal{P}$.

**Example 48** *Consider a degree 2 vertex in a framework $(G,p)$ in $(\mathbb{R}^2, \|\cdot\|_1)$ as illustrated on the left of Figure 24.1. The unit ball $\mathcal{P}$ in $(\mathbb{R}^2, \|\cdot\|_1)$ is indicated on the right. If the vertex $v$ is pinned then, under any motion of the framework which preserves the $\ell^1$ distance between adjacent vertices, the vertices $a$ and $b$ are constrained to lie on the indicated polytopes. Note that in this example the rigidity map $f_G$ is not differentiable at $p$ as the vector $p_v - p_b$ lies in the conical hull of two different facets of $\mathcal{P}$.*

**Figure 24.1**
A framework $(G,p)$ in $(\mathbb{R}^2, \|\cdot\|_1)$ where $p$ is a point of non-differentiability for the rigidity map $f_G$.

**Theorem 24.2** *Let $G$ be a graph and let $\|\cdot\|_{\mathcal{P}}$ be a polyhedral norm on $\mathbb{R}^d$. If the rigidity map $f_G$ is differentiable at $p$ then,*

*(i)* $\operatorname{rank} df_G(p) \leq d|V| - d$*, and,*

*(ii)* $(G,p)$ *is rigid in* $(\mathbb{R}^d, \|\cdot\|_{\mathcal{P}})$ *if and only if* $\operatorname{rank} df_G(p) = d|V| - d$.

The appearance of the second $d$ in the above rank formula needs some explanation as in the Euclidean context the corresponding number is $\frac{d(d+1)}{2}$. This number indicates the dimension of the space of trivial infinitesimal motions which, in Euclidean space, can be attributed to $d$ translational motions and $\frac{d(d-1)}{2}$ rotational motions. On replacing the Euclidean norm with a polyhedral norm, rotations are no longer isometric and so do not induce trivial infinitesimal motions. We are left, in this case, with only $d$ translational flexes generating the space of all trivial infinitesimal motions.

### 24.2.2 The Rigidity Matrix

In order to compute the rank of $df_G(p)$ we require a convenient matrix representation. Such a representation is provided below in which non-zero entries are determined by the facets of the polytope $\mathcal{P}$. To describe these entries note that $\mathcal{P}$ is an intersection of half-spaces and so has the form,

$$\mathcal{P} = \bigcap_{j=1}^{n} \{x \in \mathbb{R}^d : x \cdot \hat{F}_j \leq 1\}$$

for some $\hat{F}_1, \dots, \hat{F}_n \in \mathbb{R}^d$. Note that the facets of $\mathcal{P}$ have the form,

$$F_j = \{x \in \mathcal{P} : x \cdot \hat{F}_j = 1\}$$

and the vectors $\hat{F}_1, \dots, \hat{F}_n$ are the extreme points of the polar set $\mathcal{P}^\triangle$,

$$\mathcal{P}^\triangle = \{x \in \mathbb{R}^d : x \cdot y \leq 1, \forall\, y \in \mathcal{P}\}.$$

(For example, in the case of the $\ell^\infty$ norm on $\mathbb{R}^2$ we may take $\hat{F}_1 = (1,0)$, $\hat{F}_2 = (0,1)$ and their negatives).

**Theorem 24.3** *Let $G = (V,E)$ be a graph and let $\|\cdot\|_\mathcal{P}$ be a polyhedral norm on $\mathbb{R}^d$. If the rigidity map $f_G$ is differentiable at $p$ then $df_G(p)$ admits a matrix representation with rows indexed by $E$ and columns indexed by $V \times \{1,2,\dots,d\}$. The row entries for an edge $vw \in E$ are,*

$$\begin{array}{c} \\ vw \end{array} \begin{array}{c} \begin{array}{ccccccccccc} & & & v & & & & w & & & \end{array} \\ \left[\begin{array}{ccccccccccc} 0 & \cdots & 0 & \hat{F} & 0 & \cdots & 0 & -\hat{F} & 0 & \cdots & 0 \end{array}\right], \end{array}$$

*where $F$ is the unique facet of $\mathcal{P}$ with a conical hull that contains $p_v - p_w$ and $\hat{F} \in \mathbb{R}^d$ is the corresponding extreme point in the polar set $\mathcal{P}^\triangle$.*

Note that the above matrix representation of $df_G(p)$ has the same size and the same identically zero entries as the standard Euclidean rigidity matrix. It is unique up to the ordering of vertices and edges.

### 24.2.3 Framework Colors

Let $(G,p)$ be a bar-joint framework in $(\mathbb{R}^d, \|\cdot\|_\mathcal{P})$ and let $vw \in E$ be an edge of $G$. If $p_v - p_w$ is contained in the conical hull of a facet $F$ of $\mathcal{P}$ then the pair of facets $[F] = \{F, -F\}$ is referred to as an induced *framework color* for the edge $vw$. If the rigidity map $f_G$ is differentiable at $p$ then each edge $vw$ has exactly one framework color, denoted $\mathcal{P}hi(vw)$. A subgraph of $G$ is *monochrome with framework color* $[F]$ if each edge in the subgraph has induced framework color $[F]$. The *maximal* monochrome subgraph with framework color $[F]$ is denoted $G_F$.

**Example 49** *Consider the placement $p$ of the complete graph $K_3$ in $(\mathbb{R}^2, \|\cdot\|_\infty)$ illustrated in Figure 24.2. The unit ball $\mathcal{P}$ is indicated on the left with representative facets labelled $F_1$ and $F_2$. This framework is well-positioned for the $\ell^\infty$ norm as each of the three vectors $p_a - p_b$, $p_b - p_c$ and $p_a - p_c$ lies in the conical hull of exactly one facet of $\mathcal{P}$ (see Prop. 24.1). Each edge has exactly one framework color,*

$$\Phi(ab) = [F_1], \quad \Phi(ac) = [F_2], \quad \Phi(bc) = [F_2].$$

*The monochrome subgraphs $G_{F_1}$ and $G_{F_2}$ are indicated below in black and gray, respectively. The polyhedral rigidity matrix is,*

$$\begin{array}{c} \\ ab \\ bc \\ ac \end{array} \begin{array}{c} \begin{array}{cccccc} a,1 & a,2 & b,1 & b,2 & c,1 & c,2 \end{array} \\ \left[\begin{array}{cc|cc|cc} 1 & 0 & -1 & 0 & 0 & 0 \\ 0 & 0 & 0 & 1 & 0 & -1 \\ 0 & 1 & 0 & 0 & 0 & -1 \end{array}\right] \end{array}$$

*Note that the rigidity matrix has rank 3 which, by Theorem 24.2, indicates that $(K_3, p)$ is flexible. An evident flex is obtained by pinning $a$ and $b$ while translating $c$ horizontally.*

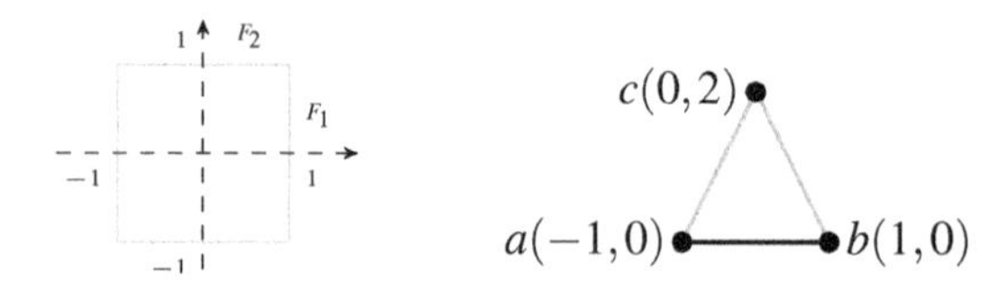


**Figure 24.2**
A placement for $K_3$ in $(\mathbb{R}^2, \|\cdot\|_\infty)$ with induced framework colors indicated in black and gray.

### 24.2.4 Connectivity

The following observation asserts that rigid frameworks satisfy a strong form of connectivity which may be expressed in terms of the framework coloring.

**Theorem 24.4** *If $(G,p)$ is rigid in $(\mathbb{R}^d, \|\cdot\|_{\mathcal{P}})$ then $G$ is connected and any subgraph obtained from $G$ by removing the edges of fewer than $d$ maximal monochrome subgraphs is connected and spanning in $G$.*

To see why this is the case, consider a two-dimensional framework $(G,p)$ with a maximal monochrome subgraph $G_F$, the removal of which disconnects the graph into two connected components. A non-trivial flex of $(G,p)$ is obtained by pinning one of these components while translating the other component in a direction orthogonal to $\hat{F}$. This argument also extends to $d$-dimensional frameworks. The following example demonstrates that the strong connectivity condition alone is not sufficient for rigidity.

**Example 50** *Consider the polyhedral norm on $\mathbb{R}^2$ given by*

$$\|x\|_{\mathcal{P}} = |x \cdot b_1| + |x \cdot b_2| + |x \cdot b_3|$$

*where $b_1 = (1,0)$, $b_2 = (0,1)$ and $b_3 = (1,1)$ and let $(K_3,p)$ be the framework in $(\mathbb{R}^2, \|\cdot\|_{\mathcal{P}})$ illustrated in Figure 24.3. The maximal monochrome subgraphs corresponding to the facets $F_1$, $F_2$, and $F_3$ are indicated by black, gray and dashed lines, respectively. The polyhedral rigidity matrix is,*

$$\begin{array}{c} \\ (ab,F_1) \\ (bc,F_3) \\ (ac,F_2) \end{array} \begin{array}{c} \begin{array}{cccccc} a,1 & a,2 & b,1 & b,2 & c,1 & c,2 \end{array} \\ \left[\begin{array}{cc|cc|cc} 2 & 2 & -2 & -2 & 0 & 0 \\ 0 & 0 & -2 & 0 & 2 & 0 \\ 0 & 2 & 0 & 0 & 0 & -2 \end{array}\right] \end{array}$$

*Note that $G$ satisfies the strong connectivity condition, however $(K_3,p)$ is not rigid. An example of a non-trivial flex, indicated by the arrows in Figure 24.3, is the continuous path $\alpha(t) = (\alpha_a(t), \alpha_b(t), \alpha_c(t))$ where $\alpha_a(t) = (t,0)$, $\alpha_b(t) = (2,2+t)$ and $\alpha_c(t) = (-1,3)$.*

### 24.2.5 Path Chasing

A collection of monochrome subgraphs $H_1, \ldots, H_k$, with framework colors $[F_1], \ldots, [F_k]$, respectively, is *independent* if $\hat{F}_1, \ldots, \hat{F}_k$ are linearly independent in $\mathbb{R}^d$.

**Theorem 24.5** *Let $(G,p)$ be a well-positioned framework in $(\mathbb{R}^d, \|\cdot\|_{\mathcal{P}})$ and let $u = (u_v)_{v \in V}$ be an infinitesimal flex of $(G,p)$. If there exist $d$ independent monochrome paths from a vertex $v$ to a vertex $w$ then $u_v = u_w$.*

Using the above theorem, rigidity can sometimes be detected by simple path chasing arguments as illustrated in the following example.

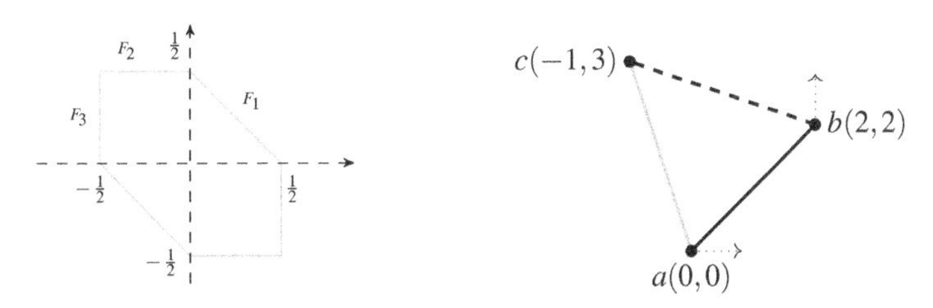


**Figure 24.3**
A flexible bar-joint framework with three maximal monochrome subgraphs which satisfy the strong connectivity condition.

**Example 51** *Consider the polyhedral norm defined for $x = (x_1, x_2) \in \mathbb{R}^2$ by,*

$$\|x\|_{\mathcal{P}} = \begin{cases} 2|x_2| & \text{if } |x_1| \leq |x_2| \\ |x_1| + |x_2| & \text{if } |x_1| \geq |x_2| \end{cases}$$

*and let $(G, p)$ be the framework in $(\mathbb{R}^2, \|\cdot\|_{\mathcal{P}})$ illustrated in Figure 24.2.5. The maximal monochrome subgraphs induced by the facets $F_1$, $F_2$, and $F_3$ are indicated by black, gray and dashed lines, respectively, and the corresponding extreme points of the polar set $\mathcal{P}^{\triangle}$ are,*

$$\hat{F}_1 = (1,1), \quad \hat{F}_2 = (0,2), \quad \hat{F}_3 = (-1,1).$$

*By computing the rank of the polyhedral rigidity matrix we see that $(G, p)$ is rigid. Alternatively, we may apply the following path chasing argument. There exist two independent monochrome paths from vertex $a$ to vertex $d$ and so if $u$ is an infinitesimal flex of $(G, p)$ then $u_a = u_d$. Similarly, $u_d = u_e$, $u_e = u_f$ and $u_f = u_c$. It follows that we may pin the vertices $a, c, d, e, f$ and it only remains to note that in this case $b$ must also be pinned. Thus the space of infinitesimal flexes consists of translational flexes only and so $(G, p)$ is rigid.*

Isostatic frameworks in $(\mathbb{R}^d, \|\cdot\|_\infty)$ can be completely characterized in terms of maximal monochrome subgraphs.

**Theorem 24.6** *Let $(G, p)$ be a well-positioned framework in $(\mathbb{R}^d, \|\cdot\|_\infty)$. The following statements are equivalent.*

*(i) $(G, p)$ is isostatic.*

*(ii) Each maximal monochrome subgraph is a spanning tree for $G$.*

The proof of the above theorem combines the strong connectivity condition for rigid frameworks and the path-chasing method.

### 24.2.6 Symmetry

For frameworks with a non-trivial symmetry group, rigidity may also be detected by considering monochrome subgraph decompositions in an associated gain graph $(G_0, \psi)$. This is particularly useful when applying constructive methods to characterize graphs which admit rigid symmetric placements. To illustrate the method, consider a framework $(G, p)$ in $(\mathbb{R}^2, \|\cdot\|_\infty)$ with half-turn rotational symmetry. The induced monochrome subgraphs of $G$ are themselves symmetric and so induce two edge-disjoint subgraphs of $G_0$. These are referred to as the maximal monochrome subgraphs of $G_0$.

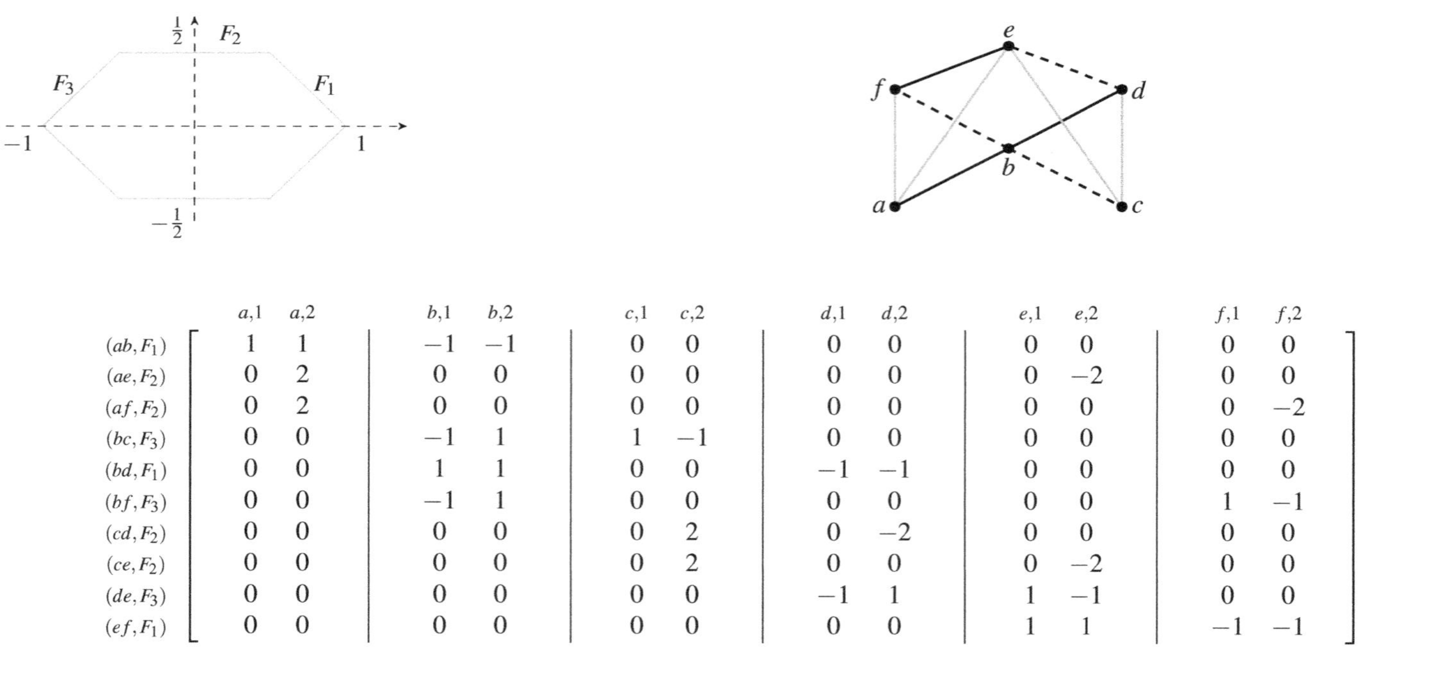


|  | $a,1$ | $a,2$ | $b,1$ | $b,2$ | $c,1$ | $c,2$ | $d,1$ | $d,2$ | $e,1$ | $e,2$ | $f,1$ | $f,2$ |
|---|---|---|---|---|---|---|---|---|---|---|---|---|
| $(ab,F_1)$ | 1 | 1 | −1 | −1 | 0 | 0 | 0 | 0 | 0 | 0 | 0 | 0 |
| $(ae,F_2)$ | 0 | 2 | 0 | 0 | 0 | 0 | 0 | 0 | 0 | −2 | 0 | 0 |
| $(af,F_2)$ | 0 | 2 | 0 | 0 | 0 | 0 | 0 | 0 | 0 | 0 | 0 | −2 |
| $(bc,F_3)$ | 0 | 0 | −1 | 1 | 1 | −1 | 0 | 0 | 0 | 0 | 0 | 0 |
| $(bd,F_1)$ | 0 | 0 | 1 | 1 | 0 | 0 | −1 | −1 | 0 | 0 | 0 | 0 |
| $(bf,F_3)$ | 0 | 0 | −1 | 1 | 0 | 0 | 0 | 0 | 0 | 0 | 1 | −1 |
| $(cd,F_2)$ | 0 | 0 | 0 | 0 | 0 | 2 | 0 | −2 | 0 | 0 | 0 | 0 |
| $(ce,F_2)$ | 0 | 0 | 0 | 0 | 0 | 2 | 0 | 0 | 0 | −2 | 0 | 0 |
| $(de,F_3)$ | 0 | 0 | 0 | 0 | 0 | 0 | −1 | 1 | 1 | −1 | 0 | 0 |
| $(ef,F_1)$ | 0 | 0 | 0 | 0 | 0 | 0 | 0 | 0 | 1 | 1 | −1 | −1 |

**Figure 24.4**
An isostatic bar-joint framework in $(\mathbb{R}^2, \|\cdot\|_{\mathcal{P}})$ as discussed in Example 51. The unit ball $\mathcal{P}$ is depicted on the left, the framework $(G,p)$ with its three induced monochrome subgraphs on the right and a rigidity matrix below. The labelling of the rows of the rigidity matrix indicates an edge of $G$ and the corresponding facet of $\mathcal{P}$ representing the framework color for that edge.

**Theorem 24.7** *Let $(G, p)$ be a well-positioned framework in $(\mathbb{R}^2, \|\cdot\|_\infty)$ with half-turn rotational symmetry. If the rotation acts freely on the vertex set then the following are equivalent.*

*(i)* *$(G, p)$ is rigid.*

*(ii)* *The maximal monochrome subgraphs of the gain graph both contain connected spanning unbalanced map graphs.*

A key idea in the proof of the above theorem is that unbalanced map graphs in the gain graph correspond to subgraphs with symmetric components in the covering graph. In particular, a connected unbalanced map graph corresponds to a connected symmetric subgraph. The result follows from this observation and Theorem 24.6.

**Example 52** *Let $(G, p)$ be the well-positioned framework in $(\mathbb{R}^2, \|\cdot\|_\infty)$ illustrated in Figure 24.5. This framework is symmetric under half-turn rotation and the rotation acts freely on the vertex set. The maximal monochrome subgraphs of G, indicated in black and gray, respectively, are both spanning trees and so $(G, p)$ is rigid by Theorem 24.6. Alternatively, note that the induced maximal monochrome subgraphs of the gain graph are both spanning and contain a single unbalanced cycle and so $(G, p)$ is rigid by Theorem 24.7.*

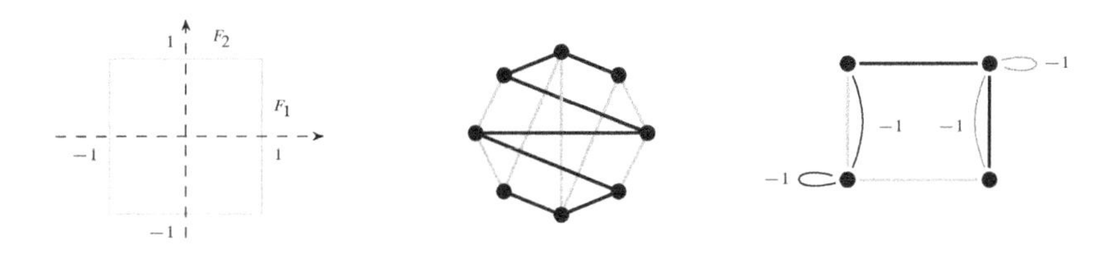


**Figure 24.5**
An isostatic framework in $(\mathbb{R}^2, \|\cdot\|_\infty)$ with half-turn rotational symmetry (center) and associated gain graph (right). The induced monochrome subgraphs are indicated in black and gray.

Consider a framework $(G, p)$ in $(\mathbb{R}^2, \|\cdot\|_\infty)$ with reflectional symmetry in a coordinate axis. The induced framework colors are preserved by the reflection and so induce a natural framework coloring on the edges of the associated gain graph $G_0$. If the reflection acts freely on the vertex set then rigidity is again characterized by considering the maximal monochrome subgraphs of $G_0$.

**Theorem 24.8** *Let $(G, p)$ be a well-positioned framework in $(\mathbb{R}^2, \|\cdot\|_\infty)$ with reflectional symmetry in a coordinate axis. If the reflection acts freely on the vertex set then the following are equivalent.*

*(i)* *$(G, p)$ is rigid.*

*(ii)* *The maximal monochrome subgraphs of the gain graph both contain connected spanning unbalanced map graphs.*

The characterization of rigidity rather than isostaticity in the previous two theorems is deliberate. If the maximal monochrome subgraphs of the gain graph are themselves connected spanning unbalanced map graphs this does not guarantee that the maximal monochrome subgraphs of the covering graph are spanning trees. The following example illustrates this point. In fact, in the next section it is observed that an isostatic framework with reflectional symmetry must contain a fixed vertex and so any reflection framework with a free action on the vertex set will fail to be isostatic.

**Example 53** *Let $(G,p)$ be the well-positioned reflection framework in $(\mathbb{R}^2, \|\cdot\|_\infty)$ illustrated in Figure 24.6. The maximal monochrome subgraphs of $G$ both contain spanning trees and so $(G,p)$ is rigid by Theorem 24.6. Alternatively, note that the maximal monochrome subgraphs of the gain graph are both connected, spanning unbalanced map graphs and so $(G,p)$ is rigid by Theorem 24.8.*

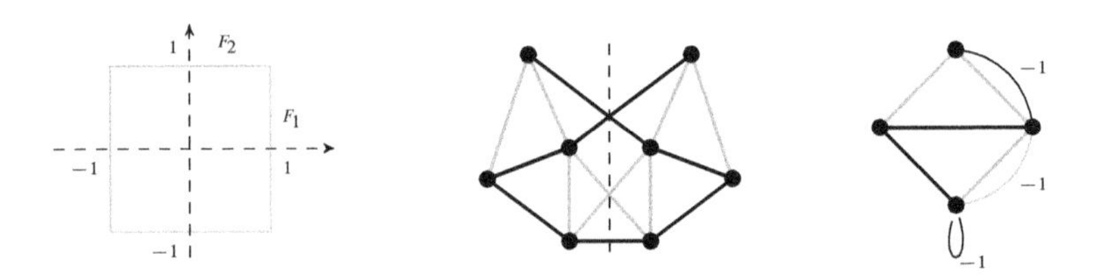


**Figure 24.6**
A rigid reflection framework in $(\mathbb{R}^2, \|\cdot\|_\infty)$ (center) and an associated gain graph (right). The maximal monochrome subgraphs of the gain graph are both spanning unbalanced map graphs.

A flex of a symmetric framework is *symmetric* if it preserves the symmetry of the framework. A symmetric framework is *symmetrically rigid* if every symmetric flex of the framework is trivial. Consider for example a $\mathbb{Z}_2$-symmetric framework $(G,p)$ in $(\mathbb{R}^d, \|\cdot\|_\mathcal{P})$ with a free action on the set of vertices. If the rigidity map is differentiable at $p$ then the rigidity operator $df_G(p)$ may be expressed as a direct sum $R_1 \oplus R_2$ where the kernel of $R_1$ consists of the symmetric infinitesimal flexes of $(G,p)$. In the following $V_0$ denotes the set of vertex orbits for $G$ and $E_0$ denotes the set of edge orbits.

**Theorem 24.9** *Let $(G,p)$ be a well-positioned reflection framework in $(\mathbb{R}^2, \|\cdot\|_\infty)$. If the reflection acts freely on the vertex set then the following are equivalent.*

*(i)* $(G,p)$ *is symmetrically rigid.*

*(ii)* $\text{rank} R_1 = 2|V_0| - 1$.

The following matrix representation for $R_1$ is known as an *orbit matrix*. To define it we must first fix a choice of vertex orbit representatives. The corresponding gain for an edge orbit $[e]$ is denoted $\psi_{[e]}$. For each vertex orbit $[v]$, let $\tilde{v}$ denote the chosen representative vertex in $G$.

**Theorem 24.10** *Let $(G,p)$ be a well-positioned and $\mathbb{Z}_2$-symmetric framework in $(\mathbb{R}^d, \|\cdot\|_\mathcal{P})$. Then $R_1$ admits a matrix representation with rows indexed by $E_0$ and columns indexed by $V_0 \times \{1,2,\ldots,d\}$.*

*If an edge orbit $[e] = [vw]$ is not a loop then the corresponding row entries are,*

$$\begin{array}{c} \\ [e] \end{array} \begin{array}{c} \begin{array}{ccccccccccc} & & & [v] & & & & [w] & & & \end{array} \\ \left[\begin{array}{ccccccccccc} 0 & \cdots & 0 & \hat{F} & 0 & \cdots & 0 & -\hat{F} & 0 & \cdots & 0 \end{array}\right], \end{array}$$

*where $F$ is the unique facet of $\mathcal{P}$ which has a conical hull that contains $p_{\tilde{v}} - p_{\psi_{[e]}\tilde{w}}$.*

*If $[e]$ is a loop at a vertex $[v]$ then the row entries are,*

$$\begin{array}{c} \\ [e] \end{array} \begin{array}{c} \begin{array}{ccccccc} & & & [v] & & & \end{array} \\ \left[\begin{array}{ccccccc} 0 & \cdots & 0 & 2\hat{F} & 0 & \cdots & 0 \end{array}\right], \end{array}$$

*where $F$ is the unique facet of $\mathcal{P}$ which has a conical hull that contains $p_{\tilde{v}} - p_{-\tilde{v}}$.*

A framework is *symmetrically isostatic* if it is symmetrically rigid and no proper symmetric spanning subframework is symmetrically rigid. It is possible to determine when a framework in $(\mathbb{R}^2, \|\cdot\|_\infty)$ with reflectional symmetry in a coordinate axis is symmetrically isostatic by considering the induced maximal monochrome subgraphs in the gain graph $G_0$. In this case the characterization is that one of the maximal monochrome subgraphs of $G_0$ is a spanning unbalanced map graph and the other is a spanning tree. Which of the two is a map graph and which is a tree will depend on the coordinate axis in which the reflection acts. This is illustrated in the example below.

**Example 54** *Consider the well-positioned reflection framework $(G,p)$ illustrated in Figure 24.7. Label the vertices of the gain graph $a,b,c,d$ moving anti-clockwise and ending with the loop at $d$. Then the polyhedral orbit matrix is,*

| | $a,1$ | $a,2$ | $b,1$ | $b,2$ | $c,1$ | $c,2$ | $d,1$ | $d,2$ |
|---|---|---|---|---|---|---|---|---|
| $(ac,F_1)$ | 1 | 0 | 0 | 0 | −1 | 0 | 0 | 0 |
| $(bc,F_1)$ | 0 | 0 | 1 | 0 | −1 | 0 | 0 | 0 |
| $(ad,F_1)$ | 1 | 0 | 0 | 0 | 0 | 0 | −1 | 0 |
| $(dd,F_1)$ | 0 | 0 | 0 | 0 | 0 | 0 | 2 | 0 |
| $(ab,F_2)$ | 0 | 1 | 0 | −1 | 0 | 0 | 0 | 0 |
| $(bc,F_2)$ | 0 | 0 | 0 | 1 | 0 | −1 | 0 | 0 |
| $(cd,F_2)$ | 0 | 0 | 0 | 0 | 0 | 1 | 0 | −1 |

*Note that the rank of the orbit matrix is 7 and so, by Theorem 24.9, the framework is symmetrically rigid. In fact, $(G,p)$ is symmetrically isostatic as the removal of any orbit of edges results in a symmetric framework with a non-trivial symmetric flex. Note that the maximal monochrome subgraphs in the gain graph consist of a spanning unbalanced map graph and a spanning tree. Also note that the strong connectivity condition is not satisfied by the induced framework coloring on $G$ and so $(G,p)$ is not rigid.*

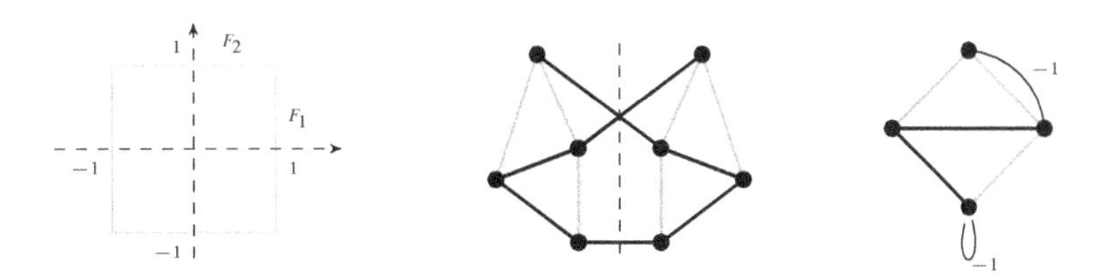


**Figure 24.7**
A symmetrically isostatic reflection framework in $(\mathbb{R}^2, \|\cdot\|_\infty)$ (center) with an associated gain graph (right).

## 24.3 Rigidity of Graphs

In this section we survey properties of graphs which are necessary, and in some cases also sufficient, for the existence of a rigid placement, with or without symmetry.

### 24.3.1 Sparsity Counts and Tree Decompositions

The rank formula for rigid frameworks implies the following counting conditions must hold for any finite simple graph which admits a well-positioned rigid or isostatic placement. These conditions are analogous to Maxwell's counting conditions for Euclidean frameworks.

**Theorem 24.11** *Let G be a finite simple graph.*

*(i) If there exists a rigid well-positioned placement of G in* $(\mathbb{R}^d, \|\cdot\|_\mathcal{P})$ *then,* $|E| \geq d\,|V| - d$.

*(ii) If there exists an isostatic well-positioned placement of G in* $(\mathbb{R}^d, \|\cdot\|_\mathcal{P})$ *then* $|E| = d\,|V| - d$ *and* $|E(H)| \leq d\,|V(H)| - d$ *for each subgraph H.*

If a graph satisfies the counting conditions in $(ii)$ then it is said to be *$(d,d)$-tight*. The following characterization provides a converse to Theorem 24.11$(ii)$ in the case of polyhedral norms on $\mathbb{R}^2$.

**Theorem 24.12** *Let G be a finite simple graph. Then, for all polyhedral norms* $\|\cdot\|_\mathcal{P}$ *on* $\mathbb{R}^2$*, the following statements are equivalent.*

*(i) There exists a well-positioned isostatic placement of G in* $(\mathbb{R}^2, \|\cdot\|_\mathcal{P})$.

*(ii) G is* $(2,2)$*-tight.*

One method of proof for the above theorem is to use a construction scheme for $(2,2)$-tight graphs consisting of four types of graph move: 0-extensions, 1-extensions, vertex splitting and vertex-to-$K_4$ moves. The base graph in this class is $K_4$. See Chapter 19 for a discussion of inductive constructions.

For all polyhedral norms on $\mathbb{R}^2$ the existence of a well-positioned isostatic placement of a graph is also characterized by the following spanning tree property.

**Theorem 24.13** *Let* $G = (V,E)$ *be a finite simple graph. The following statements are equivalent.*

*(i) There exists a well-positioned isostatic placement of G in* $(\mathbb{R}^2, \|\cdot\|_\mathcal{P})$.

*(ii) G is expressible as a union of two edge-disjoint spanning trees.*

It is not currently known whether the previous two theorems extend to $d$-dimensional frameworks. However, it is known that graphs which are $(d,d)$-tight are precisely those which are expressible as an edge-disjoint union of $d$ spanning trees. This is a result of Nash-Williams [10]. Thus if either one of the above theorems extends to $d$-dimensional frameworks then they must both extend. Considering Theorem 24.6, it is sufficient for the $\ell^\infty$ norm to show that any decomposition of $G$ into $d$ edge-disjoint spanning trees is realizable in the sense that the spanning trees are precisely the maximal monochrome subgraphs induced by some placement of the graph in $\mathbb{R}^d$. This has recently been proved for $d = 2$ (see [2]) but remains an open problem for $d \geq 3$.

### 24.3.2 Regular Points

If a graph $G$ admits an infinitesimally rigid placement in a $d$-dimensional Euclidean space then almost all placements of $G$ in this space will be rigid. For polyhedral norms, this is no longer the case and so some care must be taken in adapting any Euclidean notions of *genericity* for polyhedral norms.

A point $p$ is a *regular point* for the rigidity map if $f_G$ is differentiable at $p$ and the differential $df_G(x)$, regarded as a function of $x$, achieves its maximum rank at $p$. In this case $(G,p)$ is said to be a regular framework. If there exists an infinitesimally rigid placement for $G$ then $(G,p)$ must be rigid for all regular points $p$ and, in the Euclidean setting, these regular points form an open and dense set in $(\mathbb{R}^d)^{|V|}$.

In the case of a polyhedral norm, the set of regular points of the rigidity map $f_G$ is still an open set but it is no longer dense in $(\mathbb{R}^d)^{|V|}$. To see this note that a small perturbation of any well-positioned placement of a graph will not alter the induced framework coloring. In particular, the rigidity matrix will be unchanged. This situation is illustrated in the example below.

**Example 55** *Let $(G,p)$ be the well-positioned bar-joint framework in $(\mathbb{R}^2, \|\cdot\|_\infty)$ illustrated in Figure 24.8. The maximal monochrome subgraph $G_{F_2}$ (indicated in gray) is not a spanning subgraph of $G$ and so $(G,p)$ is flexible. The graph $G$ is $(2,2)$-tight and so, by Theorem 24.11, there exists an isostatic placement of $G$ in $(\mathbb{R}^2, \|\cdot\|_\infty)$. By Theorem 24.2, the set of regular placements of $G$ is precisely the set of well-positioned rigid placements of $G$. Thus $p$, and any sufficiently small perturbation of $p$, is a non-regular placement of $G$.*

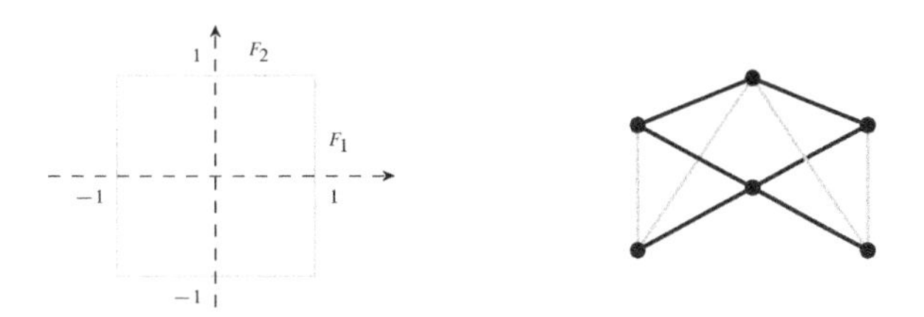


**Figure 24.8**
A non-regular, flexible placement of a $(2,2)$-tight graph in $(\mathbb{R}^2, \|\cdot\|_\infty)$.

A stronger notion of genericity sometimes used in Euclidean rigidity theory is to require not only that $(G,p)$ is regular but that every subframework of the complete framework $(K_V, p)$ is regular. Let us refer to such a placement as *completely regular*. The set of all completely regular placements of a graph in $d$-dimensional Euclidean space is always non-empty (in fact it is dense in $(\mathbb{R}^d)^{|V|}$). This is not the case for polyhedral norms as the following example shows.

**Example 56** *Consider a well-positioned placement $p$ of the complete graph $K_6$ in $(\mathbb{R}^2, \|\cdot\|_\infty)$. The induced framework coloring of the edges of $K_6$ contains a monochrome subgraph $G_F$ which itself must contain a copy of the complete graph $K_3$. The rigidity matrix for the subframework $(K_3,p)$ has rank 2. Since the regular placements of $K_3$ have rigidity matrix with rank 3, $(K_3,p)$ is not regular. Thus there does not exist a well-positioned placement of $K_6$ in $(\mathbb{R}^2, \|\cdot\|_\infty)$ for which all subframeworks are regular. More generally, it follows from Ramsey's theorem that given any polyhedral norm on $\mathbb{R}^d$ there is a complete graph for which no completely regular placements exist.*

### 24.3.3 Symmetric Isostatic Placements

In the presence of a non-trivial symmetry group, a graph which admits an isostatic symmetric placement must satisfy further counting conditions. We present below a general formula from which these conditions can be derived. The set of vertices (respectively, edges) which are fixed by a symmetry operation $\gamma \in \Gamma$ is denoted $V_\gamma$ (respectively, $E_\gamma$).

**Theorem 24.14** *If there exists an isostatic, $\Gamma$-symmetric, well-positioned placement of $G$ in $(\mathbb{R}^d, \|\cdot\|_\mathcal{P})$ then,*

$$|E_\gamma| = \mathrm{trace}(\tau(\gamma))\,(|V_\gamma| - 1),$$

*for each symmetry operation $\gamma \in \Gamma$.*

**Table 24.1**
Necessary counting conditions for graphs which admit a placement as an isostatic framework in $(\mathbb{R}^d, \|\cdot\|_{\mathcal{P}})$ where $d = 2$ or 3.

| Symmetry operation | Dimension | |
|---|---|---|
| | **2** | **3** |
| 1 | $\|E\| = 2\|V\| - 2$ | $\|E\| = 3\|V\| - 3$ |
| $s$ | $\|E_s\| = 0$ | $\|V_s\| \geq 1$, and, $\|V_s\| = 1$ iff $\|E_s\| = 0$ |
| $i$ | $\|V_i\| = 1$ and $\|E_i\| = 0$, or,<br>$\|V_i\| = 0$ and $\|E_i\| = 2$, | $\|V_i\| = 1$ and $\|E_i\| = 0$, or,<br>$\|V_i\| = 0$ and $\|E_i\| = 3$ |
| $C_2$ | (as above) | $\|V_{C_2}\| = 1$ and $\|E_{C_2}\| = 0$, or,<br>$\|V_{C_2}\| = 0$ and $\|E_{C_2}\| = 1$ |
| $C_3$ | $\|V_{C_3}\| = 1$ and $\|E_{C_3}\| = 0$, or,<br>$\|V_{C_3}\| = 0$ and $\|E_{C_3}\| = 1$ | $\|E_{C_3}\| = 0$ |
| $C_4$ | $\|V_{C_4}\| \leq 1$ and $\|E_{C_4}\| = 0$ | $\|V_{C_4}\| = 1$ and $\|E_{C_4}\| = 0$ |
| $C_n, n \geq 5$ | $\|V_{C_n}\| = 1$ and $\|E_{C_n}\| = 0$ | $\|V_{C_n}\| = 1$ and $\|E_{C_n}\| = 0$ |
| $S_3$ | n/a | $\|V_{S_3}\| = 1$ and $\|E_{S_3}\| = 0$ |
| $S_4$ | n/a | $\|V_{S_4}\| = 1$ and $\|E_{S_4}\| = 0$, or,<br>$\|V_{S_4}\| = 0$ and $\|E_{S_4}\| = 1$ |
| $S_5$ | n/a | $\|V_{S_5}\| = 1$ and $\|E_{S_5}\| = 0$ |
| $S_6$ | n/a | $\|E_{S_6}\| = 0$ |
| $S_n, n \geq 7$ | n/a | $\|V_{S_n}\| = 1$ and $\|E_{S_n}\| = 0$ |

*Note*: The left-hand column lists the possible symmetry operations. The counting conditions refer to the number of vertices and edges which are necessarily fixed by a given symmetry operation.

In the case of 2- and 3-dimensional frameworks the necessary counts are listed in Table 24.1 for all possible symmetry operations. Standard notation is used: $s$ denotes a reflection, $i$ denotes an inversion, $C_n$ denotes an $n$-fold rotation and $S_n$ denotes an improper rotation. In Example 52, a $C_2$-symmetric isostatic placement of a graph in $(\mathbb{R}^2, \|\cdot\|_\infty)$ is illustrated. Note that in this case the graph has no fixed vertices and two fixed edges (one of two possible counting conditions listed in Table 24.1). In general, these counts alone will not be sufficient to guarantee the existence of a rigid placement with a particular symmetry group. However, as demonstrated in the next section, they may be used to establish sufficient conditions.

In the presence of a free action on the vertex set, graphs which admit a symmetrically isostatic placement with reflectional symmetry can be characterized in terms of sparsity counts on its associated gain graph.

**Theorem 24.15** *Let $G$ be a graph with a group action $\theta : \mathbb{Z}_2 \to \mathrm{Aut}(G)$ which acts freely on the vertex set. The following are equivalent.*

*(i) There exists a well-positioned and symmetrically isostatic placement of $G$ in $(\mathbb{R}^2, \|\cdot\|_\infty)$ with reflectional symmetry in a coordinate axis.*

*(ii) The gain graph is $(2,2,1)$-gain-tight.*

The $(2,2,1)$-gain-tight condition states that the gain graph $G_0$ satisfies $|E(G_0)| = 2|V(G_0)| - 1$, each set $F$ of edges in $G_0$ satisfies $|F| \leq 2|V(F)| - 1$ and each balanced set of edges in $G_0$ satisfies the stronger condition $|F| \leq 2|V(F)| - 2$. Such graphs are constructible from a single unbalanced loop using four types of graph move (see [5]).

### 24.3.4 Symmetric Tree Decompositions

The isometry group for the Euclidean plane is generated by uncountably many reflections and rotations. The $\ell^\infty$ norm on $\mathbb{R}^2$, in contrast, admits just eight linear isometries. As a result, we can group the non-trivial symmetry operations for a framework in $(\mathbb{R}^2, \|\cdot\|_\infty)$ into the following four types: reflection in a coordinate axis, reflection in a diagonal line, half-turn rotation and four-fold rotation. It is possible to characterize graphs which admit a symmetric isostatic placement $(\mathbb{R}^2, \|\cdot\|_\infty)$ for each of these types of symmetry operation.

**Theorem 24.16** *Let $G$ be a finite simple graph. The following statements are equivalent.*

*(i) There exists a well-positioned isostatic placement of $G$ in $(\mathbb{R}^2, \|\cdot\|_\infty)$ with reflectional symmetry in a coordinate axis.*

*(ii) There exists a group action $\theta : \mathbb{Z}_2 \to \mathrm{Aut}(G)$ such that $G$ is an edge-disjoint union of two symmetric spanning trees and every edge orbit contains two distinct edges.*

The class of $\mathbb{Z}_2$-symmetric graphs which satisfy the conditions of the above theorem are constructible using four graph moves (see [8]). The smallest graph in this class is the *wheel graph* $W_5$.

**Example 57** *Figure 24.9 illustrates a placement of the wheel graph $W_5$ as an isostatic reflection framework in $(\mathbb{R}^2, \|\cdot\|_\infty)$. Note that the induced maximal monochrome subgraphs are edge-disjoint symmetric spanning trees and each edge orbit contains two distinct edges.*

**Figure 24.9**
A placement of the wheel graph $W_5$ in $(\mathbb{R}^2, \|\cdot\|_\infty)$ as an isostatic reflection framework.

The main reason that we need to distinguish between reflections in a coordinate axis and reflections in a diagonal line is that in the former case framework colors are preserved under the reflection while in the latter case they are reversed. This seemingly minor difference actually results in two very different classes of graph. In particular, graphs arising in the former case must contain exactly one fixed vertex while those in the latter may contain any number of fixed vertices. In the following a pair of edge-disjoint spanning trees is referred to as *anti-symmetric* if they are interchanged by the action of the group.

**Theorem 24.17** *Let G be a finite simple graph. The following statements are equivalent.*

*(i) There exists a well-positioned isostatic placement of G in* $(\mathbb{R}^2, \|\cdot\|_\infty)$ *with reflectional symmetry in a diagonal line.*

*(ii) There exists a group action* $\theta : \mathbb{Z}_2 \to \mathrm{Aut}(G)$ *such that G is an edge-disjoint union of two anti-symmetric spanning trees and every edge orbit contains two distinct edges.*

Much like a reflection in a coordinate axis, a half-turn rotation applied to a framework will preserve the framework color of each edge. However, the graphs which admit isostatic placements with half-turn symmetry may have either zero or two fixed edges and so form a strictly larger class than that in Theorem 24.16.

**Theorem 24.18** *Let G be a finite simple graph. The following statements are equivalent.*

*(i) There exists a well-positioned isostatic placement of G in* $(\mathbb{R}^2, \|\cdot\|_\infty)$ *with half-turn rotational symmetry.*

*(ii) There exists a group action* $\theta : \mathbb{Z}_2 \to \mathrm{Aut}(G)$ *such that G is an edge-disjoint union of two symmetric spanning trees and either all, or, all but two edge orbits contain two distinct edges.*

Applying a four-fold rotation to a framework will reverse the induced framework color for each edges. This is similar to the case of reflection in a diagonal line but again the class of graphs arising in these two cases are very different.

**Theorem 24.19** *Let G be a finite simple graph. The following statements are equivalent.*

*(i) There exists a well-positioned isostatic placement of G in* $(\mathbb{R}^2, \|\cdot\|_\infty)$ *with four-fold rotational symmetry.*

*(ii) There exists a group action* $\theta : \mathbb{Z}_4 \to \mathrm{Aut}(G)$ *such that G is an edge-disjoint union of anti-symmetric spanning trees and either all, or, all but two edge orbits contain four distinct edges.*

# References

[1] Victor Alexandrov. Flexible polyhedra in Minkowski 3-space. *Manuscripta Math.*, 111(3):341–356, 2003.

[2] K. Clinch and D. Kitson. Constructing isostatic frameworks for the $\ell^\infty$-plane. *Preprint*, 2017.

[3] Steven J. Gortler and Dylan P. Thurston. Generic global rigidity in complex and pseudo-Euclidean spaces. In *Rigidity and symmetry*, volume 70 of *Fields Inst. Commun.*, pages 131–154. Springer, New York, 2014.

[4] D. Kitson and S. C. Power. Infinitesimal rigidity for non-Euclidean bar-joint frameworks. *Bull. Lond. Math. Soc.*, 46(4):685–697, 2014.

[5] D. Kitson and B. Schulze. Motions of grid-like reflection frameworks. *Journal of Symbolic Computation*, To appear. arxiv.org/abs/1709.09026.

[6] Derek Kitson. Finite and infinitesimal rigidity with polyhedral norms. *Discrete Comput. Geom.*, 54(2):390–411, 2015.

[7] Derek Kitson and Bernd Schulze. Maxwell-Laman counts for bar-joint frameworks in normed spaces. *Linear Algebra Appl.*, 481:313–329, 2015.

[8] Derek Kitson and Bernd Schulze. Symmetric isostatic frameworks with $\ell^1$ or $\ell^\infty$ distance constraints. *Electron. J. Combin.*, 23(4):Paper 4.23, 23, 2016.

[9] Horst Martini, Konrad J. Swanepoel, and Gunter Weiß. The geometry of Minkowski spaces—a survey. I. *Expo. Math.*, 19(2):97–142, 2001.

[10] C. St. J. A. Nash-Williams. Decomposition of finite graphs into forests. *J. London Math. Soc.*, 39:12, 1964.

[11] F.V. Saliola and W. Whiteley. Some notes on the equivalence of first-order rigidity in various geometries. *Preprint*, 2007. arxiv.org/abs/0709.3354.

[12] Bernd Schulze and Walter Whiteley. Coning, symmetry and spherical frameworks. *Discrete Comput. Geom.*, 48(3):622–657, 2012.

[13] Hellmuth Stachel. Flexible octahedra in the hyperbolic space. In *Non-Euclidean geometries*, volume 581 of *Math. Appl. (N. Y.)*, pages 209–225. Springer, New York, 2006.

[14] A. C. Thompson. *Minkowski geometry*, volume 63 of *Encyclopedia of Mathematics and its Applications*. Cambridge University Press, Cambridge, 1996.

# Chapter 25

## *Combinatorial Rigidity of Symmetric and Periodic Frameworks*

**Bernd Schulze**

*Lancaster University, Lancaster, U.K.*

**CONTENTS**



## 25.1 Introduction

Many structures – be they man-made, such as a building, bridge or mechanical linkage, or found in nature, such as a biomolecule, protein or crystal – exhibit non-trivial symmetries. It is therefore important to study the impact of symmetry on the rigidity and flexibility properties of geometric constraint systems.

In Section 25.2, we present some fundamental methods and results for the detection of symmetry-induced infinitesimal flexes and self-stresses in frameworks which count to be isostatic without symmetry.

In Section 25.3, we then study the rigidity of *forced-symmetric* frameworks. That is, given a framework with a certain symmetry, we aim to decide whether the framework has a non-trivial motion that maintains this symmetry. This theory has undergone rapid development in recent years and is particularly useful for the detection of hidden *continuous* flexibility in structures.

The question of whether an *incidentally symmetric* framework is infinitesimally rigid, i.e., whether a symmetric framework has *any* (possibly symmetry-breaking) infinitesimal flex, is more challenging. However, extensions of the tools from the theory of forced-symmetric rigidity have recently also led to new insights into the infinitesimal rigidity of incidentally symmetric frameworks. In particular, for a number of symmetry groups, there now exist combinatorial characterizations of infinitesimally rigid frameworks which are as generic as possible subject to the given symmetry constraints. These results are summarized in Section 25.4.

Finally, in Section 25.5, we also provide some fundamental results concerning the rigidity of infinite *periodic* frameworks, both with a fixed and a flexible lattice representation.

## 25.2 Incidentally Symmetric Isostatic Frameworks

### 25.2.1 Glossary

**$\Gamma$-symmetric graph:** For a finite group $\Gamma$, a simple graph $G=(V,E)$ for which there exists a group action $\theta:\Gamma\to\text{Aut}(G)$, where $\text{Aut}(G)$ denotes the automorphism group of $G$.

**$\Gamma$-symmetric framework:** For a $\Gamma$-symmetric graph $G$ (with respect to $\theta:\Gamma\to\text{Aut}(G)$), and a homomorphism $\tau:\Gamma\to O(\mathbb{R}^d)$, a framework $(G,p)$ satisfying

$$\tau(\gamma)(p_i)=p_{\theta(\gamma)(i)} \qquad \text{for all } \gamma\in\Gamma \text{ and all } i\in V.$$

We also refer to $(G,p)$ as a **$\Gamma$-symmetric realization** of $G$ (with respect to $\theta$ and $\tau$).

**Symmetry group:** For a $\Gamma$-symmetric framework, the subgroup $\tau(\Gamma)=\{\tau(\gamma)\mid\gamma\in\Gamma\}$ of the orthogonal group $O(\mathbb{R}^d)$.

**Group representation:** For a group $\Gamma$ and a linear space $X$, a homomorphism $\rho:\Gamma\to\text{GL}(X)$. The space $X$ is called the **representation space** of $\rho$. Two representations are considered equivalent if they are similar.

**Invariant subspace:** For a representation $\rho:\Gamma\to\text{GL}(X)$, a subspace $U\subseteq X$ with the property that $\rho(\gamma)(U)\subseteq U$ for all $\gamma\in\Gamma$.

**Irreducible representation:** A group representation $\rho:\Gamma\to\text{GL}(X)$ which has no non-trivial $\rho$-invariant subspaces.

**Tensor product of representations:** For two representations $\rho_1$ and $\rho_2$ of a group $\Gamma$, the representation $\rho_1\otimes\rho_2$ of $\Gamma$ given by $\rho_1\otimes\rho_2(\gamma)=\rho_1(\gamma)\otimes\rho_2(\gamma)$ for all $\gamma\in\Gamma$.

**Character:** For a representation $\rho$ of a group $\Gamma$, the vector $\chi(\rho)$ whose $i$th component is the trace of $\rho(\gamma_i)$ for some fixed ordering $\gamma_1,\ldots,\gamma_{|\Gamma|}$ of the elements of $\Gamma$.

**Fixed vertex or edge:** For a $\Gamma$-symmetric graph $G$ (with respect to $\theta:\Gamma\to\text{Aut}(G)$) and $\gamma\in\Gamma$, a vertex $i$ is fixed by $\gamma$ if $\theta(\gamma)(i)=i$. The number of vertices of $G$ that are fixed by $\gamma$ is denoted by $|V_\gamma|$. Similarly, an edge $e=\{i,j\}$ of $G$ is fixed by $\gamma$ if $\theta(\gamma)(e)=e$, that is, if either $\theta(\gamma)$ fixes both $i$ and $j$ or $\theta(\gamma)(i)=j$ and $\theta(\gamma)(j)=i$. The number of edges of $G$ that are fixed by $\gamma$ is denoted by $|E_\gamma|$.

**Γ-generic framework:** A Γ-symmetric framework $(G,p)$ (with respect to $\theta$ and $\tau$) whose rigidity matrix has maximal rank among all Γ-symmetric realizations of $G$ (with respect to $\theta$ and $\tau$).

**Γ-symmetric isostatic graph:** A Γ-symmetric graph $G$ for which some (equivalently, almost all) Γ-symmetric realizations of $G$ are isostatic.

### 25.2.2 Symmetry-Adapted Maxwell Counts

A fundamental result for the rigidity analysis of symmetric structures is that the rigidity matrix of a symmetric framework $(G,p)$ with symmetry group $\tau(\Gamma)$ can be transformed into a block-diagonalized form via methods from group representation theory [21, 47, 48]. Using this block structure of the rigidity matrix, the (infinitesimal) rigidity analysis of $(G,p)$ can be broken up into independent subproblems, one for each irreducible representation $\rho_i$ of $\Gamma$, where each subproblem considers the relationship between external forces on the joints of $(G,p)$ and the resulting internal forces in the bars of $(G,p)$ that share the same symmetry properties described by $\rho_i$ [12, 22, 36, 48].

More precisely, for a Γ-symmetric framework $(G,p)$ (with respect to $\theta$ and $\tau$), let $P_E : \Gamma \to \mathrm{GL}(\mathbb{R}^{|E|})$ be the representation which assigns to each $\gamma \in \Gamma$ the permutation matrix of the permutation $\theta(\gamma)$ of $E$, that is, $P_E(\gamma) = [\delta_{i,\theta(\gamma)(j)})]_{i,j}$, where $\delta$ denotes the Kronecker delta symbol. Further, let $\tau \otimes P_V : \Gamma \to \mathbb{R}^{d|V|}$ be the tensor product of the representations $\tau$ and $P_V$, where (analogous to $P_E$) $P_V : \Gamma \to \mathrm{GL}(\mathbb{R}^{|V|})$ assigns to each $\gamma \in \Gamma$ the permutation matrix of the permutation $\theta(\gamma)$ of $V$. Then we have the following basic result [48, 55].

**Theorem 25.1 (Intertwining Property Of The Rigidity Matrix)** *Let $(G,p)$ be a Γ-symmetric framework with respect to $\theta : \Gamma \to Aut(G)$ and $\tau : \Gamma \to O(\mathbb{R}^d)$. Then, for the rigidity matrix $R(G,p)$, we have*

$$R(G,p)(\tau \otimes P_V)(\gamma) = P_E(\gamma)R(G,p) \quad \text{for all } \gamma \in \Gamma$$

From Theorem 25.1 and Schur's lemma, we obtain the following corollary.

**Corollary 25.2 ((Block-Diagonalization Of The Rigidity Matrix))** *Let Γ be a group, and let $\rho_0, \dots, \rho_r$ be the irreducible representations of Γ. If $(G,p)$ is a Γ-symmetric framework, then there exist invertible matrices $S$ and $T$ such that the rigidity matrix takes on the block form*

$$T^\top R(G,p)S := \widetilde{R}(G,p) = \begin{pmatrix} \widetilde{R}_0(G,p) & & \mathbf{0} \\ & \ddots & \\ \mathbf{0} & & \widetilde{R}_r(G,p) \end{pmatrix}.$$

Note that for $i \in \{0,\dots,r\}$, the size of the block matrix $\widetilde{R}_i(G,p)$ is $\dim(W_i) \times \dim(V_i)$, where $W_i$ and $V_i$ denote the $P_E$-invariant and $(\tau \otimes P_V)$-invariant subspace corresponding to the irreducible representation $\rho_i$, respectively.

Let $\mathcal{T}(G,p)$ denote the space of trivial infinitesimal motions of $(G,p)$. Then it is easy to show that $\mathcal{T}(G,p)$ is a $(\tau \otimes P_V)$-invariant subspace of $\mathbb{R}^{d|V|}$ [48]. Thus, we may form the subrepresentation $(\tau \otimes P_V)^{(\mathcal{T})}$ of $\tau \otimes P_V$ with representation space $\mathcal{T}(G,p)$, and $\mathcal{T}(G,p)$ can be written as a direct sum of $(\tau \otimes P_V)$-invariant subspaces $T_i$, $i = 0,\dots,r$. Clearly, for $(G,p)$ to be isostatic, we must have $\dim(W_i) = \dim(V_i) - \dim(T_i)$ for each $i$. These conditions can be summarized as follows [12, 36, 48].

**Theorem 25.3 (Symmetry-Adapted Maxwell Rule)** *Let $(G,p)$ be an isostatic framework which is Γ-symmetric with respect to $\theta$ and $\tau$. Then we have*

$$\chi(P_E) = \chi(\tau \otimes P_V) - \chi((\tau \otimes P_V)^{(\mathcal{T})}). \tag{25.1}$$

Each of the characters in Equation (25.1) can easily be computed for any symmetry group $\tau(\Gamma)$ [11, 47, 48]. The calculations of characters for isostatic frameworks in the plane and in 3-space are shown in Table 25.2.2 and 25.2.2, respectively (see also Theorem 25.4). In these tables (and throughout this chapter) we use the Schoenflies notation to denote the various symmetry groups and their elements, as this is one of the standard notations for symmetric structures [1, 3]. In this notation, $s$ and $C_n$ denote a reflection in a $(d-1)$-dimensional hyperplane, and a rotation by $2\pi/n$ about a $(d-2)$-dimensional axis, respectively. Moreover, $\mathcal{C}_s$ is a symmetry group generated by a single reflection, and $\mathcal{C}_n$ is a symmetry group generated by a rotation $C_n$.

Note that we may write each of the characters in Equation (25.1) uniquely as a linear combination of the characters of the irreducible representations of $\Gamma$. Therefore, if the symmetry-adapted Maxwell rule shows that a framework is not isostatic, then a comparison of coefficients in these linear combinations may be used to determine the invariant subspaces to which the detected infinitesimal flexes or self-stresses belong.

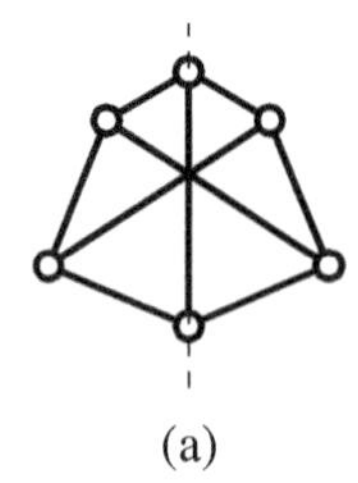

(a)

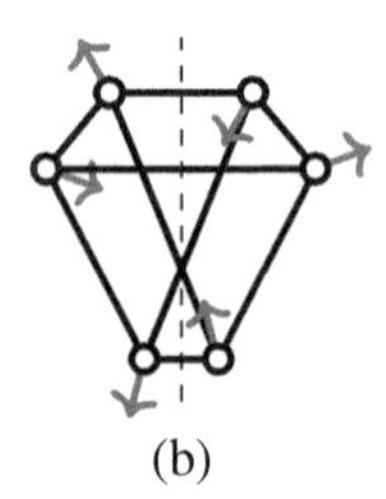

(b)

**Figure 25.1**
Realizations of the complete bipartite graph $K_{3,3}$ with reflection symmetry $\mathcal{C}_s = \{Id, s\}$ in the plane. The framework in (a) is isostatic, whereas the framework in (b) is infinitesimally flexible, as detected by the symmetry-adapted Maxwell rule.

**Example 58** *For the framework $(G,p)$ in Figure 25.1 (b), we have $\chi(P_E) = (9,3)$, $\chi(P_V) = (6,0)$, $\chi(\tau) = (2,0)$, and $\chi(\tau \otimes P_V) = (12,0)$. Moreover, $\chi((\tau \otimes P_V)^{(\mathcal{T})}) = (3,-1)$. Thus,*

$$\chi(P_E) = (9,3) \neq (12,0) - (3,-1) = \chi(\tau \otimes P_V) - \chi((\tau \otimes P_V)^{(\mathcal{T})}),$$

*and hence, by Theorem 25.3, $(G,p)$ is not isostatic.*

*Let $\rho_0$ be the trivial ("fully-symmetric") irreducible representation of $\mathcal{C}_s$ which assigns 1 to each element of $\mathcal{C}_s$, and let $\rho_1$ be the non-trivial ("anti-symmetric") irreducible representation of $\mathcal{C}_s$ which assigns 1 to Id and $-1$ to s. Then we have $(9,3) = 6\rho_0 + 3\rho_1$ and $(9,1) = 5\rho_0 + 4\rho_1$, and hence we may conclude that $(G,p)$ has an anti-symmetric infinitesimal flex and a fully-symmetric self-stress.*

By considering the vector equation (25.1) componentwise, we may obtain very simple necessary conditions for a $\Gamma$-symmetric framework $(G,p)$ to be isostatic in terms of the number of vertices and edges of $G$ that are fixed by the elements of $\Gamma$ [11].

**Theorem 25.4 (Conditions For Individual Symmetry Operations)** *Let $(G,p)$ be an isostatic framework which is $\Gamma$-symmetric with respect to $\theta$ and $\tau$. Then, for every $\gamma \in \Gamma$, we have*

$$|E_\gamma| = trace(\tau(\gamma)) \cdot |V_\gamma| - trace((\tau \otimes P_V)^{(\mathcal{T})}(\gamma)).$$

**Table 25.1**
Calculations of characters for the symmetry-adapted Maxwell rule in the plane.

| | $Id$ | $C_{n>2}$ | $C_2$ | $s$ |
|---|---|---|---|---|
| $\chi(P_E)$ | $\lvert E\rvert$ | $\lvert E_{C_n}\rvert$ | $\lvert E_{C_2}\rvert$ | $\lvert E_s\rvert$ |
| $\chi(\tau\otimes P_V)$ | $2\lvert V\rvert$ | $(2\cos\frac{2\pi}{n})\lvert V_{C_n}\rvert$ | $-2\lvert V_{C_2}\rvert$ | $0$ |
| $\chi((\tau\otimes P_V)^{(\mathcal{T})})$ | $3$ | $2\cos\frac{2\pi}{n}+1$ | $-1$ | $-1$ |

In particular, it follows from Theorem 25.4 and Table 25.2.2 that if the symmetry group $\tau(\Gamma)$ of a 2-dimensional isostatic framework contains a reflection $s$, then we must have $|E_s| = 1$. Similarly, if $\tau(\Gamma)$ contains a half-turn $C_2$, then we must have $|V_{C_2}| = 0$ and $|E_{C_2}| = 1$, and if $\tau(\Gamma)$ contains a three-fold rotation $C_3$, then we must have $|V_{C_3}| = 0$. Moreover, it follows that there cannot exist any $n$-fold rotation in $\tau(\Gamma)$ with $n > 3$. Thus, there are only 5 non-trivial symmetry groups for which we can construct an isostatic framework in the plane, namely the rotational groups $\mathcal{C}_2$ and $\mathcal{C}_3$, the reflectional group $\mathcal{C}_s$, and the dihedral groups $\mathcal{C}_{2v}$ and $\mathcal{C}_{3v}$ of order 4 and 6 [11].

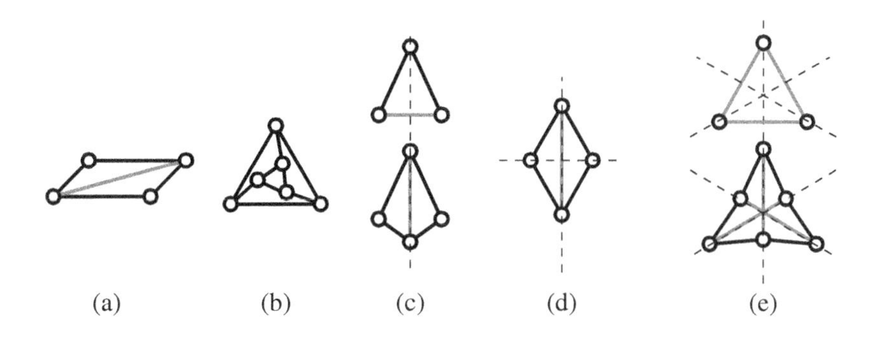


**Figure 25.2**
Symmetric isostatic frameworks in the plane (with fixed edges shown in gray color). (a) for $\mathcal{C}_2$, we have $|V_{C_2}| = 0$ and $|E_{C_2}| = 1$; (b) for $\mathcal{C}_3$, we have $|V_{C_3}| = 0$; (c) for $\mathcal{C}_s$, we have $|E_s| = 1$; (d,e) for the groups $\mathcal{C}_{2v}$ and $\mathcal{C}_{3v}$, the conditions for (a), (b), (c) must be satisfied by all rotations and reflections.

**Table 25.2**
Calculations of characters for the symmetry-adapted Maxwell rule in 3-space.

| | $Id$ | $C_{n>2}$ | $C_2$ | $s$ | $i$ | $S_{n>2}$ |
|---|---|---|---|---|---|---|
| $\chi(P_E)$ | $\lvert E\rvert$ | $\lvert E_{C_n}\rvert$ | $\lvert E_{C_2}\rvert$ | $\lvert E_s\rvert$ | $\lvert E_i\rvert$ | $\lvert E_{S_n}\rvert$ |
| $\chi(\tau\otimes P_V)$ | $3\lvert V\rvert$ | $(2\cos\frac{2\pi}{n}+1)\lvert V_{C_n}\rvert$ | $-\lvert V_{C_2}\rvert$ | $\lvert V_s\rvert$ | $-3\lvert V_i\rvert$ | $(2\cos\frac{2\pi}{n}-1)\lvert V_{S_n}\rvert$ |
| $\chi((\tau\otimes P_V)^{(\mathcal{T})})$ | $6$ | $4\cos\frac{2\pi}{n}+2$ | $-2$ | $0$ | $0$ | $0$ |

While for 3-dimensional symmetric isostatic frameworks, there are still restrictions on the number of fixed structural elements (which can be derived from the calculations in Table 25.2.2 [11]), we may use Cauchy's rigidity theorem for triangulated convex polyhedra to construct infinite families of isostatic frameworks for *every* symmetry group in 3-space [11]. Note that in Table 25.2.2, $i$ denotes inversion in the origin, and $S_n$ denotes an "improper rotation," i.e., a rotation by $2\pi/n$, followed by a reflection in a mirror perpendicular to the rotational axis.

Analogous symmetry-adapted counting rules have also been established for various other types of geometric constraint systems, such as body-bar and body-hinge frameworks [13, 17, 53], point-line frameworks [36], periodic frameworks [15] and frameworks in non-Euclidean normed spaces [24, 25] (see also Chapter 24).

### 25.2.3 Characterizations of Symmetric Isostatic Graphs

Let $G$ be a $\Gamma$-symmetric graph (with respect to $\theta : \Gamma \to \text{Aut}(G)$) and let $\tau : \Gamma \to O(\mathbb{R}^d)$ be a homomorphism. Then all $\Gamma$-generic realizations of $G$ (i.e., almost all $\Gamma$-symmetric realizations of $G$) share the same infinitesimal rigidity properties [48, 49]. Thus, $\Gamma$-generic rigidity is a purely combinatorial property.

In the following we provide combinatorial characterizations of $\Gamma$-generic isostatic frameworks for the groups $\mathcal{C}_2$, $\mathcal{C}_3$, and $\mathcal{C}_s$ in the plane, starting with the simplest case, the group $\mathcal{C}_3$ [51]. The analogous conjectures for the dihedral groups remain open [50]. Due to the well-known difficulties for the non-symmetric situtation (see [61] for example), analogous results for symmetric bar-joint frameworks in dimension 3 and higher have also not yet been established.

In the following, a graph $G$ is said to satisfy the *Laman conditions* if $|E| = 2|V| - 3$, and for every subgraph $(V', E)$ with $|V'| \leq 2$ vertices, we have $|E'| \leq 2|V'| - 3$.

**Theorem 25.5** **($\mathcal{C}_3$-Generic Isostatic Graphs)** *Let $G$ be a $\mathbb{Z}_3$-symmetric graph (with respect to $\theta : \mathbb{Z}_3 \to Aut(G)$) with at least three vertices. Further, let $\mathbb{Z}_3 = \langle \gamma \rangle$ and $\tau : \mathbb{Z}_3 \to O(\mathbb{R}^2)$ be a homomorphism so that $\tau(\mathbb{Z}_3) = \mathcal{C}_3$. Then the following are equivalent:*

*(a)* *there exists a $\mathcal{C}_3$-symmetric framework $(G, p)$, and $G$ is a $\mathbb{Z}_3$-symmetric isostatic graph (with respect to $\theta$ and $\tau$);*

*(b)* *$G$ satisfies the Laman conditions and $|V_{\mathcal{C}_3}| = 0$;*

*(c)* *$G$ has a proper 3Tree2 partition into three trees $T_0, T_1, T_2$ so that $\theta(\gamma)T_i = T_{i+1}$ for $i = 0, 1, 2$, where the indices are added modulo 3 (see Figure 25.3 (a)).*

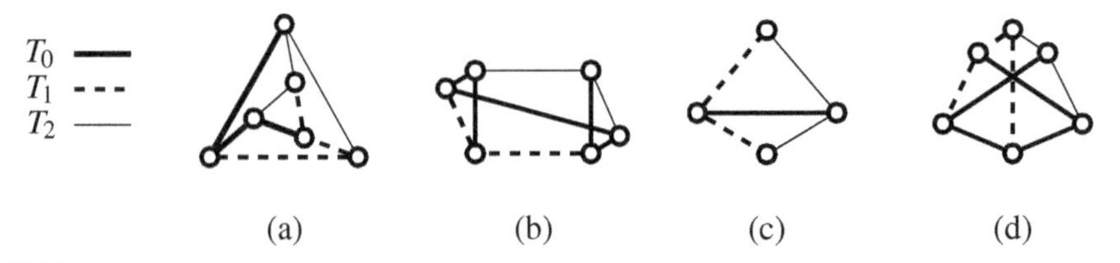


**Figure 25.3**
Proper 3Tree2 partitions (see Chapter 18, Theorem 18.25) satisfying the conditions in part (c) of Theorems 25.5, 25.6, and 25.7.

For the half-turn symmetry group $\mathcal{C}_2$ in the plane, we have the following result.

**Theorem 25.6** **($\mathcal{C}_2$-Generic Isostatic Graphs)** *Let $G$ be a $\mathbb{Z}_2$-symmetric graph (with respect to $\theta : \mathbb{Z}_2 \to Aut(G)$) with at least two vertices. Further, let $\mathbb{Z}_2 = \langle \gamma \rangle$ and $\tau : \mathbb{Z}_2 \to O(\mathbb{R}^2)$ be a homomorphism so that $\tau(\mathbb{Z}_2) = \mathcal{C}_2$. Then the following are equivalent:*

(a) *there exists a $\mathcal{C}_2$-symmetric framework $(G,p)$, and $G$ is a $\mathbb{Z}_2$-symmetric isostatic graph (with respect to $\theta$ and $\tau$);*

(b) *$G$ satisfies the Laman conditions as well as $|V_{\mathcal{C}_2}| = 0$ and $|E_{\mathcal{C}_2}| = 1$;*

(c) *$G$ has a proper 3Tree2 partition into three trees $T_0, T_1, T_2$, where $\theta(\gamma)T_1 = T_2$, and $T_0$ is a spanning tree with $\theta(\gamma)T_0 = T_0$ (see Figure 25.3 (b)).*

Finally, we provide the analogous result for the group $\mathcal{C}_s$ describing mirror symmetry in the plane.

**Theorem 25.7 ($\mathcal{C}_s$-Generic Isostatic Graphs)** *Let $G$ be a $\mathbb{Z}_2$-symmetric graph (with respect to $\theta : \mathbb{Z}_2 \to Aut(G)$) with at least two vertices. Further, let $\mathbb{Z}_2 = \langle\gamma\rangle$ and $\tau : \mathbb{Z}_2 \to O(\mathbb{R}^2)$ be a homomorphism so that $\tau(\mathbb{Z}_2) = \mathcal{C}_s$. Then the following are equivalent:*

(a) *there exists a $\mathcal{C}_s$-symmetric framework $(G,p)$, and $G$ is a $\mathbb{Z}_2$-symmetric isostatic graph (with respect to $\theta$ and $\tau$);*

(b) *$G$ satisfies the Laman conditions and $|E_s| = 1$;*

(c) *$G$ has a proper 3Tree2 partition into three trees $T_0, T_1, T_2$ so that either $\theta(\gamma)T_1 = T_2$, and $\theta(\gamma)T_0 = T_0$ (see Figure 25.3 (c)), or there exists an edge $e = \{i,j\}$ in $T_1$ whose end-vertices $i$ and $j$ are both fixed by $\theta(\gamma)$, and $\theta(\gamma)(T_1 - i) = T_2$ and $\theta(\gamma)T_0 = T_0$ (see Figure 25.3 (d)).*

For the symmetry groups $\tau(\Gamma) = \mathcal{C}_2, \mathcal{C}_3, \mathcal{C}_s$ in the plane, there are also characterizations of $\Gamma$-symmetric isostatic graphs in terms of symmetric Henneberg-type construction sequences [50, 51, 55] (see also Chapter 19).

For example, for the group $\mathcal{C}_3$, this construction sequence starts with the complete graph $K_3$ on three vertices and consists of three graph operations, namely a symmetric Henneberg 1- and a symmetric Henneberg 2-move (which consist of three standard Henneberg 1- and 2-moves, respectively, carried out simultaneously in a symmetric fashion) and a symmetric "'$K_3$-addition move," which symmetrically joins each vertex of a new $K_3$ to a vertex of the previous graph, as illustrated in Figure 25.2 (b) (see [51] for details).

For the group $\mathcal{C}_2$, appropriate symmetric versions of the Henneberg 1- and 2-moves suffice to characterize $\mathcal{C}_2$-symmetric isostatic graphs [50]. For the group $\mathcal{C}_s$, however, we also need a symmetric version of the $X$-replacement [50, 61].

For a number of conjectures, as well as some initial results, regarding sufficient conditons for $\Gamma$-generic *body-bar frameworks* in 3-space to be isostatic, we refer the reader to [17].

## 25.3 Forced-Symmetric Frameworks

### 25.3.1 Glossary

**Free action on the vertex set of a graph:** For a $\Gamma$-symmetric graph $G$, a group action $\theta : \Gamma \to \text{Aut}(G)$ with the property that $\theta(\gamma)(i) \neq i$ for all $i \in V$ and all non-trivial $\gamma \in \Gamma$. Similarly, $\theta$ is **free on the edge set of** $G$ if $\theta(\gamma)(e) \neq e$ for all $e \in E$ and all non-trivial $\gamma \in \Gamma$.

**Quotient graph:** For a $\Gamma$-symmetric graph $G$, the multigraph $G/\Gamma$ with vertex set $V/\Gamma = \{\Gamma i \mid i \in V\}$ and edge set $E/\Gamma = \{\Gamma e \mid e \in E\}$, where $\Gamma i$ and $\Gamma e$ are the vertex and edge orbits of $i$ and $e = \{i,j\}$ defined by $\{\theta(\gamma)(i) \mid \gamma \in \Gamma\}$ and $\{\{\theta(\gamma)(i), \theta(\gamma)(j)\} \mid \gamma \in \Gamma\}$, respectively.

**Quotient $\Gamma$-gain graph:** Let $G$ be a $\Gamma$-symmetric graph with respect to the group action $\theta : \Gamma \to \text{Aut}(G)$ which is free on the vertex set of $G$. The quotient $\Gamma$-gain graph of $G$ is obtained from the quotient graph $G/\Gamma$ of $G$ by orienting the edges of $G/\Gamma$ and labelling the edges of $G/\Gamma$ via the function $\psi : E/\Gamma \to \Gamma$ as follows. Each edge orbit $\Gamma e$ connecting $\Gamma i$ and $\Gamma j$ in $G/\Gamma$ can be written as $\{\{\theta(\gamma)(i), \theta(\gamma) \circ \theta(\alpha)(j)\} \mid \gamma \in \Gamma\}$ for a unique $\alpha \in \Gamma$. For each $\Gamma e$, orient $\Gamma e$ from $\Gamma i$ to $\Gamma j$ in $G/\Gamma$ and assign to it the group element $\alpha$. Note that the resulting quotient $\Gamma$-gain graph $(G_0, \psi)$ of $G$ is unique up to choices of representative vertices and that the orientation is only used as a reference orientation and may be changed, provided that we also modify $\psi$ so that if $\Gamma e$ is an edge in one direction, and $\Gamma e^{-1}$ is the same edge in the opposite direction, then $\psi(\Gamma e^{-1}) = \psi(\Gamma e)^{-1}$.

In the following, we will denote the vertex and edge set of $(G_0, \psi)$ by $V_0$ and $E_0$, respectively, and use the tilde symbol to denote the vertices and edges of $(G_0, \psi)$. We also refer to $G$ as the **covering graph** of $(G_0, \psi)$.

**Gain of a closed walk:** For a quotient $\Gamma$-gain graph $(G_0, \psi)$ and a closed walk

$$W = \tilde{i}_1, \tilde{e}_1, \tilde{i}_2, \tilde{e}_2, \tilde{i}_3, \ldots, \tilde{i}_k, \tilde{e}_k, \tilde{i}_1$$

of $(G_0, \psi)$, the group element $\psi(W) = \Pi_{t=1}^{k} \psi(\tilde{e}_t)^{\text{sign}(\tilde{e}_t)}$, where $\text{sign}(\tilde{e}_t) = 1$ if $\tilde{e}_t$ is directed from $\tilde{i}_t$ to $\tilde{i}_{t+1}$, and $\text{sign}(\tilde{e}_t) = -1$ otherwise.

**Subgroup induced by edge set:** For a quotient $\Gamma$-gain graph $(G_0, \psi)$, a subset $F \subseteq E_0$, and a vertex $\tilde{i}$ of the vertex set $V(F) \subseteq V_0$ induced by $F$, the subgroup $\langle F \rangle_{\psi,\tilde{i}} = \{\psi(W) | W \in \mathcal{W}(F, \tilde{i})\}$ of $\Gamma$, where $\mathcal{W}(F, \tilde{i})$ is the set of closed walks starting at $\tilde{i}$ using only edges of $F$.

**Balanced edge set:** For a quotient $\Gamma$-gain graph $(G_0, \psi)$, a connected edge subset $F$ of $E_0$ is called balanced if $\langle F \rangle_{\psi,\tilde{i}} = \{\text{id}\}$ for some $\tilde{i} \in V(F)$ (or equivalently, $\langle F \rangle_{\psi,\tilde{i}} = \{\text{id}\}$ for all $\tilde{i} \in V(F)$). A disconnected subset of $E_0$ is balanced if all of its connected components are balanced. A subset of $E_0$ is called **unbalanced** if it is not balanced (i.e., if it contains a cycle that is not balanced).

$(k, \ell, m)$**-gain-sparse:** For non-negative integers $k, \ell, m$ with $m \leq \ell$, a quotient $\Gamma$-gain graph $(G_0, \psi)$ satisfying

$$|F| \leq \begin{cases} k|V(F)| - \ell, & \text{for all non-empty balanced } F \subseteq E_0, \\ k|V(F)| - m, & \text{for all non-empty } F \subseteq E_0. \end{cases}$$

If we also have $|E_0| = k|V_0| - m$, then $(G_0, \psi)$ is called $(k, \ell, m)$**-gain-tight**.

**Cyclic edge set:** For a quotient $\Gamma$-gain graph $(G_0, \psi)$, a connected edge subset $F$ of $E_0$ is called cyclic if $\langle F \rangle_{\psi,\tilde{i}}$ is a cyclic subgroup of $\Gamma$ for some $\tilde{i} \in V(F)$ (or equivalently, for all $\tilde{i} \in V(F)$). A disconnected subset of $E_0$ is cyclic if all of its connected components are cyclic.

**Orbit rigidity matrix:** For a $d$-dimensional $\Gamma$-symmetric framework $(G, p)$ (with respect to $\theta$ and $\tau$), where $\theta : \Gamma \to \text{Aut}(G)$ is free on the vertex set of $G$, and its quotient $\Gamma$-gain graph $(G_0, \psi)$, the $|E_0| \times d|V_0|$ matrix $O(G_0, \psi, p)$ defined as follows. Choose a representative vertex $\tilde{i}$ for each vertex $\Gamma i$ in $V_0$. The row corresponding to the edge $\tilde{e} = (\tilde{i}, \tilde{j})$, $\tilde{i} \neq \tilde{j}$, with gain $\psi(\tilde{e})$ has the form

$$\begin{pmatrix} 0 \ldots 0 & \overbrace{p(\tilde{i}) - \tau(\psi(\tilde{e}))p(\tilde{j})}^{\tilde{i}} & 0 \ldots 0 & \overbrace{p(\tilde{j}) - \tau(\psi(\tilde{e}))^{-1}p(\tilde{i})}^{\tilde{j}} & 0 \ldots 0 \end{pmatrix}.$$

If $\tilde{e} = (\tilde{i}, \tilde{i})$ is a loop at $\tilde{i}$, then the row corresponding to $\tilde{e}$ has the form

$$\begin{pmatrix} 0 \ldots 0 & \overbrace{2p(\tilde{i}) - \tau(\psi(\tilde{e}))p(\tilde{i}) - \tau(\psi(\tilde{e}))^{-1}p(\tilde{i})}^{\tilde{i}} & 0 \ldots 0 & 0 & 0 \ldots 0 \end{pmatrix}.$$

**Fully Γ-symmetric infinitesimal motion:** For a Γ-symmetric framework $(G,p)$, an infinitesimal motion $u : V \to \mathbb{R}^d$ of $(G,p)$ satisfying

$$\tau(\gamma)u_i = u_{\theta(\gamma)(i)} \qquad \text{for all } \gamma \in \Gamma \text{ and all } i \in V.$$

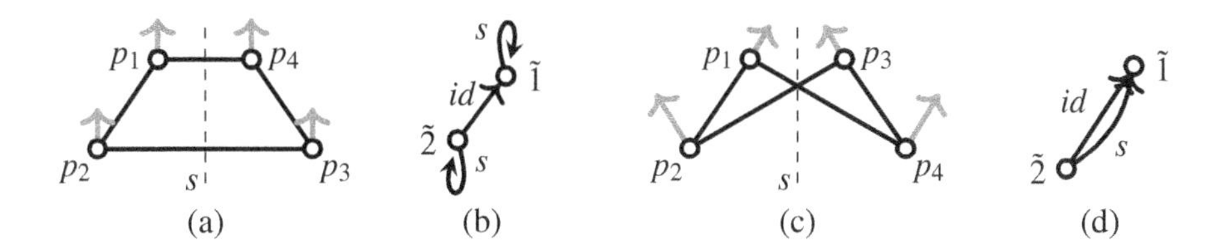


**Figure 25.4**
(a,b) A fully $\mathcal{C}_s$-symmetric infinitesimally rigid framework and its corresponding quotient $\mathcal{C}_s$-gain graph. The infinitesimal translation in (a) spans the space of fully $\mathcal{C}_s$-symmetric trivial infinitesimal motions. (c,d) A $\mathcal{C}_s$-symmetric framework with a fully $\mathcal{C}_s$-symmetric infinitesimal flex and its corresponding quotient $\mathcal{C}_s$-gain graph.

**Fully Γ-symmetric self-stress:** For a Γ-symmetric framework $(G,p)$, a self-stress $\omega : E \to \mathbb{R}$ with the property that $\omega(e) = \omega(f)$ whenever $e$ and $f$ belong to the same edge orbit $\Gamma e$ of $G$.

**Fully Γ-symmetric infinitesimally rigid framework:** A Γ-symmetric framework for which every fully Γ-symmetric infinitesimal motion is trivial. (See also Figure 25.4.)

**$\rho_0$-generic framework:** A Γ-symmetric framework $(G,p)$ (with respect to $\theta$ and $\tau$) whose orbit rigidity matrix has maximal rank among all Γ-symmetric realizations of $G$ (with respect to $\theta$ and $\tau$).

### 25.3.2 Symmetric Motions and the Orbit Rigidity Matrix

In this section we restrict attention to those motions of a symmetric framework which preserve all of the symmetries of the framework throughout the path. Our first result is a symmetric version of the Theorem of Asimov-Roth [2], which says that for almost all Γ-symmetric realizations of a graph $G$, the existence of a fully Γ-symmetric infinitesimal flex also guarantees the existence of a *continuous* symmetry-preserving flex [14, 52]. This result provides one of the key motivations for the study of forced-symmetric rigidity.

**Theorem 25.8 (Symmetry-Preserving Motions)** *A Γ-symmetric $\rho_0$-generic framework $(G,p)$ has a fully Γ-symmetric infinitesimal flex if and only if $(G,p)$ has a continuous flex which preserves the symmetry of $(G,p)$ throughout the path.*

Note that the fully Γ-symmetric rigidity properties of a Γ-symmetric framework are described by the submatrix block $\widetilde{R}_0(G,p)$ of the block-diagonalized rigidity matrix $\widetilde{R}(G,p)$ (recall Theorem 25.3), which corresponds to the trivial irreducible representation $\rho_0$ of Γ (i.e., the 1-dimensional representation which assigns 1 to each element of Γ). However, to obtain the explicit entries of this matrix, we need to go through the laborious process of block-diagonalizing the rigidity matrix. To simplify the rigidity analysis of forced-symmetric frameworks, the orbit rigidity matrix was introduced in [56]. This matrix is equivalent to the matrix $\widetilde{R}_0(G,p)$ (see Theorem 25.9), but its entries have the transparent form described in Section 25.3.1.

**Theorem 25.9 (The Orbit Rigidity Matrix)** *Let $(G,p)$ be a $\Gamma$-symmetric framework with respect to $\theta:\Gamma\to Aut(G)$ (which is free on the vertex set of $G$) and $\tau:\Gamma\to O(\mathbb{R}^d)$. The kernel of the orbit rigidity matrix $O(G_0,\psi,p)$ is isomorphic to the space of fully $\Gamma$-symmetric infinitesimal motions of $(G,p)$, and the kernel of $O(G_0,\psi,p)^T$ is isomorphic to the space of fully $\Gamma$-symmetric self-stresses of $(G,p)$.*

Let $\mathrm{triv}_{\tau(\Gamma)}$ denote the dimension of the space of fully $\Gamma$-symmetric trivial infinitesimal motions of $(G,p)$. Then $(G,p)$ is clearly fully $\Gamma$-symmetric infinitesimally rigid if and only if rank $O(G_0,\psi,p)=d|V_0|-\mathrm{triv}_{\tau(\Gamma)}$, provided that $G$ has at least $d$ vertices. This leads to the following necessary conditions for a framework to be fully $\Gamma$-symmetric rigid [20, 56].

**Theorem 25.10 (Forced-Symmetric Maxwell Counts)** *Let $G$ be a graph with at least $d$ vertices and let $(G,p)$ be a fully $\Gamma$-symmetric infinitesimally rigid framework with respect to the action $\theta:\Gamma\to Aut(G)$ which is free on the vertex set of $G$ and $\tau:\Gamma\to O(\mathbb{R}^d)$. Then the quotient $\Gamma$-gain graph of $G$ contains a spanning subgraph $(H,\psi)$ with edge set $E_0'$ which satisfies*

*(a)* $|E_0'|=d|V_0|-triv_{\tau(\Gamma)}$

*(b)* $|F|\leq d|V(F)|-triv_{\tau(\langle F\rangle_{\psi,\tilde{i}})}$ *for all* $F\subseteq E_0'$ *and all* $\tilde{i}\in V(F)$,

*where $triv_{\tau(\langle F\rangle_{\psi,\tilde{i}})}$ is the dimension of the space of fully $(\langle F\rangle_{\psi,\tilde{i}})$-symmetric trivial infinitesimal motions of the framework induced by the edges in $F$.*

Note that $\mathrm{triv}_{\tau(\Gamma)}$ can easily be computed for any symmetry group in any dimension. For all symmetry groups in the plane and in 3-space, these numbers are also recorded in the corresponding character tables for these groups [1, 3].

For example, for mirror symmetry in the plane, we have $\mathrm{triv}_{\mathcal{C}_s}=1$, as an infinitesimal translation along the mirror line forms a basis of the space of fully $\mathcal{C}_s$-symmetric trivial infinitesimal motions (see also Figure 1.4 (a)). Similarly, for half-turn symmetry in the plane, we have $\mathrm{triv}_{\mathcal{C}_2}=1$, as an infinitesimal rotation about the origin spans the space of fully $\mathcal{C}_2$-symmetric trivial infinitesimal motions.

Therefore, using Theorems 25.8 and 25.10, we may detect hidden continuous (symmetry-preserving) flexibility in frameworks by simply counting the number of vertex and edge orbits in the underlying graph.

A simple example in 3-space is the $\rho_0$-generic realization of the octahedral graph with half-turn symmetry $\mathcal{C}_2$ shown in Figure 25.5(a). Generic realizations of this graph without symmetry are isostatic in 3-space. For the numbers of vertex and edge orbits, we have $|V_0|=3$ and $|E_0|=6$. Moreover, we have $\mathrm{triv}_{\mathcal{C}_2}=2$ [1, 3, 56]. Thus, $|E_0|<3|V_0|-\mathrm{triv}_{\mathcal{C}_2}$, and we may conclude that there exists a symmetry-preserving continuous flex.

The definition of the orbit rigidity matrix has also been extended to $\Gamma$-symmetric frameworks where $\Gamma$ does not act freely on the vertex set [56]. This leads to adjusted necessary conditions for fully $\Gamma$-symmetric infinitesimal rigidity. While the overall counts on the numbers of vertex and edge orbits can easily be obtained (as demonstrated by the example below), the corresponding gain-sparsity counts become signifcantly less clear and transparent, and they have not yet been studied in detail.

Consider the $\rho_0$-generic realization of the octahedral graph with reflectional symmetry $\mathcal{C}_s$ shown in Figure 25.5 (b). For the numbers of vertex and edge orbits, we have $|V_0|=4$ and $|E_0|=6$. Moreover, we have $\mathrm{triv}_{\mathcal{C}_s}=3$ [1, 3, 56]. Since there are two vertices which are fixed by the reflection, and each of these vertices contributes only two columns to the orbit rigidity matrix (as each of them needs to remain on the mirror plane), the framework has $2\cdot 3+2\cdot 2-3=7$ fully $\mathbb{Z}_2$-symmetric non-trivial degrees of freedom. Since $|E_0|<7$, it follows that there exists a symmetry-preserving continuous flex.

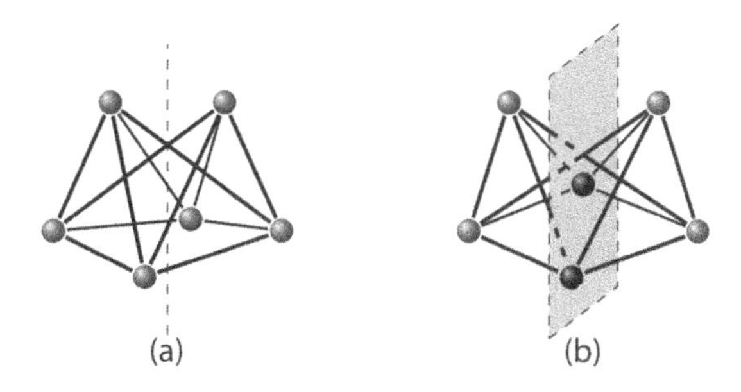


**Figure 25.5**
Flexible octahedra (also known as "Bricard octahedra") with half-turn symmetry (a) and with mirror symmetry (b).

### 25.3.3 Characterizations of Forced-Symmetric Rigid Graphs

All $\rho_0$-generic realizations of a graph share the same fully $\Gamma$-symmetric rigidity properties, and for a number of symmetry groups in the plane, Laman-type theorems for fully $\Gamma$-symmetric rigidity have been established. For the groups $\mathcal{C}_s$ and $\mathcal{C}_n$, we have the following result (see [20, 27, 28, 33, 31] for various proofs).

**Theorem 25.11 (Reflectional or Rotational Symmetry in 2D)** *Let $(G,p)$ be a $\mathbb{Z}_{n\geq 2}$-symmetric $\rho_0$-generic framework with respect to the action $\theta : \mathbb{Z}_n \to Aut(G)$ which is free on the vertex set of $G$ and $\tau : \mathbb{Z}_n \to O(\mathbb{R}^2)$. Then $(G,p)$ is forced $\mathbb{Z}_n$-symmetric infinitesimally rigid if and only if the quotient $\mathbb{Z}_n$-gain graph of $G$ contains a spanning subgraph $(H,\psi)$ which is $(2,3,1)$-gain-tight.*

The count $2|V_0|-1$ for the number of edges of $(H,\psi)$ is due to the fact that $\text{triv}_{\tau(\mathbb{Z}_n)} = 1$ for all cyclic groups $\mathbb{Z}_n$, $n \geq 2$. For a dihedral group $\mathcal{C}_{nv}$, we always have $\text{triv}_{\mathcal{C}_{nv}} = 0$. This leads to the following result [20].

**Theorem 25.12 (Dihedral Symmetry in 2D)** *Let $D_{2n}$ denote the dihedral group of order $2n$, where $n \geq 3$ is an odd integer. Further, let $(G,p)$ be a $D_{2n}$-symmetric $\rho_0$-generic framework with respect to the action $\theta : D_{2n} \to Aut(G)$ which is free on the vertex set of $G$, and $\tau : D_{2n} \to O(\mathbb{R}^2)$, where $\tau(D_{2n}) = \mathcal{C}_{nv}$. Then $G(p)$ is forced $D_{2n}$-symmetric infinitesimally rigid if and only if the quotient $D_{2n}$-gain graph of $G$ contains a spanning subgraph $(H,\psi)$ with edge set $E_0'$ which satisfies*

*(a)* $|E_0'| = 2|V_0|$

*(b)* $|F| \leq \begin{cases} 2|V(F)|-3 & \textit{for all non-empty balanced } F \subseteq E_0', \\ 2|V(F)|-1 & \textit{for all non-empty unbalanced and cyclic } F \subseteq E_0', \\ 2|V(F)| & \textit{for all non-empty } F \subseteq E_0'. \end{cases}$

Note that the gain-sparsity counts in Theorems 25.11 and 25.12 can be checked in polynomial time [5, 20]. In general, if we want to check a quotient $\Gamma$-gain graph for $(k,\ell,m)$-gain-sparsity, where $0 \leq \ell \leq 2k-1$, then we may proceed as follows. We first verify that the quotient $\Gamma$-gain gaph is $(k,m)$-sparse using a standard "pebble game algorithm" [6, 26]. We then test whether every edge set violating the $(k,\ell)$-sparsity count induces a certain subgroup of $\Gamma$. It suffices to test this for every circuit in the matroid induced by the $(k,\ell)$-sparsity count, and these circuits can be enumerated in polynomial time [58]. As we will see in the following sections, this algorithm may also be used to check for infinitesimal rigidity of incidentally symmetric frameworks and for rigidity of periodic frameworks.

For the groups $\mathcal{C}_s$, $\mathcal{C}_n$, and $\mathcal{C}_{nv}$ ($n$ odd), there also exist alternative characterizations of $\rho_0$-generic fully $\Gamma$-symmetric infinitesimally rigid frameworks in terms of symmetric Henneberg-type construction sequences for the corresponding quotient $\Gamma$-gain graphs [20].

The only remaining groups are the dihedral groups of "even order," that is, the groups $D_{2n}$, where $n$ is an even integer. For these groups, the counts in Theorem 25.12 are in general not sufficient for a $\rho_0$-generic framework to be fully $D_{2n}$-symmetric infinitesimally rigid. Two counterexamples are depicted in Figure 25.6. See [20] for further examples, as well as for some results on recursive constructions which preserve fully $D_{2n}$-symmetric infinitesimal rigidity.

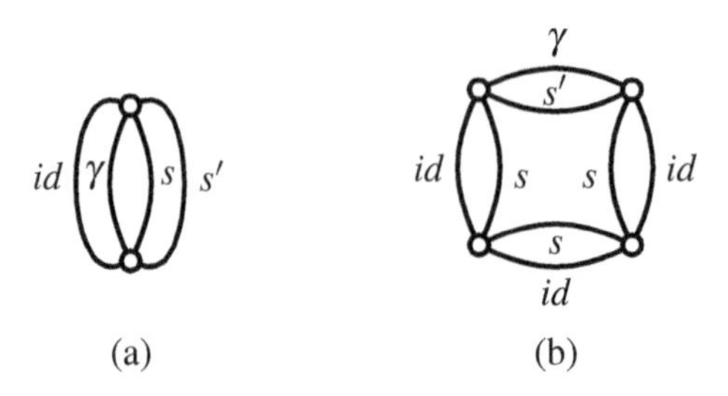


**Figure 25.6**
Quotient $D_4$-gain graphs (with the directions of the edges omitted), where $D_4$ consists of the identity id, the half-turn $\gamma$, and the reflections $s$ and $s'$: (a) a quotient $D_4$-gain graph whose covering graph, the complete bipartite graph $K_{4,4}$, is flexible for $D_4$-generic realizations in the plane (this is also known as Bottema's mechanism [10]); (b) another quotient $D_4$-gain graph whose covering graph is flexible for $D_4$-generic realizations in the plane.

For *body-bar frameworks*, there are combinatorial characterizations for fully $\Gamma$-symmetric infinitesimal rigidity for all symmetry groups in all dimensions [59]. These counts are analogous to the counts in Theorem 25.10, and are conjectured to also extend to body-hinge and molecular frameworks [38].

In [35] combinatorial characterizations for fully $\Gamma$-symmetric infinitesimal rigidity have also been obtained for Euclidean frameworks in 3-space whose vertices are constrained to lie on various surfaces (such as a cone or a cylinder).

Finally, we note that the fully $\Gamma$-symmetric rigidity properties of a $\Gamma$-symmetric framework (including continous symmetry-preserving flexibility) can also be transferred to other metrics, such as the spherical or hyperbolic metric, using the orbit rigidity matrix and a symmetrized version of the technique of "coning" [57].

## 25.4 Incidentally Symmetric Infinitesimally Rigid Frameworks

### 25.4.1 Glossary

**Phase-symmetric infinitesimal motion:** Let $\Gamma$ be the group $\mathbb{Z}_k = \{0,1,\ldots,k-1\}$ and for $t = 0,1,\ldots,k-1$, let $\rho_t : \Gamma \to \mathbb{C}\setminus\{0\}$ be the irreducible representation of $\Gamma$ defined by $\rho_t(j) = \omega^{tj}$, where $\omega$ denotes the root of unity $e^{\frac{2\pi i}{k}}$. For a $\Gamma$-symmetric framework $(G,p)$, an infinitesimal motion $u : V \to \mathbb{R}^d$ of $(G,p)$ is called **$\rho_t$-symmetric** if it lies in the $(\tau \otimes P_V)$-invariant subspace

$V_t$ corresponding to $\rho_t$, that is, if it satisfies

$$\tau(\gamma)u_i = \omega^{t\gamma}u_{\theta(\gamma)(i)} \qquad \text{for all } \gamma \in \Gamma \text{ and all } i \in V.$$

(See also Figure 25.7.) Similarly, a self-stress of $(G,p)$ is called $\rho_t$-symmetric if it lies in the $P_E$-invariant subspace $W_t$ corresponding to $\rho_t$.

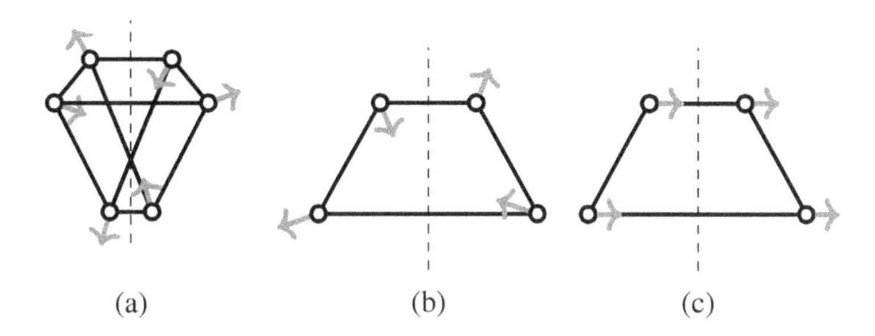


**Figure 25.7**
$\rho_1$-symmetric (or anti-symmetric) infinitesimal motions of frameworks with mirror symmetry in the plane: (a), (b) anti-symmetric infinitesimal flexes; (c) an anti-symmetric trivial infinitesimal motion.

**Phase-symmetric infinitesimally rigid framework:** For the group $\Gamma = \mathbb{Z}_k$, a $\Gamma$-symmetric framework $(G,p)$ is $\rho_t$-symmetric infinitesimally rigid if every $\rho_t$-symmetric infinitesimal motion of $(G,p)$ is trivial.

**Phase-symmetric orbit rigidity matrix:** Let $(G,p)$ be a $\Gamma$-symmetric framework with respect to $\theta : \Gamma \to \mathrm{Aut}(G)$ and $\tau : \Gamma \to O(\mathbb{R}^d)$, where $\Gamma$ is the group $\mathbb{Z}_k$ and $\theta$ is free on the vertex set of $G$. Further, let $(G_0, \psi)$ be the quotient $\Gamma$-gain graph of $G$. The $\rho_t$-symmetric orbit rigidity matrix is the $|E_0| \times d|V_0|$ matrix $O_t(G_0, \psi, p)$ defined as follows. Choose a representative vertex $\tilde{i}$ for each vertex $\Gamma i$ in $V_0$. The row corresponding to the edge $\tilde{e} = (\tilde{i}, \tilde{j})$, $\tilde{i} \neq \tilde{j}$, with gain $\psi(\tilde{e})$ has the form

$$\big(0\ldots0 \quad \overbrace{p(\tilde{i}) - \tau(\psi(\tilde{e}))p(\tilde{j})}^{\tilde{i}} \quad 0\ldots0 \quad \overbrace{\omega^{t\psi(\tilde{e})}(p(\tilde{j}) - \tau(\psi(\tilde{e}))^{-1}p(\tilde{i}))}^{\tilde{j}} \quad 0\ldots0\big).$$

If $\tilde{e} = (\tilde{i}, \tilde{i})$ is a loop at $\tilde{i}$, then the row corresponding to $\tilde{e}$ has the form

$$\big(0\ldots0 \quad \overbrace{p(\tilde{i}) - \tau(\psi(\tilde{e}))p(\tilde{i}) + \omega^{t\psi(\tilde{e})}(p(\tilde{i}) - \tau(\psi(\tilde{e}))^{-1}p(\tilde{i}))}^{\tilde{i}} \quad 0\ldots0 \quad 0 \quad 0\ldots0\big).$$

**$\rho_t$-generic framework:** A $\Gamma$-symmetric framework $(G,p)$ (with respect to $\theta$ and $\tau$) whose $\rho_t$-symmetric orbit rigidity matrix has maximal rank among all $\Gamma$-symmetric realizations of $G$ (with respect to $\theta$ and $\tau$).

**Unbalanced circuit:** A minimal edge subet $F$ of a quotient $\mathbb{Z}_k$-gain graph $(G_0, \psi)$ so that

(a) $F$ is unbalanced;

(b) $|F| > 2|V(F)| - 1$;

(c) there is a vertex $\tilde{i} \in V(F)$, an element $\gamma \in \mathbb{Z}_k$, and a labeling function $\psi' : E_0 \to \mathbb{Z}_k$ equivalent to $\psi$ such that $\psi'(\tilde{e}) = id$ for every $\tilde{e} \in F$ not incident to $\tilde{i}$, and $\psi'(\tilde{e}) \in \{id, \gamma\}$ for every $\tilde{e} \in F$ directed to $\tilde{i}$ (assuming that every edge incident to $\tilde{i}$ is directed to $\tilde{i}$). See also Figure 25.8.

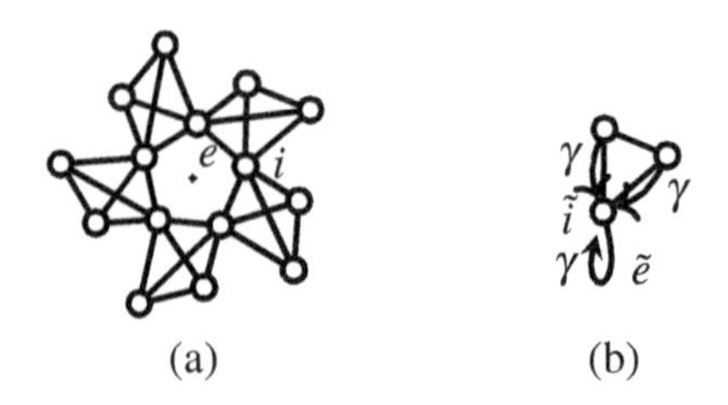


**Figure 25.8**
A $\mathbb{Z}_5$-symmetric graph (a) and its corresponding quotient $\mathbb{Z}_5$-gain graph whose edge set is an unbalanced circuit (b). The orientation and gain labeling is omitted for all edges with gain id, and $\gamma$ denotes rotation by $2\pi/5$.

### 25.4.2 Phase-Symmetric Orbit Rigidity Matrices

As we have seen in Section 25.2.2, an isostatic $\Gamma$-symmetric framework $(G, p)$ must obey certain restrictions on the numbers of vertices and edges that are fixed by the various elements of $\Gamma$. Consequently, an infinitesimally rigid symmetric framework usually does not contain a spanning isostatic subframework with the same symmetry (see also Figure 25.9). Thus, to decide whether a given $\Gamma$-generic framework is infinitesimally rigid is in general not possible using the combinatorial results of Section 25.2.3. However, in those cases further insights may be gained by analyzing the $\rho_t$-symmetric infinitesmal rigidity properties of $(G, p)$ for each irreducible representation $\rho_t$ of $\Gamma$. The basic idea is that if $(G, p)$ is $\rho_t$-symmetric infinitesimally rigid *for every* $t$, then the block-diagonalization of the rigidity matrix $\widetilde{R}(G, p)$ guarantees that $(G, p)$ is infinitesimally rigid. The $\rho_t$-symmetric infinitesimal rigidity properties of $(G, p)$ are described by the block matrix $\widetilde{R}_t(G, p)$ of $\widetilde{R}(G, p)$. However, the explicit entries of $\widetilde{R}_t(G, p)$ are quite arduous to obtain and difficult to work with. Therefore, analogous to the forced-symmmetric situation (which is associated with the trivial irreducible representation of $\Gamma$), an equivalent but more transparent "orbit rigidity matrix" has been established in [55] for each block matrix $\widetilde{R}_t(G, p)$. While there seem to be no major obstacles in extending these methods to irreducible representations of arbitrary finite groups, abelian groups are particularly simple since their irreducible representations are all 1-dimensional (over the complex numbers). Here we focus on the simplest case where $\Gamma$ is the cyclic group $\mathbb{Z}_k$.

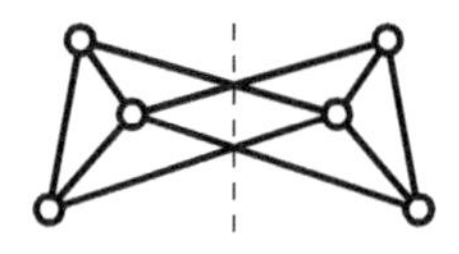

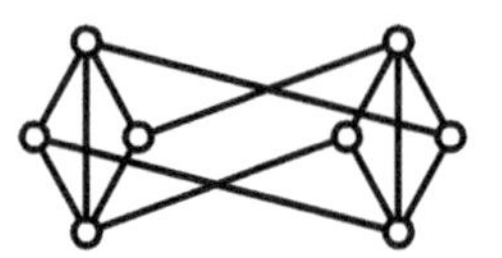

**Figure 25.9**
Infinitesimally rigid symmetric frameworks in $\mathbb{R}^2$ with respective symmetry groups $\mathcal{C}_s$ and $\mathcal{C}_2$ which do not contain a spanning isostatic subframework with the same symmetry.

**Theorem 25.13 (Phase-Symmetric Orbit Rigidity Matrices)** *Let* $\Gamma = \mathbb{Z}_k$ *and let* $(G, p)$ *be a* $\Gamma$-*symmetric framework (with respect to* $\theta : \Gamma \to Aut(G)$ *and* $\tau : \Gamma \to O(\mathbb{R}^d)$*). The kernel of the* $\rho_t$-*symmetric orbit rigidity matrix* $O_t(G_0, \psi, p)$ *is isomorphic to the space of* $\rho_t$-*symmetric infinitesimal motions of* $(G, p)$*, and the kernel of* $O(G_0, \psi, p)^T$ *is isomorphic to the space of* $\rho_t$-*symmetric self-stresses of* $(G, p)$.

Analogous to Theorem 25.10, there exist some simple necessary conditions for phase-symmetric infinitesimal rigidity for all symmetry groups in all dimensions. These conditions are obtained by replacing each number $\text{triv}_{\tau(\langle F\rangle_{\psi,\tilde{i}})}$ in the counts of Theorem 25.10 by the dimension of the corresponding space of $\rho_t$-*symmetric* trivial infinitesimal motions. In particular, for the cyclic groups in the plane we have the following result [55].

**Theorem 25.14 (Phase-Symmetric Maxwell Counts)** *Let $(G,p)$ be a $\mathbb{Z}_k$-symmetric framework with respect to the action $\theta : \mathbb{Z}_k \to Aut(G)$ which is free on the vertex set of $G$ and $\tau : \mathbb{Z}_k \to O(\mathbb{R}^2)$. If $(G,p)$ is $\rho_t$-symmetric infinitesimally rigid, then the quotient $\mathbb{Z}_k$-gain graph of $G$ contains a spanning subgraph $(H,\psi)$ so that*

*(a) for $k=2$, $(H,\psi)$ is $(2,3,1)$-gain-tight if $t=0$ and $(2,3,2)$-gain tight if $t=1$;*

*(b) for $k\geq 3$ and $t\in\{0,1,k-1\}$, $(H,\psi)$ is $(2,3,1)$-gain-tight;*

*(c) for $k\geq 4$ and $t\notin\{0,1,k-1\}$, $(H,\psi)$ is $(2,3,0)$-gain-tight.*

Since a $\Gamma$-symmetric framework $(G,p)$ can only be infinitesimally rigid if it is $\rho_t$-symmetric infinitesimally rigid for each irreducible representation $\rho_t$ of $\Gamma$, Theorem 25.14 provides necessary conditions for $(G,p)$ to be infinitesimally rigid. For the groups of order 2 and 3, the conditions in Theorem 25.14 are also sufficient for $\Gamma$-generic infinitesimal rigidity (see Theorems 25.17 and 25.19). However, for $\mathbb{Z}_k$, $k\geq 4$, there are further necessary conditions [55]. It follows from Theorem 25.14 that for $k\geq 4$, a $\mathbb{Z}_k$-generic framework can only be infinitesimally rigid if the corresponding $\mathbb{Z}_k$-quotient gain graph $(G_0,\psi)$ contains a spanning subgraph $(H,\psi)$ with $2|V_0|$ edges, which in turn contains a spanning $(2,3,1)$-gain-tight subgraph. In addition, $(H,\psi)$ cannot contain any unbalanced circuit [19, 55]. (Recall Figure 25.8.) Moreover, it follows from the following theorem that if $(H,\psi)$ contains an edge subset $F$ with $\langle F\rangle_{\psi,\tilde{i}}\simeq\mathbb{Z}_2$, then $F$ must be $(2,3,2)$-gain sparse [19, 55]. In particular, $(H,\psi)$ cannot contain a loop with gain $k/2$.

**Theorem 25.15 (Further Necessary Conditions For Infinitesimal Rigidity)** *Let $k\geq 4$, and let $(G,p)$ be a $\mathbb{Z}_k$-symmetric framework with respect to the action $\theta : \mathbb{Z}_k \to Aut(G)$ which is free on the vertex set of $G$ and $\tau : \mathbb{Z}_k \to O(\mathbb{R}^2)$. Further, let $t$ be an odd integer with $1\leq t\leq k-1$. If $O_t(G_0,\psi,p)$ is row independent, then $F$ is $(2,3,2)$-gain sparse for every $F\subseteq E_0$ such that $\langle F\rangle_{\psi,\tilde{i}}$ is isomorphic to $\mathbb{Z}_2$.*

Using Theorem 25.15, we may construct $\mathbb{Z}_k$-generic infinitesimally flexible frameworks for even $k\geq 6$ whose underlying graphs are generically rigid without symmetry [55]. The easiest example is the complete bipartite graph $K_{3,3}$ with $\mathcal{C}_6$ symmetry shown in Figure 25.10 (b).

It was conjectured in [55] that if $k$ is odd, then the existence of a spanning Laman subgraph (i.e., a spanning subgraph which is generically isostatic without symmetry) in a graph $G$ guarantees that $\mathbb{Z}_k$-generic realizations of $G$ are still infinitesimally rigid. However counterexamples to this conjecture have recently been constructed in [18]. (See Figure 25.10(c).)

### 25.4.3 Characterizations of Symmetric Infinitesimally Rigid Graphs

Using the anti-symmetric orbit rigidity matrix and recursive operations on quotient $\mathbb{Z}_2$-gain graphs, the following result was established in [55].

**Theorem 25.16 (Anti-Symmetric Infinitesimal Rigidity for $\mathcal{C}_s$ and $\mathcal{C}_2$ in 2D)** *Let $(G,p)$ be a $\mathbb{Z}_2$-symmetric $\rho_1$-generic framework with respect to the action $\theta : \mathbb{Z}_2 \to Aut(G)$ which is free on the vertex set of $G$ and $\tau : \mathbb{Z}_2 \to O(\mathbb{R}^2)$. Then $(G,p)$ is $\rho_1$-symmetric infinitesimally rigid if and only if the quotient $\mathbb{Z}_2$-gain graph of $G$ contains a spanning subgraph $(H,\psi)$ which is $(2,3,2)$-gain-tight.*

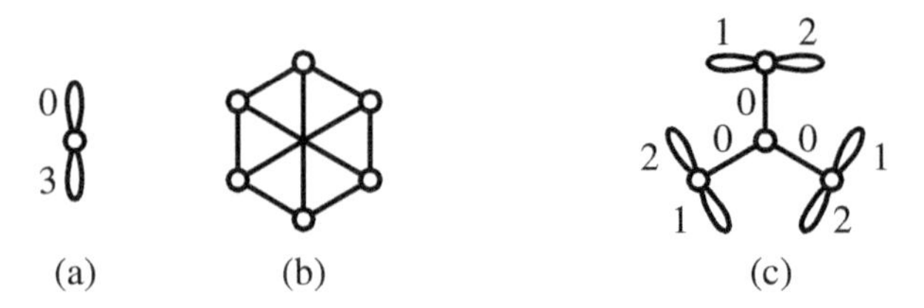


**Figure 25.10**
A quotient $\mathbb{Z}_6$-gain graph (a) and its covering graph $K_{3,3}$ (b), which is isostatic for generic realizations, but infinitesimally flexible for $\mathbb{Z}_6$-symmetric realizations; (c) a quotient $\mathbb{Z}_k$-gain graph whose covering graph is generically rigid without symmetry, but becomes infinitesimally flexible for $\mathbb{Z}_k$-symmetric realizations, where $k \geq 7$. The directions of edges are omitted in the gain graphs.

Note that the count $2|V_0| - 2$ for the number of edges of $(H, \psi)$ is due to the fact that the dimension of the space of $\rho_1$-symmetric trivial infinitesimal motions is equal to 2. As a simple consequence of Theorems 25.11 and 25.16, we obtain the following combinatorial characterizations for infinitesimal rigidity under reflection or half-turn symmetry in the plane [55].

**Theorem 25.17 ($\mathbb{Z}_2$-Generic Infinitesimal Rigidity for $\mathcal{C}_s$ and $\mathcal{C}_2$ in 2D)** *Let $(G, p)$ be a $\mathbb{Z}_2$-generic framework with respect to the action $\theta : \mathbb{Z}_2 \to Aut(G)$ which is free on the vertex set of $G$ and $\tau : \mathbb{Z}_2 \to O(\mathbb{R}^2)$. Then $(G, p)$ is infinitesimally rigid if and only if the quotient $\mathbb{Z}_2$-gain graph of $G$ contains a spanning $(2,3,i)$-gain-tight subgraph $(H_i, \psi_i)$ for each $i = 1, 2$.*

For the symmetry group $\mathcal{C}_3$, $\rho_t$-symmetric infinitesimal rigidity is described by the same $(2,3,1)$-gain sparsity count for each $t = 0, 1, 2$.

**Theorem 25.18 (Phase-Symmetric Infinitesimal Rigidity for $\mathcal{C}_3$ in 2D)** *Let $t \in \{0, 1, 2\}$ and let $(G, p)$ be a $\mathbb{Z}_3$-symmetric $\rho_t$-generic framework with respect to the action $\theta : \mathbb{Z}_3 \to Aut(G)$ which is free on the vertex set of $G$ and $\tau : \mathbb{Z}_3 \to O(\mathbb{R}^2)$. Then $(G, p)$ is $\rho_t$-symmetric infinitesimally rigid if and only if the quotient $\mathbb{Z}_3$-gain graph of $G$ contains a spanning subgraph $(H, \psi)$ which is $(2,3,1)$-gain-tight.*

Theorem 25.18 gives rise to the following characterization for $\mathcal{C}_3$-generic infinitesimal rigidity (which is equivalent to the characterization given in Theorem 25.5 (b)) [55].

**Theorem 25.19 ($\mathbb{Z}_3$-Generic Infinitesimal Rigidity for $\mathcal{C}_3$ in 2D)** *Let $(G, p)$ be a $\mathbb{Z}_3$-generic framework with respect to the action $\theta : \mathbb{Z}_3 \to Aut(G)$ which is free on the vertex set of $G$ and $\tau : \mathbb{Z}_3 \to O(\mathbb{R}^2)$. Then $(G, p)$ is infinitesimally rigid if and only if the quotient $\mathbb{Z}_3$-gain graph of $G$ contains a spanning $(2,3,1)$-gain-tight subgraph.*

As we have seen in the previous section, for $k \geq 4$, the necessary counting conditions for a $\mathbb{Z}_k$-symmetric framework to be infinitesimally rigid are more complex. However, for odd $k < 1000$, a characterization for $\mathbb{Z}_k$-generic infinitesimal rigidity has very recently been obtained in [18, 19]. In particular, for the cyclic groups $\mathbb{Z}_k$ of prime order $k < 1000$, the necessary conditions outlined in Section 25.4.2 have been shown to be sufficient for $\mathbb{Z}_k$-generic infinitesimal rigidity [18, 19].

For dimensions 3 and higher, combinatorial characterizations for infinitesimal rigidity have been obtained for *body-bar* and *body-hinge frameworks* for the groups $\mathbb{Z}_2 \times \cdots \times \mathbb{Z}_2$. These characterizations are given in terms of packings of bases of signed-graphic matroids on quotient multi-graphs [54]. Below we provide a sample result for reflection and half-turn symmetry in 3-space. For precise definitions of symmetric body-bar and body-hinge frameworks and their quotient graphs, see [54].

**Theorem 25.20 ($\mathbb{Z}_2$-Generic Infinitesimal Body-Hinge Rigidity in 3D)** *Let $(G,h)$ be a $\mathbb{Z}_2$-generic body-hinge framework in $\mathbb{R}^3$ with respect to the action $\theta : \mathbb{Z}_2 \to Aut(G)$ which is free on the vertex set and the edge set of $G$, and $\tau : \mathbb{Z}_2 \to O(\mathbb{R}^3)$. Then $(G,h)$ is infinitesimally rigid if and only if the quotient $\mathbb{Z}_2$-gain graph $(G_0, \psi)$ of $G$ has the following properties.*

*(a) for $\tau(\mathbb{Z}_2) = \mathcal{C}_s$, $(G_0, \psi)$ contains three edge-disjoint spanning trees and three subgraphs with the property that each connected component contains exactly one cycle, which is unbalanced.*

*(b) for $\tau(\mathbb{Z}_2) = \mathcal{C}_2$, $(G_0, \psi)$ contains two edge-disjoint spanning trees and four subgraphs with the property that each connected component contains exactly one cycle, which is unbalanced.*

It is conjectured that these results also extend to the special class of molecular frameworks [38].

## 25.5 Periodic Frameworks

### 25.5.1 Glossary

***d*-periodic graph:** For a group $\Gamma$ isomorphic to $\mathbb{Z}^d$, a simple infinite graph $\tilde{G} = (\tilde{V}, \tilde{E})$ with finite degree at every vertex for which there exists a group action $\theta : \Gamma \to \text{Aut}(\tilde{G})$ which is free on the vertex set of $\tilde{G}$ and such that the quotient graph $\tilde{G}/\Gamma$ is finite.

**Quotient $\mathbb{Z}^d$-gain graph:** For a $d$-periodic graph $\tilde{G}$, the directed group-labelled multi-graph $(\tilde{G}_0, \psi)$ obtained from the quotient graph of $\tilde{G}$ by orienting and labelling the edges (via the function $\psi : \tilde{E}/\Gamma \to \Gamma$) in the analogous way as described for the quotient $\Gamma$-gain graph of a finite $\Gamma$-symmetric graph in Section 25.3.1. (See also Figure 25.11.)

***d*-periodic framework:** For a $d$-periodic graph $\tilde{G}$ (with respect to $\theta$), a $d$-periodic placement of $\tilde{G}$ is a function $\tilde{p} : \tilde{V} \to \mathbb{R}^d$ and a faithful representation $L : \Gamma \to \mathcal{T}(\mathbb{R}^d)$, where $\mathcal{T}(\mathbb{R}^d)$ is the group of translations of $\mathbb{R}^d$ (which we identify with the space $\mathbb{R}^d$ of translation vectors) such that

$$\tilde{p}_i + L(\gamma) = \tilde{p}_{\theta(\gamma)(i)} \qquad \text{for all } \gamma \in \Gamma \text{ and all } i \in \tilde{V}.$$

A $d$-periodic graph $\tilde{G}$ (with respect to $\theta$) together with its $d$-periodic placement $(\tilde{p}, L)$ is called a $d$-periodic framework, which is denoted by $(\tilde{G}, \tilde{p}, L)$.

***d*-periodic infinitesimal motion:** Choose an isomorphism $\Gamma \simeq \mathbb{Z}^d$, and let $(\tilde{G}, \tilde{p}, L)$ be a $d$-periodic framework (with respect to $\theta : \Gamma \to \text{Aut}(\tilde{G})$). Let $v_1, \ldots, v_a$ be a set of representatives for the vertex orbits of $\tilde{G}$, and let the $b$ pairs $(v_i, \theta(\gamma_\beta)v_j)$ be a set of representatives for the edge orbits of $\tilde{G}$. Further, let $x_i = \tilde{p}(v_i)$ for each $i$, and let $\mu_k = L(\gamma_k)$, $k = 1, \ldots, d$, be the translation vectors in $\mathbb{R}^d$ which correspond to the standard basis $\gamma_1, \ldots, \gamma_d$ of $\Gamma$. Let $\gamma_\beta = \sum_{k=1}^d c_\beta^k \gamma_k$ for $c_\beta^k \in \mathbb{Z}$, and $\mu(\beta) = \sum_{k=1}^d c_\beta^k \mu_k$. A vector $(y_1, \ldots, y_a, \nu_1, \ldots, \nu_d) \in \mathbb{R}^{da+d^2}$ is an infinitesimal motion of $(\tilde{G}, \tilde{p}, L)$ if

$$\langle (x_j + \mu(\beta)) - x_i, (y_j + \nu(\beta)) - y_i \rangle = 0 \quad \text{for } \beta = 1, \ldots, b,$$

where $\langle \cdot, \cdot \rangle$ represents the inner product, and $\nu(\beta) = \sum_{k=1}^d c_\beta^k \nu_k$. The matrix corresponding to the linear system above is called the **periodic rigidity matrix** of $(\tilde{G}, \tilde{p}, L)$.

***d*-periodic infinitesimally rigid framework:** A $d$-periodic framework $(\tilde{G}, \tilde{p}, L)$ for which every $d$-periodic infinitesimal motion is trivial.

**$d$-periodic generic framework:** A $d$-periodic framework $(\tilde{G}, \tilde{p}, L)$ whose periodic rigidity matrix has maximal rank among all $d$-periodic realizations $(\tilde{G}, \tilde{p}', L')$ of $\tilde{G}$ (with any choice of $L'$).

**$\mathbb{Z}^d$-rank of an edge set:** For a quotient $\mathbb{Z}^d$-gain graph $(\tilde{G}_0, \psi)$, and an edge subset $F$ of $(\tilde{G}_0, \psi)$, the smallest cardinality of a generating set for the subgroup of $\mathbb{Z}^d$ induced by $F$, where the subgroup induced by $F$ is defined in the analogous way as for an edge subset of a quotient $\Gamma$-gain graph for a finite group $\Gamma$ (see Section 25.3.1).

### 25.5.2 Maxwell Counts for Periodic Rigidity

Largely motivated by practical applications in fields such as engineering, materials science, and crystallography, as well as by mathematical applications in areas such as sphere packings, the rigidity and flexibility analysis of periodic frameworks has gained significant attention in recent years. Most of this work has focused on *forced-periodic* rigidity, i.e., on the question whether a periodic framework has a flex which preserves the periodicity of the framework throughout the motion. Therefore, we will restrict attention to forced-periodic rigidity in this section. (For other types of rigidity analyses of infinite frameworks, see [4, 32, 37, 39, 41], for example.) It is important to note that while the periodicity of the framework needs to be maintained, the periodic lattice determined by the framework does not need to be preserved as the framework moves. In other words, we are interested in the flexibility of forced-periodic frameworks with a fully flexible lattice representation. A mathematical formulation of infinitesimal rigidity for such frameworks has recently been given in [7] (see also [16, 40]). A summary of the key definitions is provided in Section 25.5.1.

An equivalent alternative description of $d$-periodic infinitesimal rigidity and the periodic rigidity matrix has been given in [42, 44]. While in [7] a periodic framework is considered as a realization of an infinite graph with a periodic group action, in [42, 44] a periodic framework is considered as a realization of a finite graph on a fundamental domain of $\mathbb{R}^d$ (i.e., a flat torus). Since in the latter description, the orientation of the periodic lattice is fixed, rotations are no longer trivial motions. In other words, frameworks on a torus are equivalence classes of periodic frameworks $(\tilde{G}, \tilde{p}, L)$ under rotation. In particular, this approach has been used to extend results to periodic frameworks with a fixed or partially flexible lattice representation. This is useful to analyze motions at short time scales as well as slower deformations of material [34, 43, 45].

The following result is stated for $d$-periodic frameworks with a fully flexible lattice representation [29]. However, it can also be adapted to other types of lattice flexibility, including the fixed lattice [42].

**Theorem 25.21 (Periodicity-Preserving Motions)** *A $d$-periodic generic framework $(\tilde{G}, \tilde{p}, L)$ has a $d$-periodic infinitesimal flex if and only if $(\tilde{G}, \tilde{p}, L)$ has a continuous flex which preserves the periodicity throughout the path.*

Using the periodic rigidity matrix (and its modified versions for the various partially flexible and the fixed lattice representations), it is easy to derive some basic necessary counts for a $d$-periodic framework to be $d$-periodic infinitesimally rigid. We summarize these conditions in the following theorem (see, for example, [29, 42]). Similar counts also exist for other types of lattice flexibility, such as a distortional (volume-preserving, but shape-changing) or hydrostatic (shape-preserving, but volume-changing) lattice deformation [42], as well as for crystallographic frameworks [30, 46, 59].

**Theorem 25.22 (Periodic Maxwell Counts)** *Let $(\tilde{G}, \tilde{p}, L)$ be a $d$-periodic infinitesimally rigid framework. Then the quotient $\mathbb{Z}^d$-gain graph of $\tilde{G}$ contains a spanning subgraph $(\tilde{H}, \psi)$ with $|\tilde{V}_0|$ vertices and $|\tilde{E}_0|$ edges such that*

*(a) for $L(\mathbb{Z}^d)$ fully flexible: $|\tilde{E}_0| = d|\tilde{V}_0| + \binom{d}{2}$;*

(b) *for translation vectors of* $L(\mathbb{Z}^d)$ *allowed to scale independently:* $|\tilde{E}_0| = d|\tilde{V}_0|$;

(c) *for* $L(\mathbb{Z}^d)$ *fixed:* $|\tilde{E}_0| = d|\tilde{V}_0| - d$.

We may also derive further necessary conditions for $d$-periodic infinitesimal rigidity by considering all edge-induced subgraphs of $(\tilde{H}, \psi)$. However, as in the case of finite symmetric frameworks, these counts are more complex. In particular, for any edge subset $F$ of $(\tilde{H}, \psi)$, we need to take into account the $\mathbb{Z}^d$-rank of $F$ (see also Theorems 25.23 and 25.24 for example).

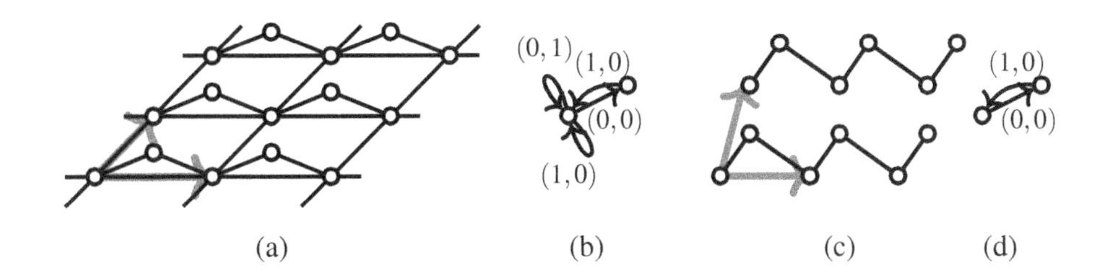


**Figure 25.11**
Parts of infinite periodic frameworks in the plane (a,c) and their respective quotient $\mathbb{Z}^2$-gain graphs (b,d). The framework in (a) is infinitesimally flexible under a fully flexible lattice representation, and the framework in (c) is isostatic under a fixed lattice representation (even though it is disconnected).

### 25.5.3 Characterizations of Periodic Rigid Graphs

As for finite forced-symmetric frameworks, the quotient $\Gamma$-gain graph plays a key role in establishing combinatorial characterizations of periodic generic infinitesimally rigid frameworks. The following characterization for periodic frameworks in the plane with a *fully flexible* lattice representation has been obtained in [29] using periodic direction networks.

**Theorem 25.23 (Periodic Rigidity For The Fully Flexible Lattice)** *Let* $(\tilde{G}, \tilde{p}, L)$ *be a 2-periodic generic framework. Then* $(\tilde{G}, \tilde{p}, L)$ *is 2-periodic infinitesimally rigid if and only if the quotient* $\mathbb{Z}^2$*-gain graph of* $\tilde{G}$ *contains a spanning subgraph* $(\tilde{H}, \psi)$ *with edge set* $\tilde{E}_0$ *and vertex set* $\tilde{V}_0$ *which satisfies*

(a) $|\tilde{E}_0| = 2|\tilde{V}_0| + 1$;

(b) $|F| \leq 2|V(F)| - 3 + 2k(F) - 2(c(F) - 1)$ *for all* $F \subseteq \tilde{E}_0$,

*where* $k(F)$ *is the* $\mathbb{Z}^2$*-rank of* $F$*, and* $c(F)$ *is the number of connected components of the subgraph induced by* $F$.

For periodic frameworks in the plane with a *fixed* lattice representation, we have the following result, which was obtained using a Henneberg-type inductive construction for the quotient $\mathbb{Z}^2$-gain graphs [45]. See also [29] for an alternative proof.

**Theorem 25.24 (Periodic Rigidity For The Fixed Lattice)** *Let* $(\tilde{G}, \tilde{p}, L)$ *be a 2-periodic generic framework, where* $L(\mathbb{Z}^2)$ *is non-singular and has to remain fixed. Then* $(\tilde{G}, \tilde{p}, L)$ *is 2-periodic infinitesimally rigid if and only if the quotient* $\mathbb{Z}^2$*-gain graph of* $\tilde{G}$ *contains a spanning subgraph* $(\tilde{H}, \psi)$ *with edge set* $\tilde{E}_0$ *and vertex set* $\tilde{V}_0$ *which satisfies*

(a) $|\tilde{E}_0| = 2|\tilde{V}_0| - 2$;

(b) $|F| \leq 2|V(F)| - 3$ *for all* $F \subseteq \tilde{E}_0$ *with* $\mathbb{Z}^2$*-rank equal to* 0*;*

(c) $|F| \leq 2|V(F)| - 2$ *for all* $F \subseteq \tilde{E}_0$.

The analogous result for periodic frameworks in the plane with a partially flexible lattice representation (with one degree of freedom) can be found in [34, 42]. Furthermore, combinatorial characterizations of infinitesimally rigid periodic frameworks in the plane with a fixed-area or fixed-angle fundamental domain are given in [31].

There also exist several initial results regarding the rigidity of periodic frameworks with additional symmetry which are forced to maintain the full crystallographic group of the framework throughout any motion [30, 43, 46, 59]. In particular, combinatorial characterizations of infinitesimally rigid crystallographic frameworks in the plane with a fully flexible lattice representation are given in [30] for the case where the group is generated by translations and rotations. For crystallographic *body-bar frameworks* with a *fixed* lattice representation in $d$-dimensional space, complete combinatorial characterizations for forced-symmetric infinitesimal rigidity are presented in [59].

One may also seek combinatorial characterizations of infinitesimally rigid periodic frameworks on a more basic level. In one of the first investigations of rigid periodic structures, it was shown in [60] that a quotient graph $\tilde{G}/\Gamma$ is the quotient graph of an infinitesimally rigid periodic framework in $d$-space with a fixed lattice representation *for some gain assignment of the edges of* $\tilde{G}/\Gamma$ if and only if $\tilde{G}/\Gamma$ contains a spanning subgraph which is the union of $d$ edge-disjoint spanning trees. This has been extended to periodic frameworks with a fully flexible lattice representation in [8].

**Theorem 25.25 (Periodic Rigidity for Generic Liftings)** *A quotient graph* $\tilde{G}/\Gamma$ *is the quotient graph of an infinitesimally rigid periodic framework in d-space for some gain assignment of the edges of* $\tilde{G}/\Gamma$ *if and only if* $\tilde{G}/\Gamma$ *contains a spanning subgraph* $\tilde{H}$ *with edge set* $\tilde{E}_0$ *and vertex set* $\tilde{V}_0$ *so that*

(a) $|\tilde{E}_0| = d|\tilde{V}_0| + \binom{d}{2}$*;*

(b) $\tilde{H}$ *has a spanning subgraph with* $d|\tilde{V}_0| - d$ *edges which has the property that every subgraph with* $m$ *edges and* $n$ *vertices satisfies* $m \leq dn - d$.

The analogous result for periodic *body-bar frameworks* in $d$-space has been established in [9].

As part of the effort to gain insights into "incidentally periodic" infinitesimally rigid frameworks, recent work has also provided algebraic (and for $d = 2$ even some combinatorial) characterizations of $d$-periodic frameworks which are "ultrarigid," that is, infinitesimally rigid *for any choice of the periodicity lattice* [32].

Finally, we note that very recent work has also established necessary conditions for a $d$-periodic generic framework to be *globally rigid* under fixed lattice representations [23]. Analogous to Hendrickon's theorem for finite frameworks, these necessary conditions consist of a graph connectivity condition and a redundant rigidity condition. In dimension 2, these conditions have also been confirmed to be sufficient, giving a combinatorial characterization of globally rigid 2-periodic generic frameworks under fixed lattice representations [23]. Extensions of this result to periodic frameworks under flexible lattice representations have not yet been obtained. There are also no analogous results yet regarding the (forced-symmetric) global rigidity of symmetric finite frameworks.

# References

[1] S.L. Altmann and P. Herzig. *Point-Group Theory Tables.* Clarendon Press, Oxford, 1994.

[2] L. Asimov and B. Roth. The Rigidity Of Graphs. *Transactions of the AMS*, 245:279–289, 1978.

[3] P.W. Atkins, M.S. Child, and C.S.G. Phillips. *Tables for group theory*. Oxford University Press, 1970.

[4] G. Badri. *Rigidity operators and the flexibility of infinite bar-joint frameworks*. Ph.D. thesis, Department of Mathematics and Statistics, Lancaster University, 2015.

[5] M. Berardi, B. Heeringa, J. Malestein, and L. Theran. Rigid components in fixed-lattice and cone frameworks. *Proc. 23rd Canadian Conference on Computational Geometry*, pages 1–6, 2011.

[6] A.R. Berg and T. Jordán. Algorithms for graph rigidity and scene analysis. *Proc. 11th Annual European Symposium on Algorithms (ESA)*, 2832 of LNCS:78–89, 2003.

[7] C. Borcea and I. Streinu. Periodic frameworks and flexibility. *Proceedings of the Royal Society A*, 466:2633–2649, 2010.

[8] C. Borcea and I. Streinu. Minimally rigid periodic graphs. *Bull. Lond. Math. Soc.*, 43(6):1093–1103, 2011.

[9] C. Borcea, I. Streinu, and S. Tanigawa. Periodic body-and-bar frameworks. *Proc. 24th ACM symposium on Computational Geometry (SoCG2012)*, pages 347–356, 2012.

[10] O. Bottema. Die Bahnkurven eines merkwürdigen Zwölfstabgetriebes. *Österr. Ing.-Arch*, 14:218–222, 1960.

[11] R. Connelly, P.W. Fowler, S.D. Guest, B. Schulze, and W. Whiteley. When is a symmetric pin-jointed framework isostatic? *International Journal of Solids and Structures*, 46:762–773, 2009.

[12] P.W. Fowler and S.D. Guest. A symmetry extension of Maxwell's rule for rigidity of frames. *International Journal of Solids and Structures*, 37:1793–1804, 2000.

[13] S.D. Guest and P.W. Fowler. A symmetry-extended mobility rule. *Mechanism and Machine Theory*, 40:1002–1014, 2005.

[14] S.D. Guest and P.W. Fowler. Symmetry conditions and finite mechanisms. *Mechanics of Materials and Structures*, 2(6), 2007.

[15] S.D. Guest and P.W. Fowler. Symmetry-extended counting rules for periodic frameworks. *Phil. Trans. of the Royal Society A*, 372(2008), 2014.

[16] S.D. Guest and J.W. Hutchinson. On the determinacy of repetitive structures. *Journal of the Mechanics and Physics of Solids*, 51(3):383–391, 2003.

[17] S.D. Guest, B. Schulze, and W. Whiteley. When is a symmetric body-bar structure isostatic? *International Journal of Solids and Structures*, 47:2745–2754, 2010.

[18] R. Ikeshita. *Infinitesimal rigidity of symmetric frameworks*. Master's thesis, Department of Mathematical Informatics, The University of Tokyo, 2015.

[19] R. Ikeshita and S. Tanigawa. Count matroids of group-labeled graphs. *Combinatorica*, doi.org/10.1007/s00493-016-3469-8, 2017.

[20] T. Jordan, V. Kaszanitzky, and S. Tanigawa. Gain-sparsity and symmetry-forced rigidity in the plane. *Discrete & Computational Geometry*, 55:314–372, 2016.

[21] R.D. Kangwai and S.D. Guest. Symmetry-adapted equilibrium matrices. *International Journal of Solids and Structures*, 37:1525–1548, 2000.

[22] R.D. Kangwai, S.D. Guest, and S. Pellegrino. An introduction to the analysis of symmetric structures. *Computers and Structures*, 71:671–688, 1999.

[23] V. Kaszanitzky, B. Schulze, and S. Tanigawa. Global rigidity of periodic graphs under fixed-lattice representations. *arXiv:1612.01379*, 2016.

[24] D. Kitson and B. Schulze. Maxwell-Laman counts for bar-joint frameworks in normed spaces. *Linear Algebra Appl.*, 481:313–329, 2015.

[25] D. Kitson and B. Schulze. Symmetric isostatic frameworks with $\ell^1$ or $\ell^\infty$ distance constraints. *The Electronic Journal of Combinatorics*, 23(4):1–13, P4.23, 2016.

[26] A. Lee and I. Streinu. Pebble game algorithms and sparse graphs. *Discrete Mathematics*, 308:1425–1437, 2008.

[27] J. Malestein and L. Theran. Generic rigidity of frameworks with orientation-preserving crystallographic symmetry. *arXiv:1108.2518*, 2011.

[28] J. Malestein and L. Theran. Generic rigidity of reflection frameworks. *arXiv:1203.2276*, 2012.

[29] J. Malestein and L. Theran. Generic combinatorial rigidity of periodic frameworks. *Advances in Mathematics*, 233(1):291–331, 2013.

[30] J. Malestein and L. Theran. Frameworks with forced symmetry II: orientation-preserving crystallographic groups. *Geometriae Dedicata*, 170(1):219–262, 2014.

[31] J. Malestein and L. Theran. Generic rigidity with forced symmetry and sparse colored graphs. *In: Proc. of Fields Workshop on Rigidity and Symmetry*, 2014.

[32] J. Malestein and L. Theran. Ultrarigid periodic frameworks. *arXiv:1404.2319*, 2014.

[33] J. Malestein and L. Theran. Frameworks with forced symmetry I: reflections and rotations. *Discrete & Computational Geometry*, 54(2):339–367, 2015.

[34] A. Nixon and E. Ross. Periodic rigidity on a variable torus using inductive constructions. *The Electronic Journal of Combinatorics*, 22(1):P1, 2015.

[35] A. Nixon and B. Schulze. Symmetry-forced rigidity of frameworks on surfaces. *Geometriae Dedicata*, 182(1):163–201, 2016.

[36] J.C. Owen and S.C. Power. Frameworks, symmetry and rigidity. *Int. J. Comput. Geom. Appl.*, 20:723–750, 2010.

[37] J.C. Owen and S.C. Power. Infinite bar-joint frameworks, crystals and operator theory. *New York Journal of Mathematics*, 17:445–490, 2011.

[38] J. Porta, L. Ros, B. Schulze, A. Sljoka, and W. Whiteley. On the symmetric molecular conjectures. *Computational Kinematics, Mechanisms and Machine Science*, 15:175–184, 2014.

[39] S.C. Power. Crystal frameworks, matrix-valued functions and rigidity operators. In *Concrete Operators, Spectral Theory, Operators in Harmonic Analysis and Approximation*, pages 405–420. Springer, 2014.

[40] S.C. Power. Crystal frameworks, symmetry and affinely periodic flexes. *New York Journal of Mathematics*, 20:665–693, 2014.

[41] S.C. Power. Polynomials for crystal frameworks and the rigid unit mode spectrum. *Philosophical Transactions of the Royal Society A*, 372(2008), 2014.

[42] E. Ross. *The Rigidity of Periodic Frameworks as Graphs on a Torus*. Ph.D. thesis, Department of Mathematics and Statistics, York University, 2011.

[43] E. Ross. The rigidity of periodic body-bar frameworks on the three-dimensional fixed torus. *Phil. Trans. of the Royal Society A*, 372(2008), 2014.

[44] E. Ross. The rigidity of periodic frameworks as graphs on a fixed torus. *Contributions to Discrete Mathematics*, 9(1), 2014.

[45] E. Ross. Inductive constructions for frameworks on a two-dimensional fixed torus. *Discrete & Computational Geometry*, 54(1):78–109, 2015.

[46] E. Ross, B. Schulze, and W. Whiteley. Finite motions from periodic frameworks with added symmetry. *International Journal of Solids and Structures*, 48:1711–1729, 2011.

[47] B. Schulze. *Combinatorial and Geometric Rigidity with Symmetry Constraints*. Ph.D. thesis, Department of Mathematics and Statistics, York University, 2009.

[48] B. Schulze. Block-diagonalized rigidity matrices of symmetric frameworks and applications. *Beiträge zur Algebra und Geometrie*, 51(2):427–466, 2010.

[49] B. Schulze. Injective and non-injective realizations with symmetry. *Contributions to Discrete Mathematics*, 5:59–89, 2010.

[50] B. Schulze. Symmetric Laman theorems for the groups $c_2$ and $c_s$. *The Electronic Journal of Combinatorics*, 17(1):1–61, R154, 2010.

[51] B. Schulze. Symmetric versions of Laman's Theorem. *Discrete & Computational Geometry*, 44(4):946–972, 2010.

[52] B. Schulze. Symmetry as a sufficient condition for a finite flex. *SIAM Journal on Discrete Mathematics*, 24(4):1291–1312, 2010.

[53] B. Schulze, P.W. Fowler, and S.D. Guest. When is a symmetric body-hinge structure isostatic? *International Journal of Solids and Structures*, 51:2157–2166, 2014.

[54] B. Schulze and S. Tanigawa. Linking rigid bodies symmetrically. *European Journal of Combinatorics*, 42:145–166, 2014.

[55] B. Schulze and S. Tanigawa. Infinitesimal rigidity of symmetric frameworks. *SIAM Journal on Discrete Mathematics*, 29(3):1259–1286, 2015.

[56] B. Schulze and W. Whiteley. The orbit rigidity matrix of a symmetric framework. *Discrete & Computational Geometry*, 46(3):561–598, 2011.

[57] B. Schulze and W. Whiteley. Coning, symmetry, and spherical frameworks. *Discrete & Computational Geometry*, 48(3):622–657, 2012.

[58] P.D. Seymour. A note on hyperplane generation. *J. Combin. Theory Ser. B*, 61:88–91, 1994.

[59] S. Tanigawa. Matroids of Gain Graphs in Applied Discrete Geometry. *Trans. Amer. Math. Soc.*, 367:8597–8641, 2015.

[60] W. Whiteley. The union of matroids and the rigidity of frameworks. *SIAM J. Discrete Math.*, 1(2):237–255, 1988.

[61] W. Whiteley. Some Matroids from Discrete Applied Geometry. *Contemporary Mathematics*, 197:171–311, 1996.

# Index